Periodic table of the elements with atomic weights*

Transition elements

1 H 1.0079																	2 He 4.00260
3 Li 6.941	4 Be 9.01218											5 B 10.81	6 C 12.011	7 N 14.0067	8 O 15.9994	9 F 18.99840	10 Ne 20.179
11 Na 22.98977	12 Mg 24.305											13 Al 26.98154	14 Si 28.086	15 P 30.97376	16 S 32.06	17 Cl 35.453	18 Ar 39.948
19 K 39.098	20 Ca 40.08	21 Sc 44.9559	22 Ti 47.90	23 V 50.9414	24 Cr 51.996	25 Mn 54.9380	26 Fe 55.847	27 Co 58.9332	28 Ni 58.70	29 Cu 63.546	30 Zn 65.38	31 Ga 69.72	32 Ge 72.59	33 As 74.9216	34 Se 78.96	35 Br 79.904	36 Kr 83.80
37 Rb 85.4678	38 Sr 87.62	39 Y 88.9059	40 Zr 91.22	41 Nb 92.9064	42 Mo 95.94	43 Tc (97)	44 Ru 101.07	45 Rh 102.9055	46 Pd 106.4	47 Ag 107.868	48 Cd 112.40	49 In 114.82	50 Sn 118.69	51 Sb 121.75	52 Te 127.60	53 I 126.9045	54 Xe 131.30
55 Cs 132.9054	56 Ba 137.34	57 La 138.9055	72 Hf 178.49	73 Ta 180.9479	74 W 183.85	75 Re 186.207	76 Os 190.2	77 Ir 192.22	78 Pt 195.09	79 Au 196.9665	80 Hg 200.59	81 Tl 204.37	82 Pb 207.2	83 Bi 208.9804	84 Po (209)	85 At (210)	86 Rn (222)
87 Fr (223)	88 Ra 226.0254	89 Ac (227)	104	105													

Lanthanides

58 Ce 140.12	59 Pr 140.9077	60 Nd 144.24	61 Pm (145)	62 Sm 150.4	63 Eu 151.96	64 Gd 157.25	65 Tb 158.9254	66 Dy 162.50	67 Ho 164.9304	68 Er 167.26	69 Tm 168.9342	70 Yb 173.04	71 Lu 174.97

Actinides

90 Th 232.0381	91 Pa 231.0359	92 U 238.029	93 Np 237.0482	94 Pu (244)	95 Am (243)	96 Cm (247)	97 Bk (247)	98 Cf (251)	99 Es (254)	100 Fm (257)	101 Md (258)	102 No (255)	103 Lw (260)

*1973 values based on $^{12}C = 12$.

CHEM ONE

CHEM ONE

Jürg Waser
Department of Chemistry
California Institute of Technology

Kenneth N. Trueblood
Department of Chemistry
University of California, Los Angeles

Charles M. Knobler
Department of Chemistry
University of California, Los Angeles

McGraw-Hill Book Company
New York   St. Louis   San Francisco   Auckland
Düsseldorf   Johannesburg   Kuala Lumpur   London   Mexico
Montreal   New Delhi   Panama   Paris   São Paulo
Singapore   Sydney   Tokyo   Toronto

CHEM ONE

Copyright © 1976 by McGraw-Hill, Inc. All rights reserved. Printed in the United States of America. No part of this publication may be reproduced, stored in a retrieval system, or transmitted, in any form or by any means, electronic, mechanical, photocopying, recording, or otherwise, without the prior written permission of the publisher.

1 2 3 4 5 6 7 8 9 0   MURM   7 9 8 7 6

Library of Congress Cataloging in Publication Data

Waser, Jürg.
    Chem one.

    Includes bibliographies.
    1. Chemistry. I. Trueblood, Kenneth N., joint author. II. Knobler, Charles M., date   joint author. III. Title.
QD31.2.W36   540   75–30507
ISBN 0-07-068420-0

This book was set in Times by York Graphic Services, Inc. The editors were Robert H. Summersgill, Janet Wagner, and Anne T. Vinnicombe; the designer was Janet Durey Bollow; the production supervisor was Joe Campanella. The drawings were done by Mark Schroeder.
The printer was The Murray Printing Company; the binder, Rand McNally & Company.

Contents

Preface

"Chem One" is intended for students planning careers in the physical and life sciences and in engineering. It had its origins in notes prepared for use in introductory courses over the last decade and almost all of the book rests directly on material that we have presented to our students in first-year courses.

We have attempted to avoid the pitfalls of half-truth and misleading oversimplification. (No student should have to discover in an advanced course that ideas dutifully learned earlier are incorrect.) However, a book with such a goal need not *begin* each topic on an advanced level. We have tried, through careful definitions, rather full explanations, and many worked examples, to provide the means by which a student who is willing to work can master the fundamental concepts and be able to apply them. Those whose prior preparation has been inadequate may find that the "Study Guide," prepared by Eleanor Siebert, will smooth their way.

Certain concepts are encountered a number of times, first in an introductory fashion in the early chapters, then again in greater depth and sophistication in the middle sections of the book, and finally still again as they are used in the discussions of descriptive chemistry that comprise the final seven chapters. In addition to the pedagogical advantages that accompany repeated exposure at increasing depth, including a better overall perspective on the unifying themes of chemistry, this scheme makes it possible for students to do significant laboratory experiments during the early stages of the course. Thus bonding, three-dimensional structure, and reactions in solution are first introduced in an elementary fashion in Chapters 1–7, and it is at this stage also that errors, stoichiometry, and gases are treated in detail.

Conventions regarding units and notation are changing as the SI recommendations come into use. We have used the SI system primarily, because those being trained now will doubtless use SI units throughout most of their working lives. However, we have not followed the SI system slavishly; we have retained the angstrom, and we have not completely given up the milliliter in favor of the exactly equivalent cubic centimeter because students will use glassware calibrated in milliliters. Table headings, labels on the axes of graphs, and some equations are written in "slash" notation. In this convention, dimensioned quantities are divided by their units to give dimensionless numbers. Thus, we write log (P/atm), which represents the logarithm of a number, as the pressure in atmospheres divided by the unit atmosphere. Each of the SI conventions is explained when it is introduced and the SI system is discussed in detail in Appendix A.

The structure of the book allows several orders of presentation in addition to assigning chapters in serial order. For courses in which laboratory experiments dealing with gases are begun at an early stage, Chapter 5 (gases) can be introduced immediately after Chapter 2 (errors and stoichiometry). On the other hand, if experiments involving solutions are begun as early as the fourth week of the first term, Chapters 6 and 7 can be assigned immediately after Chapter 3. Chapters 8 through 12 (quantum theory, atomic and molecular structure, the periodic table) can be moved to a later position in whole or in part (e.g., Chapters 11 and 12 may be postponed until just before the discussion of descriptive chemistry). The possibility of following Chapters 1 through 7 by Chapters 13 through 15 (which deal with equilibrium) may be particularly useful in courses that have early laboratory assignments involving titrations or other equilibrium experiments that are to be dealt with quantitatively.

In any book as comprehensive as this there are topics or sections that can be omitted or abridged to suit the tastes of the instructor and the students as well as the limitations of time. We have attempted to relegate the most advanced and least essential material to the latter parts of many chapters so that they may most readily be skipped. Some sections of the book can be left for the student to read and need not be discussed extensively in class. Chapter 4, for example, can stand on its own as an introduction to structure. Chapter 19 can be omitted entirely if the concept of a Latimer diagram, essential to the descriptive material in later chapters, is introduced in lecture. With selective omissions, a lecturer should have time to devote at least three weeks to organic chemistry and biochemistry, which comprise the final two chapters, if this is felt desirable. Although the latter are often found in second-year courses, there are many students who do not continue beyond the first year. It seems to us essential that any general chemistry course provide such students with some exposure to these important areas of modern chemistry.

This book has been in gestation for so long that it is difficult to express our appreciation to all those who have contributed to it in various ways. Among those whose ideas and encouragement played an early role are Alan Wingrove, J. D. McCullough, Bill Benjamin, Ed Friedrich, Julie and Judy Swain, and many other students and colleagues at UCLA and Cal Tech. Portions of the manuscript have been read critically at every stage by many people, including Jay M. Anderson, Kyle Bayes, John P. Chesick, Deirdre Devereux, Jenny Glusker, James B. Ifft, Daniel Kivelson, Caroline Lanford, Samuel Markowitz, Richard E. Marsh, Julian L. Roberts, Jr., Verner Schomaker, Raymond J. Suplinskas, Charles West, and a number of helpful anonymous reviewers. Kathleen North devoted a major portion of a summer to a detailed page-by-page critique of more than half the book from the standpoint of a student who had just been through the course, and Robert Weiss checked a large fraction of the problems. Many have labored over the typing of innumerable drafts, most notably Karen Gleason, Delna Jacobs, Marcella Hughes, Ellen Dunlevy, Glenda Grant, Edi Bierce and Judy Stewart Glazer. Finally, we have appreciated the professionalism and talent of the designer, Janet Bollow, and the illustrator, Mark Schroeder, and of Robert Summersgill, Anne Vinnicombe, Alice Goehring, Janet Wagner, and Joe Campanella who have supervised various stages of production with grace and common sense. To all, our genuine thanks.

Jürg Waser
Kenneth N. Trueblood
Charles M. Knobler

To the Student

The material in this book varies widely in difficulty. Some topics may seem essentially a review of high school chemistry, while others will be new to all students and sometimes rather abstract. Many topics are treated a number of times so that you can become familiar with them while working with them. They are first encountered at an elementary and qualitative level, are later developed in more detail, often quantitatively, and finally turn up again when they are utilized in systematizing and explaining the great body of chemical facts called "descriptive chemistry".

We do not pretend that all of the material is easy to grasp; we and most of our colleagues in chemistry had to struggle with many parts of it when we were learning it. We have, however, tried to smooth the way for you. New words are italicized when they are first defined. Many of the figures have extensive legends; these are intended to help clarify both the figures and the accompanying textual material and should be studied carefully. You will find many worked examples that show you how to apply principles to specific situations, and you will also find many problems that should help develop your ability to work with chemical concepts.

Frequent cross-references tie together related concepts and facts found in different sections of the book. Three of the four appendixes are self-contained essays on special topics that you may want to consult a number of times. Frequently used tables are located inside the front and back covers. Finally, the extensive index should be consulted liberally to locate definitions and topics encountered earlier. If you are curious, you may even want to use the index to see where and how some topic will be met again.

The *Study Guide* that accompanies the book may also be helpful, particularly in those places in the text where some background in physics or mathematics is essential to a thorough understanding. Additional elementary problems are also included in the *Study Guide*.

We suggest that when the going gets tough, as it will at times, you give the more difficult material a rest after a first cursory reading. Follow up later with a second, more careful, study, jotting down key words and concepts and frequently closing the book for quick mental reviews. Retrace the steps of derivations and check the details of the worked examples. Don't worry if at first you understand some new topic only partially and even have some wrong ideas about it. Often an initial false start that is later corrected helps to clarify something, because it gives a perspective not available to someone who hasn't thought about the topic at all.

Try to retain a critical attitude at all times. Don't accept anything stated here, or elsewhere, simply on the basis of the apparent authority of the source. Apply your powers of reasoning as much as you can; search for internal consistency. We have tried to avoid errors but it seems unlikely that we have caught them all.

Learning is a lonely pursuit and takes a good deal of discipline. Yet, all these sober words of caution and advice should not obscure the fact that chemists really *enjoy* chemistry. We know that you can too—and we hope you will.

Jürg Waser
Kenneth N. Trueblood
Charles M. Knobler

"Those sciences are vain and full of errors that are not born from experiment, the mother of all certainty, and that do not end with one clear experiment."

Leonardo da Vinci

"According to convention there is a sweet and a bitter, a hot and a cold, and according to convention there is a color. In truth there are atoms and a void."

Demokritos, Greek philosopher, fifth century B.C.

1 Introduction

1-1 Chemistry as One of the Natural Sciences

The natural sciences. The development of modern science during the last three centuries has been at once so broad and so deep that a great deal of learning and thought is required for even the best of human minds to encompass a portion of it. This is in large part because progress in science is cumulative—science builds on and extends what has been observed and understood earlier. It grows by the interplay of experimental observations, imaginative reasoning, predictions based on this reasoning, and experiments to test the predictions. As the body of systematized observations and generalizations about the natural world has increased, it has been artificially subdivided into "different" scientific fields, and these have in turn been partitioned, all because human life is too short and the mind too limited to be able to learn all that has been observed and postulated.

The different natural sciences were once considered to be physics, chemistry, biology, geology, and astronomy, but now the subdivisions and extensions of these fields have become far more numerous. Not only do they include areas in which several of these disciplines overlap, such as geophysics (the physics of the earth) and biochemistry (the chemistry of living things) but, increasingly now, they encompass areas in which the methods of many of the classical natural sciences are integrated and brought to bear on a particular portion of our world. For example, oceanography includes the related studies of the physical, chemical, biological, and geological aspects of the ocean, and planetary science involves a correlated study in all the foregoing disciplines of the members of our solar system.

Chemistry. Chemistry is concerned with the composition, structure, and properties of substances, the transformation of these substances into others by reactions, and the different kinds of energy changes that accompany these reactions. In the main, the chemist is interested in the properties and reactions of matter as it commonly exists on the earth, and most of this book deals with this aspect of chemistry. However, astronomical observations and space exploration give us every reason to believe that the general principles discussed are more widely applicable.

Since the field of chemistry covers an enormous range of activities, it is, in turn, subdivided loosely into many branches. This division is done in several distinct ways, depending on the focus of interest. It may be based on the kinds of substances that are involved; for example, organic chemistry deals with compounds containing the element carbon (of which there are literally millions) and inorganic chemistry deals with substances that do not contain this element. However, the distinctions are not clear cut. For example, the carbonate minerals are regarded as inorganic even though they contain carbon.

Another classification scheme focuses on the types of operations and reactions that one may perform. Two of the earliest major areas of chemistry were analytical chemistry, the determination of the identity and the proportions of the components of a compound or of a mixture, and synthetic chemistry, the creation of one substance from others. Still another way of subdividing the field of chemistry is on the basis of its overlap with other fields—thus one finds references to physical chemistry, biochemistry, geochemistry, cosmochemistry, and so on. Finally, the contrasting phrases "theoretical chemistry" and "descriptive chemistry" are often used to refer to the concepts, principles, and theories on the one hand and to the experimentally observed facts on the other.

Many of the experimental methods of modern chemistry were developed originally by men and women regarded as physicists—for example, the methods of spectroscopy and those of structure determination by the diffraction of X rays. Similarly, modern theoretical chemistry is based chiefly on thermodynamics, statistical mechanics, and quantum mechanics, all originally regarded as fields of theoretical physics. The interaction has not all been one-sided, however; many of the ideas important in modern physics came originally from chemistry. The atomic nature of matter was first recognized quantitatively by John Dalton (1766–1844) and was part of chemists' thinking for about a century before most physicists accepted it without reservation. The interactions of chemistry with the other nominally distinct branches of science, most notably biology and geology, have also been numerous and important.

A word of caution is needed about the vocabulary of chemistry. You will encounter many new terms, most of which have quite precisely defined meanings. Not only must you learn these meanings carefully so that you understand how to use the terms and how they differ from related ones, but sometimes you must avoid confusion with popular usage of the same words that is usually (though not always) broader and vaguer. Thus, the terms *energy, work, heat,* and *force* have more restricted meanings in physics and chemistry than in ordinary parlance; on the other hand, the words *salt* and *alcohol* have more general meanings, referring to classes of substances rather than to specific ones. A few scientific terms have multiple meanings, and while each meaning is quite precise, only the context makes it possible to decide which of them applies in any particular case. For example, the term *neutral* may refer to the absence of an excess of positive or negative charge, or to the absence of an excess of acid or base. Such possible ambiguities will often be pointed out, but you should be alert for them.

Physical and chemical properties. The simplest way to describe a substance is in terms of its properties. Its *physical properties,* such as color, density, melting point, and solubility, are those that do not depend on its reaction with other substances (or with itself). Physical properties usually describe the response of a substance to external influences, such as temperature changes or an electric field. The coefficient of thermal expansion, for example, specifies the change in volume accompanying a given temperature increase, and the electrical conductivity specifies the relation between the flow of electric current and the strength of an applied electric field. The *chemical properties* of a substance, on the other hand, are those that relate to its behavior in the presence of other substances, in particular to its reactions with other substances (or with itself) to form new substances with different properties, or its failure to react under particular conditions.

There is no sharp line between chemical and physical properties. The interaction of a substance with light may not only reveal its color but also cause chemical reaction, as it does in a photographic emulsion; the flow of electric current may cause chemical changes, and mechanical interaction may cause explosives to detonate (a chemical change). Similarly, the dissolving of a substance in water is often accompanied by chemical interaction between the dissolving particles and the water molecules.

Pure substances and mixtures. The term *substance* (or *pure substance*) is used to refer to a sample of matter that has distinct physical and chemical properties and a definite composition. A few of the materials that we deal with in everyday life are pure substances—for example, sugar, salt, copper, and silver. Pure substances may be further categorized as either elements or compounds, a distinction amplified in Sec. 1-6. Elements are represented by one- or two-letter symbols (H, O, Cl) representing important initial letters in the name that was in common use for the element when the symbol was adopted. The chemical composition of a compound is expressed by formulas (H_2O, Na_2SO_4, $C_{12}H_{22}O_{11}$) that indicate the relative numbers of atoms of different elements present in the substance.

Most familiar materials—milk, vinegar, brass, air, dirt, concrete—are mixtures of various substances.[1] As a result they do not have one or more of the well-defined properties of pure substances. For example, they may not have sharp melting or boiling temperatures, or they may appear to the naked eye to have several distinct ingredients. Even if they *appear* to be composed of a single substance, it may be possible to separate them into substances with different properties. This might be accomplished, for instance, by heating or by treatment with a suitable solvent.

Mixtures can be categorized as either homogeneous—uniform in properties throughout the sample—or heterogeneous. Homogeneous mixtures are called *solutions* (Chaps. 6 and 7). The implications of the terms *homogeneous* and *heterogeneous* are discussed further in the next section.

Macroscopic and microscopic viewpoints. Explanations and interpretations of chemical phenomena in terms of hypotheses and theories involve two distinct points of view, the macroscopic and the microscopic. The macroscopic world is that part of the physical world that is directly apparent to our senses. In common usage it includes all phenomena that can be observed by the naked eye. In chemistry and related fields, however, all those phenomena that are not on an atomic or molecular scale are considered *macroscopic,* while *microscopic* phenomena are those that occur at the atomic and molecular level.

One of the chief objectives of chemistry is to find explanations for macroscopic phenomena on the microscopic level, that is, in terms of the properties and interactions of atoms and molecules. Thus the interplay between macroscopic and microscopic viewpoints is considered throughout this text.

1-2 States of Aggregation

Phases. A piece of matter is called *homogeneous* if its macroscopic properties (e.g., density, pressure, or velocities of light and sound in it) are the same throughout. If it consists of macroscopic regions that have different properties and are separated by macroscopically sharp boundaries, it is said to be *heterogeneous.* The homogeneous parts of such a system are called *phases* (Greek: *phainein,* to appear). For example, a piece of granite consists of three principal solid phases—quartz, feldspar, and mica. A pitcher of water with ice cubes contains the liquid and solid phases of water, while the air above the water, including some water vapor, is a third, gaseous phase.

In some respects a one-phase region may not be strictly homogeneous. The pressure in a gas or in a liquid is different at points that differ in height, and properties that depend on the pressure, such as density, are also different. These differences are usually small. It is important to note that the properties in a one-phase region change continuously and slowly with position, whereas they change abruptly when a boundary to a new phase is crossed.

---

[1]Even "pure" substances—distilled water, refined sugar, electrolytically purified copper—inevitably have traces of contaminants, although perhaps only a few parts of impurity per million parts of the substance itself. In special cases even higher purities are obtainable. Germanium used in transistors is routinely produced with only 1 part in 10 million of impurity.

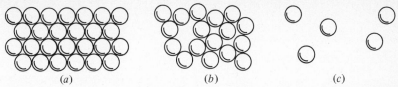

(a)   *(b)*   *(c)*

FIGURE 1-1   **Atomic aspects of (*a*) a crystalline solid, (*b*) a liquid, and (*c*) a gaseous phase.** The spheres represent molecules or atoms. In (*a*) the arrangement is regular and adjacent molecules are in contact with each other; in (*b*) it is irregular but there is still contact between adjacent molecules; in (*c*) the arrangement is irregular and most molecules are distant from each other.

States of aggregation.   The three common states of aggregation of matter are the solid, the liquid, and the gaseous states. Liquid and solid states are also called condensed states or condensed phases. Their densities are usually much greater than those of gases, and the density of a solid is usually somewhat greater than that of the corresponding liquid phase.

It is instructive to compare the macroscopic and microscopic descriptions of these three states (Fig. 1-1). A solid has the macroscopic property of resisting attempts to change its original shape and volume. A liquid assumes the shape of as much of its container as it fills, but it resists attempts to change its volume.[1] A gas fills all the space available to it uniformly, as long as effects of gravitation are negligible.

On the microscopic level we start with a description of gases. The molecules of a gas move about freely, with distances between them that are large compared to their size—for example, under ordinary conditions of temperature and pressure, about 10 molecular diameters. The forces between the molecules are small, and except for many mutual collisions, the molecules move about almost independently at high speeds, preferring no part of the container over another and filling it uniformly. The pressure exerted by the gas is the result of the numerous collisions of its molecules with the container walls.

On the other hand, the molecules[2] of a solid are packed closely and oscillate about fixed positions, moving back and forth over increasingly larger distances as the temperature rises. Only occasionally do molecules in a solid break loose and interchange places. Solids are not easily compressed because of the strong repulsion at small distances between molecules, and the resistance to other changes in shape is due to the forces that hold the molecules close together.

The packing of molecules in liquids is somewhat looser, and the molecules slip by each other continually. There is still sufficient attraction for the molecules to cling together, and strong repulsive forces arise when they are pushed even closer, as when the liquid is compressed. Thus it is understandable that a liquid resists attempts to change its volume, but not its shape.

---

[1]Drops of liquids, and larger amounts in the absence of gravitation, tend to assume a spherical shape because of *surface tension,* that is, because of forces that strive to make the surface area of the liquid a minimum.

[2]The word *molecule* will be used frequently to stand for the words *atom* and *ion* (charged atom or molecule) as well, whenever the difference among these terms is not important, as in the present discussion.

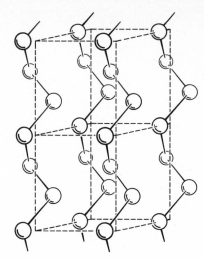

FIGURE 1-2   **The structure of β-selenium crystals.**   Each selenium (Se) atom is bonded to two others to form helical chains of indefinite extension. The distance between nearest-neighbor atoms in adjacent chains is significantly greater than that between the bonded atoms in each chain. The dashed lines enclose two "unit cells" of the crystal. The actual crystal is a repetition of such unit cells in all directions. The properties of the crystal in the direction of the chains are markedly different from the properties perpendicular to the chain directions.

When the properties of a phase are the same in all directions, the phase is said to be *isotropic;* otherwise it is *anisotropic* (Greek: *iso,* same; *tropos,* turn). Crystals are anisotropic. A typical direction-dependent property of crystals is the pronounced cleavage that often exists, as in mica, which may be split into very thin sheets. The electrical conductivity may vary with the direction of the current flow as it does in crystals of graphite or selenium (Fig. 1-2), and optical properties may differ markedly in different directions, as evidenced in the double refraction shown by calcite crystals (Iceland spar) (Fig. 1-3).

In crystalline solids the molecules are arranged with great regularity. Except for minor flaws, this regularity extends throughout the crystal and thus affects its macroscopic properties; there is long-range order. The microscopic reason why crystals are anisotropic is that any *ordered* array of molecules shows different prop-

FIGURE 1-3   **Double refraction in Iceland spar (calcite, $CaCO_3$).**   A light ray entering a crystal that exhibits double refraction is usually propagated in two different directions inside the crystal. An object viewed through such a crystal may thus be seen double.

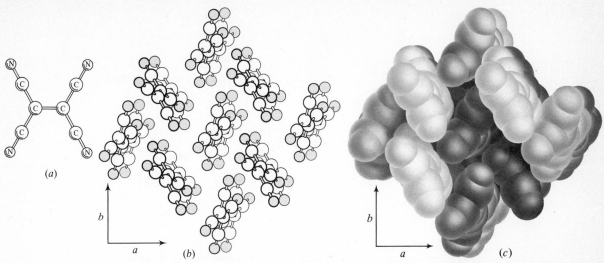

FIGURE 1-4   **The structure of a molecular crystal.**   The molecule is tetracyanoethylene, depicted in terms of a ball-and-stick model in (*a*). All atoms of the molecule lie in one plane. The arrangement of the molecules in the crystal is shown in (*b*). In (*c*) the molecules are represented with their appropriate relative sizes ("packing" radii). The molecule in the lower right of the layer closest to the viewer has been omitted in (*c*) to give a better view of the packing. The molecules in any one stack along direction *b* are parallel to one another, but the molecules in adjacent stacks are tilted relative to one another. Note also that the spacing and the sequence of the atoms vary in different directions, such as *a* and *b*, thus explaining the direction dependence (anisotropy) of these crystals.

erties in different directions. In Fig. 1-4, for example, properties in direction *a* are different from those in direction *b*. In effect, crystals are anisotropic because the regularity of the molecular arrangement persists through macroscopic regions, so that it shows up in the macroscopic properties.

There are also noncrystalline solids, called *amorphous solids*. Their molecular arrangement is more or less random (Fig. 1-5). While on the atomic level the immediate surroundings of a given atom are irregular and anisotropic, these irregularities usually average out on a macroscopic level, with the result that most amorphous solids are isotropic. For the same reason, liquids are usually, and gases are always, isotropic.

The arrangement of molecules in amorphous solids and in liquids is not completely random, however. Each atom occupies a certain volume that cannot, therefore, be occupied even in part by another atom, as would have to occur on occasion for complete randomness. Furthermore, the molecules of an amorphous solid are packed closely together, and their shapes bring about partial order. In amorphous solids that contain ions, there is a further element of order: each ion is surrounded predominantly by ions of opposite charge. This is short-range order. Long-range randomness is sufficient to make a solid or liquid isotropic on a macroscopic scale.

In some solids there may be partial long-range order without the complete three-dimensional order characteristic of a crystal. Such solids are also anisotropic. Many naturally occurring materials of biological origin such as rubber, silk, hair,

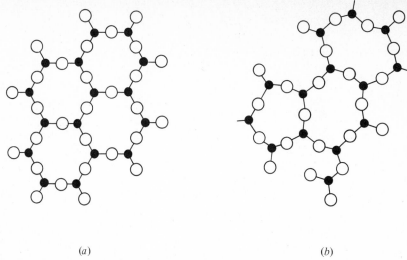

(a)   *(b)*

FIGURE 1-5   Two-dimensional models for (a) a crystalline substance and (b) an amorphous substance (glass).   In the crystalline substance the arrangement is regular; in the glass, irregular.

cellulose, and DNA (deoxyribonucleic acid, the genetic material in the cell nucleus) show one- or two-dimensional regularity when stretched. Similarly, many artificial plastics consist of long, stringlike molecules that can be aligned more or less parallel to one another, thus causing anisotropy on a macroscopic scale. For example, cellophane and similar materials can often be made anisotropic by stretching. It is, in fact, possible to produce partial ordering in many systems by applying electrical, mechanical, or other forces and to induce anisotropy in this way. Although liquids normally are isotropic, as mentioned above, there are some exceptions involving molecules with dimensions that differ greatly in different directions. Under some circumstances these molecules tend to line up parallel to one another, and the result is again anisotropy. Such liquid phases are referred to as *liquid crystals*.

1-3   Temperature, Heat, and Work[1]

Molecules and their constituent atoms are in continuous random motion. The *temperature* of a phase is a measure of the *average kinetic energy* associated with this *random* motion. *Heat,* on the other hand, describes a type of energy that is evident only when being exchanged between one sample of matter and another. *Work* also represents a form of energy that exists only while it is being transferred; for example, when a force causes a displacement, the work is the product of force and displacement.

Temperature and heat are commonly confused. The random motion of the

---

[1]Students not familiar with the concepts of force, work, and energy should read the Study Guide before studying this section.

molecules in a body at high temperature represents considerable energy, but it is incorrect to say that the body contains heat. When this body is brought into contact with another body at a lower temperature, some of its energy is transferred to the second body. The transferred energy is termed heat.

Units of energy. Units as well as numbers are needed to specify many physical quantities, such as length, time, or energy. Such quantities are said to have dimensions, while pure numbers, such as the ratio of two quantities with the same dimensions, are termed dimensionless. Most units used in this text are the internationally adopted SI units, described in more detail in Appendix A. The SI units of heat are the joule (J), which is the common unit of energy, and the kilojoule (kJ), equal to 1000 J. Other units still widely used by scientists are the calorie (cal) and the kilocalorie (kcal), 1000 cal. The calorie was for many years defined as the amount of energy required to raise the temperature of one gram of water from 14.5 to 15.5°C (see below for a definition of °C). By international agreement the calorie (or more precisely the thermochemical calorie) is now by definition 4.1840 joules:

$$1 \text{ cal} = 4.1840 \text{ J} = 4.1840 \times 10^7 \text{ ergs}$$

In terms of work the joule and the calorie are large units. This will be apparent from several examples.

EXAMPLE 1-1   **Use of a joule and a calorie to lift weight.** How many meters can 1 kg be lifted by the energies represented by 1 J and by 1 cal?

Solution. The work w required to lift an object a height h in the earth's gravity is given by

$$w = mgh$$

where m is the mass of the object and g is the acceleration of gravity, 9.80 m s^{-2}. Applying this formula, we find that to lift 1.00 kg a height of 1.00 m requires

$$w = 1.00 \text{ kg} \times 9.80 \text{ m s}^{-2} \times 1.00 \text{ m} = 9.80 \text{ kg m}^2 \text{ s}^{-2}$$

The unit $\text{kg m}^2 \text{ s}^{-2}$ is called the joule (Appendix A); that is,

$$w = 9.80 \text{ kg m}^2 \text{ s}^{-2} \times \frac{1 \text{ J}}{1 \text{ kg m}^2 \text{ s}^{-2}} = 9.80 \text{ J}$$

Note that in the preceding equations the units have been treated like algebraic quantities and can be multiplied or canceled. Such a procedure provides a check of the answer: if the calculation has been performed correctly, the answer must come out in the proper units. Since 9.80 J is sufficient energy to lift 1 kg a distance of 1 m, 1.00 J can raise 1 kg a distance of $1.00 \text{ J}/9.80 \text{ J m}^{-1} = \underline{0.102 \text{ m}}$ and 1 cal can lift 1 kg a distance of

$$\frac{1 \text{ cal}}{(9.80 \text{ J m}^{-1})(4.184 \text{ cal J}^{-1})} = \underline{0.427 \text{ m}}$$

For chemical reactions the kilojoule is more useful than the joule. For example, the heat of combustion of 1 g of wood is about 17 kJ, and since the weight of a kitchen match is about 0.5 g, the energy liberated by burning it is roughly 10 kJ. The heat of combustion of 1 g of sugar or 1 g of starch is also about 17 kJ. This is not surprising, since sugar, wood, and starch are closely related chemically.

EXAMPLE 1-2   **Utilizing the energy of sugar.** How high can a person weighing 125 lb be lifted by fully utilizing the energy in 10 g of sugar?

Solution. We rearrange the formula from Example 1-1:

$$h = \frac{w}{mg} = \frac{17 \times 10^3 \text{ J g}^{-1}(10 \text{ g})}{125 \text{ lb} \times 0.454 \text{ kg lb}^{-1} \times 9.80 \text{ m s}^{-2}}$$

$$= \frac{17 \times 10^4 \text{ kg m}^2 \text{ s}^{-2}}{556 \text{ kg m s}^{-2}} = \underline{3.1 \times 10^2 \text{ m}}$$

This is equivalent to 3.1×10^2 m (3.28 feet m^{-1}) = $\underline{1.0 \times 10^3 \text{ feet}}$.

Thus, to work off the energy in a well-sugared cup of tea, a climb of 1000 feet would be about right if all the energy from the sugar were to go into climbing, which is far from true. Even when resting, an adult human uses up some 100 to 120 kJ per hour. Moderate exercise may triple this quantity, and heavy work may raise the energy requirement by a factor of 5 or 10. In these terms 10 g of sugar would keep a resting adult going for some 30 min and he would need considerably more when working.

Confusion may arise because the "calorie" of nutritionists is the kilocalorie of chemists. Sometimes the abbreviation Cal with a capital C is used to designate kilocalories.

Temperature scales. Temperature is measured on different scales. Chemists and physicists favor the *Kelvin scale,* the *Celsius scale,* and the *centigrade scale,* the last two of which are for most practical purposes indistinguishable.

To define a temperature scale, two fixed points of reference are needed; these fix both the size of the degree and the zero of the scale. The centigrade scale is defined by the conditions that the ice point, the temperature at which water freezes when saturated with air at one atmosphere,[1] is at exactly zero degrees and the boiling point of water at one atmosphere pressure is at exactly one hundred degrees.

The SI temperature scale is the Kelvin scale, the unit of which is the *kelvin,* designated by K. This temperature unit was formerly called the degree Kelvin, abbreviated °K, a usage that is still often found. The Kelvin scale is an absolute scale, because its zero point is the lowest temperature that can be reached theoretically ("the absolute zero of temperature"). The second fixed point on this scale is the temperature at which pure air-free water freezes at the pressure of its own vapor (the triple point of water; see Fig. 3-5), to which is assigned the value 273.1600 K. With this assignment the ice point is at 273.150 K and the boiling point of water at 1 atm pressure is at 373.150 K. Another scale, closely similar to the centigrade

[1]The pressure unit atmosphere (atm) is defined in Sec. 3-1.

scale but tied by definition to the Kelvin scale, is the Celsius scale, temperatures on which are designated by °C. By definition,

$$\text{Degrees Celsius} = \text{kelvin} - 273.150 \text{ degrees}$$

The difference between the centigrade and Celsius scales is one of definition rather than of numerical values, and °C may be read "degrees centigrade" as well as "degrees Celsius". Temperature on either of these scales is designated here by t, and on the Kelvin or absolute scale it is represented by T, as is widely customary.

Two other temperature scales in wide current use are the Fahrenheit and the Rankine scales. They are shown in Fig. 1-6 together with the Celsius and the Kelvin scales.

1-4   Dalton and His Predecessors: Early Ideas about the Atom

Scientific concepts inevitably change with time. Some, like that of phlogiston, a substance believed by a number of eighteenth-century scientists to have negative weight and to escape from an object as it was burned, are abandoned and remain only of historical interest. Others, such as the corpuscular theory of light supported by Newton in the early eighteenth century but superseded by the wave theory during the nineteenth century, are apparently abandoned but then appear again in modified form—the modern photon theory of light incorporates both wave and particle (or corpuscular) aspects.

The idea that there are elementary components, or elements, common to all substances is at least as old as the fifth century B.C. Leukippos and Demokritos

FIGURE 1-6   **Different temperature scales.**   The Fahrenheit (°F) and Rankine (°R) scales are largely used by engineers. The relationships to the Celsius and Kelvin scales are $t/°F = \frac{9}{5}(t/°C) + 32$, and $T/°R = \frac{9}{5}(T/K)$. The Rankine scale is an absolute scale in the same sense that the Kelvin scale is. (*Note:* An equation relating two physical quantities must be *dimensionally* as well as *numerically* correct. One way of assuring such consistency is to cast equations in dimensionless quantities. The quantity $t/°F$, which represents the temperature in degrees Fahrenheit divided by the *unit* degree Fahrenheit, is a dimensionless quantity, as are the pure number 32 and $t/°C$, which represents the temperature in degrees Celsius divided by the unit degree Celsius.)

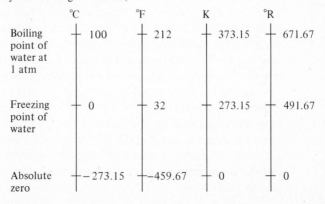

	°C	°F	K	°R
Boiling point of water at 1 atm	100	212	373.15	671.67
Freezing point of water	0	32	273.15	491.67
Absolute zero	−273.15	−459.67	0	0

developed at that time an *atomistic* (Greek: *a*, not; *tomos*, divisible) theory in which all substances were regarded as composed of indivisible and indestructible particles (atoms) which differed from each other in shape, weight, relative position, and orientation but did not differ internally. Differences in the properties of substances were considered to arise from the differing configurations and combinations of atoms. Thus, a liquid such as water would be composed of smooth and round atoms that rolled over one another readily, while a solid such as iron would be composed of atoms that differed from those of water in having a jagged, rough shape, which would permit them to cling together to form a solid body.

These early atomic ideas survive today, in greatly modified form, but they were partly in the shadow of an alternative theory for many centuries. This view of the composition of matter retained the concept of a few elements but regarded them as qualities rather than as indivisible material particles. It was initiated by Anaxagoras and Empedokles, and was developed and supported by the immense authority of Aristotle. According to this notion, all matter was continuous—that is, indefinitely divisible—and composed of the same fundamental material. Differences in properties resulted from the presence of differing amounts of two sorts of antagonistic principles: dry-wet and hot-cold. It is not clear why just these principles were regarded as fundamental; the fact that they relate only to the sense of touch and not to the other senses suggests that only the sense of touch was deemed trustworthy. The four classical elements were composed of combinations of pairs of these principles: fire (hot and dry), air (hot and wet), water (cold and wet), and earth (cold and dry).

The alchemists of the Middle Ages and later centuries adopted a similar set of elements, although their "principles", such as metallic behavior, solubility, and combustibility, were different. Their continuing efforts to transmute common metals such as lead into rare ones such as gold are often regarded now with moderate contempt or amusement, but from the standpoint of a theory that views distinct substances as differing only in the proportion of different qualities, such conversions might be accomplished merely by appropriate changes in this proportion. There is no intrinsic reason why this sort of change should be regarded as less likely than some of the remarkable syntheses done routinely in the laboratory nowadays—until one understands more about the true nature of the chemical elements and the principles governing their interconversion.

The atomic theory became established unambiguously only after the brilliant insight of John Dalton in the early nineteenth century had been supported and developed by subsequent convincing experimental evidence. Dalton's essential contribution was to emphasize the possibility of deducing by experiment the relative weights of different atoms and the combining proportions of these atoms in different compounds. Many of his conjectures about chemical formulas were incorrect—chiefly because he did not acknowledge the possibility that the fundamental "particle" of an element, which we now call a molecule, might contain more than one atom (e.g., in modern terminology, his formulas for gaseous oxygen and hydrogen were O and H rather than O_2 and H_2). However, this shortcoming was soon corrected by Avogadro and Cannizzaro, and with improved analytical techniques and data the atomic theory was eventually put on a firm quantitative basis, although not without some controversy.

We shall consider in Chap. 2 the quantitative relations among the components of substances and among the reactants and products involved in chemical changes. The present chapter is devoted to a discussion of some of our present ideas of atoms and molecules, together with an account of a few of the critical experiments that helped to establish them. The details of atomic and molecular structure are considered in later chapters.

1-5   The Emergence of Modern Concepts about the Atom

In the present-day view, atoms have a structure and are made up of constituents. While "constituents of atoms"—parts of the indivisible—is literally a contradiction in terms, this merely illustrates again the way in which the implications of scientific concepts change. Dalton's idea of atoms as the fundamental building blocks of the chemist has survived, but these atoms are now recognized to have components that can be separated from one another under appropriate conditions.

Rudimentary ideas concerning the structure of atoms developed during the nineteenth century, but it was not until the end of that century and the start of the present one that these ideas took definite form and the modern picture began to emerge. Important contributions to the development of atomic theory included the discovery of the electron by J. J. Thomson in 1897, the development of the nuclear model of the atom by Ernest Rutherford in 1911, and the application of quantum ideas to atomic structure by Niels Bohr in 1913. Our current views on atomic structure emerged only after the elaboration of modern quantum theory in the mid-1920s and the discovery of the neutron in 1932. The first of these developments is discussed in this chapter; the later ones are considered in Chap. 8.

The electron.   It has been known since ancient times that rubbing certain substances, such as amber and glass, with fur or wool makes them attract light objects such as bits of paper. Objects touched by rubbed amber or glass repel each other. These attractions and repulsions, termed electrical phenomena (Greek: *elektron,* amber), were explained in the early part of the eighteenth century by attributing them to the presence in matter of two kinds of fluid: positive and negative electricity. It was postulated that there is an attraction between portions of fluid of opposite sign and a repulsion between fluids of the same sign. A substance would normally contain the fluids in equal amounts, so that their properties canceled. The fluids could be separated by friction, however, and all the observed electrical phenomena could be explained as arising from an excess of either positive or negative electricity.

In 1827, Georg Ohm likened the flow of electrical fluid to the flow of water and used this analogy to give precise meaning to the concepts quantity of electricity, electric current, and electromotive force. By this time, a number of experimenters had shown that the passage of electric current through solutions caused chemical changes. For example, water is decomposed when electricity is passed through it (Fig. 1-7), hydrogen being produced at one electrode, i.e., at one of the wires dipping into the liquid, and oxygen at the other.

Michael Faraday coined the name *electrolysis* for the process of decomposing

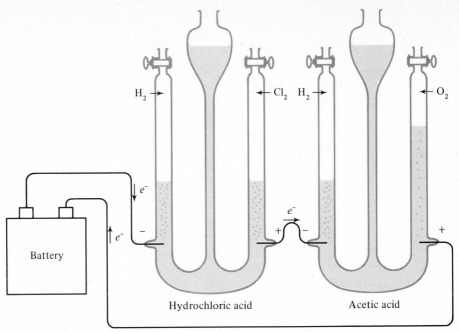

FIGURE 1-7 **An electrolysis experiment.** The same amount of current passes through the two electrolysis cells. The volume of hydrogen evolved at the electrode at which the electrons enter is the same in each cell. Similar volume relationships apply to gases evolved at the other electrode (and from other solutions by a given current). Electrolysis is discussed in Chap. 18.

substances by electricity, and his quantitative investigations of the phenomenon gave the first experimental indication that electricity might exist as discrete particles. Faraday showed that the passage of the same electric current through two electrolysis cells causes the same volume of hydrogen to be produced from different solutions (Fig. 1-7). Furthermore, this volume is directly proportional to the quantity of electricity passed through the cells. Thus, if matter was atomic in nature, a given quantity of electricity liberated a specific number of atoms. This suggested that there must be some fundamental unit of electricity associated with every atom.

Convincing experimental evidence for the particulate nature of electricity was obtained only in the last decade of the nineteenth century as the outgrowth of four decades of study of the electrical conductivity of gases. These studies were made chiefly in glass tubes of various designs, usually evacuated to low pressures. Rays were found to emanate from the cathode (the negative electrode) of a tube containing gas at a low pressure to which a sufficiently high voltage (about 10^4 V) had been applied to cause conductivity. It was shown that these *cathode rays* carry negative electricity, but there were conflicting views on whether the rays consisted of particles or waves. J. J. Thomson in 1897 hypothesized that they were particles, for which he adopted the name proposed a few years earlier by Stoney, *electrons*. Thomson designed experiments for measuring the ratio of their charge (the quantity of electricity they possess) to their mass, e/m, by two different methods, the simpler of which involved balancing opposing deflections of the rays by electrostatic and magnetic fields (Fig. 1-8). He found the same value of *e/m whatever the gas present in the tube;* this strongly suggested the presence of just one kind of negative particle.

The value of *e/m* found by Thomson was higher by about 10^3 than that for the simplest charged particles previously known, which were hydrogen ions—what we now recognize as protons. This implied either that the charge of the electron is much higher than that of the proton, or that its mass is much smaller, or some combination of these possibilities. R. A. Millikan's determination of the electronic charge in 1911 permitted separate evaluation of *e* and *m* for the electron; it was established that the mass of the electron is about $\frac{1}{1800}$ that of the proton and its charge the same in magnitude as that of the proton. Millikan's experiment also established unambiguously that electric charge occurs in discrete (that is, indivisible) units and thus seemed to "prove" that electrons are particles. However, scarcely another decade had passed before it was suggested, and demonstrated experimentally, that electrons have wave character as well—a valuable lesson in the limitations of scientific concepts and proofs. This "wave-particle duality" is discussed in Chap. 8.

X rays and radioactivity.   The next steps in the development of our modern picture of the atom were a direct outgrowth of three remarkable discoveries that

FIGURE 1-8   **A cathode-ray tube.** In an evacuated glass tube similar to a television picture tube, electrons are emitted from a hot wire, accelerated, and shaped into a beam by a set of positively charged, perforated disks (*A*). The beam may be deflected toward the reader to a larger or smaller degree (*D*) by increasing or decreasing the charges on the vertical plates (*B*) or by moving the magnet (*C*) in or out. By quantitative experiments J. J. Thomson determined an approximate value of the ratio *e/m* for electrons. Similar measurements can be made for other charged particles.

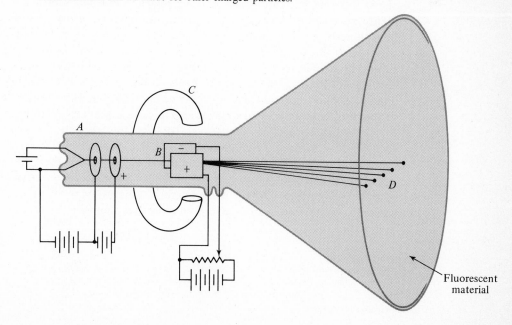

Fluorescent
material

came during the last few years of the last century. The first two were experimental—the discoveries of X rays and of radioactivity. The third was Planck's quantum hypothesis and its later development, the theoretical implications of which marked a major turning point in the history of physics and of our world. These implications were so revolutionary that they were unacceptable to many who had been trained in nineteenth-century physics.

The discovery of X rays in 1895 by W. K. Röntgen was purely accidental. Röntgen was investigating the discharge of electricity through gases in an evacuated tube, as so many had before him. In the room with the equipment he also had a paper screen that glowed when exposed to ultraviolet light. Röntgen noticed that this screen emitted light even when various objects were between it and the cathode-ray tube and sought to explain this chance observation. He found that some extremely penetrating radiation was coming from the point on the cathode-ray tube where the cathode rays struck the glass wall and caused a weak glow (fluorescence). In a very short time he established many of the important properties of this radiation, which he termed X rays, including its diagnostic value in medicine. However, it was nearly two decades before the exact nature of X rays as very short wavelength electromagnetic radiation, similar in many ways to visible light, was clearly understood.

Röntgen's discovery caused great excitement, and considerable skepticism as well, and many physicists immediately turned to a study of these mysterious rays. Antoine Henri Becquerel, a professor in Paris, supposing that there might be some connection between the fluorescence of the glass wall of the cathode-ray tube and the production of X rays, undertook a systematic study of all minerals that fluoresced or phosphoresced when exposed to sunlight.[1]

Among the substances Becquerel studied was uranium potassium sulfate, which shows a strong but short-lived phosphorescence that fades in less than a second. Hence his technique was to try to detect the possible X rays associated with this phosphorescence by placing the salt on a photographic plate that was carefully wrapped in thick black paper and exposing the combination to bright sunlight for several hours. He found that some very penetrating radiation was indeed present and would even pass through thin sheets of aluminum or copper. However, during a period of bad weather, while storing some photographic plate and salt combinations in the dark until he had an opportunity to expose them to sunlight, he found that there was no connection whatsoever between the exposure to sunlight and the emission of penetrating radiation from the salt. Further experiments showed that the radiation was characteristic of uranium in any form, whether phosphorescent or not, and persisted undiminished for months. The name *radioactivity* was applied to this phenomenon by one of Becquerel's students, Marie Curie, who (with her husband Pierre) discovered several radioactive elements, including radium and polonium.

---

[1]Fluorescence and phosphorescence both involve the emission of visible light by a substance after it absorbs radiation, usually of a different wavelength; the distinction is essentially that fluorescence does not persist after the source of absorbed radiation is removed, whereas phosphorescence does persist for at least a short time.

Becquerel's discovery, like Röntgen's, was essentially an accidental one, resulting in this instance from the pursuit of a false hypothesis. However, the way in which each of these men carefully noted and recognized the implications of his apparently anomalous initial observations illustrates well the truth of Louis Pasteur's dictum that "chance favors the prepared mind only".

Ernest Rutherford, a young New Zealand physicist trained under Thomson at Cambridge, was among many who turned their attention to radioactivity. He and others soon found three distinct types of radiation from radioactive substances, termed initially alpha, beta, and gamma rays. Alpha rays were shown by Rutherford to consist of streams of doubly charged helium atoms, He^{2+}. Beta rays were identified as streams of electrons, like cathode rays. Gamma rays were found to be similar to X rays in their properties. Rutherford and F. Soddy succeeded in establishing the fact that radioactivity was accompanied by the spontaneous change of the radioactive atoms into different atoms, an idea that was hard for many to accept because of the supposed immutability of atoms.

The nuclear atom. The model of the atom considered most plausible around 1910 was one proposed by Thomson in which the positive electricity was spread uniformly throughout a sphere inside which the electrons moved in circular orbits. From considerations of various properties of gases and the densities of solids, the effective diameters of atoms were known to be a few angstroms (Å) (1 Å $\equiv 10^{-8}$ cm); the number of electrons in any atom was approximately known from X-ray scattering.

Rutherford did not trust this model and began, with his collaborators, to probe atomic structure with alpha particles. The experimenters used atoms of metals such as silver, gold, and platinum as targets by bombarding thin foils with alpha particles (Fig. 1-9). The fraction of particles scattered was measured for each angle, and it

FIGURE 1-9 **Rutherford's scattering experiment.** A beam of α particles is made to pass through a gold foil. The particles can be observed by the light flashes they produce when striking a fluorescent screen. The screen can be rotated about the center of the gold foil so that the angular dependence of the scattering of the α particles can be determined.

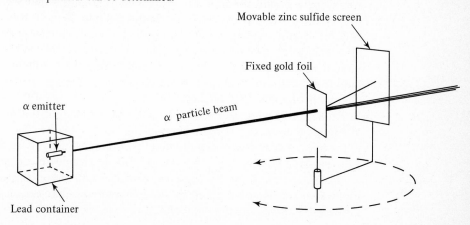

was found that while most of the alpha particles passed through the foil undeflected a few of them were deflected through large angles, some even in the backward direction (that is, through angles greater than 90°). The number of these strongly scattered alpha particles increased in proportion to the thickness of the gold foil.

Rutherford's results could not be explained on the basis of Thomson's model, which pictured the positive charge of the gold atoms as spread quite *uniformly* through the foil. The large scattering angles for a few alpha particles could be understood only if the positive charge and most of the mass of each atom were concentrated in a very small region, which Rutherford termed the *nucleus* (Latin: *nucula,* small nut). Occasional backward scattering of alphas could then be explained as a result of a more or less direct collision of each of these alphas with a nucleus; scattering of the preponderance of alphas through small or zero angles could be explained as a result of electrostatic repulsion between alphas and nuclei at greater distances. Any electrons present in the atoms would not scatter alpha particles noticeably, because the electron mass is about 7500 times smaller than that of an alpha particle. Expressed in terms of the scattering from a (hypothetical) foil consisting of a single layer of gold atoms, about 1 alpha particle in 10^8 was scattered back near the direction from which it had come. From this result, Rutherford was able to estimate that the area of the nucleus was about 10^{-8} times that of the entire atom, and thus the diameter of the nucleus was about 10^{-4} times that of the entire atom, or about 10^{-4} Å. Furthermore, he was able to deduce that the positive charge on the nucleus (expressed in terms of the electronic charge) was about half of what is called the atomic weight. Since the atom as a whole was neutral, this was also the number of electrons in the atom.

These results formed the basis of Rutherford's atom model, in which a massive, central, positively charged nucleus is surrounded by orbiting electrons, similar to a sun surrounded by planets. The nucleus contains almost the entire mass of the atom and has a positive charge of Z electronic units, exactly neutralizing the negative charge of the Z electrons. This model, however, presented formidable difficulties. It can be proved that no static arrangement of positive and negative charges is stable, and so the electrons had to be in motion around the nucleus, but it was well known from the laws of classical physics that a charged particle moving in a circular or similar path should continuously radiate energy. Thus atoms should constantly lose energy and eventually the electrons should fall into the nucleus and the atom should collapse—yet it was known that atoms did not continuously radiate and that almost all atoms were stable indefinitely. The resolution of this conflict between experimental results and theoretical predictions came with Bohr's quantum theory of atomic structure in 1913 and its modification and extension in the quantum mechanics of the mid-1920s (Chap. 9).

Rutherford's work had established the importance of the nuclear charge as an attribute of atoms. A few years later, in 1913 and 1914, the fundamental significance of the nuclear charge was consolidated by a young British physicist, H. G. J. Moseley, who was killed in his twenties in World War I. Moseley measured the wavelengths of the X rays produced by samples of nearly half the elements then known and discovered a remarkably simple relation between the charge Z on the atomic nucleus, termed the *atomic number,* and the X-ray wavelength: the product of the atomic number and the square root of the wavelength was the same for all elements. It was found that an arrangement of the elements in order of increasing atomic number, as established by Moseley's method, was not always in agreement with an ordering based on atomic mass. Chemical evidence shows, in fact, that the atomic number is more closely related to the chemical properties of an atom than is its mass.

The atomic weight and isotopes.   Dalton emphasized the possibility of determining the relative weights of different atoms from precise data on the composition and reacting proportions of different substances, and a great deal of effort was devoted to this task during the nineteenth century and much of the present one. The term *atomic weight* is customarily used to designate the relative masses or weights of the different atoms, referred to a common standard that is defined by setting the mass of the most abundant form of carbon atom at exactly 12. On this scale, which is discussed further in Sec. 1-7, the simplest element, hydrogen, has an atomic weight of very nearly 1. Since weight is proportional to mass if all measurements are made under the same conditions of gravity, there is usually no distinction made between the terms *atomic mass* and *atomic weight* referred to this dimensionless scale. We shall normally use the more common term, atomic weight.

After Thomson determined the charge-to-mass ratio for the electron, he applied a similar technique to the determination of this ratio for many positively charged atoms and molecules (positive ions). His method was extended by F. W. Aston and others, and several high-precision instruments called *mass spectrometers* have been developed for measuring the charge-to-mass ratio for almost any charged particle (ion) that can be formed in a gas. One mass spectrometer is described in Fig. 1-10. Since the charge on the particle, expressed in units of the electronic charge, is always a small integer, usually 1 or 2, the mass of the ion can thus be determined with high precision.

When the first mass spectrometer was used with a pure sample of the gaseous element neon, it was discovered that there were two distinct atoms present, with atomic weights approximately 20 and 22. It was soon found that many other chemical elements consisted of atoms with different masses but essentially identical chemical properties. The term *isotope*[1] (Greek: *isos,* equal; *topos,* place) was coined to apply to these different atoms that occupy the same place in the classification of the elements according to their chemical properties, the periodic table.

If a naturally occurring element consists of a mixture of isotopes, as most do, the atomic weight of this element as determined by chemical methods represents the average weight of the different isotopes, weighted in proportion to their abundance. The average weight is observed because typical chemical experiments involve very large numbers of atoms and molecules. On the other hand, atomic weights determined with the mass spectrometer represent the weights of the individual isotopes of which the different elements are composed. Although many chemical

[1]The term *isotope* originally implied a relationship and thus one should not speak of an element consisting of only a single isotope, any more than one should speak of an only child as a sister. Nevertheless, this usage has become common and is sometimes employed here.

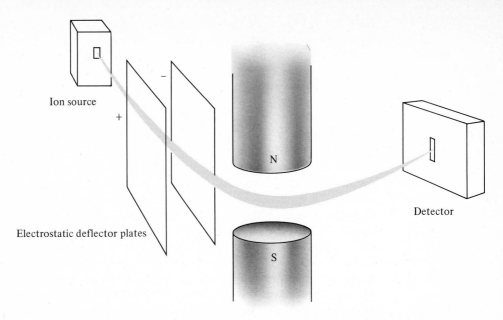

Ion source

Electrostatic deflector plates

N

S

Detector

Magnetic deflector

FIGURE 1-10   **A mass spectrometer.** The atoms or molecules of interest are ionized and accelerated by electric fields in the ion source. They are deflected from a straight-line path by an electric field and then by a magnetic field, both arranged in such a way that the total deflection does not depend on the velocity of these ions but just on the ratio e/m of their charge to their mass. The ions are finally collected and counted with a detector. Photographic plates were first used for detection, but electronic means are now much more common. A variable magnetic field is used so that ions of different e/m successively pass through the slit of the detector as the magnetic field is changed.

atomic weights (listed inside the front cover) are nonintegral, the weights of individual isotopes are very nearly integral.

In 1932, James Chadwick, a British physicist working in Rutherford's laboratory in Cambridge, demonstrated the existence of a neutral particle of atomic weight approximately 1, the neutron. The neutron had actually been suggested by Rutherford in 1920, but many experiments to detect it during the 1920s had been unsuccessful. Chadwick's discovery made possible the present view that a nucleus "contains" protons and neutrons, as described in more detail in the next section.

1-6   Atoms, Molecules, and Compounds

Atoms and chemical elements. The modern definition of a chemical element is that it is a substance that contains only atoms of the same atomic number, i.e., of the same nuclear charge. There are 106 such elements now known, each with a unique name and symbol, except for the simplest element, hydrogen, for which the three isotopes have different symbols and names, as discussed later. The names and symbols are listed inside the front cover.

The nucleus of each atom is composed of protons and neutrons (except for the isotope of hydrogen of atomic weight 1, whose nucleus consists of a single proton). The number of protons is equal to the atomic number Z because the proton charge is not lost when neutrons and protons combine to form the nucleus. Neutrons and protons are collectively called *nucleons,* and the total number of nucleons in a nucleus is called the mass number A. A given nucleus thus contains Z protons and $(A - Z)$ neutrons.

The strong interactions between the protons and neutrons of a nucleus change many of the properties these particles exhibit in the free state. For example, a free neutron is an unstable particle, with a 50 percent chance of changing into a proton and an electron within about 13 min. In a nucleus, however, the neutron may be stable; that is to say, whether or not a given nucleus is stable does not depend on the original stability or instability of the nucleons that make it up. In fact, nucleons lose their identity when they become part of the nucleus and it is impossible to tell the original protons and neutrons apart, although the nuclear charge is a reminder of the number of protons that went into the formation of the nucleus. Some properties of the electron, the proton, and the neutron are summarized in Table 1-1.

The nucleus of an atom is surrounded by Z electrons, so that the entire atom is electrically neutral. The chemical behavior of an atom is determined almost exclusively by these electrons. It is, in fact, mainly the outermost of these electrons, also known as *valence electrons,* that are responsible for the chemical properties, as discussed in later chapters. To a very small extent, the chemical properties depend on the mass number A, but this influence of A on the chemistry of a particular atom can usually be neglected. The chemical identity of an atom is thus determined almost entirely by its atomic number Z.

Nuclides and isotopes.   An atomic species of given atomic number Z and specified mass number A is called a *nuclide.* It is often represented by adding a preceding superscript A and subscript Z to the chemical symbol "Ch" of the atom:

$$_Z^A\text{Ch}$$

Thus $_6^{12}\text{C}$ is a carbon nuclide whose nucleus contains 6 protons and 6 neutrons. It

TABLE 1-1   **Some properties of elementary particles**

	Mass		Charge	
	kg	Atomic mass* units, u	Coulombs, C	Electronic charge units
Electron	9.1095×10^{-31}	5.4858×10^{-4}	-1.602×10^{-19}	-1
Proton	1.67265×10^{-27}	1.00728	$+1.602 \times 10^{-19}$	$+1$
Neutron	1.67495×10^{-27}	1.00867	0	0

*See Sec. 1-7.

is often called carbon 12. The subscript 6 is redundant, since the same information is contained in the symbol C. It is not easy to remember the atomic numbers of all elements, however, so explicit listing of Z as a subscript is often useful. For the nuclides of the lighter elements ($Z \leq 20$), A is about twice Z; for the heavier elements, A is somewhat greater than twice Z. Isotopes are nuclides with the same atomic number (Z) but different mass numbers (A).

The simplest element, hydrogen, has three isotopes, $_1^1H$, $_1^2H$, and $_1^3H$, the last of which is radioactive. The relative difference in mass among these isotopes is the largest known; that is, for no other element are the ratios of isotopic masses as large as $3:1$ and $2:1$ or even $3:2$. The corresponding chemical effects of these mass differences are thus the largest known. For this reason, separate names and symbols are sometimes used for the isotopes: hydrogen 2 is called deuterium, D; hydrogen 3, tritium, T; and hydrogen 1, occasionally protium. The nucleus of deuterium is called the deuteron and the tritium nucleus is called the triton; in addition to one proton they contain one and two neutrons, respectively.

Three isotopes of carbon, ^{12}C, ^{13}C, and ^{14}C, are found naturally, and the heaviest of these is also radioactive. The natural abundances of the isotopes of hydrogen and carbon are listed in Table 1-2. These are number percents, so that, for example, of 100,000 representative hydrogen nuclei 99,984 are $_1^1H$ and 16 are $_1^2H$.

Molecules and compounds.   Discrete uncharged groupings of atoms, tied together by specific forces in a definite arrangement, are called *molecules*. The arrangement need not be rigid, but there must be a clear distinction between atoms that belong to the molecule and those that do not. The atoms in a particular molecule may be either the same (H_2, O_2, P_4, S_8) or different.

A chemical compound contains at least two elements. The defining characteristic is that a compound contains different elements in a *definite* ratio, independent of the way in which the compound was prepared. This may be a consequence of the compound's existing as molecules, for example, H_2O or CO_2. However, other compounds are composed of ions, that is, charged atoms or molecules, and the definite composition of such a compound results from its overall neutrality, which implies that it contains appropriate numbers of ions of opposite charge to make the charges balance. Thus sodium nitrate, $NaNO_3$, contains equal numbers of Na^+ and NO_3^-

TABLE 1-2   **Natural abundances of isotopes of hydrogen and carbon**

Nuclide	Percent abundance	Atomic weight	Nuclide	Percent abundance	Atomic weight
$_1^1H$	99.984	1.0078	$_6^{12}C$	98.89	12.0000*
$_1^2H$ (=D)	0.016	2.0141	$_6^{13}C$	1.11	13.0034
$_1^3H$ (=T)	10^{-16}†	3.0160	$_6^{14}C$	10^{-10}†	14.0032

*The atomic weight of ^{12}C is by definition exactly 12.

†Tritium and carbon 14 occur naturally, despite their continuous radioactive decay at a rate that is rapid on a geologic time scale, because they are continually formed in the upper atmosphere by the reaction of neutrons with ^{14}N and ^{16}O. The neutrons are formed by the interaction of cosmic-ray protons with these same atmospheric constituents. Tritium is also formed during thermonuclear explosions, and it may be detected in atmospheric and surface waters in much larger proportions for many years after such an explosion whose products are not confined.

ions (the names and formulas of common ions are listed in Appendix B), and ammonium sulfate, $(NH_4)_2SO_4$, contains two ammonium ions (NH_4^+) for every sulfate ion (SO_4^{2-}). These compounds, which are crystalline solids at ordinary temperatures, have properties that depend on those of their component ions; no molecules are present, that is, it is not possible to associate a given sodium ion with a given nitrate ion, or some pair of ammonium ions with a given sulfate ion.

1-7 Atomic and Molecular Weights and the Mole

Atomic, molecular, and formula weights. As mentioned earlier, atomic weights are normally given on a scale defined by setting the atomic weight of carbon 12 exactly equal to 12. The unit of mass normally used for individual atoms is the atomic mass unit, u, exactly one-twelfth the mass of a carbon-12 atom.[1] It is important to remember that the atomic weights required for most chemical purposes are averages of those of the different isotopes, weighted according to their natural abundances.

EXAMPLE 1-3   **Atomic weight from natural abundances.** Silicon consists of three stable isotopes, whose atomic weights and relative abundances are as follows:

	Atomic weight	Percent abundance
^{28}Si	27.98	92.2
^{29}Si	28.98	4.7
^{30}Si	29.97	3.1

What is the (chemical) atomic weight of silicon?

Solution. The atomic weight is the weighted average of the individual values:

$$\text{Atomic weight Si} = \frac{27.98(92.2) + 28.98(4.7) + 29.97(3.1)}{92.2 + 4.7 + 3.1}$$

$$= \underline{28.09}$$

We speak of *chemical* atomic weights when this situation requires emphasis. The accuracy with which the atomic weight of a given element is known reflects not only the quality of the experimental work on which the value is based but also the fact that the natural abundances of the different isotopes vary slightly, depending on the source of the element or compound considered. It is fortunate that for most elements natural abundances do not vary enough to affect usual weight relations in chemical reactions.

---

[1]The mass in grams of a ^{12}C atom can only be determined by experiment; hence its value is subject to refinement. A change in this value will not alter the definition of the atomic mass unit, but it will change the number of atomic mass units per gram.

The molecular weight of a molecule (abbreviated MW) is the sum of the atomic weights of the atoms it contains. We normally use the term *molecular weight* only with reference to compounds that consist of molecules. For other compounds the term *formula weight* (FW) will be used, defined as the sum of the atomic weights of all atoms shown in the formula of the compound. The importance of this distinction, which is more one of principle than of practice, will become clearer when we discuss solutions. Some examples follow.

EXAMPLE 1-4   **Molecular weight.** Carbon dioxide exists as discrete CO_2 molecules in the solid, liquid, and gaseous states. The molecular weight and the formula weight are thus identical: $12.01 + 2(16.00) = \underline{44.01}$.

EXAMPLE 1-5   **Formula weight.** Sodium sulfate, Na_2SO_4, in the crystalline form or when melted, consists only of Na^+ and SO_4^{2-} ions, in the ratio $2:1$. Similarly, in aqueous solutions (solutions in water) there are no Na_2SO_4 molecules, but only Na^+ and SO_4^{2-} ions, associated fleetingly with various numbers of water molecules, as discussed in Chap. 7. Under ordinary circumstances, no Na_2SO_4 molecules exist and the term *molecular weight* is inappropriate. The formula weight of Na_2SO_4 is $2(22.99) + 32.06 + 4(16.00) = \underline{142.04}$.

EXAMPLE 1-6   **Water of crystallization.** When an aqueous solution of sodium sulfate is evaporated, crystals of $Na_2SO_4 \cdot 10H_2O^*$ (glauber salt) appear below $32.4°C$ and crystals of anhydrous (water-free) Na_2SO_4 appear above this temperature. The formula weight of $Na_2SO_4 \cdot 10H_2O$ is $142.04 + 10(18.015) = \underline{322.19}$. The water molecules in crystals of glauber salt are typical of what is called water of crystallization, and the compound exists in the crystalline state only.

EXAMPLE 1-7   **Variation in formula under different conditions.** A more complicated situation is illustrated by aluminum chloride, $AlCl_3$. In the crystalline state there are layers extending throughout the crystal, with overall composition $AlCl_3$, but containing no discrete molecules. Liquid aluminum chloride consists of Al_2Cl_6 molecules. At low temperatures the vapor also contains Al_2Cl_6 molecules, but at elevated temperatures it consists chiefly of molecules of $AlCl_3$. In certain water-free organic solvents aluminum chloride exists as $AlCl_3$ also. Aqueous solutions, however, contain mainly Cl^- ions and a species of formula $[Al(H_2O)_6]^{3+}$ called hexahydrated aluminum ion.

   The formula weight of $AlCl_3$ is $26.98 + 3(35.45) = \underline{133.33}$, whereas the molecular weight depends on the conditions; it is $\underline{133.33}$ for $AlCl_3$, $\underline{266.66}$ for Al_2Cl_6, and has no meaning when discrete molecules do not exist.

---

*The dot ($\cdot$) is used in formulas such as that of glauber salt to indicate that the components shown are associated in a definite proportion, without any implication about the nature of the exact molecular or ionic species present or the way in which they are combined. This usage of dots must not be confused with their employment in Lewis formulas to represent shared and unshared valence electrons (Chap. 3).

Avogadro's number, the mole, and the gram formula weight. Most chemical experiments involve macroscopic amounts of substances, and it is therefore extremely useful to define a macroscopic quantity that always contains the same number of microscopic particles such as atoms, molecules, or ions. This unit is called the *mole* (SI symbol, mol),[1] and the number of particles it contains is called *Avogadro's number,* for which the symbol N_A will be used. Avogadro's number N_A is *defined* as the number of atoms contained in exactly twelve grams of ^{12}C. Further, *by definition,* one mole of a chemical species (atoms, molecules, or ions) is the quantity that contains N_A particles. The *value* (as opposed to the definition) of N_A is 6.022×10^{23}. Sometimes the term *mole* is used to refer to N_A particles even when they are not all of the same kind. For example, one mole of air contains Avogadro's number of "air molecules", with appropriate contributions from the component gases, chiefly N_2 and O_2.

This *number aspect of the mole* should be carefully noted and remembered: a mole is a specific number of things, just as a dozen is, although the number in a mole is almost incomprehensibly large.[2] When we speak of a dozen objects, their shape, weight, or color is unimportant. There are 12 objects, and in 10 dozen there are 120. Similarly, 1 mol refers to 6.022×10^{23} of the objects considered, and 10 mol to 6.022×10^{24}. Although our main concern will be with moles of atoms, molecules, and ions, we shall have occasion to refer to a mole of electrons (which is called the *faraday,* a fundamental unit in electrochemistry).

A second aspect of the mole is of considerable practical value in chemical calculations. *The weight in grams of a mole of any atomic, molecular, or ionic species is numerically the same as the weight of that species on the atomic weight scale.* This is a necessary consequence of the definition of the atomic weight scale and of Avogadro's number, N_A. Since the mass of one atom of ^{12}C is defined to be exactly 12 u, where u is 1 atomic mass unit, and since, by definition, N_A atoms of ^{12}C weigh

---

[1] The unusually lengthy symbol *mol* was presumably selected because further contraction of *mole* might lead to confusion. Its usage is parallel to that of other symbols such as m or kg. No plural exists and the symbol is usually employed with numerals, as in "3 mol" or "5 kg", in contrast to "three moles" or "five kilograms". *Mol* must not be taken as an abbreviation for molecule(s).

[2] The sizes of very large and very small numbers are hard to comprehend, especially when they are expressed as powers of 10. For example, 1 Å is 10^{-10} m; this implies that it takes 250 million Å, a number well beyond everyday comprehension, to make up 1 inch. It is particularly difficult to appreciate the enormous size of Avogadro's number and the infinitesimal dimensions of atoms and atomic nuclei. The following examples are illustrative:

The volume of N_A dust grains, represented by tightly packed cubes each 0.010 mm on a side, is that of a cube more than 840 m (about a half mile) on a side.

Were we to mark the molecules in 100 ml (about $\frac{1}{3}$ cup) of water and mix them with the entire volume of the oceans (1.37×10^9 km^3), each 100 ml of the mixture would still contain over 200 of the original molecules.

Imagine trying to see with your unaided eye a large beachball and a sand grain resting on the surface of the moon. Their images at that distance are, respectively, about the same size as those of an iodine atom and its nucleus viewed from a distance of 20 cm.

exactly 12 g, it follows that

$$12N_A \, u = 12 \text{ g} \qquad \text{and} \qquad N_A \, u = 1 \text{ g}$$

or

$$1 \text{ u} = \frac{1}{N_A} \text{ g} = \frac{1}{6.022 \times 10^{23}} \text{ g} = 1.6606 \times 10^{-24} \text{ g}$$
$$= 1.6606 \times 10^{-27} \text{ kg}$$

This relation and Table 1-3 show that Avogadro's number is a *scale factor* for conversion from atomic mass units to grams. Table 1-3 also illustrates the advantage of using atomic mass units when individual atoms or molecules are considered, because the masses in macroscopic units, such as grams, are inconveniently small. For the large numbers of atoms and molecules that are normally used in chemical experiments, the mole is a convenient unit because it represents a macroscopic quantity, and more specifically because for any molecular compound the weight of *one mole of molecules,* expressed in grams, is numerically equal to the molecular weight of the molecule. Moreover, if *one formula unit* of a compound is defined as that quantity that contains exactly the numbers of different species indicated by the formula of the compound, then the weight of *one mole of formula units,* expressed in grams, is numerically equal to the formula weight of the compound.

The value of Avogadro's number necessarily depends both on the choice of the standard for the atomic weight scale and on the choice of the gram as the unit of mass for comparison. For example, the basic SI unit of mass is the kilogram (kg), and the scale factor from atomic mass units to kilograms would be 1000 times Avogadro's number. Hence, a factor 10^{-3} is required when the mass of 1 mol is to be expressed in SI mass units. For example, the masses of 1 mol each of ^{12}C and ^{1}H atoms are, respectively, 12×10^{-3} kg and 1.008×10^{-3} kg.

A mole of atoms of any element is sometimes called a *gram atom* of that element, but we shall not use this term. Note that 1 mol H *atoms* weighs 1.008 g, and 1 mol H_2 *molecules* weighs 2.016 g. If no particular molecular species is indicated, as in "a mole of hydrogen", the inference is that the stable species at room temperature is meant. This still requires you to know the chemical formula of the element; under normal conditions, hydrogen is a diatomic molecule, H_2, and so "a mole of hydrogen" normally means 6.022×10^{23} hydrogen molecules, or 2.016 g of the element. It is

TABLE 1-3   Molecular and molar quantities

Particle*	Weight of one particle		Weight of a mole of particles	
	Atomic mass units	Grams	Atomic mass units	Grams
^{12}C atom	12†	1.9927×10^{-23}	7.2264×10^{24}	12†
C atom	12.01115	1.9945×10^{-23}	7.2331×10^{24}	12.01115
O atom	15.9994	2.6568×10^{-23}	9.6348×10^{24}	15.9994
O_2 molecule	31.9988	5.3132×10^{-23}	19.2697×10^{24}	31.9988
CO_2 molecule	44.0100	7.3082×10^{-23}	26.5028×10^{24}	44.0100

*An average over the natural isotopic abundances is implied unless specific isotopes are indicated by superscripts.
† By definition.

important to learn the formulas of simple substances as quickly as possible so as to know what is implied by a mole of such substances.

It is preferable to use the term *mole* only when speaking of *well-defined* particles (charged or uncharged). When it is either *immaterial or not known* whether a given substance consists of molecules, the *gram formula weight* (gfw), the quantity of substance that has a weight in grams numerically equal to the formula weight, is used. For example, 1 gfw Na_2SO_4(142.04 g) contains 1 mol SO_4^{2-} and 2 mol Na^+, as does 1 gfw (or 322.19 g) $Na_2SO_4 \cdot 10H_2O$ (which contains, in addition, 10 mol water). Frequently the prefix *gram* in gram formula weight is dropped, so that *formula weight* has two meanings: (1) the sum of the atomic weights for all the atoms in the formula and (2) the gram formula weight. It usually follows from the context which of the two is meant, and in any case the abbreviation gfw is definite. The mole and the formula weight are in most respects parallel units to express macroscopic quantities of chemical compounds. One gfw contains N_A formula units just as one mole contains N_A molecules.

EXAMPLE 1-8   **Calculations using molecular weight and density.** Calculate the following quantities for the compound difluorohexane, $C_6H_{12}F_2$, a liquid with density 0.90 g ml^{-1}:

(*a*)   *Grams in 2.00* mol: The molecular weight is $6(12.0) + 12(1.0) + 2(19.0) = 122.0$. Hence 2.00 mol corresponds to 2.00×122.0 g = 244 g.

(*b*)   *Volume occupied by 1.00* mol: Density information must be used here. We know from (*a*) that 1 mol weighs 122.0 g. To convert to milliliters we multiply by the ratio milliliters per gram (ml/g) or, equivalently, we divide by the ratio grams per milliliter (g/ml), which is what the density represents. Thus the volume occupied by 1 mol is[1]

$$\frac{122.0 \text{ g}}{0.90 \text{ g ml}^{-1}} = 136 \text{ ml}$$

(*c*)   *Weight in grams which contains 3.0×10^{23} hydrogen atoms:* This is $\frac{1}{2}$ mol hydrogen atoms, since 1 mol contains 6.0×10^{23} of whatever is being considered. The formula of the compound shows that one molecule of it contains 12 hydrogen atoms, so 1 mol contains 12 mol hydrogen atoms. Thus, if x represents the moles of compound desired,[2]

$$\frac{x}{0.50 \text{ mol H atoms}} = \frac{1 \text{ mol compound}}{12 \text{ mol H atoms}}$$

---

[1]Note that the answer is given to three significant figures even though one factor in the preceding step is given only to two significant figures. The reason is that if the last digit in this factor, 0.90, were different by 1 this would be a change of 1 part in 90, or about 1 percent. The answer should reflect a comparable percentage uncertainty, which is the case with 136 ml, whereas retention of only two significant figures as with 1.4×10^2 ml would imply an uncertainty of 1 part in 14, or 7 percent. Significant figures are discussed in detail at the start of Chap. 2.

[2]It is convenient to let x and other symbols represent a quantity that *includes* the units, because appropriate units then automatically turn up in the answer, as they do in the present example, and may serve as a check. When x is divided by the appropriate units, the result is a pure number; examples are T/K, $V/$liter.

$x = \frac{1}{24}$ mol, which weighs

$$\tfrac{122}{24} = 5.1 \text{ g}$$

Another method of solving problems of this kind is to multiply the original quantity by quotients that represent equivalent quantities, such as (1 mol compound/12 mol hydrogen atoms), in units chosen so that in the end there is cancellation of all units except those desired for the answer. The procedure is similar to, but not identical with, that used in part (*b*). Thus

$$3.0 \times 10^{23} \text{ hydrogen atoms} \times \frac{1 \text{ mol hydrogen atoms}}{6.0 \times 10^{23} \text{ hydrogen atoms}} \times \frac{1 \text{ mol compound}}{12 \text{ mol hydrogen atoms}}$$

$$\times \frac{122.0 \text{ g compound}}{1 \text{ mol compound}} = 5.1 \text{ g compound}$$

In other words, the different factors convert the number of hydrogen atoms given initially, in turn, into moles of hydrogen atoms, moles of compound, and grams of compound.

(*d*) *Weight in grams of 10 molecules:* Since 1 mol weighs 122.0 g, one molecule weighs $(122.0/N_A)$ g and 10 molecules weigh $(1220/N_A)$ g $= (1220/6.022 \times 10^{23})$ g $= 2.026 \times 10^{-21}$ g.

(*e*) *Weight that contains 6.0 g carbon:* Since the atomic weight of C is 12.0, 6.0 g represents 0.50 mol C atoms. One mole of the compound contains 6 mol carbon atoms; thus, if x is moles of compound,

$$\frac{x}{0.50 \text{ mol C atoms}} = \frac{1 \text{ mol compound}}{6 \text{ mol C atoms}}$$

and $x = 0.083$ mol compound, which amounts to

$$0.083 \times 122 \text{ g} = 10.2 \text{ g}$$

Using the second method of part (*c*), we find

$$6.0 \text{ g carbon} \times \frac{1 \text{ mol C atoms}}{12.0 \text{ g C}} \times \frac{1 \text{ mol compound}}{6 \text{ mol C atoms}} \times \frac{122.0 \text{ g compound}}{1 \text{ mol compound}}$$

$$= 10.2 \text{ g compound}$$

The distinction between mole and gram formula weight is important in contrasting two methods of describing a solution or any other chemical system. The first method is to furnish an explicit prescription for the preparation, such as: "dissolve 2.0 gfw Na_2SO_4 (or 2.0 gfw $Na_2SO_4 \cdot 10H_2O$) and 0.5 gfw $HgCl_2$ in water and dilute it to 1 liter." This is also called an *operational* description because it permits the reconstruction of the operations that would reproduce the system. The second method is to state the kinds and numbers of moles of ions and molecules the system contains—in the example given the description is: 0.5 mol $HgCl_2$ molecules, 2.0 mol

SO_4^{2-}, and 4.0 mol Na^+ ions.[1] The gram formula weight permits a *prescriptive,* the mole a *descriptive,* account of the system.

Frequently the units *millimole* (mmol) and *milliformula weight* (mfw) are convenient. They are a thousand times smaller than the mole and the formula weight, respectively.

1-8 Experimental Determination of Avogadro's Number

The determination of the value of Avogadro's number was of great importance to the general acceptance of the atomic hypothesis, particularly by physicists. In 1865, J. Loschmidt, a noted Austrian physicist, used the then novel kinetic theory of gases (Chap. 5) to calculate an astonishingly good value for N_A, about 5×10^{23}.

Many different methods have been used to measure N_A, some of them of considerable ingenuity. They include estimates derived from the scattering of light by air molecules (the phenomenon that causes the sky to appear blue), the rate of settling of very fine particles suspended in water, the random (Brownian) motion of fine particles suspended in water, measurements of radioactive decay, comparison of the volume taken up in a crystal of known mass by one molecule (as determined with X rays) with the volume of the crystal itself, and the most precise method: comparison of the charge on a single electron with the charge on a mole of electrons.

Problems and Questions

1-1 Definitions of terms. (*a*) What is the distinction between a physical property and a chemical property? Give examples to illustrate your answer. (*b*) Can a sample of a pure substance be heterogeneous? Can a sample of an impure substance be homogeneous? Explain, citing examples.

1-2 Temperature scales. The inhabitants of the very cold planet Cryos base their temperature scale on the properties of grain alcohol, ethanol. Their temperature unit is termed the degree frigid (°f) and the zero of their scale is the melting point of ethanol, $-117°C$. The boiling point of ethanol, $78°C$, is assigned the value $100°f$. What are the values of the freezing and boiling points of water in degrees frigid?

1-3 Utilizing the energy of gasoline. The combustion of 1.00 g gasoline liberates about 480 kJ of energy, which is equivalent to about 1.23×10^5 kJ gal^{-1}. If it were possible to convert the energy released by burning 1 gal gasoline completely into work, how high could 1 metric ton (1000 kg or 2200 lb) be lifted?   *Ans.* 12.5 km or 7.8 miles

1-4 Composition of atoms and ions. How many protons, how many neutrons, and how many electrons are there in each of the following species: ^2H, $^3H^+$, ^{40}K, $^{37}Cl^-$, ^{17}O?

---

[1]In a complete description the existence and concentrations of small quantities of H^+, OH^-, and HSO_4^- ions would also have to be noted.

1-5   Definitions. Define the following terms: nucleus, nuclide, neutron, isotope, mass number, atomic number.

1-6   Definitions. (*a*) Define each of the following terms carefully: atom, molecule, element, compound, ion. (*b*) Give examples of molecules that are compounds and of molecules that are not compounds.

1-7   Abundances from isotopic weights and atomic weight. Chlorine consists of two isotopes, ^{35}Cl with atomic weight 34.97 and ^{37}Cl with atomic weight 36.97. The atomic weight of chlorine is 35.45. What are the relative abundances of the isotopes?   *Ans.* 76, 24

1-8   Isotopic abundance. Element Y has three naturally occurring isotopes of atomic weight $(A - 4)$, A, and $(A + 1)$. If the atomic weight of natural Y is just A, what are the maximum and minimum percent abundances of the heaviest isotope consistent with this information?

1-9   The size of Avogadro's number

The Walrus and the Carpenter
  Were walking close at hand:
They wept like anything to see
  Such quantities of sand:
"If this were only cleared away,"
  They said, "it would be grand!"

"If seven maids with seven mops
  Swept it for half a year,
Do you suppose," the Walrus said,
  "That they could get it clear?"
"I doubt it," said the Carpenter,
  And shed a bitter tear.

> *Through the Looking Glass,*
>   LEWIS CARROLL

The sands of the beach are usually regarded as countless, but in fact the number of grains of sand on an average beach is considerably fewer than the number of gas molecules in 1 cm^3 of ordinary air. Suppose that there are, on the average, 60 grains of sand per mm^3 on a typical beach. If the beach is 100 m wide and the sand is 10 m deep, how long a strip of beach is needed to contain N_A grains of sand?   *Ans.* 10^7 km, or more miles of beach than there are on earth

1-10   Determination of Avogadro's number. Rutherford and Boltwood found in 1911 that the steady alpha radiation that issues from a standard preparation of radium corresponds to 27.8×10^{-3} mg He for 1 g of the preparation in 1 year. It had also been found, by counting the light flashes produced by the impact on a zinc-sulfide screen of the individual alpha particles that issue from such a preparation, that 1 g of the standard preparation emits 13.8×10^{10} alpha particles per second. Calculate (*a*) the mass of one He atom (in grams), (*b*) the number of grams corresponding to 1 atomic mass unit, and (*c*) Avogadro's number.

1-11   Number of molecules in a new mole. Suppose that the atomic weight of ^{12}C were taken to be 5.000 and that a mole were defined as the number of atoms in 5.000 kg carbon 12, or more generally as the number of atoms in 1 kilogram atomic weight of any element. Calculate the number of molecules in a mole under these conditions. *Ans.* 2.51×10^{26}

1-12 **The ark-mole.** (*a*) Suppose that there is another inhabited planet, Arko, in a distant galaxy, with intelligent beings for whom a convenient unit of mass is the ark. This unit is so defined that 1.00 ark equals 70.0 g. The life of these beings is, like ours, tied to the chemistry of carbon. Since they worship the number 5, they have chosen to assign the most abundant isotope of carbon (which we call carbon 12) an atomic weight of exactly 5.00 . . . . What is the analog of Avogadro's number on Arko, that is, the number of molecules in an ark-mole? (*b*) Assume the information given in (*a*), and calculate the molecular weight of CO_2 on the scale used by the inhabitants of Arko.

1-13 **Quantitative implications of chemical formulas.** Give the following quantities for the compound $C_9H_{12}N_2$, a solid with density 1.25 g ml^{-1}: (*a*) grams in 3.00 mol; (*b*) volume occupied by 2.50 mol; (*c*) weight that contains 7.0 g nitrogen; (*d*) weight in grams that contains 1.0×10^{23} atoms of hydrogen; (*e*) weight in grams of five molecules.
Ans. (*a*) 444 g; (*b*) 296 ml; (*c*) 37 g; (*d*) 2.1 g; (*e*) 1.23×10^{-21} g

1-14 **Quantitative implications of chemical formulas.** (*a*) The compounds phosphorus trioxide and phosphorus pentoxide are often given the formulas P_2O_3 and P_2O_5. What are the formula weights corresponding to these formulas? (*b*) These compounds exist as molecules in the vapor state and in some crystals, with the formulas P_4O_6 and P_4O_{10}, respectively. Give the following with respect to these two molecules: (i) molecular weight; (ii) moles of oxygen atoms in 100 g; (iii) weight of compound containing 62 g phosphorus; (iv) weight of compound containing 1.0×10^{23} oxygen atoms.

1-15 **Quantitative implications of chemical formulas.** Chlorobenzaldehyde is a solid that melts at 47°C. Its molecular formula is C_7H_5OCl, MW = 140, and density = 1.25 g ml^{-1}. Calculate the following quantities with respect to this compound: (*a*) moles of compound in 80 g compound; (*b*) volume of 1.50 mol compound; (*c*) total number of atoms in 28 g compound; (*d*) weight of compound containing as many hydrogen atoms as 90 g water.
Ans. (*a*) 0.57 mol; (*b*) 168 ml; (*c*) 1.7×10^{24}; (*d*) 280 g

1-16 **Quantitative implications of chemical formulas.** Calculate the following quantities for cryolite, Na_3AlF_6: (*a*) grams in 2.10 gfw; (*b*) moles of fluorine atoms in 7.0 g compound; (*c*) percent of sodium in the compound; (*d*) number of atoms in 4.0 g compound; (*e*) number of gram formula weights that contain 2.00 g Al.

"It often seems that the practical strength of the scientific method is not so much that science provides the right answer as that the scientist has learnt how to calculate the margins of error that surround his tentative answer.... The scientist is not so much a man who likes to get all the answers right as a man with a very clear view of error, and what errors mean, and when they are important."

David Wilson in *The Listener*, 1967

"We must recognize an invisible hand that holds the balance in the formation of compounds. A compound is a substance to which Nature assigns fixed ratios; it is, in short, a being which Nature never creates other than balance in hand, pondere et mensura.*"*

Joseph Louis Proust, 1799

2 Stoichiometry: The Quantitative Relationships Implied by Chemical Formulas and Equations

When the basic assumptions of the atomic theory are used to interpret the changes that occur in a chemical reaction, quantitative[1] relationships among the constituents of a compound or the substances participating in the reaction can be deduced. The

---

[1] The terms *quantitative* and *qualitative* are used in a contrasting sense in many discussions in the sciences. The use of the adjective *quantitative* implies a concern with measurement of some property numerically and usually rather precisely. *Qualitative* suggests instead a concern with distinctive characteristics or qualities, but is sometimes used, as in "a qualitative estimate", to suggest a very approximate (rather than a precise) estimate of amount, perhaps good within only a factor of 2 or 3, as in a forecast of the size of a crop.

term *stoichiometry* (Greek: *stoicheion,* element; *metron,* measure) is applied to these relationships. This chapter is devoted to stoichiometry; it begins with a discussion of errors in measurement and ways of carrying out calculations so as to avoid giving a false impression of the uncertainty inherent in experimental results.

2-1 Errors and Significant Figures

All quantitative science is based on measurements, but measurements are usually afflicted by errors, which must be taken into account. Several strategies may be used to serve this purpose. One of them is to repeat the measurements many times. The results are usually not identical, and an analysis of the way they are scattered yields information about their reliability. This analysis is called the statistical treatment of data. Another approach is to measure the same quantity by several widely different methods. The closeness of the results is an indication of their trustworthiness. For most routine problems, such as the determination of the chlorine content of a sample, the first procedure is almost always used. One method executed in triplicate is often satisfactory, provided the method is well established and its limitations are known and adequate. However, for quantities of central importance, such as Avogadro's number, many elaborate and differing types of experiments are used.

An important aspect of the handling of data is that the quantity of primary interest is usually not what is measured directly, so that intermediate computations are required. This raises two questions. One concerns the relation of the uncertainty in the quantity of interest to the errors in the original measurements. In other words, how do the errors propagate through the equations that connect the quantity of interest with the data? The other question concerns the accuracy with which intermediate calculations should be carried out. To do justice to the data, the accuracy of the calculations should be sufficiently high that no new errors are introduced, but excessive accuracy is pointless and inefficient.

A related issue concerns ways of expressing the uncertainty in an experimentally determined quantity. In a rough way this is done by the conventions of significant figures, to be discussed shortly. We now consider some of these questions, beginning with definitions of the concepts of error, accuracy, and precision.

Error, accuracy, and precision. The *error* of an observation is the difference between the observation and the actual, or true, value of the quantity observed. The *accuracy* of a set of observations is the difference between the average of the values observed and the true value of the observed quantity. The *precision* of a set of measurements is a measure of the range of values found, that is, of the reproducibility of the measurements. The meanings of accuracy and precision are thus quite different. A set of observations may, for example, have high precision and low accuracy at the same time (Fig. 2-1).

Errors are often classified as either systematic or random. *Systematic errors* may be caused by fundamental flaws in the experimental equipment (e.g., the experimenter is unaware that his stopwatch runs fast) or by inadequate understanding of the theory underlying the measurement (e.g., unwarranted neglect of the buoyant

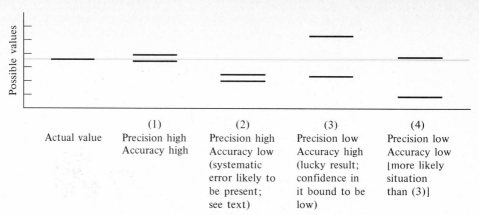

	(1)	(2)	(3)	(4)
Actual value	Precision high Accuracy high	Precision high Accuracy low (systematic error likely to be present; see text)	Precision low Accuracy high (lucky result; confidence in it bound to be low)	Precision low Accuracy low [more likely situation than (3)]

FIGURE 2-1   **Different possible combinations of high and low precision and accuracy.**

effect of air on weights and on an object being weighed). Sometimes systematic errors have their source in the observer. He may, for example, be prejudiced by prior information, as when he is repeating thermometer readings and finds it impossible to forget estimates of fractional degrees made just a moment earlier. Similarly, his estimate of a tenth of a division may be different when he is reading close to one of the etched marks and when he reads halfway between marks. Systematic errors usually do not average out, even if the observations are repeated many times.

Random errors are accidental errors that vary in a completely unreproducible way from measurement to measurement. The concept of randomness can best be explained by the use of examples: the sequence of heads and tails in tossing a perfect coin is said to be random. Random errors find their origin, at least in part, in the limited precision of instrument readings (e.g., the reading of a ruler with engraved scale marks of finite widths) and in the degree to which external conditions (e.g., the temperature of the ruler) are controlled. (Both effects may cause systematic errors as well.)

Random errors can be treated statistically, which makes it possible to relate the precision of an experimental result to the precision with which each of the experimental variables (e.g., pressure, temperature) is known. An error analysis also shows the weakest link in the chain of observations that leads to the determination of a certain quantity, the step that needs to be improved first if the precision of the result is to be improved.

Systematic errors are frequently difficult to discover. They may usually be detected by going outside the framework of the original measurements, for example, by measuring the quantity of interest in several fundamentally different ways. Close agreement of the results gives confidence in the unlikelihood of systematic errors, but the detection of such errors is never certain.

An illustration of the effects of unsuspected systematic errors is provided by the determination of the electronic charge by Millikan. His value for e was dependent on the viscosity of air in a way that he understood, and he considered carefully all the different precise measurements of the viscosity that had been made. Millikan then estimated that the value he used could "scarcely be in error by more than one part in two thousand". Taking into account this limit of error and those in his own work, he estimated that his value of *e*, 1.5924×10^{-19} C (coulomb), was in error by no more than 0.1 percent. Yet the currently accepted value is about 0.6 percent higher, 1.60219×10^{-19} C, with an estimated uncertainty of only 0.005 percent. Almost all the discrepancy between Millikan's value and that currently accepted is attributable to an error in the viscosity of air, which was about 0.5 percent low, although it was believed to be about 10 times more precise than that. Millikan estimated his precision quite correctly; it was a systematic error that made his value far less accurate than he thought.

Significant figures and relative precision. In general, results of observations should be reported in such a way that the last digit given is the only one whose value may be in doubt. The digits that constitute the result, excluding *leading* zeros, are then termed *significant figures*. The number of significant figures is the same in 23.5 mg and 0.0235 g; this is reasonable, for the uncertainty in a result should not depend on the units in which the answer is expressed. Note that there is some uncertainty about the number of significant figures in a weight reported to be 2350 kg, because it is unclear whether the final zero is significant or not. The matter is clarified by reporting the weight either as 2.350×10^3 kg or 2.35×10^3 kg, whichever applies.

The concept of significant figures does not apply to quantities known to be integers; for example, in a molecular formula, such as C_6H_{14}, the number of figures in each subscript does not imply that the composition is uncertain. We can assume that there are exactly 6 mol carbon atoms per mole of compound, even though the formula might to the unsophisticated seem to imply that this ratio is known only to low precision. Similar considerations apply to integral numbers of objects; the integers are regarded as infinitely precise.

There is no general agreement about the degree of uncertainty associated with the least significant digit given, that is, with the last digit on the right. It is sometimes assumed that the implied uncertainty is one unit of the decimal position of the least significant digit if no more definite statement of the uncertainty is indicated. According to this convention, a weight given only as 2.350 kg is uncertain by 0.001 kg. This practice is rather restrictive, and we recommend that only a general impression of uncertainty should be associated with the least significant digit.

The number of significant figures with which a quantity is given expresses roughly its *relative precision*. Thus, if we assume an uncertainty of one or two units in the least significant digit, the quantities 201 and 0.0198 are each precise to 1 or 2 parts in about 200, or about 1 percent. However, it must be noted that when a value such as 98.3 mm is given with three significant figures, its relative precision is comparable to that of a value such as 103.7 mm, given with four significant figures. If we associate an uncertainty of 1 in the last place with each quantity, each is given to a precision of about 1 part in 1000, although the number of significant figures differs. Despite minor shortcomings of this kind, the system of significant figures is often adequate and useful.

In statements of experimental procedures or of numerical problems, more significant figures are sometimes implied than are given. For example, the meanings of "1 liter" and "10 g" *may* well be those of "1.000 liter" and "10.000 g". Such implications can usually be established from the context, and confusion should seldom arise once it is understood that under such circumstances the "1" in "1 liter" does not necessarily imply a one-digit significance.

2-2   Precision in Calculations

There are several useful rules for carrying out calculations so that the precision of the answer properly reflects the precision of the data used.

Rounding off.   This operation consists of discarding digits that are not significant and adjusting the residue appropriately, as deduced from the simple rules given below:

1.   Discard the unwanted digits.

2.   (*a*)   Increase the last retained digit by one unit (*round up*) if the most significant discarded digit is larger than 5 or if it is equal to 5 but followed by other digits some of which are different from zero (for example, 94.3507 rounds to 94.4 if only three significant digits are warranted).
(*b*)   Leave the last retained digit unchanged (*round down*) if the most significant digit discarded is less than 5 (for example, 94.348 rounds to 94.3).
(*c*)   If the digits discarded are a 5 followed only by zeros (or by nothing), round to the nearest even digit. This last convention provides that, in the long run, the total number of rounding-up operations is the same as the number of rounding-down operations so that averages are unaffected. Thus, 4.55 and 4.65 would both be rounded to 4.6 if the final 5 were not a significant digit.

Addition and subtraction.   First look for the term of lowest absolute precision. This determines the precision warranted for the result, and it is meaningless and misleading to give the result with significant figures implying an absolute precision greater than that of this term. An illustration is given in Table 2-1. Note that rounding all terms to the lowest absolute precision before performing the calculation is undesirable because rounding errors may accumulate in the subsequent computations and affect the figures finally retained. Instead, before adding or subtracting, round in a preliminary fashion the terms given with higher precision, retaining one or two places beyond the precision warranted for the result. Then perform the additions or subtractions, and finally round the result.

Multiplication and division.   The situation here is like that discussed for addition and subtraction except that *relative* rather than absolute precisions are involved. The term *relative* here implies expressing the uncertainty as a fraction of the quantity of interest (or as a percentage of it). For example, if the length of a rod is given as 5.1 m with an uncertainty of 0.1 m, the absolute precision is 0.1 m and the relative precision is 0.1 m/5.1 m, or 0.02, or 2 percent. The result of any multiplication or division must not be expressed with a relative precision higher than

TABLE 2-1 **Combination of terms with different absolute precision in addition and subtraction**

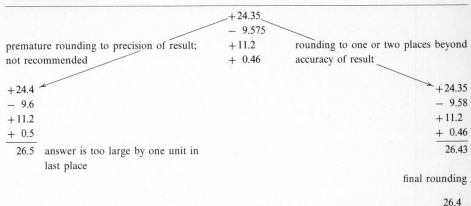

	+24.35	
	− 9.575	
premature rounding to precision of result; not recommended	+11.2	rounding to one or two places beyond accuracy of result
	+ 0.46	

+24.4		+24.35
− 9.6		− 9.58
+11.2		+11.2
+ 0.5		+ 0.46
26.5 answer is too large by one unit in last place		26.43

final rounding

26.4

that of the term involved that has the lowest relative precision. However, to prevent the accumulation of rounding errors, terms given with higher relative precision should be rounded in a preliminary fashion only. During the intermediate calculations, retain one or two figures beyond the relative precision warranted by the least precise quantity and perform the rounding as the last step.

EXAMPLE 2-1 **Multiplication and division.** Find $(437.689 \times 5.2)/4.03$.

Solution. Multiplying first and retaining four significant figures, we obtain $\frac{2276}{4.03} = 564.8 \approx 5.6 \times 10^2$. The end result can be given to two significant figures only, in keeping with the relative precision of the term given with least relative accuracy, 5.2. Note that retaining three significant figures in the multiplication would have led to the result $2.28 \times 10^3/4.03 = 5.66 \times 10^2 \approx 5.7 \times 10^2$, whereas retaining only two significant figures in the intermediate calculations yields $2.3 \times 10^3/4.0 = 5.75 \times 10^2 \approx 5.8 \times 10^2$, provided the rounding-off procedure described earlier is followed.

General intermediate computations. In computations that are more involved than those just illustrated, calculate intermediate results to one or two places beyond the precision warranted by the data and round off only the final results. There are some circumstances in which one must be extremely careful about premature rounding off. The following discussion illustrates the danger of incorrectly neglecting small terms in intermediate steps because these terms seem insignificant. Whenever you are tempted to neglect a "small" quantity, remember always to ask: "Small compared with what? What effect does neglecting this quantity have on the final answer?"

EXAMPLE 2-2 **Small terms in solutions of quadratic equations.** A certain problem, cast into algebraic form, calls for the positive root of the quadratic equation

$$x^2 + 0.1x - 1.0 \times 10^{-14} = 0 \qquad (2\text{-}1)$$

The formal solution to the equation is

$$x = \frac{-0.1 + \sqrt{(0.1)^2 + 4.0 \times 10^{-14}}}{2} \qquad (2\text{-}2)$$

Reduce this expression.

Solution. It may be argued that the significant figures of $(0.1)^2$ in the radicand suggest that the term 4.0×10^{-14} is of no concern and that to keep it would give the proceedings an undesirable aura of false precision. Deletion of this term leads, however, to $x = 0$, which is wrong. The term must therefore be retained. The square root may be evaluated by use of the approximation

$$\sqrt{1 + x} \approx 1 + \frac{x}{2} \qquad \text{for } |x| \ll 1$$

$$\sqrt{(0.1)^2 + 4.0 \times 10^{-14}} = \sqrt{(0.1)^2(1 + 4.0 \times 10^{-12})}$$
$$= 0.1(1 + 2.0 \times 10^{-12} + \cdots)$$
$$\approx 0.1 + 2.0 \times 10^{-13}$$

(Note that a typical small electronic calculator does not retain enough decimal places to allow the square root in (2-2) to be evaluated directly and that to find this square root in a table would have required a table of much higher accuracy than is usually available.) The 0.1 from the square root exactly cancels the 0.1 preceding it in (2-2), and the final answer is

$$x = \underline{1.0 \times 10^{-13}}$$

If extensive algebraic developments are required, they should be carried through to the final result symbolically and computations should be performed only at the end. Computation of intermediate results may cause errors, because terms that should be identical may become slightly different through rounding off and may no longer cancel in the end as they should. No special symbols need be invented for such algebraic manipulations, because terms such as the 0.1 and the 10^{-14} in (2-1) can be used as algebraic symbols themselves.

2-3 Chemical Formulas

Chemical formulas have already been used in Chap. 1. Several types of formulas serve to designate chemical compounds, each type conveying different kinds of information. We consider here some of the most important types of formulas, and in the following section we represent chemical reactions by equations between such formulas. The quantitative implications of formulas and equations are examined in Secs. 2-6 and 2-7.

Empirical formulas (also called *simplest formulas*) express the results of elemental analysis in the simplest way possible. For stoichiometric compounds the atoms involved carry the smallest integral subscripts possible, so that the different subscripts have no common factor. If a substance consists of molecules, there is no implication

TABLE 2-2   **Different types of chemical formulas**

Name	Empirical formula	Molecular or ionic formula	Structural formula (topological)
Phosphorus	P	P_4 (white P) P_∞ (red, black P)	See Fig. 2-2
Sodium chloride	NaCl	Na^+Cl^- or NaCl	Ionic; see Fig. 3-7
Phosphorus pentoxide	P_2O_5	P_4O_{10}	See Fig. 2-2
Ammonium peroxydisulfate	H_4NO_4S	$(NH_4^+)_2S_2O_8^{2-}$ or $(NH_4)_2S_2O_8$	See Fig. 2-2
Dichloroethylene	CHCl	$C_2H_2Cl_2$	ClHC=CHCl (two distinct compounds; geometric isomers)
Lactic acid	CH_2O	$C_3H_6O_3$	CH_3—CHOH—COOH (two distinct compounds; optical isomers)

that the empirical formula corresponds to the correct number of atoms contained in one molecule. The molecule may correspond to the empirical formula or any integral multiple of it. Examples are given in the second column of Table 2-2.

Molecular formulas give the correct atomic composition of molecules. No information about the structure of the molecule is implied, except in very simple cases such as He, H_2, or NO (nitric oxide). To be consistent, ionic compounds should be given *ionic formulas,* for example, Na^+Cl^- and $Ca^{2+}SO_4^{2-}$, but the plus and minus signs are awkward and are usually left out. Just as with molecules, the total number of atoms in an ion need not be that implied by the empirical formula. For example, sodium peroxydisulfate has the empirical formula $NaSO_4$, but the ionic formula is $(Na^+)_2S_2O_8^{2-}$ or $Na_2S_2O_8$ (see also Fig. 2-2).

FIGURE 2-2   **Examples of three-dimensional structural formulas.** The four P atoms in P_4 are at the corners of a tetrahedron. This is also true for P_4O_{10}, but here the P atoms are linked by O atoms that are positioned just outside the centers of the tetrahedral edges. An additional O atom is linked to each P atom. In red phosphorus and black phosphorus the P atoms are bonded together in three-dimensional networks.

$(NH_4)_2S_2O_8$ consists of tetrahedral NH_4^+ ions and an $S_2O_8^{2-}$ ion in which two tetrahedral SO_4 groups are joined by an O-O bond. The lines between atoms represent covalent bonds (Sec. 3-2).

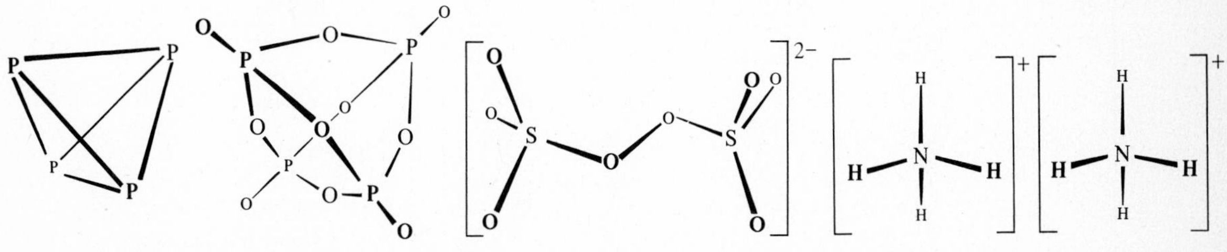

P_4 P_4O_{10} $(NH_4)_2S_2O_8$

The relation between the molecular formula and the empirical formula can easily be found, if the molecular weight is known, by considering the ratio of the molecular weight to the formula weight corresponding to the empirical formula. The molecular weight may be determined, as discussed in later chapters, from measurements of the density of the vapor of the compound or from measurements of certain properties of its solutions, such as the lowering of the freezing point. Direct structure determinations, for example by X-ray diffraction, also provide molecular and ionic formulas. Examples of molecular and ionic formulas are given in the third column of Table 2-2.

When desired, *states of aggregation* may be indicated by letters in parentheses, as in $Br_2(g)$, $Br_2(l)$, and $Br_2(s)$, referring to gaseous bromine, liquid bromine, and solid bromine, respectively. Similarly, the symbol (*aq*) may be used, as in $Al^{3+}(aq)$ or $Cl_2(aq)$, to indicate a species dissolved in water to form what is termed an aqueous solution (Chap. 7).

Structural formulas give partial or complete information on the detailed atomic arrangement in a compound. There are several types of structural formulas, some of which are discussed here. All of them reveal, at least by implication, what might be called the connectivity or topology of the molecule or ion, that is, the way in which the atoms are linked. Some show no more than that, while others reveal the detailed three-dimensional spatial arrangement of the atoms.

The formula

$$\begin{array}{ccc} H & & H \\ \diagdown & & \diagup \\ H-C-C-H \\ \diagup & & \diagdown \\ H & & O-H \end{array}$$

shows how the atoms in the ethanol (ethyl alcohol) molecule are linked. It also implies some features of the three-dimensional arrangement to one who knows that four single bonds to a carbon atom are directed approximately toward the corners of a regular tetrahedron and two single bonds to an oxygen atom make an angle of about 109° with one another. This structural formula is often given in the abbreviated form CH_3CH_2OH (chiefly for typographical convenience), which indicates the linkage of the atoms but says little about the spatial arrangement. Other examples of the same types of formulas are

$$\begin{array}{ccc} H & & O \\ \diagdown & & \diagup\!\!\!\!= \\ H-C-C \\ \diagup & & \diagdown \\ H & & O-H \end{array} \qquad \text{or} \qquad CH_3COOH$$

for acetic acid and

$$\begin{array}{ccccc} H & & O & & H \\ \diagdown & & & & \diagup \\ & C & & C & \\ \diagup & | & & | & \diagdown \\ H & & & & H \\ & H & & H & \end{array} \qquad \text{or} \qquad CH_3OCH_3$$

for dimethyl ether.

Ethyl alcohol and dimethyl ether have different molecular structures and, as a result, they have very different properties. Note, however, that these compounds have the same molecular formula, C_2H_6O. This phenomenon is called *isomerism* (Greek: *iso,* same; *meros,* part), and compounds with the same elemental composition and molecular weight but different structures are called *isomers* of one another. There are many forms of isomerism, some of which will be encountered in later chapters. One of them is geometric isomerism or stereoisomerism, in which the isomeric species have the same connectivity or topology but differing arrangements of the atoms in space and hence different geometries. An even more subtle form is optical isomerism. Optical isomers have the same connectivity and the same geometry but are related to one another as a right hand is to a left hand; that is, they are nonsuperimposable mirror images of one another.

2-4 Chemical Equations

The reactions of chemical substances may be represented by chemical equations that show the formulas and relative numbers of the reactants and products involved. Equations should, by definition, be balanced: they should show the same number of atoms of a given kind on each side, and the net charge should be the same on each side. The condition, sometimes slavishly insisted upon, that all coefficients in an equation must be integral is not essential; coefficients may be fractional or may even have a common factor. Experienced chemists sometimes do not bother to write balanced equations but simply indicate the pertinent reactants and products because they are often not concerned with the quantitative relationships among reactants and products that are correctly implied only by a balanced equation.

Many equations are easy to balance, once the important reactants and products are known. For example, when dissolved in water, silver nitrate, $AgNO_3$, and sodium sulfate, Na_2SO_4, react to form solid silver sulfate, Ag_2SO_4, which precipitates from the solution. An unbalanced equation for the reaction is

$$AgNO_3 + Na_2SO_4 \longrightarrow Ag_2SO_4 + \cdots \qquad (2\text{-}3)$$

which may be balanced to

$$2AgNO_3 + Na_2SO_4 \longrightarrow Ag_2SO_4 + 2NaNO_3 \qquad (2\text{-}4)$$

The arrow is often replaced by an equal sign for typographical reasons, but this should be done only when the equation is balanced. Throughout this book, we use arrows. Methods of balancing equations in less obvious situations are discussed in Appendix C.

The reaction considered in (2-3) and (2-4) is actually one between the ions Ag^+ and SO_4^{2-} rather than between $AgNO_3$ and Na_2SO_4. It is therefore usually written as an *ionic equation:*

$$2Ag^+ + 2NO_3^- + 2Na^+ + SO_4^{2-} \longrightarrow Ag_2SO_4 + 2Na^+ + 2NO_3^- \quad (2\text{-}5)$$

or
$$2Ag^+ + SO_4^{2-} \longrightarrow Ag_2SO_4 \qquad (2\text{-}6)$$

The ions NO_3^- and Na^+ remain unchanged and may thus be omitted, in the same way that the H_2O molecules present in the solution are also not shown in the equation. The same basic reaction would occur between any two water-soluble substances that contain, respectively, Ag^+ and SO_4^{2-}; for example, Na_2SO_4 could be replaced by K_2SO_4.

In (2-5) and (2-6) the total of the charges on each side is zero, but it need not be, as long as it is the same on each. For example, in the balanced equation

$$8H^+ + MnO_4^- + 5Fe^{2+} \longrightarrow Mn^{2+} + 5Fe^{3+} + 4H_2O$$

the total of the charges on either side is $+17$ $[= +8 - 1 + (5 \times 2) = +2 + (5 \times 3)]$.

While certain ions may not be shown in a reaction equation, it must be remembered that their presence is nevertheless required to keep the solution electrically neutral. Thus, SO_4^{2-} ions may not be introduced into a solution without some positive ions such as Na^+, K^+, or Mg^{2+}. Such extraneous ions must, of course, not participate in the reaction considered (otherwise they would have to be shown in the reaction equation).

The balanced chemical equation shows the overall results of a chemical reaction and provides quantitative information about the relative amounts of reactants and products concerned. It is often referred to as the *stoichiometric equation* for the reaction. Such an equation, however, provides no information about the detailed sequence of ionic and molecular processes that actually occur as the reactants combine to form the products.

2-5   General Principles of Stoichiometry

The term *stoichiometry* denotes the quantitative relationships among the constituents of a chemical formula or a balanced equation for a chemical reaction. The principles of stoichiometry are no more complicated than those of everyday calculations based on proportions in shopping and elsewhere. The following facts are of principal importance:

1.   All stoichiometric calculations involve *whole units* on the atomic level: atoms, molecules, ions, or atomic groups.

2.   As discussed in Chap. 1, for practical reasons these units are scaled up to macroscopic dimensions by multiplication by Avogadro's number. One therefore deals with moles and gram formula weights instead of with individual atoms, molecules, and ions.

3.   As a consequence, the proportions of the different constituents in a compound or of the different reactants and products in a particular reaction are definite and characteristic of the compound or the reaction at issue.

Our concern here is to illustrate and clarify the principles involved in stoichiometric calculations rather than to develop a set of procedures for solving different types of problems. Consequently, no formal rules are given; instead, the working of stoichiometric problems is demonstrated by examples. It is possible, and tempting, to systematize many procedures of stoichiometry and offer "plug-in" methods, but overreliance on such rote methods is dangerous when new situations are encountered. A comprehensive grasp of the overall principles—which involve no more than is stated in (1), (2), and (3) above—will prove far more useful. One way to develop this mastery is to consider alternative methods of solution; a variety of approaches is possible for every example discussed.

The solution of any problem can be developed in a series of steps. For a beginner, this sequence of explicit steps is often somewhat lengthy, but as insight, experience, and confidence develop, some steps can usually be combined and the whole solution considerably shortened. However, this process should not be rushed. Among the common difficulties of the beginner are:

1. Failure to appreciate the fact that scientific terms have precise definitions, and failure to learn and to use these definitions properly, even for such common terms as fraction and percent.

2. Failure to appreciate the implications of both the number aspect and the weight aspect of the mole (Sec. 1-7).

3. Carelessness in the use of significant figures to express the precision of an answer. In most of the examples and problems presented here, "slide-rule accuracy" (0.2 to 0.5 percent) is adequate and is all that is needed for an understanding of the principles involved. However, the accuracy of the customary methods of chemical analysis is usually such that higher precision is often justified and indeed is necessary if the full implications of the analysis are to be appreciated.

Extensive practice with problems of various kinds is essential for the development of an understanding of chemical principles. We proceed to a discussion of some typical problems, which may be classified into two general categories: those dealing with the interconversion of chemical formulas and percentage compositions, and those dealing with weight relations in chemical reactions. A number of problems overlap both these categories.

2-6 Chemical Formulas and Percentage Composition

One of the simplest applications of stoichiometry is the calculation of the composition of a compound in weight[1] percent from its formula and atomic weight information.

EXAMPLE 2-3   **Composition in weight percent.** Find the composition of nitric acid, HNO_3, on a weight basis.

Solution. The method used to compute the answer is indicated by the steps shown in Table 2-3. The atomic weights of H, N, and O are, respectively, 1.01, 14.01, and 16.00. Therefore, the weights of the components in 1 formula weight HNO_3 are as

---

[1]We are not differentiating between the terms *mass* and *weight;* see Study Guide.

TABLE 2-3 **Weight-percentage composition of HNO_3**

Number of moles of atoms	Weight/g	Weight fraction	Weight percent
1 H	1.01	$\frac{1.01}{63.02} = 0.016$	1.6
1 N	14.01	$\frac{14.01}{63.02} = 0.222$	22.2
3 O	$3 \times 16.00 = 48.00$	$\frac{48.00}{63.02} = 0.762$	76.2
SUMS:	63.02	1.000	100.0
	(formula weight)		

indicated in the second column of the table. The total of these entries is equal to the formula weight of HNO_3. The percentage composition is given in the fourth column.

The deduction of an empirical (or simplest) formula from percentage composition information is the reverse of the calculation illustrated in Example 2-3. However, it usually presents appreciably greater difficulties to the beginner and so, instead of starting with a chemical example, we present first an illustration involving something more familiar than atoms.

EXAMPLE 2-4   **Empirical formula of postage.** The postage on a package consists of 24 percent 15¢ stamps (designated Fn), 16 percent 20¢ stamps (Ty), and 60 percent 50¢ stamps (Fy). How many stamps of each kind are there?

Solution. The solution is found in two steps, with results collected in Table 2-4. The information provided is shown in the first two columns. The first step is to find the numbers of stamps that would correspond to a convenient assumed total postage, such as 100¢. These relative numbers are in the third column; for example, the number of 15¢ stamps that corresponds to 24¢ is $\frac{24}{15}$. The final step is to scale the relative numbers until they are all integers, since no fractional numbers of stamps can be used. This is done by trial. For example, as a first step we may divide all the relative numbers by the smallest: $\frac{1.6}{0.8} = 2.0$, $\frac{0.8}{0.8} = 1.0$, $\frac{1.2}{0.8} = 1.5$. Multiplication by 2 then yields the smallest integral numbers of stamps, given in the last column.

The "empirical formula" that corresponds to the specified postage is $Fn_4Ty_2Fy_3$. It represents a monetary value of $(4 \times 15) + (2 \times 20) + (3 \times 50)$¢ $= 250$¢ that corresponds to the formula weight of a compound. Note that any integral multiple of the

TABLE 2-4 **Postage on parcel**

Kind of stamp	Percent of value	Relative numbers of stamps	Smallest integral numbers of stamps
15¢ (Fn)	24	$\frac{24}{15} = 1.6$	4
20¢ (Ty)	16	$\frac{16}{20} = 0.8$	2
50¢ (Fy)	60	$\frac{60}{50} = 1.2$	3

formula ($Fn_8Ty_4Fy_6$, $Fn_{24}Ty_{12}Fy_{18}$, and so on) and of the total value (500¢, 1500¢, and so on) would lead to the same percentage distribution of the stamps. On the basis of the information given it is impossible to determine the actual postage used.

The example just worked is parallel to the problem faced by a chemist who wishes to find the chemical formula from the analysis of a compound:

EXAMPLE 2-5   **Chemical formula from elemental composition.** The elemental composition of a substance is as indicated by the first and second columns of Table 2-5. What is the empirical formula of the compound?

Solution. If a quantity of 100 u of substance is considered, the numbers in the second column of the table represent the weights of the different kinds of atoms in atomic mass units. The *relative* numbers of atoms are found by dividing these weights by the atomic weights, as shown in the third column. The smallest *integral* numbers of atoms, given in the fourth column, are found by appropriate scaling—that is, dividing by the largest common factor among the relative numbers of atoms, here 1.31, which gives the smallest integers for the numbers of the different kinds of atoms. Note that because of experimental uncertainties in the percentage composition data, some numbers may vary by as much as about 1 percent from integral values.

The *empirical* formula of the compound is thus N_2H_4CS. The actual formula cannot be established from the data given. It could be any integral multiple of the empirical formula. A compound of composition that fits the data given is ammonium thiocyanate, NH_4SCN, containing the ions NH_4^+ and SCN^-.

The same problem can also be worked in macroscopic units. If the scaling is by Avogadro's number, then 100 g rather than 100 u of substance is assumed and the numbers in the third column of Table 2-5 give the number of *moles* of atoms rather than the number of atoms in the assumed sample. However, the sample might be 100 ounces, 100 tons, or any arbitrary number of any size unit; the procedure of finding the simplest integers relating the numbers of different kinds of atoms present remains the same.

In this example, the starting point is the elemental composition, which is information that generally does not follow from chemical analysis without computations. However, to find the formula of a compound it is not necessary to work out the elemental composition; the analytical data may be used directly as the starting point. Before considering an example to illustrate this point (Example 2-11 below), we first discuss some simple examples of weight relations in chemical reactions.

TABLE 2-5   **Empirical formula of chemical compound**

Kind of atom	Weight percent	Relative numbers of atoms	Smallest integers
N	36.8	$\frac{36.8}{14.0} = 2.63$	2
H	5.3	$\frac{5.3}{1.0} = 5.3$	4
C	15.8	$\frac{15.8}{12.0} = 1.32$	1
S	42.1	$\frac{42.1}{32.1} = 1.31$	1

2-7 Quantities of Substances Participating in Reactions

One of the simplest applications of stoichiometry to chemical reactions is to find out the quantities of substance that react and that are produced. These quantities may be expressed in moles, weights, volumes, or in still other ways. It is always assumed, unless explicitly stated otherwise, that the reactions considered go to completion, that is, that they proceed until one or more of the starting materials is used up. Even though this may not always be true, the assumption is a reasonable and simplifying one, and corrections for incomplete reaction can be made.

EXAMPLE 2-6

Quantities of reactants and products. When carbon monoxide is heated with water to a high temperature, hydrogen and carbon dioxide are produced:

$$CO + H_2O \longrightarrow H_2 + CO_2 \qquad (2\text{-}7)$$

How many grams of H_2 and how many moles of CO_2 would be formed if 56 g CO reacted with an excess of water? How many grams of water would be consumed?

Solution. The balanced equation (2-7) indicates that 1 mol CO reacts with 1 mol H_2O to form 1 mol H_2 and 1 mol CO_2. Perhaps the easiest way to solve this problem is to calculate the number of moles of CO at the start and then use the simple mole relationships implied by the equation. The molecular weight (MW) of CO is 12.0 + 16.0 = 28.0, and thus 56 g CO is

$$\frac{56 \text{ g CO}}{28.0 \text{ g CO/mol CO}} = \underline{2.0 \text{ mol CO}} \qquad (2\text{-}8)$$

Hence if all the CO reacts, it will use up 2.0 mol H_2O and form 2.0 mol H_2 and 2.0 mol CO_2. To calculate the weights of hydrogen and water involved we need their molecular weights, 2.0 for H_2 and 18.0 for H_2O. Hence the weight of H_2 formed is $(2.0 \text{ mol } H_2)(2.0 \text{ g } H_2/\text{mol } H_2) = \underline{4.0 \text{ g}}$, and the weight of water consumed is $(2.0 \text{ mol})(18.0 \text{ g } H_2O/\text{mol } H_2O) = \underline{36 \text{ g}}$.

Note how the units in (2-8) are treated like algebraic quantities. The "g CO" in the numerator and denominator on the left cancel one another, and the "mol CO", because of its position in the denominator of the denominator of the term on the left, is thereby transferred into the numerator on the right. Similarly, in the calculation of the weights of H_2 and H_2O from the numbers of moles of each and the molecular weights, the units are again manipulated as if they were algebraic quantities, the moles in numerator and denominator canceling.

EXAMPLE 2-7

Quantities of substances produced. When crystals of potassium permanganate ($KMnO_4$) are heated to several hundred degrees, oxygen, solid potassium manganate (K_2MnO_4), and solid manganese dioxide (MnO_2) are formed by the reaction

$$2KMnO_4(s) \longrightarrow K_2MnO_4(s) + O_2(g) + MnO_2(s) \qquad (2\text{-}9)$$

How many grams of MnO_2 and how many moles of O_2 result when 10.0 g $KMnO_4$ is heated?

Solution.   The balanced equation (2-9) shows that in this reaction 2 gfw $KMnO_4$ produces 1 gfw MnO_2. This problem may be solved in a manner parallel to that used for Example 2-6, but we shall use an alternative, though equivalent, approach. Let x represent the grams of MnO_2 that are formed. Then, by (2-9), the following proportion exists:

$$\frac{FW\ MnO_2}{2(FW\ KMnO_4)} = \frac{x}{10.0\ g}$$

so that
$$x = 10.0\ g \times \frac{FW\ MnO_2}{2(FW\ KMnO_4)}$$

The formula weights are 86.9 for MnO_2 and 158.0 for $KMnO_4$. Inserted,

$$x = 10.0\ g \times \frac{86.9}{2(158.0)} = \underline{2.75\ g}$$

Note again that the units work out properly, to grams, as they should.

Finally, to find the moles of O_2 formed, note that, by (2-9), 2 gfw $KMnO_4$ forms 1 mol O_2. If we let y be the moles of O_2 produced, then we can set up the proportion

$$\frac{1\ mol\ O_2}{2(gfw\ KMnO_4)} = \frac{y}{10.0\ g\ KMnO_4}$$

Solving for y,

$$y = 10.0\ g \times \frac{1\ mol\ O_2}{2(gfw\ KMnO_4)} = \frac{10.0\ g\ (1\ mol\ O_2)}{2(158.0\ g)} = \underline{0.0316\ mol\ O_2}$$

Note that the units again work out.

More involved considerations are needed when one or several of the reactants are not completely used up at the end of the reaction. The reactant that runs out first and thereby limits the extent of the reaction must first be identified.

EXAMPLE 2-8   **Extent of a reaction that is limited by a reactant.**   Iron is produced in a blast furnace by the reaction of hematite (Fe_2O_3, ferric oxide) with carbon monoxide gas (CO) made by the partial oxidation of carbon. The idealized reaction equations are

$$2C + O_2 \longrightarrow 2CO \tag{2-10}$$
$$3CO + Fe_2O_3 \longrightarrow 2Fe + 3CO_2 \tag{2-11}$$

How much iron can be produced ideally by the reaction of 1.00 ton C and 5.00 tons Fe_2O_3?

Solution.   The first step is to find which (if either) of the two starting materials is exhausted first and thus limits the reaction. By (2-11) each Fe_2O_3 requires 3CO and by (2-10) each CO requires 1C, so that for the complete reaction each Fe_2O_3 requires 3C. It is convenient to summarize this result by the relationship

$$Fe_2O_3 \leftrightharpoons 3CO \leftrightharpoons 3C$$

where the symbol $\simeq$ can be read as "corresponds to" or "is equivalent to". The formula weights of the substances involved are $C = 12.0$ and $Fe_2O_3 = 159.7$. Thus, by these reactions,

$$159.7 \text{ g } Fe_2O_3 \simeq 3 \times 12.0 \text{ g } C = 36.0 \text{ g } C$$

which implies that carbon can reduce about 4.4 times its weight of Fe_2O_3. However, the weight of Fe_2O_3 actually present in this blast furnace is five times the weight of the carbon present. Thus, some Fe_2O_3 remains unreacted at the end; the carbon is used up completely in the ideal case [that is, according to reactions (2-10) and (2-11)] and is the limiting reactant.

The second step is to find the quantity of Fe that can be produced by complete reaction of the carbon present. Equations (2-10) and (2-11) show that

$$2Fe \simeq 3CO \simeq 3C$$

and since the formula weight of Fe is 55.85,

$$2 \times 55.85 \text{ g } Fe = 111.7 \text{ g } Fe \simeq 3 \times 12.0 \text{ g } C = 36.0 \text{ g } C$$

The proportion of iron to carbon in this reaction—111.7:36.0—is fixed. If the number of tons of Fe produced is symbolized as x, then

$$\frac{x}{1.00 \text{ ton}} = \frac{111.7 \text{ g}}{36.0 \text{ g}}$$

Solving this for x leads to

$$x = \frac{(1.00 \text{ ton})(111.7)}{36.0} = \underline{3.10 \text{ tons Fe}}$$

Note that x comes out in tons, as it should.

Stoichiometry may be used to determine the extent to which a reaction has been completed. The idea is to find a relationship that contains the information given as well as the unknown extent of the reaction. The next example illustrates this situation.

EXAMPLE 2-9   **Partially completed reaction.** The iron ore siderite, $FeCO_3$ (ferrous carbonate), may be transformed into solid ferric oxide, Fe_2O_3, by heating in air:

$$4FeCO_3(s) + O_2(g) \longrightarrow 2Fe_2O_3(s) + 4CO_2(g) \tag{2-12}$$

A quantity of $FeCO_3$ was treated in this way, some of it being converted into Fe_2O_3 and CO_2 according to the equation above. The CO_2 escaped as a gas. The original sample showed a weight loss of 24.6 percent. What fraction of the $FeCO_3$ was converted into Fe_2O_3?

Solution. The first step is to analyze (2-12) to see why there should be any weight loss as the reaction proceeds, and then find the relation of this weight loss to the fraction of $FeCO_3$ that has reacted. One might at first suppose naively that if the weight loss is 24.6 percent, then 24.6 percent of the sample of $FeCO_3$ had reacted, but this ignores the fact that as it reacts it is converted into solid Fe_2O_3 that remains in the sample.

One approach to this problem is to suppose that *all* the $FeCO_3$ had decomposed and calculate what the loss in weight would then be. The loss in weight is proportional to the fraction of $FeCO_3$ that has decomposed, so the ratio of the actual loss to the loss that would have occurred if all had decomposed will give the fraction desired. Since no specified quantity of $FeCO_3$ has been given, it is permissible to work with a quantity that makes the calculation simple. Exactly 1 gfw $FeCO_3$ (FW = 115.86) would be converted, by (2-12), into just $\frac{1}{2}$ gfw Fe_2O_3 (FW = 159.70). Thus complete conversion would result in a weight loss of 115.86 g $- \frac{1}{2}(159.70$ g$) = 115.86$ g $- 79.85$ g $= 36.01$ g, or a fractional loss of $\frac{36.01}{115.86} = 0.3108$, or 31.08 percent. Since the actual percentage loss was 24.6 percent, the fraction decomposed must have been $\frac{24.6}{31.08} = \underline{0.792.}$

An alternative approach is to assume for convenience that the original sample of $FeCO_3$ weighed just 100.0 g. Since the final mixture weighed 24.6 percent less, it must have weighed 75.4 g. If we let x represent the grams of siderite that are converted into Fe_2O_3, the final mixture must contain 100.0 g $- x$ of $FeCO_3$ as well as the weight of Fe_2O_3 that would be formed from the conversion of $FeCO_3$. That weight is, by the argument of the preceding paragraph, $x(\frac{79.85}{115.86}) = 0.6892x$. Thus, expressing the weight of the mixture in terms of the unknown x, we have

$$75.4 \text{ g} = (100.0 \text{ g} - x) + 0.6892x$$

or

$$x = \frac{24.6}{0.3108} = 79.2 \text{ g}$$

Since the initial weight of $FeCO_3$ was 100.0 g, of which 79.2 g was converted to Fe_2O_3, the fraction converted is found to be, as before, $\underline{0.792.}$

The next example concerns a mixture of two compounds that have at least one element in common. As will be seen, it is possible to find the percentages of the two compounds once it is known how much of the common element the mixture contains.

EXAMPLE 2-10 **Mixture of oxides and consideration of error.** A mixture of the oxides CuO and Cu_2O (cupric and cuprous oxides) contains 13.4 ± 0.3 percent oxygen. The plus or minus quantity is an error estimate. What weight fraction of the mixture is Cu_2O and what is the error estimate of the answer?

Solution. Let p be the percentage of oxygen in the sample and q the percentage of Cu_2O. A 100-g sample of the mixture thus contains p g oxygen, distributed over the two oxides as follows: q g Cu_2O contains q (FW O/FW Cu_2O) g oxygen, and $(100 - q)$ g CuO contains $(100 - q)$(FW O/FW CuO) g oxygen. Hence

$$p = q \times \frac{\text{FW O}}{\text{FW Cu}_2\text{O}} + (100 - q)\frac{\text{FW O}}{\text{FW CuO}}$$

Solved for q, this gives

$$q = \frac{100 \times \dfrac{\text{FW O}}{\text{FW CuO}} - p}{\dfrac{\text{FW O}}{\text{FW CuO}} - \dfrac{\text{FW O}}{\text{FW Cu}_2\text{O}}} \qquad (2\text{-}13)$$

The formula weights of CuO and Cu_2O are 79.55 and 143.09, respectively, so that FW O/FW CuO = 0.2012 and FW O/FW Cu_2O = 0.1118. Inserted into (2-13), these lead to

$$q = \frac{20.12 - p}{0.2012 - 0.1118}$$

and with $p = 13.4$,

$$q = \frac{20.12 - 13.4}{0.0894} = \frac{6.7}{0.0894} = 75 \qquad (2\text{-}14)$$

Thus, the sample contains 75 percent Cu_2O (by weight). Note that in taking the difference in the numerator in (2-14), $20.12 - 13.4$, the result is properly expressed as 6.7 and not 6.72, because the precision given for p provides no information whatsoever about the second digit beyond the decimal point; if the known precision of p warranted its being expressed as 13.40, the zero would be explicitly written. In fact, we are told that there is an uncertainty of 0.3 in p; to find the error estimate in q corresponding to this stated uncertainty in p, we consider that if p were decreased or increased by 0.3, q would be changed by $\pm \frac{0.3}{0.0894} = 3$ percent. Combining this with the foregoing result, we see that $q = 75 \pm 3$ percent.

As mentioned earlier, analytical results or appropriate information about the products of reactions may be used directly to determine the empirical formula of a compound without working out its elemental composition as an intermediate step. This procedure is illustrated in the next example.

EXAMPLE 2-11   **Chemical formula from analytical results.**   A compound containing only P and S was analyzed by converting the sulfur to $BaSO_4$ and the phosphorus to $Mg_2P_2O_7$ (magnesium pyrophosphate). The weights of these solids were determined by weighing, and in a typical analysis the results were, per gram of compound, 3.1254 g $BaSO_4$ and 2.0443 g $Mg_2P_2O_7$. What is the empirical formula of the compound?

Solution.   Instead of computing the percentages of S and P in the compound and proceeding as in Example 2-5, we calculate the moles of S atoms and of P atoms in 1.000 g compound directly from the data given. Thus, because 1 gfw $BaSO_4$ (233.40 g) contains 1 mol S, 3.1254 g $BaSO_4$ contains the following number of moles of S:

$$\text{mol S} = 3.1254 \text{ g} \times \frac{1 \text{ mol}}{\text{gfw } BaSO_4} = \frac{3.1254 \text{ mol}}{233.40} = 1.339 \times 10^{-2} \text{ mol}$$

Similarly, 1 gfw $Mg_2P_2O_7$ (222.56 g) contains 2 mol P and thus the number of moles of P atoms contained in $Mg_2P_2O_7$ is

$$\text{mol P} = 2.0443 \text{ g} \times \frac{2 \text{ mol}}{\text{gfw } Mg_2P_2O_7} = \frac{2.0443 \times 2 \text{ mol}}{222.56} = 1.837 \times 10^{-2} \text{ mol}$$

The ratio of moles of P to S is thus $1837:1339 = 1.372$. The smallest integers with a ratio near this number are 4 and 3 ($4:3 = 1.333$). The integers 11 and 8 are in a ratio that is even closer to 1.372 ($11:8 = 1.375$), but these integers are suspiciously large, and it is likely that the difference between 1.333 and 1.372 is caused by experimental inaccuracy. The empirical formula would thus be $\underline{P_4S_3}$.

To check the accuracy of the analysis, it is useful to find the actual weights of P and S and to see whether they add up to the sample weight:

$$g\,S = 3.1254\,g \times \frac{FW\,S}{FW\,BaSO_4} = 3.1254\,g \times \frac{32.06}{233.40} = 0.4293\,g$$

$$g\,P = 2.0443\,g \times \frac{2(FW\,P)}{FW\,Mg_2P_2O_7} = 2.0443\,g \times \frac{2 \times 30.97}{222.56} = 0.5689\,g$$

$$\text{(sum of weights)} \quad \overline{0.9982\,g}$$

The weights of sulfur and phosphorus found by analysis do not quite add to the 1.000-g sample weight. Thus, the analysis is demonstrably not very accurate and the simpler interpretation of the analytical results, leading to the formula P_4S_3 (rather than $P_{11}S_8$), is the more reasonable. A clearer distinction between these two possibilities could be provided by more accurate analyses.

EXAMPLE 2-12   **Chemical formula by determination of quantities of products formed in combustion.**   An unknown compound contains only carbon, hydrogen, and oxygen. When 100.0 g of the compound is burned with an excess of oxygen, 39.2 g water and 95.6 g carbon dioxide are formed. What is the simplest formula of the unknown substance?

Solution.   When the compound reacts with oxygen to form CO_2 and H_2O, all the carbon and hydrogen must come from the compound itself. We can use the data given to calculate the numbers of moles of carbon atoms and hydrogen atoms in the original sample, in a manner analogous to that used to find the moles of phosphorus and sulfur atoms in the preceding example. However, to determine the oxygen content of the original sample we must also know the *weight* of the carbon and hydrogen in it; the weight of original compound not accounted for as C or H must be oxygen. The weight of oxygen is then converted to moles of oxygen atoms, and finally the empirical formula of the compound is found from the relative numbers of moles of hydrogen, carbon, and oxygen atoms that the sample contains.

The calculation of moles of hydrogen atoms and carbon atoms is summarized in the following table. The number of moles of hydrogen atoms is equal to twice the number of moles of water formed, while the number of moles of carbon atoms is equal to that of carbon dioxide. An excess of oxygen is assumed to be present during combustion, since if there were insufficient oxygen, some of the carbon would have been converted to CO rather than CO_2.

Element	Combustion product	Weight of combustion product	Moles of combustion product	Moles of atoms in 100.0-g sample	Weight of element
H	H_2O	39.2 g	$\frac{39.2}{18.02} = 2.18$	4.36 mol H	4.4 g
C	CO_2	95.6 g	$\frac{95.6}{44.01} = 2.17$	2.17 mol C	26.1 g

The 100.0-g sample contains a total of 30.5 g carbon and hydrogen, and thus must contain 69.5 g oxygen, the only other constituent. This corresponds to $\frac{69.5}{16.00} = 4.34$ mol oxygen atoms. Finally, then, we know that the sample contains C, H, and O atoms in the ratio 2.17:4.36:4.34; thus, the empirical formula of the compound is $\underline{CH_2O_2}$. The molecular formula may be the same, or any integral multiple of it. (The molecule might be, for example, formic acid, HCOOH.)

Partial knowledge of the composition of a compound may yield information about its formula weight, as shown by the following example.

EXAMPLE 2-13   **Formula weight and percentage of one component.** A compound of bromine contains 83.6 weight percent Br. Give possible values of the formula weight.

Solution. The formula of the compound is Br_nX, where n is an integer and X symbolizes all atoms other than Br. In terms of the atomic weight of Br, 79.9, the compound contains $100(79.9n)/FW$ percent Br. Equating this expression with the percentage given and solving for the formula weight yields $FW = \frac{7990}{83.6}n = \underline{95.6n}$. (The compound is $POBr_3$ with $FW = 286.7$ and $n = 3$.)

Problems and Questions

2-1   **Precision of calculations.** Express the results of the following arithmetic operations with the appropriate precision:
(a) 429×0.0020
(b) $3.7 \times 0.025 - 0.2370$
(c) $(4.30 + 0.1031)(101 - 88.1)$
(d) $(1037 - 952)/(0.2790 - 0.2687)$
(e) $(5300)(0.014000)$

Ans. (a) 0.86;   (b) −0.144;   (c) 57;   (d) 8.3×10^3;   (e) 74.20

2-2   **Simplest formula.** A certain oxide of phosphorus contains 56.3 percent oxygen and an even number of oxygen atoms. What is its simplest formula?

2-3   **Simplest formula.** A 10.0-g sample of a compound of C, H, and N contains 17.7 percent nitrogen and 3.8×10^{23} atoms of hydrogen. What is its simplest formula?   *Ans.* C_5H_5N

2-4   **Simplest formula.** A certain compound contains only carbon, hydrogen, and oxygen. If it contains 47.4 percent carbon and if there is one oxygen atom present for every four hydrogen atoms, what is its simplest formula?

2-5   **Simplest formula and molecular formula.** A certain solid compound contains 31.8 percent nitrogen and 13.6 percent hydrogen; the rest is carbon. (a) What is the simplest formula of this compound? (b) Suppose that you know that 2 mol of the compound contain 56 g nitrogen. What is the molecular formula of the compound?   *Ans.* (a) C_2H_6N;   (b) $C_4H_{12}N_2$

2-6   **Simplest formula and molecular formula.** A certain oxide of carbon is analyzed and found to contain 50.0 percent carbon. A very rough determination of the molecular weight shows it to be 300, within a possible error of about 10 percent. (a) What is the simplest formula of the oxide consistent with the analysis (without considering the molecular weight)? (*b*) What is the correct formula of the oxide?

2-7   Simplest formula. Water may be shown by analysis to consist of 11.2 percent hydrogen and 88.8 percent oxygen. Suppose that you believed that an atom of hydrogen weighed one-twelfth as much as an atom of oxygen. What would be the simplest formula for water consistent with these supposed relative weights and with the analytical data?   *Ans.* H_3O_2

2-8   Atomic weights from analytical data. Let X and Y represent two unknown elements that form a compound with oxygen of formula $X_8Y_4O_6$. If 53.6 g of this compound contains 19.2 g X and 7.2×10^{23} atoms of oxygen, what are the atomic weights of X and Y?

2-9   Formula weight from analytical data. A certain stable, nonvolatile chemical substance, X, may readily be crystallized from solution as a dihydrate $X \cdot 2H_2O$. When 5.00 g of the dihydrate is heated gently, it loses all its water and leaves a residue of 4.00 g pure X. What is the formula weight of X?   *Ans.* 144

2-10   Simplest formula from analytical data. When 24.0 g of an unknown gaseous compound containing only carbon and hydrogen was burned in excess oxygen, 36.0 g water and 73.4 g CO_2 were formed. What is the simplest formula of the compound? Show how the answer could have been obtained from the weight of only one product, water or CO_2.

2-11   Simplest formula and molecular formula. A certain compound, X, contains only C, H, and N. Analysis shows that it contains 60.0 percent carbon. When 2.00 g of the compound is burned, all the hydrogen is converted to water weighing 0.90 g. (*a*) What is the weight of each element in a 2.00-g sample of X? (*b*) What is the simplest formula of X? (*c*) Suppose an experiment shows that, within a precision of 5 percent, the molecular weight of X is 155. What is the molecular formula of X?
                    Ans. (*a*) 1.20 g C, 0.10 g H, 0.70 g N;   (*b*) C_2H_2N;   (*c*) $C_8H_8N_4$

2-12   Atomic weight from chemical composition. A certain metal forms a compound with bromine that contains 71.4 percent Br and 28.6 percent metal. What conclusions can be drawn about the atomic weight of the metal if it is assumed that one atom of the metal combines with *n* atoms of bromine, where *n* is some positive integer?

2-13   Atomic weight from chemical composition. An oxide of element X contains 24.2 percent by weight oxygen. (*a*) Find at least one possible value of the atomic weight of X. (It would be preferable to find the general set of values.) (*b*) The atomic number of X is 33. What is a likely value for its atomic weight?
                    Ans. (*a*) 50.1(*n*/*m*), where *n* and *m* are positive integers;   (*b*) 75.2

2-14   Rule of Dulong and Petit. The *rule of Dulong and Petit,* which holds approximately for solid elements with atomic weights above about 30, states that the heat capacity of a mole of atoms of a solid element is about 26 J mol^{-1} K^{-1}. (The heat capacity of a substance is defined as the energy that must be absorbed by the substance to raise its temperature by one kelvin.) This rule proved of great value during the first half of the nineteenth century since it permitted choosing correct atomic weights among the several that were possible on chemical grounds. For example, the chemical composition of a metallic oxide can be used to calculate the atomic weight of the metal only if the empirical formula of the oxide is known; different possible formulas yield different possible atomic weights for the metal that are related by the ratios of simple integers, such as $2:1$ or $3:2$. Thus a rule for choosing an approximate value of the atomic weight, such as that of Dulong and Petit, was of great help; it permitted choosing among different possible formulas and thus finding the precise atomic weight.

A pure red oxide of iron contains 69.9 weight percent iron and 30.1 percent oxygen. The *specific heat* of iron (i.e., the heat capacity per gram of iron) is $0.48 \text{ J g}^{-1} \text{ K}^{-1}$. Assume that the atomic weight of oxygen is known to be 16.00 and find the simplest formula of the oxide and the precise atomic weight of iron.

2-15   Rule of Dulong and Petit.   An element A has a heat capacity per gram of $1.02 \text{ J g}^{-1} \text{ K}^{-1}$. The chloride of A contains 25.53 percent A. With the help of the rule of Dulong and Petit (Prob. 2-14), calculate the precise atomic weight of A.   *Ans.* 24.31

2-16   Atomic weight from analytical data.   When 5.50 g of a certain metal M reacts with 1.50 g fluorine, a compound MF_3 is formed. (*a*) What is the atomic weight of M? (*b*) What weight of M is needed to prepare 2 mol of the fluoride MF_3?

2-17   Atomic weight and simplest formula.   A certain metal M forms a bromide of formula MBr_4. When 1.15 g MBr_4 is treated with an excess of silver nitrate, all the bromide is converted to AgBr weighing 1.88 g. (*a*) What is the atomic weight of M? (*b*) If M forms an oxide containing 85.4 percent M, what is the empirical formula of the oxide?

Ans.   (*a*) 1.4×10^2;   (*b*) M_2O_3

2-18   Composition of fertilizer.   A fertilizer contains 73 percent of its total nitrogen in the form of KNO_3 and 27 percent as $(NH_4)_2SO_4$. What is the weight ratio of the two salts in this fertilizer?

2-19   Quantities of reactants and products.   How many moles of H_3PO_4 are needed to convert 112 g CaO into $Ca_3(PO_4)_2$ if no other product containing calcium is formed?

Ans. 1.33 mol

2-20   Quantities of reactants and products.   How many moles of H_2SO_4 are needed to convert 51 g Al_2O_3 into $Al_2(SO_4)_3$ if no other product containing aluminum is formed?

2-21   Quantities of reactants and products.   Cyclopentane, C_5H_{10}, is a liquid with density 0.75 g ml^{-1}. How many moles of CO_2 can be formed by burning 200 ml cyclopentane with excess oxygen?   *Ans.* 10.7 mol

2-22   Quantities of reactants and products.   Ethylene bromide, $C_2H_4Br_2$, is a liquid with density 2.18 g ml^{-1}. At high temperatures all the bromine in the molecule will react with lead (Pb) to form $PbBr_4$. What is the maximum weight of $PbBr_4$ that can be formed from 20.0 ml $C_2H_4Br_2$?

2-23   Quantities of reactants and products.   Some fuel oils contain significant quantities of sulfur. When the oil is burned, the sulfur is oxidized to SO_2:

$$S + O_2 \longrightarrow SO_2$$

In a major city, 465 tons SO_2 are emitted by power plants each day. If the SO_2 comes from the combustion of fuel oil that contains 3.0 percent sulfur by weight, how many tons of fuel oil are burned per day?   *Ans.* 7.8×10^3 tons

2-24   Maximum yield of product.   Calculate the maximum amount of silver that could be obtained from 10.0 g of a mixture that contains 60 percent AgCl, 20 percent AgI, and 20 percent Na_2SO_4.

2-25 **Quantities of reactants and products: limiting reagent.** Magnesium reacts with oxygen to form MgO but does not react with neon. Suppose that 0.180 g magnesium is allowed to react completely with 0.250 g of a mixture of neon and oxygen that contains 60.0 percent oxygen by weight. Calculate the weight of each substance that will be present after reaction has occurred. *Ans.* 0.100 g Ne, 0.299 g MgO, 0.031 g O_2

2-26 **Proportions of reactants and products: limiting reagent.** Zinc and iodine form only one compound, ZnI_2. If 13 g zinc and 60 g iodine are sealed in a tube and heated until reaction is complete, what weights of what substances will be in the tube?

2-27 **Proportions of reactants and products.** Sodium nitrite, $NaNO_2$, can be prepared by the reduction of sodium nitrate with lead (Pb), the reaction being

$$NaNO_3 + Pb \longrightarrow NaNO_2 + PbO$$

Suppose that you wish to use a 10 percent excess of lead over the theoretical quantity needed. How much lead and how much sodium nitrate would you take to make 10.0 lb $NaNO_2$? *Ans.* 33.0 lb Pb; 12.3 lb $NaNO_3$

2-28 **Composition of a mixture.** Potassium hydrogen sulfate, $KHSO_4$, when heated strongly, forms potassium pyrosulfate, $K_2S_2O_7$, and water. The water is volatile and escapes. (*a*) Write a balanced equation for this reaction. (*b*) Calculate the percentage of $KHSO_4$ in a mixture if a 50-g sample of the mixture lost 1.8 g when heated strongly. Assume that no other reaction takes place and that no other components of the mixture are volatile.

2-29 **Composition of a mixture.** Ethanol, C_2H_5OH, reacts with oxygen by burning to form carbon dioxide and water according to the following equation:

$$C_2H_5OH + 3O_2 \longrightarrow 2CO_2 + 3H_2O$$

Sodium hydroxide, NaOH, reacts with CO_2 as follows:

$$2NaOH + CO_2 \longrightarrow H_2O + Na_2CO_3$$

When a 200-g sample of a mixture containing ethanol was burned with an excess of air and the resulting CO_2 was allowed to react with NaOH, 212 g Na_2CO_3 was formed. The mixture contained no other substance that could form Na_2CO_3. What was the percentage of ethanol in the mixture? *Ans.* 23 percent

2-30 **Compensation of weight loss.** To counterbalance the weight loss of fuel used in the operation of a dirigible, the water contained in the exhaust gases is partly condensed whereas the carbon dioxide is allowed to escape. Exactly what percentage of the H_2O produced need be condensed to achieve weight balance if the chemical compositon of the fuel is represented by the formula C_nH_{2n}, with n any (positive) integer?

2-31 **Mixture of silver halides.** A mixture of pure AgCl and pure AgBr contains 60.4 percent silver. What is the percentage of bromine? What is the error estimate of the answer if the error estimate of the percentage of silver is 0.1 (absolute) percent? *Ans.* 35.5 ± 0.2 percent

2-32   **Composition of a mixture.**   When calcium carbonate, $CaCO_3$, is heated to about 900°C, it decomposes to CaO and CO_2. The gaseous CO_2 escapes; the CaO is not further changed. Calcium sulfate, $CaSO_4$, is stable and unchanged under these conditions.

A mixture of $CaCO_3$ and $CaSO_4$ that initially weighed 300 tons was heated at 900°C until all the CO_2 had been driven off and the weight had become constant. The loss in weight was 99 tons. (*a*) What was the percentage of $CaCO_3$ in the original mixture? (*b*) What was the percentage of Ca in the final product?

2-33   **Composition of a mixture.**   A mixture of calcium carbonate and carbon is heated strongly in air. The calcium carbonate decomposes completely to calcium oxide and carbon dioxide, and the carbon is oxidized by the air to carbon dioxide. If the total weight of the carbon dioxide formed is equal to the weight of the original mixture, what is the percentage of carbon in the original mixture?   *Ans.* 17.3 percent

2-34   **Erroneous analysis.**   A certain mineral contains just 20.0 percent calcium. A student was given this mineral as an unknown and was asked to determine the content of calcium. The procedure used was to dissolve the sample, precipitate the calcium as calcium oxalate monohydrate, $CaC_2O_4 \cdot H_2O$, and heat this until it decomposed entirely to CaO, with attendant formation of CO, CO_2, and H_2O. The CaO was then to be weighed. However, the student did not heat the sample sufficiently and succeeded only in converting the salt to calcium carbonate; unfortunately he did not know this and assumed he was weighing the oxide. If he made no other errors, what was the percentage of Ca he reported?

"There are therefore Agents in Nature able to make the Particles of Bodies stick together by very strong Attractions. And it is the Business of Experimental Philosophy to find them out."

Isaac Newton, 1717

"We shall say that there is a chemical bond between two atoms or groups of atoms in case that the forces acting between them are such as to lead to the formation of an aggregate with sufficient stability to make it convenient for the chemist to consider it as an independent molecular species."

Linus Pauling,[1] 1939

3 Phase Changes and Interatomic Forces

We considered in Chap. 1 the three common states of matter—solid, liquid, and gaseous—and introduced the concepts of temperature and heat, as well as some fundamental ideas about atoms, molecules, and compounds. We continue that discussion in the present chapter, beginning with the transitions that occur among the solid, liquid, and gaseous states of a substance. This topic is followed by a consideration of interatomic and intermolecular forces, both the comparatively strong forces that lead to the formation of the aggregates referred to as molecules and compounds, and the weaker forces that are responsible, at low enough temperatures, for the condensation of all gases into liquids and solids.

[1]From Linus Pauling, "The Nature of the Chemical Bond", 3d ed., Cornell University Press, Ithaca, N.Y., 1960. Copyright 1939 and 1940, third edition © 1960, by Cornell University.

3-1   **Phase Transitions and Vapor Pressure**

Phase transitions.   At temperatures very close to absolute zero, all substances are solid, with the exception of helium, which remains liquid at the lowest temperatures unless sufficient pressure is applied (about 26 times normal atmospheric pressure). When the temperature of a solid is raised, the average energy of molecular motion increases, the molecules effectively take up more space as a consequence of this motion, and the solid expands. Eventually the motion becomes sufficiently extensive to destroy the close association that keeps the molecules near fixed positions, and the solid *melts.* For crystalline solids this transition occurs at a very definite temperature, the *melting point* (mp), while for amorphous solids the transition to the liquid state is gradual. It begins with a softening of the solid at a temperature that cannot be measured or specified precisely. As the temperature is raised, the solid gradually begins to flow as a thick liquid. With further increase in temperature the *viscosity* of the liquid, that is, its resistance to flow, usually decreases because the increasing agitation of the molecules more readily overcomes their mutual attractions.

Conversely, when the temperature of a liquid is lowered sufficiently it freezes. If the resulting solid is crystalline, the transition occurs at a well-defined temperature, the *freezing point,* identical with the melting point. The liquid may also turn into an amorphous solid, the transition taking place in a temperature interval. In this second case the resulting solids are often called *glasses,* because ordinary glass behaves in this way. Other examples of glasses are lava, pitch, resin, and many common plastics. The fact that the transition between glasses and their melts is continuous rather than sharp has been used as an argument to classify glasses as liquids rather than as solids—liquids with a viscosity so great that they do not change shape readily. The mechanical properties of many glasses are, however, so much like those of crystalline solids that it seems preferable to call them solids. For example, the breaking strength and elastic properties of fibers of ordinary glass make them competitive with steel in many applications.

It is possible to cool some liquids below their freezing points by cooling them in a smooth and absolutely clean vessel protected against vibrations. Water may be *supercooled* in this way to about $-20°C$. Jarring the vessel or introducing a speck of impurity causes sudden freezing. Crystallization is also aided greatly by impurities or wall irregularities on which the phase change may begin to take place. Best of all is the presence of some crystals of the actual substance to be crystallized; these are called *seed crystals,* and only a very few small ones are needed. The properties of a supercooled liquid are a smooth continuation of the properties the liquid possesses at higher temperatures. If freezing does not occur, there are no discontinuities when the temperature passes through the normal freezing point (see Figs. 3-5 and 3-6).

Many solid compounds can occur in several distinct crystalline forms, and transitions between these different solid phases can be observed under appropriate conditions. Transitions between solid phases are usually very slow.

Vapors and gases.   All liquids and solids are to some extent volatile; that is, they tend to evaporate to form gases.[1] For most solids, the pressure of the vapor so formed is extremely small at ordinary temperatures, but it is significant for some solids as it is for most liquids. The term *vapor* is frequently used to refer to a gas that can readily be condensed to a liquid or solid; thus, one usually speaks of water vapor, sodium vapor, naphthalene vapor, and so on. However, there is *no difference between a vapor and a gas;* every vapor is a gas and every gas can be condensed under appropriate conditions. It is merely an accident of our terrestrial environment that we happen to refer to gaseous water, sodium, and naphthalene as *vapors* and to gaseous oxygen and argon as *gases*. A key to the understanding of the relation between a liquid and its vapor lies in consideration of the pressure exerted by the vapor, a topic that will be considered after a brief discussion of pressure units.

Pressure units.   Pressure is defined as force per unit area. A device to measure the pressure of a gas is the manometer, shown in Fig. 3-1. The difference between the levels of the liquid in the two branches of the U-tube can be related to the pressure exerted by the gas. A liquid often used in the manometer is mercury, and

---

[1]As the temperature of a substance is raised, the individual motions of atoms *within* molecules become more vigorous and can become violent enough to break the molecule apart. For some substances this decomposition occurs below the boiling point or even below the melting point. Such substances may therefore decompose in the solid or liquid state before they vaporize significantly.

FIGURE 3-1   **Two manometers.** In either of the two versions shown, the U-shaped part of the glass tube contains a liquid, usually mercury, and the level difference h is an indication of the pressure of the gas in a vessel to which the left side of the manometer is connected. The *open* manometer (*a*) is open to the atmosphere on the right and measures the pressure difference between vessel and outside atmosphere. This difference is also called the *gauge pressure*. In the example shown, P_1 is greater than atmospheric pressure, the difference being measured by h_1.

In the *closed* manometer (*b*) the space above the liquid at the closed end is evacuated and the instrument shows the actual pressure; P_2 is thus measured directly by h_2. It is assumed that the pressure of the vapor of the manometer fluid in the evacuated space is negligible. This is a reasonable assumption for mercury at room temperature, for at 25°C the pressure exerted by the mercury vapor is only about 0.002 torr.

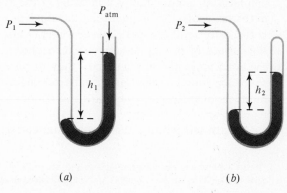

(*a*)   (*b*)

one millimeter of level difference (at 0°C and standard gravity) is used directly as a pressure unit. This unit, formerly called mmHg or just mm, is now called the *torr* after Torricelli, the inventor of the barometer. Another common pressure unit in chemistry is the *standard atmosphere* (atm), equal to 760 torr. It is the average pressure exerted by the atmosphere of the earth at sea level.

The pressure unit in the SI system is the newton per square meter ($N\,m^{-2}$ or $kg\,m^{-1}\,s^{-2}$) (Appendix A). To find its relation to the standard atmosphere, we consider the pressure exerted by a column of mercury exactly 760 mm (= 0.760 m) high at standard gravity (9.80665 $m\,s^{-2}$) and 0°C. The density of mercury at this temperature is 13.5995 $g\,cm^{-3}$ = 13.5995 $\times$ $10^3\,kg\,m^{-3}$. Thus,

$$1 \text{ atm} = 0.760000 \text{ m} \times 9.80665 \text{ m s}^{-2}\,(13.5995 \times 10^3\,kg\,m^{-3})$$

$$= 1.01325 \times 10^5\,kg\,m^2\,s^{-2}\,m^{-3} = 1.01325 \times 10^5\,N\,m^{-2}$$

(This is, in fact, the present *definition* of the standard atmosphere.) In the cgs (centimeter-gram-second) system the unit of pressure is the dyne per square centimeter ($dyn\,cm^{-2}$), equal to $0.10\,N\,m^{-2}$. Thus 1 atm = $1.01325 \times 10^6\,dyn\,cm^{-2}$. A pressure unit approximately equal to the atmosphere but more directly related to the newton per square meter is the bar: 1 bar = $10^5\,N\,m^{-2}$ = 0.987 atm.

A pressure of one atmosphere is very nearly equal to the force on an area of one square centimeter caused by the weight of one kilogram; that is, 1 atm corresponds to[1]

$$1.0332 \text{ kg cm}^{-2} = 1.0332 \times 10^4 \text{ kg m}^{-2}$$

A technical unit often used is the pound per square inch, $lb\,in^{-2}$, the pressure exerted by one pound on an area of one square inch at the earth's surface; 1 atm corresponds to[1]

$$14.696 \text{ lb in}^{-2} \approx 14.7 \text{ lb in}^{-2}$$

The vapor pressure of liquids.   To discuss the transition of a liquid into the vapor state, we consider a closed vessel containing only a liquid and its vapor (Fig. 3-2). The vessel is surrounded by a constant-temperature bath. As already emphasized, temperature is a measure of the average energy of the random motion of the molecules. Some molecules have just this average energy, but molecules with both greater and less than average energy are also present. Some molecules near the surface of the liquid always have sufficient energy to escape the attraction of the other molecules in the liquid—they *evaporate*. Conversely, vapor molecules that strike the liquid surface are usually retained; that is, they *condense*. When equal numbers of molecules evaporate and condense per unit time, we say that there is equilibrium.

At each temperature this equilibrium corresponds to a definite pressure $P_{vap}(T)$

---

[1]In the present context, kilogram and pound are units of weight (i.e., of the force of attraction by the earth) rather than of mass: $(1.0332 \times 10^4 \text{ kg}) (9.80665 \text{ m s}^{-2}) = 1.0132 \times 10^5 \text{ N}$.

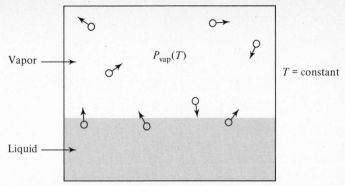

FIGURE 3-2 **Liquid and vapor in closed container.** A closed vessel is partly filled by a liquid, and the temperature is maintained at a value T by a surrounding bath of constant temperature. Given enough liquid and sufficient time, a vapor phase establishes itself above the liquid, with a pressure that is a property of the liquid and depends upon the temperature. Molecules in the liquid and near the surface may break loose and escape into the vapor phase if they have sufficient energy. Molecules in the vapor phase that strike the liquid surface may be captured and stay with the liquid. Another gas may be present in the vapor space without affecting the liquid-vapor equilibrium.

of the vapor.[1] It is called the *vapor pressure* of the liquid, and it increases with increasing temperature. When the pressure of the vapor phase is below this equilibrium value, more molecules evaporate than condense, continually increasing the pressure in the gas phase. When the gas pressure is above $P_{vap}(T)$, more molecules condense than evaporate, thus lowering the gas pressure. Any heat liberated or absorbed during such condensation or evaporation is taken up or furnished by the constant-temperature bath. At equilibrium, the rates of evaporation and condensation are in balance and no further net change occurs; all visible aspects of condensation and evaporation cease. This balance of opposing rates is the essential feature of all *dynamic* equilibria, of which this is a typical example.

The system illustrated in Fig. 3-2 is at constant volume and temperature. What happens if the volume of the vapor space is changed slowly while the pressure is kept substantially constant? This situation is discussed in Fig. 3-3. Because of the dynamic nature of the equilibrium between the liquid and its vapor, the system adjusts to the volume change (as long as both phases are present) while maintaining the pressure of the vapor at the equilibrium value $P_{vap}(T)$ that corresponds to the temperature of the bath.

The vapor pressure of a pure liquid does *not* depend on the amount of liquid or vapor present, nor on the area of the interface between the two phases. It is necessary, however, that some of each phase be present to ensure that there is an equilibrium between them. The vapor pressure is an inherent property of the liquid.

Figure 3-4 shows some typical curves representing the vapor pressure as a function of temperature. Of course, the concept of vapor pressure is not tied to the presence

[1] We use the symbol T in parentheses here to stress the fact that the vapor pressure is strongly dependent on the temperature.

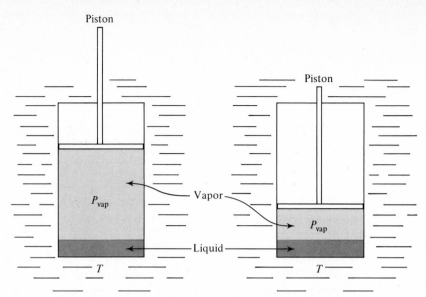

FIGURE 3-3 **Liquid and vapor at constant pressure.** A cylinder is partly filled by liquid, and the remaining volume is taken up by the vapor of the same substance. The temperature is maintained by an external bath at a constant value T, and a piston that forms a tight seal exerts a constant pressure P. If P is larger than the vapor pressure $P_{vap}(T)$, vapor keeps condensing and the piston moves down until it is in direct contact with the liquid. If P is less than $P_{vap}(T)$, liquid keeps evaporating and the piston moves up until all liquid is gone (provided the cylinder is large enough). If P is equal to $P_{vap}(T)$, a state of dynamic equilibrium exists no matter what the position of the piston is, and the piston does not move in or out. If the force on the piston is increased or decreased very slightly, the piston can be slowly moved in or out and the system adjusts by evaporation of liquid or condensation of vapor, the pressure remaining constant (except for the small temporary decrement or increment of pressure required for the actual motion of the piston). These considerations apply as well to an equilibrium between a solid and its vapor.

of a closed vessel. In a vessel open to the outside, some vapor molecules escape continually and an equilibrium can no longer be established—the liquid keeps evaporating. When the temperature of the liquid is raised to the point where the vapor pressure equals the outside pressure, vapor bubbles begin to form and the liquid boils. The vapor now has sufficient pressure to push the air away. The *normal boiling point* (bp) of a liquid is defined as the temperature at which its vapor pressure equals one atmosphere (760 torr; Fig. 3-4).

If air or other gases are also present in the vapor phase, the total pressure P_{tot} in the gas phase is the sum of two contributions: the pressure caused by the vapor molecules and that due to the other gas molecules. The individual contributions of the different gases are called their *partial pressures* (see Chap. 5 for a more precise definition).

In the presence of other (nonreacting) gases the equilibrium between a liquid and its vapor leads to the same vapor pressure as in the absence of such gases, but $P_{vap}(T)$ is now a partial pressure rather than the total pressure. When the partial pressure of the vapor is less than the vapor pressure P_{vap} of the liquid, evaporation exceeds condensation; if it is larger than P_{vap} (e.g., because the temperature has just been lowered), condensation exceeds evaporation. Equilibrium exists when the partial pressure of the vapor in the gas phase equals P_{vap}.

The vapor pressure of solids. A solid also has a vapor pressure, as mentioned earlier. Usually there are molecules near the surface of a solid that have sufficient energy to escape and form a vapor phase, and in a closed container a dynamic equilibrium is achieved. The vapor pressure of a solid is an inherent property, depending only on the temperature and the kind of substance. It increases with the temperature in a manner also analogous to that for the vapor pressure of a liquid. A solid may thus evaporate directly, without melting, a process called *sublimation*. For many solids the vapor pressure is so small at normal temperatures that effectively no evaporation takes place. However, some solids have measurable vapor pressures, that of ice at 0°C being 4.6 torr or about 0.006 atm. Although this might at first seem a very low pressure, those who are familiar with cold climates know that snow often disappears from the ground and frozen wash on the clothesline dries by sublimation even when the temperature never rises above the melting point of ice.

A few solids have vapor pressures that reach 1 atm at temperatures below their melting points. The temperature for which the vapor pressure of a solid is 1 atm is called the normal sublimation point. At this temperature, the vapor has sufficient pressure to push away the surrounding atmosphere, but of course evaporation of such a solid also occurs below this temperature, just as it does for a liquid below

FIGURE 3-4 **Vapor-pressure curves for ethyl ether, water, and mercury.** Note the increase of the vapor pressure with temperature, which becomes ever more substantial as the temperature increases. The abbreviation bp stands for the normal boiling point, i.e., the boiling temperature at 1 atm pressure. The quantities P_{vap}/torr and $t/°C$ are dimensionless, and, for example, $P_{vap}/\text{torr} = 500$ implies that $P_{vap} = 500$ torr. The axes of graphs are usually labeled by such dimensionless quantities, as are entries in tables.

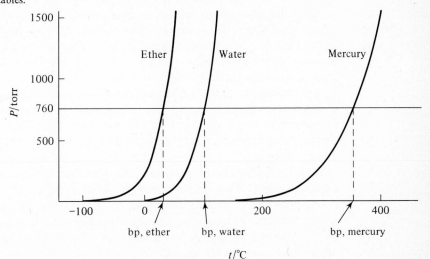

its normal boiling point. Solid carbon dioxide (dry ice) is an example of a substance that normally sublimes rather than melts. Its sublimation point is $-78\,^{\circ}\mathrm{C}$.

An important distinction.   It is essential to recognize that, in the absence of the liquid or solid, the pressure exerted by the molecules of a vapor at a given temperature T may have any value from zero up to the vapor pressure at that temperature, $P_{\mathrm{vap}}(T)$. As has been stressed, the vapor pressure is a property characteristic of an equilibrium between a vapor and some other phase (or phases) at a particular temperature. In the presence of vapor alone or under nonequilibrium conditions, the *pressure of the vapor* of any substance is not necessarily the same as, and is usually less than, the *vapor pressure*. For example, on a relatively dry day at (say) $25\,^{\circ}\mathrm{C}$ ($77\,^{\circ}\mathrm{F}$), the pressure of water vapor in the atmosphere may be only 10 torr while the vapor pressure of water at that temperature is 24 torr, so that the relative humidity is $\frac{10}{24}(100) = 42$ percent. When the relative humidity is very high, the pressure of water vapor in the atmosphere approaches the vapor pressure. The pressure of the vapor may also be above the vapor pressure; such vapor is called supercooled or supersaturated and eventually condenses, as we discuss later.

Stability of phases.   When equilibrium exists between any two condensed phases, for example, between a liquid and a solid phase, the vapor pressure of the two phases must be identical. Otherwise the phase with the higher vapor pressure would evaporate continually and condense into the other phase and no equilibrium would be reached. In the end only the phase with the lower vapor pressure would exist. These considerations apply equally well when there are two different solid phases in equilibrium with each other, as happens under appropriate conditions for a number of substances (for example, different forms of ice under high pressure, or sulfur at about $96\,^{\circ}\mathrm{C}$ and 1 atm).

When more than one condensed phase of a substance can exist at a given temperature and the phases are not in equilibrium with each other, that is, they do not have the same vapor pressure, then the phase with the lower or lowest vapor pressure is called the *stable* phase, all others being termed *metastable* phases. For example, supercooled water (that is, liquid water below $0\,^{\circ}\mathrm{C}$) has a higher vapor pressure than ice at the same temperature (see Fig. 3-5). The term metastable implies that the unstable phase exists long enough to be observed, because the rate at which it changes into the stable phase is very low.

The boiling of liquids is often associated with a delay effect, *superheating*, similar to the supercooling of liquids. The smooth boiling of liquids is enhanced by specks of impurities or rough spots on the container walls, both of which further the formation of vapor bubbles. In the absence of either, vapor bubbles may not form at once, although the boiling point has been reached and even passed. If the temperature is further increased, however, vapor bubbles may form suddenly with almost explosive vigor. This phenomenon is called bumping; it can be minimized by avoiding excessive heating rates and by making the formation of vapor bubbles easier. The latter objective is often accomplished in the laboratory by adding "boiling chips"—porous, unreactive materials containing many air-filled cavities, which can act as nuclei for bubble formation.

Conversely, vapor may be supercooled in the absence of suspended solid or liquid particles, the pressure of the vapor being above the vapor pressure. If a charged high-energy particle traverses such supercooled vapor, electrons are knocked off the molecules in its path and the vapor tends to condense in tiny droplets on the ions created. This phenomenon is used in the *Wilson cloud chamber* to make the tracks of high-energy particles visible. In a similar way the modern *bubble chamber* is based on the fact that in a superheated liquid tiny vapor bubbles tend to form on the ions created along the path of the particle, which is thus made visible by the trail of bubbles formed in its wake (Fig. 8-1).

Phase diagrams for pure substances. The ranges of temperature and pressure in which the different phases of a pure substance are stable may be represented graphically by a *phase diagram*. Each substance has its own characteristic diagram; those for water and carbon dioxide are shown in Figs. 3-5 and 3-6. These diagrams show whether the most stable state at any specified temperature and pressure is a gas, a liquid, or a solid, or whether several phases may be equally

FIGURE 3-5 **Schematic phase diagram for water.** This diagram describes the stable states of water at various combinations of temperature (T) and pressure (P), and makes it possible to predict what changes will occur as T and P are changed from any initial conditions. Consider, for example, point 1, corresponding to conditions ($-25°C$, 30 torr) under which water is solid (ice). Suppose that T is increased at constant P; this change in conditions corresponds to moving horizontally toward the right in the diagram. When the boundary between the solid and liquid phases is crossed, at point 2, the ice is transformed into liquid water. Similarly, at point 3 the boundary between the liquid and gaseous phases is crossed and the water is vaporized. The negative slope (exaggerated in the diagram) of the liquid-solid boundary is unusual and indicates that a pressure increase causes a lowering of the melting point. This is a consequence of the fact that the density of liquid water is greater than that of ice.

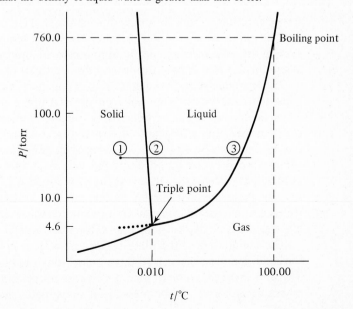

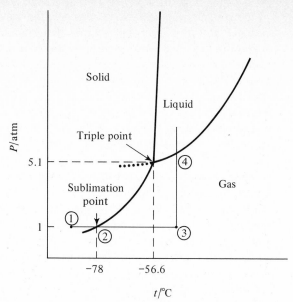

FIGURE 3-6   **Schematic phase diagram for carbon dioxide.** A noteworthy aspect of this diagram is that the triple point lies above atmospheric pressure. It takes at least 5.1 atm to liquefy carbon dioxide, and at atmospheric pressure only solid and gaseous carbon dioxide can exist. For example, when carbon dioxide at $-90°C$ and 1 atm (point 1) is heated at constant pressure, it sublimes at $-78°C$ (point 2). Gaseous carbon dioxide at $-50°C$ and 1 atm (point 3) may be liquefied at constant temperature by an increase in pressure (point 4). The positive slope (exaggerated here) of the liquid-solid boundary corresponds to an increase of melting temperature with increasing pressure, a much more common circumstance than that depicted in Fig. 3-5.

stable and thus in equilibrium with each other. The area in which the gaseous state is most stable extends to low pressures and high temperatures, whereas that representing the solid state extends to high pressures and low temperatures. The stability range for the liquid state forms a wedge between the other two regions.

The boundary line between any two areas gives the temperatures and pressures consistent with equilibrium between the two phases corresponding to these areas. Thus, the boundary line between the liquid and gaseous regions is the vapor-pressure curve for the liquid, like those shown in Fig. 3-4, and the boundary curve between the solid and gaseous regions is the vapor-pressure curve for the solid. The equilibrium curve between liquid and solid is almost vertical; the deviations from the vertical have been exaggerated in Figs. 3-5 and 3-6 to make them perceptible. This steep slope means that melting temperatures (corresponding to solid-liquid equilibrium) are not greatly affected by pressure changes; this is true not only for water and carbon dioxide, but for other substances as well.

The boundaries of the three regions meet at a point; this *triple point* indicates the conditions under which all three phases may exist in equilibrium with each other. The dotted continuation of the liquid-vapor curve into the area of the solid (in each diagram) shows the vapor pressure of the supercooled liquid. The vapor pressure of a supercooled liquid is greater than that of the corresponding crystalline solid because the solid is the more stable phase below the triple point. There is no corresponding extension of the vapor-pressure curve for the solid above the triple point; pure crystals always melt sharply at the melting point and cannot be super-heated.

Energy effects accompanying phase changes. When a liquid or solid evaporates, energy is always absorbed. Unless energy is supplied rapidly enough from outside the evaporating substance, the energy must come from the substance and hence its temperature decreases. Conversely, when a vapor condenses to a liquid or solid, energy is released and the temperature of the system increases unless the energy is passed on to the surroundings. These thermal effects are of enormous practical importance—for example, in the mechanism of cooling of organisms by evaporation and in the large-scale meteorological phenomena accompanying atmospheric storms in which much water vapor condenses as rain.

Energy is always absorbed when a solid melts ("fuses") and is always released when a liquid freezes. Heats of fusion and heats of vaporization are considered further in Chap. 16.

3-2 Interatomic and Intermolecular Forces

Background. The existence of condensed phases suggests that there are attractive forces between molecules. Similarly, it follows from the existence of distinct chemical compounds containing definite proportions of different kinds of atoms that there are forces holding these atoms together in stable aggregates. This led nineteenth-century chemists to the idea of valency, an expression of the number of other atoms with which a given atom can combine in a given compound. Much of the chemical research during that century was devoted to establishing the formulas of compounds in an effort to understand the tendency of atoms to combine with one another, and many theories were evolved about the nature of the bonds—i.e., the forces—between atoms and thus the nature of combining power or valency. By the end of the century it was recognized that some substances (for example, sodium chloride) were composed of oppositely charged ions with mutual electrostatic attraction, but it was only after the development of the Rutherford-Bohr model of the atom that a clear idea of the nature of the forces in nonionic compounds began to develop, initially through the work of G. N. Lewis.

One of the great triumphs of nineteenth-century chemists was the recognition of the fact that when the elements are arranged in order of increasing atomic weight (or better, as we now know, atomic number), there are regular (*periodic*) recurrences of remarkably similar chemical and physical properties. This periodic classification and the structures of atoms are discussed in detail in Chap. 10, but some of the simpler conclusions about the relation of the electronic structure of atoms and the periodicity of chemical properties, which you perhaps recall from an earlier chemistry course, are used as a preliminary to the present discussion of types of compounds.

A modern form of the periodic table is given inside the front cover; the columns of that table include elements with closely similar chemical properties.

During the last half of the nineteenth century and the early years of this one, extensive studies of the light absorbed and emitted by atoms (Chap. 8) gave detailed information about the relative energies associated with the electrons in an atom. It was found that atomic electrons can be grouped according to the energy needed to remove them. The removal of an electron from a neutral atom results in the formation of a positively charged ion and hence this process is called ionization. The electrons for which the energy of ionization is relatively small in any particular atom are referred to as *outer* electrons (a term whose spatial connotations are in some situations misleading). It was also noticed that atoms whose outer electrons are hardest to remove (helium, neon, argon, and the other "inert gases") are the least reactive elements, while those elements whose outer electrons are most easily removed (lithium, sodium, potassium, and the other *alkali metals*) are among the most reactive. Similarly, for other atoms that occupy a given column of the periodic table and therefore have closely related chemical properties, the energies required for removal of the outer electrons are comparable. These and many other observations led to a picture of *chemical interaction* involving the *outer electrons* of atoms and to a cardinal generalization that there are certain particularly stable arrangements of these electrons that are found again and again. These arrangements correspond to those of the inert gases and are called *completed shells* or *closed shells*.

Chemical and physical interactions.   As our understanding of the nature of the forces between atoms and molecules has increased, the term *valence* has generally been abandoned as an expression of the specific "combining power" of individual atoms because the interactions of atoms are too varied and subtle to be described with one general term. The word *valence* is still used, however, to refer to the general phenomenon of chemical affinity, the power of atoms to form compounds.

The forces that hold atoms together in molecules, and molecules together in larger aggregates, are all electrical in origin, but they can be fully understood only with the aid of quantum mechanics. These forces vary widely in strength; some are referred to as chemical bonds or chemical interactions, others as physical interactions. The distinction is, however, somewhat artificial, and because it is not sharply defined, there are some differences in usage. We use the term *chemical interaction,* or *chemical bond,* to refer to an interaction leading to a molecular or ionic species that persists sufficiently long to be detected, at least by instrumental methods. These interactions, discussed at length below and in later chapters, include ionic and covalent bonds, hydrogen bonds, and some dipole interactions. Attractive interactions that do not lead to species that persist long enough for detection are termed *physical*. The distinction between chemical and physical interactions is to some extent a reflection of the character of our normal terrestrial environment, particularly the temperature. At high temperatures, for example, many interactions normally regarded as chemical become ineffective compared to the prevailing high thermal energy, and familiar compounds may decompose. The fragments formed may themselves be stable at these high temperatures although they would react with each other extremely rapidly at ordinary temperatures. When discussing properties of substances it is thus often convenient to refer to *normal conditions,* which imply a temperature not far from 20°C and a pressure of about 1 atm. They must not be confused with standard conditions, which refer to much more precisely prescribed circumstances. Several sets of standard conditions are used in later chapters.

The forces that hold together atoms and groups of atoms in the aggregates we refer to as compounds are called *chemical bonds.* There are two types of strong chemical bonds, the ionic bond and the covalent bond. Typical energies of interaction—energies required to separate atoms held together by such bonds—are in the range 150 to 400 or more kJ mol^{-1}. A much weaker but very important chemical bond is the hydrogen bond, with an interaction energy of some 5 to 40 kJ mol^{-1}.

Ionic bonds. The force of attraction between oppositely charged ions, that is, anions and cations,[1] is commonly referred to as the *ionic bond.* This force is due to the mutual electrostatic (coulomb) attraction of the opposite charges (Appendix A). Consequently the energy ϵ of such a bond varies directly as the product of the charges $\mathbf{Q}_1$ and $\mathbf{Q}_2$ of the ions concerned, and inversely as the distance r between their centers:[2]

$$\epsilon = \frac{\mathbf{Q}_1\mathbf{Q}_2}{\alpha r} \tag{3-1}$$

Since the distance between an anion and a cation is approximately equal to the sum of their radii, the energy required to break a bond between two small ions is considerably greater than that required to break a bond between two larger ions with charges of the same magnitude. For example, the energy of the bond between Li$^+$ and F$^-$, two relatively small ions, is about 570 kJ mol^{-1}, which is about 70 percent larger than the energy of the bond between the significantly larger Cs$^+$ and I$^-$ ions, about 340 kJ mol^{-1}.

Oppositely charged ions may form ionic crystals, in which the ions are packed together in highly regular three-dimensional arrays that are held together chiefly by ionic bonds. For example, crystals of sodium chloride contain equal numbers of Na$^+$ and Cl$^-$ ions in a regular pattern (Fig. 3-7) in which each Na$^+$ has six Cl$^-$ neighbors and each Cl$^-$ has six Na$^+$ neighbors. While electrostatic forces are attractive between ions of opposite charge, they are repulsive between ions of the same charge. Thus, in the sodium chloride crystal the Na$^+$ ions repel each other and the Cl$^-$ also repel each other. In the interlaced arrangement of Na$^+$ ions and Cl$^-$ ions, however, the sum of all attractive forces is larger than that of the repulsive forces, so that the solid is stable.

In addition to these electrostatic forces that cause long-range attractions and

---

[1] The names *anion* for a *negatively* charged ion and *cation* for a *positively* charged ion are derived from the names of the electrodes to which the respective ions move in an electrolysis cell (Fig. 1-7). Anions are attracted to the positive electrode, the anode, and cations to the negative electrode, the cathode.

[2] The α has to do with units. In the SI system the units for r are meters (m) and those for $\mathbf{Q}_1$ and $\mathbf{Q}_2$ are coulombs (C), which with $\alpha = 1.113 \times 10^{-10}$ C^2 J^{-1} m^{-1} yields joules (J) as energy units (Appendix A).

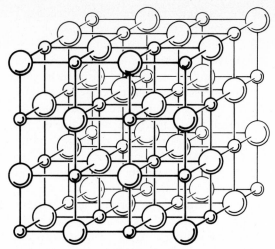

FIGURE 3-7 **The arrangement of sodium and chloride ions in a crystal of sodium chloride.**
The larger spheres represent Cl⁻, the smaller ones Na⁺. This is a skeletal rather than a packing model of the structure (the contrast between these two ways of representing structures is shown in Fig. 1-4). The only purpose of the lines connecting the spheres is to aid in the visualization of geometric relationship, and they must not be mistaken for covalent bonds. The radii of the spheres representing the ions have been made relatively small so that the arrangement of the ions in the structure can be visualized. This arrangement is harder to see in a packing model, such as that in Table 4-1.

repulsions, there are also forces of repulsion that become noticeable only when the distance between the two ions (or indeed two neutral atoms) becomes less than a critical value (Fig. 3-8). These forces increase rapidly the closer the ions get pushed together. Without this repulsion ionic crystals would collapse.

It is important to note that sodium chloride crystals do not contain individual NaCl molecules, but only Na⁺ and Cl⁻ ions. Similarly, a melt or aqueous solution of sodium chloride contains only Na⁺ and Cl⁻ ions, and no NaCl molecules. Only NaCl vapor, existing at high temperatures, contains NaCl "molecules", a "molecule" actually being a tightly associated combination of an Na⁺ and a Cl⁻ ion, more appropriately called an *ion pair*. It might thus be better to write Na⁺Cl⁻ rather than NaCl, but the second of these forms is preferred for simplicity.

Sometimes it is asserted that an ionic compound is formed by the transfer of electrons from one atom to another. This is indeed one possibility, but ionic compounds can also be formed by starting with ions that already exist, for example by crystallization from a solution containing ions. Thus there is no necessity that electrons be transferred during the formation of an ionic bond.

The names and formulas of many common ions are given in Appendix B.

Covalent bonds. Uncharged atoms with incomplete outer electron shells often attract one another strongly. The chemical bond formed under these circumstances results from the *sharing* of one or more pairs of electrons by the two atoms, usually in such a fashion that each atom then has a completed outer electron shell (counting the shared electrons with each atom). The forces involved are electrostatic in origin,

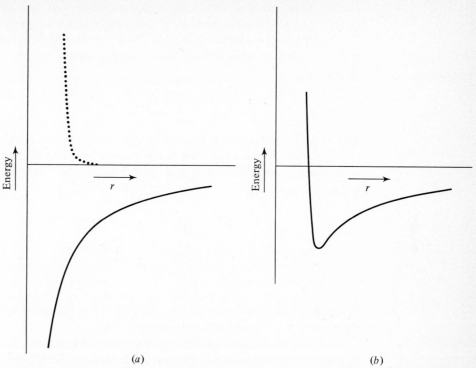

(a) (b)

FIGURE 3-8   **Interaction energies of oppositely charged ions as a function of the distance between their centers** (a) Electrostatic energy of attraction (solid) and energy of repulsion (dotted); (b) net interaction energy, the sum of the two curves in (a).

The lower curve in (a) is a hyperbola, represented analytically by Equation (3-1). The energy varies inversely as $1/r$ and becomes increasingly negative with decreasing r because the ions have charges of opposite sign. The slope of this curve is everywhere positive. Zero energy is associated with an infinite distance between charges. The upper curve in (a) has a negative slope everywhere and becomes extremely steep at short distances.

Curve (b) represents the actual interaction energy of a pair of ions such as Na^+ and Cl^-. The force acting on an ion is in the direction of downward slope, i.e., in the direction of lower potential energy. In the absence of kinetic energy, the minimum of the potential energy curve represents a stable state. At this distance, attraction and repulsion just balance—the force acting on the ion is zero. This separation corresponds to the average distance (about which the ions would oscillate) in an ion pair, Na^+Cl^-. This is not quite the same as the average distance in an NaCl crystal because of the modifying effects of the other surrounding ions.

but the principle of sharing electrons can be fully understood only with the help of quantum mechanics, and further discussion is deferred until Chaps. 11 and 12.

A shared pair of electrons constitutes a *covalent bond*. In diagrams depicting the structure of molecules, each shared pair of electrons in a covalent bond is conventionally represented by a line between the two bonded atoms, for example, H—H or H—Cl. There may be two or three such pairs shared between two atoms, giving rise to what is called a *double* or *triple* bond. The *unshared* or *lone* electron pairs on each atom are often of chemical importance as well, because they can partake in chemical reactions to form new covalent bonds with approaching atoms and because they have a strong influence on the geometric arrangement of molecules. It is customary to represent unshared outer electrons by dots, as in the formulas

$$\text{H}-\overset{\cdot\cdot}{\underset{|}{\text{O}}}: \quad \text{H}-\overset{\cdot\cdot}{\underset{\cdot\cdot}{\text{F}}}: \quad \text{H}-\overset{\cdot\cdot}{\underset{|}{\text{N}}}-\text{H} \quad :\overset{\cdot\cdot}{\text{O}}=\text{C}=\overset{\cdot\cdot}{\text{O}}: \quad \text{H}-\text{C}\equiv\text{N}: \tag{3-2}$$

$$\overset{|}{\text{H}} \qquad\qquad\qquad\qquad \overset{|}{\text{H}}$$

The bonds that hold together the atoms in polyatomic ions are predominantly covalent also—for example, the bonds between the nitrogen and hydrogen atoms in the ammonium ion, NH_4^+, and those between the sulfur and oxygen atoms in the sulfate ion, SO_4^{2-}:

$$\left[\begin{array}{c} \text{H} \\ | \\ \text{H}-\text{N}-\text{H} \\ | \\ \text{H} \end{array}\right]^+ \qquad \left[\begin{array}{c} :\overset{\cdot\cdot}{\text{O}}: \\ | \\ :\overset{\cdot\cdot}{\text{O}}-\text{S}-\overset{\cdot\cdot}{\text{O}}: \\ | \\ :\overset{\cdot\cdot}{\text{O}}: \end{array}\right]^{2-} \tag{3-3}$$

Formulas of the kind illustrated in (3-2) and (3-3) are called electron-dot or Lewis formulas and are explained more fully in Chap. 11.

Ionic and covalent bonds represent the two simplest and most important kinds of chemical forces. A number of contrasts between them are worthy of emphasis. Because the ionic bond is the result of simple electrostatic attraction between charged particles, it is nonspecific; that is, it does not depend on the chemical nature of the ions involved but only on their charges and the distance between them. There is no limit to the number of such bonds a given ion can form, provided only that there is room around it for all the ions of opposite charge.

On the other hand, covalent bonds are highly specific. They vary considerably in strength and in other respects with the detailed nature of the atoms involved and even to some extent with the nature of the other bonds formed by these atoms. Furthermore, covalent bonds are highly directional; if two atoms are bonded co-valently to a third atom, the angle between them depends strongly, in a quite predictable way, on the electronic structure of the atom to which they are bonded. In contrast, since the electric field around a spherical ion is spherically symmetrical, ionic bonds are not directional. Finally, the number of covalent bonds that can be formed by a given atom is limited. For example, hydrogen normally forms only one covalent bond and carbon normally forms no more (nor less) than four. Although there are some apparent exceptions to this generalization, most notably among "electron-deficient compounds", it is valid for the great majority of covalently bonded substances. This limit to the number of covalent bonds an atom can form is some-times described by saying that the covalency of the atom can be saturated. This is in marked contrast to the situation with ionic bonding.

Pure ionic and pure covalent bonds represent ideal extremes. Most ionic bonds (e.g., the Na^+Cl^- bonds in sodium chloride crystals) have some covalent character, and most covalent bonds (e.g., the H-O bonds in H_2O) have some ionic character.

One speaks of *partial ionic character* of a covalent bond in this connection, pure covalent and pure ionic bonds corresponding, respectively, to 0 and 100 percent partial ionic characters. The partial ionic character of a bond is largely governed by the *electronegativity* of the two atoms bonded. This term refers to the tendency of a bonded atom to attract electrons from other atoms in the same molecule or ion. A large electronegativity difference between two atoms causes a large partial ionic character in a bond between them, the more electronegative atom acquiring some negative charge and the other a positive charge of equal magnitude (unless additional circumstances further modify these charges). When atoms have equal or nearly equal electronegativities, bonds between them are almost purely covalent. This topic is taken up in more detail in Chap. 11.

Van der Waals forces. In addition to ionic and covalent bonds, two other types of forces are of great importance in the chemical behavior of molecules. These are van der Waals forces, which are nonspecific, and hydrogen bonds, which are sufficiently specific and directional in character that they are often termed chemical bonds (and are so regarded in this text), although they are far weaker than most ionic and covalent bonds.

Weak attractive forces exist between uncharged atoms or molecules even in the absence of any tendency for formation of covalent bonds. Perhaps the most obvious evidence is the fact that all gases condense to liquids and eventually solidify at sufficiently low temperature and sufficiently high pressure. Accurate measurements show that even in the gaseous state all substances behave in a way that indicates the presence of weak intermolecular forces (Chap. 5).

These weak forces have long been known collectively as *van der Waals forces* because the Dutch physicist J. D. van der Waals was the first to obtain an estimate of their magnitude in gases by modification of the ideal gas law (Chap. 5). Although hydrogen bonds are just one specific kind of weak attractive force, they are sufficiently unique that modern usage places them in a special category that is not included under the general heading van der Waals forces.

One type of van der Waals force arises when the center of the positive charges of a molecule does not coincide with the center of the negative charges. Such molecules are said to have electric dipole moments (Fig. 3-9). The forces between molecules with permanent dipoles (*polar* molecules) depend strongly on the relative orientations and positions of these molecules (Fig. 3-10). They vary as the inverse fourth power of the distance between the dipoles and are thus of very short range compared to the range of electrostatic forces between ions. The dipolar forces play an important role in determining the packing of ordered arrays of molecules in crystals. They also contribute to the physical interactions in liquids and gases, but in these cases (and in some solids) the relative orientation of the dipoles is largely random, and this leads to mutual repulsion as well as attraction. The balance of attraction and repulsion can be shown to amount to a weakly attractive and very short-range force, proportional to the inverse seventh power of the distance between the dipoles. The energy of attraction varies as the inverse sixth power of this distance.

Even individual atoms and molecules with no permanent dipoles or other permanent charge distributions are subject to interactions quite apart from covalent bond

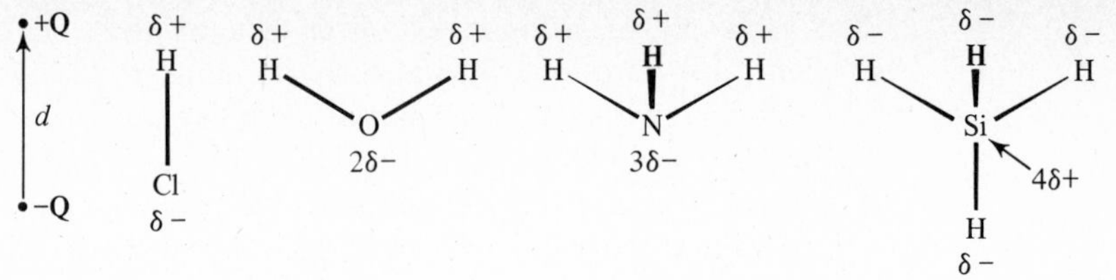

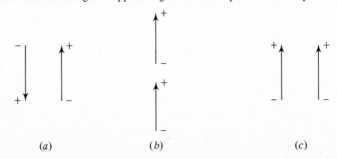

FIGURE 3-9   **Dipoles.** A simple dipole (left) and several molecular examples are shown. The symbols $\delta+$ and $\delta-$ indicate positive and negative charges associated with the atoms in question. They are equal to each other in magnitude for any given molecule here, but they may differ from molecule to molecule. In HCl the separated positive and negative charges are associated with the two atoms. In H_2O the center of the positive charges is halfway between the two H atoms. In NH_3 the H atoms form the base of a relatively flat pyramid with N at the top (shown here inverted); the center of the positive charges is at the center of the base. In SiH_4 (silane) the H atoms surround the Si tetrahedrally and the center of the positive charges coincides with that of the negative charges. The molecule of silane is therefore nonpolar, even though each bond in the molecule is a dipole. This situation is not uncommon. In all the other molecules shown, the directions of the dipoles are the same as in the dipole on the far left consisting of the two point charges $+Q$ and $-Q$ separated by the distance d. The dipole moment μ is the product of Q and d.

formation. These interactions occur chiefly because other atoms and molecules, when close enough, can deform the electronic charge distribution of the given atom or molecule, *inducing* dipole moments in them (Fig. 3-11). The attractive force between a pair of such induced dipoles is sometimes called the *dispersion force* or the *London force,* after F. London, the man who first explained it quantitatively. London forces increase with the number of electrons in the atoms or molecules concerned and with the ease with which an electric field can deform the electronic charge distribution. The composite of these two qualities is termed *polarizability.*[1]

[1]In the simplest case the polarizability of an atom or molecule is a proportionality constant that relates the dipole moment induced in an atom or molecule by an electric field to the electric field strength (see Study Guide). The name *dispersion force* has its roundabout origin in the circumstance that polarizability is the basis not only of dispersion forces but also of the refractive index of transparent substances, which in turn governs the separation of white light into its colored spectral components by prisms. The last phenomenon was called *dispersion* of light by Isaac Newton.

FIGURE 3-10   **Dipoles with different relative orientations.** In (*a*) the positive end of one dipole is close to the negative end of the other while like-charged ends are further removed from each other. The overall result is attraction between the dipoles. A similar situation holds for (*b*). In (*c*) charges of equal sign are close to each other and charges of opposite sign are farther apart. The two dipoles repel each other.

(*a*)   (*b*)   (*c*)

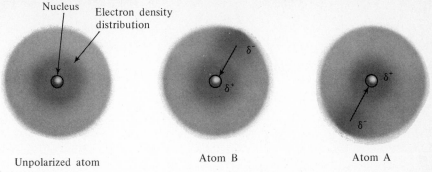

Unpolarized atom   Atom B   Atom A

Mutually polarized atoms

FIGURE 3-11   **London dispersion forces (highly schematic drawing).** On the left is an unpolarized atom with a spherically symmetric electronic charge distribution. On the right, one of the two atoms, such as atom A, has become polarized through a momentary fluctuation that causes the electronic charge distribution to deviate from spherical symmetry, creating a momentary dipole, symbolized by the arrow. This dipole affects the charge distribution of atom B and induces a dipole in atom B in such a direction that the two dipoles attract each other. This situation is rapidly followed by a sequence of similar situations, with dipoles that are always oriented so as to attract each other. The net result is weak attraction, the London dispersion force. (In these schematic drawings, the relative size of the nucleus has been greatly exaggerated.)

Since the boiling point of a liquid is a rough measure of the attractive forces between the molecules it contains, the magnitudes of the London forces between molecules without permanent dipole moments can be assessed roughly by comparison of their normal boiling points (Fig. 3-12). London forces are a general property of matter because of its electronic makeup. The energy of attraction is large between substances with large polarizability and depends inversely on the sixth power of the distance between atoms, which makes the attraction effective only over a very short range of distance. One consequence of this inverse sixth-power dependence is that the energy due to London forces between molecules or parts of molecules is strongly shape-dependent because the number of close atomic approaches between the molecules considered depends on their shapes.[1]

At sufficiently short distances all atoms repel one another because of interpenetration of their outer electronic shells. Without this or some other repulsive force, solid and liquid matter as we know it would collapse to a state of much higher (or infinite) density.[2] The balance between attractive and repulsive forces implies that a graph of the energy of interaction of a pair of atoms as a function of their distance apart shows a minimum. Figure 3-13 shows a comparison among the energies characteristic of typical covalent, ionic, and London interactions. One important

---

[1] With a $1/r^6$ law, an increase of only 12 percent in the distance halves the energy and a twofold increase in the distance reduces the energy by a *factor* of 64.

[2] Matter with much higher than normal density does exist in certain dwarf stars, which represent advanced stages in stellar evolution. White dwarfs have densities about 10^5 greater than that of the earth or of typical stars like the sun. The density of matter in so-called neutron stars is about 10^{14} times that of the earth; this is also the approximate density of matter in atomic nuclei.

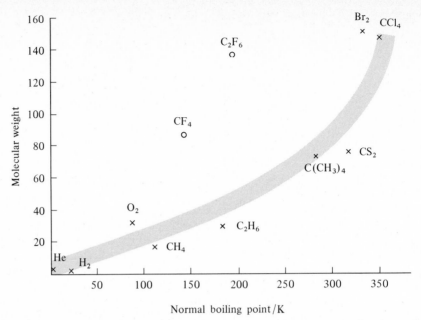

FIGURE 3-12   **Boiling points and molecular weights of some nonpolar molecules.**   Only an approximate correlation of boiling points and molecular weights is to be expected, since the intermolecular forces depend strongly on the distances between pairs of atoms in different molecules and on the ease with which the electronic structure of these atoms can be deformed (the polarizability of the atoms). The polarizability of fluorine compounds is relatively low because fluorine atoms bind their electrons particularly tightly and thus nonpolar fluorine compounds have lower boiling points and higher volatility than other substances of similar molecular weight. A plot of boiling point against *numbers of electrons* is very similar to the curve shown here, since the number of electrons present increases regularly with increasing molecular weight.

point is that the repulsive portions of these curves are so steep as to represent a dependence on interatomic distance r of approximately r^{-10} or even r^{-12}. This means that when atoms or ions approach one another more closely than a minimum distance, they behave as though they were almost incompressible, since a small decrease in distance leads to a great increase in energy. The position of the minimum in the energy depends on whether the interaction is of the covalent, ionic, or van der Waals type.

In considering the van der Waals interaction of two atoms that are not joined by a chemical bond, it is a good approximation to regard the atoms as nearly hard and slightly sticky spheres, with characteristic radii called the van der Waals radii (Fig. 3-14). For example, molecules in crystals pack in such a way that atoms of different molecules rarely are closer to each other than the sum of their van der Waals radii. One speaks of a *van der Waals contact* when the distance between two nonbonded atoms is approximately equal to the sum of their van der Waals radii. These radii are in the range 1 to 2 Å (1 Å = 10^{-10} m). Typical values are H, 1.1 Å; O, 1.4 Å; N, 1.5 Å; C, 1.6 Å; Cl, 1.8 Å; Br, 2.0 Å; I, 2.2 Å. Characteristic radii for

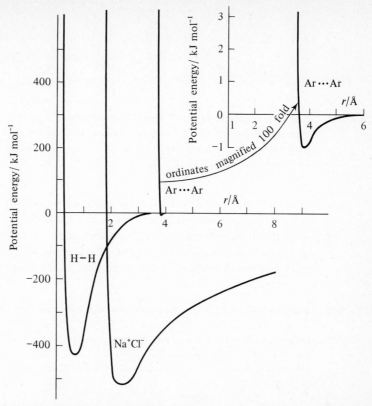

FIGURE 3-13 **Interaction energies between atoms.** Typical energy versus distance curves are shown for three situations, a covalent bond (H—H), an ionic bond (Na^+Cl^-) (compare Fig. 3-8), and a van der Waals interaction (Ar $\cdots$ Ar). All curves shown have a positive slope at large distances, signifying attraction, a steep negative slope at short distances, corresponding to strong repulsion, and a minimum, representing the position of balance between attraction and repulsion. The curves for covalent and ionic interactions are similar to each other except that the ionic interaction has a much larger range, which extends appreciably further than the covalent interaction. The minimum of the van der Waals curve is very shallow and can be clearly seen only in the inset on the upper right, in which the ordinates have been magnified by 100. The minimum of the van der Waals curve is usually too shallow to permit the formation of molecules. Furthermore, it occurs at a larger distance than do the minima for covalent or ionic interactions for atoms of similar atomic number. The strongly repulsive portions of all three curves are similar because basically they have the same origin, the mutual interpenetration of the electron clouds of the two atoms concerned. The dots used to show the dispersion interaction (Ar $\cdots$ Ar) should not be confused with electrons.

covalent bonding and ionic bonding can also be assigned to different atoms (Chap. 11).

Hydrogen bonds. In contrast to the nonspecific van der Waals forces, which depend only on the relative positions and polarizabilities of the atoms involved but not otherwise on their individual nature, the *hydrogen bond* is a specific kind of

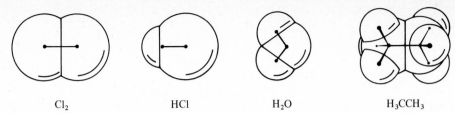

Cl_2 HCl H_2O H_3CCH_3

FIGURE 3-14 **Van der Waals models of simple molecules.** Centers of atoms are indicated by dots. The spheres approximately represent the space taken up by the atoms, in the sense that there is strong repulsion when nonbonded atoms approach each other so closely that the spheres begin to interpenetrate. A van der Waals representation of molecules is also shown in Fig. 1-4c.

attractive chemical interaction. Its existence was first postulated in 1920 by M. L. Huggins (in an undergraduate thesis) and by W. M. Latimer and W. H. Rodebush to explain the surprisingly high boiling point of water, a molecule with only 10 electrons and a molecular weight of 18 (Fig. 3-15). Evidently there are strong attractive forces between water molecules. Many other properties of water are also

FIGURE 3-15 **Boiling points of some compounds of hydrogen and the inert gases.** The curves for the inert gases and the compounds of the formula XH_4 are similar, the compounds of formula XH_4 boiling some 80 to 100 K higher than the inert gases with the same total number of electrons. The situation for the compounds of formula XH_3, H_2X, and HX is similar, with the exception of NH_3, H_2O, and HF, each of which has an abnormally high boiling point (particularly H_2O), signifying strong inter-molecular forces.

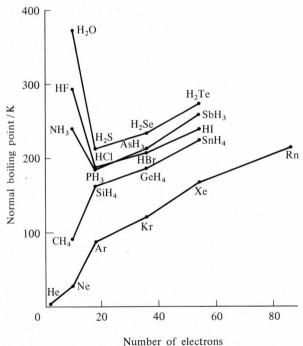

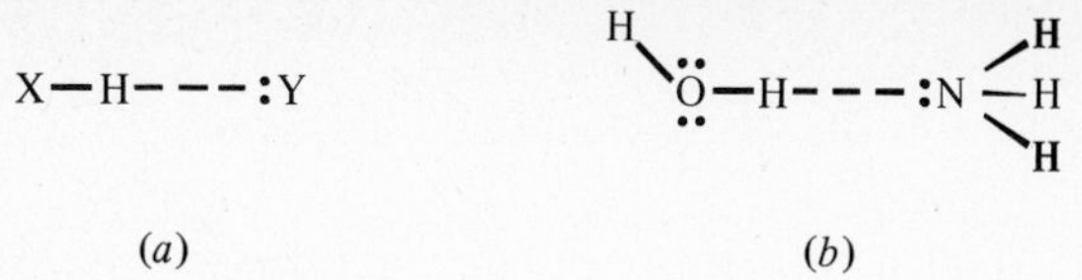

$$X—H----:Y$$

$$\begin{array}{c} H \\ \diagdown \\ \ddot{O}—H----:N \\ \diagup \\ \ddot{} \end{array} \begin{array}{c} H \\ \diagup \\ \diagdown \\ H \end{array}$$

(a)   (b)

FIGURE 3-16   **The hydrogen bond.** The hydrogen atom attached to atom X in (a) points to and is closer to atom Y than would correspond to a normal van der Waals contact. Atoms X and Y may be identical but need not be. A specific example of a hydrogen bond is shown in (b), where the oxygen atom of the water molecule is hydrogen bonded to the nitrogen atom of the ammonia molecule, which has an unshared electron pair (see also Fig. 3-17).

unusual and can be explained by the existence of such forces, as can similar unusual properties of some other substances, all of which contain hydrogen atoms. Studies of these substances, by a variety of methods, in the solid, liquid, and gaseous states and in solution have provided detailed information about the unique interaction called the hydrogen bond.

A hydrogen bond involves an arrangement of three atoms X—H- - -Y (Fig. 3-16), with the symbol - - - representing this bond. Other atoms are usually bonded to X and Y, which may be identical and must be strongly electronegative (that is, they must exhibit a strong tendency to attract electrons). Atom Y must have an unshared pair of electrons. The hydrogen atom is covalently bonded to atom X and points in the general direction of atom Y. As a result of its electronegative character, atom X pulls the shared pair in the X-H bond toward itself, leaving the hydrogen atom with a residual positive charge and relatively unshielded by electrons. Atom Y is endowed with a residual negative charge because of its strong electronegativity. The electrostatic attraction between the positive hydrogen atom and the negative atom Y, particularly its unshared electron pair, is what constitutes the hydrogen bond. There may also be a *small* contribution of covalent bonding between atoms H and Y. Atom X is referred to as the donor atom since it donates the hydrogen atom for the hydrogen bond; atom Y is called the acceptor atom.

It might be thought that other kinds of atoms, as for example a Li atom covalently bonded to atom X, might exert similar attraction for the lone pair of atom Y, forming (for Li) a "lithium bond". This does not happen. The hydrogen atom behaves uniquely because it possesses only one electron. A hydrogen atom involved in a hydrogen bond is able to come very close to the unshared pair on atom Y, and in this respect it acts like a bare atomic nucleus. Thus a strong coulombic attraction can develop. The hydrogen isotopes deuterium and tritium are capable of forming similar bonds since they too have but a single electron.

The most effective donors and acceptors for hydrogen bonds are the atoms F, O, and N, which may be covalently bonded to still other atoms to form molecules or ions (for example, CH_3OH, OH^-, OH_3^+, NH_4^+, NH_2^-, SO_4^{2-}). An outright negative charge enhances the power of an atom to act as an acceptor, which not only makes F^- an excellent acceptor but also permits Cl^- and Br^- to act in the same capacity. A few other atoms, for example, S, may also be involved in weak hydrogen bonds.

The energy of hydrogen bonds is between 5 and 40 kJ mol^{-1}, or about 5 to 10

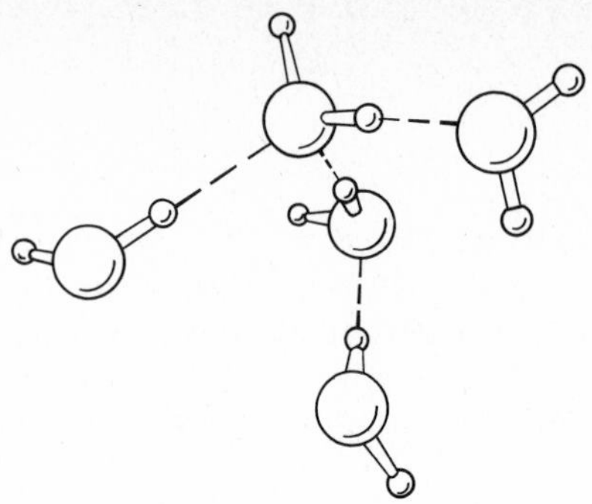

FIGURE 3-17   **Hydrogen-bonded water molecules in liquid water or ice.**   Each oxygen atom is covalently bonded to two hydrogen atoms at about 1 Å to form a water molecule. The two hydrogen atoms are in general directed toward and close to the oxygen atoms of two neighboring molecules (at about 1.8 Å), forming hydrogen bonds. In ice, two additional hydrogen bonds are formed to a given oxygen atom by hydrogen atoms of neighboring water molecules, so that each water molecule is surrounded tetrahedrally by four others, to each of which it is tied by a hydrogen bond. (The H—O—H angles in the drawing appear to vary because of differing viewing angles.) When ice melts, some of the hydrogen bonds are broken, more or less randomly, and the average number of hydrogen bonds formed by each water molecule decreases somewhat.

percent of that of a typical covalent bond. The H- - -Y distance is usually about 0.5 to 1 Å smaller than the sum of the van der Waals radii of H and Y, attesting to the strength of the interaction.

Figure 3-17 is a schematic representation of hydrogen bonding in water and ice. As shown, the hydrogen atoms lie closer to the oxygen atoms to which they are covalently bonded. This is the usual situation, but there are a few examples of very strong hydrogen bonds between fluorine atoms, and between oxygen atoms, in which the hydrogen atom is halfway between the other two atoms. The roles of donor and acceptor atoms then become indistinguishable.

Hydrogen bonds are of great importance in every area of chemistry. For example, they play a dominant role in determining the physical behavior of ice and liquid water, and they exert a major influence in the detailed three-dimensional architecture of protein molecules and of the strands of DNA, deoxyribonucleic acid. Hence the details of many of the most essential processes occurring in organisms are controlled by hydrogen bonds.

The preceding discussion of van der Waals forces and hydrogen bonds has been presented in terms of interactions between different molecules (*inter*molecular interactions). It applies equally well to interactions between different parts of the same molecule when these parts are in appropriate relative positions. *Intra*molecular hydrogen bonds and van der Waals forces are particularly significant in the chemistry of large molecules, and the important interactions in proteins and DNA just mentioned are almost entirely intramolecular.

Problems and Questions

3-1   **Pressure units.**   In some experiments at high temperature, the metal gallium has been used as a manometer fluid because it has a lower vapor pressure than mercury. The density of gallium is about 6.1 g cm^{-3}. How high is a column of gallium equivalent to 1 standard atmosphere?

Ans. 169 cm

3-2   **Pressure measurement.**   The gravitational acceleration decreases by 3.09 × 10^{-6} m s^{-2} for every meter of altitude above the earth's surface. What is the height of a column of mercury (at 0°C) equivalent to 1 standard atmosphere when the manometer is at an altitude of 60,000 feet? (1 foot = 0.305 m)

3-3   **Vapor pressure and stability.**   The vapor pressure of liquid benzene at a temperature *t* can be computed from the empirical formula

$$\log \frac{P}{\text{torr}} = \frac{-1784.8}{t/°C + 273.2} + 7.962$$

and that of solid benzene from

$$\log \frac{P}{\text{torr}} = \frac{-2309.7}{t/°C + 273.2} + 9.846$$

(*a*) Which of the phases of benzene, solid or liquid, is the more stable at 10°C? (*b*) At what temperature are the solid and liquid in equilibrium?   *Ans.* (*a*) liquid;   (*b*) 5.5°C

3-4   **Definitions.**   (*a*) Give concise definitions of the following terms: melting point, sublimation, superheating, boiling point, vapor pressure. (*b*) Distinguish clearly between *vapor pressure* and *pressure of a vapor*.

3-5   **Phase diagram of CO$_2$.**   Refer to Fig. 3-6. Suppose that a sample of carbon dioxide at −80°C and 0.1 atm is (1) compressed at constant temperature to a pressure of 7 atm, then (2) is heated at constant pressure to a temperature of −50°C, and then (3) is held at constant temperature while the pressure is lowered to 3 atm. Describe any phase changes that will occur during each of the stages 1, 2, and 3. Indicate whether they come about at some particular temperature and pressure (if so, what?) or over a range.

3-6   **Phase diagram.**   Figure 3-18 is a phase diagram for a hypothetical substance X that has two solid forms, 1 and 2; solid 2 is stable at higher pressures, as indicated. (*a*) Explain the significance of point *A* (each letter refers to the point of intersection of the lines adjacent to it) and point *B*. (*b*) Describe any phase changes that would occur, giving the approximate temperature and pressure of each, if the following sequence of steps were followed: (i) Some vapor at 60°C and 0.001 atm is compressed at constant temperature to 100 atm and cooled to −10°C; the pressure is then gradually reduced until it is 10^{-4} atm. (ii) Some solid 1 is gradually heated at a constant pressure of 0.01 atm from −20°C until the temperature is 60°C.

3-7   **Interpretation of phase diagram of water.**   Imagine that a beaker of water supercooled to −5°C is placed next to a beaker of ice, also at −5°C, and the two are isolated in a closed box. If the temperature of the box is kept at −5°C and the water does not freeze, what change *will* take place eventually? (Under proper conditions water can be kept supercooled for a long time.) Explain, with the help of the phase diagram for water, Fig. 3-5.

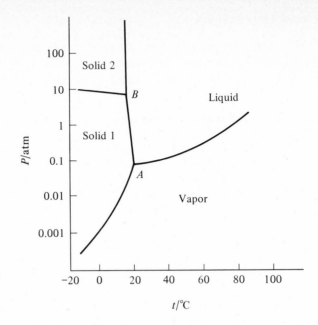

FIGURE 3-18

3-8   Interpretation of phase diagram.   The schematic phase diagram for water, Fig. 3-5, shows that at room temperature and a pressure of 1 atm water is in its liquid form. We know from experience, however, that there is water vapor above the surface of water in a glass at these conditions of temperature and pressure. Explain this apparent contradiction.

3-9   Cooling by evaporation.   Explain in terms of heat effects during phase changes why a hot humid day is usually far more uncomfortable than a dry day at the same temperature.

3-10   Comparison of physical properties.   In each of the following pairs of substances, choose the one that you would expect to have the *higher* value of the property mentioned. Explain briefly why this should be so.
(a)   Solubility in water: $C_2H_4(OH)_2$; C_4H_{10}
(b)   Melting point: SiH_4; LiF
(c)   Boiling point: CCl_4; CBr_4

3-11   Types of bonding.   Distinguish between ionic and covalent bonds, and give examples of compounds held together by each type of bond.

3-12   Polar molecules.   What is meant by the term *polar molecule?* Give an example and show how it fits your definition.

3-13   Forces between nonpolar molecules.   What evidence is there that forces exist between *non*polar molecules? On what do these depend? Why are they so sensitive to molecular shape?

3-14   Intermolecular forces.   Coulomb interactions are often referred to as *long-range* forces, and dispersion interactions are termed *short-range* forces. Compare the falloff with distance of these two types of interactions by computing for each the ratio of the energy of attraction at 2, 3, 4, and 5 molecular diameters to that at 1 molecular diameter.

Ans. coulomb: 0.50, 0.33, 0.25, 0.20;

dispersion: 1.6×10^{-2}, 1.4×10^{-3}, 2.4×10^{-4}, 6.4×10^{-5}

3-15 Hydrogen bonding. Would you expect the hydrogen-bonding tendencies of deuterium fluoride to be appreciably greater than, about the same as, or appreciably less than those of hydrogen fluoride? Explain.

3-16 Hydrogen bonds. Ammonia is both a donor and an acceptor of hydrogen bonds. Indicate by a sketch an interaction of two ammonia molecules that illustrates this fact. Make it clear what is implied by the term *hydrogen bond*.

4 A Classification of Chemical Substances

This chapter describes a classification of chemical substances that is based on the nature of the interactions between the atomic or molecular units of which the substance is composed. A brief description of what are called *nonstoichiometric* compounds is also presented.

4-1 Elements and Compounds Composed of Molecules

Many elements are composed of isolated molecules that contain one or more atoms. Examples include the monatomic inert gases (He, Ne, Ar, Kr, Xe, and Rn); various diatomic species such as H_2, O_2, and the halogens (F_2, Cl_2, Br_2, I_2, and At_2); and polyatomic molecules such as P_4 and S_8. Similarly, many compounds consist of

*Quoted from A. F. Wells, "The Third Dimension in Chemistry", Oxford University Press, New York, 1956.

84

molecules that are all the same kind, the overall composition of the compound then being the same as that of one of its molecules.[1]

We can, somewhat arbitrarily, divide this category into *volatile* and *nonvolatile* substances according to the magnitudes of their vapor pressures under ordinary conditions. There is of course no sharp boundary between these groups; a normal vapor pressure of 0.1 torr ($\sim 10^{-4}$ atm) or greater might be taken as a criterion of volatility, but the exact value chosen is unimportant. Typical volatile substances include sulfur dioxide (SO_2), ethyl alcohol (C_2H_5OH), and iodine, which are respectively gaseous, liquid, and solid under normal conditions. Figure 1-4 shows the crystal structure of another such compound (tetracyanoethylene). Typical nonvolatile molecular substances are sucrose (ordinary sugar, $C_{12}H_{22}O_{11}$) and substances with high molecular weight (e.g., polyethylene, C_nH_{2n} with n in the thousands), as well as more complicated molecules such as proteins.

In the solid state, substances composed of molecules are often soft; in the liquid or solid form they are usually transparent or translucent and often colorless. They show no metallic luster and do not conduct electricity. In both the solid and the liquid states, the molecules are held together by van der Waals forces and hydrogen bonds. The degree of volatility simply reflects the energy characteristic of the interactions among neighboring molecules. If there are many weak van der Waals attractions per molecule (as in polyethylene) or several strong hydrogen bonds (as in sugar), the energy required to remove a molecule from the solid or liquid is substantially greater than the thermal energies available under ordinary conditions and the substance is nonvolatile. On the other hand, if there are comparatively few and weak intermolecular forces, the molecules can easily break loose and the substance is volatile. Since there are no ions present and the molecules do not transfer electrons to one another readily, no charge carriers are present; consequently these substances do not conduct electricity.

Crystals that contain only discrete molecules (and not, for example, ions) are called *molecular crystals*. The molecules in the crystal need not all be identical; sometimes two or more kinds of molecules may pack together in an ordered and energetically favorable array. For example, urea, $(NH_2)_2CO$, and hydrogen peroxide, H_2O_2, crystallize together to form a stable crystal containing a 1:1 ratio of the components, $(NH_2)_2CO \cdot H_2O_2$ (Fig. 4-1), and iodine forms crystals with various planar hydrocarbons, again usually containing the components in a 1:1 ratio. Frequently these *molecular compounds* exist only in the crystalline state. This may be a consequence of some accidentally favorable mode of packing—the shapes may be complementary so that mutually attractive portions of different molecules fit together readily. Hydrogen bonds are frequently involved. If the mutual attraction of the different components is strong enough, the molecular compound may persist in the liquid state, in solution, and even, rarely, in the gaseous state.

---

[1] Or better, of one of its *average* molecules, taking the existence of isotopes of the elements into account. On a weight basis the elemental composition of such a compound therefore varies slightly if the isotopic composition of the elements is changed. The composition is exactly constant when considered on the basis of *numbers* of atoms of different elements involved.

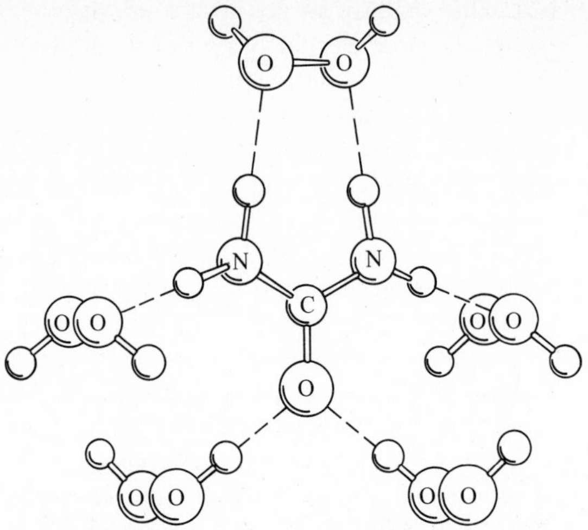

FIGURE 4-1   A portion of the structure of a crystal of (NH₂)₂CO·H₂O₂. The urea [(NH₂)₂CO] molecule in the center is involved in six hydrogen bonds to five hydrogen peroxide molecules, four of which are shown almost end-on (the small unlabeled spheres represent hydrogen atoms). Four of the six hydrogen bonds are of the type N—H- - -O and two are of the type O—H- - -O. All other urea molecules in the crystal have the same environment. Each hydrogen peroxide molecule is in turn involved in six hydrogen bonds of the same types to five different urea molecules (only one of which is shown here for each H₂O₂ molecule).

4-2   Covalent Solids

In many crystalline substances, the atoms are linked by covalent bonds into one-dimensional, two-dimensional, and three-dimensional arrays, which are often referred to as chains, networks, and frameworks, respectively. Each of these extends, at least ideally, throughout the structure in one, two, or three dimensions. Because these arrays do extend throughout the crystal, and thus normally involve millions of atoms, they are often said to be infinite in extent, since the only limitation on their size is that crystals are finite in size. In other words, they would continue to grow in size in a suitable medium.

What is the minimum number of neighboring atoms to which each atom must be bonded in the different kinds of arrays? If atoms of a particular kind form no bonds, the resulting crystals necessarily contain single atoms and thus are simple molecular crystals. If an atom can have only one bonded neighbor, it necessarily forms diatomic molecules, which also yield molecular crystals (Fig. 4-2*a*). Hydrogen and the halogens fall in this class. With two bonded neighbors per atom, there are two possible alternatives: an infinite chain or a ring (Fig. 4-2*b*). Sulfur and the elements below it in the periodic table belong, in their elemental state, in this category and exhibit both possibilities. Networks or layer structures become possible when each atom is bonded to three neighbors, but isolated molecules can also exist (e.g., the tetrahedron of Fig. 4-2*c*). Both possibilities illustrated in Fig. 4-2*c* occur

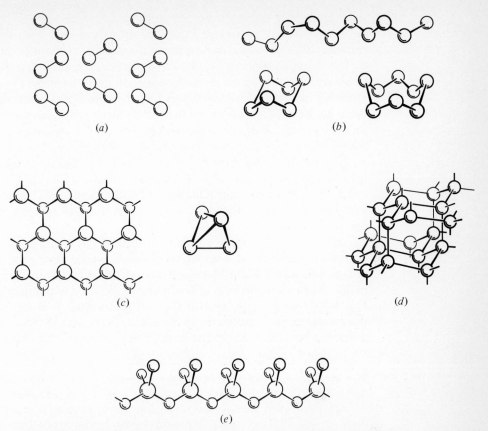

FIGURE 4-2   Some different kinds of zero-, one-, two-, and three-dimensional arrays. In (*a*), each atom forms bonds to only one neighbor; a molecular crystal containing diatomic molecules results (for example, I_2). In (*b*), each atom forms bonds to two neighbors; infinite chains or rings are produced (for example, S_n or Se_n, S_8 or Se_8, S_6) (see Fig. 1-2). In (*c*), each atom is bonded to three neighbors; a two-dimensional network or layer structure results (for example, As_n) or a tetrahedron (P_4 or As_4). In (*d*), each atom is bonded to four neighbors and a three-dimensional framework structure is formed [for example, C (diamond), ZnS, SiC, or BN]. In (*e*), another type of chain structure is shown, with some of the atoms in the chain forming more than two bonds but without covalent cross-links between different chains [for example, $(SiO_3)_n^{2n-}$, $(SO_3)_n$, $(CH_2O)_n$].

for arsenic and phosphorus. Their vapors and crystalline white phosphorus contain tetrahedral molecules, while crystals of arsenic, black phosphorus, and the elements below them in the periodic table consist of puckered layers. The layer is puckered rather than planar because there is an unshared electron pair on each atom; this effect is discussed further in Chap. 12.

When all atoms can form bonds to four neighbors, a three-dimensional (framework) structure is the usual result (Fig. 4-2*d*). If all atoms are assumed to be identical, Fig. 4-2*d* depicts the structure of diamond, a form of carbon; if it is assumed that there are equal numbers of two different kinds of atoms (Fig. 4-6), Fig. 4-2*d* represents the structure of a great variety of substances, including the abrasive carborundum (SiC) and the common mineral sphalerite (a form of ZnS).

The examples given in Fig. 4-2*b* to *d* involve no bonds other than those required to link the one-, two-, and three-dimensional arrays together. Figure 4-2*e* illustrates a chain with some additional atoms joined to the atoms composing the chain, but these additional atoms form no further covalent bonds, so that the chains are not linked together covalently to form a layer or a three-dimensional structure. Silicate chains of this type exist with composition $(SiO_3)_n^{2n-}$, with n a large integer. They occur, for example, in the mineral diopside, $CaMg(SiO_3)_2$, in which the chains are held together in the crystal by electrostatic attraction to the Ca^{2+} and Mg^{2+} ions interspersed in the structure. One form of sulfur trioxide, $(SO_3)_n$, also exists as chains like those depicted in Fig. 4-2*e*; these chains are held together in crystals by van der Waals forces.

An example of a crystal structure built up from separate chains weakly linked to one another ($CuCl_2$) is shown in Fig. 4-3. Typical layer structures are depicted in Figs. 4-4 and 4-5. Figure 4-4 illustrates individual layers and their mode of stacking in the structures of one form of boron nitride, BN, and of graphite, a form of carbon; Fig. 4-5 shows a single layer of the $CdCl_2$ structure. It is instructive to compare the arrangement of the atoms in the $CdCl_2$ layer with that in Fig. 4-2*c*. The only difference between them is that in the $CdCl_2$ structure, which is common to many

FIGURE 4-3   **The structure of crystals of copper chloride, $CuCl_2$.** Chains containing copper atoms surrounded by four chlorine atoms at the corners of an approximate square constitute the primary structural feature. Each chlorine atom is joined to two copper atoms at an angle of nearly 90°. The chains run vertically in the diagram. They are tied together by weaker bonds between copper and chlorine atoms in neighboring chains, in a direction nearly perpendicular to the page. Each copper atom is thus surrounded by an elongated octahedron of chlorine atoms, two of which, on opposite sides of the copper atom, are about 25 percent more distant and much more weakly bonded than the others. The illustration shows this bonding for the chain in the middle only. The structure of $CuCl_2$ illustrates the difficulties of rigid classification; while it is primarily a chain structure, it definitely shows aspects of a three-dimensional framework as well.

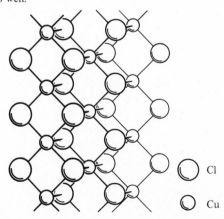

Cl

Cu

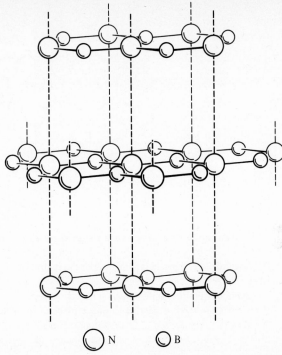

N   B

FIGURE 4-4   **The crystal structures of boron nitride (BN) and graphite.** Crystals of the common form of BN consist of layers in which boron and nitrogen atoms are covalently linked into a planar arrangement of regular hexagons, the corners of which are alternately occupied by boron and nitrogen atoms. The layers are held together in a parallel array by van der Waals forces. A common type of graphite, a form of carbon, has this same structure, the corners of the hexagons being occupied by covalently linked carbon atoms. In other specimens of graphite the stacking of the layers is less regular than that shown here; the structures of these graphites are thus termed partially disordered. The structures of other forms of BN and carbon are illustrated in Figs. 4-2*d* and 4-6.

salts with the general formula MX_2, an additional atom has been added below the center of each of the six-membered rings in the layers of Fig. 4-2*c*. This kind of relationship between two structures—interstices in one being filled systematically to form the other—is a common one, especially in ionic crystals. Figure 4-6 shows the three-dimensional framework of the sphalerite structure.

In all these crystals the chains or the layers may be considered to be molecules, and the entire crystal may be so considered in the case of a three-dimensional network. However, these molecules are infinitely large in the sense that they are of limited size only because of limitations in material available. It is better to restrict the term *molecule* to particles of a *finite size,* a size *that is always the same* no matter how much material is available, as is the case for the molecules H_2O and $C_{12}H_{22}O_{11}$ and even for giant protein molecules.

Crystals containing the kinds of arrangements of atoms discussed in this section are referred to as *covalent crystals* because predominantly covalent forces are opera-

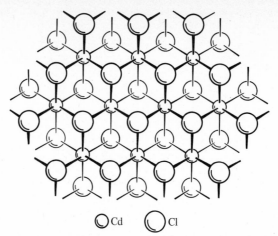

◯ Cd   ◯ Cl

FIGURE 4-5   **The structure of a layer in a crystal of cadmium chloride, CdCl₂.** Cadmium chloride crystals are composed of layers held to each other by van der Waals forces. One of these layers is represented here. Each cadmium atom in the central plane of the layer is bonded to six chlorine atoms, three above the central plane and three below it. The chlorine atoms are at the six corners of an octahedron with the cadmium atom at the center. Each chlorine atom is bonded to three cadmium atoms. The layers are stacked parallel to one another to form the crystal and are held together by van der Waals forces. Other substances exhibiting the same crystal structure include NiI_2, $FeCl_2$, $CoCl_2$, $NiCl_2$, $ZnBr_2$, $MnCl_2$, and $MgCl_2$. The bonds holding the layers together in these substances have various amounts of ionic character ranging from about one-eighth in NiI_2 to somewhat more than one-half (or predominantly ionic) in $MgCl_2$.

tive throughout the crystal, in one dimension for chains, in two for layers, and in three for frameworks. Such compounds are characterized by high melting and boiling points, and they frequently decompose before they melt or sublime. Electrical conductivity is normally extremely low in the solid state, and solubility in water is also very low unless there is chemical reaction with water. For example, the infinite chains in $(SO_3)_n$ react with water, the chains breaking apart to form individual molecules of sulfuric acid, H_2SO_4.

Covalent crystals are usually translucent and show no metallic luster. Those based on a three-dimensional framework are often very hard and brittle; diamond and boron nitride (BN, borazon), with the structure shown in Fig. 4-6, are the hardest substances known. Examination of the structure suggests the reason: to fracture the crystal, it is necessary to break many strong covalent bonds no matter in what direction one applies a force. Almost as hard is silicon carbide (SiC, carborundum), which exists in many forms with closely related structures, one of which is that of Fig. 4-6.

Crystals based on two-dimensional networks tend to cleave in layers (e.g., graphite, $CdCl_2$, mica). Crystals composed of chainlike structures cleave in strings (e.g., asbestos, a silicate that contains chains like those of Fig. 4-2e, joined together in pairs).

Covalent solids need not be crystalline; often they are amorphous (Sec. 1-2). Most natural and synthetic fibers contain atoms covalently linked in chains with only weak forces between the chains (nylon, cellulose, DNA). In amorphous silica, SiO_2, each Si atom is tetrahedrally surrounded by four O atoms and each O links two Si together, which accounts for the composition. The substance is not crystalline, however, because there is no long-range order. It is, in fact, a typical glass.

4-3   High Polymers

An important class of covalent compounds is that of high polymers. These consist of extremely large molecules, with molecular weights of 10^4 to 10^5 or even higher. They are made by combination of many small molecules, called *monomers*. The monomer molecules used to make a given polymer may all be identical or may be of two kinds that will suitably react with one another. Most polymers are mixtures of molecules that vary chiefly in size. Consequently in many ways polymers behave like pure compounds, but they seldom have sharp melting points. The common name *plastics* is often applied to many of these substances. Typical synthetic polymers include synthetic rubber, nylon, polystyrene, polyethylene (polythene), polytetra-fluoroethylene (Teflon), and polymethylmethacrylate (Lucite or Plexiglas) (Chap. 26).

Even elements may exist in polymeric form. An example is lambda sulfur (plastic sulfur), which consists of long chains of sulfur atoms and has rubberlike properties. Many natural substances are polymeric, including cellulose, starch, and rubber.

FIGURE 4-6   **The crystal structure of sphalerite, ZnS.**   Sphalerite is a mineral of composition ZnS. Each zinc atom is surrounded tetrahedrally by four sulfur atoms, and conversely, so that a three-dimensional framework is created. A number of other compounds, among them BeS, BeSe, BeTe, ZnSe, CdTe, HgS, HgSe, HgTe, CuCl, CuBr, CuI, and SiC, have the same structure, with the ionic character of the bonds varying from very small in HgTe to about one-quarter in CuCl. By application of very high pressure and temperature, boron nitride can be transformed from the structure shown in Fig. 4-4 to the present structure. Diamond, a form of carbon, has a similar structure except that all atoms are carbon atoms. Graphite (Fig. 4-4) can be transformed into diamond by applying very high pressure and temperature.

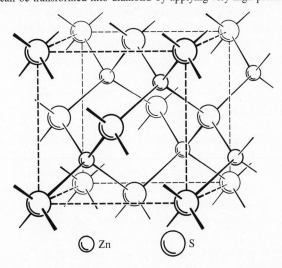

Zn   S

Furthermore, the myriad of different kinds of protein molecules and nucleic acid molecules that exist in nature can be regarded in many ways as polymers, although there are more different kinds of monomers from which any individual large molecule is constructed than for most other polymers, and these individual units are arranged in highly definite sequences in each protein or nucleic acid.

4-4 Metals

About 70 percent of all elements are metallic, and many mixtures and compounds also show certain *metallic properties*. They have a characteristic luster and are opaque in all but the thinnest layers. They are good conductors of electricity and heat. There are no chemical changes when metals conduct electricity, in contrast to the behavior of ionic compounds (see below). These properties are associated with the presence of very mobile electrons, which not only move through the metal easily but can also be made to pass into and out of it readily. Many metallic substances, especially when pure, are soft and can be worked easily—rolled or hammered into sheets (malleable) and drawn into wires (ductile). Mercury is the only metallic element that is liquid at ordinary temperature; it freezes at $-39°C$.

There are several distinct and useful points of view concerning the nature of the bonding in metals, considered in some detail in Chap. 23. In one view, the atoms are linked into three-dimensional arrays by bonds that are similar to covalent bonds but also permit the shifting of electrons throughout the crystal. In this sense metallic solids are related to the three-dimensional covalent solids just discussed. In the other predominant picture, the metal is regarded as consisting of a framework of positive ions surrounded by a sea of very mobile electrons.

An *alloy* is a metallic substance containing at least two elements.[1] It may be liquid or solid, homogeneous or heterogeneous. If it is homogeneous, it may be an intermetallic compound such as $NaZn_{13}$, Al_9Co_2, or Fe_5Zn_{21}. However, some homogeneous crystalline alloys are of variable composition because some of the atoms can easily be replaced by others without separation of a second phase; such crystals are typical solid solutions. Other solid alloys are polycrystalline; that is, they are mixtures containing small crystals of several different metallic substances in various proportions and therefore have no fixed composition or fixed properties.

4-5 Ionic Compounds

Many crystalline substances contain ions, and the chemical composition of such crystals is usually definite and constant. There are two reasons for this: electrical neutrality requires a fixed ratio between the numbers of ions of different charge, and the packing requirements in the crystal usually preclude substitution of one ion by another of the same charge. This is not an invariable rule (Sec. 4-8), but it is a very good one for most substances.

[1] See Sec. 23-2.

Ionic compounds lose their identity when dissolved in water. For example, a solution of potassium nitrate, KNO_3, contains no KNO_3 particles but rather only K^+ and NO_3^- ions; its characteristics are due to the ions K^+ and NO_3^-, not to molecules of KNO_3. A solution made by dissolving amounts of NaCl and KNO_3 chosen to make the numbers of K^+, Na^+, Cl^-, and NO_3^- ions equal cannot be distinguished from a solution in which appropriate amounts of $NaNO_3$ and KCl are dissolved. This solution contains only Na^+, K^+, Cl^-, and NO_3^- ions. Which of the four possible compounds NaCl, KCl, $NaNO_3$, or KNO_3 will crystallize first when a solution containing these ions is evaporated depends on the relative numbers of ions present and on the temperature. For example, with equal numbers of ions present, KNO_3 crystallizes first at temperatures near 0°C.

The physical characteristics of ionic crystals are high melting points and very high boiling points. Ionic substances show no metallic luster, and they are brittle and relatively hard. They conduct electricity when molten or dissolved, because of the movement of ions, and chemical reaction (electrolysis) takes place at the electrodes where the current enters and leaves. Ionic substances are at least slightly, and often appreciably, soluble in water. The solubility may be very small but it is almost always measurable, while the solubility in water of metals and covalent solids (but *not* molecular crystals) is with few exceptions immeasurably small.

Crystalline ionic solids may contain molecules in addition to ions. A case in point is the frequent occurrence of water of crystallization (Example 1-6, Sec. 1-7). To illustrate, evaporation of a solution of sodium carbonate below 32°C leads to crystals of the formula $Na_2CO_3 \cdot 10H_2O$. This compound exists because it is possible to pack Na^+, CO_3^{2-}, and H_2O in the correct proportions in a favorable way. Another such solid is the mineral carnallite, which is sometimes given the formula $MgCl_2 \cdot KCl \cdot 6H_2O$. It is called a double salt, but this name and the foregoing formula are misleading because the compound does not contain $MgCl_2$ and KCl but rather Mg^{2+}, K^+, Cl^-, and H_2O in the ratio $1:1:3:6$. Carnallite owes its existence solely to the accident that a favorable mode of packing exists for these particles in these proportions. The formula $MgKCl_3 \cdot 6H_2O$ is therefore preferable.

Many compounds of this type exist. Some are only slightly soluble and therefore can be used in analytical chemistry to precipitate certain ions from a solution. An example is $MgNH_4PO_4 \cdot 6H_2O$, a crystalline compound that contains the species Mg^{2+}, NH_4^+, PO_4^{3-}, and H_2O. It is used for the precipitation of either Mg^{2+} or PO_4^{3-}. For the determination of Mg^{2+}, excess NH_4^+ and PO_4^{3-} are added under appropriate conditions to the solution containing an unknown quantity of Mg^{2+}. In the determination of PO_4^{3-}, excess Mg^{2+} and NH_4^+ are similarly added to the solution containing an unknown quantity of PO_4^{3-}. In both situations the amount of the ion originally present in the solution can be determined from the weight of $MgNH_4PO_4 \cdot 6H_2O$ that precipitates.

A number of the compounds mentioned contain ions that are made up of several atoms linked by covalent bonds, such as NO_3^-, PO_4^{3-}, and NH_4^+. Another important polyatomic ion of this sort is present in the substance K_2PtCl_6, which contains K^+ but not Pt^{4+} and Cl^- as might be expected. Instead, Pt and Cl are linked by covalent bonds to form ions, $PtCl_6^{2-}$ (Fig. 4-7). This hexachloroplatinate ion is an example of a *complex ion,* of which a great many exist (Chaps. 7 and 25).

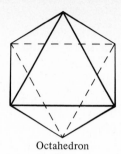

Octahedron

Tetrahedron

$$\left[\begin{array}{c} \text{Cl} \\ \text{Cl} \underset{\text{Cl}}{\overset{\text{Cl}}{\underset{|}{\overset{|}{\text{Pt}}}}} \text{Cl} \\ \text{Cl} \end{array} \right]^{2-} \qquad \left[\begin{array}{c} \text{Al(H}_2\text{O)}_6 \end{array} \right]^{3+} \qquad \left[\begin{array}{c} \text{Zn(OH)}_4 \end{array} \right]^{2-}$$

FIGURE 4-7   **Examples of complex ions: PtCl$_6{}^{2-}$, Al(H$_2$O)$_6{}^{3+}$, and Zn(OH)$_4{}^{2-}$.** In PtCl$_6{}^{2-}$, the chlorine atoms surround the platinum atom octahedrally, and in a similar way the aluminum atom in Al(H$_2$O)$_6{}^{3+}$ is surrounded octahedrally by the oxygen atoms of the six water molecules. In other complexes a central atom may be surrounded by different numbers of atoms in different arrangements. Thus, in Zn(OH)$_4{}^{2-}$ the zinc atom is surrounded tetrahedrally by four oxygen atoms of OH groups; in most Ni^{2+} complexes in which the nickel atom has four near neighbors, they occupy the corners of a square about the nickel atom. Many other complexes, with a variety of structural arrangements, are known.

Frequently part or all of the water of crystallization is covalently bound to a central metal atom with the formation of a complex ion; such a hydrated ion is [Al(H$_2$O)$_6$]$^{3+}$ (Fig. 4-7). In crystals of CuSO$_4 \cdot$5H$_2$O, each copper atom (Fig. 4-8) is at the center of a square formed by the oxygen atoms of four water molecules and is tied to these oxygen atoms by covalent bonds. (In addition, there is an oxygen atom of a sulfate group directly above and another sulfate oxygen directly below each copper atom.) The fifth water molecule in the formula lies between two oxygen atoms of sulfate groups and two oxygen atoms of water molecules, tying them together with hydrogen bonds. The structures of many other salt hydrates show similar features. The combination Cu(H$_2$O)$_4{}^{2+}$ is known to persist even in aqueous solution, as do other hydrated ions.

4-6   Clathrate Compounds

In recent years many examples of a novel type of solid compound have been discovered. In these compounds, called *clathrates* (Latin: *clathratus,* encaged, enclosed by crossbars), ions or molecules form a three-dimensional array with regularly spaced cavities in which still other molecules are trapped. In most of the clathrates

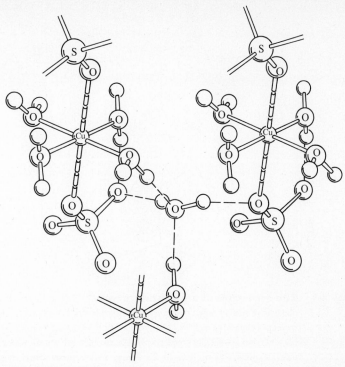

FIGURE 4-8 **The structural arrangement of ions and molecules in a crystal of $CuSO_4 \cdot 5H_2O$.** The crystal structure is built up by a periodic repetition of the components indicated in the formula; a portion of it is depicted here.

so far studied, the framework is held together by hydrogen bonds, but in a few it is covalently bonded at least in two dimensions. Among the most common clathrates are those in which hydrogen-bonded water molecules make up the cages, enclosing molecules like Cl_2, CH_4, and even inert gas atoms such as Xe. Gas hydrates of definite composition are formed, for example, $Cl_2 \cdot 8H_2O$, $Xe_4 \cdot 23H_2O$, and $(CH_4)_4 \cdot 23H_2O$ (Fig. 4-9). Clathrates are stabilized by the mutual van der Waals attraction of the molecules enclosed in the cage and the molecules forming the cage. If many of the cavities are vacant, the structure is usually unstable, and most gas hydrates are at best rather unstable because of the ease with which the hydrogen bonds linking the framework together can be broken.

4-7 Summary of the Different Types of Substances

Table 4-1 is a compilation of the different types of substances that have been discussed. It lists characteristic properties and gives examples. While this classification is valuable, it must be realized that there are no sharp boundaries between the categories and many substances do not fit into the scheme.

For example, crystals of pyrite (FeS_2) and of galena (PbS) are exceptions. They

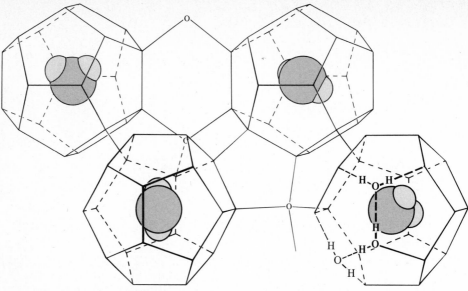

FIGURE 4-9   **The structure of a clathrate crystal.** The framework of hydrogen-bonded water molecules consists of nearly regular pentagonal dodecahedra linked together by additional water molecules, some oxygen atoms of which are shown. The polygons that border the spaces between linked dodecahedra have two hexagonal and six pentagonal faces; they are not shown explicitly. All hydrogen bonds, indicated either explicitly or by lines in the figure, are about the same length as in ordinary ice. The two kinds of polyhedra act as chambers, entrapping atoms (such as Ar, Kr, and Xe) or small molecules (such as CH_4 or H_2S). The figure shows H_2S molecules in random orientation. Not all H_2S molecules entrapped are shown. The ratio of H_2S to water corresponds to the formula $(H_2S)_4 \cdot 23H_2O$. Clathrates involving many different frameworks of water molecules are known.

show metallic luster but are poor conductors of electricity and are hard; they are based on three-dimensional, largely covalent arrays of atoms. Many compounds formed by certain metals (such as titanium, vanadium, and tungsten) with boron, carbon, or nitrogen (compounds called borides, carbides, or nitrides) are metallic in character, showing luster and high electrical conductivity; yet they are extremely hard and brittle, and have very high melting points and low chemical reactivity. Oxide crystals like quartz (a form of SiO_2) often show ionic as well as covalent character.

Other substances may belong to several categories, depending on the temperature and other circumstances. Aluminum chloride as a solid has a covalent, layered structure. It exists as $Al(H_2O)_6^{3+}$ and Cl^- ions in solution but is molecular in the melt (Al_2Cl_6), in the vapor $(Al_2Cl_6$ and $AlCl_3)$, and in organic solvents, as discussed in Example 1-7.

4-8   Nonstoichiometric Compounds

Compounds consisting of identical molecules have a fixed composition and are called *stoichiometric* compounds. In solid compounds not made up of identical molecules it sometimes happens that enough of the constituent ions or molecules are of the

TABLE 4-1
Types of chemical substances

Types	Macroscopic properties	Microscopic properties	Examples
Covalent molecules	Gases or liquids and solids of mp and bp below ~400°C; electrical insulators if solid; often soft, no luster, translucent (at least in thin layers), often soluble	Finite molecules of small or intermediate size 	*Elemental:* He, Ne, Ar, Kr, Xe, Rn, H_2, N_2, O_2, F_2, Cl_2, Br_2, I_2, P_4, S_8 *Compounds:* CO_2, H_2O, NH_3, HCl, CH_4 (methane), C_6H_6 (benzene), P_4O_{10} (phosphorus pentoxide)
Covalent solids	Crystalline or amorphous; high mp and bp; insulators; brittle and hard, translucent, no luster, insoluble	Atoms held in one-, two-, or three-dimensional arrays by covalent forces Diamond	*Elemental:* C (diamond), C (graphite), Si, P (black), Se (chains) *Compounds:* SiC, BN (borazon), AlB, ZnS, $CdCl_2$, $CuCl_2$
High polymers	Solid or viscous liquids; relatively soft and flexible; no luster; when soluble, solution viscous; plastics	Long chains or networks of covalently bonded atoms; usually mixtures of molecules of same type but different size	*Elemental:* S (chains, amorphous), Se *Compounds* (often mixtures): starch, cellulose, rubber, nylon, Teflon, polystyrene, polyethylene
Metals	Highly conducting in solid or liquid state; metallic luster, opaque even in thin layers, insoluble	Highly mobile electrons 	*Elemental:* Na, Cu, Fe, Cr *Compounds:* $NaZn_{13}$, Al_9Co_2, Fe_5Zn_{21}

TABLE 4-1 (Continued)

Types	Macroscopic properties	Microscopic properties	Examples
Ionic compounds	Hard and brittle crystals; high mp and bp; water-soluble although at times in trace amounts; electrically conducting in melts and solutions; translucent, no luster	Crystals consisting of ions—for example, NaCl: Na^+, smaller spheres; Cl^-, larger spheres*	*Elemental:* none *Compounds:* NaCl, CaF_2, KNO_3, $(NH_4)_2SO_4$, K_2PtCl_6
Clathrates	Solid; insulators; translucent, no metallic luster	Cages filled by molecules and formed by a three-dimensional lattice of atoms, molecules, or ions	*Elemental:* none *Compounds:* $Cl_2 \cdot 8H_2O$, $Xe_4 \cdot 23H_2O$, $SO_2 \cdot 3C_6H_4(OH)_2$, $Ni(CN)_2 \cdot NH_3 \cdot C_6H_6$

*The illustration is taken from a paper published by William Barlow in 1898. Barlow proposed the sodium chloride structure and four others from considering the highly symmetric packing of spheres of equal and unequal sizes. His proposal came before the discovery of X rays and before it could therefore be proved that the structures he proposed actually occurred in nature.

wrong kind or are missing, that is, there are enough vacancies or other irregularities in the crystal lattice (lattice defects), to affect the macroscopic properties. If there is an excess of ions of one kind, the crystal gains or loses as many electrons as are necessary to maintain electrical neutrality. The composition of such crystals depends on the details of the irregularities and may lie within a certain range rather than remain constant. It has proved useful to employ the term *compound* for these substances as long as they retain properties characteristic of the stoichiometric substance. Such compounds of variable composition[1] are called *nonstoichiometric* compounds.

Stoichiometric compounds are sometimes called *daltonides* and nonstoichiometric compounds *berthollides,* after the English chemist Dalton and the French chemist Berthollet. Dalton was a strong proponent of the idea that compounds always have a constant composition; Berthollet opposed this idea equally strongly, in part by arguments later shown to be in error by the French chemist Proust. Berthollet believed that the composition of compounds depended on their history, that is, on their mode of preparation.

Many metal oxides and sulfides are berthollides, including the oxides of iron. Stoichiometric compounds exist with composition Fe_2O_3 and Fe_3O_4, which can be

[1]We are not concerned here with slight variations in weight composition caused by different isotopic composition. What matters is the composition by numbers of atoms of different elements.

written $Fe_{0.67}O$ and $Fe_{0.75}O$. However, no sample with the composition FeO has ever been found. The material called "ferrous oxide" is not FeO but rather exists with a range of compositions between about $Fe_{0.84}O$ and $Fe_{0.94}O$. The fact that there is a slightly smaller number of iron ions than of oxide ions means that some of the iron ions must be Fe^{3+} instead of Fe^{2+} in order to maintain charge balance.

Intermetallic compounds are also usually berthollides. Their nonstoichiometric composition is due mainly to replacement in the crystal lattice of atoms of the appropriate kind by wrong atoms. For example, the ideal composition of gamma brass is Cu_5Zn_8 or $CuZn_{1.60}$, but the gamma-brass phase, as it is called, extends over the range $CuZn_{1.58}$ to $CuZn_{1.65}$. The boundaries of such ranges depend on the temperature, and the numbers are quoted just to give an idea of magnitudes.

Problems and Questions

4-1 **Volatility.** What is a *volatile* substance? What is the relation between volatility and the strength of intermolecular forces?

4-2 **Molecular crystals.** Define the term *molecular crystal* and give two examples. What are some characteristic physical properties of molecular crystals?

4-3 **Molecular crystals.** Benzene molecules consist of covalently bonded carbon and hydrogen atoms, yet benzene forms molecular crystals. Explain.

4-4 **Covalent solids.** Explain the relation between the number of bonds that can be formed by a typical atom in a crystal and the possibility of forming linear, two-dimensional, and three-dimensional network structures.

4-5 **Covalent crystals.** Define the term *covalent crystal*. What are the characteristic physical properties of covalent crystals?

4-6 **Metallic properties.** Many familiar household items are made of metal. Which characteristic properties make metals the materials of choice for the fabrication of the following items: pots; foil wrapper; jewelry; electrical wiring; light-bulb filaments?

4-7 **Distinguishing ionic and molecular crystals.** Although large crystals of sugar (rock candy) and large crystals of salt (rock salt) have different geometric shapes, they look much the same to the untrained observer. What physical tests other than taste might be performed to distinguish between these two crystalline substances?

4-8 **Distinguishing types of solids.** You are presented with samples of ionic, covalent, molecular, and metallic solids. The samples are unlabeled and you must distinguish them. Describe a series of physical tests you could perform to categorize the samples.

4-9 **Distinguishing types of solids.** Describe the chief differences in structure and properties between molecular crystals and covalent network solids. Indicate the reasons for these differences.

4-10 **Reasons for volatility.** Give a structural interpretation of the fact that all the halogens are relatively volatile—that is, have an appreciable vapor pressure at ordinary temperatures—while carbon and silicon are not.

"The particles of air are in contact with each other; yet they do not fit closely in every part, but void spaces are left between them, as between the grains of sand on a seashore. (These grains must be imagined to correspond to the particles of air, and the air between the grains of sand to the void spaces between the particles of air.) Hence, when any force is applied to it, the air is compressed and—contrary to its nature—is forced into the vacant spaces by the pressure exerted on its particles. When the force is withdrawn, however, the air returns again to its former position on account of the elasticity of its particles; in this, it resembles horn shavings and sponge which, if compressed and then released, return to their original position and volume."

Hero of Alexandria, "Pneumatika", around A.D. 60*

5 The Behavior of Gases

5-1 Introductory Remarks

In this chapter we shall be concerned with the nature and properties of gases. To a very considerable degree one can discuss the gaseous state both qualitatively and quantitatively without considering the specific nature of the molecules of which the gas is composed. Our concern is primarily with an "ideal" or "perfect" gas instead of with any particular real gas. However, the properties of some actual gases are considered in Sec. 5-3.

Speculation about the nature of the gaseous state occupied philosophers of science

*Quoted by Toulmin and Goodfield, "The Architecture of Matter", Harper & Row, Publishers, Incorporated, New York, 1962.

for many centuries. Some of the ancient speculations had considerable merit, but it was only after the quantitative experiments of Robert Boyle in the seventeenth century, and those of his successors during the ensuing two centuries, that the modern kinetic-molecular picture of a gas evolved. As the name implies (Greek: *kinesis, motion*), this picture is a microscopic-level view of the consequences of the independent motion of molecules in gases. During the nineteenth century a quantitative and remarkably successful theory based on this simple model was developed that permits quite precise quantitative predictions of many different properties of gases (Sec. 5-4). Furthermore, to the extent that some properties of real gases deviate from the predictions of the theory, the deviations are often understandable in terms of the assumptions used in formulating the model. Thus, appropriate modifications of the model can be made and the theory can be improved in a systematic way. This kind of interplay of model, theory, and experimental observation is the essence of modern science.

The interrelations of the properties of any phase, be it solid, liquid, or gas, may be expressed quantitatively in its *equation of state*. This is often written as a relation between the pressure, the volume, the temperature, and the mass of material involved, although other variables may be used. The ideal gas has a particularly simple equation of state, a fact that is of considerable practical importance since the behavior of most real gases under ordinary conditions approximates the ideal within a few percent. Better equations for real gases exist but are more complicated. There is no generally applicable ideal equation of state for liquids or for solids.

We begin with a discussion of ideal gases and apply the ideal equation of state to numerous situations. This discussion is followed by an exposition of real-gas behavior, including condensation. The kinetic-molecular theory, which explains the properties of gases, is considered in Secs. 5-4 and 5-5.

5-2   Ideal Gases

The ideal-gas law.   The equation of state for an ideal gas is

$$PV = nRT \tag{5-1}$$

where
P = pressure of the gas
V = volume
T = absolute temperature
n = number of moles
R = universal constant called the *gas constant*

The numerical value of the gas constant depends on the choice of units for the variables in the equation—P, V, n, and T.

Common units of pressure have been discussed in Sec. 3-1, and unless stated otherwise, gas pressures will be given in atmospheres. Volumes will usually be expressed in liters, and absolute temperatures in kelvins. The units of n are moles, but this does not imply that the gas considered must be pure; as discussed in Sec. 1-7, a mole of air contains a total of N_A molecules, where N_A is Avogadro's number, with the different components present in proportion to their relative abundances. Many gas mixtures, including air, behave like nearly ideal gases under some conditions. For a gas mixture, n is simply the total number of gas molecules divided by N_A, or, what amounts to the same thing, the sum of the numbers of moles of the component gases.

The value of the gas constant appropriate to the most common units is

$$R = 0.082057 \text{ liter atm mol}^{-1} \text{K}^{-1}$$

The units are necessarily the same as those of PV/nT. If some of the variables in the ideal-gas equation are in other units, the numerical value of R must be adjusted appropriately.

EXAMPLE 5-1　　**Change of units of the gas constant** R.　What is the value of R when the volume is expressed in cubic centimeters and the pressure is in torr?

Solution.　Multiply or divide by conversion factors in such a way that cancellation yields the units sought:

$$R = 0.082057 \text{ liter atm mol}^{-1} \text{K}^{-1} \times \frac{1000 \text{ cm}^3}{\text{liter}} \times \frac{760 \text{ torr}}{\text{atm}}$$

$$= 6.236 \times 10^4 \text{ cm}^3 \text{ torr mol}^{-1} \text{K}^{-1}$$

Conversion factors can always be written as quotients in which numerator and denominator express the same physical quantity in different units.

All real gases become more nearly ideal as the pressure is lowered at a given temperature or the temperature is raised at a given pressure,[1] that is, as the volume occupied by a given quantity of gas increases. As shown in Sec. 5-4, ideal-gas behavior corresponds on the molecular level to a situation in which there are no forces between the gas molecules and in which the molecules themselves take up no space. Thus all gases become more nearly ideal as their volume is increased, because the effects of intermolecular forces decrease as the distances between the molecules increase and at the same time the volume of the gas molecules themselves becomes a smaller fraction of the total volume occupied by the gas.

We now turn to a consideration of some applications of the ideal-gas law under special conditions, with different variables in (5-1) held constant. Historically, these special cases were discovered before the ideal-gas law and are associated with the names of their discoverers. In the mid-nineteenth century the information contained in these separate laws was embodied into the generalized law given in (5-1).

Boyle's law.　In the seventeenth century, Robert Boyle, one of the first men to appreciate the importance of quantitative experimentation, established the fact that the product of the pressure and the volume of a fixed quantity of gas is constant

---

[1] A few gases tend to dissociate with decreasing pressure or increasing temperature; that is, the gas molecules tend to break into smaller molecules or atoms. This dissociation may lead to *apparent* deviations from ideality if corrections for the change in the total number of molecules and thus the number of moles n are not made (Sec. 13-4).

at a given temperature; that is,

$$PV = \text{const} \qquad (\text{if } T = \text{const}, n = \text{const}) \tag{5-2}$$

This relationship, Boyle's law, is seen to be merely a special case of the ideal-gas law. Boyle himself stated the law in terms of a constant weight of gas (which is equivalent to a constant number of moles in the absence of dissociation or its reverse, association) rather than in terms of constant n, inasmuch as the development of the molecular hypothesis was still a century and a half in the future. For the same reason, Boyle did not make the generalization that PV is the same for *different* gases if n and T are the same. Boyle's law is illustrated in the following example and in Fig. 5-1.

EXAMPLE 5-2   **Boyle's law.** A sample of gas occupies 100 liters at 3.0 atm pressure. What pressure is needed to bring it to a volume of 20 liters at the same temperature?

Solution. Using the subscripts 1 and 2 for the initial and final states, respectively, we have $P_1 V_1 = P_2 V_2$. Thus

$$P_2 = \frac{P_1 V_1}{V_2} = \frac{100 \text{ liters} \times 3.0 \text{ atm}}{20 \text{ liters}} = \underline{15 \text{ atm}}$$

The law of Gay-Lussac and Charles. Another special case results when the quantity of gas and the pressure are kept constant, so that, in terms of (5-1),

$$\frac{V}{T} = \text{const} \qquad (\text{for } P = \text{const}, n = \text{const}) \tag{5-3}$$

This law was discovered in an alternative form by Gay-Lussac and Charles in the late eighteenth century, and it played a key role in the later establishment of an absolute temperature scale. They found that the volume of a fixed quantity of a

FIGURE 5-1   **Volume and pressure of an ideal gas at constant temperature.** For a fixed amount of an ideal gas, the relationship between volume and pressure at constant temperature is represented by a hyperbola. Each of the curves shown represents Boyle's law at a given temperature.

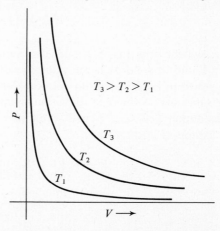

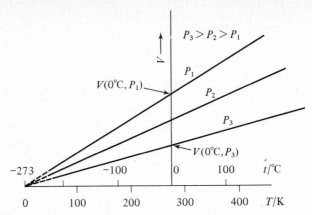

FIGURE 5-2 **Volume of a fixed amount of an ideal gas at constant pressure.** Each of these lines represents the law of Gay-Lussac and Charles at a particular pressure. At constant pressure, the volume of a fixed amount of an ideal gas is linearly related to its temperature and is proportional to its Kelvin temperature. Thus the volume of an ideal gas can be used as a "thermometer" to indicate absolute temperature. As T goes toward zero, so does the volume V of an ideal gas. There is no known substance that behaves like an ideal gas near 0 K, because all real gases eventually condense. Nevertheless, most gases exhibit the linear relation of volume and temperature shown here with reasonably high precision when the temperature is not too low and the pressure not too high. Temperature measurements near 0 K would be difficult or impossible if ideal-gas behavior were the only basis of the absolute scale. However, this has no bearing on the concept of an absolute zero of the temperature scale.

gas at constant pressure was a linear function of its temperature (Fig. 5-2). With temperature expressed in degrees Celsius, symbolized by $t/°C$,

$$V = a + b(t/°C) = a\left[1 + \frac{b}{a}(t/°C)\right] \qquad (5\text{-}4)$$

The constant a is merely the volume of the gas when $t = 0°C$, which we shall call V_0, and the constant b/a turns out to be approximately the same for all gases, about 1/273. Thus (5-4) may be rewritten as

$$V = V_0\left(1 + \frac{t}{273°C}\right) \qquad (5\text{-}5)$$

It is evident from Fig. 5-2 that the volume of a gas (provided it did not condense to a liquid or solid as it was cooled) would become zero at about $-273°C$. This suggested to Lord Kelvin the introduction of a new temperature scale with degrees the same size as those of the Celsius scale but with a different zero point. By the substitution $T/K = t/°C + 273$ into (5-5), the law of Gay-Lussac and Charles can be expressed in the form

$$V = V_0\left(1 + \frac{T - 273\ K}{273\ K}\right) = V_0\frac{T}{273\ K}$$

or

$$\frac{V}{T} = \frac{V_0}{273\ K} = \text{const}$$

which is just (5-3). In this context, the "absolute" aspect of T lies in the fact that its introduction permits an (idealized) description of the thermal expansion of any gas by the same simple proportionality, expressed in (5-3). The relationship between (5-3) and (5-4) is illustrated in Fig. 5-2, where the abscissa shows kelvins as well as degrees Celsius.

The law of Gay-Lussac and Charles shows that the volume and absolute temperature of a perfect gas are directly proportional to each other when P and n are kept constant. Indeed, this proportionality may be used as the basis for measuring absolute temperature.[1] The one aspect of the Kelvin scale that is not absolute is the size of its degree, but this has recently been defined by international agreement (as discussed in Sec. 1-3) in conjunction with the Celsius and the centigrade scales. While the exact relationship between the Kelvin temperature T and the Celsius temperature t is given by

$$t = T - 273.150 \text{ K} \tag{5-6}$$

we usually use the approximation

$$t \approx T - 273 \text{ K}$$

Constancy of PV/T for a fixed number of moles of gas. If only the number of moles of gas remains constant in (5-1), then we have

$$\frac{PV}{T} = \text{const} \qquad \text{(when } n = \text{const)}$$

or

$$\frac{P_1 V_1}{T_1} = \frac{P_2 V_2}{T_2} \qquad \text{(when } n = \text{const)} \tag{5-7}$$

This is a very useful general relationship when one has a fixed number of moles of gas; its utility is illustrated in the following example.

EXAMPLE 5-3 **Effect of simultaneous change of volume and temperature.** A sample of air at 20°C is obtained by filling a 100-ml tube to a pressure of 1.00 atm. The sampling procedure is repeated with a 150-ml tube on a day when the air temperature is 5°C. To what pressure should the second sample tube be filled to provide the same quantity of air as taken in the first sampling?

Solution. The problem is equivalent to the calculation of the effect of changes in volume and temperature on the pressure of a fixed quantity of gas; that is, $n = P_1 V_1/RT_1 = P_2 V_2/RT_2$, or

$$P_2 = P_1 \frac{V_1}{V_2} \frac{T_2}{T_1}$$

---

[1] The proportionality between the volume and the absolute temperature of a perfect gas is not the only means of establishing the absolute temperature; the concept has other and deeper roots in thermodynamics, as we shall see in Chap. 17.

The required pressure is given by

$$P_2 = 1.00 \text{ atm} \times \frac{278 \text{ K}}{293 \text{ K}} \times \frac{100 \text{ ml}}{150 \text{ ml}} = \underline{0.63 \text{ atm}}$$

Note that this expression may be considered as involving the multiplication of the initial pressure by two factors: (1) the temperature ratio, applied in such a way that a reduced temperature corresponds to a decreased pressure (in other words, this factor must be less than unity here); (2) the ratio of volumes, applied in such a way that a larger volume corresponds to a lower pressure, so that this factor must also be less than unity here. The units in the ratios cancel; hence the answer is dimensionally correct. This is not a complete check of the correctness of the answer, because the dimensions would still be correct if one or both ratios had inadvertently been turned upside down.

It must be kept in mind that each of the special laws so far discussed is *valid only if n is constant*. There are two reasons why n might vary when the conditions of a system are changed: (1) gas may be added or removed without chemical reaction, as by vaporization or condensation; (2) gas molecules may be added or removed by chemical reaction. For example, nitrogen tetroxide, N_2O_4, dissociates partially into NO_2 when the pressure on it is decreased:

$$N_2O_4(g) \longrightarrow 2NO_2(g)$$

Since each molecule of N_2O_4 is replaced by two molecules of NO_2 upon dissociation, there is a change in n. This second contingency is discussed in Chap. 13 in Example 13-2 and Sec. 13-4.

Avogadro's law.   When P and T are constant, V is proportional to n,

$$V = \frac{nRT}{P} = n \times \text{const} \tag{5-8}$$

with the constant the same for all ideal gases. This means that *at the same T and P equal volumes of two ideal gases contain the same number n of moles* and therefore also the *same number nN_A of molecules*. This is called Avogadro's law.

In discussions of gas volumes, the standard temperature 0°C and pressure 1 atm, abbreviated STP, are often used. The *standard molar volume,* i.e., the volume of 1 mol of ideal gas at STP,[1] is

$$\tilde{V}_{\text{STP}} = \frac{V}{n} = \frac{RT}{P}$$

$$= 0.082057 \text{ liter atm mol}^{-1} \text{ K}^{-1} \times \frac{273.15 \text{ K}}{1.00000 \text{ atm}} \tag{5-9}$$

$$= 22.414 \text{ liters mol}^{-1}$$

For real gases, the molar volume at STP may deviate from this value. The following

---

[1]A tilde ($\sim$) is used to indicate values that pertain to 1 mol; thus $\tilde{V}$ is the volume of 1 mol of substance.

values are illustrative. The molar volumes given apply at standard conditions; the percentage deviations are from the ideal value:

Gas	He	H_2	N_2	O_2	CH_4	CO_2	NH_3
$\widetilde{V}$/liter mol^{-1}	22.425	22.431	22.402	22.392	22.366	22.258	22.076
Deviation/%	+0.05	+0.08	−0.05	−0.09	−0.21	−0.70	−1.5

As is evident from the table, deviations in either direction may occur. They reflect both molecular size and intermolecular forces. Small nonpolar gas molecules with low polarizability, such as He, H_2, and Ne, deviate at most a few tenths of a percent over a wide range of pressures at ordinary temperatures. As the molecular size and polarizability increase, the deviations may be larger. Nonetheless, the ideal-gas law is a remarkably good approximation for real gases in many situations.

An important consequence of Avogadro's law is that volume composition and mole composition of an ideal gas are identical.

EXAMPLE 5-4   **Volume and mole composition of a gas.** A 22.4-liter sample of a gas at STP contains by volume 20 percent ethylene (C_2H_4), 50 percent methane (CH_4), and 30 percent nitrogen (N_2). How many moles of carbon atoms are contained in this gas sample?

Solution. Since 22.4 liters at STP corresponds to 1.00 mol gas, the quantities of component gases are 0.20 mol C_2H_4, 0.50 mol CH_4, and 0.30 mol N_2. Therefore, the number of moles of carbon atoms is equal to

$$2 \times 0.20 + 0.50 = \underline{0.90 \text{ mol}}$$

Historical importance of Avogadro's law. John Dalton postulated that atoms were the smallest particles of elements and wrote formulas for water and ammonia that corresponded to HO and NH. He considered the hypothesis that equal volumes of gases contain equal numbers of atoms but rejected it because it did not appear to fit certain observed relationships between volumes of reacting gases. For example, two volumes of hydrogen and one of oxygen combine (above 100°C) to give two volumes of water vapor whereas one volume of oxygen and one of nitrogen form two volumes of nitric oxide (NO). Dalton did not consider that in elemental substances several atoms might be combined into one molecule, as in H_2 and O_2. Amedeo Avogadro recognized this possibility and postulated in 1811 that equal volumes of gases contained equal numbers of *molecules,* not atoms. He saw, for example, that $2x$ hydrogen atoms (with x = any integer) might constitute a hydrogen molecule H_{2x} and similarly $2x$ oxygen atoms might constitute a molecule O_{2x}. If water had the formula $H_{2x}O_x$, reaction between hydrogen and oxygen would then be described by

$$2H_{2x} + O_{2x} \longrightarrow 2H_{2x}O_x$$

which would, by his law of volumes, explain the experimental observation that two volumes of hydrogen and one of oxygen result in two volumes of water vapor. It would also give oxygen the atomic weight 16 relative to an atomic weight of 1 for hydrogen, rather than 8 as implied by the formula HO for water. There was no way at that time to determine the value of x, which was taken to be 1 for simplicity's sake. In a similar way the equation

$$N_2 + O_2 \longrightarrow 2NO$$

would reproduce correctly the volumes of nitrogen and oxygen that combine to form NO, and again a common factor, here set equal to 1, remained undetermined.

The idea of molecules composed of atoms of the *same kind* was unpalatable to Avogadro's contemporaries and it took many decades before it was accepted. In the meantime many new elements were discovered and their compounds analyzed, particularly by Jöns Jakob Berzelius, who also introduced our system of chemical symbols. Avogadro's principle was applied to the new elements and their compounds by Stanislao Cannizzaro, who successfully obtained in this way a consistent set of atomic weights and set the stage for the discovery of the periodic system in 1869 by Dmitri Mendeleev and, independently, by Lothar Meyer.

Molecular weight of gas molecules. The ideal-gas law can be given a form that shows an important connection between the density ρ (Greek letter rho) of a gas and the molecular weight of its molecules. Let m be the mass of a sample of ideal gas and let $\widetilde{M}$ be the mass per mole of gas, so that the molecular weight is $\text{MW} = \widetilde{M}/(\text{g mol}^{-1})$. The number of moles of the gas considered is

$$n = \frac{m}{\widetilde{M}} \tag{5-10}$$

so that

$$PV = \frac{mRT}{\widetilde{M}} \tag{5-11}$$

We note that it is possible to isolate on one side the combination m/V, which is just the gas density ρ:

$$\rho = \frac{m}{V} = \widetilde{M}\frac{P}{RT} \tag{5-12}$$

The density of a gas at a given T and P is thus proportional to the molecular weight.

EXAMPLE 5-5   **Density of a gas.** What is the approximate density of CO at 700 torr and 30°C?

Solution. The molecular weight of CO is $12.0 + 16.0 = 28.0$. Using the approximation that CO behaves like an ideal gas, we have, by (5-12),

$$\rho = \frac{P\widetilde{M}}{RT} = \frac{\frac{700}{760}\text{ atm} \times 28.0\text{ g mol}^{-1}}{0.0821\text{ liter atm mol}^{-1}\text{K}^{-1} \times 303\text{ K}}$$

$$= \underline{1.037\text{ g liter}^{-1}}$$

with the least significant digit uncertain. Note that the factor $\frac{700}{760}$ in the third part of the equation is present to convert 700 torr into atmospheres.

We may for convenience rewrite (5-12) by solving for $\widetilde{M}$,

$$\widetilde{M} = \rho \frac{RT}{P} \tag{5-13}$$

and calculate the molecular weight from the density of a gas at a given temperature and pressure.

Neither (5-12) nor (5-13) should be memorized. They can always be derived from the ideal-gas equation, $PV = nRT$. The following two examples are applications of (5-13).

EXAMPLE 5-6   **Molecular weight from vapor density.**   The density of ethyl alcohol vapor, C_2H_5OH, at 76°C and 625 torr is found to be 1.373 g liter^{-1}. What molecular weight corresponds to this vapor density?

Solution.   By (5-13),

$$\widetilde{M} = \frac{\rho RT}{P} = \frac{1.373 \text{ g liter}^{-1} \times 0.0821 \text{ liter atm mol}^{-1}\text{K}^{-1} \times 349 \text{ K}}{\frac{625}{760} \text{ atm}}$$

$$= 47.8 \text{ g mol}^{-1}$$

The sum of the atomic weights is 46.1, smaller than the value just found by about 3.5 percent. The difference is due to gas imperfection, which is particularly to be expected for a vapor close to the pressure at which it condenses. The vapor pressure of ethanol at 76°C is 670 torr and the normal boiling point is 78°C.

EXAMPLE 5-7   **Molecular aggregation from vapor density.**   Arsenious trioxide, of composition As_2O_3, forms a vapor of density 5.66 g liter^{-1} at 743 torr and 571°C. What is the approximate molecular weight, and what is the molecular formula of this compound?

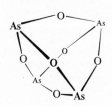

Solution.   We proceed as in the previous example and find

$$\widetilde{M} = \frac{5.66 \text{ g liter}^{-1} \times 0.0821 \text{ liter atm mol}^{-1}\text{K}^{-1} \times 844 \text{ K}}{743 \text{ torr}/760 \text{ torr atm}^{-1}}$$

$$= 401 \text{ g mol}^{-1}$$

The sum of the atomic weights for As_2O_3 is 197.8, which is about half the value found. Arsenious oxide thus exists as the dimer As_4O_6, with molecular weight 395.6, at least under these conditions in the vapor phase. This has been confirmed by electron diffraction studies of the vapor, which also yield information about the structure.

Stoichiometry involving gases.   When gases are involved in stoichiometric considerations, the best procedure is to find the number of moles of gas directly rather than to calculate weights.

EXAMPLE 5-8   **Volume of gas by stoichiometry.**   Iron reacts with oxygen to form Fe_2O_3. What volume of oxygen at 800 torr and 25°C reacts quantitatively with exactly 1 g iron?

Solution. The reaction equation is

$$4Fe + 3O_2 \longrightarrow 2Fe_2O_3$$

Therefore, 4 mol iron reacts with 3 mol oxygen; hence 1 mol iron reacts with $\frac{3}{4}$ mol O_2. Since 1 g iron represents $1 \text{ g}/(55.8 \text{ g mol}^{-1}) = 0.01792$ mol, the number of moles of O_2 reacting with 1 g (or 0.01792 mol) Fe is $\frac{3}{4}(0.01792 \text{ mol}) = 0.01344$ mol. The volume of this number of moles of oxygen at 25°C and 800 torr is

$$V = \frac{nRT}{P} = \frac{0.01344 \text{ mol} \times 0.0821 \text{ liter atm mol}^{-1} \text{ K}^{-1} \times 298 \text{ K}}{800 \text{ torr}/760 \text{ torr atm}^{-1}}$$

$$= \underline{0.312 \text{ liter oxygen}}$$

In the next example the volume of a gas produced in a reaction is given and the problem posed is to interpret this volume stoichiometrically.

EXAMPLE 5-9   **Extent of reaction from volume of gas.** A 5.00-g sample of crude ZnS ore was heated in air, a process called roasting. At the completion of the reaction 1.560 liters SO_2 gas at 200°C and 750 torr pressure had been produced and all ZnS converted to ZnO. What is the minimum volume of air (containing 20 vol percent oxygen) at STP needed for the reaction? What is the percentage of ZnS in the ore?

Solution. First, we find the number of moles of SO_2 formed:

$$n = \frac{PV}{RT} = \frac{(750 \text{ torr}/760 \text{ torr atm}^{-1})1.560 \text{ liters}}{0.0821 \text{ liter atm mol}^{-1} \text{ K}^{-1} \times 473 \text{ K}}$$

$$= 39.6 \times 10^{-3} \text{ mol}$$

To answer the first part of the question we consider the reaction equation

$$ZnS + \tfrac{3}{2}O_2 \longrightarrow ZnO + SO_2$$

and note that $\frac{3}{2}$ mol O_2 is used up for each mol SO_2 formed. Thus, the number of moles of O_2 needed equals $1.5 \times 0.0396 \text{ mol} = 0.0594 \text{ mol } O_2$, or in terms of air,

$$\frac{0.0594 \text{ mol } O_2}{0.20 \text{ mol } O_2/\text{mol air}} = 0.30 \text{ mol air}$$

At STP the air takes up a volume of $0.30 \times 22.4 = \underline{6.7 \text{ liters}}$. For the second part of the question we note that according to the reaction equation 1 mol SO_2 corresponds to 1 gfw ZnS, or $(65.4 + 32.1) \text{ g} = 97.5 \text{ g ZnS}$. Thus, the quantity of gas calculated earlier corresponds to

$$(39.6 \times 10^{-3} \text{ mol } SO_2)\frac{1 \text{ gfw ZnS}}{1 \text{ mol } SO_2} \times \frac{97.5 \text{ g ZnS}}{1 \text{ gfw ZnS}} = 3.86 \text{ g ZnS}$$

In a 5.00-g sample this is $3.86 \times \frac{100}{5.00} = \underline{77.2 \text{ percent}}$.

In problems that are concerned with the volumes of reacting gases, it is simplest to remember Avogadro's law: total numbers of moles and gas volumes are proportional to each other at constant T and P.

EXAMPLE 5-10   **Reaction of gases.** To 50 ml of a mixture of CO and CO_2 is added 40 ml O_2, all at 20°C and 740 torr. The mixture is heated until reaction has converted the carbon monoxide completely to carbon dioxide. When the original temperature and pressure are reestablished, the gas mixture is found to occupy 75 ml. What was the volume composition of the original gas?

Solution. Let there be x ml CO and $(50 - x)$ ml CO_2 in the original mixture. The reaction equation is

$$CO + \tfrac{1}{2}O_2 \longrightarrow CO_2$$

and thus x ml CO uses up $(x/2)$ ml O_2 and produces x ml CO_2. At the end there is 50 ml CO_2—that is, $(50 - x)$ ml unchanged plus x ml produced by oxidation of CO—and $(40 - x/2)$ ml O_2, which combined must total 75 ml: $50 + 40 - (x/2) = 75$. Thus, $x = 30$. The original gas mixture therefore consists of 30 ml CO and 20 ml CO_2.

Dalton's law of partial pressures. As indicated at the start of the chapter, the ideal-gas law applies as well to mixtures as it does to pure gases. Indeed, it was with a mixture, ordinary air, that Boyle, Gay-Lussac, and Charles did their early experiments. Suppose we have a mixture of ideal gases that is itself ideal, occupying a volume V at a temperature T. If we denote the different components of the mixture by subscripts $i = 1, 2, 3, \ldots$, we may write the ideal-gas law, (5-1), for each:

$$P_1 = n_1 \frac{RT}{V}$$

$$P_2 = n_2 \frac{RT}{V} \tag{5-14}$$

and generally

$$P_i = n_i \frac{RT}{V}$$

with the pressure P_i being the pressure that each component would exert if it occupied the total volume V individually. Suppose that we now add these pressures, symbolizing this addition by the symbol ΣP_i, with Σ, the Greek letter sigma, implying summation:

$$\Sigma P_i = \Sigma n_i \frac{RT}{V} = \frac{RT}{V} \Sigma n_i = \frac{nRT}{V} \tag{5-15}$$

where we have taken RT/V as a common factor in front and have used the fact that the total moles, n, in the gas mixture is the sum of the moles of component gases, $n = \Sigma n_i$. Since, by (5-1), the total pressure $P = nRT/V$, we have

$$\Sigma P_i = P \tag{5-16}$$

For an ideal-gas mixture, *the total pressure is thus equal to the sum of the pressures that each component gas would exert if it were present alone.* The pressures P_i are called the *partial pressures* of the different component gases of the mixture, and the law just expressed is called *Dalton's law.* An alternative way of stating this law, and one that is of considerable utility, can be deduced by combining (5-14), (5-15), and

(5-16) to give

$$\frac{P_i}{P} = \frac{n_i}{n}$$

(5-17)

The partial pressure of component i is thus related to the total pressure in the following simple way:

$$P_i = \frac{n_i}{n} P = x_i P$$

(5-18)

where

$$x_i = \frac{n_i}{n}$$

(5-19)

is called the *mole fraction* of component i. It represents the fraction of the total moles (or molecules) present that consists of component i.

While we have just seen that in an ideal-gas mixture each component gas behaves as if it were present by itself, this is only approximately true for a mixture of *real* gases. Here the partial pressure is again defined as the pressure each component would exert if it were present by itself, but the total pressure may no longer be equal to the sum of the partial pressures. In other words, while the total pressure may be considered to be the sum of contributions by the component gases, these contributions may be different from the pressures the components would exert when alone. The reason is, of course, that there are interactions between the components of a mixture. The components no longer behave as if they were present alone. However, such deviations from Dalton's law are usually very small and we shall neglect them.

FIGURE 5-3   **Collecting a gas over water.**   To find the pressure due to the gas, a correction must be applied to allow for the partial pressure of the water vapor in the gas collected. The gas is usually saturated with water vapor, and the total pressure of the gas collected (P_{tot}) is equal to the sum of the pressure (P_{gas}) the gas would exert at the same volume and temperature if the water vapor were removed and the vapor pressure of water (P_{vap, H_2O}) at the temperature considered. An additional correction may have to be applied for the hydrostatic pressure difference caused by the liquid column of height h. This correction is $(h/mm)(\frac{1.00}{13.6})$ torr, where 1.00 and 13.6 are the respective densities of water and mercury in g cm^{-3}. In the situation illustrated here, this correction must be subtracted from the outside pressure; if the liquid level inside the cylinder were below that outside, it would have to be added.

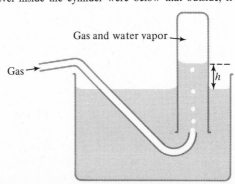

Gas and water vapor

Gas

h

Applications of Dalton's law. When a gas is collected over water (Fig. 5-3), the total pressure includes a contribution from the water vapor. Dalton's law can be used to calculate the pressure that would be exerted by the gas alone.

EXAMPLE 5-11

Collecting a gas over water. A gas Y was collected above water (Fig. 5-3) at 18°C and an outside pressure of 745 torr. Its volume was 234.0 ml and there was no hydrostatic pressure difference between the inside and outside of the cylinder. How many moles of Y were collected?

Solution. According to Dalton's law, the total pressure is given by

$$P_{tot} = P_Y + P_{H_2O}$$

and here P_{H_2O} is the vapor pressure of water at 18°C because the space above the water is saturated with water vapor. The vapor pressure of water at 18°C is 15.5 torr (Table D-1, Appendix D), so that the pressure of Y is 745 − 15.5 = 729.5 torr. (Even though it is not significant, the final digit is retained during the calculation, as discussed in Sec. 2-2.) The number of moles of Y is then given by

$$n_Y = \frac{P_Y V}{RT}$$

$$= \frac{729.5 \text{ torr} \times 234.0 \text{ ml} \times \frac{1 \text{ liter}}{1000 \text{ ml}}}{62.4 \text{ liter torr mol}^{-1} \text{K}^{-1} \times 291 \text{ K}}$$

$$= 9.40 \times 10^{-3} \text{ mol}$$

A second example of the application of Dalton's law concerns the density of a gas mixture.

EXAMPLE 5-12

Density of a gas mixture. What is the density at 700 torr and 27°C of a gas containing, by volume, 30.0 percent ethane (C_2H_6) and 70.0 percent carbon dioxide?

Solution. Since we are free to consider whatever quantity of gas is most convenient, we choose 1.000 mol of gas, consisting of 0.300 mol C_2H_6 and 0.700 mol CO_2. The molecular weight of C_2H_6 is 30.0 and that of CO_2 is 44.0. The total weight is therefore $m = (0.300 \times 30.0 + 0.700 \times 44.0) \text{ g} = 39.8 \text{ g}$. To find the volume we assume that the mixture behaves like an ideal gas, so that

$$V = \frac{nRT}{P} = \frac{1.000 \times 0.0821 \times 300}{\frac{700}{760}} \text{ liters} = 26.7 \text{ liters}$$

The density is

$$\rho = \frac{m}{V} = \frac{39.8}{26.7} \text{ g liter}^{-1} = 1.49 \text{ g liter}^{-1}$$

Note that the mass of 1 mol of the gas mixture is 39.8 g, so that the gas behaves as if its molecular weight were 39.8. This value is called the *apparent molecular weight* of the gas mixture.

Another application shows how volume percent in a gas mixture may be converted into weight percent.

EXAMPLE 5-13   **The composition of dry air.**   The average composition by volume of dry air is, restricting ourselves to the three major constituents, N_2, 78.0 percent; O_2, 21.0 percent; and Ar, 1.0 percent. What is the composition by weight?

Solution.   Again we consider 1 mol of gas. The first row of Table 5-1 shows the moles of components in 1 mol of air. Their sum is 1.000 as shown in the last column. On the second line are the weights of the components, obtained by multiplication by the respective molecular weights, 28.0, 32.0, and 39.9. The sum of these weights, 28.9 g, is the weight of 1 mol of dry air. The apparent molecular weight of dry air is thus 28.9. On the third line are the weight fractions, obtained by division by 28.9. Multiplication by 100.0 converts to the percentages shown on the last line.

Despite popular belief that on a humid day the air is "heavier" because it seems more oppressive, air containing water vapor has a lower density and apparent molecular weight than dry air at the same pressure, if the pressure is kept constant, because H_2O has molecular weight 18, considerably less than that of the other constituents. On the other hand, the presence of appreciable CO_2, with molecular weight 44, will raise the density of air.

5-3   Real Gases

The behavior of real gases.   Real gases obey the ideal-gas law only approximately. Large deviations may occur at high pressure or at low temperature, particularly when a gas is close to condensation.

The product $P\widetilde{V}$ is constant for an ideal gas at constant temperature and has the value RT. Thus the ratio $P\widetilde{V}/RT$ has the value 1 for an ideal gas at any temperature. Figure 5-4 shows a graph of $P\widetilde{V}/RT$ versus P for some *real* gases, and it is apparent that there are sizable deviations from ideal-gas behavior when the pressure exceeds a few atmospheres.

The condensation of a gas can be thought of as a catastrophic deviation from ideality. Consider carbon dioxide, for which large negative deviations are shown (Fig. 5-4) at 40°C for moderate pressures, for example, 50 to a few hundred atm.

TABLE 5-1   **Composition of 1 mol of dry air**

	N_2	O_2	Ar	Σ
n_i/mol	0.780	0.210	0.010	1.000
m_i/g	0.780×28.0 $= 21.8$	0.210×32.0 $= 6.72$	0.010×39.9 $= 0.40$	28.9
Weight fraction	$\frac{21.8}{28.9} = 0.754$	$\frac{6.72}{28.9} = 0.232$	$\frac{0.40}{28.9} = 0.014$	1.000
Weight percent	75.4	23.2	1.4	100.0

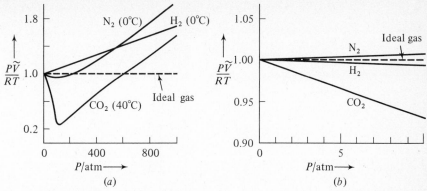

FIGURE 5-4   **Behavior of real gases.** The ratio $P\widetilde{V}/RT$ plotted against P approaches 1 as P goes to zero because at very low pressure gases behave ideally, so $P\widetilde{V} = RT$. The horizontal dashed line corresponds to an ideal gas. The deviations of the curves from this horizontal line are a measure of the deviations from ideal-gas behavior. Graph (*a*) extends from 0 to 1000 atm and exhibits large deviations. For pressures that do not exceed a few atmospheres the deviations are small, as shown in graph (*b*), which shows pressures up to 10 atm.

The behavior of this gas is shown in more detail in Fig. 5-5, which contains a number of curves of P versus V, all at different but constant temperatures. Such curves are called *isotherms* (Greek: *isos,* equal; *therme,* heat). At temperatures below 31.1°C the isotherms have horizontal straight-line portions that correspond to condensation of the CO_2 to a liquid. During condensation the pressure stays constant and the volume may vary arbitrarily within certain limits, depending only on the relative amounts of gas and liquid. The isotherms at 13.1°C and 21.5°C are seen to have three portions: the first (at low pressures) where only gas is present; the second (horizontal, with the pressure equal to the vapor pressure at the temperature of the isotherm) where gas and liquid are both present, at equilibrium; and the third (at high pressures) where only liquid is present. As the temperature is raised, the straight-line horizontal portions of the isotherms become shorter; for the isotherm at 31.1°C this portion is reduced to a point, for which the curve has a horizontal tangent. Isotherms at higher temperatures have no horizontal portions or horizontal tangents. The 31.1°C isotherm is called the *critical isotherm* because at this temperature or above carbon dioxide does not condense to a liquid, no matter how high the pressure. The temperature of this isotherm is called the *critical temperature* T_c, and the pressure at which it has a horizontal tangent (point *A* in Fig. 5-5) is called the *critical pressure* P_c (the vapor pressure at the critical point). As seen in the diagram, there is no condensation phenomenon at pressures above P_c, just as there is none for temperatures above T_c. It should not, however, be inferred that at 31.1°C or higher CO_2 cannot have liquidlike properties. At sufficiently high pressures, for example, it has a high density that changes little with pressure, a behavior characteristic of liquids. The point is, however, that there is no *discontinuous* transition between the gaseous and the liquid state, as there is below 31.1°C. When the gas is compressed above 31.1°C, there is a continuous increase in density and viscosity but no phase boundary appears until at high pressure ($\sim 6 \times 10^3$ atm) solid begins to form.

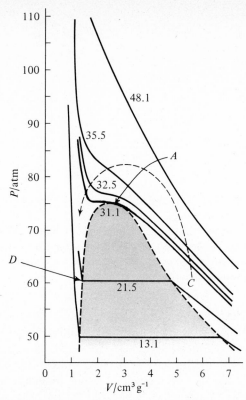

FIGURE 5-5   *P,V* **diagram for carbon dioxide.**   Each line (isotherm) represents a plot of pressure against volume at constant temperature. The dotted region is a two-phase region in the diagram, which represents the coexistence of liquid and vapor at an equilibrium characterized by a volume independent of the pressure and thus by horizontal portions of the isotherms. The horizontal tangent of the critical isotherm at $31.1°C$ is also tangent to the two-phase region. Along isotherms that do not pass through the two-phase region there is no discontinuous transition between liquid and gas. It is possible to convert any gas into a liquid by choosing a path that avoids the two-phase region. For example, to convert C into D in this way we may proceed along a path that is partly indicated by the dashed line with the arrow. To emphasize this continuity in properties and the similarity between gases and liquids in other regards, they are both referred to as *fluid* states.

 All substances have similar *P,V* diagrams. Each is characterized by its values of the critical temperature, critical pressure, and critical density. The critical temperature is a measure of the strength of the forces acting between molecules. Substances such as helium and hydrogen, which have weak attractive forces, have low critical temperatures. Stronger interactions, like those in water, lead to correspondingly higher critical temperatures. Table 5-2 gives critical data for a few common substances, including densities of the fluid at the critical point, called *critical densities*.

The *P,V,T* diagram.   The equation of state for a fixed quantity of a pure substance is a relation between three variables: pressure, temperature, and volume. Equation-of-state data may be represented on two-dimensional diagrams by keeping one of

TABLE 5-2   **Critical temperature, pressure, and density for some common substances**

Substance	Critical temperature/K	Critical pressure/atm	Critical density/g cm^{-3}	Critical molar volume/cm^3
He	5.3	2.26	0.0693	57.8
H$_2$	33.3	12.8	0.0310	65.0
N$_2$	126.1	33.5	0.3110	90.1
O$_2$	153.4	49.7	0.430	74.4
CO$_2$	304.2	73.0	0.460	95.6
NH$_3$	405.6	111.5	0.235	26.4
H$_2$O	647.2	217.7	0.325	55.4

the variables constant. In Fig. 5-5, lines of constant temperature have been shown. Graphs showing the relation between P and T at constant V, or that between T and V at constant P, can also be constructed. These three sets of graphs can be looked upon as planar slices through a three-dimensional P,V,T diagram like that depicted in Fig. 5-6.

An examination of the P,T projection, Fig. 5-6*a*, shows it to be like the phase diagrams encountered in Sec. 3-1 (Figs. 3-5 and 3-6). The projection on the P,T plane of the points that define the (two-phase) liquid-vapor region is the vapor-pressure curve. It ends at the critical temperature and pressure. Similarly, the melting curve is the P,T projection of points that define the region in which solid and liquid can coexist, and the sublimation curve is the projection of the solid-vapor coexistence region. No solid-liquid or solid-vapor critical points are observed.

There is no practical equation of state that can accurately represent the behavior of any substance in all regions of the P,V,T diagram, although equations that are applicable for specific regions have been developed. For example, the properties of real gases can be described by the virial equation of state

$$\frac{P\widetilde{V}}{RT} = 1 + \frac{B}{\widetilde{V}} + \frac{C}{\widetilde{V}^2} + \cdots \tag{5-20}$$

The quantities B and C are called virial coefficients, and $\widetilde{V}$ is the molar volume. The values of the virial coefficients depend on the substance and the temperature. They are usually determined by experiment and can be related by exact theory to intermolecular forces.

A relatively simple equation that approximates the behavior of real gases and simulates their condensation to liquids was put forward by J. D. van der Waals in 1873. This equation, which can be justified in terms of corrections to the ideal-gas equation, has the form

$$\left(P + \frac{a}{\widetilde{V}^2}\right)(\widetilde{V} - b) = RT \tag{5-21}$$

The term $a/\widetilde{V}^2$ is added to account for the attraction between molecules, and the molar volume is reduced by the quantity b to correct for the volume occupied by the molecules. For N$_2$, a has the value 1.38 atm liter^2 mol^{-2} and b is 0.0394 liter

FIGURE 5-6 *P,V,T* **surface for a substance that contracts on freezing.** (*a*) Three-dimensional model; (*b*) projection of *P,V,T* surface on the *P,T* and *P,V* planes; (*c*) and (*d*) details of the projections (*b*). (*Adapted from F. W. Sears: "Introduction to Thermodynamics, the Kinetic Theory of Gases and Statistical Mechanics," 2d ed. Addison-Wesley, Reading, Mass., 1953. By permission of the publisher.*)

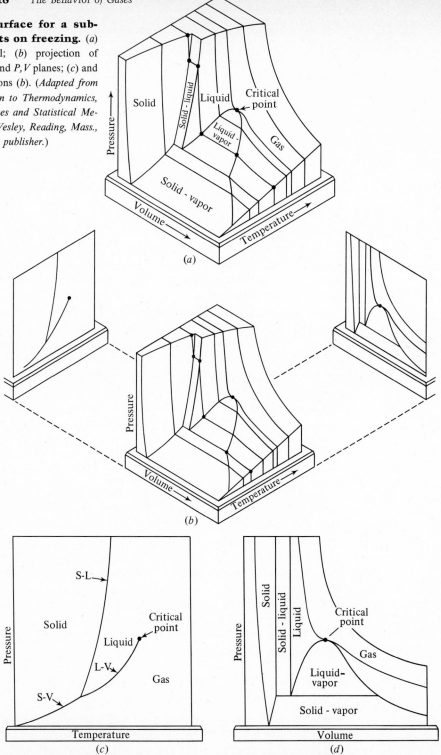

mol^{-1}. Thus, at STP the $a/\widetilde{V}^2$ term represents a correction to the pressure of about 0.3 percent and b reduces the volume by 0.2 percent. Because larger molecules occupy more volume and have stronger interactions, their van der Waals constants are characteristically higher. For example, a and b for SO_2 are 6.7 atm liter^2 mol^{-2} and 0.056 liter mol^{-1}, respectively.

5-4   Kinetic Theory and the Ideal-gas Law

Introductory remarks.   We turn now from the macroscopic properties of gases to the microscopic picture of a gas—the kinetic theory of gases, developed during the nineteenth century chiefly by Rudolf Clausius, James Clerk Maxwell, and Ludwig Boltzmann. The aim of kinetic theory is to explain the detailed behavior of gases on the basis of elementary principles of physics. The theory is convincingly successful; we consider here how this theory leads to the ideal-gas law and what it implies about molecular velocities.

The molecular velocity and its components.   Recall that the molecules of a gas are in rapid and continuous motion. In order to develop a microscopic theory of gases it is necessary to describe this motion mathematically. The position of any molecule can be given by its cartesian coordinates, x, y, and z. When the molecule moves, any or all of these coordinates can change, and it is possible to describe the motion in terms of the simultaneous speeds of the molecule in the x, y, and z directions.

Consider a molecule moving with speeds v_x, v_y, and v_z in the three cartesian directions. Its motion can be represented by the three points plotted on the coordinate axes shown in Fig. 5-7. The arrow drawn in the figure is called the velocity vector. The velocity vector points in the direction of travel and its length v, that is, its magnitude, is a measure of the speed of the molecule. The quantities v_x, v_y, v_z are called, respectively, the x, y, and z components of the velocity. Application of Pythagoras' theorem (see legend to Fig. 5-7) shows that

$$v^2 = v_x{}^2 + v_y{}^2 + v_z{}^2 \qquad (5\text{-}22)$$

The molecules of a gas are, on the average, widely separated. However, despite the fact that a gas is "mostly empty space", there are about 3×10^{19} molecules per cm^3 at ordinary temperature and pressure, and any given molecule collides about 10^{10} times per second with other molecules in the gas. Molecules also collide with the walls of the container, and these collisions are responsible for the pressure exerted by the gas.

As a result of collisions with other molecules, the motion of any given molecule consists of a zigzag sequence of many short straight-line segments. It might appear that because of this complicated motion any microscopic theory of gases must be hopelessly complex. This is not so. We can to a very good approximation neglect the intermolecular collisions and depict the molecules as moving in straight paths interrupted only by collisions with the *walls*. We are able to neglect collisions

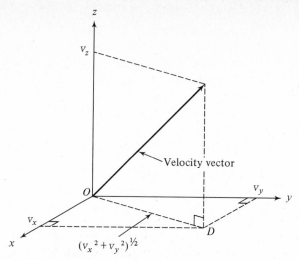

FIGURE 5-7   **Velocity vector and its components.** By applying Pythagoras' theorem to the horizontal right triangle, we find for the square of the distance $\overline{OD}$ that $\overline{OD}^2 = v_x{}^2 + v_y{}^2$. For the vertical triangle $v^2 = \overline{OD}^2 + v_z{}^2$, so that $v^2 = v_x{}^2 + v_y{}^2 + v_z{}^2$.

between molecules because the role of a molecule whose velocity has been changed by such a collision is in essence assumed by another molecule whose velocity has, in turn, just been suitably changed. This results from the fact that the number of molecules in even a small gas sample is enormous, and the number of intermolecular collisions in even a short time interval is very large.

Molecular collisions with the wall. We consider next the collisions between the gas molecules and the container walls. These collisions with the wall, which are responsible for the pressure exerted by the gas, are assumed to be *elastic*. Elastic collisions leave the velocity component parallel to the wall unchanged, while they reverse the sign of the component perpendicular to the wall.

For simplicity we assume the gas to be enclosed in a cube of edge e and introduce a cartesian coordinate system having its origin at a corner and its axes along the cube edges. Figure 5-8 depicts the path of one molecule as it moves through the cubic container and depicts the projection of the path on one face of the cube. The velocity components before the first collision considered, at A, are v_x, v_y, and v_z. The collision with the wall at A changes the z component of the *velocity* from v_z to $-v_z$ while the other velocity components remain unchanged. Similarly in the next wall collision, at B, the y component of the *velocity* is changed from v_y to $-v_y$ while the other velocity components remain unchanged.

Consider now the collisions of the molecule with the cube face at $y = e$, which we shall call wall W. Between any two such collisions the molecule travels all the way to the face at $y = 0$ and back to the wall W. How long does it take the molecule to travel that distance? We can calculate the time it takes a molecule to travel any

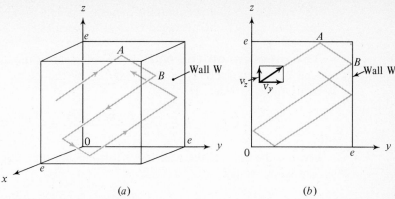

FIGURE 5-8   (a) **Idealized path of a molecule in a cubic container; (b) projection of path on the** y,z **face of the container.**

distance in a given direction if we know its speed in that direction. The *speed* of the molecule in the direction perpendicular to the wall W (and the wall at $y = 0$) is always v_y regardless of whether collisions with other walls intervene, because collisions with other walls do not change the speed. The distance the molecule travels between collisions with W, measured in a direction perpendicular to W, is $2e$, twice the cube edge. Thus the time between collisions with W is the distance traveled divided by the speed, $2e/v_y$. The number of collisions with the wall W in a given time τ is then τ/(time between collisions). Thus, in 1 second and for the one molecule considered,

$$\frac{\text{Collisions}}{\text{Second}} = \frac{1}{2e/v_y} = \frac{v_y}{2e} \tag{5-23}$$

In each collision the momentum (Study Guide) of the molecule changes from mv_y to $-mv_y$, a total change in momentum of $-2mv_y$, where m is the molecular mass. This change of momentum corresponds to a force on the wall, the magnitude of which can be obtained by the following argument. By Newton's second law of mechanics, force equals the rate of change of the momentum. The force exerted on the wall by each molecule per second is equal to the momentum transferred per collision, $2mv_y$, times the number of collisions per second, $v_y/2e$, or

Force per molecule = momentum transferred per second per molecule

$$= 2mv_y \frac{v_y}{2e} = \frac{mv_y^2}{e} \tag{5-24}$$

Let the cube be of a size such that it contains just 1 mol of gas and thus N_A gas molecules. The total average force on wall W is given by

$$\text{Total average force on } W = \frac{N_A m \langle v_y^2 \rangle}{e} \tag{5-25}$$

The symbol $\langle \ \rangle$ indicates that the quantity inside it is to be averaged—in the present case that $v_y{}^2$ is to be averaged over all molecules.[1]

The pressure on the wall. The collisions of the molecules with the wall are seen macroscopically as pressure exerted by the gas on the wall. This pressure P, being force per area, is

$$P = \frac{F}{A} = \frac{N_A m \langle v_y{}^2 \rangle / e}{e^2} = \frac{N_A m \langle v_y{}^2 \rangle}{e^3}$$

because the area of the wall is e^2. Introducing the cube volume, $V = e^3$, and re-arranging, we get

$$PV = N_A m \langle v_y{}^2 \rangle \tag{5-26}$$

The quantity $\langle v_y{}^2 \rangle$ may be related to $\langle v^2 \rangle$, the average of the squares of the molecular velocities. When both sides of (5-22) are averaged, it follows that

$$\langle v^2 \rangle = \langle v_x{}^2 \rangle + \langle v_y{}^2 \rangle + \langle v_z{}^2 \rangle \tag{5-27}$$

because the average of a sum equals the sum of the averages. Furthermore, since all directions of velocities are equally probable,

$$\langle v_x{}^2 \rangle = \langle v_y{}^2 \rangle = \langle v_z{}^2 \rangle \tag{5-28}$$

Thus $\langle v_y{}^2 \rangle = \langle v^2 \rangle / 3$ and

$$PV = \frac{N_A m \langle v^2 \rangle}{3} \tag{5-29}$$

Finally, the average translational kinetic energy[2] $\langle \epsilon_k \rangle$ of a gas molecule is

$$\langle \epsilon_k \rangle = \frac{m \langle v^2 \rangle}{2} \tag{5-30}$$

and therefore also

$$PV = \tfrac{2}{3} N_A \langle \epsilon_k \rangle \tag{5-31}$$

The same equation may be derived by much more refined and rigorous methods. Its validity does not depend on the size or shape of the container or on the assumptions of elastic collisions. It does, however, depend on the following two assumptions:

1. The volume of the molecules themselves is zero. The space accessible to each molecule is the entire volume V of the container and is thus not decreased by the presence of the other molecules.

---

[1] The operations of averaging and squaring may not be interchanged. In fact, $\langle v_y \rangle = 0$, because the velocity components $+|v_y|$ and $-|v_y|$ are equally likely. Thus $\langle v_y \rangle^2 = 0$ while $\langle v_y{}^2 \rangle$ is necessarily positive if there is any motion in the $\pm y$ direction.

[2] Translational energy is energy associated with the motion of the center of mass of the object. In translational motion all parts of an object move along parallel paths, which is not true of rotational or vibrational motion.

2.   There are no forces between the molecules.

These two assumptions are interrelated, because if the volume of the molecules themselves is not zero, this implies that a repulsive force arises when two molecules threaten to occupy the same space. (If molecules are represented by "hard" spheres, i.e., spheres with no "give" whatever, the force would arise on contact only but would then be infinite.)

In deriving (5-31) we emphasized that molecular motion is random and that we could relate the pressure only to the *average* of the momentum transferred per second to the wall. In air at ordinary temperature and pressure there are about 10^{24} collisions with each square centimeter of the walls during each second, but there are fluctuations as a result of the random motions of the molecules. The reason that these fluctuations are not noticeable at ordinary pressures is that random fluctuations in a large number N usually vary as $N^{1/2}$, so the *relative* fluctuation is $N^{1/2}/N = 1/N^{1/2}$. Thus if the pressure-measuring device has an area of 1 cm^2, the expected fluctuation in pressure from one second to the next is of the order of 1 part in 10^{12}, far too little to detect. If on the other hand the area is greatly reduced, the fluctuations are noticeable. Brownian motion of fine dust particles in air results from this kind of effect—the collisions of gas molecules on opposite sides of the dust particle do not balance out and hence the particle is buffeted about and moves in a jerky, random fashion. Relative fluctuations can also be magnified by reducing the pressure or, with delicate instrumentation, by reducing the time required to measure it.

The ideal-gas equation.   Equations (5-29) and (5-31) look very much like the ideal-gas equation for 1 mol of gas, $P\widetilde{V} = RT$. It is known *experimentally* that the ideal-gas equation describes approximately the behavior of a real gas (the lower the density of the gas, the better the approximation). This is precisely the condition under which the assumptions leading to (5-29) and (5-31) are expected to be correct, because at low densities the molecules are far apart so that both the forces between molecules and the molecular volumes are of little account.

If (5-31) is to be identified with the ideal-gas equation, its right side must be equal to RT since we have chosen our cube to contain just 1 mol of gas. Thus,

$$\tfrac{2}{3}N_A\langle\epsilon_k\rangle = RT \tag{5-32}$$

This means that the average kinetic energy of one gas molecule must equal

$$\langle\epsilon_k\rangle = \frac{m\langle v^2\rangle}{2} = \frac{\tfrac{3}{2}RT}{N_A} = \tfrac{3}{2}kT \tag{5-33}$$

where

$$k = \frac{R}{N_A} \tag{5-34}$$

is the gas constant per molecule, also called the *Boltzmann constant*. The average translational kinetic energy per gas molecule is thus proportional to the temperature, and the proportionality factor is *independent of the kinds of molecules* that are being considered. Thus, at a given temperature the average translational kinetic energy of the molecules H_2, O_2, Hg, and C_2H_6 in the gaseous state is the same in spite of differences in their masses and sizes. While this conclusion has been reached here by a combination of theoretical and experimental results, the more refined kinetic theory of gases furnishes this result by theoretical arguments alone.

The kinetic theory of gases also provides a basis for Avogadro's law. Molecules of an ideal gas fill the space available to them by virtue of their rapid motion and not by taking up the space themselves. The volume filled by the gas at a given temperature and pressure is thus not affected by the nature of the molecules but only by their number.

Equation (5-32) makes it apparent that RT is an energy per mole, a fact that is perhaps surprising, considering that the units we have so far favored for the gas constant R have been liter atmospheres per mole per kelvin (liter atm mol^{-1} K^{-1}). In other words, liter atmospheres are energy units—the product of volume and pressure represents an energy. Consider the relation

$$\text{Volume} \times \text{pressure} = \text{volume} \times \text{force/area}$$

Since volume/area has the dimensions of a length or a distance, volume $\times$ pressure has the same dimensions as distance $\times$ force. These are the dimensions appropriate to mechanical work, so that liter atmospheres are indeed energy units.

To convert liter atmospheres to joules, we note that 1 liter is by definition exactly 1000 cm^3 while 1 atm is $1.01325 \times 10^5 \text{ N m}^{-2}$. Thus

$$1 \text{ liter atm} = 1000.000 \text{ cm}^3 \times \frac{1 \text{ m}^3}{10^6 \text{ cm}^3} \times 1.01325 \times 10^5 \text{ N m}^{-2}$$

$$= 101.325 \text{ J} \tag{5-35}$$

Division by $4.1840 \text{ J cal}^{-1}$ converts this into calories:

$$1 \text{ liter atm} = 24.217 \text{ cal} \tag{5-36}$$

The most useful units for the gas constant R depend on the circumstances. They may be liter atmospheres, ergs, joules, or calories—all multiplied by mol^{-1} K^{-1}. In SI units, J mol^{-1} K^{-1},

$$R = 0.082057 \text{ liter atm mol}^{-1} \text{K}^{-1} \times 101.325 \text{ J liter}^{-1} \text{atm}^{-1}$$

$$= 8.314 \text{ J mol}^{-1} \text{K}^{-1}$$

Table 5-3 gives R in four different units. Additional energy-conversion factors are given in Appendix A, Tables A-4 and A-5.

The Boltzmann constant k may be obtained in different units by dividing any of these values by Avogadro's number. Thus,

TABLE 5-3   **The gas constant R in several units**

0.08206	liter atm mol^{-1} K^{-1}
8.314	J mol^{-1} K^{-1}
8.314×10^7	erg mol^{-1} K^{-1}
1.987	cal mol^{-1} K^{-1}

$$k = \frac{R}{N_A} = \frac{8.314 \text{ J mol}^{-1} \text{ K}^{-1}}{0.6022 \times 10^{24} \text{ mol}^{-1}}$$
$$= 1.3807 \times 10^{-23} \text{ J K}^{-1} = 3.300 \times 10^{-24} \text{ cal K}^{-1} \tag{5-37}$$

where the units mol^{-1} have been associated with N_A.

In the foregoing all molecules were considered to be of the same kind, but generalization to gas mixtures is straightforward and leads to the result that in a gas mixture the *average translational kinetic energy of all molecules is the same*, as given by (5-33), irrespective of their masses.

The root-mean-square velocity.   Equation (5-33) contains information about the average (or mean) speed of gas molecules. The quantity directly involved is $\langle v^2 \rangle$, the average of the square of the velocity, for which

$$\langle v^2 \rangle = \frac{3kT}{m} \tag{5-38}$$

Multiplying numerator and denominator on the right by Avogadro's number and replacing $N_A k$ by the gas constant, R, and $N_A m$ by the molecular weight, $\widetilde{M}$, we find

$$\langle v^2 \rangle = \frac{3RT}{\widetilde{M}} \tag{5-39}$$

All that remains to be done to find $\langle v \rangle$ is taking the square root—or so it may seem. However, the square root of an average value is always greater than the average of the square roots (except in the trivial case that all the numbers being averaged are the same). The reason is that in averaging squares there is always a bias favoring the larger numbers. Consider, for example, the numbers 1, 4, and 9, with an average of $\frac{14}{3} = 4.67$, the square root of which is 2.16. The square roots of 1, 4, and 9 are 1, 2, and 3, with an average of 2, somewhat less than the square root of the average of the original numbers.

The square root of the average (or mean) of the square of any quantity is called the *root-mean-square* value, abbreviated rms; thus $\langle v^2 \rangle^{1/2}$ is the rms speed, v_{rms}, and is always greater than $\langle v \rangle$, the average speed of the molecules. More detailed considerations, discussed in Sec. 5-5 below, show that for gas molecules v_{rms} is always only about 8 percent larger than $\langle v \rangle$ (a relationship that is, coincidentally, also true of the numbers quoted in the simple example in the preceding paragraph). Thus the value of the root-mean-square velocity,

$$v_{\text{rms}} = \sqrt{\frac{3RT}{\widetilde{M}}} \tag{5-40}$$

gives a good indication of the average speed of gas molecules.

EXAMPLE 5-14   **The root-mean-square velocity of H$_2$ at 1000 K.**   What is v_{rms} of H$_2$ at 1000 K?

Solution. To apply (5-39) we use R in SI units, $R = 8.31 \text{ J mol}^{-1} \text{ K}^{-1}$, and note that the mass per mole for H_2 is $2.016 \times 10^{-3} \text{ kg mol}^{-1}$. Thus

$$\langle v^2 \rangle = \frac{3RT}{\widetilde{M}} = \frac{3 \times 8.31 \times 1000}{2.02 \times 10^{-3}} \text{ J kg}^{-1}$$

$$= 12.34 \times 10^6 \, (\text{m s}^{-1})^2$$

Here we used the relationship $J = \text{kg m}^2 \text{ s}^{-2}$ so that $\text{J kg}^{-1} = (\text{m s}^{-1})^2$. Taking the square root yields

$$v_{\text{rms}} = 3.51 \times 10^3 \text{ m s}^{-1}$$

$$= 3.51 \text{ km s}^{-1} = 12.64 \times 10^3 \text{ km hour}^{-1}$$

$$= \frac{12.64 \times 10^3 \text{ km hour}^{-1}}{1.609 \text{ km/mile}} = 7.9 \times 10^3 \text{ mile hour}^{-1}$$

Table 5-4 contains additional values of v_{rms}. These rms molecular velocities may seem surprisingly large, but it must be remembered that under ordinary conditions a gas molecule does not move in a straight line for more than a very short time because it suffers innumerable collisions with other gas molecules. Thus the mixing of two gases at ordinary pressures solely as a result of their intrinsic molecular motion (rather than as a consequence of bulk stirring by convection or by mechanical means) is a comparatively slow process despite the high velocities of individual molecules.

The average of the distances through which a gas molecule moves between collisions is called the *mean free path.* It is of the order of 10^{-7} m for oxygen at 1 atm. At a given temperature the mean free path is inversely proportional to the pressure, so that for O_2 at 10^{-7} atm, a moderately good vacuum easily obtainable in the laboratory, it has increased to 1 m.

Inspection of (5-40) shows that v_{rms} is directly proportional to $\sqrt{T}$ and inversely proportional to $\sqrt{\widetilde{M}}$. The same dependence on T and $\widetilde{M}$ is shown by phenomena that are related to the motion of gas molecules, such as *diffusion,* the motion of one kind of molecule through an assemblage of others as a result of random motion, and *effusion,* the streaming of a gas into a vacuum through a fine hole in a thin

TABLE 5-4   **Root-mean-square velocities for some gases**

Gas	Temperature/K	Molecular weight	$v_{\text{rms}}/\text{km s}^{-1}$
H_2	273 1000	2.02	1.84 3.51
O_2	273 1000	32.00	0.461 0.883
H	10^7 (in stars)	1.01	497
Hg	273 1000	200.6	0.184 0.353

wall. These two phenomena obey laws of the same form, even though the basic molecular mechanisms are quite different. Both laws were discovered by Thomas Graham, his law of diffusion in 1831 and his law of effusion in 1846. These laws state that for two gases, 1 and 2, the *times* required for passage of equal numbers of molecules past some reference point under the same conditions are in the ratio

$$\frac{t_1}{t_2} = \sqrt{\frac{\widetilde{M}_1}{\widetilde{M}_2}} \tag{5-41}$$

It is thus possible to compare the molecular weights of two gases by comparing their rates of effusion.

EXAMPLE 5-15   **Molecular weight from effusion time.** The rate of effusion of a gas through a certain orifice at STP is 1.41 that of oxygen under the same conditions. What is the molecular weight of the gas?

Solution. Since the *rate* of effusion is 1.41 times as great, the *time* for effusion must be 1.41 times smaller. Thus, with $\widetilde{M}_1 = 32.0$ g mol^{-1} (for O_2) and $t_2/t_1 = 1/1.41$ in (5-41), we have $\widetilde{M}_2 = \widetilde{M}_1(t_2{}^2/t_1{}^2) = (32.0$ g mol$^{-1})/(1.41)^2 = \underline{16$ g mol$^{-1}}$. (The gas in question was methane, CH_4. Note that the first step in almost any problem is to see what one can infer qualitatively so that one can check the reasonableness of the answer. The fact that the gas effuses more rapidly than oxygen implies at once that it must have a lower molecular weight, which is consistent with the answer. It is all too easy to get ratios inverted when doing problems like this one.)

Diffusion can, under appropriate conditions, be used to separate gases of different molecular weight. If a mixture of oxygen and methane is allowed to diffuse through a porous wall, the methane passes through more rapidly because of its smaller molecular weight. Thus the mixture that has penetrated the wall is enriched in methane and there is a depletion of methane in the original mixture. There are actually better methods for the separation of these two gases, based on their chemical properties or their different solubilities in liquids, but the present method is useful for one class of mixtures difficult to separate in other ways: mixtures of isotopic molecules. The classical case is the separation of ^{235}U and ^{238}U in the form of their fluorides, UF_6, which are gaseous at ordinary conditions. Since the masses of $^{235}UF_6$ and $^{238}UF_6$ are almost identical, the effect is very small. In fact the ratio of the two diffusion times is

$$\frac{t_1}{t_2} = \left(\frac{235 + 6 \times 19}{238 + 6 \times 19}\right)^{1/2} = \left(\frac{349}{352}\right)^{1/2} = 0.9957$$

To be effective, the separation procedure has to be repeated many times, which is achieved by diffusion through a sequence of thousands of porous barriers. To a first approximation, the effect of s barriers is to change the ratio 0.9957 to $(0.9957)^s$.

EXAMPLE 5-16   **Separation of ^{235}U and ^{238}U.** What is the effect of 1000 porous barriers on the separation of $^{235}UF_6$ and $^{238}UF_6$ by diffusion?

Solution. The ratio 0.9957 is changed to $(0.9957)^{1000}$. To find the value of this number (call it f) we use logarithms: $\log f = \log(0.9957)^{1000} = 1000 \log 0.9957 = 1000(-0.00187) = -1.87 = 0.13 - 2$. Thus $f = 1.35 \times 10^{-2} \simeq \frac{1}{74}$. Having 1000 barriers is thus seen to be a great improvement over having only one barrier since the ratio of diffusion times is changed by a factor of 74 (as though the molecular weights differed by a factor of 74^2 or almost 5500!). An elaborate experimental design, first used at Oak Ridge, Tennessee, during World War II, is needed to put this method into effect.

It should be noted that Graham's law of diffusion applies only when certain conditions regarding the porous wall are satisfied; the sizes of the molecules concerned are also of importance. The sizes of isotopic molecules such as $^{235}UF_6$ and $^{238}UF_6$ are virtually identical.

5-5 The Boltzmann Factor and the Maxwell-Boltzmann Distribution

The kinetic theory of gases does not consider the behavior of individual molecules but rather is a statistical treatment of the immense number of molecules contained even in small samples of gas. An important question solved first with the help of this theory concerns the fraction of molecules that have energies in a given range. This molecular energy distribution may be understood in terms of the *Boltzmann factor,* which is discussed briefly in the following paragraphs. We shall encounter the Boltzmann factor again because it is very generally applicable and thus extremely important in all problems of constant-temperature energy distributions in chemistry and physics.

The Boltzmann factor. The Boltzmann factor governs the population of energy states at a given temperature. It can be derived by purely theoretical arguments, but we shall only present the results.

Consider a system at constant temperature, in which different energy states are possible—for example, gas molecules with varying kinetic energy (the application of chief interest in the present chapter), electrons with different possible potential energy states in an atom, polar molecules with varying orientation and thus varying potential energy in an electric field, or small magnets (for example, certain nuclei) with varying orientations in a magnetic field. The Boltzmann factor, which is applicable to all these systems, relates the numbers of particles in each energy state to the differences in energy between the states. If n_1 and n_2 are the numbers of particles per unit volume, with energies ϵ_1 and ϵ_2, respectively, then

$$\frac{n_2}{n_1} = \frac{e^{-\epsilon_2/kT}}{e^{-\epsilon_1/kT}} = e^{-(\epsilon_2-\epsilon_1)/kT} = e^{-\Delta\epsilon/kT} = e^{-\Delta\tilde{E}/RT} \tag{5-42}$$

where $\Delta\epsilon = \epsilon_2 - \epsilon_1$ is the difference in molecular energies, $\Delta\tilde{E}$ is the corresponding energy difference per mole, and $e = 2.718\ldots$ is the base of natural logarithms.

The exponential factor $e^{-\Delta\epsilon/kT} = e^{-\Delta\widetilde{E}/RT}$ is the Boltzmann factor; for typographical convenience it is sometimes written on one line as $\exp(-\Delta\widetilde{E}/RT)$, where $\exp(a)$ is equivalent to e^a. The law expressed in (5-42) may be stated alternatively by saying that the number of particles of energy ϵ is proportional to $\exp(-\epsilon/kT)$:

$$n = \text{const} \times \exp\frac{-\epsilon}{kT} \qquad (5\text{-}43)$$

The equivalence of (5-43) and (5-42) is seen by writing (5-43) for two different energies and taking the ratio, whereupon the constant terms cancel and (5-42) results.

EXAMPLE 5-17   **The Boltzmann factor.**   The first vibrational energy state of CO is found at an energy of 43.1×10^{-21} J per molecule above the ground state. At what temperature is the ratio of molecules in the first excited state to that in the ground state just $1:100$?
Solution.   By (5-42),

$$0.0100 = \exp\frac{-43.1 \times 10^{-21}}{1.381 \times 10^{-23}\, T/\text{K}}$$

Taking natural logarithms yields

$$-\ln 100 = \frac{-3121}{T/\text{K}}$$

or $T = \underline{678\text{ K.}}$

Our present interest in the Boltzmann factor is its application to the distribution of speeds in gas molecules at a given temperature, that is, the relative numbers of molecules with speeds in given ranges. Since the speed is directly related to the kinetic energy, and the average kinetic energy is proportional to the temperature [Equation (5-33)], this distribution may be deduced by appropriate application of the Boltzmann factor.

The Maxwell-Boltzmann distribution of speeds of gas molecules.
Although the distribution of molecular speeds was first derived in 1860 from the kinetic theory of gases, it has become possible with modern techniques to determine the distribution experimentally. The results of some measurements on thallium vapor are shown in Fig. 5-9. Molecular speed is plotted along the horizontal axis; the quantity plotted in the vertical direction is proportional to the fraction of molecules with a given speed. The general shape of this curve is typical of the speed distributions of all gases. Relatively small fractions of the molecules have very low or very high speeds, although such distributions are unsymmetric and have tails that extend to high speeds. The line drawn through the experimental points has been calculated from kinetic theory. The agreement is well within experimental error, and many similar experiments have confirmed the accuracy of the theory.

The calculated curves for N_2 shown in Fig. 5-10 illustrate the striking effect of temperature on the speed distribution. Most important is the increase of the fraction of fast molecules as the temperature is raised. Since the total area underneath each curve must be unity (i.e., the sum over all speeds of the fractions with given speeds

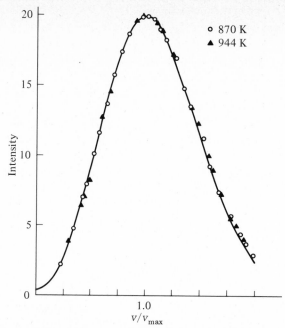

FIGURE 5-9   **Speed distribution of thallium atoms obtained by molecular beam measurements.**   The data represented by circles and triangles were measured at different temperatures. However, when v/v_{max} is used as horizontal variable rather than v (where v_{max} is the speed observed most often), the same curve, calculated from the kinetic theory of gases, fits all data points.

must total 1), a flattening of the curves must accompany the rise in the high-temperature tail with an increase in temperature. The maximum in the curve (corresponding to the most probable speed) therefore becomes lower. For chemical purposes, the molecules that have a speed greater than a certain minimum are of special interest because usually only molecules with an excess of energy may react upon collision with other molecules. The differences in the areas underneath the curves to the right of the vertical line at 730 m s^{-1} in Fig. 5-10 illustrate the increase with temperature of the fraction of molecules with speeds equal to or in excess of 730 m s^{-1}. The implications of this increase are discussed further in Sec. 20-4.

The function plotted in Fig. 5-10 is the *Maxwell-Boltzmann speed distribution,* which has the mathematical form

$$B(v) = \text{const} \times v^2 \exp \frac{-mv^2}{2kT} \tag{5-44}$$

This equation can be derived from kinetic theory; here we attempt no more than to show that a function of this form is reasonable.

Every speed v has associated with it a kinetic energy $mv^2/2$, so that a description of the distribution of molecular speeds is also a description of the distribution of molecular kinetic energy. As we have just seen, energy distributions are related to the Boltzmann factor, hence the term exp $(-mv^2/2kT)$ in (5-44).

There are many velocities that correspond to the same speed because, as can be seen from (5-22), different combinations of the squares of velocity components can sum to the same v^2. Each velocity must be counted in the distribution. How does the number of possible combinations of the velocity components depend on v? Consider the velocity vectors shown in Fig. 5-11. They are of equal length; hence they all correspond to the same speed. The points at the tips of all such vectors form the surface of a sphere of radius v, so the surface area $4\pi v^2$ is a measure of the number of these vectors. It is for this reason that the factor v^2 appears in the Maxwell-Boltzmann distribution.

The proportionality constant in (5-44) can be evaluated by requiring the total of all fractions to be unity. The final result for $B(v)$ is

$$
\begin{aligned}
B(v) &= \left(\frac{2}{\pi}\right)^{1/2}\left(\frac{m}{kT}\right)^{3/2} v^2 \exp\left(-\frac{mv^2}{2kT}\right) \\
&= \left(\frac{2}{\pi}\right)^{1/2}\left(\frac{\widetilde{M}}{RT}\right)^{3/2} v^2 \exp\left(-\frac{\widetilde{M}v^2}{2RT}\right)
\end{aligned}
\tag{5-45}
$$

The two formulas are equivalent because $m/k = N_A m/N_A k = \widetilde{M}/R$. The details of the proportionality constant in front of the two expressions for $B(v)$ are relatively

FIGURE 5-10   **Maxwell-Boltzmann distribution for N$_2$ at three temperatures.** The hatched area indicates the fraction of nitrogen molecules at 450 K with speeds in the range v to $v + \delta v$ and is, for small δv, just $B(v)\,\delta v$ with $B(v)$ given by (5-44). The total area under any curve must equal unity since it represents the fraction of molecules having all speeds from zero to "infinity", which includes all molecules. The areas underneath the curves to the right of the vertical dashed line at 730 m s^{-1} represent the fractions 0.007, 0.035, and 0.261 of the entire areas, in the order of increasing temperature. These fractions represent the fractions of molecules with speeds equal to or above 730 m s^{-1}.

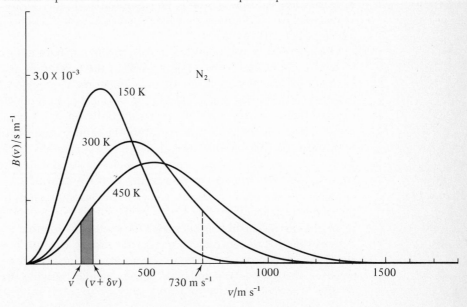

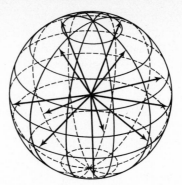

FIGURE 5-11   **Velocity vectors with the same speed** v.   The arrows represent velocity vectors from the origin of the sphere to various points on its surface. Since the velocity vectors correspond to the same speed, they all have the same length and the points of their tips all lie on the spherical surface, represented in the figure by longitude and latitude lines. The area of the surface is $4\pi v^2$ because v is the radius of the sphere.

unimportant; what is important to remember is the way in which $B(v)$ depends on the speed v, both by a factor v^2 and by an exponential, $\exp[-(mv^2/2kT)]$.

It may at first seem strange that the Boltzmann factor, which indicates an exponential decrease in population with increasing energy, gives rise to a distribution function with a maximum at an energy (and thus speed) far from zero. A negative exponential function, such as that in the Boltzmann factor, has its maximum value when the exponent is zero, corresponding here to a velocity of zero. The explanation of the maxima in the curves of Figs. 5-9 and 5-10 is directly related to the fact that the distribution contains two conflicting factors: the Boltzmann factor, giving an exponential decrease with v, and the v^2 factor. At first the latter dominates and $B(v)$ increases with increasing v, but soon (more rapidly at lower temperature) the exponential term dominates and $B(v)$ decreases, with the tail of the function at high v being essentially exponential.

This interplay of two opposing factors, giving a maximum in a function that falls to zero or approaches zero on either side of the maximum, is a common situation in chemistry and physics. A simple two-dimensional analogy, which is relevant also to the electron distribution in atoms (Sec. 9-4), may be useful. Consider a hypothetical city for which the population density falls off exponentially with increasing distance from the center of the city toward the suburbs and surrounding countryside. A problem of considerable importance for city planners is the population living within a certain radius of the center of the city. If we consider the number of people, $N(r)$, within different 0.1-km ranges, 0.0 to 0.1, 0.1 to 0.2, ..., 5.1 to 5.2, and so on, we find again that the number goes through a maximum (Fig. 5-12). Although the number of people per unit area of the city, the population density, decreases regularly with increasing distance from the center, the number of people living within a given range at distance r, $N(r)$, increases at first because the area of the ring in which they live increases in proportion to the distance from the center. Eventually, however, the very low population density of the outlying areas (the factor e^{-r})

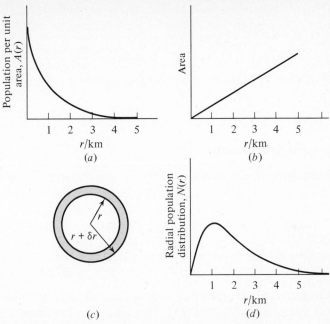

FIGURE 5-12   **Two-dimensional analogy to the Maxwell-Boltzmann distribution.**   (*a*) The population density $A(r)$ (population per unit area) in a hypothetical city is assumed to decrease exponentially with the increasing radius r, i.e. $A(r) = ce^{-r}$, with c a constant. (*d*) The radial population distribution $N(r)$ is so defined that the number of people living in a ring lying between r and $r + \delta r$, shown in (*c*), is approximately $N(r)\,\delta r$ and is equal to the product of $A(r)$ and the area of the ring. This area, which is approximately $2\pi r\,\delta r$, increases linearly with r as shown in (*b*). Both approximations improve as δr decreases and apply exactly in the limit as δr approaches zero. Thus, $N(r) = 2\pi r A(r) = 2\pi r c e^{-r}$ and the curve in (*d*) is the product of those in (*a*) and (*b*).

becomes dominant and overcomes the factor $2\pi r$ so that $N(r)$ decreases. $N(r)$ is called a radial distribution function because it describes the distribution as a function of the radial distance from the center. A similar function in three dimensions is used to describe the electron density in atoms (Sec. 9-4) and the distribution of molecules in liquids (Sec. 6-1).

EXAMPLE 5-18   **Kinetic energy of gas molecules.**   Suppose that in a sample of argon gas all atoms move in the same direction at a speed of 100 m s^{-1}. This state of affairs can last for only a brief moment because collisions of the atoms with the walls of the container and with other atoms redistribute the energy until a Maxwell-Boltzmann distribution is reached. What is the temperature of the gas at this final state of equilibrium, assuming that none of the energy of the gas atoms is transferred to the wall or to the surroundings?

Solution.   The average kinetic energy of an atom is $\langle \epsilon_k \rangle = \frac{3}{2}kT$ for the equilibrium distribution, which must be equal to the kinetic energy of each atom at the beginning.

Therefore, $mv^2/2 = \frac{3}{2}kT$, or, multiplying by $2N_A$, $\tilde{M}v^2 = 3RT$. Rearrangement yields $T = \tilde{M}v^2/3R$. Using the SI system and $\tilde{M} = 39.9 \times 10^{-3}$ kg mol^{-1} for Ar, we find

$$T = \frac{(39.9 \times 10^{-3} \text{ kg mol}^{-1})(10^4 \text{ m}^2 \text{ s}^{-2})}{3 \times 8.31 \text{ J mol}^{-1} \text{ K}^{-1}}$$

$$= \underline{16.0 \text{ K}}$$

Problems and Questions

5-1 **Application of gas laws.** A 224-liter steel reaction vessel is filled with hydrogen at 100°C and 1.00 atm. Calculate (*a*) the density of the gas; (*b*) the volume of the same quantity of hydrogen at 0°C and 1.50 atm; (*c*) the pressure of an equal weight of neon, Ne, in the same vessel at 100°C; (*d*) the moles of H_2 that must be removed if the pressure is to remain 1.00 atm when the vessel is heated to 200°C; (*e*) the volume of oxygen, measured at 20°C and 740 torr, needed to react with all the hydrogen to form water. *Ans.* (*a*) 0.066 g liter^{-1}; (*b*) 109 liters; (*c*) 0.100 atm; (*d*) 1.55 mol; (*e*) 90 liters

5-2 **Heating air.** Suppose that air were not soluble in body fluids and that you inhaled 2.00 liters of air at 1.00 atm and −20°C and then held your breath so that the air was warmed to body temperature (37°C). If there were no change in the volume available to the gas, what would be the change in pressure? Conversely, if the pressure of the gas were held constant, what would be the increase in volume?

5-3 **Application of gas laws.** A tank is filled with 1000 g nitrogen at 0°C and 16.0 atm. The tank is then heated to 50°C and the valve is opened. What is the total weight of the nitrogen that escapes if the external pressure is 1.0 atm and the temperature is kept at 50°C?
Ans. 9.5×10^2 g

5-4 **Molar volume.** One of the commonest errors of beginning students of chemistry is to assert that the volume occupied by a mole of a substance is 22.4 liters, with no further qualification. Under what conditions is this true? (Be neither too general nor too specific in your answer.)

5-5 **Volume and cost of natural gas.** Natural gas is a mixture of low-molecular-weight hydrocarbons such as methane (CH_4) and propane (C_3H_8). This mixture of gases is an excellent fuel and is stored prior to use in large natural underground reservoirs of fixed volume and relatively unchanging temperature or in large tanks on the surface of the ground. (*a*) Suppose that a supply of natural gas is stored in an underground reservoir of volume 6.0×10^5 m^3 at a pressure of 3.6 atm and a temperature of 17°C. How many storage tanks of volume 2.0×10^4 m^3 could be filled with this gas at 7°C and 1.20 atm? (*b*) Natural gas is normally sold to consumers by volume. Suppose a consumer obtains some gas in the winter, at an average temperature of 3°C and a pressure of 1.00 atm, and additional gas in the spring, at an average temperature of 17°C and 1.00 atm. If the price per unit volume is the same in winter and spring, will a given amount of money buy more, fewer, or the same number of moles of gas in winter as in spring? Explain your reasoning clearly, and indicate the percentage difference, if any.
Ans. (*a*) 87; (*b*) more in winter by 5 percent

5-6 **Change of units.** What numerical value would you have to assign the gas constant to get correct results when directly using the following units in the ideal-gas equation: pressure in torr, volume in cubic inches, and temperature in degrees Rankine (°R)?

5-7 **Limestone analysis.** A 1.00-g sample of a limestone produces 219 ml CO_2 at 30°C and 750 torr pressure when treated with HCl. What percentage of the mineral is $CaCO_3$ if this is the only carbonate in the mineral?   *Ans.* 87 percent

5-8 **Molecular formula.** A certain compound contains 60.0 percent carbon, 5.0 percent hydrogen, and 35.0 percent nitrogen. At 149 torr and 21°C its vapor has a density of 1.30 g liter^{-1}. What is the molecular formula of the compound?

5-9 **Molecular formula and atomic weight.** A 9.8-g sample of the gaseous oxide of a (hypothetical) element X occupies a volume of 2.00 liters at 27°C and a pressure of 1.00 atm. (*a*) What is the approximate molecular weight of the gaseous oxide? (*b*) If the oxide contains 40.0 percent oxygen, how many atoms of oxygen are there in each molecule of the oxide? (*c*) Give two values of the atomic weight of X consistent with the information given.
$$Ans. \quad (a) \quad 121; \quad (b) \quad 3; \quad (c) \quad 72/n, \, n = \text{integer}$$

5-10 **Molecular formula.** (*a*) An unknown gaseous hydrocarbon (that is, a compound containing only C and H) has a density of about 3.5 g liter^{-1} at 746 torr and 100°C. What is its approximate molecular weight? (*b*) When this gas is burned with an excess of oxygen, the products being carbon dioxide and water, the weight of water formed is just the same as the weight of hydrocarbon at the start. What is the empirical formula of the unknown hydrocarbon? (*c*) What is the molecular formula of the compound and what is its exact molecular weight? If there is a discrepancy from the value in (*a*), explain it.

5-11 **Octane and air mixture.** Octane (C_8H_{18}) is vaporized at 25°C and 1.00 atm into 1.00 liter air (containing 21 vol percent oxygen). What is the maximum weight of octane that can be present if it is to be burned completely to H_2O and CO_2? What is the volume of the mixture of air and octane vapor at 25°C and 1.00 atm?   *Ans.* 0.078 g; 1.02 liters

5-12 **Volume composition of a gas.** Given a gas mixture of the following volume composition: 20 parts He, 20 parts N_2, 50 parts NO, 50 parts N_2O. How many grams of nitrogen are there in 200 liters of the mixture at 0°C and 760 torr?

5-13 **Number of atoms in molecules.** Two elements A and B in the gaseous state may react in the following proportions. Three volumes of A react completely with two volumes of B to produce six volumes of a new gas that is definitely one compound, not a mixture. What can you say from these data about the number of atoms contained (*a*) in the molecules of element A, (*b*) in the molecules of element B, (*c*) in the molecules of the new compound?
$$Ans. \quad (a) \quad 2n; \quad (b) \quad 3m; \quad (c) \quad \text{at least two}$$

5-14 **Gas composition from its density.** A mixture of CO and CO_2 has a density of 1.82 g liter^{-1} at STP. What weight fraction of the gas is CO?

5-15 **Stoichiometry of a gas reaction.** A 100-ml sample of O_2 is added to 50 ml of a mixture of CO and C_2H_6 (at the same temperature and pressure) and the entire mixture is heated. The CO and C_2H_6 burn completely to CO_2 and H_2O; some of the oxygen may be left unreacted. After the initial temperature and pressure are restored and the water has been removed entirely, the gas volume is 85 ml. Find the volume fractions of CO and C_2H_6 in the original mixture (before the oxygen was added). *Ans.* CO, 0.60

5-16 Gas collected over water. A sample of nitrogen is collected over water at 30°C at which temperature the vapor pressure of water is 32 torr. The total pressure of the gas is 656 torr, and the volume is 606 ml. How many moles of nitrogen does the sample contain?

5-17 Vapor pressure of unknown liquid. A volume of just 4.00 liters oxygen was collected over a certain liquid at 25°C and a pressure of 750 torr. When the vapor of the liquid was removed from the oxygen by appropriate treatment and the oxygen volume was again measured, this time at STP, it was found to be 3.00 liters. Calculate the vapor pressure of the liquid; assume that the oxygen was initially saturated with vapor. *Ans.* 128 torr at 25°C

5-18 Mixture of oxygen and water vapor. A mixture of oxygen and water vapor at 90°C and 738 torr was cooled to 20°C. Some liquid water condensed from the sample during the cooling; at 20°C, the vapor pressure of water is 18 torr. The pressure of the gas after cooling was the same as that at the start; the volume of the gas had decreased to 0.500 of what it had been originally. What was the final pressure of the oxygen, and what was it initially (at 90°C)? (If you need this information, the vapor pressure of water at 90°C is 526 torr.)

5-19 Humidity of air. On a humid day in the tropics when the air was at 35°C, a 20.0-liter air sample was collected at a pressure of 1.00 atm. The air sample was dried by passing it over a salt that would remove all the moisture; the weight of the water removed was 0.72 g. (*a*) How many moles of water were removed from the air? (*b*) What was the pressure of water in the sample of air before it had been dried? (*c*) What relative humidity does this correspond to?
Ans. (*a*) 0.040 mol; (*b*) 38 torr; (*c*) 90 percent

5-20 Mixture of NO_2 and N_2O_4. Under ordinary conditions the chemical substance with simplest formula NO_2 is actually a mixture of two gaseous species, NO_2 and N_2O_4. At 45°C and 1.00 atm total pressure, the density of a sample containing only these substances is 2.56 g liter^{-1}. (*a*) What is the partial pressure of each component? (*b*) What is the percentage by weight of each component?

5-21 Deviation from ideal-gas law. At 50 atm and 50°C, 10.0 g CO_2 occupies 85 ml. What is the percentage deviation from the ideal-gas law? *Ans.* −29 percent

5-22 Real gases. Sulfur dioxide shows significant deviations from the ideal-gas law at STP. At 0°C its molar volume is 22,414 cm^3 at a pressure of 0.976 atm, rather than 1 atm. (*a*) Use the van der Waals equation (5-21) and the constants given in the text for SO_2 to calculate the pressure at which a mole of SO_2 occupies a volume of 22,414 cm^3 at 0°C. (*b*) Perform the same calculation using the more realistic virial equation of state [Equation (5-20)]. (At 0°C the second virial coefficient B for SO_2 is −530 cm^3 mol^{-1}.)

5-23 Isotherms of a real gas. Draw several typical isotherms, showing the relationship between P and V at constant T, for a real gas. These isotherms should be at temperatures for which (*a*) the gas behavior is approximately ideal, (*b*) the gas condenses to a liquid, and (*c*) the gas exhibits critical behavior. (*d*) Describe condensation in terms of one of the isotherms drawn. (*e*) Indicate on your diagram a path by which it is possible to change a gas into a liquid without going through a two-phase region, that is, a region in which the liquid phase and the gas phase are separated by a meniscus.

5-24 Helium-nitrogen mixture. A mixture of helium and nitrogen occupies a certain volume at a temperature of 27°C and a pressure of 4.00 atm. The same sample is allowed to expand to a volume exactly five times the original volume, and the temperature is then changed until the pressure is just 2.00 atm. (*a*) What is the final temperature (°C)? (*b*) If the gas mixture contains just 33 percent helium by weight, what is the final partial pressure of nitrogen? (*c*) What is the ratio of the average molecular velocity of helium to that of nitrogen in the mixture?

5-25 Mixture of gases. A flask of volume 8.2 liters contains 4.0 g hydrogen, 0.50 mole oxygen, and sufficient argon so that the partial pressure of argon is 2.00 atm. The temperature is 127°C. (*a*) What is the density of the mixture in the flask? (*b*) What is the total pressure in the flask? (*c*) What is the mole fraction of hydrogen? (*d*) Suppose that a spark is passed through the flask so that the following reaction occurs until one reactant is entirely used up:

$$2H_2(g) + O_2(g) \longrightarrow 2H_2O(g)$$

What will the pressure be when the temperature returns to 127°C?

   Ans. (*a*) 4.9 g liter^{-1}; (*b*) 12.0 atm; (*c*) 0.67; (*d*) 10.0 atm

5-26 Comparison of two gases. Suppose that you have two flasks of the same volume and at the same temperature, one containing gas A and the other gas B. The density of A is just four times that of B in these flasks. Assume that the gases are ideal. What can you conclude about the ratio of the pressures of A and B in the flasks? (If additional information is needed, state precisely what it is and show just how you would use it.)

5-27 Gaseous diffusion and molecular velocities. (*a*) A certain molecule X is observed to diffuse 0.50 times as rapidly at 300 K as does oxygen at 400 K. What is the molecular weight of X? (*b*) At what temperature will the rms speed of a hydrogen molecule be the same as that of a molecule of Cl_2 at 100°C?   *Ans.* (*a*) 96; (*b*) 10.6 K

5-28 Comparison of oxygen and helium. Imagine two boxes of equal size at the same temperature and pressure, one containing oxygen and the other helium. Compare the contents of the boxes as quantitatively as possible in the following respects: (*a*) number of molecules, (*b*) total mass, (*c*) average kinetic energy of molecules, (*d*) average number of impacts per second on walls of box, (*e*) average force exerted on wall by each collision.

5-29 Boltzmann factor and thermal excitation of nitrogen. At 1000 K, 3.50×10^{-2} as many molecules in a sample of N_2 are in the first excited vibrational-energy state as in the ground state. How far above the ground state is the first excited state, in kilojoules per mole?
   Ans. 27.9 kJ mol^{-1}

5-30 Boltzmann distribution. The relative energies of four states are 0, 500, 2000, and 3000 J mol^{-1}. The relative equilibrium populations at a temperature T of the lower two states are 100.0 and 50.0. (*a*) What are the populations of the third and fourth states relative to a population of 100.0 for the lowest state? (*b*) What is T?

"While the behavior of molecules in a gas can be described mathematically with very great precision, leading to elegant and simple laws, the picture of molecular motion in liquids is unpleasant and untidy. The molecules forming a liquid can be compared to Japanese beetles caught in abundance in a patented trap, or to the worms in a fisherman's can. They crawl and wiggle around each other, they constantly come to clinches like two inexperienced boxers, and they do not permit a physicist to say anything simple or reasonable about them. We can build simple theories and formulate simple mathematical laws only when things themselves are simple, and the motion of molecules in liquids is certainly not that."

George Gamow[1]

6 Liquids and Liquid Solutions

Liquids, and especially solutions in liquid solvents, the concerns of the present chapter, play a central role in our lives and in the world about us. The most familiar liquid, water, is essential to all forms of life—virtually all biochemical reactions occur in aqueous media. The geochemical processes that continually shape the earth involve molten solids or aqueous solutions, and the great majority of the important industrial chemical processes used to prepare synthetic plastics and fibers, medicinal chemicals, photographic supplies, petroleum products, ultrapure materials for transistors, and myriad other products of modern civilization involve liquids and solutions

[1]G. Gamow, "Matter, Earth, and Sky", p. 204, Prentice-Hall, Inc., Englewood Cliffs, N.J., 1958.

at various stages. The solvent properties of many liquids, and especially the remarkable properties of water, make possible homogeneous conditions for carrying out chemical reactions. Such conditions are enormously advantageous because they ensure uniform and intimate contact of the reactants.

6-1 The Liquid State

Although the nature and structure of liquids are rather well understood qualitatively, no good quantitative theories of the liquid state have been developed, despite intensive efforts over many years by able theoreticians. The reasons are that (1) the molecules of a liquid are close together, so that the essential simplifying assumptions of the kinetic theory of gases—negligible intermolecular forces and negligible molecular volumes—are totally inapplicable, and (2) the molecules are not packed in an ordered array, so that the symmetry that characterizes the crystalline state and makes it susceptible to precise quantitative treatment is also lacking.

Structure of liquids. It is extremely difficult to design experiments that will give information about the microscopic structure of liquids, and many of our ideas about this subject come from calculations made with computers. In essence the computer is used to imitate or simulate nature in order to permit a detailed microscopic picture of a model of a liquid to be obtained at any instant. In a typical computer simulation experiment, the trajectories of a few hundred molecules in a cubic volume about 20 molecular diameters on a side are calculated. At the start of the calculation the molecules are given random kinetic energies whose average is consistent with the desired temperature. The positions and velocities of all the molecules are then calculated as they change with time because of collisions and less direct interactions. Throughout the study, the interactions of each molecule with all other molecules in the volume are computed according to some specified intermolecular potential-energy function. When a molecule leaves the volume, the computer program generates an identical molecule that enters from the opposite side, so as to avoid spurious effects that would result from collisions with the walls.

The basic reliability of the simulation technique can be checked in a number of ways. For example, in a properly designed simulation the system reaches an equilibrium condition in which the molecular velocities follow the Maxwell-Boltzmann distribution, which holds for liquids as well as gases. The temperature of the system is the T in the Maxwell-Boltzmann expression [Equation (5-44)], const $\times v^2 \exp(-mv^2/2kT)$. There is little doubt that the results obtained from computer studies exhibit the essential characteristics of real liquids and can therefore be used to gain insight into the microscopic properties of the liquid state. The basic ideas that have emerged are these: to a good approximation, the structure of a simple liquid such as argon or methane is determined primarily by the repulsive forces between molecules; the attractive forces act only as a "glue" that holds the fluid together. Thus, when structure is considered, one might picture liquid argon as a collection of equal-size marbles. At high densities (i.e., when the marbles are confined to a small volume) the marbles pack into an ordered solid structure and have

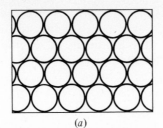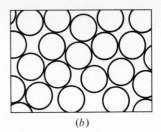

(a)   (b)

FIGURE 6-1   **Effect of density change on regularity of packing.** (*a*) Ordered structure at high density; (*b*) disordered structure at slightly lower density.

little mobility (Fig. 6-1*a*). However, if the density is decreased slightly, by about 10 percent, the highly ordered structure breaks down and the marbles can move past each other (Fig. 6-1*b*). For almost all real substances, melting is accompanied by a similar decrease in density and increase in mobility, in agreement with this "hard sphere" picture.

There are some exceptions to this generalization, however, the most notable being water, which at the melting temperature is about 8 percent *more* dense than ice. This is a consequence of the fact that at ordinary pressures the structure of ice is quite open (Fig. 21-4), with only four nearest-neighbor water molecules about any given molecule instead of the maximum of twelve that is possible for a close-packed ordered array. With such loose packing, liquid water has a structure with sufficient vacancies to be fluid, even though it is more dense than ice.

Any structural view of a liquid must take into account that liquids are dynamic systems. The relative positions of the molecules in a liquid are constantly changing. If we could film the molecular motion (with an ultramicroscopic and ultra-high-speed camera), we could examine each frame of the film to determine at a given instant the geometric arrangement of the molecules—the liquid structure. This information is available from computer simulations.

A typical "frame" for a monatomic liquid is shown in Fig. 6-2. Note that there

FIGURE 6-2   **Computer simulation of a liquid.** Note that collisions of atoms with the walls have been eliminated by designing the program so that atoms that leave the boundary on one side reenter on the other side.

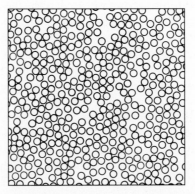

are regions in which the ordered structure typical of a solid is seen, but in many places the order is destroyed because one or more atoms are missing. The order in the liquid is of short range. Many atoms are surrounded by a first shell of neighbors in a nearly regular arrangement, but the regularity of packing with respect to a given "central" atom rapidly decreases as the distance from that atom is increased. Remember, also, that the frame shown is an instantaneous view. In another instant the positions of the atoms will have altered and the arrangement around any given atom will have changed.

One way in which the structure of a liquid can be represented is to construct a plot of the number of atoms at a distance r from an average atom. Such a curve, called a radial distribution function, can be computed from simulation calculations, and one such distribution function is shown in Fig. 6-3a. Radial distribution functions for real liquids can be obtained by X-ray or neutron diffraction studies. The data in Fig. 6-3b to d are typical and represent three different kinds of liquids: (b) a monatomic liquid, Ar, with only comparatively weak, nondirectional forces between the atoms; (c) water, for which the primary intermolecular attraction results from hydrogen bonding, which has some directional character; and (d) an ionic liquid, molten LiCl, with strong (but also nondirectional) ionic forces between the components of the melt. As in the simulation results, there is a continuous range of interatomic distances above a certain minimum that corresponds approximately to the sum of the van der Waals (or ionic) radii of the components. This range is in marked contrast to the distribution of distances in the corresponding solids, for which only certain discrete distances occur, represented in Fig. 6-3b to d by the dotted peaks.

Energy requirements in phase transitions. There is molecular[1] motion in all solids, which consists of vibrations of the molecules about fixed positions and is present even at very low temperatures. This vibrational motion in solids has been simulated in computer studies (Fig. 6-4a). As the temperature of the solid is increased, the amplitude of the motion increases but, at least in the simplest cases, the average positions of the molecules remain substantially unchanged and the solid retains its integrity.

When the temperature of a solid reaches a certain value, the motion becomes sufficiently great that the solid melts to a liquid (Fig. 6-4b) and an equilibrium can be established between solid and liquid. Like all equilibria, this equilibrium represents a state of balance. Here the balance is between a tendency for the energy to decrease (corresponding to the stronger molecular interactions in the solid) and a tendency toward disorder (corresponding to the more random arrangement in the liquid). The disordering tendency becomes more important with increasing temperature, as we shall see when we consider the thermodynamics of equilibria (Chap. 17).

The melting of a solid is invariably accompanied by absorption of energy. This fact is easily understood when the volume of the liquid is somewhat greater than

---

[1]In this discussion the terms *molecule* and *molecular* refer to any of the basic units that comprise a solid—atoms, ions, or molecules.

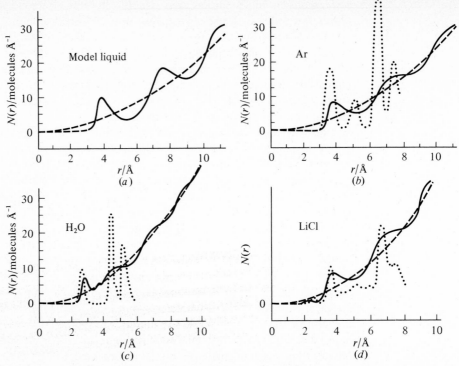

FIGURE 6-3   **Radial distribution curves for four liquids.**   The first curve is an adaptation of the results found for a computer-simulated monatomic liquid, of which Fig. 6-2 showed an instantaneous view. Curves (*b*), (*c*), and (*d*) show X-ray diffraction results. The vertical coordinate for the curves, $N(r)$, is defined so that $N(r) \, \delta r$ is the number of neighbor molecules in a spherical shell of thickness δr at a distance r from an average molecule. The area under the distribution curve out to a distance r' is the total number of neighbors at distances less than or equal to r'.

The dotted peaks represent the first few shortest interatomic distances in the corresponding crystalline solids, broadened by the effects of thermal motion. The areas of these peaks are proportional to the number of atom pairs at the corresponding distances.

All distribution curves for the liquids are flat for very small interatomic distances, indicating zero neighbors at distances below the closest possible approach. A sequence of broad peaks follows, which are attributed to preferred distances of separation, separated by minima, indicating less likely distances of separation. At distances large compared to atomic dimensions, the curves merge with a parabolic curve (shown dashed) that corresponds to the average number of neighbors expected for a liquid without distinct structure.

that of the solid. For such substances, the average interatomic distance increases on melting, so that some energy must be expended in increasing the average separation of the molecules slightly and thus raising their average potential energy. This explanation is oversimplified, however, for energy is also absorbed during the melting of those few solids, such as ice, that *contract* on melting. This paradox can be resolved by noting that when these substances melt, some of the shortest intermolecular distances lengthen. These increases in length are more than offset by decreases in

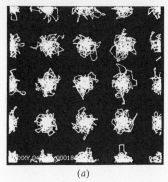
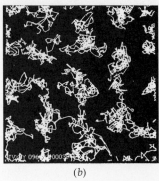

(a)   (b)

FIGURE 6-4   **Atomic motion in solids and liquids.** The two figures show computer-simulated projections of the trajectories of 32 rigid spheres during the time it takes them to make 3000 collisions inside a cubic box. In (a) a solid is simulated; in (b), a liquid. In the solid each sphere moves in a well-defined region; in the liquid the regions of individual motion are much less well defined and a sphere may move from one region to another. The reason that in the liquid there are regions of individual motion at all is that the space taken up by one sphere cannot also be occupied by another; the spheres are impenetrable, as is essentially true also of atoms and molecules. (*B. J. Alder and T. E. Wainwright, Scientific American, p. 113, October 1959.*)

some of the longer distances, so the average distance decreases and the substance becomes more dense. However, the shape of a typical intermolecular potential-energy diagram (Fig. 3-13) shows that much more energy is required to separate nearby molecules by a given amount than is gained when more distant ones move the same distance closer. There may well be other specific effects that also increase the potential energy. For example, stabilization by hydrogen bonding may decrease when ice melts because of increasing distortion of hydrogen-bonding arrangements.

We now turn to the phase transition between liquid and vapor. Since the distribution of kinetic energy among the molecules of a liquid follows the Maxwell-Boltzmann law, Equation (5-45), the average kinetic energy is $\frac{3}{2}kT$ per molecule or $\frac{3}{2}RT$ per mole, just as in the vapor. The fact that at a given temperature the average *kinetic* energy is the same in the vapor and the liquid does not mean that the average potential energy is the same. Were this so there would be no energy change during the transition from liquid to vapor. The process of vaporization corresponds, at ordinary temperature and pressure, to an increase in the average intermolecular separation by a factor of about 10, since the volume increases by about 10^3. Because this increased separation is associated with an increase in the potential energy, energy must be supplied to vaporize a liquid. The amount of energy required varies widely for different liquids, depending on the strength of the interactions among the particles of which the liquid is composed.

6-2   Solutions

Solutions are homogeneous mixtures. They are intimate mixtures on the molecular level and may exist in all three states of aggregation. Liquid solutions are the most familiar and are our primary concern here. However, since all gases mix homogeneously with one another in any proportions, all mixtures of gases can be classified as solutions. Solid solutions also exist, not only for amorphous solids but for crystals as well, because in many crystals some of the original atoms, ions, or molecules may be replaced randomly by foreign atoms, ions, or molecules without destroying the regular structure characteristic of the original substance.

Composition of solutions. The predominant component of a solution is referred to as the *solvent,* the minor components as the *solutes.* These terms are somewhat arbitrary when the components are present in equal or nearly equal amounts, but this is not a common situation. The proportions of the components may always be varied over at least a narrow range, with an accompanying continuous variation in the properties of the solution. Any particular solution has a definite composition, which may be specified in many different ways. For example, dry air contains 21.0 percent oxygen by volume, which means that 100.0 liters of air can be reproduced by combining 21.0 liters of oxygen with 79.0 liters of the remaining components of air (chiefly nitrogen), all at the same temperature and pressure. Note the difference between weight and volume percent (Example 5-13). There is 23.2 wt percent oxygen in air, so that 100.0 g air contains 23.2 g oxygen. A percentage composition without further specification usually implies weight percent, because the weights of the components of a solution are additive whereas the volumes often are not. Composition by weight is therefore easier to establish.

EXAMPLE 6-1   **Weight and volume percent.** A solution of alcohol (ethanol) in water containing 33.36 wt percent alcohol is said to contain 40.00 vol percent alcohol at 4°C. Its density at this temperature is 0.9510 g ml^{-1}; that of pure alcohol is 0.7932 g ml^{-1} and that of water is 1.0000 g ml^{-1}, both at 4°C. How many milliliters of pure alcohol and how many milliliters of water must be mixed (at 4°C) to yield 100.00 ml of the mixture?

Solution. The 100.00 ml of mixture weighs 95.10 g and contains 95.10 g × 0.3336 = 31.73 g alcohol and 63.37 g water. The pure alcohol has a volume of (31.73 g)/(0.7932 g ml^{-1}) = 40.00 ml, and the 63.37 g water has a volume of 63.37 ml. The "40.00 vol percent alcohol" thus refers to the volumes of pure alcohol and final solution; the solution could with equal justification be described as containing 63.37 percent water by volume. The two volume percentages do not add to 100.00, because there is a substantial contraction when alcohol and water are mixed.

Saturated solutions. The *concentration* of a solution is a quantitative measure of the relative amounts of solute and solvent present—for example, moles of solute per liter of solution. Different concentration units are discussed below and in the following chapter.

When a solution of a substance is in contact with additional pure solute, the dissolving of more solute depends upon the concentration of the solute already in solution. A solution is said to be *saturated* when there is equilibrium between dissolved solute and any excess solute. This equilibrium is a dynamic one, like those involving vapor pressure discussed in Sec. 3-1. At equilibrium the rates at which the solute goes into the solution and comes out of it are equal, so that there is no net change in the amount of dissolved material. The concentration of solute in a saturated solution is called the *solubility* of the substance; the amount of excess undissolved solute present does not affect the solubility. Two liquids may be mutually soluble (*miscible*) in all proportions, as, for example, are ethanol and water. For such solutions, the term *solubility* has no meaning since the solution cannot be saturated.

Supersaturation. It is sometimes possible to prepare solutions that contain more solute than they would under equilibrium conditions. For example, the temperature of a saturated solution of a solid may be changed carefully to a temperature at which the solubility is smaller. When this is done without agitation in a vessel with smooth walls and in the absence of specks of solid impurities, the excess of the solute often fails to crystallize. Such solutions are called *supersaturated*. The phenomenon is in many respects similar to those of superheating and supercooling. Scratching the container walls or introducing a tiny crystal of the solute (or occasionally even a speck of *any* solid) usually causes the excess solute to crystallize. To give an example, a saturated solution of $Na_2SO_4 \cdot 10H_2O$ contains 92.8 g of this substance per 100 g water at 30°C and 13.8 g at 5°C, but when a filtered solution saturated at 30°C is carefully cooled, no solid precipitates and the solution is supersaturated. If a tiny seed crystal of the solute is dropped into the supersaturated solution, however, crystallization begins immediately.

Solutions of gases in liquids can also be supersaturated. The solubility of all gases increases with increasing pressure and decreases with increasing temperature. Consequently when the temperature of such a solution is raised, or the pressure of gas above the solution is decreased, the amount of dissolved solute exceeds the solubility under the new conditions. If the solution is not disturbed, it may remain supersaturated for some time—but if it is shaken, it usually releases the excess gas almost explosively, as anyone knows who has ever shaken a warm, freshly opened bottle of a carbonated drink.

Volatile solutes. When the solute is a volatile solid or liquid, the partial pressure of the solute must be the same for all phases in equilibrium with each other. For example, its partial pressure above a saturated solution in equilibrium with pure solute must be the same as that of the pure solute (Fig. 6-5). The reason is that solute and saturated solution may interact through the vapor phase as well as by direct contact. Equilibrium can exist only if it covers all possible paths of interaction. If the vapor pressure of the pure solute were higher than its partial pressure above the saturated solution, the solute would evaporate and condense in the solution.

For example, the vapor pressure of solid iodine at 25°C is 0.31 torr and this is also the partial pressure of iodine above a saturated solution of iodine in water, or in any other solvent, at 25°C. It is even more instructive to consider a saturated solution of ether in water, which may be obtained by shaking the two liquids together. When equilibrium has been reached and the shaking is stopped, two layers separate (Fig. 6-6). The aqueous layer at the bottom is saturated with ether, and the ether layer on top is saturated with water. This does not imply that the proportion of ether to water in the bottom layer is the same as that of water to ether in the top

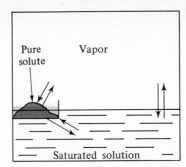

FIGURE 6-5   **A saturated solution of a volatile substance.**   The wire basket contains excess solid that is partly immersed in the solution. At equilibrium the solution is saturated and no further solute dissolves. The existence of equilibrium implies that the vapor pressure of the solute above the *pure* solute and above the *saturated solution* must be the same.

layer, but these two ratios are such that (1) the partial pressure of ether is the same over each phase and (2) the partial pressure of water is the same over each phase.

6-3   Concentration Units

One significant advantage of working with solutions in carrying out chemical reactions is that precise and accurate measurements of liquid volumes can readily be made. Thus it is often easy, with concentration units expressed in terms of the amount of solute per unit *volume* of solution, to study the stoichiometry of chemical reactions or to make measurements of reaction rates or positions of equilibrium. Concentration

FIGURE 6-6   **Saturated solutions of water in ether (above) and ether in water (below) in a separatory funnel.**   When equal volumes of water and ether [more properly diethyl ether, $(C_2H_5)_2O$] are shaken together until each of these liquids is saturated with the other, two layers separate. The less dense is a solution of water in ether containing, at 20°C, about 1.2 wt percent water. The more dense layer is a solution of ether in water containing about 6.5 wt percent ether at 20°C. The separatory funnel permits convenient separation of two immiscible liquids.

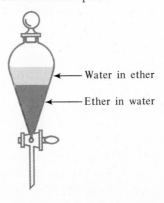

units based on the *weight* of the solution or its components are also sometimes used, particularly in the study of what are called the colligative properties of solutions (Sec. 6-4).

Volume concentration units. Among the most common units of concentration expressed as amount of solute per unit *volume* of *solution* are molarity and formality, defined as follows:

1. *Molarity,* symbolized by square brackets around the chemical formula of the solute:

$$[A] = \frac{\text{number of moles of molecular or ionic species A}}{\text{volume of } solution \text{ in liters}} \tag{6-1}$$

The units of molarity are thus moles of *solute* per liter of *solution,* abbreviated mol liter^{-1} or just *M*. For example, a solution that contains 0.1000 mol A liter^{-1} is said to be 0.1000 molar in A, or 0.1000 *M* in A.

2. *Formality,* equal to

$$\frac{\text{Number of gram formula weights of solute}}{\text{Volume of } solution \text{ in liters}} \tag{6-2}$$

The units of formality are thus gram formula weights of *solute* per liter of *solution,* abbreviated gfw liter^{-1} or just *F* (read as "formal").

Another volume concentration unit, *normality,* is of considerable importance in solution stoichiometry. It is discussed in the following chapter. A few other volume concentration units are also occasionally used, the most common being g solute per 100 ml of *solvent,* a unit used in many handbooks to express solubilities (especially in water). To convert information given in this way into formalities, it is necessary to know the density of the solvent and the density of the solution or to have information from which these densities can be calculated.

EXAMPLE 6-2   **Interconversion of concentration units.** The solubility of sodium chloride is listed in a handbook as 36 g per 100 ml water at 0°C. The density of the resulting solution is 1.21 g ml^{-1} and the density of water at this temperature is 1.00 g ml^{-1}. What is the formality of the solution?

Solution. To calculate the formality, one must know the number of gram formula weights per liter of *solution*. One liter of solution weighs (1000 ml) (1.21 g solution/ml solution) = 1210 g. However, this weight includes both solute and solvent. The fraction of it attributable to NaCl must be known. To calculate this, we note that the solution is prepared by dissolving 36 g NaCl in 100 ml, or 100 g water, yielding 136 g solution. Since 136 g solution contains 36 g NaCl, 1210 g solution contains $(1210 \times \frac{36}{136})$ g = 320 g NaCl. Finally, the formula weight of NaCl is 35.5 + 23.0 = 58.5, and the number of formula weights of NaCl in 1 liter is $\frac{320}{58.5}$ gfw = 5.5 gfw. This is, by definition, the formality, and the solution is 5.5 *F*.

EXAMPLE 6-3   **Formal and molar composition.** What is the composition in terms of formalities and molarities of a solution, 500 ml of which contains 0.200 gfw $CaCl_2 \cdot 2H_2O$
and 0.150 gfw HCl?

Solution. Since 1000 ml solution contains twice the quantities stated, the solution is
<u>0.400 *F*</u> in $CaCl_2 \cdot 2H_2O$ and <u>0.300 *F*</u> in HCl. An alternative description would be in
terms of the anhydrous salt $CaCl_2$, 0.200 gfw of which could be substituted (per 500 ml
solution) for the dihydrate; the solution is 0.400 *F* in $CaCl_2$ and 0.300 *F* in HCl. In terms
of molarities, both $CaCl_2$ and HCl are essentially fully ionized, so that
$[Ca^{2+}] =$ <u>0.400 *M*</u>, $[Cl^-] = 2 \times 0.400 + 0.300 =$ <u>1.100 *M*</u>, and $[H^+] =$ <u>0.300 *M*</u>. The
water of hydration has no bearing on the molar composition of the solution.

Volume concentration units have many practical advantages, but one significant
disadvantage is that they are temperature-dependent. As the temperature changes,
the volume of solution changes while the quantity of solute remains constant, so that
their ratio, the concentration, also changes. Consequently, careful temperature
control is needed in precise quantitative volumetric work.

Weight concentration units. Concentrations that are expressed as a ratio of
weights (or as numbers proportional to weights, such as numbers of moles) do not
vary with temperature, because weights are essentially temperature-independent.
Among useful weight concentration units are:

1. Weight molality $= \dfrac{\text{moles solute}}{\text{kg } solvent} = c_w$ (6-3)

Concentrations expressed in this way will be denoted by c_w. They are sometimes
referred to as mola*l* concentrations, the final letter distinguishing them from mola*r*
concentrations, defined in (6-1) above. However, we usually use the phrase *weight
molal* to help emphasize the distinction.

2. Weight formality $= \dfrac{\text{gfw solute}}{\text{kg } solvent}$ (6-4)

Here the prefix *weight* is essential, since the term *formality* alone refers to the unit
defined in (6-2) above.

3. Weight percent $= \dfrac{\text{g solute}}{\text{g } solution} \times 100$ (6-5)

It is important to note that in the definitions of weight molality and weight formality
the denominator is the kilograms of solvent, not the kilograms of the entire solution.
   A useful feature of weight concentrations is that the weights of solute and solvent
are additive, their sum being the weight of the solution. It is only rarely and accidentally true that the volume of a solution is the sum of the volumes of solute and solvent.
No knowledge of densities is required to find weight formality or weight molality
from the weight percent of a solution, but to transform a weight concentration to a
volume concentration does require such knowledge.

Mole fraction. The mole fraction, defined in Equation (5-19) for gas mixtures, is also a useful concentration unit for solutions. Let the molecular or ionic species in a solution be designated by the index i, this index running over all species present, including the solvent. The solution then consists of n_1 moles of species 1, n_2 moles of species 2, and so on if there is more than one solute. Then

Total number of moles present: $\qquad n = \Sigma\, n_i$

$$\text{Mole fraction of component } i\text{:} \qquad x_i = \frac{n_i}{n} \qquad\qquad (6\text{-}6)$$

The sum of all mole fractions is necessarily unity:

$$\Sigma\, x_i = 1 \qquad\qquad (6\text{-}7)$$

EXAMPLE 6-4   **Interconversion of concentration units.** A certain solution of HNO_3 in water contains 30.0 percent by weight of HNO_3 and has a density of 1.180 g ml^{-1}. Calculate (*a*) its formality, (*b*) its weight formality, and (*c*) its composition in terms of molarities, weight molalities, and mole fractions, assuming that in this solution the acid is 84 percent dissociated,[1] i.e., that 16 percent exists as HNO_3 and 84 percent as H^+ and NO_3^-.

Solution

(*a*) To calculate the formality, we proceed much as in Example 6-2. To find the number of gfw HNO_3 (FW = 63.0) in 1000 ml of solution, we first use the density of the solution to calculate its weight,

$$\text{Weight of 1000 ml solution} = 1000 \text{ ml} \times 1.180 \text{ g ml}^{-1} = 1180 \text{ g} \qquad (6\text{-}8)$$

and then use the percentage composition to calculate the weight of HNO_3 in this weight of solution,

$$\text{Weight of } HNO_3 \text{ in 1000 ml} = \frac{\text{wt percent } HNO_3}{100} \times \text{wt of solution}$$

$$= 0.300 \times 1180 \text{ g} = 354 \text{ g} \qquad (6\text{-}9)$$

Finally, gfw HNO_3 per liter $= \frac{354}{63.0}$ gfw $= \underline{5.62 \text{ gfw}}$, which is numerically equal to the formality.

(*b*) To calculate the weight formality, we must know the number of gfw HNO_3 per kg H_2O (the solvent in this solution). From part (*a*) above we know that for each 1180 g solution there are 354 g HNO_3, and thus, by difference, 826 g H_2O. Thus there are $(1000 \times \frac{354}{826})$ g $= 428.6$ g HNO_3 per 1000 g of water, or $\frac{428.6}{63.0}$ gfw $= 6.80$ gfw HNO_3 per kilogram of solvent, and hence the weight formality is $\underline{6.80.}$

(*c*) The molar composition is described by $[H^+] = [NO_3^-] = 0.84 \times 5.62 = \underline{4.7\ M,}$ and $[HNO_3] = 0.16 \times 5.62 = \underline{0.9\ M.}$ In terms of weight molalities, the concentrations of both H^+ and NO_3^- are $0.84 \times 6.80 = \underline{5.7 \text{ weight molal}}$ while HNO_3 is $0.16 \times 6.80 = \underline{1.1 \text{ weight molal}}.$ To calculate the mole fractions of all constituents, H^+,

---

[1] In less concentrated solutions, for example, 1 F or less, HNO_3 is almost completely dissociated.

NO_3^-, HNO_3, and H_2O, we note that 1.000 kg water contains $\frac{1000}{18.02} = 55.5$ mol water; this kilogram of water contains, from the foregoing, 5.7 mol H^+, 5.7 mol NO_3^-, and 1.1 mol HNO_3, a total of 68.0 mol. The individual mole fractions are $x_{H^+} = x_{NO_3^-} = \frac{5.7}{68.0} = \underline{0.084}$; $x_{HNO_3} = \frac{1.1}{68.0} = \underline{0.016}$; $x_{H_2O} = \frac{55.5}{68.0} = \underline{0.816}$. The sum of the mole fractions is 1.000, as it must be.

6-4   Colligative Properties of Solutions

There are certain properties of any given solution that are quantitatively interrelated and are therefore known as *colligative properties* (Latin: *colligare,* to tie together). They include the lowering of the solvent vapor pressure, the elevation of the boiling point, the lowering of the freezing point, the osmotic pressure, the pressure dependence of gas solubilities, and the distribution law that describes the distribution of solute between two immiscible solvents. The magnitude of each of these effects depends on the concentration and temperature of the solution, the pressure, and other variables. Once the detailed dependence of any one colligative property on some variable has been determined for a given solvent-solute system, the quantitative dependence of all other colligative properties on this same variable follows. The theoretical framework that permits this correlation of different solution properties is *thermodynamics* (Greek: *therme,* heat; *dynamis,* power), a theory that was a product of investigations on properties of heat and temperature and thus might not be expected to have a bearing on the topics discussed here. Nevertheless, the development of thermodynamics showed that the consequences of its laws are far-reaching and penetrate almost every branch of natural science. It is a characteristic of thermodynamics that its application *correlates* different phenomena *quantitatively.* To illustrate, neither the osmotic pressure law nor the law of the lowering of the vapor pressure upon dissolving a substance follows from thermodynamics, but once either of them has been established, the other can be derived quantitatively from the first by the application of thermodynamic laws.

The colligative properties of solutions are described and illustrated in the following pages. In general, the quantitative laws stated below are only approximately valid for most solutions, but they become increasingly applicable as solute concentrations decrease, that is, for *dilute* solutions.

Raoult's law.   The vapor pressure of a solution usually lies between the vapor pressures of its pure components. An approximate relation describing the partial pressure of the solvent P_1 is Raoult's law

$$P_1 = P_1^{\bullet} x_1 \tag{6-10}$$

where $P_1^{\bullet}$ is the vapor pressure of the pure solvent at the temperature in question and x_1 is the mole fraction of the solvent. A solution for which Raoult's law holds exactly, over the entire concentration range from solute mole fraction zero to unity and at all temperatures and pressures, is said to be *ideal.* No change in volume occurs upon dissolving the solute in an ideal solution, nor is there any release or absorption of energy.

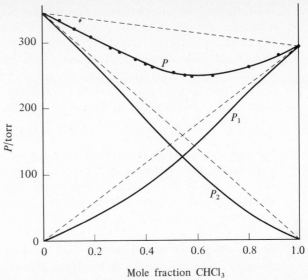

FIGURE 6-7   Vapor pressure of a solution. The graph shows the partial and total pressures for solutions of chloroform and acetone as a function of mole fraction at 35°C. Full lines signify measured pressures: P_1 = partial pressure of chloroform, P_2 = partial pressure of acetone, P = total pressure. Dashed lines represent the partial and total pressures calculated from Raoult's law [Equation (6-10)]. Note that for high concentrations of chloroform P_1 approaches the Raoult's law value. Similarly, Raoult's law is a good approximation for P_2 at high concentrations of acetone. When, as in this case, the total pressure is less than the ideal-solution value, we speak of negative deviations from Raoult's law. Positive deviations from Raoult's law, that is, $P > P_{ideal}$, are also found.

If the solute in an ideal solution is also volatile, Raoult's law applies to it as well, and indeed if there are several volatile solutes, it applies to all of them. Even for nonideal solutions (6-10) applies to the solvent reasonably well, at least when the solution is dilute, but there are often marked deviations in more concentrated solutions (Fig. 6-7).

The vapor pressure of a solution is the sum of the partial pressures of the components. This relationship is important when one is considering the boiling point of a solution. A solution will boil when the *total* pressure of its vapor equals the applied pressure. If a dissolved solute is nonvolatile, its partial pressure is zero and the vapor pressure of the solution is simply the partial pressure of the solvent. The vapor pressure of the solvent has effectively been lowered by the addition of a nonvolatile solute. Sometimes this lowering of the vapor pressure, that is, $P_1^\bullet - P_1$, is of interest. In a binary ideal solution, that is, a solution with two components,

$$P_1^\bullet - P_1 = P_1^\bullet - x_1 P_1^\bullet = P_1^\bullet(1 - x_1) = P_1^\bullet x_2 \qquad (6\text{-}11)$$

since $x_2 = 1 - x_1$. Thus the lowering of the vapor pressure of the solvent by a dissolved solute is proportional to the mole fraction of the solute in an ideal solution, and approximately so in a real dilute solution. In the presence of several solute species, $1 - x_1$ is the *sum* of the mole fractions of all solute species and x_2 in the last expression in (6-11) must therefore be replaced by this sum.

Boiling-point elevation. The boiling point of a solution of a nonvolatile solute lies above that of the pure solvent. The reason is that, by Raoult's law, at any given temperature the partial pressure of the solvent in a solution is lower than that of the pure solvent. Consequently it is necessary to heat the solution to a somewhat higher temperature than the normal boiling point of the solvent to raise its vapor pressure to 1 atm (Fig. 6-8). For a dilute solution, the increase in the boiling point, ΔT_b, is proportional to the weight molality of the solute, c_w:

$$\Delta T_b = k_b c_w \qquad (6\text{-}12)$$

The constant k_b is called the *molal boiling-point elevation;* typical values are given in Table 6-1. We shall treat (6-12) as an empirical law, but thermodynamics shows how k_b is related to the boiling temperature of the solvent, its heat of vaporization, and its molecular weight. When several solute species are present, their effect on the boiling point is additive.

Freezing-point depression. The freezing point of a solution is below that of the pure solvent provided that, as is usually true, the crystals formed as the solution freezes are pure solvent and not a solid solution containing solute. The lowering of the freezing point is proportional to the weight molality of the solute, c_w:

$$\Delta T_f = k_f c_w \qquad (6\text{-}13)$$

The constant k_f is called the *molal freezing-point lowering;* typical values are given in Table 6-1. As with boiling-point elevation, the effect is additive when several solute species are present.

FIGURE 6-8 **Boiling-point elevation of a solution of a nonvolatile solute.** The solid curve represents the vapor pressure of the pure solvent as a function of temperature; the solvent boils at T_b. The dotted curve, somewhat lower at any given temperature, represents the vapor pressure of the solution. If the solute is nonvolatile, the vapor pressure above the solution necessarily reaches 1 atm (and thus the solution boils) only at a temperature above T_b. The new boiling point is $T_b + \Delta T_b$, with ΔT_b given by (6-12).

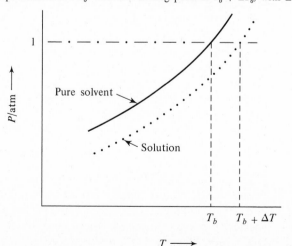

TABLE 6-1 Some boiling-point-elevation and freezing-point-lowering constants

BOILING-POINT ELEVATION [EQUATION (6-12)]		FREEZING-POINT LOWERING [EQUATION (6-13)]	
Substance	$k_b/(\text{K kg mol}^{-1})$	*Substance*	$k_f/(\text{K kg mol}^{-1})$
H_2O	0.52	H_2O	1.86
Ethanol (C_2H_5OH)	1.23	Naphthalene ($C_{10}H_8$)	6.8
Carbon disulfide (CS_2)	2.34	Benzene (C_6H_6)	5.1
Benzene (C_6H_6)	2.64	Phenanthrene ($C_{14}H_{10}$)	12.0
Toluene ($C_6H_5CH_3$)	3.37	Camphor ($C_{10}H_{16}O$)	40

Figure 6-9 illustrates schematically the reason for the freezing-point depression of a solution when the solid solvent that separates is pure. If the solute and the solvent form a solid solution (also called mixed-crystal formation), the situation is more complex.[1]

EXAMPLE 6-5 **Molecular weight from boiling-point elevation.** When 5.0 g of an unknown organic compound is dissolved in 100 g benzene, the boiling point of the benzene rises 0.65 K above its normal value. What is the approximate molecular weight of the unknown substance?

Solution. This solution is sufficiently dilute that (6-12) should be a reasonable approximation. Since k_b for benzene is 2.6 K kg mol^{-1}, the weight-molal concentration of the unknown solution is

$$c_w = \frac{\Delta T_b}{k_b} = \frac{0.65 \text{ K}}{2.6 \text{ K kg mol}^{-1}}$$

$$= 0.25 \text{ mol kg}^{-1} \quad \text{or } 0.25 \text{ weight molal}$$

Since there is 0.25 mol per kilogram of solvent, or 0.025 mol per 100 g solvent, we know that the 5.0-g sample represents 0.025 mol. Hence the mass per mole is

$$\widetilde{M} = \frac{5.0 \text{ g}}{0.025 \text{ mol}} = 2.0 \times 10^2 \text{ g mol}^{-1}$$

no higher precision being warranted.

EXAMPLE 6-6 **Molecular weight from freezing-point depression.** When 25.0 mg of a naturally occurring organic substance of unknown structure, just isolated from the bark of a West African tree known to possess unusual medicinal properties, was dissolved in 1.00 g camphor, the melting point of the camphor fell by 2.0 K. What is the approximate molecular weight of the unknown material?

---

[1]The freezing point of a solid solution may be lower than, the same as, or even higher than that of the pure solvent. This may be understood by considering the curve of the vapor pressure of the solvent above the solid solution. This curve would lie below that for the pure solid solvent in Fig. 6-9. The freezing point is determined by the intersection of the new curve with the dotted curve in the figure. Thus the freezing point depends on the position of the new curve and could lie anywhere in a range of temperatures above or below the normal freezing point T_f (but necessarily above $T_f - \Delta T_f$).

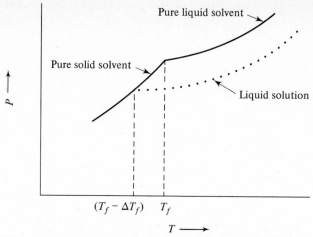

FIGURE 6-9 **Freezing-point lowering of a solution from which pure solid solvent separates.**
The solid curve represents the vapor pressure of the pure solvent as a function of temperature. The
portions of the curve representing the vapor pressure of the liquid and that of the solid meet at the
normal freezing point, since at that temperature these two phases are in equilibrium and their vapor
pressures must thus be the same (see Fig. 3-5 or Fig. 3-6). The freezing point of the solution is also
necessarily the temperature at which the vapor pressure of solvent above the solution is the same as
that of the solvent above the crystallized solid phase. Since the vapor pressure of the solution (represented
by the dotted curve) is lower than that of the pure solvent, it is necessary to go to a lower temperature
to make it equal to that of the solid solvent.

Solution. Applying (6-13) with the value of k_f for camphor equal to 40, we have

$$c_w = \frac{\Delta T_f}{k_f} = \frac{2.0 \text{ K}}{40 \text{ K kg mol}^{-1}}$$

$$= 0.050 \text{ mol kg}^{-1}$$

Since the solution contains 25 mg of unknown per gram of camphor, it contains 25 g
per kilogram of camphor. Thus 25 g of the unknown is 0.050 mol, so that the mass
per mole is

$$\tilde{M} = \frac{25 \text{ g}}{0.050 \text{ mol}} = 5.0 \times 10^2 \text{ g mol}^{-1}$$

no higher precision being warranted.

Henry's law. In a dilute solution, the equilibrium concentration (that is, the
solubility) of a dissolved gas A is proportional to its partial pressure:

$$[A] = k_H P_A \tag{6-14}$$

The Henry's law proportionality constant, k_H, depends on the nature of the gas
and the solvent. The increase of gas solubility with increasing pressure is responsible
for the difficulties that deep-sea divers experience with "the bends" if they rise too
rapidly to the surface. The atmosphere they breathe at appreciable depths is at

relatively high pressure, and if they rise too rapidly the excess dissolved gas does not equilibrate fast enough with the gradually lowered pressure in their lungs. Consequently the fluids in their tissues are supersaturated with gas, which separates as fine (and usually very painful and dangerous) bubbles throughout their body tissues. One way to minimize this effect is to dilute the essential oxygen in a diver's atmosphere with a gas that has very low solubility even at high pressure, such as helium (rather than with nitrogen as in air).

Henry's law applies, at least approximately, to dilute solutions not only of gases but also of most volatile solutes. For the latter it is more natural to say that the partial vapor pressure of the solute A is proportional to its concentration, rather than the other way around as in (6-14). Furthermore it is often more satisfactory to use the mole fraction of A rather than its molarity, thus leading to

$$P_A = k'_H x_A \tag{6-15}$$

Equation (6-15) is approximately equivalent to (6-14) for dilute solutions, for which the mole fraction and the molarity are approximately proportional to one another.[1]

Since Henry's law may be stated in several ways, using different concentration units, it is especially important to note the units of the Henry's law constants in any tabulation before using such constants.

Distribution law. Suppose that a molecular or ionic species A is soluble in each of two immiscible (that is, mutually insoluble) solvents, I and II. Let a quantity of A be shaken with any mixture of the two solvents and be thus distributed between them. When equilibrium is established the ratio of the concentrations of A in the two solvents, $[A]_I$ and $[A]_{II}$, is always the same at a given temperature, at least if the solutions are relatively dilute:

$$\frac{[A]_I}{[A]_{II}} = K \tag{6-16}$$

The constant K depends on the temperature and the nature of solute and solvents; it is independent of the relative volumes of the two solvents. For gases or volatile solutes, (6-16) can be derived from (6-14), but it is applicable to nonvolatile solutes as well.

Osmotic pressure. Suppose a solution is separated from the pure solvent by a membrane that is permeable to solvent molecules but not to solute species; such a membrane is called *semipermeable*. If both solution and pure solvent are under the same (usually atmospheric) pressure, it is observed that more solvent molecules pass through the semipermeable membrane in the direction of the solution than in the reverse direction, so that the solution becomes continuously more dilute. This may be interpreted as resulting from the fact that the vapor pressure of the pure

---

[1]Although (6-15) resembles Raoult's law, Equation (6-10), it is *not* in general true that k'_H is the same as the constant in Raoult's law; k'_H depends on the nature of solute and solvent, as well as on the temperature, and may vary widely for different solutes in the same solvent as well as for the same solute in different solvents. The two constants are necessarily the same for an ideal solution.

solvent is greater than its partial pressure above the solution. For the two phases to be in equilibrium, the solvent vapor pressure must be the same in each. The vapor pressure of the solvent in the solution can be raised without changing the temperature by applying a higher external pressure to the solution, because it happens that the vapor pressure of a liquid is raised by applied pressure, as can be shown by thermodynamics. The excess pressure on the solution, at which there is equilibrium so that there is no net passage of solvent through the diaphragm in either direction, is called the *osmotic pressure*, here designated Π (Greek capital pi). For an ideal solution, or any sufficiently dilute solution,

$$\Pi V = n_2 RT \tag{6-17}$$

where
$$V = \text{volume of solution}$$
$$n_2 = \text{total number of moles of solute present}$$
$$R = \text{gas constant}$$
$$T = \text{absolute temperature}$$

If the osmotic pressure Π is expressed in atmospheres, V in liters, and T in kelvins, then the appropriate value of R is 0.08206 liter atm mol^{-1} K^{-1}. Because (6-17) is identical in form with the ideal-gas law, a great deal of effort was expended during the last century in trying to explain the osmotic pressure law in terms of some sort of ideal-gas behavior of the solute molecules, but such interpretations have little significance. The osmotic pressure is not a pressure on the membrane; it is a pressure that must be applied to the solution to achieve equilibrium with pure solvent.

Measurements of osmotic pressure are of considerable value in providing estimates of the molecular weights of very large molecules, including both natural and synthetic polymers.

EXAMPLE 6-7   **Molecular weight from osmotic pressure.** The osmotic pressure of a solution containing 5.0 g polystyrene per liter of benzene was found to be 7.6 torr at 25°C. What is the approximate molecular weight of this sample of polystyrene?

Solution. This solution is sufficiently dilute (about 0.5 percent) that (6-17) should be a good approximation. Substituting $\Pi = 7.6$ torr/(760 torr atm^{-1}) = 0.0100 atm, $V = 1.00$ liter, $n_2 = 5.0$ g/$\tilde{M}$ (where $\tilde{M}$ is the mass per mole of the sample), and $T = 298$ K,

$$\tilde{M} = \frac{5.0 \text{ g} \, (0.0821 \text{ liter atm mol}^{-1}\text{ K}^{-1}) \, (298 \text{ K})}{(0.0100 \text{ atm})(1.00 \text{ liter})}$$

$$= \tfrac{122}{0.0100} \text{ g mol}^{-1} \qquad = \underline{12.2 \times 10^3 \text{ g mol}^{-1}}$$

Osmotic phenomena are responsible for maintaining a balance of fluids in the body, particularly in the functioning of the kidney. The contents of body cells, such as red blood corpuscles, are normally at osmotic equilibrium with the surrounding blood plasma. If red blood cells are put in distilled water, they swell and eventually burst. This does not happen in a salt solution that has the same osmotic pressure as blood plasma. Such a solution is known as *isotonic* (Greek: *isos*, equal; *tonos*, tension) saline solution.

Osmosis also plays a role in one method of desalting seawater and other brines. When salt water is separated from fresh water by a membrane that permits only the passage of water, an excess pressure on the salt water equal to its osmotic pressure Π is required to prevent its dilution by the fresh water. If a pressure larger than Π is applied to the solution (e.g., by a piston), pure water is squeezed out. The semipermeable membrane acts in effect as a molecular filter. This procedure is called reverse osmosis. The average salt content of ocean water corresponds to an osmotic pressure of about 22 atm, and reverse osmosis of ocean water therefore requires pressures in excess of this value. Lower pressures suffice for less salty water.

Comparison of magnitudes of colligative properties.   The relative magnitudes of the effects manifested in the different colligative properties are instructive. The relative vapor-pressure lowering is usually very small. For example, in a 1.00 wt percent aqueous solution of sucrose ($C_{12}H_{22}O_{11}$, MW = 342) the mole fraction of solute is 5.3×10^{-4}, so that the vapor pressure of water is lowered by 5.3×10^{-2} percent. At 25°C this means a pressure drop of 0.013 torr out of a total pressure of 23.8 torr. The calculated boiling-point increase is 0.015 K and the calculated freezing-point lowering, 0.054 K. Both temperature changes can easily be measured, although special thermometers are needed. The osmotic pressure is calculated to be 0.71 atm at 25°C, a pressure that can easily be measured without special equipment.

6-5   Colloidal Solutions

Very small particles suspended in a liquid or in a gas may settle to the bottom only very slowly or may not even settle at all over long periods of time. Suspensions of fine particles that do not settle are called *colloidal suspensions* or *colloids;*[1] if the medium in which they are suspended is a liquid, they are often referred to as colloidal solutions. At ordinary temperatures the suspended particles execute brownian motion, a random, rapid, vibratory motion that never ceases and prevents the particles from settling.

The particles in many colloidal solutions are so small that they are not even visible with a high-powered microscope and special means must be used to observe them. One such method is to use the Tyndall effect. A beam of light passing through a colloidal solution forms a trace that is visible from the side. This effect is observed when a beam of strong light is visible in hazy or smoke-filled air. The beam can be seen because the suspended particles, even though they are too small to be observed directly as individual particles, scatter some of the light sideways. Information about the size and shape of colloidal particles can be obtained by careful studies of the pattern of light scattered by a colloidal solution.

Because of their size, colloidal particles cannot pass through the pores of some

---

[1]The term *colloid* comes from the Greek: *kolla,* glue. Many glues are made from the gelatinlike material collagen, which is the main constituent of fibrous connective tissue in animals. Gelatin is a typical colloid.

membranes, such as cellophane, collodion, or parchment paper. This fact is often used to free a colloidal suspension of small molecules and ions that may happen to be dissolved in the suspending liquid. The suspension is placed in a cellophane or other suitable bag immersed in a bath of solvent that is frequently renewed. The dissolved small ions and molecules pass through the pores into the solvent and the colloidal solution remains in the bag. This process is called *dialysis* (Greek: *dia,* through; *lyein,* to loosen). Both natural and artificial kidneys function essentially by dialysis, small molecules and some salts being eliminated from the body in the urine and large molecules being retained.

The behavior of colloidal solutions is in many respects similar to that of ordinary solutions, and there is no sharp boundary between the two. The suspended particles may be solid or liquid aggregates of many molecules that are themselves insoluble or are soluble to a very small degree in the suspending liquid. They may also be single large molecules. In this case the suspension is in effect an intimate mixture of solute and solvent particles on the molecular level, just like any other solution. However, the fact that the solute particles are so much larger than the solvent molecules makes it sometimes desirable to classify the solution as colloidal. In other words, the distinction between ordinary and colloidal solutions is best based on the criterion of *size* of solute particles rather than on whether the solute particles are molecules or aggregates of molecules, even though the size distinction is not sharp and clear-cut. Particles with dimensions in the range 10^2 to 10^5 Å (10^{-5} to 10^{-2} mm) are often considered colloidal.

Many proteins and other large molecules form colloidal solutions, as do some simpler compounds and even some elements. For example, a colloidal suspension of gold in water can be prepared. At a particle size that has been found to yield long-lasting colloidal suspensions, this solution has a beautiful red color. One such suspension prepared by Michael Faraday has been kept apparently unchanged for more than 100 years. At other particle sizes, the suspension may be yellow, blue, or violet. The color is caused by interference (Sec. 8-1) between the light waves scattered from different parts of the particles and therefore depends on the particle size. Colloidal suspensions of gold in glass cause a deep ruby color also, at a dilution of about 1 part by weight of gold to 50,000 parts of glass. Colloidal metal sulfides and other precipitates are often obtained inadvertently (and to the despair of the beginning analyst) when the sulfides or other compounds of metal ions are precipitated.

The term *colloid* is used to refer to more than colloidal solutions. The characteristic that all colloidal systems have in common is that some substance, finely divided in at least one dimension, is dispersed through a second one. Water is a common medium for dispersion, but some colloids are solid, such as the ruby glass mentioned above. The dispersion medium may also be a gas; fogs and smokes are often colloidal suspensions of fine particles of liquids and solids, respectively, in air. Foams and other thin films are properly regarded as colloidal too, when one of their dimensions is in the colloidal range. Surface effects play a major role in colloidal phenomena because any finely divided substance has a surface area greatly increased over that which it would have if not so divided. Molecules and ions are attracted to any surface by van der Waals and Coulomb forces, and if the surface is very large, the numbers of particles so attracted may also be large. Thus, this phenomenon of fixation of particles on surfaces (*adsorption*) plays an important part in many colloidal systems.

6-6 Separation of Substances

Highly purified substances are essential in many different fields of research and in industry and commerce as well, since traces of impurities sometimes significantly alter physical and chemical properties. The essence of purification is the separation of a substance from others with similar properties. The most important methods for separation include filtration, fractional crystallization, distillation, and chromatography.

Filtration. A filter permits the separation of material suspended in a fluid by virtue of differences in particle size. Macroscopic particles larger than a certain minimum size do not pass through the filter. The role of the filter may also be played by a semipermeable membrane, considered in the discussions of osmosis and colloids (Sec. 6-5).

Fractional crystallization. This term implies a separation of one of several solutes from a solution by crystallization in relatively pure form. Separation by fractional crystallization may be accomplished by slowly lowering the temperature of a solution or by evaporating the solvent until crystals form. Which of the different possible crystalline substances will form first depends upon the temperature of the solution, upon the concentrations of the various solutes present, and upon their solubilities at different temperatures. The temperature variation of the solubility often differs markedly from one compound to another, even for substances that are very similar chemically. For example, the solubility of KCl is more than twice as great at 100°C as at 0°C while that of NaCl increases by only about 11 percent over this same temperature interval.

Crystallization is a highly selective process, so that even in the presence of many impurity molecules or ions only a particular species usually enters a particular growing crystal while the impurities remain in solution. The reason for this is structural: each molecule or ion has a particular size and shape, and sometimes also particular ways by which it may interact strongly or specifically with its environment, for example, by hydrogen bonding or by ionic forces. Other molecules or ions cannot normally fit into the spaces in the crystal that are tailored for this particular species, nor can they interact in the specific way it does. On the other hand, if another molecule or ion is very similar in size and shape to one of the structural units of the crystal, a solid solution or *mixed crystal* may form.

EXAMPLE 6-8   **Separation of KBr and KBrO₃.** Bromide and bromate (BrO_3^-) ions are formed by the reaction of bromine with hydroxide ion:

$$3Br_2 + 6OH^- \longrightarrow 5Br^- + BrO_3^- + 3H_2O \qquad (6\text{-}18)$$

Suppose that an excess of bromine reacts with 0.18 gfw KOH in 100 g water and the excess bromine is removed by evaporation of the solution until only 50 g water is left. Information about the solubilities of KBr and $KBrO_3$ is given in Table 6-2. Devise a scheme for obtaining relatively pure crystals of KBr and $KBrO_3$ from the solution.

Solution. If it is assumed that reaction (6-18) has occurred, then the solution after evaporation contains $\frac{0.18}{6} = 0.030$ gfw $KBrO_3$, or 5.0 g, and $5(0.18)/6 = 0.15$ gfw KBr, or 18 g, in 50 g water. Table 6-2 shows that at temperatures near 100°C, 50 g water can dissolve 52 g KBr and 25 g $KBrO_3$. At 0°C, however, this quantity of solvent can hold only $0.50(3.1) = 1.6$ g $KBrO_3$, although it can still dissolve $0.50(53.5) = 27$ g KBr.

Thus, if the solution is cooled in an ice-water bath at 0°C, about $(5.0 - 1.6)$ g $= 3.4$ g $KBrO_3$ should crystallize, while no KBr should form. The crystalline $KBrO_3$ can be removed by filtration. The solution that passes through the filter (the filtrate) can then be heated and evaporated carefully until only about 10 g of water remains and can then be cooled to, say, 60°C. At this temperature, 10 g water can dissolve 2.3 g $KBrO_3$, so none of the 1.6 g $KBrO_3$ still in the solution should crystallize. On the other hand, only about 8.6 g KBr can remain dissolved under equilibrium conditions in 10 g water at 60°C, so nearly 10 g of the 18 g KBr initially present should have crystallized during this second stage.

Both the KBr and the $KBrO_3$ formed under the different conditions of crystallization should be nearly pure, with only small amounts of the other material present. Impurities may be present, however, if the surrounding solution (the *mother liquor*) was not effectively washed from the spaces between the crystals or if some of this solution was trapped within a crystal that grew too rapidly. Alternatively some foreign ions may have been adsorbed (Sec. 6-5) on the surface of the crystals. Mixed-crystal formation is unlikely with ions as different in size and shape as Br^- and BrO_3^-. The first batches of crystals can be further purified by recrystallization from fresh solvent.

Distillation and related methods. Among the most effective and widely used separation methods are those based upon differences in vapor pressure. The term *distillation* implies converting a liquid or solid phase into a gas and then condensing it again by cooling. With a mixture of nonvolatile and volatile substances for which only the nonvolatile materials are of interest, a separation may be achieved

TABLE 6-2   **Solubilities of KBr and $KBrO_3$ at various temperatures**

	Solubility	
	g/(100 g water)	
Temperature/°C	KBr	$KBrO_3$
0	53.5	3.1
20	65.2	6.9
40	75.5	13.2
60	85.5	22.7
80	95.0	34.0
100	104.0	50.0

by evaporation alone (without recondensation). For example, salt is obtained by evaporation of seawater and other brines, and in the technique of freeze-drying, a solvent (usually water) is removed by sublimation from a solution frozen in a thin layer.

It often happens that two or more desired components of a mixture to be separated are volatile and may even have similar vapor pressures. Separation by distillation can still be achieved as long as the vapor is richer in one of the components than is the liquid. Consider, for example, a liquid mixture of N_2 and O_2 at a pressure of 1 atm. Figure 6-10 depicts the compositions of the vapor and the liquid in equilibrium with each other at different temperatures. Note that the vapor is always richer in N_2, so that boiling a sample of the liquid of any composition and condensing the resulting vapor always yields a condensate richer in N_2. This condensate can be distilled again, leading to still further enrichment in N_2. Such a procedure of successive redistillations is called fractional distillation. There are ways of doing it in one continuous step with an appropriately designed *distillation column,* rather than in many distinct steps.

Since the vapor is richer in N_2 than is the boiling liquid from which it is removed,

FIGURE 6-10 **Liquid and vapor compositions for a mixture of O_2 and N_2.** The curves labeled "Vapor" and "Liquid" show the compositions of the respective phases in equilibrium with each other at 1 atm and any given temperature, as specified by the appropriate horizontal line. For example, at 84 K liquid of composition 69 percent O_2 and 31 percent N_2 (point *A*) is in equilibrium with vapor containing 40 percent O_2 and 60 percent N_2 (point *B*). If this liquid mixture is boiled and the vapor is removed, the liquid remaining becomes progressively richer in O_2; that is, its composition moves to the right in the diagram and its boiling temperature increases. At 86 K, liquid of composition *C* is in equilibrium with vapor of composition *D*. For mixtures of some other substances, the corresponding curves display a maximum or a minimum, and in a certain range of compositions the vapor is richer in the *less* volatile component.

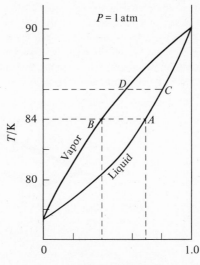

Weight fraction O_2

the liquid left behind must necessarily be richer in O_2 than it was initially. As distillation continues, the remaining liquid becomes progressively richer in O_2; that is, its composition moves to the right in Fig. 6-10 and its boiling temperature increases, as indicated by the curve labeled "liquid". At the same time, the composition of the vapor (and hence of the initial liquid condensed, the distillate) changes as indicated by the vapor curve.

Distillation is an extremely important industrial and laboratory method with many highly sophisticated variations. One of its most important applications is as a stage in the refining of petroleum. In 1969 about 500 billion liters of crude oil were fractionated into various grades of gasoline, kerosene, oils, and other products in the United States alone.

Chromatography.   While the operations of filtration, crystallization, and distillation have been known and used for millennia, many other important separation methods are relatively modern. Perhaps the most versatile of these is chromatography. The method was originated in 1906 by the Russian botanist Tswett to separate and isolate plant pigments; his term *chromatography* (Greek: *chroma,* color) is somewhat misleading because the method is not limited to colored substances. The great potential of chromatographic methods first began to be realized in the 1930s and 1940s.

In one form of chromatography, essentially that originated by Tswett, a solution of the mixture to be separated is poured into a vertical glass tube packed with a powdered solid, such as aluminum oxide (Al_2O_3) or silica gel (especially prepared SiO_2), capable of adsorbing the substances to be separated. Initially the components of the mixture are adsorbed in the top portion of the packed column. Then a suitable

FIGURE 6-11   **Paper chromatography.**   A drop of the mixture to be separated is placed near one end of a strip of filter paper, which is then suspended in a jar in such a way that the mixture to be separated is at the lower end of the strip. Enough liquid solvent of an appropriate composition is placed in the bottom of the jar so that it just touches the lower edge of the paper strip, and the jar is then closed. The solvent rises in the paper by capillary action and the components of the mixture, migrating at different rates up the paper strip, gradually separate. If they are colored, their positions are visible. If not, they can be detected in various ways; for example, they might be visible when viewed in ultraviolet light or they might react with an appropriate reagent to give a colored product.

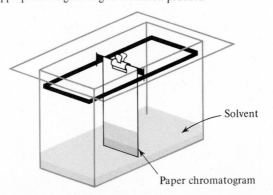

Solvent

Paper chromatogram

solvent is allowed to flow slowly through the column. The substances to be separated move gradually down the column because they are at times dissolved in the flowing solvent and at other times adsorbed on the column. The rate at which a given solute moves depends on a delicate balance between its interactions with the solvent and with the column. With appropriately chosen solvent and adsorbent, the components of the mixture emerge in well-separated portions at the bottom of the column and their solutions can be collected in different containers.

An important variant of the general method is paper chromatography, in which the role of the solid adsorbent is played by a strip of filter paper (Fig. 6-11). Some disadvantages of using paper, such as the fact that it is hard to be sure that different batches have the same properties, can be overcome by using *thin-layer chromatography* (TLC), in which a glass plate coated with adsorbent is used instead of paper. Both paper chromatography and TLC are of great value in the analysis of a great variety of complex mixtures; they are rapid methods that require only very modest equipment and are applicable to microgram quantities. One of the most sensitive current methods of separation and detection is *vapor-phase chromatography* (VPC) (Fig. 6-12). Vapor-phase chromatography can be used in the quantitative analysis of a sample of a mixture weighing as little as 10^{-6} or 10^{-7} g. The volatility of the

FIGURE 6-12 **Vapor-phase chromatography.** The mixture to be separated flows as a vapor in an inert *carrier gas,* e.g., helium, over a column of finely divided solid that has been coated with a film of a liquid such as a high-boiling, essentially nonvolatile, hydrocarbon or silicone (Sec. 22-3) oil. As the mixture of vapor and carrier gas passes slowly through the column, a gradual separation of the components of the mixture occurs. The separated fractions are detected and may be quantitatively estimated as they emerge from the column. One common detector is a device that ionizes each component and measures the resulting current; it is capable of detecting as little as 10^{-10} g s^{-1}.

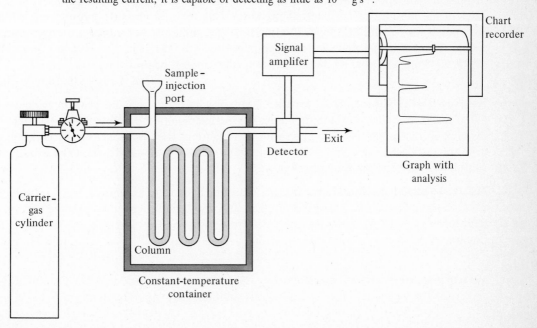

sample to be analyzed need not be high, and it can be increased by carrying out the experiment at an elevated temperature.

The common feature of all chromatographic methods is that a fluid mobile phase containing the sample to be analyzed passes over a stationary phase with which the components of the mixture can interact reversibly. As long as different components are distributed differently between the mobile and immobile phases at any moment, they will migrate with different speeds and hence can be separated. Chromatography has played a major role in many of the spectacular advances in chemistry and biochemistry in recent decades just because it provides convenient, versatile, sensitive, and highly specific methods of analysis.

Problems and Questions

6-1   Differences between solids and liquids.   (*a*) Describe in detail the differences between the microscopic structure of a typical substance just below and just above the melting point. Include in your description the average distance between molecules, the long- and short-range order, the degree of molecular motion, and the average potential energy per molecule. (*b*) Explain how the differences in microscopic properties of solids and liquids are reflected in differences in the macroscopic properties of solids and liquids.

6-2   Differences between liquids and gases.   Compare and contrast the microscopic properties of a liquid and its vapor. Include in your discussion those properties specified in Ques. 6-1. Explain how these differences are reflected in the macroscopic properties.

6-3   Internal pressure in liquids.   The van der Waals equation (5-21) describes the behavior of a substance in the liquid as well as in the gaseous state. In this equation the pressure P is increased by $a/\widetilde{V}^2$, a term that represents an internal pressure that would prevail in the fluid even in the absence of an external pressure. This internal pressure is quite large for liquids. Estimate it for N_2 and SO_2 by evaluating $a/\widetilde{V}^2$ for these two substances at their normal boiling points. The respective densities and temperatures are 0.81 g cm^{-3} ($-196°C$) and 1.43 g cm^{-3} ($-73°C$). Values of a are 1.38 and 6.7 atm liter^2 mol^{-2}, respectively.

Ans. 1.15×10^3 atm; 3.3×10^3 atm

6-4   Radial distribution functions.   Explain why (*a*) there is no peak in the curves of Fig. 6-3 at distances below a minimum value; (*b*) the average of the number of molecules at a given distance increases parabolically, i.e., with the square of the distance; and (*c*) as the interatomic distance increases, there is a general flattening of the peaks and the peaks eventually blend completely with the parabolic background.

6-5   Hydrobromic acid solution.   (*a*) Gaseous hydrogen bromide is very soluble in water. What volume of the gas, measured at 20°C and 750 torr, is needed to prepare 300 ml of 4.0 *F* hydrobromic acid? (*b*) If the density of the resulting solution is 1.10 g ml^{-1}, what is the mole fraction of water in the solution? Note that HBr is completely dissociated into H^+ and Br^-.

Ans. (*a*)   29 liters;   (*b*)   0.84

6-6   Solubility of sodium carbonate.   A 14.0 percent solution of Na_2CO_3 in water has a density of 1.146 g ml^{-1}. Eighty-five ml of this solution was evaporated to dryness. (*a*) What weight of dry Na_2CO_3 was obtained? Suppose that the residue had the composition $Na_2CO_3 \cdot 10H_2O$; what weight of it would be present? (*b*) What are the volume and weight formalities of the solution?

6-7 **Density of ethanol solution.** A 9.9 *M* solution of ethanol (C_2H_5OH) in water contains just 50.0 percent ethanol by weight. What is the density of the solution? *Ans.* 0.91 g ml^{-1}

6-8 **Fructose solution.** A certain aqueous solution of fructose ($C_6H_{12}O_6$) contains 40.0 percent by weight of fructose and has density 1.180 g ml^{-1} at 20°C. (*a*) What is the concentration of fructose in grams per liter of solution? (*b*) What is the molarity of fructose in this solution? (*c*) What is the weight molality of fructose? (*d*) What is the mole fraction of water?

6-9 **Perchloric acid solution.** Suppose that you need some pure perchloric acid and wish to prepare it by mixing ordinary commercial concentrated aqueous $HClO_4$ (density 1.67 g ml^{-1}, 70.6 percent perchloric acid by weight) with 300 g of 6.1 percent (by weight) solution of Cl_2O_7 in $HClO_4$ that contains no water. Cl_2O_7 reacts with water to form $HClO_4$. What volume of the concentrated aqueous $HClO_4$ should you add, and how many grams of pure perchloric acid would you obtain? *Ans.* 3.7 ml, 306 g

6-10 **Glycerol solution.** The mole fraction of water is 0.884 in a 4.8 *M* aqueous solution of glycerol, $C_3H_5(OH)_3$. (*a*) What is the percentage by weight of glycerol in the solution? (*b*) What is the approximate freezing point of the solution?

6-11 **Molecular weight by freezing-point lowering.** The molal freezing-point-lowering constant for benzene is 5.1 K kg mol^{-1}. When 0.196 g compound A is dissolved in 20.0 g benzene, the freezing point decreases by 0.51 K. (*a*) What is the approximate molecular weight of A? (*b*) If compound A contains just 25.0 percent carbon (by weight), what is the exact molecular weight of A? Explain any discrepancy from the value obtained in (*a*). *Ans.* (*a*) 98; (*b*) 96, nonideal-solution behavior

6-12 **Automobile radiator.** Ethylene glycol, CH_2OHCH_2OH, is a common antifreeze liquid for use, mixed with water, in automobile radiators. (*a*) What should be the approximate molality of solute in an aqueous solution if the solution is to freeze at a temperature no higher than −20°C (−4°F)? (*b*) What volume of ethylene glycol (density 1.11 g cm^{-3}) should be added to 30 liters water (density 1.00 g cm^{-3}) to give the molality calculated in (*a*)? (*c*) What would be the approximate boiling point of this solution at 1 atm?

6-13 **Temperature variation of solubility.** The solubility of $(NH_4)_2SO_4$ in water is 706 g liter^{-1} of water at 0°C and 1033 g liter^{-1} at 100°C. Suppose that 500 g of the salt is added to 200 ml water at 100°C and that the mixture is stirred until no more of the salt dissolves and is then filtered at 100°C. Assume that no water evaporates and that equilibrium conditions are attained. (*a*) How much $(NH_4)_2SO_4$ is retained by the filter? (*b*) If the solution passing through the filter is chilled to 0°C, how much of the dissolved salt crystallizes out? (*c*) If you did not take account of the difference in density of water at 0°C and 100°C (1.000 and 0.958 g cm^{-3}, respectively), by how much is your answer to (*b*) in error? *Ans.* (*a*) 293 g; (*b*) 71 g; (*c*) −8 percent

6-14 **Solubility of a gas.** The solubility of carbon dioxide in water is 335 mg per 100 g water at 0°C when the partial pressure of carbon dioxide is 760 torr. How many milligrams of carbon dioxide will dissolve in 1.00 kg water at 0°C when the partial pressure of carbon dioxide is 5.0 atm? What is the freezing point of this solution?

6-15 Vapor-pressure lowering. A 4.0-g sample of a substance X is dissolved in 156 g benzene (MW = 78) and found to lower the vapor pressure of benzene from 200.0 to 196.4 torr. (*a*) What are the mole fractions of benzene and of X? (*b*) What is the molecular weight of X?

Ans.   (*a*) $x_X = 0.018$;   (*b*) 1.1×10^2

6-16 Molal boiling-point elevation. When 5.12 g naphthalene ($C_{10}H_8$) is dissolved in 100 g CCl_4, the boiling point of the CCl_4 is raised 2.00 K. What is the molal boiling-point-elevation constant for CCl_4?

6-17 Molecular weight from freezing-point lowering. A 6.0-g sample of a solute dissolved in 100 g H_2O lowers the freezing point of water by 1.02 K. Calculate the apparent molecular weight of the solute.   *Ans.* 109

6-18 Molecular weight of a protein. The osmotic pressure at 25°C of a solution containing 1.35 g protein in 100 ml water was found to be 9.9×10^{-3} atm. What molecular weight would account for this datum?

6-19 Relationship between osmotic pressure and freezing-point lowering. Calculate the approximate osmotic pressure, at 25°C, of an aqueous solution that freezes at −0.035°C.   *Ans.* 0.46 atm

6-20 Raoult's law for a mixture of volatile liquids. At 80°C the vapor pressure of benzene (C_6H_6) is 753 torr and that of toluene (C_7H_8) is 290 torr. What is the total pressure and what is the composition of the vapor over a solution of exactly $\frac{1}{3}$ mol benzene and exactly $\frac{2}{3}$ mol toluene? Assume Raoult's law for both components of the solution.

6-21 Vapor pressure of the solution of a volatile solute. The following data apply at 25°C. Solid iodine has a vapor pressure of 0.31 torr. Chloroform (liquid), $CHCl_3$, has a vapor pressure of 199.1 torr. In a saturated solution of iodine in chloroform, the mole fraction of iodine is 0.0147. (*a*) What is the partial pressure of iodine vapor at equilibrium with such a saturated solution? (*b*) Assuming Raoult's law, what is the total vapor pressure of this solution?

Ans.   (*a*)   0.31 torr;   (*b*)   196.5 torr

6-22 Heptane-octane mixture. The liquids *n*-heptane (MW = 100) and *n*-octane (MW = 114) form a nearly ideal solution. At 40°C, heptane has a vapor pressure of 100 torr and octane a vapor pressure of 40 torr. Suppose that a solution is prepared by mixing equal weights of these two liquids. (*a*) What is the mole fraction of each component in this solution? (*b*) What is the partial pressure of each component over this solution at 40°C? (*c*) Suppose that some of the vapor described in (*b*) is condensed to a liquid. What would be the mole fraction of each component in this liquid, and what would be the vapor pressure of each component above this liquid at 40°C?

6-23 Distillation of an ideal solution. Liquids A and B form an ideal solution. The vapor pressures of pure A and pure B at 100°C are 300 and 100 torr, respectively. Suppose that the vapor above a solution composed of 1.00 mol A and 1.00 mol B at 100°C is collected and condensed. This condensate is then heated to 100°C and the vapor is again condensed to form liquid X. What is the mole fraction of A in X?   *Ans.* 0.90

6-24 **Equilibration of two solutions through the vapor.** Two open 1-liter beakers, A and B, containing aqueous solutions of NaCl are placed together in a box, which is then sealed tightly. Initially there are 600 ml solution in A and 300 ml in B. Consider two situations: (*a*) the concentration of the solution in A is initially twice that in B; (*b*) the concentration in A is initially half that in B. Assume that equilibrium is established by waiting sufficiently long (perhaps several days). What will be the relation between the volumes of the solutions in A and B for case (*a*) and for case (*b*)? Explain clearly.

"Acids, are all those things which taste sour upon being reduced to a proper degree of strength; as vinegar, juice of lemons, oil of vitriol, etc., and which effervesce with chalk, and the salts of vegetables prepared by incineration; and with such like substances form a neutral salt. They are likewise distinguished from alkalies, by turning syrup of violets red.

"Alkali, is any substance, which being mixed with an acid, ebullition and effervescence ensue thereon, and which afterwards forms a neutral salt. They are likewise distinguished from acids by turning syrup of violets green.

"Neutral salts are a sort of intermediate salts between acid and alkali, and though composed of both, yet upon trial, exhibiting the marks of neither one nor the other."

William Lewis, 1746

7 Acids and Bases, Ionic Solutions, and Oxidation-Reduction

Aqueous solutions merit special attention. Water is a good solvent for a great variety of compounds, ranging from ionic substances like common salt to covalent compounds such as sugar, ammonia, and hydrogen chloride, and even to giant protein molecules. An understanding of the chemical properties of aqueous solutions is essential to an understanding of many different fields of chemistry. Among the most common reactions in solution are those in which protons are exchanged between the reactants—acid-base reactions. Another fundamental class of reactions is that

168

in which electrons are exchanged—oxidation-reduction reactions. In this chapter we introduce these fundamental chemical systems and concepts: acids and bases, solutions of ionic substances in water, and oxidation and reduction. The chemical equilibria in these systems are considered quantitatively in Chaps. 14, 15, 17, 18, and 19.

7-1 Acids and Bases

Early views. The alchemists and earliest chemists used the term *acid* (Latin: *acidus,* pungent, sharp) to refer to any of a variety of substances that had certain qualities and chemical properties in common, such as a characteristic sour, slightly sharp (acidic) taste, reaction with some metals to produce hydrogen, and the ability to cause certain color changes in various vegetable dyes. The first modern chemist, Lavoisier, noted two centuries ago that the oxides of sulfur, phosphorus, nitrogen, and carbon all produced acidic solutions. After his discovery of the element common to all these compounds, he chose the name *oxygen* for this element on the basis of the misconception that all acids contained oxygen (Greek: *oxys,* sharp, pungent, acid; *gennan,* to form). However, it was soon recognized that acids containing no oxygen existed and that the acid property was related to hydrogen rather than oxygen. In 1815, Davy and Dulong defined an acid as a substance containing hydrogen that could be replaced by metal, a point of view later emphasized by Liebig as well.

As the concept of acids was developing, several other important classes of compounds were also recognized, including bases and salts. Solutions of typical bases (or alkalies) could be obtained by soaking the ashes left after burning various plants and filtering out the insoluble material. "Potash" crystallized from such a solution was soon recognized as potassium carbonate, K_2CO_3, and the element potassium owes its name to this process. Bases have characteristic properties in common, just as acids do; they have a somewhat bitter taste, produce particular colors with various vegetable dyes (different from the colors produced by acids), and react with acids to form salts, the distinctive properties of both the acid and the base thereby disappearing or being *neutralized.*

The modern view of the nature of solutions of acids, bases, and salts in water was first proposed in 1887 by Svante Arrhenius (and was greeted with much skepticism before the evidence in favor of it was recognized as overwhelming). This theory of *ionic dissociation* asserts that acids, bases, and salts are partially or completely dissociated into ions in aqueous solution. When dissolved in water, an acid HX yields H^+ and a charged ion, X^-. Thus, for example, a solution of hydrochloric acid contains H^+ and Cl^- and a solution of nitric acid contains H^+ and NO_3^-. The bases NaOH, KOH, and $Ca(OH)_2$ yield OH^- ions and the positive ions Na^+, K^+, and Ca^{2+}, respectively. Arrhenius therefore defined an acid as a substance that yields H^+ ions when it is dissolved in water and defined a base as a substance that yields OH^- ions when it is dissolved in water. Acids and bases react to form a *salt* and water, as, for example,

$$2HNO_3 + Ca(OH)_2 \longrightarrow Ca(NO_3)_2 + 2H_2O \tag{7-1}$$

or, writing this as an ionic equation balanced with respect to charge as well as species,

$$2H^+ + 2NO_3^- + Ca^{2+} + 2OH^- \longrightarrow Ca^{2+} + 2NO_3^- + 2H_2O \tag{7-2}$$

This characteristic reaction between an acid and a base is called *neutralization*. Since the Ca^{2+} and NO_3^- ions remain unchanged, they may be omitted and the equation for the neutralization reaction can be written simply as

$$2H^+ + 2OH^- \longrightarrow 2H_2O \tag{7-3}$$

Strong and weak electrolytes. Substances whose solutions conduct electric current by migration of ions are called *electrolytes* (Fig. 7-1). Arrhenius recognized that the reason some acids and bases are much poorer electrolytes than others is that they are only slightly dissociated into ions whereas the others are more completely dissociated. The terms *strong* and *weak* are used to distinguish qualitatively between electrolytes that are almost completely dissociated and those that are dissociated only to a small extent. As with other qualitative terms, there is no sharp boundary between strong and weak electrolytes, but in practice this causes little difficulty. More quantitative ways of describing the strength of acids and bases are discussed in Sec. 14-2.

It is important not to confuse the relative solubility of an electrolyte with its strength. For example, acetic acid is extremely soluble in water but is a weak electrolyte. On the other hand, barium hydroxide is not very soluble but what does dissolve dissociates almost completely so that barium hydroxide is a strong electrolyte in water. Similarly, the present use of the terms strong and weak should not be confused with popular usage; hydrofluoric acid, HF, is a weak acid but it is extremely corrosive and will even dissolve glass.

Pure water is a very weak electrolyte, as it contains relatively few ions and is a poor conductor of electricity. Very sensitive instruments are needed to detect its conductivity. If an acid, a base, or a salt is dissolved in water, the conductivity becomes greater, and it increases with increasing concentration and with the extent of dissociation of the electrolyte involved.

FIGURE 7-1 **Equipment for demonstrating the conductivity of solutions.** When the electrodes are immersed in the liquid, the brightness of the bulb is an indication of the number of ions present in solution. (*a*) Strong electrolyte, (*b*) weak electrolyte, (*c*) nonelectrolyte.

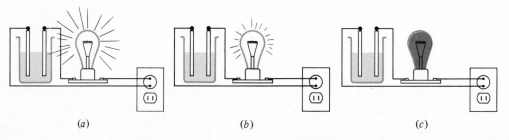

(*a*)   (*b*)   (*c*)

Strong and weak acids. In aqueous solutions, HCl, HBr, HI, HNO_3, and $HClO_4$ (perchloric acid) are essentially completely dissociated, according to the reaction

$$HA \longrightarrow H^+ + A^- \tag{7-4}$$

and are thus strong acids. Sulfuric acid, H_2SO_4, is also strong insofar as the dissociation to give one hydrogen ion is concerned:

$$H_2SO_4 \longrightarrow H^+ + HSO_4^-$$

However, the HSO_4^- is not completely dissociated except in solutions of very low concentration, so that an aqueous solution of H_2SO_4 contains both undissociated HSO_4^- and SO_4^{2-} ions in addition to H^+ ions. No H_2SO_4 molecules are detectable in an aqueous solution. Among the commonest weak acids are phosphoric (H_3PO_4), nitrous (HNO_2), and hydrofluoric (HF), which are of intermediate strength, and carbonic (H_2CO_3) and hydrocyanic (HCN), which are quite weak.

Acidic hydrogen atoms. Many important acids containing carbon are classified as organic acids. An example is acetic acid, which is somewhat weaker than nitrous acid and has the formula

$$\begin{array}{c} H \\ | \\ H-C-C \\ | \qquad\ \ \diagdown \\ H \qquad O-H \end{array} \ \ \diagup\!\!\!\!O$$

As mentioned in Sec. 2-3, this formula is frequently written CH_3COOH for typographical convenience; note, however, that there is no O-O bond in the molecule. The hydrogen atom that ionizes is the one joined to the oxygen atom, as is true for almost all oxygen-containing acids. The group of atoms —COOH is present in many organic acids and is called the *carboxyl group*. We abbreviate acetic acid by the symbol HOAc, Ac thus indicating the atomic grouping CH_3CO. In abbreviated form, the ionic dissociation of acetic acid is indicated by the equation $HOAc \longrightarrow H^+ + OAc^-$, which in detail corresponds to

$$\begin{array}{c} H \\ | \\ H-C-C \\ | \qquad\ \ \diagdown \\ H \qquad O-H \end{array}\diagup\!\!\!\!O \quad\longrightarrow\quad \begin{array}{c} H \\ | \\ H-C-C \\ | \qquad\ \ \diagdown \\ H \qquad O^- \end{array}\diagup\!\!\!\!O \ + H^+$$

Although acetic acid contains four hydrogen atoms, only the one attached to the oxygen atom comes off easily or is, in other words, an *acidic hydrogen atom*. This is a general feature of organic compounds: hydrogen atoms bonded to carbon atoms are seldom noticeably acidic in aqueous solution. The rare exceptions involve molecules in which the carbon atom has a much higher positive charge than usual.

There are many acids, both organic and inorganic, that contain more than one ionizable proton, including H_2SO_4, H_3PO_4, H_2CO_3, H_2SO_3 (sulfurous acid), and HOOC—COOH (oxalic acid). In each of these *polyprotic* acids (sometimes, somewhat confusingly, called polybasic acids), the ionizable hydrogen atoms are attached to oxygen atoms, so the formulas might better be written $(HO)_2SO_2$, $(HO)_3PO$, and so on. Some polyprotic acids do not contain oxygen; an example is H_2S (hydrogen sulfide), which is a very weak acid in aqueous solution. Conversely, there are many substances, both organic and inorganic, that contain the —OH group but are not acids in aqueous solution. Examples include ethanol (CH_3CH_2OH) and other alcohols, a class of compounds containing a hydroxyl group attached to a carbon atom that is in turn linked to hydrogen or other carbon atoms. Typical alcohols do not ionize in aqueous solution, and they show neither acidic nor basic properties.

Strong and weak bases.   It has perhaps occurred to the reader that there is another class of compounds containing O-H bonds that does not show acidic properties in aqueous solutions: the metallic hydroxides such as NaOH and $Ca(OH)_2$. These compounds give OH^- ions in aqueous solution and thus are typical bases. The corresponding pure solid hydroxides are ionic, containing Na^+ and OH^- ions in one case and Ca^{2+} and OH^- in the other. Many weaker bases, as well as many acids, have the covalent structure X-O-H, where X represents the rest of the molecule. In some of these compounds the O-H bond is cleaved in solution to form H^+, whereas in others the X-O bond breaks to form OH^-. In still others neither bond breaks and no ionization at all occurs.

Typical strong bases include NaOH, KOH, and the hydroxides of the other alkali metals (Li, Cs, and Rb), as well as the hydroxides of the alkaline-earth elements other than beryllium (Be)—for example, $Mg(OH)_2$, $Ca(OH)_2$, and $Ba(OH)_2$. These alkaline-earth hydroxides are not very soluble. Only one of the two hydroxide ions separates completely from the positive ion, so solutions of the alkaline-earth hydroxides contain significant concentrations of MOH^+ ions (such as $MgOH^+$ and $CaOH^+$) in addition to M^{2+} ions and OH^- ions.

The commonest weak base is ammonia, NH_3, which reacts to a small extent with water to form ammonium ions and hydroxide ions:

$$H_2O + NH_3 \longrightarrow OH^- + NH_4^+ \qquad (7\text{-}5)$$

The fraction of ammonia molecules that undergo this reaction is small, and thus ammonia is considered a weak base. The reaction is not strictly a dissociation, but is rather a *proton-transfer* or *proton-exchange* process. As discussed in the next section, all acid-base reactions in aqueous solution can be viewed as proton-exchange reactions. Although for many years chemists considered that ammonium hydroxide, NH_4OH, was the predominant species responsible for the properties of a solution of ammonia in water, it has recently been shown that little or no ammonium hydroxide is present in such solutions or, indeed, exists under any circumstances.[1] In aqueous solution NH_3 forms hydrogen-bonded aggregates with water molecules.

---

[1]If the species NH_4OH were present, the products in (7-5) would be formed by its dissociation. Undoubtedly the authority of Arrhenius' dissociation theory provided the original rationale for the belief in the species NH_4OH. Evidence against its existence has been provided only by modern physical methods, including various types of spectroscopy and diffraction.

Both O-H- - -N and N-H - - -O bonds are present, as in

$$(7\text{-}6)$$

When ammonia reacts with an acid, a proton is transferred to the NH_3 molecule, forming the ammonium ion, NH_4^+, as in (7-5). The unshared pair of electrons on NH_3 is responsible for its basicity:

$$(7\text{-}7)$$

The ammonium ion is hydrogen bonded to water molecules through N-H - - -O bonds in a manner similar to that shown for NH_3 in (7-6).

The neutralization reaction between H^+ and OH^-, Equation (7-3), goes nearly to completion, leaving only a very small concentration of unreacted H^+ ions and OH^- ions. Correspondingly, pure water is dissociated into H^+ and OH^- only to a very small extent:

$$H_2O \longrightarrow H^+ + OH^- \qquad (7\text{-}8)$$

Water is thus a very weak base as well as an equally weak acid. A precise quantitative relation exists between the concentrations of the hydrogen and hydroxide ions as a result of (7-8); it is considered later [Equations (7-18) and (7-19)].

7-2 The Hydronium Ion and Proton-Transfer Reactions in Aqueous Solutions

The hydronium ion. The true hydrogen ion is a hydrogen atom stripped of its electron. In other words, it is the bare nucleus of a hydrogen atom—a proton when we are dealing with the nuclide ^1H. This is a very small particle compared to the size of a hydrogen atom—about 100,000 times smaller in diameter. It is also very small compared to the smallest other positive ions, such as Li^+ or Be^{2+}, which are comparable in size to atoms. The electrons in the shell of another atom can therefore approach the positive charge of a bare hydrogen ion much more closely than that of any positive ion that has not been stripped of all its electrons. For this reason the bare hydrogen ion exists in reasonable concentrations only in hydrogen gas that has been subjected to an electric discharge or is at very high temperatures.

No significant concentration of free H^+ exists in aqueous solutions because H^+ immediately combines with a water molecule to form the *hydronium ion*,[1] H_3O^+:

$$\left[\begin{array}{c} \quad H \\ :O-H \\ \quad H \end{array} \right]^+$$

This ion has been identified spectroscopically when an electric discharge takes place in water vapor, and it has also been found by X-ray and neutron diffraction to exist in solids such as perchloric acid monohydrate, $HClO_4 \cdot H_2O$, crystals of which are found to contain H_3O^+ and ClO_4^- ions. Recent evidence suggests that in aqueous solutions the hydronium ion is tied by hydrogen bonds to three more water molecules,

$$(7\text{-}9)$$

corresponding to the formula $H_9O_4^+$. As illustrated in Fig. 3-9, the water molecule is a dipole in which each hydrogen atom carries a residual positive charge and the oxygen atom a residual negative charge. The positive charge on the $H_9O_4^+$ ion therefore attracts the oxygen atoms of additional water molecules, so that the complex is almost certainly even larger. The size of this complex is unknown, and it is in fact difficult to decide just where the influence of the original H_3O^+ ion on the surrounding water molecules ends. We may call this entire complex the *hydrated hydronium ion* and use the symbols $H_3O^+(aq)$ or just H_3O^+. However, in general we shall simply speak of the hydrogen ion and use the symbol H^+, the understanding always being that in an aqueous medium it represents a hydrated hydronium ion. Because of the extreme instability of bare protons in water, there are no grounds for confusion; it has been estimated that it would take an incredibly large amount of water, 10^{130} liters, to contain an average of one bare proton at all times. Only when the situation requires it do we specifically refer to the hydronium or H_3O^+ ion.

Hydrogen ions may be transferred from one molecule to the next. In such a transfer the H_2O losing the hydrogen ion would, however, become an OH^- ion and the H_2O gaining a hydrogen ion an H_3O^+ ion, if the transfer were not followed by others that involve still further water molecules. At any given time the concentrations of the ions OH^- and H_3O^+ are very small in pure water, and their identities

---

[1]The officially approved name for this ion is *oxonium*, but the name hydronium is more common.

change rapidly because of extremely rapid proton exchanges. Proton exchange is one of the fastest chemical reactions known, the average lifetime of a given H_3O^+ ion being only about 10^{-12} s.

The Brønsted definition of acids and bases. The fact that in aqueous solutions the hydrogen ion exists only as aquated hydronium ion means that the acid dissociation reactions discussed earlier are actually proton-transfer reactions, a new bond being formed as well as one being broken. Thus, representing an acid generally as HA, the reaction of "dissociation" (7-4) is really[1]

$$HA + O\!\!\begin{array}{c} \diagup H \\ \diagdown H \end{array} \longrightarrow A^- + \left[O\!\!\begin{array}{c} \diagup H \\ \diagdown H \end{array}\!\!\!-H \right]^+ \tag{7-10}$$

Comparison of (7-10) with (7-5) shows that they involve the same fundamental process, a transfer of a proton from one species to another. This process is indeed the common feature of all acid-base reactions in aqueous media, a feature that led the Danish chemist Brønsted to propose in 1923[2] that an *acid be defined as a molecule or ion capable of losing a proton—a proton donor—and a base be defined as a proton acceptor.* Thus, in reaction (7-5) water behaves as an acid, donating a proton to the acceptor ammonia. In reaction (7-10), water acts as a base to which the acid HA donates a proton.

This way of viewing acids and bases leads to a natural definition of a *conjugate acid-base pair* (Latin: *conjugatus,* joined together). The conjugate base of any acid is the species obtained by removing a proton from the acid; thus Cl^- is the conjugate base of HCl, OAc^- is the conjugate base of HOAc, OH^- is the conjugate base of H_2O, and NH_3 is the conjugate base of the ammonium ion, NH_4^+. Conversely, the conjugate acid of any base is the acid obtained by adding a proton to the base; for example, H_3O^+ is the conjugate acid of the base H_2O, HCl is the conjugate acid of Cl^-, and so on. Some representative conjugate acid-base pairs are listed in Table 7-1. As these examples show, and as the definition implies, the charge on the conjugate acid is always more positive by one unit than that on the base.

A general formulation of the relation between the members of a conjugate acid-base pair is

$$(H_nA)^m \longrightarrow H^+ + (H_{n-1}A)^{m-1} \tag{7-11}$$

where n is some positive integer and the charge on the acid, m, is any integer, positive, negative, or zero. If the emphasis is on the acid member of the pair, however, it is convenient to replace this somewhat awkward equation by the simpler formulation

$$HA \longrightarrow H^+ + A^- \tag{7-12}$$

---

[1]In (7-10) and similar equations, the unshared electron pairs (on H_2O, H_3O^+, and other species) are omitted for typographical convenience.

[2]The acid-base definition given here is frequently attributed to both Brønsted and the British chemist T. M. Lowry, but Lowry did not give the definition explicitly and, indeed, stated later that Brønsted's view of the anion of an acid as a base was novel.

TABLE 7-1   **Typical conjugate acid-base pairs**

Acid	Base	Acid	Base
HOAc	OAc^-	H_2CO_3	HCO_3^-
HCl	Cl^-	HCO_3^-	CO_3^{2-}
H_3O^+	H_2O	H_3PO_4	$H_2PO_4^-$
H_2O	OH^-	$H_2PO_4^-$	HPO_4^{2-}
NH_4^+	NH_3	HPO_4^{2-}	PO_4^{3-}
NH_3	NH_2^-		

Here HA represents the acid regardless of its actual charge or the charge on its conjugate base. If the emphasis is on the basic member of the pair, it is convenient to use the similar relation

$$HB^+ \longrightarrow H^+ + B \qquad (7\text{-}13)$$

Here the base is symbolized generally by B, again regardless of what the actual charges might be.

With this definition, the general reaction between an acid and a base is one in which the accompanying proton-transfer results in the formation of a new acid and a new base, the conjugates of those present originally:

$$\underset{\text{Acid 1}}{HA} + \underset{\text{Base 2}}{B} \longrightarrow \underset{\text{Base 1}}{A^-} + \underset{\text{Acid 2}}{HB^+} \qquad (7\text{-}14)$$

Equations (7-5) and (7-10) are seen to represent reactions that may be formulated in just this way.

A number of species in Table 7-1 (for example, HCO_3^-) may act either as an acid or a base, losing or gaining a proton. This behavior is termed *amphiprotic* (Greek: *amphi*, both). An important example of amphiprotic behavior is that shown by water; one molecule reacts with another, the first acting as base and the second as acid:

$$H_2O + H_2O \longrightarrow H_3O^+ + OH^- \qquad (7\text{-}15)$$

This *autoprotolysis* reaction (Greek: *auto*, self; *lysis*, loosening, splitting) is fundamentally the same reaction as the dissociation reaction (7-8) of water,

$$H_2O \longrightarrow H^+ + OH^- \qquad (7\text{-}16)$$

Comparison of (7-15) with (7-16) shows that (7-16) is properly considered to be a proton-transfer reaction of the general acid-base type (7-14). The same is true of the neutralization reaction (7-3), which is the reverse of the reaction just considered. Reactions completely analogous to (7-15) and its reverse occur in some other solvents that are capable of sustaining acid-base reactions among solutes dissolved in them. Examples are pure ammonia (in which the products are NH_4^+ and NH_2^-) and pure sulfuric acid (in which the products are $H_3SO_4^+$ and HSO_4^-). There are significant differences between acid-base reactions in such media and those in water. Our primary concern in this and subsequent chapters, except where specifically noted, will be with aqueous media.

7-3   pH and the Concept of Equivalence in Acid-Base Reactions

The pH.   The hydrogen-ion concentration of an aqueous solution is one of its important properties. Since this concentration may range over many powers of 10, and since these powers are mainly negative when the concentration is expressed in moles of hydrogen ion per liter of solution, as is common, it proves convenient to work with another quantity, the pH, which is the negative logarithm to the base 10 of the hydrogen-ion concentration. Symbolizing this concentration in moles per liter (or M) by $[H^+]$, we have[1]

$$pH = -\log[H^+] = -\log[H_3O^+] \qquad (7\text{-}17)$$

Note that the larger the H^+-ion concentration, the lower the pH. When $[H^+] = 1.00\ M$ the pH is zero, and for H^+-ion concentrations larger than this the pH becomes negative. For pure water at $24°C$ the pH is 7.00, corresponding to $[H^+] = 1.00 \times 10^{-7}\ M$. Since hydroxide and hydronium ions are produced in equal numbers in pure water by reaction (7-15), the concentration of hydroxide ions in pure water is always equal to that of hydrogen ions. Thus, at $24°C$, the concentration of OH^- ions is $[OH^-] = 1.00 \times 10^{-7}\ M$ and pOH (meaning $-\log[OH^-]$) is 7.00. Thus pH + pOH = 14.00 for water at $24°C$.

In acidic and basic aqueous solutions, the concentrations of hydrogen and hydroxide ions are no longer equal, since acidic solutions contain excess H^+ and basic (or alkaline) solutions contain excess OH^-. However, as discussed at length in Chap. 14, in *any* aqueous solution at room temperature it is approximately true that

$$[H^+][OH^-] = 1.0 \times 10^{-14}\ M^2 \qquad (7\text{-}18)$$

or, expressed alternatively,

$$pH + pOH = 14.0 \qquad (7\text{-}19)$$

The number of significant figures with which a pH is expressed must be appropriately related[2] to the number with which the corresponding $[H^+]$ is given so that each can be calculated from the other without loss of precision.

EXAMPLE 7-1   **pH of acidic solutions.**   Give the pH of solutions with the following $[H^+]$ values in moles per liter: (*a*) 0.047; (*b*) 1.0; (*c*) 5.0; and (*d*) 3×10^{-6}.

Solution.   (*a*) We have $[H^+] = 0.047\ M$ so that pH $= -\log 0.047 = -\log(10^{-2} \times 4.7) = 2.00 - \log 4.7 = 2.00 - 0.67 = \underline{1.33}$. Note that two significant figures of $[H^+]$ correspond here to two significant figures beyond the decimal point for the pH, that

---

[1]Strictly speaking pH $= -\log([H^+]/M)$, but the units mole per liter or M from $[H^+]$ are usually dropped, resulting in expressions (7-17). This is often done when the logarithm of a physical quantity such as a length or a pressure is taken (see also Appendix A).

[2]The *characteristic* of a logarithm (the number preceding the decimal point) merely denotes the power of 10 involved and is thus not a significant figure in the normal sense. Significant figures apply only to the remaining portion of the logarithm, the *mantissa*.

is, in the mantissa of the logarithm. Thus, a value $[H^+] = 0.048\ M$ would yield pH $= +2.00 - \log 4.8 = 2.00 - 0.68 = 1.32$. (b) When $[H^+] = 1.0\ M$, the pH $= -\log 1.0 = \underline{0.0.}$ (c) Here pH $= -\log 5.0 = \underline{-0.70.}$ (d) pH $= 6 - \log 3 = 6 - 0.5 = \underline{5.5.}$

EXAMPLE 7-2   **H^+-ion concentration from the pH.**   What is $[H^+]$ for a solution in which (a) pH $= 2.0$; (b) pH $= 3.75$?

Solution. (a) Here we have $[H^+] = 10^{-2.0}\ M = \underline{1.0 \times 10^{-2}\ M}$. (b) It is best to consider that $3.75 = 4.00 - 0.25$, so that $[H^+] = 10^{0.25} \times 10^{-4}\ M$. Since the antilogarithm of 0.25 is 1.8, we have $[H^+] = \underline{1.8 \times 10^{-4}\ M}$.

Equivalence and normality. Stoichiometric calculations with acids and bases can often be simplified with the concept of chemical equivalence. *One equivalent* (abbreviated eq) *of an acid is defined as that amount of acid that donates one mole of protons in a specified reaction. One equivalent of a base is that amount of base that accepts one mole of protons in a specified reaction.*

The *equivalent weight* (EW) of an acid or a base is the (relative) weight of one equivalent, expressed on the atomic-weight scale. It can be defined for a particular substance only with respect to a particular reaction, unless only one reaction is possible. Thus the equivalent weight of HCl, acting as an acid, is necessarily the same as the formula weight (or the molecular weight), since there is only one proton that can be donated by this acid. However, for an acid with more than one acidic hydrogen atom, such as H_2SO_4 or H_3PO_4, the equivalent weight cannot be specified until one knows *how many protons react per molecule in the reaction under consideration*. If H_2SO_4 reacts only to form HSO_4^-, there is one equivalent per mole of acid and the equivalent weight is the same as the molecular weight or formula weight. On the other hand, if H_2SO_4 reacts to form SO_4^{2-}, there are two equivalents per mole of H_2SO_4, so the equivalent weight is half the molecular weight.

It is extremely important to recognize this dependence of the equivalent weight EW on the nature of the reaction. It is meaningless to ask "What is the equivalent weight of this compound?" unless the context of the question, or better an explicit statement, clarifies the reaction that the questioner has in mind. The most common difficulty of the beginner with the concept of equivalence is failure to recognize this interdependence.

Equivalents are particularly useful in considering the stoichiometry of reactions in solution. As indicated in Chap. 6, it is convenient to express concentration in terms of the amount of solute per unit volume of solution because volumes of solution can readily be measured precisely. The volume concentration unit for equivalents, defined in a fashion analogous to molarity and formality (Sec. 6-3), is normality:

$$\text{Normality} = \frac{\text{number of equivalents of solute}}{\text{volume of } solution \text{ in liters}} \qquad (7\text{-}20)$$

The units of normality are thus equivalents of solute per liter of solution, abbreviated eq liter^{-1}, or just N (read as "normal").

EXAMPLE 7-3   **Equivalents and normalities.** A number of reactions are listed in Table 7-2, together with the equivalent weight and the normality of a $0.2\ F$ solution of one reactant. For clarity in the present context, the equations are written with molecular formulas although all the reactions are in fact ionic.

In reaction (1), 1 gfw KOH reacts with 1 mol protons. Reactions (2) and (3) present an important contrast and illustrate the point stressed earlier, that the equivalent weight of a given reagent depends on the reaction under consideration. In reaction (2), each gram formula weight of sulfuric acid furnishes 2 mol protons and hence the equivalent weight is half the formula weight. In (3), on the other hand, only one of the two hydrogen atoms in the sulfuric acid molecule is replaced and consequently the equivalent weight is the same as the formula weight. Reactions (4) and (5) illustrate the same point regarding sodium carbonate; in (4), Na_2CO_3 reacts with two protons, in (5) with only one. Note especially that a given solution [e.g., the $0.2\ F\ H_2SO_4$ of reactions (2) and (3) or the $0.2\ F\ Na_2CO_3$ of reactions (4) and (5)] can have different normalities, depending on the reaction of the solute.

Reaction (6) is another acid-base reaction, the base being silver oxide, with two equivalents per gram formula weight.

The advantage of the equivalence concept is that, by definition, one equivalent of reactant A reacts with exactly one equivalent of reactant B. This eliminates the need for balancing equations explicitly when dealing with equivalents, but this apparent gain is in part illusory since most of the same considerations required in balancing an equation are needed to find the appropriate equivalent weights. As indicated in Example 7-3, to define an equivalent of a given reagent in a particular reaction one must know the reactants and products and must, in essence, balance the equation—at least until one has had a good deal of experience with the commonly occurring reactions. When that stage is reached, the concept of equivalence does save time and effort.

Normalities are frequently employed for expressing concentrations of solutions used in quantitative volumetric analysis because they simplify calculations. The experimental procedure consists in taking a known volume V_1 of a solution contain-

TABLE 7-2   **Examples of equivalents and normalities**

Reaction					EW	Normality of a $0.2\ F$ solution
					Quantities pertaining to underlined reactants	
(1)	HCl	+ $\underline{KOH}$	$\longrightarrow$ KCl	+ H_2O	FW KOH	$0.2\ N$
(2)	$\underline{H_2SO_4}$	+ 2KOH	$\longrightarrow$ K_2SO_4	+ $2H_2O$	FW $H_2SO_4/2$	$0.4\ N$
(3)	$\underline{H_2SO_4}$	+ KOH	$\longrightarrow$ $KHSO_4$	+ H_2O	FW H_2SO_4	$0.2\ N$
(4)	2HCl	+ $\underline{Na_2CO_3}$	$\longrightarrow$ 2NaCl	+ H_2O + $CO_2(g)$	FW $Na_2CO_3/2$	$0.4\ N$
(5)	HCl	+ $\underline{Na_2CO_3}$	$\longrightarrow$ $NaHCO_3$	+ NaCl	FW Na_2CO_3	$0.2\ N$
(6)	$2HNO_3$	+ $\underline{Ag_2O(s)}$	$\longrightarrow$ $2AgNO_3$	+ H_2O	FW $Ag_2O/2$	*

*Since the Ag_2O is an insoluble solid, no concentration is given.

ing solute 1 and measuring by careful addition *that volume V_2 of a second solution*, containing solute 2, that is completely equivalent to the quantity of solute 1 present in V_1. This kind of measurement is called a *titration*. It establishes the equivalence of the quantities of reacting solutes, that is, the equality of the numbers of equivalents of each so that

$$\text{No. of equivalents of solute 1} = \text{no. of equivalents of solute 2}$$

Then, if the normalities are, respectively, N_1 and N_2 (in equivalents per liter),

$$N_1 V_1 = N_2 V_2 \tag{7-21}$$

If the volumes are expressed in liters, $N \times V$ is the number of equivalents of solute in the volume V; if the volumes are expressed in milliliters, $N \times V$ is the number of *milliequivalents* (meq), where 1 meq $= 10^{-3}$ eq.

EXAMPLE 7-4   **Solution stoichiometry.** A 20.00-ml sample of a 0.400 N solution of a certain acid was found to react with exactly 25.00 ml of a basic solution of unknown concentration. What was the normality of the base?

Solution. Because the number of equivalents of acid must be the same as the number of equivalents of base, (7-21) is applicable and

$$N_{\text{base}} = N_{\text{acid}} \frac{V_{\text{acid}}}{V_{\text{base}}}$$
$$= 0.400 \times \tfrac{20.00}{25.00} = \underline{0.320\ N}$$

Note that nothing need be known about the nature of the acid or the base.

EXAMPLE 7-5   **Equivalent weight and molecular weight.** A 0.300-g sample of an unknown water-insoluble acid was dissolved in 50.00 ml of 0.200 N NaOH. The resulting solution was then neutralized by addition of 15.00 ml of 0.400 N HCl. What is the equivalent weight of the unknown acid, and what can you say about its molecular weight?

Solution. The total number of equivalents of acid (unknown acid + HCl) must be the same as that of base.

$$\text{Equivalents of HCl} = (0.0150 \text{ liter})(0.400\ N) = 0.00600 \text{ eq}$$
$$\text{Equivalents of unknown acid} = \frac{0.300 \text{ g}}{\text{EW g/eq}}$$
$$\text{Total equivalents of acid} = 0.00600 + \frac{0.300}{\text{EW}}$$
$$\text{Equivalents of base} = (0.0500 \text{ liter})(0.200\ N) = 0.0100$$

Thus

$$0.00600 + \frac{0.300}{\text{EW}} = 0.0100$$

$$\frac{0.300}{\text{EW}} = 0.0100 - 0.00600 = 0.0040$$

$$\text{EW} = \tfrac{0.300}{0.0040} = \underline{75}$$

The molecular weight must be some integral multiple of the equivalent weight, the integer being the number of acidic hydrogen atoms reacting per mole of acid. Thus,

$$MW = \underline{75\,n}$$

where n is some positive integer.

The concept of equivalence is not restricted to acid-base reactions. More generally, one equivalent of any substance is that amount that corresponds in its reaction either directly or indirectly to one mole of hydrogen atoms or protons. The term *indirectly* refers to the possibility that the substance in question does not itself react with or form hydrogen atoms or protons but can be made to react with another reagent that, in turn, does react with or form hydrogen atoms or protons. Additional links may be included in this chain of equivalent quantities until one mole of hydrogen atoms is finally referred to. In the final section of this chapter the equivalence concept is applied to the important class of reactions involving oxidation and reduction.

7-4   Acidic, Basic, and Amphoteric Properties of Oxides

Although it was believed in Lavoisier's day that substances were acidic because they contained oxygen, it was later seen that many oxygen compounds, such as the oxides of the alkali and alkaline-earth metals, are basic in character. It is possible and instructive to classify oxides on the basis of their acidic or basic properties. A water-soluble oxide is considered acidic if its aqueous solution is acidic. Examples include NO_2, SO_2, SO_3, $P_2O_3(P_4O_6)$, and $P_2O_5(P_4O_{10})$. Oxides that do not readily dissolve in water are said to be acidic if their solubility increases as the solution is made more basic—that is, with increasing $[OH^-]$—and decreases as the solution is made more acidic. Examples of water-insoluble acidic oxides are SiO_2 and SnO_2, the dioxides of silicon and tin.

Oxides are called basic if their aqueous solutions are alkaline or if their solubility increases with increasing H^+ concentration (decreasing pH) and decreases with increasing OH^- concentration (increasing pH). Among the insoluble basic oxides are those of many metals, for example, FeO, Fe_2O_3, CoO, and NiO.

The solubility of some oxides is increased both by an increase in the concentration of hydrogen ions and by an increase in the concentration of hydroxide ions. These oxides therefore show both acidic and basic character. Such behavior is termed *amphoteric*. The solubility of amphoteric oxides in pure water is usually limited. Examples are BeO and Al_2O_3, with the reactions

$$BeO + 2H^+ \longrightarrow Be^{2+} + H_2O \tag{7-22}$$

$$BeO + 2OH^- + H_2O \longrightarrow Be(OH)_4{}^{2-} \tag{7-23}$$

$$Al_2O_3 + 6H^+ \longrightarrow 2Al^{3+} + 3H_2O \tag{7-24}$$

$$Al_2O_3 + 2OH^- + 3H_2O \longrightarrow 2Al(OH)_4{}^- \tag{7-25}$$

In each of these reactions, it is assumed that suitable oppositely charged ions are present—for example, Cl^- or NO_3^- in (7-22) and (7-24), and K^+ or Na^+ in (7-23) and (7-25). Their source could be solutions of HCl or HNO_3 for (7-22) and (7-24), or NaOH or KOH for the other two reactions. The ions $Be(OH)_4^{2-}$ and $Al(OH)_4^-$ are examples of complex ions, which are discussed in Sec. 7-6. Their -2 and -1 charges correspond to the sum of the charge on the related metal cation (Be^{2+}, Al^{3+}) and the four -1 charges from the OH^- that are part of each of these complex ions.

Other amphoteric oxides are ZnO, PbO, and SnO, which react with OH^- in H_2O to form the ions $Zn(OH)_4^{2-}$, $Pb(OH)_4^{2-}$, and $Sn(OH)_4^{2-}$ and react with H^+ in H_2O to form the (hydrated) cations Zn^{2+}, Pb^{2+}, and Sn^{2+}.

7-5 Aqueous Solutions of Salts

In an aqueous solution many salts are completely dissociated into ions. This is particularly true of salts resulting from reactions of strong acids with strong bases, salts such as $NaNO_3$ or KBr. Aqueous solutions and also melts of these salts contain no $NaNO_3$ or KBr molecules, only the ions Na^+ and NO_3^-, or K^+ and Br^-.

When solutions of electrolytes are mixed, reactions may take place with resulting changes in the quantities of substances in solution. For example, acid-base reactions may remove some of the acids and bases, or insoluble solids may precipitate, removing some ions from the solution. In calculating the concentrations of the different species present in the final solution, one must take into account any such reactions and note as well which substances present are strong electrolytes and are therefore essentially completely dissociated into ions.

EXAMPLE 7-6   **Mixture of solutions of electrolytes.** The following solutions are mixed together: 50 ml of 0.250 F potassium hydroxide, 50 ml of 0.100 F barium nitrate, and 100 ml of 0.060 F sulfuric acid. Enough water is added to make the total volume 500 ml. The only insoluble substance formed is barium sulfate; assume it to be completely insoluble. Calculate the final concentrations of K^+, OH^-, Ba^{2+}, NO_3^-, H^+, and SO_4^{2-}.

Solution. Both KOH and $Ba(NO_3)_2$ are strong electrolytes; H_2SO_4 dissociates completely to H^+ and HSO_4^-. A straightforward approach is to calculate the number of moles of each ion present before any possible reactions occur and then consider the stoichiometry of any reactions that do take place. The moles of each ion present initially are

K^+:   (0.050 liter)(0.250 mol K^+ liter^{-1}) = 0.0125 mol

OH^-:   same as K^+, 0.0125 mol

Ba^{2+}:   (0.050 liter)(0.100 mol Ba^{2+} liter^{-1}) = 0.0050 mol

NO_3^-:   two present for every Ba^{2+}, or 0.0100 mol

H^+:   (0.100 liter)(0.060 mol H^+ liter^{-1}) = 0.0060 mol

HSO_4^-:   same as H^+, 0.0060 mol

The base OH^- will react with H^+ to form H_2O; any excess OH^- will react with the other acid present, HSO_4^-, to form SO_4^{2-} and H_2O. Insoluble $BaSO_4$ will be formed by Ba^{2+} with SO_4^{2-} or HSO_4^-. Since all possible compounds of K^+ and NO_3^- that might be formed are soluble and strong electrolytes, these ions remain in the solution unchanged.

There are initially 0.0125 mol OH^- and 0.0060 mol H^+; these ions react to produce 0.0060 mol H_2O, leaving 0.0065 mol OH^-, which reacts with the 0.0060 mol HSO_4^- to form 0.0060 mol H_2O and 0.0060 mol SO_4^{2-}, leaving finally 0.0005 mol OH^- unreacted.

There is 0.0050 mol Ba^{2+}, which then removes 0.0050 mol SO_4^{2-} from the solution, leaving 0.0010 mol SO_4^{2-}.

Since the final volume of the solution is 0.500 liter, the final concentrations are

$$K^+: \quad \frac{0.0125 \text{ mol}}{0.500 \text{ liter}} = 0.0250 \ M \qquad NO_3^-: \quad \frac{0.0100 \text{ mol}}{0.500 \text{ liter}} = 0.0200 \ M$$

$$OH^-: \quad \frac{0.0005 \text{ mol}}{0.500 \text{ liter}} = 0.0010 \ M \qquad H^+: \quad \frac{1.0 \times 10^{-14}}{[OH^-]} = 1.0 \times 10^{-11} \ M$$

$$Ba^{2+}: \quad \text{zero} \qquad\qquad\qquad SO_4^{2-}: \quad \frac{0.0010 \text{ mol}}{0.500 \text{ liter}} = 0.0020 \ M$$

One check on the calculations is to be sure the answers are consistent with electrical neutrality:

Concentrations of positive charges		Concentrations of negative charges	
On K^+:	0.0250 M	On NO_3^-:	0.0200 M
On H^+:	negligible	On OH^-:	0.0010 M
		On SO_4^{2-}:	0.0040 M*
Total	0.0250 M	Total	0.0250 M

*Note that the molarity of SO_4^{2-} is multiplied by 2 in counting charges because each ion is doubly charged.

The hydrolysis reaction. Salts formed from weak acids or weak bases also dissociate completely into ions in aqueous solutions. Examples of such salts are sodium acetate, NaOAc, the salt of the weak acid HOAc and the strong base NaOH, and ammonium chloride, NH_4Cl, the salt of the weak base NH_3 and the strong acid HCl. Unlike solutions of salts formed from strong acids and strong bases, however, solutions of these salts contain some undissociated molecules of weak electrolytes formed by what is termed a *hydrolysis* reaction.

For example, a solution of NaOAc contains essentially no undissociated NaOAc molecules, but rather the ions Na^+ and OAc^-, which also exist in crystalline sodium acetate. In addition the solution contains a small concentration of HOAc molecules, produced by the reaction of OAc^- ions with water to form HOAc and OH^-:

$$H_2O + OAc^- \longrightarrow HOAc + OH^- \tag{7-26}$$

This is a typical hydrolysis reaction, the reaction of a salt ion with water; this solution becomes basic because of the OH^- formed. Reaction (7-26) occurs only to a small, but not a negligible, extent. Comparison with the general acid-base reaction (7-14) shows that (7-26) is simply another proton-transfer reaction. A small fraction of the water molecules transfer protons to the acetate ions present, producing small concentrations of HOAc and OH^-. In essence, OAc^- is trying to pull a proton from a water molecule, that is, it competes with OH^- for the proton. In this competition acetate ions finish a poor second, but they do succeed in removing some protons and thereby form some HOAc molecules.

A solution of ammonium chloride contains no undissociated NH_4Cl molecules, but rather chiefly NH_4^+ and Cl^- ions. However, it also contains a very small concentration of free ammonia formed by hydrolysis of NH_4^+:

$$NH_4^+ + H_2O \longrightarrow H_3O^+ + NH_3 \qquad (7\text{-}27)$$

This hydrolysis reaction produces a slightly acidic solution instead of the slightly alkaline one formed by sodium acetate. The essential difference here is that the NH_4Cl is formed from a weak base and a strong acid rather than a weak acid and a strong base. In each case, the hydrolysis reaction produces a small quantity of the weak electrolyte from which the salt was formed and a corresponding quantity of either H_3O^+ or OH^-.

Reaction (7-27) is another proton-exchange process, NH_4^+ being the acid (proton donor) and water the base (proton acceptor) on the left side. Water is a far weaker base than ammonia so that most of the protons remain attached to ammonia molecules in the form of ammonium ions, NH_4^+.

All acid-base reactions in water may be viewed as a competition between bases. The relative strengths of the bases and acids involved can be assessed from measurements of the extents of different reactions. This topic is discussed in some detail in Chap. 14.

Hydration of ions. We have already discussed the nature of the hydrogen ion in aqueous solutions (Sec. 7-2). Most other cations are also hydrated, that is, water molecules are bound to them more or less firmly. The strength of attachment of the water molecules increases with increasing positive charge and decreasing size of the ion, which suggests that electrostatic attraction of the ion for the negative end of the water dipole (Fig. 3-9) plays a significant role. Structural studies of many solids containing hydrated cations show that the oxygen atoms of the water molecules are indeed oriented toward the cation. This arrangement also permits formation of covalent bonds involving the unshared electron pairs on the oxygen atom of the water molecule.

The exact number of water molecules associated with any given cation in solution is difficult to specify. For example, there is evidence that Mg^{2+} is closely associated with six H_2O molecules, arranged octahedrally around it, corresponding to a formula $Mg(H_2O)_6^{2+}$. There is an additional layer of less strongly held water molecules. Should the magnesium ion in water be looked upon as a bare Mg^{2+}, as $Mg(H_2O)_6^{2+}$, or perhaps even as an ion containing more water molecules? The same question arises for practically all metal ions in aqueous solution. The present consensus is to assume the existence of definite species, such as $Mg(H_2O)_6^{2+}$, $Be(H_2O)_4^{2+}$ (tetrahedral), and $Fe(H_2O)_6^{3+}$; in each case the number of water molecules is assumed

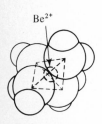

Be^{2+}

$Be(H_2O)_4^{2+}$

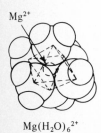

Mg^{2+}

$Mg(H_2O)_6^{2+}$

to be that in best agreement with available evidence, including arguments based on analogy with related compounds. Less firmly bound water of hydration is not counted. In all probability its extent fluctuates statistically, and the attraction of water molecules by a cation weakens so slowly with distance that it is difficult to assess the precise extent of the hydration.

When referring to ionic species in aqueous solutions, we adopt the convention already chosen for H^+: symbols such as Mg^{2+}, Be^{2+}, and Fe^{3+} are abbreviations that represent whatever species actually exist in solution. More explicit formulas such as $Mg(H_2O)_6^{2+}$ are used when this is particularly desirable—for example, in discussion of reactions that specifically involve the water molecules.

Anions such as SO_4^{2-} and Cl^- are also hydrated to some extent, but less is known about this than about the hydration of cations. However, both the number of water molecules involved and their strength of attachment are probably considerably less for anions than for cations. The available evidence on hydration of anions in solids indicates that it is not very common. When specific interaction of water molecules and anions is found, it usually involves hydrogen bonds, e.g., from a water molecule to a sulfate ion. We use symbols such as Cl^- and NO_3^- to represent anions as they actually exist in aqueous solution, with the understanding that water molecules may be loosely attached.

Many uncharged molecules in aqueous solution are hydrogen bonded to water molecules—for example, NH_3 as shown in (7-6)—but the water molecules are rarely shown in formulas. Solutes in nonaqueous solvents may also be associated with solvent molecules; this phenomenon is called *solvation*.

Partially dissociated salt molecules. Some salts are dissociated only partially in aqueous solution. An example is mercuric chloride, $HgCl_2$, solutions of which contain mainly the species $HgCl_2$, with only small concentrations of $HgCl^+$, Hg^{2+}, and Cl^-. Moreover, on addition of Cl^- ions to a solution of $HgCl_2$ the ions $HgCl_3^-$ and $HgCl_4^{2-}$ are formed:

$$HgCl_2 + Cl^- \longrightarrow HgCl_3^- \tag{7-28}$$

$$HgCl_3^- + Cl^- \longrightarrow HgCl_4^{2-} \tag{7-29}$$

These ions are typical complex ions as discussed in the following section, and reactions like (7-28) and (7-29) are of great importance in inorganic chemistry. Note that all species mentioned here may be considered to be composed of Hg^{2+} and various numbers of Cl^-.

Most salts of the alkali and alkaline-earth metals are usually completely ionized in aqueous solution. Salts involving transition metals[1] often exist at least partly as molecules, and there is also a tendency for the formation of charged complexes similar to those in reactions (7-28) and (7-29). However, it is difficult to state general rules about which salts are ionized completely and which salts are not. In the absence of contrary statements we shall always assume that there is complete dissociation of salts into the anions and cations concerned. This does not, of course, preclude

---

[1] Transition metals are the elements in the middle of the long periods of the periodic table, as illustrated inside the front cover.

partial hydrolysis of these cations and anions, as exemplified by reactions (7-26) and (7-27).

7-6 Complex Ions

Some complex ions have already been described in the preceding paragraphs and in Figs. 4-7 and 4-8. The term *complex ion* is used generally to refer to a polyatomic ion that consists of a central atom, usually of a metallic element, to which one or more atoms or groups of atoms are bonded. The attached atoms or groups of atoms are called *ligands* (Latin: *ligandus,* to be bound). Typical complex ions are listed in Table 7-3; the central atom is listed as a positive ion and the ligands as anions or neutral molecules. It is often useful to think of a complex ion as made up from components like those in Table 7-3, but this should not be taken to imply that the ligands and the central ion are linked by ionic bonds; usually the bonding is largely covalent (Chap. 25). The charge assigned to the central atom in Table 7-3 is determined by considering the charges assigned to the ligands and the overall charge on the complex ion.

The definition of a complex ion is not rigid or precise and there are minor variations in usage, but this causes no difficulties. Ions containing oxygen and only one other element—such as permanganate, MnO_4^-, sulfate, SO_4^{2-}, and nitryl, NO_2^+, which might be thought of as containing the oxide ion O^{2-} as the ligand, are not usually regarded as complex ions. These ions are so stable that substitution of ligands is much rarer than in more typical "complex ions" and there is no advantage in considering them as consisting of a central atom and ligands.

The geometric arrangement of the ligands about the central atom is an important characteristic of complex ions. When there are six atoms joined to the central atom, they are usually arrayed at the six corners of an octahedron about the central atom (Fig. 4-7); four groups are usually arranged at the corners of an approximately regular tetrahedron or at the corners of a square. If there are only two ligands, the bonds joining them to the central atom are often at $180°$, giving a linear arrangement as in $Ag(NH_3)_2^+$, $(H_3N-Ag-NH_3)^+$ or as in $Ag(CN)_2^-$, $(NC-Ag-CN)^-$. The relation of the various observed geometric arrangements to the number of ligands

TABLE 7-3 **Typical complex ions**

Complex ion	Central ion	Ligands	Geometry
$CoCl_4^{2-}$	Co^{2+}	Cl^-	Tetrahedral
$Ag(NH_3)_2^+$	Ag^+	NH_3	Linear
$Mg(H_2O)_6^{2+}$	Mg^{2+}	H_2O	Octahedral
$Al(H_2O)_5OH^{2+}$	Al^{3+}	H_2O, OH^-	Octahedral
AlH_4^-	Al^{3+}	H^-	Tetrahedral
$Fe(CN)_6^{4-}$	Fe^{2+}	CN^-	Octahedral
$Zn(OH)_4^{2-}$	Zn^{2+}	OH^-	Tetrahedral
$Pt(NH_3)_4Cl_2^{2+}$	Pt^{4+}	NH_3, Cl^-	Octahedral
$PtCl_4^{2-}$	Pt^{2+}	Cl^-	Square

and to the electronic structures of the central atom and of the complex ion as a whole is discussed in Chap. 12.

For historical reasons, compounds containing complex ions are called *coordination compounds* and the number of atoms directly bonded to the central atom is referred to as the *coordination number*. The most common coordination numbers are 4 and 6, but some atoms frequently display coordination number 2 (especially Ag^+ and Hg^{2+}) and compounds representing each possible coordination number up to 9 are known.

Hydrated metal ions may be viewed as complex ions in which the ligands are water molecules; consequently, the formation of other complex ions in solution usually involves displacing the ligand H_2O. For example, a solution of cupric sulfate, $CuSO_4$, in water has a light blue color that is due to hydrated Cu^{2+} ions, chiefly $Cu(H_2O)_4^{2+}$. When ammonia is added to the solution, an intense deep blue color appears, due chiefly to $Cu(NH_3)_4^{2+}$, the tetrammine cupric ion. Since, by convention, water of hydration attached to ions in aqueous solution is often not shown in formulas, this reaction is sometimes written as

$$Cu^{2+} + 4NH_3 \longrightarrow Cu(NH_3)_4^{2+} \tag{7-30}$$

However, it might preferably be written as

$$Cu(H_2O)_4^{2+} + 4NH_3 \longrightarrow Cu(NH_3)_4^{2+} + 4H_2O \tag{7-31}$$

to emphasize the fact that it is a *ligand-exchange reaction*.

Some hydrated cations, especially those with a high positive charge, behave as weak acids in solution. The reaction involves the transfer of a proton from a coordinated water molecule to a free water molecule—for example,

$$Fe(H_2O)_6^{3+} + H_2O \longrightarrow Fe(H_2O)_5OH^{2+} + H_3O^+ \tag{7-32}$$

The hydrated ferric ion is in fact a slightly stronger acid than acetic acid [that is, it dissociates, by (7-32), slightly more than does acetic acid at a comparable concentration].

Most compounds of the transition metals are colored because of selective absorption of certain wavelengths of visible light. The colors of the complexes of a given transition metal depend upon the ligands that are attached. For example, $Fe(H_2O)_6^{3+}$ is pale violet. If a proton is removed, as in (7-32), the resulting $Fe(H_2O)_5OH^{2+}$ is yellow; if one of the water molecules is replaced by thiocyanate to form $Fe(H_2O)_5SCN^{2+}$, a deep red color appears. Similarly, the color of Cu^{2+} compounds is strongly dependent upon the ligands present. As already mentioned, $Cu(H_2O)_4^{2+}$ is blue and $Cu(NH_3)_4^{2+}$ is an intense deep blue. Crystalline $CuCl_2$ (Fig. 4-3), which may be made by direct combination of copper and chlorine, is yellow.

Some ligands contain more than one atom capable of coordinating a central atom. An example is ethylenediamine,[1]

$$H_2N—CH_2—CH_2—NH_2 \tag{7-33}$$

---

[1] Note that the $—NH_2$ group is called the amine group, with one *m*, while NH_3, when occurring in complexes such as those in (7-30) or (7-35), is called ammine, with two *m*'s.

often abbreviated by the letters *en*. The nitrogen atoms of its two amine groups are capable of coordinating a metal ion in very much the same way as does the nitrogen atom of ammonia. It forms complexes such as $[Co(NH_3)_4(en)]^{3+}$, $[Co(NH_3)_2(en)_2]^{3+}$, and $[Co(en)_3]^{3+}$, in which each *en* is attached to the central cobalt atom by its two nitrogen atoms, taking the place of two NH_3. These ions all have octahedral structures, the first one, for example, being represented by the diagram

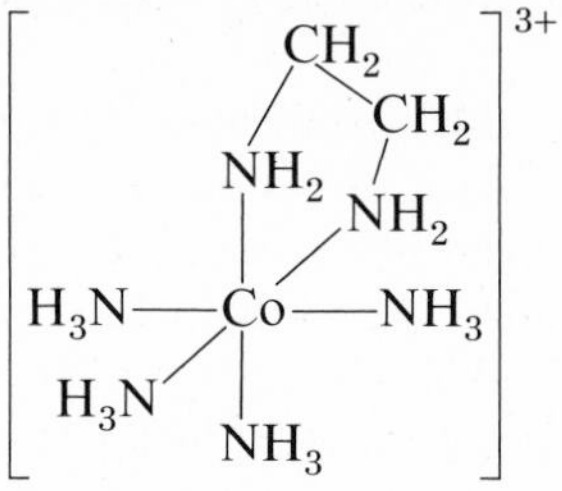

$$
\begin{bmatrix}
 & & & CH_2 & & \\
 & & & \diagup \; \searrow CH_2 & & \\
 & & NH_2 & \; \diagup & & \\
 & & | & \quad NH_2 & & \\
 H_3N\!-\!\!\!& -Co- & \!\!\!-NH_3 & & & \\
 H_3N\diagup & | & & & & \\
 & NH_3 & & & &
\end{bmatrix}^{3+}
\tag{7-34}
$$

Ligands such as ethylenediamine are called *multidentate* (many-toothed), and a complex in which a multidentate ligand is attached in several places to the same central atom is called a *chelate* (Greek: *chele,* claw).

Finally, complexes exist with more than one metal atom. These are called *polynuclear* complexes. An example is

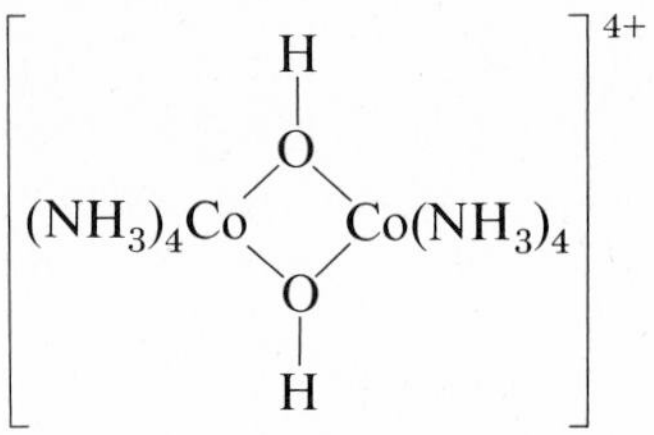

$$
\begin{bmatrix}
 & H & \\
 & | & \\
 & O & \\
 & \diagup \; \searrow & \\
 (NH_3)_4Co & & Co(NH_3)_4 \\
 & \searrow \; \diagup & \\
 & O & \\
 & | & \\
 & H &
\end{bmatrix}^{4+}
\tag{7-35}
$$

The formation of polynuclear complexes sometimes leads to very large aggregates, of essentially infinite size, which may be insoluble. Thus, in the precipitation of $Zn(OH)_2$ by reaction of Zn^{2+} and OH^-, a possible schematic sequence is

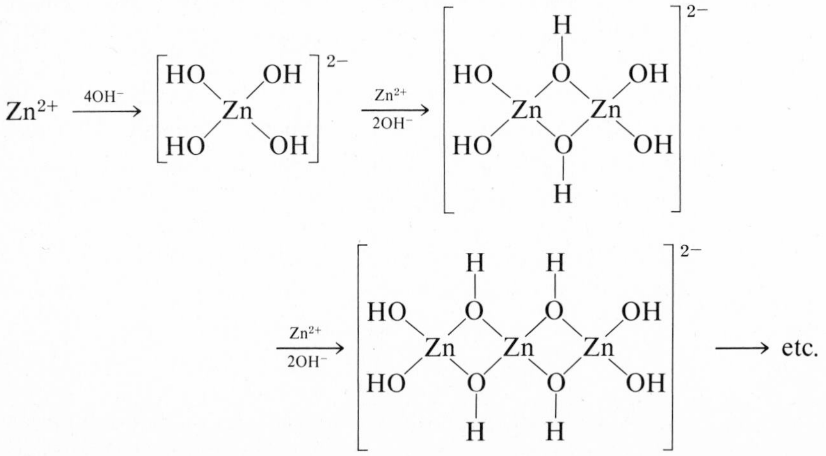

$$
Zn^{2+} \xrightarrow{4OH^-}
\begin{bmatrix}
HO & \quad & OH \\
 \searrow & Zn & \diagup \\
 \diagup & & \searrow \\
HO & \quad & OH
\end{bmatrix}^{2-}
\xrightarrow[2OH^-]{Zn^{2+}}
\begin{bmatrix}
 & H & \\
 & | & \\
HO & O & OH \\
 \searrow \; \diagup \;\; \searrow \; \diagup & & \\
 Zn \qquad Zn & & \\
 \diagup \; \searrow \;\; \diagup \; \searrow & & \\
HO & O & OH \\
 & | & \\
 & H &
\end{bmatrix}^{2-}
$$

$$
\xrightarrow[2OH^-]{Zn^{2+}}
\begin{bmatrix}
 & H & H & \\
 & | & | & \\
HO & O & O & OH \\
 \searrow \diagup \; \searrow \diagup \; \searrow \diagup & & & \\
 Zn \quad Zn \quad Zn & & & \\
 \diagup \searrow \; \diagup \searrow \; \diagup \searrow & & & \\
HO & O & O & OH \\
 & | & | & \\
 & H & H &
\end{bmatrix}^{2-}
\longrightarrow \text{etc.}
\tag{7-36}
$$

This aggregation does not necessarily lead to crystalline structures. Hydroxides often precipitate in a gelatinous form that usually incorporates (hydrogen-bonded) H_2O molecules also, in amounts that depend on the conditions of precipitation.

7-7 Oxidation and Reduction

Oxidation and reduction reactions are common in every area of chemistry—biological oxidations to supply energy, corrosion of metals, flames, electric batteries, and many chemical syntheses. This class of reactions differs from that of acid-base reactions in many ways, but there are also important analogies and similarities between the two classes.

Originally *oxidation* meant combination with oxygen, as in the reactions

$$Cu + \tfrac{1}{2}O_2 \longrightarrow CuO \tag{7-37}$$

$$Fe + \tfrac{1}{2}O_2 \longrightarrow FeO \tag{7-38}$$

$$2Fe + \tfrac{3}{2}O_2 \longrightarrow Fe_2O_3 \tag{7-39}$$

Similarly, the term *reduction* was first used in the context of reduction of a metallic oxide to the metal. This may be achieved, for example, by heating the oxide with carbon, with hydrogen, or with other metals:

$$2FeO + C \longrightarrow 2Fe + CO_2 \tag{7-40}$$

$$CuO + H_2 \longrightarrow Cu + H_2O \tag{7-41}$$

$$Fe_2O_3 + 2Al \longrightarrow 2Fe + Al_2O_3 \tag{7-42}$$

Substances that combine readily with oxygen, such as carbon, hydrogen, and aluminum, are called reducing agents. They are oxidized to CO_2, H_2O, and Al_2O_3 in the reactions given.

It was only natural to extend the terms oxidation and reduction to reactions similar to those described but not involving oxygen. For example, metallic lithium reacts vigorously not only with oxygen but also with fluorine and chlorine:

$$2Li + \tfrac{1}{2}O_2 \longrightarrow Li_2O \tag{7-43}$$

$$Li + \tfrac{1}{2}F_2 \longrightarrow LiF \tag{7-44}$$

$$Li + \tfrac{1}{2}Cl_2 \longrightarrow LiCl \tag{7-45}$$

and there is little that is special about the reaction with oxygen. All three of these reactions are therefore called oxidations. In the resulting crystals of Li_2O, LiF, and $LiCl$, the lithium exists as the Li^+ ion, so that as far as lithium is concerned all three reactions may be described by the equation

$$Li \longrightarrow Li^+ + e^- \tag{7-46}$$

that is, the lithium atom loses an electron. By the same token the other reactants, O_2, F_2, and Cl_2, gain electrons:

$$\tfrac{1}{2}O_2 + 2e^- \longrightarrow O^{2-} \tag{7-47}$$

$$\tfrac{1}{2}F_2 + e^- \longrightarrow F^- \tag{7-48}$$

$$\tfrac{1}{2}Cl_2 + e^- \longrightarrow Cl^- \tag{7-49}$$

These similarities lead to the following generalizations:

Oxidation is defined as a *loss of electrons.*

Reduction is defined as a *gain of electrons.*

An *oxidizing agent* or *oxidant* ("ox") is an ion or molecule that *takes electrons away from* another ion or molecule, thereby oxidizing it. A *reducing agent* or *reductant* ("red") is an ion or molecule that *supplies electrons to* another ion or molecule, thereby reducing it. In these reactions, the oxidizing agent is reduced:[1]

$$ox + ne^- \longrightarrow red \quad \text{(reduction)} \tag{7-50}$$

The oxidizing agent is thus transformed into a reduced form that is potentially capable of losing electrons and could therefore be a reducing agent. Similarly, the reducing agent is oxidized, losing electrons and being transformed to a form capable of acting as an oxiding agent:

$$red \longrightarrow ox + ne^- \quad \text{(oxidation)} \tag{7-51}$$

Two species related as in (7-50) or (7-51) are called a *redox couple.* In referring to redox couples in this text, the format *oxidized form/reduced form* will be used—for example, Na^+/Na, O_2/O^{2-}, or F_2/F^-.

A reaction that relates the oxidized and reduced forms of a redox couple is called a *half-reaction.* It may involve atomic or molecular species and ions, as in (7-46) to (7-49), or just ions, as for the couple Fe^{3+}/Fe^{2+}:

$$Fe^{3+} + e^- \longrightarrow Fe^{2+} \tag{7-52}$$

Species other than oxidizing and reducing agents may also participate in redox reactions, as they do in the couples MnO_4^-/Mn^{2+} and $Cr_2O_7^{2-}/Cr^{3+}$:

$$MnO_4^- + 8H^+ + 5e^- \longrightarrow Mn^{2+} + 4H_2O \tag{7-53}$$

$$Cr_2O_7^{2-} + 14H^+ + 6e^- \longrightarrow 2Cr^{3+} + 7H_2O \tag{7-54}$$

Note that charges as well as atoms are balanced.

If oxidation occurred on a macroscopic scale without simultaneous reduction, electrons released during the oxidation would accumulate in large numbers and the resulting large charge separations would be energetically very unfavorable. Consequently any macroscopic oxidation is always accompanied by a simultaneous reduction. The two half-reactions involved may be coupled electrically, as in a battery or in electrolysis (Chap. 18), each half-reaction occurring separately at an electrode that is able to furnish or take up the electrons being exchanged. A separate path is provided (e.g., a wire) to permit the electrons to flow readily from one electrode to the other. On the other hand, the oxidizing and reducing agents may exchange

---

[1]The **n** in (7-50) is dimensionless, in contrast to the n in $PV = nRT$, for which the customary units are mol.

electrons directly, with no intervening electrical circuits. The general *redox reaction,* as an oxidation-reduction reaction is often called, is an electron-transfer reaction involving two redox couples, ox_1/red_1 and ox_2/red_2:

$$ox_1 + red_2 \longrightarrow red_1 + ox_2 \qquad (7\text{-}55)$$

Note that by our generalizations oxygen is reduced in reactions (7-37) to (7-39). Similarly the production of oxygen by the electrode reaction

$$2H_2O \longrightarrow O_2(g) + 4H^+ + 4e^- \qquad (7\text{-}56)$$

is called an oxidation of water since each water molecule loses two electrons. Such statements surely would have struck the chemists of a few generations ago as odd points of view. This is a not infrequent consequence of generalizations: they make the concepts that were the starting point seem odd and out of place. Indeed this is one reason why it takes exceptional and unconventional minds to make such generalizations, minds capable of transcending the tenets of their day.

Comparison of (7-12) with (7-51) and of (7-14) with (7-55) suggests some useful analogies between acid-base and redox reactions. An acid furnishes protons, while a reducing agent furnishes electrons; conversely a base accepts protons, while an oxidizing agent accepts electrons. A redox couple is analogous to a conjugate acid-base pair, although redox couples frequently involve the exchange of more than one electron whereas practically all acid-base couples can be formulated as involving conjugate pairs, the members of which exchange a *single* proton.

Equivalence in oxidation and reduction.   The equivalence concept can be extended to redox reactions by defining one equivalent of a reducing agent as the amount that releases one mole of electrons (one faraday) in a particular reaction. One equivalent of an oxidizing agent is defined, for a given reaction, as the amount of the oxidizing agent that accepts one mole of electrons. It is important to recognize that the equivalent weight for a given oxidizing or reducing agent may vary from one reaction to another, just as the equivalent weights of some acids and bases may vary in different reactions. For example, permanganate ion, MnO_4^- (FW = 119), is a good oxidizing agent. It may take up one electron, being then reduced to MnO_4^{2-}, or three electrons, thereby forming MnO_2, or five electrons, in which case it is reduced to Mn^{2+}. The equivalent weight of MnO_4^- differs in the three processes cited, being, respectively, 119, $\frac{119}{3}$, and $\frac{119}{5}$.

Normality is defined for solutions of oxidants and reductants just as it is for acids and bases: the normality of a solution is the number of equivalents of solute per liter of solution.

Oxidation states.   The oxidation equation

$$Co^{2+} \longrightarrow Co^{3+} + e^- \qquad (7\text{-}57)$$

for the reaction in aqueous solution is shorthand for

$$Co(H_2O)_6^{2+} \longrightarrow Co(H_2O)_6^{3+} + e^- \qquad (7\text{-}58)$$

Similar reactions relate the ammonia complexes $Co(NH_3)_6^{2+}$ and $Co(NH_3)_6^{3+}$:

$$Co(NH_3)_6^{2+} \longrightarrow Co(NH_3)_6^{3+} + e^- \qquad (7\text{-}59)$$

and the cyanide complexes $Co(CN)_6^{4-}$ and $Co(CN)_6^{3-}$:

$$Co(CN)_6^{4-} \longrightarrow Co(CN)_6^{3-} + e^- \qquad (7\text{-}60)$$

While in all three cases the oxidation concerns the entire complex ion, it is convenient to consider that it is the Co atom that is oxidized, because all three reactants are related to Co^{2+} and the products to Co^{3+}. Thus, the ion $Co(CN)_6^{4-}$ is formed by the reaction of $Co(H_2O)_6^{2+}$ with $6CN^-$:

$$Co(H_2O)_6^{2+} + 6CN^- \longrightarrow Co(CN)_6^{4-} + 6H_2O \qquad (7\text{-}61)$$

and $Co(CN)_6^{3-}$ is related to $Co(H_2O)_6^{3+}$ in a similar way. We say that in all three reactions cobalt is oxidized from the $+2$ to the $+3$ state. It is customary to use roman numerals to indicate oxidation states of metal atoms, so that we may write

$$Co(II) \longrightarrow Co(III) + e^- \qquad (7\text{-}62)$$

Similarly, in reactions (7-43) to (7-45) we say that in each reaction lithium is oxidized from oxidation state zero to the $+I$ state, as expressed in (7-46) or by

$$Li(0) \longrightarrow Li(I) + e^- \qquad (7\text{-}63)$$

In the same fashion, oxygen in reactions (7-37) to (7-39) is said to be reduced from the zero state to the $-II$ state. By convention, oxygen is assigned oxidation state $-II$ in all its compounds except peroxides and others containing O-O bonds, and certain compounds with fluorine. This leads to the oxidation state $+I$ for hydrogen in water and in OH^-, and by convention this oxidation state is assigned to hydrogen in all compounds except metallic hydrides such as NaH.

It is possible to assign oxidation states to the different atoms of a molecule or ion by adoption of a set of self-consistent rules, one of which is that the sum of the oxidation states of all atoms of a molecule or ion must equal the charge on the molecule or ion. Other rules are that, with few exceptions (Appendix C), fluorine, oxygen, and hydrogen are assigned the oxidation states $-I$, $-II$, and $+I$, respectively. This procedure has already been used in earlier discussions, e.g., that following (7-23) and (7-25) relating to $Be(OH)_4^{2-}$ and $Al(OH)_4^-$, that concerning $HgCl_2$ and the related complexes in (7-28) and (7-29), and that concerning the complex ions of Table 7-3. It also leads to the assignment of oxidation states $+VI$ for sulfur in SO_4^{2-}, $+VII$ for manganese in MnO_4^-, $-III$ for nitrogen in NH_3, $+IV$ for silicon in SiO_2, and so on. Oxidation states assigned according to a consistent set of rules make it possible to define the oxidation or reduction of atoms within molecules or ions, so that *oxidation is an increase in oxidation state and reduction is a decrease.* However, a certain arbitrariness adheres to any such rules and they sometimes lose their usefulness. No classification of chemical reactions exists that works in every case and never leads to contradictions. Oxidation states are used in the precise specification of many compounds by name (see Appendix B). A generalization, oxidation numbers, is useful in balancing equations for oxidation-reduction reactions, as described in Appendix C.

Problems and Questions

7-1 **pH values and concentrations.** (*a*) Convert the following pH values into pOH and H^+ concentrations: 4.7, 8.4, 11.22, 0.52, -0.30. (*b*) Convert the following molar H^+ concentrations into pH values: 6.0, 1.5×10^{-6}, 7×10^{-10}, 3×10^{-2}, 5×10^{-14}.
   Ans.   (*a*)   pOH: 9.3, 5.6, 2.78, 13.48, 14.30; $[H^+]$: 2×10^{-5}, 4×10^{-9}, 6.0×10^{-12}, 0.30, 2.0;
(*b*) -0.78, 5.82, 9.15, 1.5, 13.3

7-2 **Solutions of electrolytes.** (*a*) Distinguish clearly between the terms *strong* (as applied to electrolytes) and *concentrated* and between the terms *weak* and *dilute*. (*b*) Which of the following substances are strong electrolytes when dissolved in water, which are weak, and which are nonelectrolytes: KOH, sugar, NH_3, $Ca(NO_3)_2$, HCl, methyl alcohol, LiBr, Na_2SO_4, HOAc (acetic acid), NaOAc (sodium acetate), NH_4Cl?

7-3 **Freezing point of solutions.** Arrange the following solutions in order of decreasing freezing point, placing that with the highest freezing point first: (*a*)   0.20 *F* KOH;   (*b*)   0.10 *F* sugar;   (*c*)   0.25 *F* NH_3;   (*d*)   0.04 *F* $BaCl_2$;   (*e*)   0.04 *F* HNO_3.   *Ans.   e, b, d, c, a*

7-4 **Composition of solutions.** (*a*) How many moles of magnesium ions and how many moles of nitrate ions are present in 150 ml of 2.0 *F* magnesium nitrate? Assume that $Mg(NO_3)_2$ is a strong electrolyte, but indicate how your answers would differ (if at all) if it were not. (*b*) How many grams of HNO_3 are needed to prepare 2.5 liters of 0.120 *F* nitric acid?

7-5 **Brønsted acids and bases.** Define the terms *acid* and *base* in the Brønsted sense. Then identify the acid and the base on each side of the following equations:

(*a*) $HOAc + CN^- \longrightarrow HCN + OAc^-$
(*b*) $HSO_4^- + HPO_4^{2-} \longrightarrow SO_4^{2-} + H_2PO_4^-$
(*c*) $NH_3 + O^{2-} \longrightarrow NH_2^- + OH^-$

Ans.   acids: HOAc, HCN; HSO_4^-, $H_2PO_4^-$; NH_3, OH^-;
bases: CN^-, OAc^-; HPO_4^{2-}, SO_4^{2-}; O^{2-}, NH_2^-

7-6 **Conjugate acid-base pairs.** (*a*) Define the term *conjugate acid-base pair,* and illustrate it with two examples not included in Table 7-1. (*b*) Like water, ammonia can act as both a base and an acid, although in aqueous solutions it normally behaves only as a base. Indicate the appropriate conjugate form when it acts as an acid and when it acts as a base. (*c*) Give the conjugate base of each of the following acids: H_3O^+, HS^-, HSO_4^-, $CH_3NH_3^+$, HF, H_3AsO_3. (*d*) Give the conjugate acid of each of the following substances acting as a base: OH^-, HSO_4^-, NO_3^-, HOAc, CO_3^{2-}.

7-7 **Equivalent weight.** Citric acid, MW = 192, can be symbolized as H_3X, that is, it has three potentially acidic hydrogen atoms, one or more of which can react, depending on the reaction conditions. What are possible values of the equivalent weight of citric acid in acid-base reactions?
Ans. 192; 96; 64

7-8 **Equivalent weight.** Hydrazine, N_2H_4, is a base that can react with either one or two protons, depending on the reaction conditions. What are possible values of the equivalent weight of hydrazine in acid-base reactions?

7-9   Equivalent weight. (*a*) A 1.07-g sample of a certain acid was dissolved in 100 ml water and the resulting solution was titrated with 0.500 *N* Ca(OH)$_2$. Exactly 30.0 ml of the base was needed to neutralize the acid. What was the equivalent weight of the acid? (*b*) The acid was H$_3$AsO$_4$. How many protons in each molecule of acid were neutralized in the reaction described in (*a*)?   *Ans.* (*a*) 71;   (*b*) two

7-10   Equivalent weight and molecular weight. Suppose that 30.0 ml of 0.224 *N* HCl is needed to neutralize 0.538 g of an unknown base. What is the equivalent weight of the base? Give two possible values of its formula weight and explain their relation to the equivalent weight.

7-11   Analysis of a sodium carbonate sample. Sodium carbonate exists in various crystalline forms with different amounts of water of crystallization, including Na$_2$CO$_3$, Na$_2$CO$_3 \cdot$ 10H$_2$O, and others. The water of crystallization can be driven off by heating; the amount of water removed depends on the temperature and duration of heating.

A sample of Na$_2$CO$_3 \cdot$ 10H$_2$O had been heated inadvertently and it was not known how much water had been removed. A 0.200-g sample of the solid that remained after the heating was dissolved in water, 30.0 ml of 0.100 *N* HCl was added, and the CO$_2$ formed was removed. The solution was acidic; 6.4 ml of 0.200 *N* NaOH was needed to neutralize the excess acid. What fraction of the water had been driven from the Na$_2$CO$_3 \cdot$ 10H$_2$O?   *Ans.*   30 percent

7-12   Equivalent weight and molecular weight. When 2.00 g of an unknown organic acid A is dissolved in 20.0 ml of 0.400 *N* NaOH, the resulting solution is basic, but the solution can be neutralized by the addition of 10.0 ml of 0.200 *N* HCl. (*a*) What is the equivalent weight of the acid A under these circumstances? (*b*) Give three possible values of the molecular weight of A. (*c*) The acid A is known to contain 5.9 percent cobalt (Co). How, if at all, does this alter your answer to (*b*)?

7-13   Solution of sodium carbonate. An aqueous solution of sodium carbonate, Na$_2$CO$_3$, is titrated with strong acid to an equivalence point corresponding to 2 eq mol^{-1} for carbonate ion. (*a*) If 20.0 ml of the carbonate solution reacts with just 40.0 ml of 0.50 *N* acid, what is the formality of the carbonate solution? (*b*) If the solution contains 5.0 wt percent sodium carbonate, what is the density of the solution? (*c*) Suppose that you wanted to prepare a liter of an identical solution by starting with crystalline sodium carbonate decahydrate, Na$_2$CO$_3 \cdot$ 10H$_2$O, rather than with solid Na$_2$CO$_3$ itself. How much of this material would you need? (*d*) What is the equivalent weight of carbonate ion in the reaction that occurs here?
Ans.   (*a*) 0.50 *F*;   (*b*) 1.06 g ml^{-1};   (*c*) 143 g;   (*d*) 30

7-14   Solution of hydrogen iodide. (*a*) What volume of gaseous HI at 25°C and 750 torr would be needed to form 250 ml of 6.0 *N* hydriodic acid? (*b*) What weight of magnesium carbonate would this solution react with if all the carbonate were converted to CO$_2$? What if all carbonate were converted to bicarbonate?

7-15   Analysis of a vinegar sample. Suppose you are an analytical chemist for a large food concern and have obtained the following data pertaining to the analysis of vinegar for acetic acid content (acetic acid is CH$_3$COOH). When 17.0 ml of 1.00 *N* sodium hydroxide was added to a 15.0-g sample of vinegar, the solution became basic and 24.0 ml of 0.208 *N* sulfuric acid was needed to neutralize it. What is the percentage of acetic acid in the vinegar? Assume that vinegar contains no other acidic compounds.   *Ans.*   4.8 percent

7-16 Titration of diphenylamine in acetic acid. Weak organic bases are sometimes dissolved in pure acetic acid so that they may be more readily titrated. What volume of 0.0100 *N* solution of perchloric acid ($HClO_4$) in acetic acid would be needed to neutralize 10.0 ml of a 1.00 wt percent solution of diphenylamine, $C_{12}H_{11}N$, in acetic acid? Diphenylamine has 1 eq FW^{-1} in this reaction. Use the following densities, measured at 25°C, if needed:

	Density/(g ml^{-1})
Pure diphenylamine (solid)	1.159
Pure acetic acid (liquid)	1.049
0.0100 *N* $HClO_4$ in acetic acid	1.051
1.00 percent diphenylamine in acetic acid	1.050

7-17 Properties of an unknown acid. When 1.50 g of an unknown acid A is dissolved in 25.0 g naphthalene ($C_{10}H_8$), the melting point of the naphthalene is depressed 2.55°C below its usual value. The molal freezing-point-depression constant for naphthalene is 6.8 K kg mol^{-1}. (*a*) What is the approximate molecular weight of A? (*b*) A 0.500-g sample of A is just neutralized by 25.0 ml of 0.500 *N* NaOH. How many protons reacted for each molecule of A in this reaction?

Ans. (*a*) 160; (*b*) four

7-18 Analysis of a gas mixture. A 3.00-liter flask contains a mixture of helium and gaseous HCl at a total pressure of 1.00 atm and 27°C. When the entire contents of the flask are passed into 500 ml of 0.200 *F* NaOH, the solution remains basic. However, addition of 100 ml of 0.100 *F* aqueous HCl to this resulting solution neutralizes it. (*a*) Calculate the total number of moles of gas in the flask. (*b*) Calculate the number of moles of HCl in the flask initially. (*c*) Calculate the partial pressures of HCl and of He in the flask initially.

7-19 Mixing solutions of electrolytes. Calculate the concentrations of the principal ions present in a solution prepared by mixing 150 ml of 0.200 *F* $AgNO_3$ with 350 ml of 0.100 *F* $BaCl_2$. Assume that all substances present are strong electrolytes, that AgCl is insoluble, and that volumes are additive.　　　*Ans.*　Ba^{2+}, 0.070 *M*; Cl^-, 0.080 *M*; NO_3^-, 0.060 *M*; Ag^+, 0.000 *M*

7-20 Mixing solutions of electrolytes. Suppose that 150 ml of 0.200 *F* K_2CO_3 and 100 ml of 0.400 *F* $Ca(NO_3)_2$ are mixed together. Assume that the volumes are additive, that $CaCO_3$ is completely insoluble, and that all other substances that might be formed are soluble. Calculate the weight of $CaCO_3$ precipitated, and calculate the concentrations in the final solution of the four ions that were present initially.

7-21 Concentrations of ions in a mixture after reaction. Suppose that 15.0 liters of gaseous SO_3 at 0.40 atm and 20°C is passed into 500 ml of an aqueous solution that is 0.48 *F* in NH_3 and 0.40 *F* in $Ba(OH)_2$. Assume that the volume of the solution does not change and that barium sulfate is insoluble. Calculate the final concentrations of the principal ionic and molecular species present (other than water) and the weight of barium sulfate precipitated (if any).　　　*Ans.*　NH_3, 0.28 *M*; NH_4^+, 0.20 *M*; SO_4^{2-}, 0.10 *M*; 47 g $BaSO_4$

7-22 Percentage of ammonia in a mixture. A 5.0-liter flask contains a mixture of ammonia and nitrogen at 27°C and a total pressure of 3.00 atm. The sample of gas is allowed to flow from the flask until the pressure in the flask has fallen to 1.00 atm. The gas that escapes is passed through 1.50 liters of 0.200 *F* acetic acid. All the ammonia in the gas that escapes is absorbed by the solution and turns out to be just sufficient to neutralize the acetic acid present. The volume of the solution does not change significantly. (*a*) Will the electrical conductivity of the aqueous solution change significantly as the gas is absorbed? Give equations for any reactions, and calculate the final concentrations of the principal ions present (if any) at the end. (*b*) Calculate the percentage by weight of ammonia in the flask initially.

7-23   **Analysis of chloride samples.**   Suppose that you are determining chloride in many solid samples by titration of a weighed amount of sample with 0.2000 *F* silver nitrate. To simplify calculations you decide to weigh out an amount of sample each time such that the volume of silver nitrate used, in milliliters, will just equal the percentage of chloride in the sample. What weight of sample should you take?   *Ans.*   0.709 g

7-24   **Analysis of a mixture of KCl and HBr.**   It was desired to neutralize a certain solution X that had been prepared by mixing solutions of potassium chloride and hydrobromic acid. Titration of 10.0 ml X with 0.100 *N* silver nitrate required 50.0 ml of the latter. The resulting precipitate, containing a mixture of AgCl and AgBr, was dried and found to weigh 0.762 g. How much 0.100 *N* sodium hydroxide should be used to neutralize 10.0 ml solution X?

"There have probably been few events in the history of science that have had such extraordinary consequences within the brief span of one generation as Planck's discovery of the elementary quantum of action. This discovery not only has formed, to an ever increasing degree, the basis for bringing order into our knowledge of atomic events . . . , but at the same time it has caused a complete change in our fundamental ways of describing natural phenomena. . . . This new knowledge has shattered the conceptual foundations not only of classical physics but also of our ordinary modes of thinking. It is to just the liberation gained in this way that we owe the marvelous progress that has been made in our insight into natural phenomena during the past generation."

Niels Bohr, 1929*

8 Particles, Waves, and Quantization

The foundations for an understanding of atomic structure, the periodic classification of the elements, and the binding of atoms together to form molecules are provided by the concepts and methods of quantum theory, which have revolutionized physics and chemistry during this century. This chapter introduces some of the fundamental ideas needed to understand the applications of quantum theory to problems of atomic and molecular structure, which are discussed in the following chapters.

8-1 Properties of Particles and Waves

During the first quarter of this century it was discovered that electromagnetic radiation such as light behaves in many circumstances like a stream of particles and that particles such as electrons and protons behave in many circumstances like waves.

Naturwissenschaften, vol. 17, pp. 483 and 486, 1929.

Both concepts—that of particles and that of waves—have their roots in classical physics, a description of the familiar macroscopic world. Their application to phenomena in the microscopic realm, phenomena involving individual atoms and molecules and their interaction with radiation, is a considerable extrapolation. The inadequacy of this extrapolation is presumably the reason that neither the wave nor the particle concept alone is sufficient to explain events on an atomic scale, where the laws of classical physics must be replaced by those of quantum mechanics. Before explaining what is implied by these statements, more needs to be said about particles and waves.

Particles. Particles can be *counted* and follow well-defined paths (*trajectories*) when in motion. They have attributes such as *mass* and (often) *charge* and, if moving, *kinetic energy* and *momentum*. Sometimes the observation of these properties is difficult and may require ingenious experimentation. For example, elaborate pieces of equipment such as a Wilson cloud chamber or a bubble chamber (Fig. 8-1) may be needed to observe the trajectories of subatomic particles.

The tracks of some kinds of particles have never been observed, yet these particles are still believed to exist. Sometimes detailed analyses of events that involve particles with visible tracks show discrepancies in the expected balances of kinetic energy, momentum, and angular momentum (Study Guide), all of which can be explained by assuming the existence of a particle with an invisible track. To postulate the existence of a particle on such evidence is not as outlandish as it seems. Remember that it is in fact impossible to *prove* the existence of *any* particle (or of anything else). The postulation of the existence of *anything* is always done in order to explain a set of observed events in a natural and unlabored way. No way has been found to make visible the trajectories of neutral particles such as neutrons and photons. However, the initial and final points of such trajectories can often be located because the particles that create a neutral particle upon collision or by radioactive decay may leave visible tracks. Moreover, when new particles are created from a neutral particle they also may have detectable tracks (Fig. 8-1*b*). In summary, then, we speak of particles when we observe events that can be described in terms of at least some typical attributes of particles.

Waves. Some of the measurable attributes of waves are *frequency* ν ("nu"), the number of vibrations per second; *amplitude* A_0; and *wavelength* λ ("lambda") (Fig. 8-2). The *velocity* or *speed of propagation* of the wave, u, is equal to the distance traveled by the wave in one second. This distance must contain λ exactly ν times; hence,

$$u = \lambda\nu \tag{8-1}$$

Another quantity that can be used to characterize a wave is the wave number $\tilde{\nu}$, the reciprocal of the wavelength,[1] $\tilde{\nu} = \lambda^{-1}$. The *intensity* (energy per second) associated with a wave is proportional to the square of its amplitude. The *phase* γ of the wave specifies the position of the crest with respect to a fixed point. As

[1]Note that here the tilde ($\sim$) does not indicate a molar quantity as it usually does in this text.

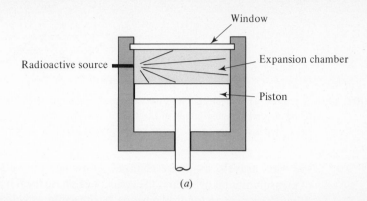

(a)

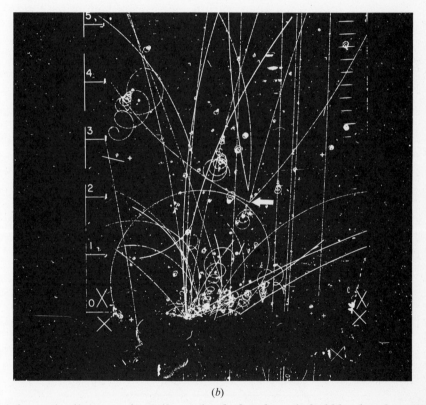

(b)

FIGURE 8-1 **(a) Schematic diagram of a Wilson cloud chamber; (b) bubble chamber tracks.** In (a) the expansion chamber contains air saturated with the vapor of water or ethanol. When the piston is moved downward quickly, the air expands and its temperature falls. The vapor becomes supersaturated and tends to condense in tiny droplets on the ions created when charged particles of high energy traverse the cloud chamber. The droplets make the particle trajectories visible.

In (b) in a bubble chamber tiny vapor bubbles form along the trajectories of charged particles passing through a liquid, such as liquid H_2, the pressure of which has just been reduced to slightly below the boiling pressure. The arrow indicates the appearance of tracks produced by charged particles that originate from the decay of a neutral species.

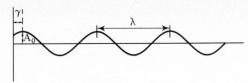

FIGURE 8-2   **A wave.** The wavelength λ of the wave shown is the distance between neighboring crests or troughs or other corresponding points of the wave train. The phase γ is the position of a crest relative to a fixed point such as the origin. It may be specified as a fraction of λ, γ/λ being about 0.12 in this example.

the wave travels, its phase changes, and since the choice of fixed point and time of measurement is arbitrary, the concept of phase itself may seem arbitrary. However, the *difference in phase* of two waves of the same wavelength, *traveling with the same speed,* is *not* arbitrary; it is independent both of the time of measurement and of the choice of reference point for measuring the phase of either wave separately. Phase differences are extremely important in the interaction of waves. Various phase differences between two waves are illustrated in Fig. 8-3.

In certain experiments, waves show the phenomenon of *interference.* This means

FIGURE 8-3   **Interference of two waves.** Three examples are shown of what occurs when two waves of the same wavelength and equal amplitude add. The two separate waves are shown on the left and their sum, or resultant wave, on the right. The three examples are characterized, respectively, by (*a*) a phase difference zero—total reinforcement; (*b*) a phase difference λ/4—partial reinforcement; (*c*) a phase difference λ/2—complete destruction. The resultant wave has the same wavelength λ [except in case (*c*), where the term *wavelength* has no meaning].

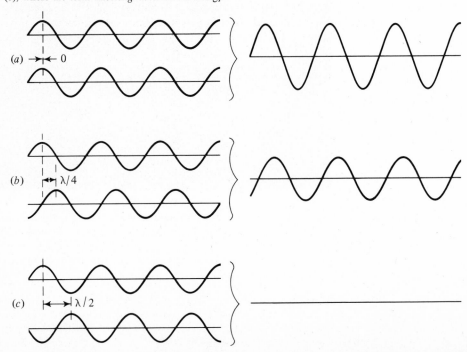

that the amplitudes of waves of the same wavelength, which may come from different directions, may add or subtract when they meet (Fig. 8-3). When the crest of one wave coincides with the crest of the other (which means that troughs also coincide), there is reinforcement or addition of the amplitudes: the two waves are *in phase* and undergo *constructive interference.* When the crest of one wave coincides with the trough of the other, the two waves cancel. They are *out of phase;* there is *destructive interference.*

8-2   Light and Electromagnetic Radiation

The electromagnetic spectrum.   In 1672 Newton published his first scientific paper, an account of experiments in which sunlight was passed through a glass prism and was spread out into a rainbow-colored band (Fig. 8-4*a*). Further experiments convinced him that white light was a mixture of light of all colors of the rainbow, and he suggested as an improvement of a theory of light proposed by Robert Hooke that each color was associated with different-wavelength vibration of an all-pervading, intangible fluid called the "ether".

The wave nature of light was not established until the early years of the nineteenth century when Thomas Young, A. J. Fresnel, and others showed that it offered an explanation for all the properties of light that had been observed, particularly interference phenomena such as those shown in Fig. 8-5. When light of a single color passes through two slits in a card and falls on a screen, a characteristic interference pattern is produced: areas of brightness alternate with areas of darkness. Their explanation was that each slit may be considered as a source of waves. Dark locations on the screen are those where the crest of one wave arrives at the same

FIGURE 8-4   **Schematic spectrograph and a typical atomic emission spectrum.**   (*a*) In the optical range, where glass prisms, lenses, and photographic plates can be used, the spectrum may be examined after the light has passed through a fine slit and been spread out by a prism or grating. (*b*) Atomic spectra are characterized by spectral lines, each of which consists of radiation of a particular wavelength or frequency also called monochromatic radiation. Each kind of atom has its own characteristic sets of spectral lines in absorption and in emission. The spectrum of H is typical. The diagram shows both the wavelengths and the customary symbols for the four most common lines in the spectrum of H (the *Balmer* lines).

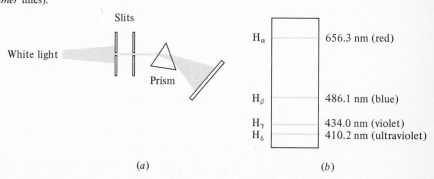

(*a*)   (*b*)

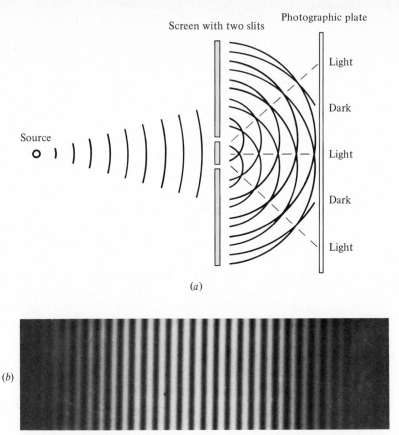

(a)

(b)

FIGURE 8-5 **An interference experiment.** (*a*) Radiation of a particular color (wavelength) is emitted by the point source at left. The crests of the waves at a given instant are represented by the arcs emanating from the point source. Two slits in a screen (middle) allow passage of the radiation, each slit representing the source of a new wave that is in phase with the original wave. The crests of the new waves are represented by concentric circles centered at the slits. The dashed lines represent locations for which the crests of the slit-centered waves coincide and indicate the portions of the screen (at right) that receive maximum illumination.

(*b*) Photograph of an interference pattern. (*From F. A. Jenkins and H. E. White, Fundamentals of Optics, 3rd ed., p. 235, McGraw-Hill, New York, 1957. Copyright © 1957 by McGraw-Hill Book Company, Inc. Used with permission of McGraw-Hill Book Company.*)

time as the trough of another. The brightest locations are those where crests arrive at the same time and therefore reinforce. In between the darkest and brightest regions there is partial cancellation.

In 1864, Maxwell showed on theoretical grounds that the properties of light were those expected of electromagnetic waves. Associated with such waves are a magnetic field and an electric field that oscillate in a plane perpendicular to the direction of travel. All electromagnetic waves have the same speed in empty space, called the speed of light and symbolized by *c*. The whole range of electromagnetic waves, with wavelengths from zero to infinity, constitutes the *electromagnetic spectrum*. Since, as discussed below, visible light is part of the electromagnetic spectrum, the term *light* is often used to designate electromagnetic radiation in general.

Planck's radiation law. Matter may absorb or emit electromagnetic radiation. The intensity of the radiation emitted by an object at each wavelength (its *emission spectrum*) or the amount of radiation absorbed at each wavelength (its *absorption spectrum*) can be measured by instruments related to the simple prism used by Newton (Fig. 8-4).

Absorption and emission spectra were first intensively investigated during the last decades of the nineteenth century. Absorption was recognized as representing a gain in energy, emission a loss. Although the energy received by absorption at certain wavelengths may lead to specific phenomena such as the emission of radiation at a new wavelength, the general result of absorption is a temperature increase of the body absorbing the radiation. Emission of radiation correspondingly causes a temperature decrease. In such a situation there is always the possibility of equilibrium—the radiation intensity may be such that emission and absorption of radiation just balance. It is found that when there is equilibrium between radiation and matter at a given temperature, the relative intensities of light of each wavelength are *independent* of the chemical or physical nature of the body (Fig. 8-6). Radiation that has this particular spectral composition is called *blackbody radiation*.

As shown in Fig. 8-6, when the temperature increases, the maximum in the

FIGURE 8-6 **The spectral distribution of blackbody radiation at three temperatures.** The intensity (energy per second) of blackbody radiation is shown as a function of the wavelength at three different temperatures. For example, the shaded area shows the intensity of the radiation in the wavelength range λ to $\lambda + \delta\lambda$ at 2000 K. Only a small part of the total wavelength range shown constitutes visible light. Note that (1) the total intensity (total area under a curve) increases rapidly as the temperature is increased; (2) the maximum of the intensity curve shifts toward shorter wavelengths as the temperature is increased; (3) only a small fraction of the total intensity falls into the visible range, even at 2000 K. The formula that describes these curves was derived by Planck.

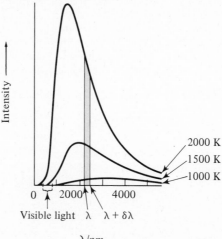

spectral distribution of blackbody radiation shifts to shorter wavelengths and the total intensity increases rapidly. Many unsuccessful efforts were made to derive from theory the experimentally well-established spectral distribution of blackbody radiation. Finally, in 1900, Planck succeeded in finding by trial a mathematical formula to fit the curves, and a few weeks later he derived this formula—now known as Planck's radiation law—by theoretical arguments. Planck showed that one revolutionary assumption was needed: his radiation law could be derived only if it was assumed that absorption and emission of light by matter involve discrete units of energy $h\nu$, where h is called Planck's constant. He called these packages of energy *quanta* (the plural of *quantum*). Their size is proportional to ν, the frequency of the radiation.

Planck was cautious and said only that the absorption and emission of light were quantized but not that the light itself must therefore exist as particles. The quantization of energy that he postulated was a new phenomenon and was accepted only hesitantly by many; even the quantization of matter, in the form of discrete atoms, was not accepted by all the leading physicists and chemists of that day and hence it is not surprising that Planck's quantum hypothesis met with some skepticism. However, the real test of any hypothesis is that it can be used to explain more than one independent set of facts, more than those it was tailored to fit. Within a very few years the quantum hypothesis had passed this test magnificently: Einstein had applied it to radiation and Bohr to atomic structure, and even the skeptics were convinced.

The photoelectric effect.   When light shines on a metal surface, the surface may emit electrons, a phenomenon called the *photoelectric effect* (Fig. 8-7). In 1905 Einstein, by an analysis of this effect, showed unequivocally that under certain conditions electromagnetic radiation exhibits the characteristics of particles. It had been determined that the maximum kinetic energy of the electrons emitted by the metal depends not upon the *intensity* of the light—that is, not upon the *total energy* impinging on the surface per second—but only upon the *frequency* of the light (Fig. 8-8). This is in contradiction to the supposed wave nature of light, for if light behaved

FIGURE 8-7   **The photoelectric effect.**   When a metal surface is irradiated by light of sufficiently short wavelength (for some metals ultraviolet radiation may be required), electrons are emitted by the metal, causing a current to flow even when the collecting electrode (at the top) is made slightly negative relative to the metal surface.

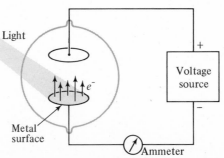

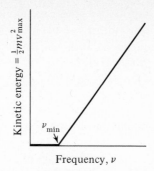

FIGURE 8-8   **Maximum kinetic energy of photoelectrons.** The maximum kinetic energy of photoelectrons emitted by a metal irradiated by light of frequency ν is shown. The inclined portion of the curve can be represented by

$$\tfrac{1}{2}mv_{max}^2 = h(\nu - \nu_{min})$$

The slope of this straight line is Planck's constant h. Electrons are not emitted unless $\nu \geq \nu_{min}$; hence the minimum energy to eject an electron is $h\nu_{min}$.

like ordinary waves, the electrons emitted would be expected to have less and less kinetic energy as the intensity of the light is decreased. What happens instead is that if the intensity of the light is cut in half, only half as many electrons are emitted but the kinetic energy of each electron remains at the original value provided that the color of the light—that is, its frequency—is not changed. As the intensity is decreased further, fewer and fewer electrons emerge, but at *any* level of intensity some electrons are emitted, and their maximum kinetic energy does not depend on the level of illumination.

   Einstein saw that the *one* assumption that explained this effect was that in some way light had particle nature and that the energy of these light quanta or *photons* was constant for electromagnetic radiation of a given frequency. A decrease in light intensity would thus result in a decrease in the number of photons involved, but not in a change of their energy. At low intensity levels the number of photons would be so small that just *one* photon would strike the metal surface every few seconds or hours, but these photons would have the same energy and each would cause the emission of one electron. As long as there was *any* light, it would exist as photons and cause electron emission. Furthermore, the energy of the electrons emitted increases as the frequency of the light used for irradiation increases, and the energy of each photon is found to be just that demanded by Planck's earlier considerations.

Physical characteristics of photons.   The *energy* of a photon is found to be

$$\epsilon = h\nu = \frac{hc}{\lambda} \tag{8-2}$$

where ν and λ are the frequency and wavelength of the radiation and c is the velocity of light, with both λ and c pertaining to radiation in a vacuum. The quantity h is Planck's constant, $h = 6.6262 \times 10^{-34}$ J s in SI units. The mass of a photon of a given frequency is obtained by dividing (8-2) by c^2, using Einstein's relationship[1] $\epsilon = mc^2$ (Study Guide):

$$m = \frac{\epsilon}{c^2} = \frac{h\nu}{c^2} = \frac{h}{c\lambda} \tag{8-3}$$

The momentum of a photon is obtained by multiplying its mass by its velocity:

$$\text{Momentum of photon} = \frac{h\nu}{c^2} c = \frac{h\nu}{c} = \frac{h}{\lambda} \tag{8-4}$$

These expressions for photon mass and momentum have also been established experimentally. In an important set of experiments, A. H. Compton showed that when short-wavelength radiation (X rays and γ rays) is scattered by atoms, some of the scattered radiation has an increased wavelength (the *Compton effect*). The shift in wavelength can be explained quantitatively if the radiation is assumed to consist of photons and the collisions between the photons and the components of atoms are assumed to occur with conservation of energy and momentum, just as for collisions between entities more commonly regarded as "particles," such as protons or baseballs.

The statement that light may behave like particles as well as waves means that light may exhibit some of the attributes of particles as well as those of waves. Which attributes are exhibited depends on the experimental situation. When light passes through slits, interference phenomena occur that are characteristic of waves; when light interacts with matter, phenomena such as energy and momentum transfer bring out its particle nature. In interactions with matter, the particle aspects of electromagnetic radiation become more important as the frequency becomes larger, because the mass of the photon increases with frequency. Thus, particle aspects of radio waves are almost never observed, and the Compton effect was established only by using short-wavelength X rays and γ rays.

Energy relationships. Although the energy of electromagnetic radiation is quantized, the individual energy packages vary in direct proportion to the frequency (and thus inversely with the wavelength) by (8-2). It is very helpful to have a feeling

---

[1]It should be noted that a photon has mass only by virtue of the fact that its velocity is the velocity of light, c. By the special relativity theory, if a particle has the mass m_0 when at rest, it has the mass

$$m = \frac{m_0}{\sqrt{1 - v^2/c^2}} \tag{1}$$

when it is moving with velocity v. At the limit where v approaches the velocity of light c, the denominator approaches zero and hence the mass m approaches infinity unless the rest mass m_0 is also zero. Two conclusions follow: (1) only particles of rest mass zero can move at the speed of light; (2) when such particles do move at the speed of light, Equation (1) yields the indeterminate expression $\frac{0}{0}$ for their mass. The mass can be found if the energy is known, as shown for photons with Equation (8-3).

for the energy of the individual photons associated with radiation of a given frequency or wavelength. Consider, for example, blue-green light of wavelength 5.0×10^2 nm (or 5.0×10^{-7} m), which corresponds to the wave number[1] $\tilde{\nu} = 1/\lambda = 2.0 \times 10^4$ cm^{-1} and the frequency $\nu = c/\lambda = c\tilde{\nu} = (2.0 \times 10^4$ cm$^{-1})$ $(3.0 \times 10^8$ m s$^{-1})(100$ cm/m$) = 6.0 \times 10^{14}$ s^{-1}. A photon of this frequency has the energy $h\nu = (6.63 \times 10^{-34}$ J s$)(6.0 \times 10^{14}$ s$^{-1}) \approx 4.0 \times 10^{-19}$ J. It is often advantageous to be able to express energies in different units so that comparisons of different kinds of energy quantities can be made; the conversion factors in Tables A-4 and A-5 may be used. Two of the most common energy units in chemistry and chemical physics are (1) the *electronvolt,* eV, which is the kinetic energy of an electron that has been accelerated, for example, by being placed between two parallel metal plates, with an electrical potential difference of exactly 1 V; and (2) the *kilojoule per mole,* kJ mol^{-1}, the energy in kilojoules of 1 mole of whatever is being considered.

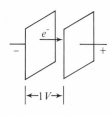

The conversion from joules to electronvolts is made by using the relationship 1 eV $= 1.6022 \times 10^{-19}$ J (Table A-5), which gives the photon energy considered in the previous paragraph, 4.0×10^{-19} J, a value in electronvolts of about $\frac{4.0}{1.6} = 2.5$ eV. The energy units electronvolts permit easy comparison with the energy stored in familiar chemical sources of electrical energy, such as storage batteries and flashlight batteries. A 2.5-V battery can provide electrons with a kinetic energy equal to the energy of a 500-nm photon. In this sense such a battery and blue-green light are energetically equivalent. It is easy to remember that the energies of ordinary chemical reactions correspond to a few electronvolts because common batteries, which operate by means of chemical reactions, furnish potentials of a few volts per cell and can thus supply electrons with energies of a few electronvolts.

To find the energy in kilojoules of a *mole* of photons of wavelength 500 nm, we multiply by N_A. The energy of N_A photons of this blue-green light is thus

$$(6.0 \times 10^{23} \text{ photons/mol})(4.0 \times 10^{-19} \text{ J/photon}) = 24 \times 10^4 \text{ J mol}^{-1}$$
$$= 2.4 \times 10^2 \text{ kJ mol}^{-1}$$

This energy (or, more precisely, 239 kJ mol^{-1}) is just below the bond energy of Cl$_2$, the energy required to convert 1 mol Cl$_2$ molecules into atoms, which is 242 kJ mol^{-1}. Thus, a photon of slightly shorter wavelength than 500 nm (for example, 490 nm) has sufficient energy to split one Cl-Cl bond.

It is important to recognize the special nature of the unit kilojoules per mole, and in particular the difference between it and the kilojoule. While the kilojoule is a plain energy unit, the unit kilojoules per mole implies that the energy quantity considered pertains to exactly 1 mol of whatever is being discussed, each of the N_A individuals in the mole having on the average $1/N_A$ of the total energy. For example, the statement that the dissociation energy of the Cl-Cl bond is 242 kJ mol^{-1} means that to break the bonds of N_A *molecules* of Cl$_2$ requires 242 kJ. The statement that the dissociation energy is 2.5 eV means that it takes 2.5 eV to break *one* Cl-Cl bond. It also takes $242/(6.02 \times 10^{23}) = 4.02 \times 10^{-22}$ kJ to break this one bond.

---

[1]The customary units for $\tilde{\nu}$ are cm^{-1}.

TABLE 8-1
The electromagnetic spectrum

Range			Energy/eV	Designation	Mode of generation
$\tilde{\nu}/cm^{-1}$	ν/s^{-1}	λ/m			
$(0)-10^{-1}$	$(0)-3 \times 10^9$	$(\infty)-10^{-1}$	$0-10^{-5}$	Radio waves	Electrical oscillations
$10^{-1}-10^6$	$3 \times 10^9-3 \times 10^{16}$	$10^{-1}-10^{-8}$	$10^{-5}-10^2$	Optical radiation	Atomic, molecular, and ionic events
10^6-10^9	$3 \times 10^{16}-3 \times 10^{19}$	$10^{-8}-10^{-11}$	10^2-10^5	X rays	Transitions of inner atomic electrons
$10^9-(\infty)$	$3 \times 10^{19}-(\infty)$	$10^{-11}-(0)$	$10^5-(\infty)$	γ rays	Nuclear phenomena

EXAMPLE 8-1   **Thermal energy at room temperature.**[1] What is the average thermal energy for a gas at 25°C in joules per mole and electronvolts per molecule?

Solution. As shown in Sec. 5-4, this energy is $\frac{3}{2} RT$ for 1 mol gas, or

$$1.5 \times 8.314 \text{ J K}^{-1} \text{ mol}^{-1} \times 298 \text{ K} = 3716 \text{ J mol}^{-1}$$
$$= \underline{3.72 \text{ kJ mol}^{-1}}$$

This is equivalent (see Table A-5) to

$$3.72 \text{ kJ mol}^{-1} \times 0.01036 \frac{\text{eV}}{\text{kJ mol}^{-1}} = \underline{0.0385 \text{ eV per molecule}}$$

Regions of the electromagnetic spectrum. The electromagnetic spectrum may be divided into different regions, depending on the modes of production and detection of the radiation, as well as on other criteria. Table 8-1 illustrates one such scheme; the optical and "visible" regions are shown in Fig. 8-9. There are, of course, no sharp boundaries; adjacent regions blend into one another smoothly.

Radio waves, the region of the spectrum with wavelengths above 0.1 m, had few chemical applications until recently because the corresponding energies are very low. Radio waves have become important chemically during the last two decades because of the development of nuclear magnetic resonance (nmr), a powerful tool that involves radio-frequency spectroscopy and is used for studying problems of molecular structure (Chap. 21).

The portion with wavelengths from about 10^{-8} to 10^{-1} m is called *optical radiation.* Visible light (400–750 nm) represents only a minor part of the region, but the term *optical* is applied to this entire region of the spectrum because the radiation can be manipulated by the same types of instruments that are used for light, such as diffraction gratings, interferometers, lenses, and prisms. However, at the two extremes of the region (Fig. 8-9)—the shortest ultraviolet and the longest infra-red wavelengths (particularly in the microwave region)—special techniques are often far more convenient and indeed necessary. Even between these extreme limits,

[1] It is sometimes customary to make "per molecule" part of the units of a quantity (e.g., as in 2.5 eV per molecule). This practice is followed in Example 8-1, but we shall not do it generally.

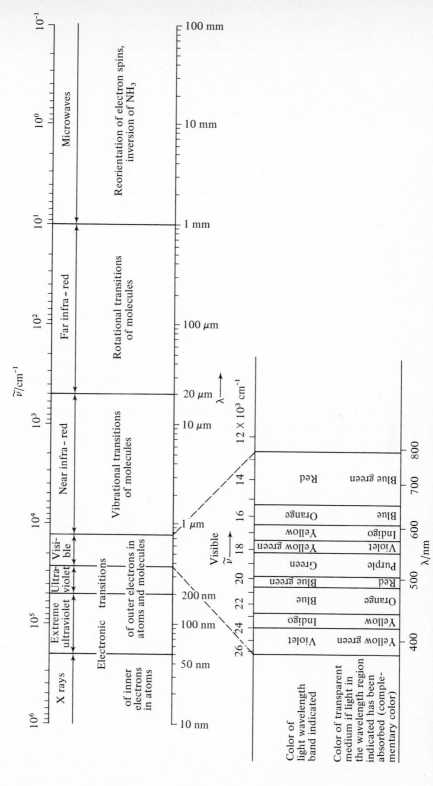

FIGURE 8-9 The optical region of the electromagnetic spectrum. Note the complementary colors indicated at the bottom of the enlarged visible portion of the spectrum.

different materials have to be used to make lenses and prisms suited to different wavelengths because the transparency of a substance to the radiation varies from wavelength to wavelength. Glass transmits light in the visible region, but quartz (SiO_2) must be substituted on the ultraviolet (short-wavelength) side of the visible spectrum. Quartz is opaque to wavelengths shorter than 200 nm and must be replaced by fluorite (CaF_2). On the long-wavelength or infra-red side of visible light, glass is satisfactory to about 4000 nm; prisms of sodium chloride, potassium chloride, or other materials are needed to work with still longer wavelengths.

Radiation in the visible and nearby ultraviolet and infra-red regions of the spectrum is emitted by atoms, molecules, and ions that have been excited thermally or in some other way, as by an electric discharge. Such radiation can also be emitted by species formed in a high-energy state by a chemical reaction. Conversely, many important reactions, known generally as photochemical reactions, are initiated by the absorption of photons in these regions of the spectrum. As indicated in the upper portion of Fig. 8-9, radiation in different parts of the optical region can be used to study changes in different kinds of molecular energy. In a very rough way, one can generalize that radiation in the far infra-red region corresponds to changes in rotational energy of molecules, that in the near infra-red to changes in vibrations of atoms relative to one another, and that in the visible and near ultraviolet to changes in the arrangement of the outer electrons in atoms and molecules.

X rays are generated by energy changes involving the inner electrons of atoms, and thus study of these changes yields information about the inner electronic structure of the atom. X rays are also produced when electrons or other particles are stopped rapidly. There are no practical lenses available for X rays because the wavelengths are of the order of atomic dimensions, so that arrays of atoms act like gratings and diffract the radiation rather than simply altering its speed.[1] This behavior can be turned to advantage, and X-ray diffraction is an important tool in the determination of molecular structure.

Gamma rays are produced by changes (transitions) from high-nuclear-energy states to lower ones, states that involve different "arrangements" of the components of nuclei in the same sense that photons in the visible and ultraviolet regions of the spectrum correspond to transitions between states involving different arrangements of the outer electrons in atoms. The fact that γ rays have energies some 10^6 times those of photons in the visible region (Table 8-1) implies that the energy changes accompanying the rearrangement of nuclear components (nuclear reactions) are of the order of a million times those of ordinary chemical reactions, which involve the rearrangement of the outer electrons of atoms. The enormous potential of nuclear reactions as energy sources—for peaceful uses such as electric power generation and propulsion of ships and space vehicles as well as for destructive purposes—is a direct consequence of this fact.

The absorption and emission spectra of any atomic or molecular species (Fig. 8-4) are very characteristic and are often used for identification and quantitative

---

[1] These diffraction effects are not important for lenses and prisms used in the optical region because the wavelength in that region is large compared to atomic dimensions. The lens materials behave as if they were continuous in nature and not made of discrete atoms.

analysis. These spectra depend greatly on the conditions under which they are obtained, such as state of aggregation, temperature, and solvent present (if any). A spectrum that consists of a continuous range of frequencies (and thus wavelengths) is called a *continuous spectrum* (Fig. 8-10), whereas in a *line spectrum* (Fig. 8-11) only certain very narrow regions of frequencies, called spectral lines, are obtained.

8-3   The Wave Nature of Particles

The de Broglie wavelength.   The wave nature of particles was discovered some two decades later than the particle nature of light. It was in 1923 that Louis de Broglie first published a satisfactory argument (later amplified in his doctoral thesis) that there might be a wave nature to particles just as there is a particle nature to electromagnetic waves. De Broglie showed that if particles had wave properties, the wavelength associated with them should be

$$\lambda = \frac{h}{p} = \frac{h}{mv} \tag{8-5}$$

This is called the *de Broglie wavelength* of a particle of mass m, velocity v, and momentum p. The relationship given in (8-5) is the same as that which applies to photons, Equation (8-4).

The first particles for which a wave nature was demonstrated experimentally were

FIGURE 8-10   **A continuous absorption spectrum.**   This graph represents the absorption spectrum of a solution of $Ti(H_2O)_6^{3+}$ in water. The relative intensity of absorption of radiation is plotted as a function of the frequency. The sample shows maximum absorption in the blue-green region of the spectrum, and consequently the transmitted light is primarily in the violet and orange-red regions. Thus the solution appears purple when viewed by transmitted light.

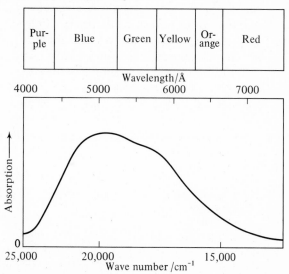

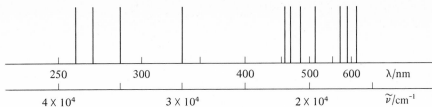

FIGURE 8-11 **Line emission spectrum of atomic sodium.** The lines represent the wavelengths at which atomic sodium emits radiation when in an electric discharge or a flame. The different lines occur with very different intensity, the most prominent feature being the two lines at 589.6 nm and 589.0 nm (not separated in the figure). This radiation is the cause of the intense yellow-orange color of sodium lamps or of flames containing sodium. The intensities are not represented here because of the difficulty of varying the shading in a line drawing. Each line is an image of the slit system of the spectrometer [see Fig. 8-4(a)].

electrons. For nearly three decades after their discovery, electrons had been regarded only as particles. They had been recognized as the carriers of electricity in metals, their mass and charge had been measured, their tracks had been seen in cloud chambers, and they had been identified as the once mysterious "beta rays" from radioactive substances. It did not occur to anyone to perform experiments that might demonstrate that particles had wavelike properties until de Broglie's thesis appeared. However, soon afterward it was shown experimentally by G. P. Thomson that the passage of electrons through thin films of crystalline material was accompanied by diffraction effects similar to those exhibited by X rays. C. Davisson and L. H. Germer showed independently that reflection of electrons by crystalline surfaces also produced diffraction effects. The wavelength found in all cases was the de Broglie wavelength. Since that time the diffraction of electrons by solids as well as by gas molecules has become an important tool for the examination of the atomic structure of matter. (Electrons used for this purpose often have kinetic energies of about 50,000 eV, which corresponds to a wavelength of 0.055 Å.) Other particles have also been shown to have wave character; indeed, the diffraction of neutrons has been developed during the last two decades into a powerful method for investigating the structures of crystals, amorphous solids, and liquids.

Entire atoms and molecules may show wave character as well. For example, Stern and others have observed diffraction when helium atoms or hydrogen molecules are reflected by the surface of a lithium fluoride crystal.

EXAMPLE 8-2 **De Broglie wavelength.** What is the de Broglie wavelength of an electron of speed 100 km s^{-1}?

Solution. The momentum is $p = mv = (9.1 \times 10^{-31}$ kg$)(100 \times 10^3$ m s$^{-1}) = 9.1 \times 10^{-26}$ kg m s^{-1}. Inserting this value into de Broglie's formula for the wavelength, we obtain

$$\lambda = \frac{6.6 \times 10^{-34} \text{ kg m}^2 \text{ s}^{-1}}{9.1 \times 10^{-26} \text{ kg m s}^{-1}}$$

$$= 0.73 \times 10^{-8} \text{ m} = \underline{73 \text{ Å}}$$

The uncertainty principle. An important aspect of the wave-particle duality of matter and radiation is the uncertainty relation, discovered first by W. Heisenberg in 1927 through a careful analysis of the procedures required to determine the location and the momentum of a particle.

Consider, for example, a particle whose momentum is precisely known. Suppose that we wish to determine its position at a given time. To do this we must *observe* the particle, which requires a beam of light. The most precise optical measurements of distance give results in terms of numbers of wavelengths of the light beam used for observation. Thus, as the wavelength of the light used is shortened, the precision of the measurement of the position improves because the comparison units on the measurement scale (wavelengths of light) become smaller. However, light consists of photons, each with the momentum h/λ, and the observation of the particle requires that it collide with at least one photon and deflect this photon toward the observer. This collision necessarily transfers momentum to the particle observed, so that the actual momentum of the particle is different from that known precisely before collision, and the measurement is uncertain by this amount. The amount of momentum transferred to the particle increases as light of shorter and shorter wavelength is used. Hence the more accurately the position of the particle is measured, the less precise is the final knowledge of its momentum. Conversely, if the disturbance of the particle momentum is diminished by choosing radiation of long wavelength, the precision with which the particle location is known decreases. Careful analysis of this situation leads to the *Heisenberg uncertainty relation*

$$\delta x \, \delta p_x \gtrsim h \qquad (8\text{-}6)$$

which relates δx, the uncertainty in position of the particle in a direction, and δp_x, the uncertainty in the component of its momentum in the same direction. The same relationship emerges again and again upon analysis of different methods of measurement. It thus appears to be independent of the particular method employed to measure x and p_x, and Bohr and Heisenberg postulated that the relationship was part of the nature of matter and radiation. Similar relationships are obtained for the y and z directions. Perfect knowledge of the position of a particle thus precludes *any* knowledge of its momentum and thus its velocity. Conversely, perfect knowledge of the momentum and velocity of a particle precludes any knowledge of its position.

By analysis of the measurement of the energy of a system, Heisenberg and Bohr discovered that the uncertainty $\delta\epsilon$ in this measurement depends on the time δt taken for the measurement in such a way that[1]

$$\delta\epsilon \, \delta t \gtrsim h \qquad (8\text{-}7)$$

The variable pairs (x,p_x), (y,p_y), (z,p_z), and (t,ϵ) are called pairs of *conjugate variables,* and the uncertainty principle states that conjugate variables may not be known *simultaneously* with a precision that would violate (8-6) or (8-7). There are other pairs of conjugate variables for which analogous relationships exist.

The uncertainty principle is manifested in many different ways. For example,

---

[1]Note that the dimensions of h are mass $\times$ length^2/time, which are the same as the dimensions of energy $\times$ time.

vibrational motion of atoms in solids persists even at the lowest possible temperatures. If it did not, the momentum of any atom would be zero (except for possible motion of the crystal), the atomic positions could be measured quite precisely by diffraction methods, and the product of the uncertainties in position and momentum would be far less than h. Similarly, the energies of states with very short lifetimes can never be measured precisely. They inevitably turn out to be "fuzzy", that is, repeated measurements give a spread of values of the energy rather than one sharply defined value. When the lifetime is very short, the uncertainty in it is correspondingly low and hence, by (8-7), the uncertainty in the energy must increase. We discuss a number of other manifestations of this most fundamental principle in later chapters; one of the most illuminating is its bearing on the sizes of atoms.

Problems and Questions

8-1   Characteristic properties of waves and particles.   Describe as precisely as possible some of the characteristics of (macroscopic) waves and of (macroscopic) particles.

8-2   Photoelectric effect.   Explain what is meant by the *photoelectric effect*. Why did it play an important role in the development of the particle-wave interpretation of the nature of electromagnetic radiation?

8-3   Frequency and wavelength.   (*a*) Calculate the frequency in s^{-1} of electromagnetic radiation of wavelength 300 m, 1 cm, 500 nm, 0.1 nm. Indicate in what region of the spectrum each of these wavelengths would be found. (*b*) Calculate the wavelength of a photon of frequency $10^{12}\,s^{-1}$, $10^{15}\,s^{-1}$, $10^{19}\,s^{-1}$. Indicate in what region of the spectrum each of these wavelengths would be found.
   Ans.   (*a*) $10^6\,s^{-1}$ (1 MHz) (radio); $3 \times 10^{10}\,s^{-1}$ (microwave); $6 \times 10^{14}\,s^{-1}$ (visible);
$3 \times 10^{18}\,s^{-1}$ (X ray);   (*b*) $3 \times 10^{-4}\,m$ (0.3 mm) (far infra-red);
$3 \times 10^{-7}\,m$ (300 nm) (ultraviolet); $3 \times 10^{-11}\,m$ (0.03 nm) (X ray)

8-4   Photoelectric effect.   Photons of wavelength 273 nm or less are needed to cause ejection of electrons from a surface of electrically neutral tungsten. (*a*) What is the energy barrier, in joules and in electronvolts, that electrons must overcome to leave an uncharged piece of tungsten? (*b*) What is T such that kT is equal to this energy, with k the Boltzmann constant? (*c*) What is the maximum kinetic energy, in joules and in electronvolts, of electrons ejected from tungsten by photons of wavelength 150 nm?

8-5   Photoelectric effect.   A minimum of 2.26 eV is required to knock electrons from the surface of an uncharged potassium electrode. What maximum kinetic energy do such electrons have when the wavelength of the incident light is 300 nm?   *Ans.* 3.00×10^{-19} J

8-6   Wave nature of particles.   Indicate what experimental evidence there is for the fact that some "particles" show wavelike properties.

8-7   De Broglie wavelengths of neutrons and atoms.   Thermal neutrons are neutrons with speeds characteristic of the translational energies of gases at room temperature. What is the wavelength of neutrons of a speed that corresponds to v_{rms} at 298 K? What are the wavelengths of Ar atoms and of Xe atoms under the same circumstances?

Ans. 1.46×10^{-10} m; 2.32×10^{-11} m; 1.28×10^{-11} m

8-8 De Broglie wavelengths of macroscopic objects. What is the de Broglie wavelength of a 1-mg weight moving at a speed of 1 cm s^{-1}? What is it for the earth (6.0×10^{24} kg) moving at an average orbital velocity of 30 km s^{-1}?

8-9 Uncertainty principle. State the uncertainty relations between the coordinates and the momentum components of a particle and between the energy and the lifetime of a state.

8-10 Uncertainty principle. (*a*) Assume that a 1.0-mg weight is positioned with an accuracy of 1.0×10^{-3} mm. What is the minimum momentum of the weight consistent with this positional uncertainty? What velocity and what kinetic energy correspond to this momentum? (*b*) Repeat the calculation for a hydrogen atom positioned with an accuracy of 1.0×10^{-3} Å.

8-11 Uncertainty principle. The excited state of an atom responsible for the emission of a photon typically has an average life of 10^{-10} s. What energy uncertainty corresponds to this value? What is the corresponding uncertainty in the frequency associated with the photon?

Ans. 6.6×10^{-24} J; 10^{10} s^{-1}

8-12 Wavelengths of a neutron and an electron. The neutron has a mass 1839 times that of the electron. What is the relation between the wavelength of an electron and that of a neutron that has the same kinetic energy as the electron?

"I am reading your paper in the way a curious child eagerly listens to the solution of a riddle with which he has struggled for a long time, and I rejoice over the beauties that my eye discovers, but which I must study in much greater detail to comprehend fully."

Max Planck, in letter to Erwin Schrödinger, April 2, 1926

"But everything resolves itself with unbelievable simplicity and unbelievable beauty, everything turns out exactly as one would wish, in a perfectly straightforward manner, all by itself and without forcing."

Erwin Schrödinger, in letter to Max Planck, June 11, 1926

9 Atomic Structure

Before we consider how the particle-wave duality of matter and radiation is manifested in current views of the structure of matter, it is instructive to examine earlier theories of atomic structure, particularly that developed by Niels Bohr in 1913. This chapter begins with a brief description of the classical picture of Rutherford's nuclear atom, the difficulties with that picture, and Bohr's resolution of these difficulties by his bold and imaginative application of quantum theory. It then continues with a discussion of the more modern quantum theory (or *wave mechanics*), developed during the 1920s, and its application to the structure of atoms.

9-1 Rutherford's Atom Model

As discussed in Sec. 1-5, atoms consist of a tiny, positively charged nucleus with a diameter of the order of 10^{-12} Å and an appropriate number of electrons surrounding the nucleus. Rutherford, basing his ideas largely on the results of his scattering experiments with alpha particles, had conceived of the atom as a planetary system with the nucleus as central sun and the electrons as orbiting planets.

By the laws of classical electrodynamics, an electron moving in a circular orbit emits electromagnetic radiation of a frequency corresponding to its number of revolutions per second. Since radiation is a form of energy, the energy of the system decreases. This lowering of the energy means a decrease in the radius of the orbit of the electron and a simultaneous increase in its speed. The electron is thus expected to spiral ever faster and ever closer to the nucleus, until the radius of its orbit reaches the value zero and the atom collapses or some other catastrophic event occurs. This train of events is quite in contradiction with experience. First, atoms are stable; their collapse is not observed. Second, this model also requires that the frequency of the orbital motion should change as the radius of the orbit changes. The radiation emitted by a collection of many atoms should therefore cover a whole range of frequencies; that is, according to this model, atomic spectra are expected to be *continuous*. In actuality, however, atoms have *discrete spectra* or *line spectra*, and for each kind of atom the observed radiation shows only certain specific frequencies, as illustrated for atomic hydrogen in Fig. 8-4 and for sodium in Fig. 8-11.

Regularities in these spectra had been sought and found during the last two decades of the nineteenth century. It was determined empirically, for example, that the wave number of any line in the spectrum of hydrogen can be described with high accuracy by the formula

$$\tilde{\nu} = 1.09678 \times 10^5 \left(\frac{1}{n_1{}^2} - \frac{1}{n_2{}^2} \right) \text{cm}^{-1} \tag{9-1}$$

where n_1 and n_2 are integers, n_2 always being greater than n_1. For $n_1 = 2$ and $n_2 = 3$, 4, 5, and 6 the wave numbers obtained from (9-1) are those of the four Balmer lines in the H spectrum (Fig. 8-4b).

9-2   Bohr's Model of the Hydrogen Atom

Bohr's postulates.   As we have just seen, the predictions of classical theory applied to the Rutherford atom model are in direct contradiction to experimental facts. Bohr succeeded in saving the model by boldly adding two novel postulates to the rules of classical physics, chosen simply because they gave correct results for the spectrum of the hydrogen atom. The postulates were:

1.   Only a special set of electron orbits is permitted, for which the angular momentum is an integral multiple of $h/2\pi$, where h is Planck's constant. For a circular orbit of radius r, the angular momentum of a particle with mass m and *speed* v is mvr. Thus,

$$mvr = \frac{nh}{2\pi} \tag{9-2}$$

The integer n is a characteristic of the orbit called its *quantum number*. For these orbits the laws of classical electrodynamics are suspended, so that a circling electron is forbidden to emit electromagnetic radiation. The system therefore loses no radiant energy and the orbits are circles and not spirals. The atom is said to be in a *stationary state.*

2.   When an electron changes orbits, the energy difference $\Delta\epsilon$ between the two orbits shows up as electromagnetic radiation. If the new orbit is of lower energy, radiation is emitted by the system; if it is of higher energy, radiation is absorbed. The frequency ν_{rad} of the radiation involved has nothing to do with the frequency of the orbiting electron but obeys the fundamental relationship

$$|\Delta\epsilon| = h\nu_{\text{rad}} \qquad (9\text{-}3)$$

This is an extension of Planck's relationship [Equation (8-2)], applied now to the transitions[1] between stationary states of atoms. An atomic transition causes the emission, or is caused by the absorption, of a photon with energy $h\nu_{\text{rad}} = |\Delta\epsilon|$.

The energy ϵ_n of an electron in a hydrogen atom orbit of quantum number n can be derived by combining Bohr's postulates with the classical expressions for the potential energy and the kinetic energy of an electron moving in a circular orbit about a proton. These different energy values are called *energy levels;* the state of lowest energy is called the *ground state,* the others *excited states.* Since there is an attraction between the proton and the electron, the energy is lowest when the electron is closest to the nucleus; energy must be put into the atom to excite the electron and move it further from the nucleus. A few energy levels of the hydrogen atom are shown in Fig. 9-1. It is customary to assign the energy zero to the state in which electron and nucleus are completely separated and the separated particles possess no kinetic energy. This implies that negative energy values must be assigned to all other states shown in Fig. 9-1.

The energy change for the transition from state n_1 to state n_2 is

$$\Delta\epsilon = \epsilon_{n_2} - \epsilon_{n_1} \qquad (9\text{-}4)$$

If the level ϵ_{n_2} is lower than the level ϵ_{n_1}, $\Delta\epsilon$ is negative and the transition is associated with the emission of a photon of frequency $|\Delta\epsilon|/h$. Conversely, if ϵ_{n_2} is above ϵ_{n_1}, a photon of frequency $\Delta\epsilon/h$ must be available and be absorbed by the atom to make the transition occur.[2]

Bohr's theory successfully predicts the frequencies of the lines in the spectrum of atomic hydrogen. It leads to an equation identical in form to (9-1) in which the integers n_1 and n_2 are identified as the quantum numbers of the lower and upper levels, respectively. The theory also predicts the frequencies of the lines in the spectra of hydrogenlike ions, that is, of one-electron ions such as He^+, Li^{2+}, and Be^{3+}. These spectra are similar to that of hydrogen. An equation like (9-1) also applies to them,

---

[1] The term *transition* is used to refer to a change from one definite energy state to another.

[2] Under appropriate conditions, the transition may also be brought about by supplying the same energy in another form—for example, by collision with another atom or molecule that has high kinetic or electronic energy.

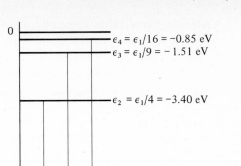

0

$\epsilon_4 = \epsilon_1/16 = -0.85$ eV
$\epsilon_3 = \epsilon_1/9 = -1.51$ eV

$\epsilon_2 = \epsilon_1/4 = -3.40$ eV

ϵ

$\epsilon_1 = -13.60$ eV

FIGURE 9-1 **Four energy levels of the hydrogen atom.** The energy zero corresponds to complete separation of electron and nucleus of the hydrogen atom. In other words $|\epsilon_1|$ is the minimum energy required to ionize a hydrogen atom in the ground state without providing the separated proton and electron with kinetic energy: $|\epsilon_1|$ is the *ionization energy*. When the electron of a hydrogen atom is in the excited state with $n = 2$, it may drop into the ground state, with the energy difference $\epsilon_2 - \epsilon_1$ appearing as a photon. Conversely, a hydrogen atom in the ground state may absorb a photon of this same energy and be excited to the state with the electron in the level $n = 2$. The situation is similar for other excited states.

except that the proportionality constant varies directly as the square of the atomic number, Z^2, as Bohr's theory predicts it should.

EXAMPLE 9-1 **Energy corresponding to a spectral line.** What is the energy in joules, in kilojoules per mole, and in electronvolts corresponding to the Balmer line H$_\alpha$ that arises from the transition between the states $n_2 = 3$ and $n_1 = 2$?

Solution. By (9-1), $\tilde{\nu} = 1.09678 \times 10^5 (\frac{1}{4} - \frac{1}{9})$ cm^{-1} = 1.5233×10^4 cm^{-1}. From $\tilde{\nu} = \lambda^{-1}$ and $\nu = c/\lambda$, it follows that the energy equivalent to the wave number $\tilde{\nu}$ is $h\nu = hc\tilde{\nu}$. For the wave number just given, this energy is $(1.5233 \times 10^4$ cm$^{-1}) (2.998 \times 10^8$ m s$^{-1})(100$ cm/m$)(6.626 \times 10^{-34}$ J s) = 3.026×10^{-19} J. (See inside back cover for the values of c and h.) From Table A-5 we find that 1 eV = 1.6022×10^{-22} kJ = 96.48 kJ mol^{-1}. The energy is therefore also 1.889 eV and 182.22 kJ mol^{-1}.

EXAMPLE 9-2 **Ionization energy of the hydrogen atom.** What is the ionization energy of the hydrogen atom?

Solution. This is the energy needed to lift an electron from the ground state of the atom to the state in which the electron is at an "infinite" distance—that is, to the state representing the zero of the energy scale, a state associated with the quantum number $n_2 = \infty$ (Fig. 9-1). The energy desired is $-\epsilon_1$, which by (9-1) with $n_1 = 1$ and $n_2 = \infty$ corresponds to $\tilde{\nu} = 1.09678 \times 10^5$ cm^{-1}. The energy relationships used in the preceding example lead to the following equivalent values: 1312 kJ mol^{-1}, 13.60 eV, and 2.179×10^{-18} J.

The old quantum theory.   During the decade after Bohr's first application of quantum theory to the atom, a set of rules was developed for treating many different systems, rules now referred to collectively as the *old quantum theory*. Other phenomena on the atomic level were treated by imposing upon the laws of classical theory rules of quantization similar to those used by Bohr. Results were usually in agreement with experiments; however, there were some notable failures of the old quantum theory—for example, its inability to explain certain features of the infra-red spectra of diatomic molecules such as HCl or to explain so fundamental a chemical fact as the formation of H_2 molecules. Furthermore the theory was unsatisfactory in that Bohr's atom model was two-dimensional, the orbits being planar. The theory was also objectionable on aesthetic grounds: it was a patchwork of arbitrary rules, justified only by the fact that they usually worked. There was no set of central postulates from which the entire theory was derived by a sequence of logical steps.

All this was changed around 1926 by the advent of three theories: Schrödinger's wave mechanics, Heisenberg's matrix mechanics, and a more abstract theory by P. A. M. Dirac. These three theories proved to be mathematically equivalent, even though they represented seemingly very different approaches. Bohr's atom model and other work had provided important guidelines in their creation. Only a descriptive treatment can be given here of this *modern quantum theory*. The theory that lends itself best to such a description is that of Schrödinger's wave mechanics.

9-3   Wave Mechanics

Some consequences of the existence of matter waves.   De Broglie's matter waves of wavelength $\lambda = h/mv$ (Sec. 8-3) gave a partial clue as to why Bohr's circular orbits worked so well. According to Bohr's postulate, the only allowed orbits are those for which $mv = nh/2\pi r$ [Equation (9-2)]. This result can be used to calculate λ_n, the wavelength of an electron in the nth orbit:

$$\lambda_n = \frac{h}{mv} = h\frac{2\pi r_n}{nh} = \frac{2\pi r_n}{n} \tag{9-5}$$

The wavelength of the orbiting electron is thus a simple fraction of the circumference of the orbit, so that the circumference is an integral multiple of the wavelength and in successive revolutions the waves are in phase and therefore reinforced (Fig. 9-2). Orbits of any other circumference do not exist—are forbidden—because the electron wave would no longer be in phase with itself after going around once and would thus destructively interfere with itself. The permitted orbits are those for which waves do not die out. Waves corresponding to such orbits are called standing waves and are well known in physics. For example, when a piano string is struck it vibrates in a standing-wave pattern (see Example 9-3 below).

The Schrödinger equation.   While de Broglie's investigations provided a flash of insight into the inner workings of the atom, they furnished only the wavelengths associated with particles. It was several years before Schrödinger proposed his

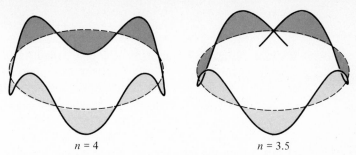

$$n = 4 \qquad\qquad n = 3.5$$

FIGURE 9-2 **De Broglie waves for permitted and prohibited orbits.** For the allowed orbit illustrated schematically here ($n = 4$), the de Broglie wave remains in phase with itself upon successive revolutions and does not destroy itself by interference. In the forbidden orbit ($n = 3.5$) there is a phase difference of $\lambda/2$ and therefore complete destruction by interference after only 1 revolution. For other nonintegral values of n, destruction by interference may take more than 1 revolution but is still inevitable.

celebrated equation that expresses the law to which matter waves are subject and provides a systematic approach to problems of quantum theory. We present here only a simple version of the Schrödinger equation, together with a description of its properties and those of some of its solutions. There will be no attempt to provide a mathematical derivation of these properties; hence most of the features of wave mechanics will have to be accepted on faith. This is a traditional cause of uneasiness for many students, but it must be lived with until the mathematical apparatus needed for a more complete understanding can be assimilated.

By using symbolic notation, the Schrödinger equation can be written in a deceptively simple-looking way; it is

$$\mathcal{H}\psi = \epsilon\psi \tag{9-6}$$

where ϵ is the total energy of the system and ψ ("psi") is its *wave function,* the meaning of which is discussed in the next section. The wave function is the unknown in the equation, and once it has been found by solving the equation, it can be used to calculate many properties of the system, as discussed below. The part of the equation that is written symbolically as $\mathcal{H}$ is called the *Hamiltonian* of the system, an algebraic conglomerate of multiplicative constants and differential operators[1] (like d/dx), all of which are to be applied to the wave function ψ in (9-6). For example, $\mathcal{H}$ for a particle of mass m that can move only in the x direction and whose potential energy $\epsilon_p(x)$ depends only on x is

$$\mathcal{H} = -\frac{h^2}{8\pi^2 m}\frac{d^2}{dx^2} + \epsilon_p(x) \tag{9-7}$$

The Schrödinger equation for this one-dimensional situation is obtained by substituting (9-7) into (9-6):

---

[1]An appreciation of these formulas requires calculus (and, for complete understanding, on an advanced level at that). The formulas are given to provide you a glimpse of their nature and form, not with the idea that you will understand all their implications.

$$\mathcal{H}\psi = -\frac{h^2}{8\pi^2 m}\frac{d^2\psi}{dx^2} + \epsilon_p(x)\psi = \epsilon\psi \qquad (9\text{-}8)$$

Because $\mathcal{H}$ has to be applied to ψ it is called an *operator* and is said to operate on the function ψ. The form taken by $\mathcal{H}$ changes from system to system. It has a certain form for the hydrogen atom, another for the methane molecule, and so on. A Hamiltonian, and therefore a Schrödinger equation, can be written for virtually any system, no matter how complex—it need not be limited to atoms or molecules.

The right side of the Schrödinger equation is much more straightforward in its meaning than the left. There are no symbolic implications: $\epsilon\psi$ is simply the product of the energy ϵ and the wave function ψ. The energy ϵ is a constant, but ψ is a function of the coordinates of the system.

The wave function ψ. Wave functions that are solutions of the Schrödinger equation describe, in a way explained below, possible *stationary* states of the system, which means states that do not vary with time. (For the description of quantum phenomena that vary with time, a slightly different form of the Schrödinger equation is needed.) The Schrödinger equation is of a general type often referred to as wave equations because waves of all kinds—ocean waves, sound waves, electromagnetic waves—obey equations of this general form. This is also why ψ is called the wave function.

The Schrödinger equation is a differential equation for ψ, meaning that it involves derivatives of ψ. When one attempts to solve a differential equation, that is, to find values of the function involved (here, ψ) that satisfy the equation, it is usually necessary to impose certain conditions upon the acceptable solutions. If we are to apply the Schrödinger equation to a particle confined in a box, for example, we require that the wave function vanish outside the boundaries of the box, since the particle must be inside the box. Furthermore, we must also impose the condition that ψ be zero *at* the boundaries. Imposition of *boundary conditions* of this kind invariably places restrictions on the wave functions ψ that are permitted to exist.

Consider now a system of *one* particle whose potential energy depends on its position coordinates (x,y,z). An equation like (9-8) can be written, and if solutions to it can be found, the way in which the wave function for this system depends on the position coordinates of the particle will be known. What then is the physical significance of ψ? In other words, with what physical quantities can ψ be identified and how can it be used? The answer to the first part of this question is that although ψ itself cannot be identified with any observable property, its square is related to the probability of finding the particle in different parts of space. The probability of finding the particle in the volume element $dV = dx\,dy\,dz$ extending from x to $x + dx$, from y to $y + dy$, and from z to $z + dz$ is

$$[\psi(x,y,z)]^2\,dV \qquad (9\text{-}9)$$

The probability is thus proportional to the square[1] of ψ and to the size of dV, so that the quantity ψ^2 is a probability per unit volume, or a *probability density.*

---

[1] In many situations ψ may be complex (i.e., contain expressions with $i = \sqrt{-1}$); for these ψ^2 must be replaced by $|\psi|^2$.

A knowledge of the wave function for a system permits the calculation not only of the probability of finding that system in any particular state (for example, of finding the aforementioned single particle in a particular region of space) but also the calculation of many other quantities of interest—for example, the permitted energy levels, the average values of the positions and momenta, and, for certain systems in excited states, the probability that the system will return to the most stable state and emit radiation within a given time. The details of these calculations are often tedious and complicated, but the simplest wave functions may be no more than ordinary exponential or trigonometric functions such as e^{-x} or $\sin x$, and calculations with them often involve only simple operations of differential and integral calculus.

One of the most important characteristics of the Schrödinger equation is that the imposition of boundary conditions inevitably means that the equation has solutions only for a set of special values of ϵ, the possible energies of the system. The discrete energy levels of atoms and molecules are thus explained in a natural way and no arbitrary quantization postulate such as that in the Bohr theory need be invoked. Values of ϵ for which the Schrödinger equation has a solution are called *eigenvalues*[1] of the system; the wave functions ψ are also called *eigenfunctions*.

The different energy levels for which the Schrödinger equation has solutions are the stationary states of the system; they correspond to different standing waves of the wave function. This is analogous to the different standing waves that can be sustained by a piano string, which occur at different but specific frequencies that represent the pitch of the string and all its harmonics. The different frequencies of the string are the eigenvalues of the wave equation that describes its motion; they correspond to the different energy levels that are the eigenvalues of a wave-mechanical system.

EXAMPLE 9-3   **Permitted frequencies of a string.** When a string is stretched and clamped tightly at both ends, it can be made to vibrate in a direction perpendicular to its length. The velocity of waves associated with such transverse vibrations is given by the formula $u = \sqrt{t_0/m_l}$, where t_0 is the tension of the string and m_l is the mass of the string per unit length. What are the permitted frequencies of vibrations for a string of length l?

Solution. Boundary conditions require that there be no motion at the clamped ends, so that l must be equal to an integral multiple of $\lambda/2$, $l = n\lambda/2$. Thus only wavelengths $\lambda_n = 2l/n$ are permitted. The values of the permitted frequencies follow from the relationship $u = \nu\lambda$ [see Equation (8-1)], so that $\nu_n = u/\lambda_n = (n/2l)\sqrt{t_0/m_l}$, with $n = 1, 2, 3, \ldots$. Anyone familiar with a stringed musical instrument—for example, a guitar, violin, harp, or piano—will recognize that this result is in accord with experience: increasing the tension on a string increases the pitch (frequency), and increasing the mass per unit length or the length decreases the pitch.

---

[1]The use of the prefix *eigen* in these words represents a direct German translation of the word *characteristic,* used by the great Irish mathematician Sir William Hamilton, who first discussed general methods of dealing with systems of equations of the kind involved here. His work dates from the first half of the nineteenth century, long before quantum theory. It is ironic that the words *eigenvalue* and *eigenfunction* are now much more common in English than *characteristic value* and *characteristic function.*

The vibrations of a drumhead, which are discussed in the next subsection, can be represented by the solutions to a wave equation. Such two-dimensional systems exhibit an important phenomenon not found in one-dimensional cases: several eigenfunctions may happen to correspond to the same energy ϵ. Such an energy level is said to be *degenerate,* and the number of different independent eigenfunctions corresponding to this energy is called the degeneracy of the level.

To distinguish the different eigenfunctions of a system, they are labeled by integers called quantum numbers. In many systems, several independent quantum numbers are needed to characterize a given state, as shown in the examples that follow.

An analogy: the vibrations of a drumhead. A two-dimensional analogy illustrates several of the statements made in the preceding pages. It concerns the possible vibrations of a circular diaphragm, such as the head of a drum, clamped along its rim. This diaphragm may exhibit standing waves that are the solutions of a differential equation quite similar to the Schrödinger equation for the one-electron atom. Several such standing waves are shown in Fig. 9-3. Each diagram represents an instantaneous picture of a standing wave. A plus sign means an excursion of the diaphragm above the plane of the drawing; a minus sign signifies an excursion below that plane. The situation is constantly changing—an instant later the reverse signs may apply. Between the areas designated by plus and minus signs there are regions of no motion. These areas where the amplitude of the wave is zero and remains zero are called *nodes.* The rim of the diaphragm must be a node also, because that is where the diaphragm is clamped and hence motionless.

To each standing wave there corresponds a frequency that is the analog of the energy in the Schrödinger equation. The different standing waves are classified in Fig. 9-3 by the numbers n and l, where n is the total number of nodes and l is the number of nodes that are straight lines. When straight-line nodes are present, their orientations are indicated by letters such as $x, y,$ and xy (to be explained below). Consider, for example, the case $n = 2, l = 1$, which contains one straight-line node. The frequency does not depend on the orientation of this line, which may be in a vertical direction, in a horizontal direction, or in any other direction. There exists an infinite number of standing waves with $n = 2, l = 1$, differing only in the orientation of the straight-line node. However, all of these may be represented as the sum of the two standing waves shown in Fig. 9-3 that have *either* a vertical *or* a horizontal nodal line. For example, a straight-line node at 45° from the x axis, in the positive quadrant, corresponds to equal contributions from the two standing waves that have a vertical and a horizontal node (see bottom of Fig. 9-3).

The standing wave with the vertical node is designated by an x because the equation of this nodal line (the y axis) is $x = 0$. The other standing wave is designated by a y because its nodal line (the x axis) has the equation $y = 0$. The frequencies of these two standing waves are identical, and this frequency is thus said to be doubly degenerate because it corresponds to two independent modes of vibration.

In the case $n = 3, l = 2$ (Fig. 9-3) there are two straight-line nodes at right angles. Again the frequency of the standing wave is independent of the orientation of these two lines, and again all possible cases can be described as the sum of just two standing

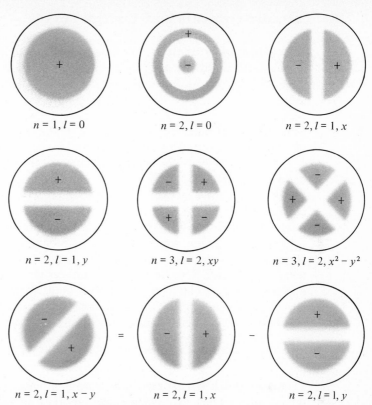

$n = 1, l = 0$ $n = 2, l = 0$ $n = 2, l = 1, x$

$n = 2, l = 1, y$ $n = 3, l = 2, xy$ $n = 3, l = 2, x^2 - y^2$

$n = 2, l = 1, x - y$ $=$ $n = 2, l = 1, x$ $-$ $n = 2, l = 1, y$

FIGURE 9-3   **Standing waves on a drumhead.** A plus sign indicates a temporary excursion of the diaphragm toward the reader; a negative sign means an excursion in the opposite direction. The number n indicates the number of regions, called nodes, that remain motionless, including the fixed rim of the diaphragm. The number l counts those nodes that are in straight lines; the letters x, y, xy, and $x^2 - y^2$ indicate their orientation relative to the horizontal x axis and the vertical y axis, as explained in the text. All possible cases for $n = 3$ are not shown.

waves. One of them has the node lines vertical and horizontal; the other has the node lines at 45° to the x and y axes. The first of these is labeled xy because the equations of the vertical and horizontal node lines are $x = 0$ and $y = 0$ and these two equations can be combined into one, $xy = 0$. The nodal lines of the other standing wave, at 45°, are $x - y = 0$ and $x + y = 0$ or, combined, $(x - y)$ $(x + y) = x^2 - y^2 = 0$; this mode of vibration is labeled $x^2 - y^2$. Again the frequency of these two modes is doubly degenerate. More complicated standing waves also exist and are labeled in a similar way.

There are many points of similarity between the standing waves on a drumhead and the standing waves of the Schrödinger wave function representing the different possible states of an atom. These points are brought out in the next section. The numbers n and l as used above are in certain ways analogous to the quantum numbers used to designate atomic states.

9-4 Atomic Orbitals and Energy Levels

The allowed energies. The Schrödinger equation for an atom consisting of a nucleus with charge Ze and one electron has solutions only for energy values[1]

$$\epsilon_n = -\frac{2\pi^2 mZ^2e^4}{n^2h^2\alpha^2} \tag{9-10}$$

with the zero of potential energy chosen as the state with the electron completely removed; m is the mass of the electron and n may have any positive integral value, 1, 2, 3, ... These are the energies that permit the existence of standing waves for the wave function ψ.

For all but the lowest energy value (where $n = 1$) there exists more than one possible eigenfunction. These energy values are therefore degenerate and quantum numbers in addition to n are needed to designate the different solutions ψ of the Schrödinger equation, as in the two-dimensional example of the preceding pages. Nodes occur in the wave functions for atoms; these nodes are surfaces in three-dimensional space. Some of the nodal surfaces are spherical, centered at the atomic nucleus. Other possible nodes are planes that pass through the nucleus. One node always present is a sphere of infinite radius, because the wave functions of atoms go to zero at distances far from the nucleus. This corresponds to the fact that the probability of finding an electron, which depends on ψ^2, approaches zero at large distances from the nucleus.

The quantum numbers. Electrons in atoms are characterized by four quantum numbers. Only two of these are important in specifying electron energies in atoms and they are discussed in the following pages: the *principal* quantum number n and the *azimuthal* quantum number l. The other two are related to the angular momentum of the electron in an orbital and to the spin of the electron (Sec. 9-5).

The principal quantum number n is identical with the n used in (9-10). It is also the total number of nodal surfaces of the wave function, including the spherical node of infinite radius. Thus the ground state of the hydrogen atom ($n = 1$) corresponds to a wave function with only one node—that at infinity.

While n may assume the value of any positive integer, $n = 1, 2, 3, 4, \ldots$, the quantum number l is restricted to the range $l = 0, 1, 2, \ldots, n - 1$. Thus for the state with $n = 1$, only the value $l = 0$ is possible, but for $n = 2$ there are two permissible values of l, 0 and 1. In general, the value of l indicates the number of nodal planes in the wave function.

For one-electron (hydrogenlike) atoms, the energy is essentially independent of the value of l. This is not true when there is more than one electron, however, and hence for most atoms and ions the energy depends both on n and on l. As a result, two subscripts are needed to denote the energy of each level, ϵ_{nl}. This dependence on l removes part of the degeneracy of the energy levels of the hydrogen atom (or of He$^+$, Li^{2+}, . . .), but some degeneracy remains: for each combination of values of n and l, there exist $2l + 1$ solutions of the wave equation for the many-electron

---

[1] $\alpha = 1.113 \times 10^{-10}$ C^{2}J^{-1} m^{-1} (see Appendix A, Equation A-4).

atom that correspond to the same energy ϵ_{nl}. When $l = 0$, there is only one energy state, but when $l = 1$, $2l + 1 = 3$ and there are thus three states identical in energy (a threefold degeneracy). When $l = 2$, there is a fivefold degeneracy.

It is customary to denote states with different values of the quantum number l by letters as follows:[1] s for $l = 0$, p for $l = 1$, d for $l = 2$, and f for $l = 3$. Values of l beyond 3 are used only rarely. Figure 9-4 shows the number of different states corresponding to different n,l pairs and also indicates, very schematically, the relative energies of the different states in many-electron atoms. When n is 1, the only permitted value of l is zero. This state is labeled $1s$ and is nondegenerate. When $n = 2$, l may be 0 or 1, corresponding to one $2s$ and three $2p$ states. For $n = 3$, l may be 0, 1, or 2, and there are these states: one $3s$, three $3p$, and five $3d$. When

---

[1] The letters s, p, d, and f were once used by spectroscopists to characterize certain lines. They are abbreviations of *s*harp, *p*rincipal, *d*iffuse, and *f*ine.

FIGURE 9-4   **Schematic energy-level diagram for a many-electron atom.** The degeneracies of the s, p, d, and f levels are seen to be 1, 3, 5, and 7, respectively, regardless of the value of n. The relative heights of levels that are very close together (such as $4s$ and $3d$) may shift in different atoms for reasons discussed in the following pages. The energy scale in this schematic diagram is not linear, being especially compressed at the lower end (see text).

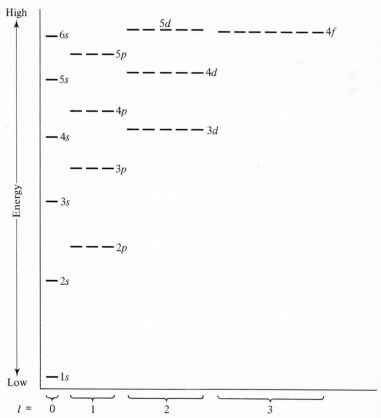

$n = 4$, l may be 0, 1, 2, or 3, corresponding to one 4*s*, three 4*p*, five 4*d*, and seven 4*f* states. The states corresponding to higher values of *l* for $n = 5$ and 6 (and more) are not shown. This important pattern of energy levels will be considered again in the following chapter. The $(2l + 1)$-fold degeneracy of the levels, and their relative positions, are of great importance for an understanding of the periodic system of the elements.

It must be emphasized that diagrams such as that in Fig. 9-4 are only schematic. The relative positions of levels that are very close, such as 3*d* and 4*s*, or 4*d* and 5*s*, are not the same in all atoms; sometimes one is slightly higher, sometimes the other. Furthermore, the vertical scale in this schematic diagram is greatly distorted. For example, the separation of the 1*s* and 2*s* levels is tens, hundreds, or even thousands of times as great as the separation of, say, the 4*s* and 5*s* levels, the relative difference increasing with increasing atomic number. Nevertheless, the diagram serves as a useful guide to the distribution of electrons among energy levels, those of lowest energy being filled first according to rules we discuss in Chap. 10. The principal rule is that no more than two electrons can occupy any of the individual states depicted in Fig. 9-4.

Description of the different atomic orbitals. The solutions ψ of the Schrödinger equation for an atom are called *atomic orbitals* because they represent the wave-mechanical counterpart of the orbits of the Bohr model. They are the products of two functions. One of these functions, denoted here by $R(r)$, depends only on the distance *r* from the center of the atom and not on the direction; it is thus spherically symmetric. This function $R(r)$ is called the *radial* part of ψ. The second function, denoted by $\Phi(\theta,\varphi)$, depends only on the direction (that is, on θ and φ; see Fig. 9-5) and is independent of *r*; it is known as the *directional* or *angular*

FIGURE 9-5 **Spherical polar coordinates.** The angles θ (colatitude) and φ (azimuth or longitude) are defined as illustrated. Consideration of the relation of *x*, *y*, and *z* to *r*, θ, and φ shows that

$$x = r \sin \theta \cos \varphi$$
$$y = r \sin \theta \sin \varphi$$
$$z = r \cos \theta$$

and also that

$$r^2 = x^2 + y^2 + z^2$$

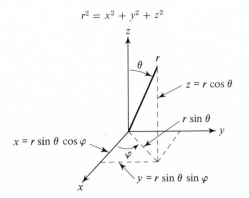

part of ψ. Thus

$$\psi(r,\theta,\varphi) = R(r)\Phi(\theta,\varphi) \qquad (9\text{-}11)$$

The functional form of $R(r)$ and $\Phi(\theta,\varphi)$ for some of the commonest atomic orbitals is shown in Table 9-1, and various graphical representations of these and other orbitals are given in Figs. 9-6, 9-8 to 9-11, and 25-6.*

Radial part. The shape of the radial part of the orbitals, $R(r)$, depends on the quantum numbers n and l. Examples of different radial parts of orbitals $R_{nl}(r)$ are shown in Fig. 9-6 and in Table 9-1. They all fall off exponentially to zero for distances large compared to atomic dimensions. However, of greater interest than $R(r)$ itself is the *radial probability distribution* for the electron in an orbital, that is, the probability of finding the electron within a narrow range of distance, δr, at a distance r from the nucleus. This problem is the three-dimensional analog of the two-dimensional radial population distribution function illustrated in Fig. 5-12 and

---

*The orbitals discussed here have been obtained by solving the wave equation for a one-electron atom and are thus strictly applicable only for atoms in which the potential energy is spherically symmetric and coulombic, that is, proportional to $-(1/r)$. However, the angular parts, which are most important for bonding considerations because of their directional character, are valid for all (isolated) atoms, in the approximation in which each electron may be considered to be in the average field of all the others. The radial parts shown here apply, at least qualitatively, after appropriate scaling.

TABLE 9-1   **Radial and angular dependence*,† of some orbital wave functions,**
$\psi(r,\theta,\varphi) = R(r)\Phi(\theta,\varphi)$ **for one-electron atoms**

Orbital	$R(r)$	$\Phi(\theta,\varphi)$
$1s$	e^{-kr}	1
$2s$	$(2 - kr)e^{-kr/2}$	1
$2p_x$	$kre^{-kr/2}$	$\sin\theta\cos\varphi = x/r$
$2p_y$	$kre^{-kr/2}$	$\sin\theta\sin\varphi = y/r$
$2p_z$	$kre^{-kr/2}$	$\cos\theta = z/r$
$3s$	$(27 - 18kr + 2k^2r^2)e^{-kr/3}$	1
$3p_x$	$kr(6 - kr)e^{-kr/3}$	$\sin\theta\cos\varphi = x/r$
$3p_y$	$kr(6 - kr)e^{-kr/3}$	$\sin\theta\sin\varphi = y/r$
$3p_z$	$kr(6 - kr)e^{-kr/3}$	$\cos\theta = z/r$
$3d_{xy}$	$k^2r^2e^{-kr/3}$	$\sin^2\theta\cos\varphi\sin\varphi = xy/r^2$
$3d_{yz}$	$k^2r^2e^{-kr/3}$	$\sin\theta\cos\theta\sin\varphi = yz/r^2$
$3d_{zx}$	$k^2r^2e^{-kr/3}$	$\sin\theta\cos\theta\cos\varphi = xz/r^2$
$3d_{x^2-y^2}$	$k^2r^2e^{-kr/3}$	$\sin^2\theta\,(\cos^2\varphi - \sin^2\varphi) = (x^2 - y^2)/r^2$
$3d_{z^2}$	$k^2r^2e^{-kr/3}$	$3\cos^2\theta - 1 = (3z^2 - r^2)/r^2$

* See Fig. 9-5 for the relation of r, θ, φ to x, y, z. The constant $k = Z/r_B$, where Z is the atomic number of the atom and r_B is the Bohr radius, 0.53 Å.

† Certain constant terms in each function have been omitted here since our present concern is the dependence of the functions on direction and distance from the nucleus. The relative magnitudes of these constants are reflected in the relative magnitudes of the different functions, which are depicted in some of the accompanying figures.

   The angular functions $\Phi(\theta, \varphi)$ can be expressed quite simply in terms of the ratios $x/r, y/r$, and z/r, which depend on θ and φ only and not on r (see Fig. 9-5). These ratios appear on the right-hand side of the equations in col. 3 of the table. The subscripts of the different orbitals are simplified versions of these expressions.

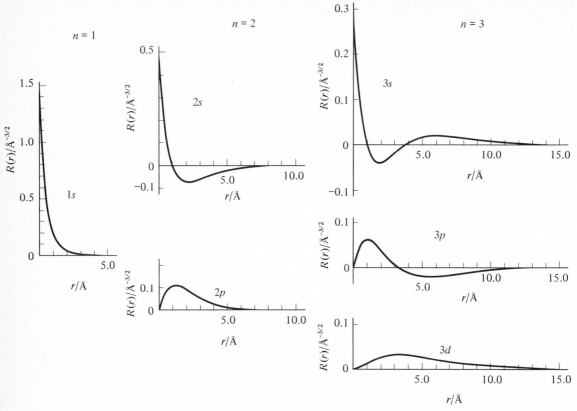

FIGURE 9-6   **Examples of radial parts of an orbital** $R_{nl}(r)$. The important features of the radial part $R(r)$ of an atomic orbital ψ are that (1) it assumes large values only for values of r that correspond to atomic dimensions and goes to zero as r becomes large on this scale and (2) it has $n - l$ nodes including the one at $r = \infty$. Negative values of $R(r)$ are permissible because it is the square $\psi^2(r,\theta,\varphi) = R^2(r)\Phi^2(\theta,\varphi)$ that is the probability density for electrons. The units of $R(r)$ are $\text{Å}^{-3/2}$, which makes Å^{-1} the units of $4\pi r^2 R^2(r)$, as is appropriate for a radial probability distribution (see Fig. 9-8). Note, however, that the vertical scales for the different $R_{nl}(r)$ are different, although they are equal for functions of the same n.

the accompanying text. If the electron is to be found between r and $r + \delta r$ from the nucleus, it must lie within a spherical shell of approximate volume $4\pi r^2\,\delta r$ (Fig. 9-7). Since the probability density is proportional to the square of the wave function, the radial probability of finding the electron within this shell can be obtained by averaging $\psi^2(r,\theta,\varphi)$ over all directions (i.e., over all values of θ and φ). The result is independent of the angular part of ψ, $\Phi(\theta,\varphi)$, and may be written as the product of the square of the radial part $R(r)$ of ψ, R^2, and the volume of the shell, $4\pi r^2\,\delta r$. The result, $4\pi r^2 R^2\,\delta r$, is proportional to the thickness δr of the shell, and the function $4\pi r^2 R^2$ is called the radial probability distribution (see also the discussions of the similar distribution shown in Fig. 5-12). Figure 9-8 shows radial probability distributions, $4\pi r^2 R^2$, for the orbitals of Table 9-1. The maximum in the distribution for the $1s$ orbital is at 0.53 Å, the radius of the smallest possible orbit in the Bohr theory,

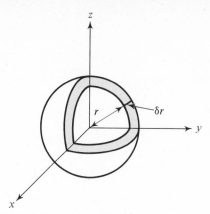

FIGURE 9-7 **Volume of a spherical shell.** One octant of the shell has been removed for purposes of exposition. The surface of the inner sphere is $4\pi r^2$ and the thickness of the shell is δr so that the shell volume is approximately equal to $4\pi r^2 \, \delta r$. The approximation improves as δr is decreased.

r_B. Thus this radius retains a fundamental significance even in the wave-mechanical picture of the hydrogen atom, but instead of being the precise radius of the $1s$ orbit of the electron, it is only the radius at which the probability of finding the electron is greatest. For example, the probability of finding the electron at about 0.20 Å or about 1.10 Å is half as great as that of finding it at about the Bohr radius.

The curves in Fig. 9-8 show that as n increases, the electron is on the average further from the nucleus, in agreement with the increase of potential energy with increasing n. A more subtle feature, which is of great importance in understanding the relative energies of the states in diagrams like Fig. 9-4 and thus in interpreting the periodic table, may be discerned by comparing the curves in Fig. 9-8 for electrons with the same n but different l—for example, $2s$ and $2p$, or $3s$, $3p$, and $3d$. Although the average distance from the nucleus is comparable for all the orbitals of a given n, it can be seen that an s electron is more likely to be found near the nucleus than is a p electron, and a p electron in turn more likely than a d electron. The relative energies of the s, p, and d electrons of a given n (Fig. 9-4) are consistent with this pattern of decreasing penetration toward the nucleus with increasing l.

Angular part. The function $\Phi(\theta,\varphi)$ does not depend on the kind of atom or ion involved or on the principal quantum number n. It does, however, depend on the azimuthal quantum number l, and for a given l there are $2l + 1$ angular functions—e.g., one s function, three independent p functions, five independent d functions, and so forth. The shapes of these functions are important because many aspects of molecular structure depend on them. Their squares, $\Phi^2(\theta,\varphi)$, represent an angular probability distribution; that is, they contain information about the probability of finding an electron in a given direction. They are the counterpart of the non-directional, spherically symmetric function $R^2(r)$ that describes the radial dependence of the function $\psi^2(r,\theta,\varphi) = R^2(r)\,\Phi^2(\theta,\varphi)$. As mentioned, the functions $\Phi(\theta,\varphi)$ are identical for all atoms and monatomic ions, a statement that is not true for the functions $R(r)$.

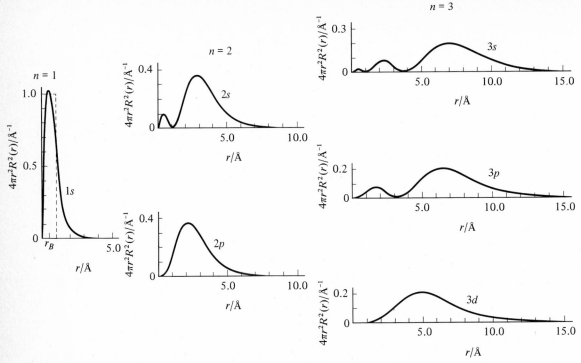

FIGURE 9-8   **Examples of radial probability distributions.**   As discussed in the text and in Fig. 9-6, the ordinate is equal to $4\pi r^2 R^2(r)/\text{Å}^{-1}$; all probability distributions are on the same scale. The horizontal scale applies for H-atom orbitals and is increasingly contracted as the atomic number Z increases (because the negative exponent in each of these functions is proportional to Z; see Table 9-1). The area under any curve between any two values of r (say, r_1 and r_2) equals the probability that an electron occupying the corresponding orbital is at a distance from the nucleus between r_1 and r_2. The total area under any of the curves is unity, since there is unit probability (i.e., there is certainty) that the electron lies between $r = 0$ and $r = \infty$. The dotted rectangle in the figure illustrating the $1s$ curve is of unit area, and it is apparent that the area of this rectangle is comparable to the area under the curve.

Examination of Table 9-1 verifies that the angular functions $\Phi(\theta,\varphi)$ for the s orbitals, regardless of the principal quantum number n, are equal to 1 and are thus constant, independent of direction. Similarly, the set of three angular functions $\Phi(\theta,\varphi)$ for all p orbitals is the same for every value of n, and the situation is analogous for the five functions $\Phi(\theta,\varphi)$ for d orbitals and the seven functions for f orbitals.

Complete atomic orbitals.   Let us now consider the complete atomic orbitals resulting from the multiplication of the radial parts $R(r)$ by appropriate angular parts $\Phi(\theta,\varphi)$. Examples of important atomic orbitals are discussed in the next few pages and various representations of them are illustrated in Figs. 9-9 to 9-11. These schematic diagrams can be very illuminating provided that the reader has a clear idea of just what is (and what is *not*) represented in them. We urge that you carefully study each figure, with its legend, in sequence as it is referred to in the text.

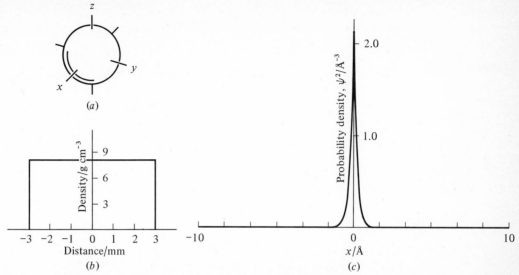

FIGURE 9-9   **Representations of spherical density distributions.** (*a*) Representation of a spherical surface, which might be that of a steel ball [see (*b*)] or might enclose most of the region in which the electron is likely to be found in a hydrogen atom with its electron in a 1*s* orbital [see (*c*)]. (*b*) Mass density variation along a line (e.g., the *x* axis) through the center of a steel ball of radius 3.0 mm. (*c*) Electron probability density variation along a line (e.g., the *x* axis) through the center of a hydrogen atom with its electron in a 1*s* orbital. (This is the square of the function shown for the 1*s* orbital in Fig. 9-6.)

The representation in (*a*) is a conventional way of illustrating an object with spherical symmetry. For a typical macroscopic hard sphere of uniform density (for example, a steel ball, a marble, or a billiard ball), the surface shown represents the "outer surface", at which the mass density (mass per unit volume) drops sharply to zero from the constant value it has within the sphere [see (*b*)].

On the other hand, for illustrating the electron distributions in atoms, there is no "outer surface" that can be represented. If we plot the square of the wave function, $\psi^2 (r,\theta,\varphi)$, which measures the probability per unit volume of finding the electron, as a function of the distance from the center of the atom, as in (*c*), it is obvious that (on this microscopic scale) there is no sharp boundary to the atom. If we draw a surface at some constant value of ψ^2, this surface will be spherical because ψ, and hence ψ^2, is spherically symmetric. Suppose that we are considering a hydrogen atom with its electron in the 1*s* orbital and we draw this surface at a radius of 2 Å. The probability of finding the electron within the surface is then greater than 95 percent, but it is not 100 percent, for as Fig. 9-8 shows, there is some probability of finding the electron beyond 2 Å.

Surfaces of constant ψ, or constant ψ^2, are spherical for 2*s*, 3*s*, and all other *ns* orbitals. However, density is nonuniform within these surfaces also, and indeed falls to zero wherever ψ changes sign (see Fig. 9-6). Still, the kind of representation used in (*a*) is often used to depict any *s* orbital.

***s* orbitals.**   Since $\Phi(\theta,\varphi) = 1$ for all *s* states, the complete *s* orbitals are spherically symmetric and identical to the appropriate radial part of the wave function, $R_{n0}(r)$. A few of these have been portrayed in Fig. 9-6 [and the functions $4\pi r^2 R^2(r)$ in Fig. 9-8]. Figure 9-9 illustrates some similarities and differences in the representation of macroscopic spherical objects and of spherically symmetric electron distributions

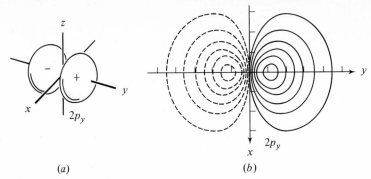

(a)   *(b)*

FIGURE 9-10   **Representations of the $2p_y$ orbital.** *(a)* Surfaces of constant ψ (and thus constant ψ^2). *(b)* A contour diagram corresponding to a section cut through the diagram shown in *(a)*; dashed contours indicate negative values of ψ.

A common convention for representing the distribution of ψ or ψ^2 in three-dimensional space is to make perspective drawings (or three-dimensional models) of surfaces on which ψ, and thus ψ^2, is constant. Such surfaces are illustrated in *(a)* above, as well as in Fig. 9-9a. The *sign* of ψ on the surface is shown by placing a plus or a minus sign nearby; the sign of ψ^2 is always positive. It is essential to remember that these signs have *nothing to do with electric charge.*

The surfaces of constant ψ represent the locus of points with the *same* value of ψ. To represent the way in which ψ *varies* with position in three-dimensional space by this kind of diagram requires somehow representing several different surfaces, each one corresponding to some particular constant value of ψ, one surface within the next, like the successive layers in a (not necessarily spherical) onion. Because this is extremely difficult to draw or model, a slice or section through such a figure is often shown, as in *(b)* above. The different lines (contours) then represent cross sections of surfaces corresponding to differing constant values of ψ. The section shown is the plane with the equation $z = 0$ (and thus with x and y varying). Because the function $2p_y$ has rotational symmetry about the y axis, the cross section cut by any other plane containing the y axis would be identical to that shown here.

Occasionally, only the angular part of ψ, $\Phi(\theta,\varphi)$, rather than all of ψ, is illustrated by a diagram that superficially resembles the representation of surfaces of constant ψ. For a p orbital, for example, the *angular* variation is sometimes shown as two spheres, tangent at the origin of the diagram. The uninitiated viewer of such a diagram is tempted to imagine that the origin of the diagram represents the center of the atom in ordinary three-dimensional space, and consequently that distances from the origin of the diagram represent distances from the center of the atom (the quantity we have denoted by r). However, this is incorrect; the distance from the origin of such a diagram represents instead the *magnitude of the angular portion Φ (θ,φ) of the wave function* ψ, rather than any distance. Thus, such a diagram is *not* a representation of the distribution of ψ or ψ^2 in three-dimensional space. Rather it is a representation in a space in which the variables θ and φ have their regular meaning, while the variable r is replaced by the *relative* values of ψ for any given *constant* value of the radius [for which $R(r)$ is constant]. To appreciate the three-dimensional shape of the orbital, one must include the effect of varying the radius, which is appreciable (as indicated in Fig. 9-6).

in s orbitals. It also shows the variation of ψ^2 for the $1s$ hydrogen-atom orbital along a line through the center of the atom.

p orbitals.   For these orbitals, there are three independent angular functions $[l = 1, 2l + 1 = 3]$. It is convenient, and indeed customary in most applications in

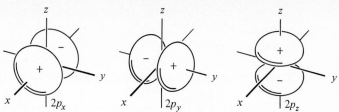

FIGURE 9-11 **The orbitals $2p_x$, $2p_y$, and $2p_z$.** These orbitals are identical except for orientation. The surfaces represent constant values of ψ and thus correspond to those of Fig. 9-10a. Each of these orbitals has a single nodal plane, that is, a plane in which the wave function vanishes. These are the plane $x = 0$ for $2p_x$, the plane $y = 0$ for $2p_y$, and the plane $z = 0$ for $2p_z$.

chemistry, to choose these angular functions to be those indicated in Table 9-1. The results of multiplying one of them by an appropriate radial part $R(r)$ to give the $2p_y$ orbital are shown in Fig. 9-10, which should be studied carefully. The value of the wave function for the $2p_y$ orbital is constant on each of the two surfaces shown in Fig. 9-10a, positive[1] for that on the right and negative for that on the left. The two surfaces enclose regions of space for which ψ (and thus ψ^2) for the $2p_y$ orbital has relatively large values. Hence, an electron occupying a $2p_y$ orbital has a large probability of being inside the space enclosed by the two lobes and a correspondingly small probability of being outside these regions.

The three usual p orbitals for any specific value of n are identical in shape but differ in orientation, being aligned along three mutually perpendicular directions. Those for $n = 2$ are illustrated in Fig. 9-11. The subscripts x, y, and z refer to the fact that, for example, the angular portion of any orbital np_x ($n = 2, 3, 4, \ldots$) is proportional to x/r (see Table 9-1). This portion is positive when x is positive and negative when x is negative, and the plane $x = 0$ is a nodal plane; that is, any np_x orbital is equal to zero in the y,z plane.

d **orbitals.** When $l = 2$, there are five independent orbitals. They are shown schematically in Fig. 25-6. All these orbitals are obtained by multiplying the appropriate angular function $\Phi(\theta,\varphi)$ by the radial function $R(r)$ for $3d$ orbitals (corresponding to $n = 3$, $l = 2$; see Table 9-1 and Fig. 9-6). As shown in Table 9-1, these angular functions can be simply expressed in terms of the ratios x/r, y/r, and z/r; such expressions also explain the subscripts used for d orbitals.

Orbits and orbitals. What is left of the orbits of the Bohr theory? For each there is a corresponding orbital or wave function, the square of which gives a probability density for finding the electron, and associated with an electron in any given orbital is a certain energy and angular momentum. It is also possible to

---

[1]It is very important to note that the signs associated with the lobes in Fig. 9-10 and elsewhere have nothing to do with electric charge; they represent the sign of ψ and arise in a natural way from the expressions in Table 9-1. The only physically meaningful aspect of a wave function is its square and this is *always* positive. However, the relative signs of different ψ's are of great importance when orbitals of different atoms overlap or orbitals of the same atom are *hybridized,* as we shall see in the discussions of bonding and hybridization in Chap. 12.

calculate for each orbital the probability that the momentum, and hence the velocity, of the electron lies in any particular range of values. While each orbital can thus be given particlelike attributes, it would be a futile proposition to attempt a detailed description of an electron *orbit*. All that can be said is that if we were to locate the electron of a hydrogen atom in a given state and were to repeat the experiment many times, we would find distributions of location, momentum, and certain other measurable properties as predicted by wave mechanics. What the electron does in the absence of the disturbances that are an inextricable part of making the observation is an undecidable question.

9-5 Angular Momentum and Spin

In wave mechanics the angular momentum associated with the motion of an electron in an atomic orbital is quantized and is determined by the azimuthal quantum number l. The magnitude of this angular momentum depends on the value of l, although instead of being an integral multiple of $h/2\pi$ as in the Bohr model, it equals $\sqrt{l(l+1)}\,h/2\pi$. Furthermore, angular momentum is a vector quantity. In a magnetic field or an electric field, the component of the angular momentum vector parallel to the direction of the field may assume only certain discrete values, $2l + 1$ in number. It is no coincidence that $2l + 1$ is the degeneracy of the energy levels corresponding to a given value of l (for any value of the principal quantum number n), as discussed in Sec. 9-4. Application of an electric or magnetic field not only causes the angular momentum vector to be oriented in specific ways but also removes the degeneracy of the energy levels ϵ_{nl}, causing each to split into $2l + 1$ separate levels, each of which is associated with a different orientation of the angular momentum.

When the external field is removed, the $2l + 1$ different energy levels coalesce again into the single degenerate level ϵ_{nl}. It then becomes meaningless to discuss the orientation of the angular momentum, because there is no longer a natural axis of reference—all directions in space have become equivalent; that is, space has become isotropic. The quantization of orbital angular momentum and consequent splitting of degenerate energy levels in the presence of an applied field is not of importance for most of our later discussions of atomic structure and chemical bonding. Nevertheless, these phenomena do play key roles in many areas of chemistry and physics, especially those involving the bonding of transition metals and the magnetic properties of matter.

The electron also has a quantized intrinsic angular momentum called *spin*. In a simple picture, electron spin is associated with rotation of the electron. A rotating charge behaves like a magnet and is affected by external magnetic fields. The spin can have two, and only two, possible orientations relative to the direction of a magnetic field. These two directions are commonly described as *parallel* and *antiparallel* to the external field. Although this description is not quite accurate, it is unambiguous and therefore useful. Two spin states are available to an electron even in the absence of an external field, and it is convenient to describe these two states by terms such as "spin up" and "spin down". These two states of "opposite spins" have the same energy in the absence of an external field and somewhat different energies in the presence of one.

9-6   The Pauli Exclusion Principle

It is possible to understand many features of the periodic system of the elements by considering two factors: (1) the positions and degeneracies of the atomic energy levels and (2) the *Pauli exclusion principle*. A common statement of Pauli's principle for atomic electrons is that at most two electrons may occupy the same orbital and that these two electrons must have opposite spins. Such electrons are said to be paired. An electron that occupies an orbital singly is said to be unpaired, since there is no second electron of antiparallel spin in the same orbital. The exclusion principle is of great importance not only to the understanding of the periodic table, as we shall see in the next chapter, but also to a comprehension of the behavior of matter in general—it plays a fundamental role, for example, in the behavior of electrons in metals, in semiconductors, and in plasmas.

9-7   Wave Mechanics and the Uncertainty Principle

We return to one of the more general questions of quantum theory—the uncertainty relation

$$\delta x \, \delta p_x \gtrsim h \qquad (9\text{-}12)$$

(where p_x again has the meaning of the x component of the momentum).

From the point of view of wave mechanics the uncertainty relation can be understood as follows. If a particle moving in the x direction has the exact momentum p_x, it is represented by a wave of the exact wavelength $\lambda = h/p_x$. However, by its very nature such a wave extends uniformly from minus infinity to plus infinity. The probability of finding the particle is thus the same everywhere; that is, there can be no information available as to the whereabouts of the particle. On the other hand, the description of a particle that is known to be in a certain region of space requires a wave function that is nonzero *only in this region*. It is a mathematical fact that such a highly localized wave can be constructed only by the superposition of many waves of *different wavelengths,* waves that represent a whole spread of λ's. But different λ's mean different values of p_x, so that an uncertainty δp_x exists. This is how δx and δp_x are inextricably interwoven.

One might ask whether the uncertainty relations are an attribute of the act of measuring or whether they represent a fundamental property of matter and radiation. If particles exist, is it that they do not *possess* precise location and precise momentum at the same time, or is it just impossible *to know* their values? These are questions that have no answer as yet. There are some scientists, for example, who believe that particles such as electrons exist, that their behavior is described by the solutions of the Schrödinger equation, and that any description other than one on a probability basis is impossible. Others think that the solutions of the Schrödinger equations describe not the behavior of particles themselves but only *the state of knowledge* that is accessible to physical measurements. There are other possible interpretations of the equations of wave mechanics. A choice among these interpretations cannot be made on any rational basis at this time. It belongs to the realm of philosophy or metaphysics and is a matter of personal preference. At least part of the problem appears to stem from the fact that both the particle concept and the wave concept are of macroscopic origin. It should not be surprising that difficulties arise when these concepts are applied in the atomic domain.

One illuminating application of the uncertainty principle is in providing an estimate of the minimum size of atoms consistent with the known binding energies of electrons. The diameter of the region within which an electron moves in a hydrogen atom is of the order of a few angstroms. Why can't it be much smaller? Suppose that it were 0.01 Å. This would imply that the uncertainty in the position of the electron could be no greater than 0.01 Å   or 10^{-12} m, and hence, by (9-12), the uncertainty in its momentum could be no less than about 7×10^{-34} J s$/10^{-12}$ m $\approx 7 \times 10^{-22}$ kg m s^{-1}. Thus the momentum itself might be this large, which means that the kinetic energy of the electron might be as large as

$$\epsilon_k = \frac{p^2}{2m} = \frac{49 \times 10^{-44}}{18 \times 10^{-31}} \text{ J} \approx 3 \times 10^{-13} \text{ J}$$

which is about 10^6 eV. However, the binding energy (the potential energy) of an electron in a hydrogen atom is only 13 eV; hence, an electron with a kinetic energy of 10^6 eV could not be contained in a hydrogen atom. Noting that the kinetic energy varies as the square of the momentum and that the uncertainty in the momentum varies inversely with the uncertainty in the position, we see that if we *increase* the size of the region within which the electron is confined by a factor of 300, to about 3 Å, the uncertainty in the momentum is *decreased* by a factor of 300 and the maximum kinetic energy is then *decreased* by 10^5 to the order of magnitude of 10 eV. This energy is consistent with the known binding energy. Thus, the electron cannot be confined in a region of dimensions much smaller than 3 Å unless the binding energy is appreciably greater than a few electronvolts. This general inverse relation between the kinetic energy of a particle and the size of the region within which it is confined in a system of atomic dimensions is of wide applicability.

Problems and Questions

9-1   **The Bohr atom.**   Discuss the difficulties with the Rutherford nuclear model of the atom according to classical physics, and state the postulates that Bohr made in deriving the formula for the spectrum of a hydrogen atom from this model.

9-2   **Spectra of atoms and ions with only one electron.**   (*a*) How are the frequencies ν, wave numbers $\tilde{\nu}$, and wavelengths λ of the spectra of the species He$^+$, Li^{2+}, and Be^{3+} related approximately to those in the spectrum of the hydrogen atom? (*b*) Calculate ν and λ for the spectral lines of H and He$^+$ when $n_1 = 1$ and $n_2 = 2, 3, 4$ and when $n_1 = 2$ and $n_2 = 3, 4, 5$. In which range of the spectrum do these lines lie? Give approximate colors if they are in the visible range. [Observed wavelengths for H are, in nanometers: 121.6 (1-2), 102.6 (1-3), 97.2 (1-4), 656 (2-3), 486 (2-4), 434 (2-5).

9-3   **Ionization energies of one-electron ions.**   Calculate the energy required to remove (to infinity) the electron in the lowest possible state (the ground state) of the ions He$^+$, Li^{2+}, and Be^{3+}.   *Ans.*   54.4 eV, 122.4 eV, 217.6 eV

9-4 **Vibrations of a drumhead.** Describe the types of nodes in the possible vibrations of a circular diaphragm that is clamped along its rim. Describe their analogies with nodes in atomic orbitals.

9-5 **Consequences of the uncertainty principle.** Indicate the implications of the uncertainty principle with regard to the sharply defined orbits of the Bohr atom.

9-6 **Energy states.** The yellow-orange light of sodium lamps consists primarily of two wavelengths: 589.59 and 589.00 nm. What is the energy of each excited state involved relative to the ground state of Na? What is their energy difference?

9-7 **Dissociation energy of a molecule.** Photochemical dissociation of a molecule can occur if the molecule absorbs a photon that has sufficient energy to dissociate that molecule. The dissociation of a *mole* of iodine requires about 150 kJ. Would a photon of yellow light of wavelength 600 nm have sufficient energy to dissociate a *molecule* of iodine? What wavelength would be just sufficient?   *Ans.*   yes; 798 nm

9-8 **Relationship between wavelengths.** Suppose an atom in an excited state can return to the ground state in two steps. It first falls to an intermediate state, emitting radiation of wavelength λ_1, and then to the ground state, emitting radiation of wavelength λ_2. The same atom can also return to the ground state in one step, with the emission of radiation of wavelength λ. How are λ_1, λ_2, and λ related? How are the frequencies of the three radiations related?

9-9 **The Franck-Hertz experiment.** When atoms in a gas are bombarded with electrons of known energy, they absorb no energy unless the electron energy is above a minimum value. As the atoms absorb energy from the electrons, they begin to emit photons. Explain. If the experiment is performed on mercury vapor, the threshold for absorption is 4.90 eV. What wavelength do you expect for the photons emitted by Hg atoms bombarded with electrons of this energy?   *Ans.*   253 nm (observed wavelength is 253.7 nm)

9-10 *p* **states and** *s* **states.** (a) State clearly what is implied by *p orbital* in terms of quantum numbers and "shape". (b) The energy difference between the most stable state of the Li atom and the lowest excited state, in which the outer electron is in a 2*p* orbital, is 1.85 eV, or 2.96×10^{-19} J. However, the energy difference between the 2*s* and 2*p* levels in the ion Li^{2+} is less than 0.0002 of this value. Explain.

9-11 **Radial portion of atomic orbitals.** Study Fig. 9-6 and deduce what graphs of $R(r)$ for $n = 4$, $l = 0$, 1, 2, 3 must look like. Sketch these graphs.

9-12 **Atomic orbitals.** Consider an *s* orbital and a *p* orbital of the same principal quantum number. What can you say about: (a) the principal quantum number; (b) the angular momentum quantum number for each orbital; (c) the angular variation of the electron distribution for the *s* orbital and (one of) the *p* orbitals; (d) the relative energy of an electron in the *s* orbital and in (one of) the *p* orbitals? Assume the atom is polyelectronic, and discuss the reason for any difference in energy in terms of the shapes of the orbitals.

9-13 **Radial nodes.** Find the positions of the radial nodes in the 2*s*, 3*s*, and 3*p* orbitals for the hydrogen atom by finding the values of *r* for which $R(r)$ (Table 9-1) goes to zero. Check your results by comparison with Fig. 9-6, noting that $r_B = 0.53$ Å.
Ans.   2*s*, $r = 1.06$ Å; 3*s*, $r = 1.01$ Å, 3.76 Å; 3*p*, $r = 3.18$ Å

9-14 Angular nodes. (*a*) Consider the definition of the angles θ and φ (Fig. 9-5) and the angular dependence of a p_y orbital wave function (Table 9-1). Show that this orbital has a nodal plane (i.e., the wave function goes to zero) at $y = 0$. (*b*) Similarly, show that a d_{zx} orbital has nodal planes at $z = 0$ and $x = 0$. Show also that a $d_{x^2-y^2}$ orbital has nodal planes at $x = y$ and $x = -y$, that is, at $\varphi = 45°$ and $135°$.

9-15 Comparison of energy-level spacings. For the hydrogen atom, the transition from the 2p state to the 1s state is accompanied by the emission of a photon with an energy of 16.2×10^{-19} J. For an iron atom, the same transition (2p to 1s) is accompanied by the emission of X rays of wavelength 0.193 nm. What is the energy difference between these states in iron? Comment on the reason for the variation (if any) in the 2p-1s energy-level spacing for these two atoms.

"The majority of English chemists represent the atomic weight of carbon by 6, that of oxygen by 8, and that of sulphur by 16. Dr. Frankland would double the atomic weight of carbon, but would retain the old atomic weights of oxygen and sulfur. Mr. Griffin, who lays claim to priority in doubling the atomic weights of carbon and oxygen, ridicules the notion of doubling that of sulphur. Dr. Williamson, Mr. Brodie, and myself have for a long time advocated the doubling of all three.

For silicon Thomson took 7.12, assuming the chloride to be $SiCl$; Gmelin prefers $SiCl_2$ and 14.25, and Berzelius 21.37 and $SiCl_3$; whereas in my opinion the balance of argument is in favour of $SiCl_4$ and 28.50."

W. Odling, 1859

". . . I have tried to base a system on the magnitudes of the atomic weights of the elements. My first attempt in this respect was the following: I chose the substances with the smallest atomic weights and arranged them according to the sizes of their atomic weights. This showed that there existed a periodicity in the properties of these simple substances and that even according to their atomicity[1] the elements followed one another in the arithmetical sequence of their atomic weights."

D. I. Mendeleev, 1869

10 The Periodic System

When the known elements are arranged in order of increasing atomic number (which is also, with only a few exceptions, the order of increasing atomic weight), it is found that there are regular recurrences of many different physical and chemical properties. The discovery of this *periodic law* was of enormous importance to the progress of

---

[1]The word *atomicity*, a literal translation from the Russian, was used by many chemists at that time for the concept later termed *valence*.

chemistry. It permitted the systematization of an immense body of hitherto apparently unrelated information and provided the impetus to a search for an underlying atomic structure that could explain this regularity in nature.

After some historical background, we proceed in this chapter to a discussion of the modern form of the periodic table, an interpretation of it in terms of atomic structure and energy levels, and a consideration of some typical examples of periodicity and other trends in chemical and physical properties.

10-1   Background

The idea of searching for patterns among the numerical values of atomic weights and combining weights was widespread in many chemists' minds during the early and middle years of the nineteenth century. In fact, even as early as the 1790s the German chemist J. B. Richter had prepared what seems to have been the first table of combining weights. It contained values for 8 bases and 13 acids, all referred to the combining weight of sulfuric acid as a standard. Richter spent some effort in trying to deduce a mathematical relationship among the values he found, and suggested that the weights fell in a series some members of which were missing. However, this line of reasoning did not prove fruitful at the time and indeed was generally discredited. Nonetheless, three-quarters of a century later, when much more precise and extensive values of atomic weights were available, this approach provided the key to the periodic classification of the elements.

Many of the early attempts at finding a quantitative systematization of the chemical properties of the elements were discarded, and even ridiculed, because they seemed too limited and often even contradictory to the sets of apparently accurate atomic weights then being determined. In fact, as some chemists of the time recognized, it was *equivalent* weights that were being established quite accurately, but the integral factors necessary to convert these to atomic weights were often very uncertain.[1] One of the earliest speculations, Prout's hypothesis, was that all atoms were composed of hydrogen atoms held together by some very strong force. When it became apparent that some atomic weights were not integral multiples of the weight of a hydrogen atom, Prout's idea seemed without value. Still later, however, a century after Prout's proposal, the discovery of isotopes showed that in fact his suggestion was considerably sounder than it had once appeared.

After the discrediting of Prout's hypothesis in the 1830s, the climate of opinion, which always plays a significant role in the evolution of science, was such that the search for regularities among chemical properties of the elements was viewed with skepticism by many influential chemists. They regarded atoms of different elements as totally distinct and unrelated to one another. Nonetheless, some of their contemporaries pursued the search for order, and gradually the existence of different groups of elements with similar properties—the halogens, the alkaline-earth metals, the family containing oxygen, sulfur, selenium, and tellurium, and other similar groups

---

[1] The state of confusion in the late 1850s with regard to the atomic weights of even the most common elements is indicated clearly in the quotation from Odling that opens this chapter.

—was recognized. After Cannizzaro had in 1858 clarified once and for all the distinction between atoms and molecules and had, with the help of Avogadro's law (Sec. 5-2), established approximately correct atomic weights for the majority of the elements then known, renewed attempts were made to find some pattern when the elements were arranged in order of increasing atomic weight. One of the most significant of these was the proposal in 1863 by the 25-year-old English chemist, John Newlands, of the *law of octaves*. Newlands noticed that when the elements were arranged in order of ascending atomic weight, there was a regular recurrence of properties among the elements of lower weight, every element resembling the eighth element following it, like the eighth note in an octave of music. The law broke down completely when applied to the heavier elements and was derided and ridiculed as too fantastic to be worthy of serious consideration. However, only 6 years later, Dmitri Mendeleev in Russia and Lothar Meyer in Germany independently and conclusively showed that there was indeed a periodicity in chemical and physical properties, although a more elaborate one than that suggested by Newlands.

Mendeleev's papers were especially convincing; he boldly disregarded some of the atomic weights then accepted, because they led to positions for the elements concerned that were not in accord with his periodic system, and he also left certain gaps in his table for elements still to be discovered. By considering carefully the chemical properties of related elements, he was able to predict with uncanny accuracy the chemical and physical properties of several of these "missing" elements and their compounds. Within two decades, three of these elements (scandium, gallium, and germanium) had been discovered and found to behave just as he had predicted (Table 10-1). Later, additional elements that he had predicted were discovered or were prepared artificially, including technetium, rhenium, and polonium.

TABLE 10-1 **Comparison of some of Mendeleev's predictions for ekasilicon with observed properties of germanium**

Ekasilicon (Es) (predicted in 1871)	Germanium (Ge) (discovered in 1886)
Atomic weight 72 (average of the atomic weights of the neighbors, Ga and As)	Atomic weight 72.6
Dark gray metal, mp higher than that of Sn, perhaps 800°C; density 5.5 g cm^{-3}	Gray metal, mp 958°C; density 5.36 g cm^{-3}
Heating in air will yield a white powder, EsO$_2$, of density 4.7 g cm^{-3}; lower oxides may also exist	Heating in air yields GeO$_2$, white; density 4.70 g cm^{-3}; GeO is also known
Action of acids such as HCl will be slight; Es will resist attack by alkalies such as NaOH	Neither HCl nor NaOH dissolves Ge, but concentrated HNO$_3$ does
Hydrated EsO$_2$ will be soluble in acids and easily reprecipitated	Ge(OH)$_4$ dissolves in dilute acid and reprecipitates upon dilution or addition of base
Like Sn, Es will form a volatile, liquid chloride, EsCl$_4$; bp somewhat below 100°C and density 1.9 g cm^{-3}	GeCl$_4$ is a volatile liquid; bp 83°C and density 1.88 g cm^{-3}
Like Sn, Es will form a yellow sulfide, EsS$_2$, insoluble in water but soluble in ammonium sulfide	GeS$_2$ is white, insoluble in water and dilute acids but readily soluble in ammonium sulfide

Mendeleev's periodic law was accepted almost at once because he presented the evidence in favor of it so convincingly, but he and others regarded it merely as an empirical generalization based on observation rather than arising from a knowledge of some underlying principles. Since the existence of atomic nuclei, and thus of nuclear charge and atomic number, was unsuspected, the initial basis for the ordering of the elements was their atomic weights rather than their atomic numbers. While atomic weights are indeed fundamental quantities that could even then be measured with precision, they are valuable in relation to the periodic system only because elements with larger atomic numbers *usually* also have larger atomic weights. However, for a few pairs of elements (Te and I, Co and Ni, and, after the discovery of argon in the 1890s, Ar and K), the order of atomic weights was (and still is) such that the element of higher atomic weight belongs first in the periodic table. The wisdom of considering the chemical properties of the elements as more fundamental than their atomic weights when these inconsistencies occurred should be particularly noted, for this bold step was at first regarded as unsound by some persons with less imagination and insight than Mendeleev, Meyer, and their followers. It is not uncommon in the development of science that the recognition of new connections and relationships requires at first disregarding some of what at the time seems relevant evidence. It was only after Moseley in 1913 and 1914 found a way to measure nuclear charge that this quantity, the atomic number, was recognized and accepted as the true basis for the ordering of the elements in the periodic table. For each of the pairs Te–I, Co–Ni, and Ar–K, the element listed first and placed first in the periodic table has indeed the lower atomic number despite the fact that it has the higher atomic weight. Shortly thereafter, the discovery of isotopes made it clear that atoms of a given element might vary in mass and thus that the atomic weight could not be of primary significance in determining chemical properties, which are nearly identical for all the different isotopes of a given element.

10-2   The Modern Periodic Table

Descriptive comments.   Many schemes have been proposed for effectively displaying the periodicity in chemical and physical properties. One modern form of the periodic table is shown inside the front cover, and a slightly altered version, with a few added features intended to emphasize the interrelationships of the elements that compose the different groups or families, is presented in Fig. 10-1.

One noteworthy feature of the arrangement in Fig. 10-1 is that the elements in the column headed 0 are given twice, once at the extreme left and again at the extreme right. This is done to emphasize the continuity of the system, a continuity that could be shown effectively by wrapping a similar chart around a cylinder and displacing the right edge down by one row relative to the left edge so that corresponding elements would just match.[1] Whatever the form of the periodic table, it

---

[1] In fact, such helical forms of the periodic table have often been proposed. One of the earliest suggestions for ordering the elements by atomic weight was made by de Chantcourtois more than a century ago, even before Mendeleev and Newlands, on the basis of a spiral arrangement that he had made.

0	IA	IIA	IIIA	IVA	VA	VIA	VIIA	0
				H 1		He 2		
He 2	Li 3	Be 4	B 5	C 6	N 7	O 8	F 9	Ne 10
Ne 10	Na 11	Mg 12	Al 13	Si 14	P 15	S 16	Cl 17	Ar 18

0	IA	IIA	IIIB	IVB	VB	VIB	VIIB	VIII			IB	IIB	IIIA	IVA	VA	VIA	VIIA	0
Ar 18	K 19	Ca 20	Sc 21	Ti 22	V 23	Cr 24	Mn 25	Fe 26	Co 27	Ni 28	Cu 29	Zn 30	Ga 31	Ge 32	As 33	Se 34	Br 35	Kr 36
Kr 36	Rb 37	Sr 38	Y 39	Zr 40	Nb 41	Mo 42	Tc 43	Ru 44	Rh 45	Pd 46	Ag 47	Cd 48	In 49	Sn 50	Sb 51	Te 52	I 53	Xe 54
Xe 54	Cs 55	Ba 56	La 57 *	Hf 72	Ta 73	W 74	Re 75	Os 76	Ir 77	Pt 78	Au 79	Hg 80	Tl 81	Pb 82	Bi 83	Po 84	At 85	Rn 86
Rn 86	Fr 87	Ra 88	Ac 89 †	104	105													

	Ce 58	Pr 59	Nd 60	Pm 61	Sm 62	Eu 63	Gd 64	Tb 65	Dy 66	Ho 67	Er 68	Tm 69	Yb 70	Lu 71
*Lanthanides														
†Actinides	Th 90	Pa 91	U 92	Np 93	Pu 94	Am 95	Cm 96	Bk 97	Cf 98	Es 99	Fm 100	Md 101	No 102	Lr 103

FIGURE 10-1   One form of the periodic table of the elements. The solid slanting lines connect groups of elements with similar properties and closely related outer electronic structure (congeners). The dashed slanting lines connect groups of elements whose similarity is less obvious, consisting principally (but not exclusively) in their having certain oxidation states in common. Elements in shaded fields are called metalloids.

is designed to group together elements with similar properties. No arrangement is entirely satisfactory, since the lengths of the periods (that is, the number of elements included before similar properties recur) are not constant, becoming greater as atomic number increases. However, the arrangements shown in Fig. 10-1 and inside the front cover reveal many of the important relationships.

The horizontal rows in these charts represent the *periods*. Each period ends with one of the inert gases (noble gases), the atoms of which are so stable that the gases are monatomic and form compounds only with difficulty if at all. Their atomic numbers are 2, 10, 18, 36, 54, and 86; their names are helium (He), neon (Ne), argon (Ar), krypton (Kr), xenon (Xe), and radon (Rn). The lengths of the periods beyond He thus represent the differences in the atomic numbers of these unusually stable elements and are, successively, 8, 8, 18, 18, and 32. The last period, starting after Rn, would presumably have 32 elements also if it were not incomplete; at present, the element with highest known atomic number is that with $Z = 106$, whereas the last element in that period would have $Z = 86 + 32 = 118$. If the first very short period (which consists just of hydrogen and helium) is disregarded, every period begins with an alkali metal.

The columns in the table contain elements with related properties, and several columns in the longer periods may be related to a given column in the shorter ones.

The columns in the two short eight-element periods (Li to Ne, $Z = 3$ to 10; Na to Ar, $Z = 11$ to 18) are numbered IA to VIIA, with 0 for the inert-gas column. In the 18-element periods, the column heading 0 is repeated for the inert gases and so are the headings IA to VIIA for the groups of elements most closely related to those in the short periods (connected to them by the solid slanting lines in Fig. 10-1). The elements in column 0 and in the A columns are often called the *representative elements*. In addition, there are ten more columns, seven of which are headed IB to VIIB and are connected by dashed slanting lines to the columns with the same roman numeral in the two shorter periods above. These lines are dashed to suggest that the similarity in properties is much less marked than for those connected by the solid lines; it is often at best only very weak. The three remaining columns in the 18-element periods are grouped as one triple column, designated VIII.

Following Pauling, we use the term *congeners* for those elements that form a closely related group, such as those in column 0 or in any column A in Fig. 10-1—as, for example, the alkali metals, the inert gases, or the halogens. In addition, the elements in each column B (such as Cu, Ag, and Au) or those in any one of the three columns of the triple column VIII (such as Ni, Pd, and Pt) are also termed congeners.

Inspection of the 18-element periods shows that after columns IA and IIA come the 10 columns not connected by solid lines to the short-period elements above them—IIIB to VIIB, the three columns VIII, and then IB and IIB. The elements in the first nine of these ten columns are often referred to as the *transition elements;* the elements in the last column, IIB, have been called the *posttransition elements*. The last six columns of the long periods contain congeners of the elements in the short periods.

The two series of 14 elements grouped separately at the bottom of the chart comprise portions of the two 32-element periods, the second of which is still incomplete. Neither of these groups has congeners among the lighter elements. The *lanthanides* (elements 58 to 71) follow the element lanthanum (La) and resemble it in many ways, as their name implies; these elements are also sometimes called, especially in the older literature, the *rare-earth* elements. The *actinides* (elements 90 to 103) occupy a similar position in the next period, following actinium (Ac). Somewhat surprisingly, the actinides do not closely resemble the lanthanides (Chap. 24). While most of the lanthanides are very similar to one another in chemical properties, the actinides differ from each other in significant ways, just as the transition elements do. Most of the actinides—all those beyond uranium (U)—occur on earth only as man-made elements, having been synthesized only during the last four decades. The nuclei of these elements decompose too rapidly for them to have survived since the creation of the earth, even if they were present then.[1]

---

[1] These elements almost certainly are continuously produced (and continuously decay) in many stars. It has, for example, been speculated that californium 254 is involved in the flare into brilliance of certain supernova stars whose initial huge brightness falls to half its value in about 55 days. It is known that the nucleus of ^{254}Cf breaks into two or more smaller nuclei spontaneously, delivering a large amount of energy, with a similar half-life (Secs. 20-3 and 24-4) of 61 days under normal terrestrial conditions.

General relationships. Several broad generalizations may be made about the chemical properties of elements on the basis of the region of the periodic table in which they are found. Perhaps the most useful of these concerns the classification of elements as metals,[1] nonmetals, and metalloids (elements with intermediate properties). Since about three-fourths of all known elements are metallic, it is easiest to consider the regions where elements without distinctly metallic properties occur. The typical nonmetals occupy a triangular region on the right side of the table, with the long side of the triangle extending from the vicinity of boron (B, $Z = 5$) and carbon (C, $Z = 6$) to that of astatine (At, $Z = 85$) and radon (Rn, $Z = 86$). All elements to the right of this line are nonmetallic. The elements lying along a strip one or two elements wide and adjacent to this line show some metallic properties and some properties typical of nonmetals. They sometimes exist in several different crystalline forms, some of which may exhibit more metallic properties than others. These are the *metalloids;* they include boron (B), silicon (Si), germanium (Ge), arsenic (As), antimony (Sb), tellurium (Te), and polonium (Po), which lie in fields indicated by shading in Fig. 10-1. All the other elements, lying to the left of this strip, are metals (except of course for the duplicated inert-gas column on the extreme left in Fig. 10-1).

Metallic character increases not only as one goes to the left in the usual form of the periodic table but also as one proceeds downward within any group. Thus the most metallic elements are cesium (Cs) and francium (Fr), which are in the lower left corner of the table; the least metallic (other than the inert gases) is fluorine (F), in the upper right corner. Since one prominent characteristic of metals is their tendency to form positive ions by losing electrons, it is not surprising that Cs and Fr lose electrons most easily, nor that, except for the inert gases He and Ne, it is most difficult to remove an electron from F. The *ionization energy* for neutral atoms—that is, the energy needed to remove the first electron from an atom X

$$X(g) \longrightarrow X^+(g) + e^- \tag{10-1}$$

(the *first ionization energy*)—shows a very marked periodicity. Figure 10-2, a plot of the first ionization energy against atomic number, illustrates the regular recurrence of certain trends. There are similar general correlations of many other properties with position in the periodic table.

A brief survey of congeners. The *inert gases* (group 0: He, Ne, Ar, Kr, Xe, and Rn) are almost completely inert chemically. Only in the last decade have attempts to make inert-gas compounds been successful, and then only for the inert gases of higher atomic number. The forces between inert-gas atoms are very small, because of their low polarizability (Sec. 3-2), and are exceptionally low for He. These gases are therefore difficult to condense, and their behavior at ordinary pressures and temperatures is closely described by the ideal-gas equation.

The *alkali metals* are all soft, low-melting metals of low density, with a strong tendency to form singly charged positive ions M^+. They react rapidly with atmos-

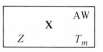

	X	AW
Z		T_m

X Chemical symbol

Z Atomic number

AW Atomic weight (except in parenthesis, when mass number of most stable or best known isotope is given)

T_m Melting temperature

		4.0
	He	
2		2 K*
		20.2
	Ne	
10		25 K
		39.9
	Ar	
18		84 K
		83.8
	Kr	
36		116 K
		131.3
	Xe	
54		161 K
		(222)
	Rn	
86		202 K

[1]Characteristic properties of metals are described in Sec. 4-4 and Chap. 23.

*A minimum pressure of about 25 atm is required.

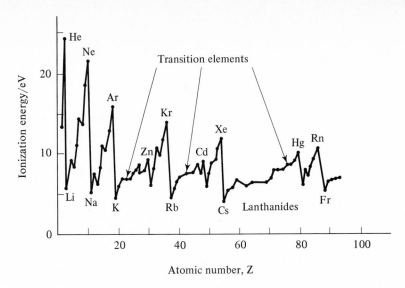

FIGURE 10-2   **The first ionization energy of the elements as a function of the atomic number.**
The ionization energy in a given period is lowest for the alkali metals and increases in a general way
as Z is increased. Deviations from this rule can be understood by considering the details of the electron
configurations of the atoms involved, as explained in Sec. 10-4. In a given group of congeners such
as the inert gases or the alkali metals, the ionization energy decreases slowly as Z increases.

pheric oxygen, the reactivity increasing from Li to Cs. The reaction of the alkali
metals with water is violent and the resulting alkali hydroxides are strong bases.

    The *alkaline-earth metals* are harder and denser than the alkali metals. They easily
form dipositive ions M^{2+}, but their reactivity is notably smaller than that of the
alkali metals. Their hydroxides are strong bases and are only slightly soluble in water,
with the exception of the moderately soluble $Ba(OH)_2$ and $Ra(OH)_2$.

    The *congeners of boron* have the tendency to form ions M^{3+}, and although the
B^{3+} ion is not stable, many boron compounds corresponding to oxidation state
(Appendix C) +III are known. The oxides of the elements in this group show acidic
as well as basic character—that is, they are amphoteric, as discussed in Sec. 7-4.

    In the *carbon group* there is a tendency to assume the oxidation state +IV. It
is also apparent that carbon is the least, and lead the most, metallic element in the
group. This exemplifies the trend toward increasing metallic character with higher
atomic number within a column in the system. In groups I to III the same trend
expresses itself as increasing reactivity when going down a column, because all
elements in these groups, with the exception of the metalloid boron, are metals. In
group IVA carbon is a nonmetal and shows only slight metallic characteristics in
the form of graphite. Silicon and germanium are metalloids, and tin shows nonmetallic properties in one of its different forms or *allotropes* (Greek: *allo*, other;
trope, way, manner)—gray tin. The oxides of the lighter elements in this column
are weakly acidic; those of the heavier are amphoteric.

IA		
		6.9
	Li	
3		454 K
		23.0
	Na	
11		371 K
		39.1
	K	
19		336 K
		85.5
	Rb	
37		312 K
		132.9
	Cs	
55		302 K
		(223)
	Fr	
87		—

IIA		
	Be	9.0
4		1560 K
	Mg	24.3
12		922 K
	Ca	40.1
20		1112 K
	Sr	87.6
38		1043 K
	Ba	137.3
56		1002 K
	Ra	(226)
88		973 K

IIIA		
	B	10.8
5		2340 K
	Al	27.0
13		933 K
	Ga	69.7
31		303 K
	In	114.8
49		430 K
	Tl	204.4
81		577 K

IVA		
	C	12.0
6		4100 K*
	Si	28.1
14		1685 K
	Ge	72.6
32		1210 K
	Sn	118.7
50		505 K
	Pb	207.2
82		600 K

VA		
	N	14.0
7		63 K
	P	31.0
15		317 K
	As	74.9
33		885 K*
	Sb	121.8
51		904 K
	Bi	209.0
83		545 K

VIA		
	O	16.0
8		54 K
	S	32.1
16		388 K
	Se	79.0
34		494 K
	Te	127.6
52		723 K
	Po	(210)
84		527 K

VIIA		
	F	19.0
9		54 K
	Cl	35.5
17		172 K
	Br	79.9
35		266 K
	I	126.9
53		387 K
	At	(210)
85		—

IB		
	Cu	63.5
29		1356 K
	Ag	107.9
47		1234 K
	Au	197.0
79		1336 K

IIB		
	Zn	65.4
30		693 K
	Cd	112.4
48		594 K
	Hg	200.6
80		234 K

IIIB		
	Sc	45.0
21		1812 K
	Y	88.9
39		1799 K
	La	138.9
57		1193 K
	Ac	(227)
89		1323 K

IVB		
	Ti	47.9
22		1943 K
	Zr	91.2
40		2125 K
	Ce	140.1
58		1071 K
	Th	232.0
90		2028 K

The *nitrogen group* contains only one element with metallic properties: bismuth, the heaviest of the group. Nitrogen and phosphorus are nonmetals; arsenic and antimony are metalloids. Their oxidation states range from +V to −III. The oxides of nitrogen, phosphorus, and arsenic form acids of decreasing strength, while the oxides of antimony and bismuth are amphoteric.

The *oxygen group* contains three nonmetals, O, S, and Se, and the two metalloids Te and Po. The oxidation states range from +VI to −II. Sulfur and selenium form acidic oxides, while the oxides of tellurium are amphoteric. This group of elements is occasionally referred to as the *chalcogens*.

The *halogens* are all strongly nonmetallic. All of them are very reactive, with fluorine taking the lead. Their hydrides are all acids; HCl, HBr, and HI are strong acids, while HF is of intermediate strength. The oxides of Cl, Br, and I are all acidic.

The *transition metals* form the B columns or B groups (excepting the posttransition

*Sublimation temperature at 1 atm. mp = 1090 K at 28 atm.

metals in column IIB, Zn, Cd, and Hg, but including the three columns VIII, the lanthanides and the actinides). The relationships between corresponding A and B groups are rather tenuous. For example, the elements in column IB (Cu, Ag, and Au) resemble the alkali metals (column IA) insofar as they tend to form compounds in the $+I$ oxidation state, such as CuI, AgBr, and AuCl, but these elements are far less reactive than the alkali metals. Furthermore, Cu forms many stable compounds containing Cu^{2+} (for example, $CuSO_4$) and Au forms compounds containing Au^{3+} (for example, $AuCl_3$). The relationships of the other pairs of corresponding A and B groups are similarly weak. The ions of the IIB metals Zn, Cd, and Hg are doubly positive, but many compounds of mercury exist in which mercury has the oxidation state $+I$, and there are other exceptions. The IIIB metals Sc, Y, La, and Ac form chiefly $+3$ ions, as do all rare-earth metals or lanthanides. In group IVB (Ti, Zr, Hf, and Th) the $+IV$ state predominates, and this group is perhaps more closely related to the elements in the corresponding A group than is the case for any other A and B groups. In the succeeding groups VB, VIB, VIIB, and VIII, the situation becomes more complicated, although the highest oxidation state is $+V$ in group VB, $+VI$ in VIB, and $+VII$ in VIIB. It is characteristic of transition elements that they are all metallic, that they can exist in many oxidation states, and that many of them form colored salts.

Diagonal relationships.   The interesting "diagonal" relationship expressed by the locations of the metalloids in the periodic table (Fig. 10-1) also exists among some elements of the first and second short periods. Thus Li is in some respects similar to Mg (e.g., both form carbonates, phosphates, and fluorides that are sparingly soluble in water, and both form nitrides with ease). Be shows similarity to Al (e.g., both show a strong tendency to form covalent compounds, and both form amphoteric oxides and volatile halides) and there is some relationship between B and Si (both form acidic oxides and volatile hydrides and halides). However, no pronounced relationship exists between C and P, between N and S, or between O and Cl.

10-3   Atomic Structure and the Periodicity of Chemical Properties

The Aufbau principle.   The key to an understanding of the periodic system in terms of atomic structure was furnished by the Pauli exclusion principle (Sec. 9-6), in combination with the system of atomic energy levels discussed in Chap. 9 and portrayed schematically in Fig. 9-4 and again in Fig. 10-3. Pauli and Bohr were the first to recognize this fact, and A. Sommerfeld pioneered in establishing the details. The fundamental ideas are these:

1.   The system of relative energy levels, reproduced schematically in Fig. 10-3,

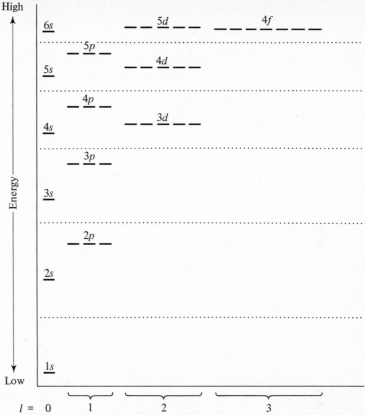

FIGURE 10-3   Schematic diagram of the relative energies of atomic orbitals. The diagram shows *relative* energies only, with an energy scale that is not meant to be uniform. The relative heights and even the sequence of levels that are very close together (such as 3d and 4s or 6s, 5d, and 4f) may shift in different atoms. Particularly important are the energy gaps (indicated by dotted lines) that separate certain sets of levels. These gaps are very pronounced at low energies, less so at high energies.

applies to all atoms. The *relative* energies of the orbitals remain approximately the same from element to element, with only rare interchanges of relative heights.

2. By the Pauli principle, each orbital may hold at most two electrons, which must have opposite spin.

3. The orbitals are filled in order with the electrons available, starting with the orbital of lowest energy and assigning at most two electrons to any orbital. This procedure is called, after Sommerfeld, the *Aufbau principle* (German: *Aufbau,* building up).

The first ten elements. Starting with hydrogen ($Z = 1$) we see that the one electron occupies the lowest available orbital, the 1s orbital. In helium ($Z = 2$) both electrons are able to occupy the 1s orbital provided that their spins are opposed. It is customary to describe the electron configuration of these two atoms by the following notation:

$$\text{H} \quad 1s$$

$$\text{He} \quad 1s^2 \quad \text{(read as ``one s squared'' or as ``one s two'')}$$

The superscript indicates the number of electrons that occupy the orbital in question when that number is larger than 1.

The next element, lithium, has three electrons. By the Pauli principle the $1s$ orbital is now filled and the third electron must be placed in the $2s$ orbital. The fourth element, beryllium, has four electrons, which fill the $1s$ and $2s$ orbitals in pairs with antiparallel spins. The configurations of Li and Be are thus $1s^2\, 2s$ and $1s^2\, 2s^2$, respectively. In all subsequent elements the $1s$ and $2s$ orbitals are filled by two pairs of electrons.

The next orbitals are the three $2p$ orbitals, all of which have the same energy in an isolated atom. These orbitals can hold a total of six electrons. Together with the $1s$ and $2s$ orbitals, they are thus able to accommodate the electrons of the next six elements, B, C, N, O, F, and Ne. A total of four electrons always occupy the $1s$ and $2s$ orbitals. In boron the additional electron is in one of the $2p$ orbitals, leading to the configuration $1s^2 2s^2 2p$. An interesting phenomenon occurs in the next two elements: the extra electrons go into separate $2p$ orbitals, leading to the configuration $1s^2 2s^2 2p2p$ for carbon and $1s^2 2s^2 2p2p2p$ for nitrogen. The spins of the electrons in the separate $2p$ orbitals are all parallel, which is permitted by the exclusion principle and leads to two and three parallel spins for C and N, respectively (Table 10-2).

This pattern of filling the degenerate $2p$ energy levels depicted in Table 10-2 has been established experimentally by interpretation of atomic spectra. It follows a generalization known as *Hund's rule:* the order of filling degenerate orbitals in an atom (or molecule) is such that as many electrons remain unpaired as possible. The nitrogen atom, having three p electrons, has the maximum number of unpaired electrons in this period. For the next three elements there is a pairing of the electrons in the $2p$ orbitals. In oxygen, the configuration is $1s^2 2s^2 2p^2 2p2p$, so that two electrons with parallel spin remain. The electron configuration of fluorine is $1s^2 2s^2 2p^2 2p^2 2p$ and that of neon is $1s^2 2s^2 2p^2 2p^2 2p^2$. The symbols involving the $2p$ orbitals, or other sets of degenerate orbitals, are usually abbreviated by ignoring any differences between the different $2p$'s and writing, for example, $1s^2 2s^2 2p^3$ for N and $1s^2 2s^2 2p^5$ for F, as in Table 10-2.

TABLE 10-2   **Electron configuration for the elements B to Ne**

Atomic number	Element	Electron configuration	Occupancy of the 2p orbitals
5	B	$1s^2 2s^2 2p$	↑ — —
6	C	$1s^2 2s^2 2p^2$	↑ ↑ —
7	N	$1s^2 2s^2 2p^3$	↑ ↑ ↑
8	O	$1s^2 2s^2 2p^4$	↑↓ ↑ ↑
9	F	$1s^2 2s^2 2p^5$	↑↓ ↑↓ ↑
10	Ne	$1s^2 2s^2 2p^6$	↑↓ ↑↓ ↑↓

Some definitions. Before we consider the normal electronic structures of atoms of the remaining elements, the implications of a few terms that are used repeatedly and sometimes in slightly varying contexts must be examined.

The word *orbital* is, as discussed in Chap. 9, used to refer to a particular solution of the Schrödinger equation for the electronic structure of an atom (or more generally a molecule). However, at times the word *orbital* is also used to allude to the *region of space* in which the corresponding value of ψ^2 has its largest values—that is, the region of space in which an electron occupying that orbital is likely to spend most of its time. This geometric implication of the word is reinforced in many people's minds by "orbital shapes" such as those shown in Chap. 9. Finally, the term *orbital* occasionally has energetic, as well as spatial, implications.

The word *subshell* is used to refer to a group of orbitals with a given n and l. Thus, for example, the three $2p$ orbitals are said to form a subshell, as are the three $4p$ orbitals, the seven $4f$ orbitals, and so forth.

Finally, the word *shell* is used in several different ways, of which two are most common: (1) the first usage is to refer to a group of orbitals with a given value of n (and varying values of l). Thus, all the orbitals with $n = 3$ are in this sense said to constitute one shell, and so forth. (2) The second usage is less well defined. It designates groups of subshells of comparable energy as constituting a shell, regardless of whether they all have the same n or not. The orbitals lying between the dotted horizontal lines in Fig. 10-3 constitute shells in this sense, a sense that is usually more useful for chemists because the closeness of the energy levels involved markedly influences the relation of the chemical properties of the elements concerned.

Since the word *shell* normally has specific geometric implications, implying a spherical or ellipsoidal distribution that is thin relative to its radius, it is important to realize that its use in discussing atomic structure does *not* usually have this implication. The term *outer shell,* although it seems to imply something about shape and position, really refers to relative energy. An outer-shell electron (or more simply, an outer electron) is one that is relatively easy to remove from an atom because its ionization energy is comparatively low, whereas an inner-shell electron is much harder to remove.[1] There is *some* implication of position as well, since the inner-shell electrons are harder to remove just because they are on the average much closer to the nucleus than the outer ones, but as we saw in Chap. 9, the shapes and relative penetrations of different orbitals vary considerably.

The remaining elements. The electron configurations of the elements beyond neon are summarized in Table 10-3, which also includes the first 10 elements for

---

[1] The terms *outer electron, inner electron, d electron,* and the like must not be construed to imply that any such electron has a distinct identity, other than that of being associated with a certain quantum state. Two electrons cannot be distinguished unless they are confined to completely separate regions of space. Electrons in atoms are not spatially separated and are completely indistinguishable. Nonetheless, if an electron is removed from an atom and a certain orbital is then unoccupied, it is common to say that *the* electron that occupied this orbital has been removed, even though the occupancy of that orbital cannot be attributed to any particular electron before the electron is removed from the atom.

TABLE 10-3
The electron configurations of the atoms of the elements

Element	Atomic number	Populations of subshells											Element
		1s	2s	2p	3s	3p	3d	4s	4p	4d	5s	4f	
H	1	1											H
He	2	2											He
Li	3	He core	1										Li
Be	4		2										Be
B	5		2	1									B
C	6	He core	2	2									C
N	7		2	3									N
O	8		2	4									O
F	9		2	5									F
Ne	10	2	2	6									Ne
Na	11				1								Na
Mg	12				2								Mg
Al	13				2	1							Al
Si	14	Ne core			2	2							Si
P	15				2	3							P
S	16				2	4							S
Cl	17				2	5							Cl
Ar	18	2	2	6	2	6							Ar
K	19							1					K
Ca	20							2					Ca
Sc	21						1	2					Sc
Ti	22						2	2					Ti
V	23						3	2					V
Cr	24						5	1					Cr
Mn	25						5	2					Mn
Fe	26						6	2					Fe
Co	27	Ar core					7	2					Co
Ni	28						8	2					Ni
Cu	29						10	1					Cu
Zn	30						10	2					Zn
Ga	31						10	2	1				Ga
Ge	32						10	2	2				Ge
As	33						10	2	3				As
Se	34						10	2	4				Se
Br	35						10	2	5				Br
Kr	36	2	2	6	2	6	10	2	6				Kr

TABLE 10-3 (Continued)

Element	Atomic number	Populations of subshells											Element
		4d	5s	5p	4f	5d	6s	6p	5f	6d	7s	
Rb	37		1									Rb
Sr	38		2									Sr
Y	39	1	2									Y
Zr	40	2	2									Zr
Nb	41	4	1									Nb
Mo	42	5	1									Mo
Tc	43	6	1									Tc
Ru	44	7	1									Ru
Rh	45	8	1									Rh
Pd	46	10										Pd
Ag	47	10	1									Ag
Cd	48	10	2									Cd
In	49	10	2	1								In
Sn	50	10	2	2								Sn
Sb	51	10	2	3								Sb
Te	52	10	2	4								Te
I	53	10	2	5								I
Xe	54	{Kr}* 10	2	6									Xe
Cs	55						1					Cs
Ba	56						2					Ba
La	57					1	2					La
Ce	58				2		2					Ce
Pr	59				3		2					Pr
Nd	60				4		2					Nd
Pm	61				5		2					Pm
Sm	62				6		2					Sm
Eu	63				7		2					Eu
Gd	64				7	1	2					Gd
Tb	65				9		2					Tb
Dy	66				10		2					Dy
Ho	67				11		2					Ho
Er	68				12		2					Er
Tm	69				13		2					Tm
Yb	70				14		2					Yb
Lu	71				14	1	2					Lu
Hf	72				14	2	2					Hf
Ta	73				14	3	2					Ta
W	74				14	4	2					W
Re	75				14	5	2					Re

Kr core (elements 37–53)

Xe core (elements 55–75)

TABLE 10-3 (Continued)

Element	Atomic number	Populations of subshells										Element
		4d	5s	5p	4f	5d	6s	6p	5f	6d	7s	
Os	76				14	6	2					Os
Ir	77				14	9						Ir
Pt	78				14	9	1					Pt
Au	79				14	10	1					Au
Hg	80		Xe		14	10	2					Hg
Tl	81		core		14	10	2	1				Tl
Pb	82				14	10	2	2				Pb
Bi	83				14	10	2	3				Bi
Po	84				14	10	2	4				Po
At	85				14	10	2	5				At
Rn	86	{Kr}* 10	2	6	14	10	2	6				Rn
Fr	87										1	Fr
Ra	88										2	Ra
Ac	89									1	2	Ac
Th	90									2	2	Th
Pa	91								2	1	2	Pa
U	92								3	1	2	U
Np	93				Rn				5		2	Np
Pu	94				core				6		2	Pu
Am	95								7		2	Am
Cm	96								7	1	2	Cm
Bk	97								9		2	Bk
Cf	98								10		2	Cf
Es	99								11		2	Es
Fm	100								12		2	Fm
Md	101								13		2	Md

* {Kr} symbolizes the configuration of the krypton atom, $1s^2 2s^2 2p^6 3s^2 3p^6 3d^{10} 4s^2 4p^6$, which is also termed the *krypton core* in atoms of higher atomic number.

completeness. The configurations of the inert-gas atoms are particularly important because of the unusual stability of these elements (manifested by their high ionization energies, representing the highest successive peaks of the curve in Fig. 10-2). These successive inert-gas configurations correspond to filling all the levels up to the successive horizontal dotted lines in Fig. 10-3. The periodicity of chemical properties reflects the fact that similar outer-electron arrangements recur at regular intervals, with successively larger values of the principal quantum numbers of the orbitals involved; inspection of Table 10-3 reveals these regular recurrences. For convenience the inner electrons are represented by reference to the electronic arrangement of the corresponding inert gas, designated by the terms "He core", "Ar core", and the like or by the symbols {He}, {Ar}, and so forth.

Table 10-3 shows that the characteristic outermost electron configuration of the inert-gas atoms (excepting that for He, which is $1s^2$) is $ns^2 np^6$, with $n = 2, 3, 4, 5,$

and 6. The element immediately following each of the inert gases in the periodic table is an alkali metal, which easily loses one electron and forms a positive ion of unit charge. This is readily understood because, as implied by the dotted lines in Fig. 10-3, in each of the alkali-metal elements this outermost electron is in a substantially higher energy level than all the others. For similar reasons, atoms of the elements that are two places beyond each inert gas—the alkaline-earth metals, Be, Mg, Ca, Sr, Ba, and Ra—are prone to lose their two outermost electrons to form ions. All these ions—M^+ for the alkali metals and M^{2+} for the alkaline-earth metals —have the electron configuration of the inert-gas atoms just before them in the table. It is very hard to remove more electrons from these ions, not only because of the stability of the inert-gas configurations but also because of the coulomb attraction of the positive ion for the negatively charged electrons. Ions such as Na^{2+} and Ca^{3+} are formed only under extreme conditions, in high-energy electric discharges, stellar atmospheres, and the like.

Atoms that have three electrons in excess of an inert-gas configuration may form +3 ions, but removal of electrons becomes harder as the charge on the resulting ion increases. This effect becomes less important as the quantum number n increases, because the increased size of atoms and ions (and hence the larger electron-nucleus separation) makes removal of electrons easier. Thus Al^{3+} is common, but not B^{3+}.

Atoms of elements that *lack* just one electron of a filled shell may acquire an electron (if available) so readily that energy is liberated. This attests again to the unusual stability of the inert-gas configurations. Thus, gaseous hydrogen atoms and halogen atoms all liberate energy when combining with electrons to form −1 ions, each of which has an inert-gas configuration; this energy is a measure of the *electron affinity* of these atoms, a property examined again in Sec. 10-4.

Study of Table 10-3 reveals that the order of filling the levels in successive atoms follows a quite regular pattern, with only occasional fluctuations. We shall discuss the reasons for these exceptions shortly. The approximate order in which the levels are filled can easily be remembered with a mnemonic scheme such as that depicted in Table 10-4.

TABLE 10-4   **Mnemonic diagram of approximate order of filling orbitals.**   The atomic orbitals are arranged in rows s, p, d, f, beginning with the s orbitals at the bottom and starting each higher row one place further to the right than the preceding row, so that orbitals of the same principal quantum number are above each other. The approximate sequence in which orbitals are filled is indicated by the slanted arrows, proceeding from left to right. Minor deviations from this sequence are discussed in the text.

At first glance the periodic table may seem formidable, because the systematization provided by considering it in terms of the electronic structures of the atoms is not apparent. However, when one is armed with a knowledge of the pattern of orbital energy levels and the principles governing the way in which these orbitals become filled with electrons, the systematic interrelationships among the 106 elements begin to emerge. We urge you to examine the general features of Table 10-3 and relate them to the periodic chart depicted in Fig. 10-1 and the schematic energy-level diagram of Fig. 10-3. Among the features to note are:

1.   The A columns, and column 0, correspond to filling s and p subshells.

2.   The B columns and column VIII, 10 columns in all, correspond to filling d subshells.

3.   The lathanides, or rare-earths, and the actinides correspond to filling f subshells, which explains why there are 14 of each.

4.   The outer-electron configuration (ignoring the principal quantum number) is the same within each of the A columns and column 0; it is nearly the same within each of the columns containing a transition element,[1] as well as in column IIB. (There are a few variations among congeners in the transition elements; for example, Ni, Pd, and Pt are congeners and have many similarities in properties, but their most stable outer-electron configurations are, respectively, $3d^8 4s^2$, $4d^{10}$, and $5d^9 6s$. These variations are not easy to explain in simple terms and reflect the fact that any d level is almost identical in energy with the next higher s level so that very subtle effects can alter relative electron populations.)

Interpretation of some details.   Why should there be unusual stability for the inert-gas configurations? How can one rationalize the fluctuations of ionization energy within any period, illustrated in Fig. 10-2? How can one explain the general trend of ionization energy within any group of congeners? What is the significance of the twofold change in subshell population in going from V ($Z = 23$) to Cr ($Z = 24$), $d^3 s^2$ to $d^5 s$, or from Ni ($Z = 28$) to Cu ($Z = 29$), $d^8 s^2$ to $d^{10} s$? Some of the effects can be understood at least qualitatively. To do so, we must bear in mind many of the principles and results discussed in earlier chapters, most notably the coulomb attraction of unlike charges and repulsion of like charges, the Pauli principle, and the varying shapes and extent of approach to the nucleus of different orbitals, depicted and discussed in Sec. 9-4.

---

[1] There is no unanimity among chemists on a definition of the term *transition element*, but a common usage, which we have adopted, is to denote in this way any element that has a partly filled d or f subshell in any of its commonly occurring oxidation states, including the elemental state [oxidation state (0)]. The elements of column IIB (Zn, Cd, and Hg) have a filled d subshell (and either an empty or filled f subshell) in both the elemental state ($d^{10} s^2$) and the other common oxidation states, including that corresponding to the +2 ion (d^{10}); this is why they are usually considered not to be transition elements. Many of the characteristics that the transition elements have in common, including variable oxidation state and color of many of their compounds, are closely associated with the presence of incomplete d or f subshells (Chaps. 24 and 25). More than half of all known elements (59 out of 106) are transition elements.

We begin by seeking to explain the variation of ionization energy of the first dozen elements, for in so doing we shall consider most of the effects that provide the answers to our questions. These ionization energies are depicted graphically in Fig. 10-2 and are listed in Table 10-5.

The ionization energy of the hydrogen atom can be calculated exactly, as discussed in Chap. 9, and both the Bohr theory and the modern quantum theory give the experimental value, 13.6 eV. The spacing of the energy levels for one-electron atoms and ions varies directly with Z^2, so that the energy required to remove the *second* electron from a helium atom (that is, the ionization energy of He$^+$ when its single electron is in the $1s$ orbital) is $(2^2 \times 13.6)$ eV = 54.4 eV. Since both electrons of a helium atom are in the $1s$ orbital, one might at first imagine that the energy needed to remove the first would be the same as that for the second. It is actually far smaller, 24.6 eV, only about 1.8 times that of hydrogen instead of 4 times as great. Does this indicate a failure of the theory? Not at all, for a neutral helium atom is not a one-electron atom, and it is just because of the presence of two electrons in the $1s$ orbital that the first ionization energy is much smaller than the second. Each electron diminishes the effective nuclear charge felt by the other electron as long as both are present, because of the electron-electron repulsion resulting from their like charges; each electron is thus said to "screen" or "shield" the nucleus, in part, from the other. Because the electrons tend to avoid one another, this shielding is not complete; the effective nuclear charge is about 1.69, as compared with the actual charge of 2 and the value of 1 that would be appropriate if the shielding by the other electron were complete. Because the radius of the $1s$ orbital is small, the first ionization energy of helium is high—indeed, the highest for any neutral atom.

The first ionization energy of the next atom, lithium, involves removing an electron from the $2s$ rather than the $1s$ orbital, the third electron being excluded from the $1s$ orbital in accord with the Pauli principle. Not only is the radius of the $2s$ orbital (that is, the position of maximum probability density, illustrated in Fig. 9-8) significantly greater than that for a $1s$ orbital for the same atom, but the nucleus is very effectively screened by the two inner-shell electrons so that the effective nuclear charge for the $2s$ electron is nearer 1 than 3. Thus the ionization energy of lithium is far smaller than that of He and only about 40 percent that of H. The screening is not, of course, perfect. As Fig. 9-8 shows, the $2s$ electron has a significant, although small, probability of being at a distance comparable to the radius of the $1s$ orbital, where the effect of the nucleus is much greater and where, at the same time, the

TABLE 10-5 **First ionization energies of the first twelve elements***

Element	eV	kJ mol^{-1}	Element	eV	kJ mol^{-1}
H	13.6	1310	N	14.5	1400
He	24.6	2370	O	13.6	1310
Li	5.4	520	F	17.4	1680
Be	9.3	900	Ne	21.6	2080
B	8.3	800	Na	5.1	500
C	11.3	1090	Mg	7.6	740

*Rounded to the nearest 10 kJ mol^{-1}.

$1s$ electrons will tend to avoid being near the $2s$ electron because of coulomb repulsion.

The electron configuration of Be, $1s^2 2s^2$, is related to that of Li much as that of He is to that of H, and the relation of the ionization energies is similar, that of Be being a little more than 1.7 times that of Li. Inspection of Fig. 10-3 or Table 10-5 shows that the first ionization energy of the next atom, B, is somewhat smaller than that of Be; the added electron is in a $2p$ orbital, $2s$ being fully occupied. As Fig. 10-3 indicates, the $2p$ orbital has somewhat higher energy than the $2s$ and thus less energy is required to remove an electron from it. The reason that a $2s$ electron is more stable is apparent when one considers orbital shapes. Any p orbital wave function becomes zero at the nucleus, while those for s orbitals do not. There is thus a higher likelihood for a $2s$ electron than for a $2p$ electron to be very near the nucleus and hence be in a more stable state. The resulting difference in ionization energy is small but perceptible, as noted in Fig. 10-2 and Table 10-5.

In the ground state of a carbon atom, the two $2p$ electrons have parallel spins and occupy different orbitals, in accord with Hund's rule. This configuration is more stable than one with the electrons paired in the same $2p$ orbital because electron-electron repulsions are minimized when the electrons occupy different $2p$ orbitals. As illustrated in Fig. 9-11, the different p orbitals are oriented at 90° to one another and thus electrons in them are in quite different regions of space. As long as the electrons in a degenerate set of orbitals have parallel spins, they must occupy different orbitals; hence the state with parallel spins ensures that the electron-electron repulsion will be minimized. The ionization energy of carbon is higher than that of boron because the electrons in the different $2p$ orbitals do not effectively shield the nucleus from one another.

Hund's rule applies as well to the configurations of N and O atoms, as mentioned earlier, for just the same reason—the minimization of the coulomb interelectron repulsions. The ionization energy of N is higher than that of C by about the same amount by which the ionization energy of C exceeds that of B, and the interpretation of the difference is similar. However, as Fig. 10-2 and Table 10-5 show, the ionization energy of the next element, O, is below that of N. Can this be rationalized in terms of the principles we have been using? The nuclear charge has again increased, but now there is no p orbital available in which there is not already an electron. Consequently, the electron-electron repulsion is increased. Evidently this increased electronic repulsion is slightly more important than the increased attraction by the nucleus, so that removal of one electron is easier. Without detailed calculations, it is difficult to predict which of these effects should be more important. With fluorine and neon, the experimental ionization energies indicate that the effect of still further increases in the nuclear charge apparently dominates.

The next atom, Na, with 11 electrons, must have its outer electron in a $3s$ orbital, the $2p$ subshell now holding six electrons, the maximum number permitted by the Pauli principle. Just as for Li, this outer s electron is quite effectively shielded from the nucleus by the completed inner shells, and since the effective radius of the $3s$ orbital is greater than that of the $2s$, the ionization energy of Na is somewhat lower than that of its congener Li. Similarly, the ionization energy of Mg is somewhat less than that of Be. In general, ionization energy decreases within a group of congeners as the atomic number increases just because the outer-electron configurations of any group of congeners involve similarly shaped orbitals with gradually increasing radii.

Hund's rule applies to the filling of any degenerate levels, such as d or f; furthermore, unpaired electrons in different orbitals will have parallel spins.

EXAMPLE 10-1

Hund's rule. How many unpaired electrons (that is, electrons of a given spin that are not matched by electrons of opposite spin) are associated with the ground state of $_{45}$Rh? Use the information given in Table 10-3.

Solution. The only orbitals that are not completely filled with electrons are the five $4d$ and the one $5s$ orbitals, which are occupied by eight and one electrons, respectively. The scheme of these states, occupied by as many electrons of parallel spin as possible (Hund's rule), is

showing that there are three unpaired electrons.

The general arguments used above in analyzing the trends in ionization energy among the first dozen elements can be applied throughout the periodic table, although it is often difficult to predict the details of electronic structure. Several points are worthy of attention. The special stability of half-filled subshells, illustrated by the fact that N has a higher ionization energy than either of its neighbors, C and O, is noticeable many times. Similarly, the fact that filled subshells represent relatively stable configurations is often evident, whether it is the ns^2 characteristic of the alkaline-earth metals or, even more strikingly, the configurations $3d^{10}4s^2$, $4d^{10}5s^2$, and $5d^{10}6s^2$ characteristic of Zn, Cd, and Hg (column IIB). These effects are particularly noticeable in Fig. 10-2.

With very close energy levels, such as $3d$ and $4s$, or $4f$, $5d$, and $6s$, the exact order of filling and the relative stability may change from one atom or ion to another, as the effects of nuclear charge, interelectronic repulsion, and shielding of the nucleus vary. These variations cannot be simply rationalized. Note, for example, in Table 10-3 that one $5d$ orbital (as well as $6s$) is occupied in La ($Z = 57$) before the first $4f$ orbital receives an electron and that two $6d$ orbitals (as well as $7s$) in Th ($Z = 90$) contain electrons before the first $5f$ orbital is occupied. It is noteworthy too that when transition-metal (and column IIB) ions are formed, the s electrons are lost first rather than the d electrons, which have a lower principal quantum number. Thus, the d orbitals of these elements appear to gain slightly in relative stability when positive ions are formed. Typical configurations observed for some of these ions are shown in Table 10-6.

Some aspects of theory. The positions of atomic energy levels can be worked out quantitatively by a detailed examination of the electron distributions in the different orbitals. The results of calculations of this kind are quite satisfactory. Two points must be kept in mind: (1) If electrons were placed in the available orbitals one at a time, each new electron would cause a shift in the entire level structure.

TABLE 10-6   **Configuration of outer electrons of some ions of transition and related elements**

Ion	Configuration	Ion	Configuration
Ti^{3+}	d	Fe^{2+}	d^6
V^{3+}	d^2	Co^{2+}	d^7
Cr^{3+}, V^{2+}	d^3	Ni^{2+}	d^8
Mn^{3+}, Cr^{2+}	d^4	Cu^{2+}	d^9
Fe^{3+}, Mn^{2+}	d^5	Zn^{2+}, Cu^+	d^{10}

In the Aufbau we really consider the level systems for the different elements with the electrons already in place and mutually interacting, with electrons that are closer to the nucleus shielding those in outer orbitals. (2) The concept of atomic orbitals that are occupied by electrons is part of an approximate model. The model is extremely valuable and has yielded many illuminating correlations between theory and experimental observation of atoms, particularly relating to their spectra, and in this sense it has provided a great deal of insight into the nature and workings of atoms. However, it is impossible to associate actual orbitals with actual atoms, because in the orbital model the mutual interactions of the electrons are taken into account only on the average rather than in detail. Orbitals are an important ingredient of modern atomic theory, but they should not be taken too literally.

10-4   Orbital Energies and Chemical Properties

The arrangement of atomic energy levels described earlier not only provides a basis for understanding the general structure of the periodic table but explains in addition many chemical properties of individual elements. These two subjects are closely intertwined. We have already discussed briefly the interpretation of the general properties of the inert gases, alkali metals, halogens, and other groups. In this final section we consider a few aspects of the general behavior of *valence electrons,* as the electrons in the incompletely filled shell are called. The shell that is partially occupied by these electrons is termed the *valence shell.* We begin with two important physical quantities that characterize the electrons and orbitals of the valence shell, the ionization energy and the electron affinity, and then proceed to a brief consideration of some other significant atomic properties.

Ionization energy.   The first ionization energy has been illustrated (Fig. 10-2) and discussed earlier. It is important to note that it is the energy needed to remove an electron from a neutral *atom* of the element in the *gas* phase; thus the ionization energy of hydrogen is the energy needed to ionize an H atom rather than an H_2 molecule, and that of sodium refers to monatomic sodium vapor, not to metallic sodium, a solid at ordinary temperatures. When ionization energy is defined in this specific way, the effects of interaction with other atoms in molecules or in the larger aggregates characteristic of condensed phases are removed and the striking periodicity displayed in Fig. 10-2 is revealed.

In one important way the previous discussion of variations and trends in ionization energy has been oversimplified, because we have focused attention only on the electronic structures of the neutral atoms and not on those of the resulting ions. The ionization energy involves the *difference* in energy between the ionized atom and the original atom, and thus one must consider the energy of each. For example, an increase in ionization energy in progressing from one element to the next might result from an increase in stability of the neutral atom, from a decrease in stability of the ion formed, or from some combination of the two (Fig. 10-4). The fact that the earlier analysis sufficed for explaining most of the observed trends indicates that the energy levels of the ions generally change less than those of the neutral atoms, but the levels of the ions must nonetheless be examined if one wishes to understand the details of Fig. 10-2.

The energies needed to remove a second, then a third, and then still further electrons from an atom are called, respectively, the second, third, and higher ionization energies. Their variations with the atomic number and electron configuration of the atoms and ions involved can be analyzed in a fashion very similar to the one we have used for the first ionization energy, taking into account also the effect of the increased ionic charge as more and more electrons are removed. For example, a graph of second ionization energy against atomic number would resemble in many ways Fig. 10-2. However, the vertical changes would be larger because more energy is required to remove an electron from a positive ion than from a neutral atom, and the curve would be shifted one place to the right because a singly charged positive ion is isoelectronic[1] with the neutral atom of atomic number smaller by 1.

[1]Two species with the same number of electrons are said to be *isoelectronic;* thus H^-, He, and Li^+ are isoelectronic, as are O^{2-}, F^-, Ne, Na^+, Mg^{2+}, and Al^{3+}, and even molecular species such as N_2, CO, and NO^+.

FIGURE 10-4 **Variation of ionization energy with relative stabilities of atoms and ions.** The first ionization energy represents the difference in the energies of the gaseous neutral atom and its singly charged positive ion. Three hypothetical cases are shown, for atoms X, Y, and Z. The length of each vertical double arrow is proportional to the corresponding ionization energy. Atom Y is less stable than atom X and yet they have the same ionization energy, because Y^+ is less stable than X^+ by the same amount. The ionization energy of Z is greater than that of Y because the neutral atom Z is appreciably more stable than Y while the ion Z^+ is only slightly more stable than Y^+. Other possibilities also exist. It is clear that the energetics of both the neutral atom and the ion should be considered in analyzing trends in ionization energies. This same principle applies in the interpretation of energy effects accompanying *any* change in state.

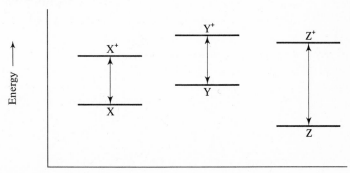

Electron affinity. The energy liberated when an electron is added to a gaseous atom, converting it into an ion with a single negative charge, is called the electron affinity. Thus in (10-2),

$$X(g) + e^- \longrightarrow X^-(g) + \epsilon_1 \qquad (10\text{-}2)$$

the electron affinity corresponds to the energy ϵ_1. It is important to note that, for historical reasons, the sign convention for defining electron affinity is opposite to that used for defining ionization energy: a positive ionization energy signifies that energy is *required* to remove an electron, whereas a positive electron affinity indicates that energy is *liberated* when an electron is added. The electron affinity of an atom is equal to the ionization energy of the related anion, as indicated by (10-3), which is just the reverse of (10-2):

$$\epsilon_1 + X^-(g) \longrightarrow X(g) + e^- \qquad (10\text{-}3)$$

Comparison of (10-3) with (10-1) shows that ϵ_1 is indeed the ionization energy of the anion X^-, as well as being [by (10-2)] the electron affinity of X.

Table 10-7 gives electron affinities for some elements; those shown here are positive, but electron affinities are by no means positive for all the elements. Precise values of electron affinities are usually hard to obtain, but the values given in Table 10-7 are consistent with our previous discussions: they are large for the halogens, whose -1 ions have the stable inert-gas configurations, are smaller for oxygen and sulfur, and rather small (but still positive) for hydrogen, corresponding to formation of the hydride ion, H^-, which is isoelectronic with He. The values for Li and Na are similar to that for H; the remaining alkali metals have even smaller electron affinities. Be and Mg have *negative* electron affinities, around -0.5 to -0.8 eV. All known *second* electron affinities of atoms [corresponding to adding an electron to a singly charged negative ion, a reaction such as $O^-(g) + e^- \longrightarrow O^{2-}(g)$] are also negative, as a consequence of the repulsion between the negative ion and the electron to be added to it.

Electronegativity. Electronegativity is a measure of the tendency of a bonded atom to attract electrons from other atoms in the same molecule or ion. As discussed briefly in Sec. 3-2 and in more detail in Sec. 11-5, a large electronegativity difference between two bonded atoms results in significant partial ionic character of the bond. Because electronegativity is a property of an atom in a bonded state, it depends to a small extent on the nature of the orbitals that the atom uses in bonding (which may vary from one compound to the next) and on the nature and number of its

TABLE 10-7   **Approximate electron affinities***

Element	eV	kJ mol^{-1}	Element	eV	kJ mol^{-1}
H	0.75	70	Br	3.4	320
Li	0.5	50	I	3.1	300
F	3.4	330	O	1.5	140
Cl	3.6	350	S	2.1	200

*Rounded to the nearest 10 kJ mol^{-1}.

bonded neighbors. Thus the electronegativity of a given atom may vary slightly from one compound to the next. Attempts to establish a quantitative scale of electronegativity involve correlating it with properties that can be measured precisely. Several different scales have been proposed. The two most common are that due to Pauling, who obtained electronegativities from a study of the energies needed to break covalent bonds, and that suggested by Mulliken, who equated the electronegativity of an atom with the average of its first ionization energy and its electron affinity. These scales give quite similar values when adjusted so that the electronegativity of fluorine is 4.0 on each. Representative values are shown in Fig. 10-5, in which the elements are arranged in the same rows as in the periodic table except that the horizontal scale is one of electronegativity.

FIGURE 10-5 **Diagram of electronegativities.** The diagram resembles a distorted periodic table, the elements of each A column being displaced toward the left with increasing atomic number. This displacement is strongest for the halogens, weakest for the alkali metals. There is little leftward displacement in the B columns (the transition metals, including the lanthanides and actinides); for these columns the electronegativities average about 1.7, varying from about 1.1 for most of the lanthanides to around 2.3 for the less reactive metals of higher atomic number, such as Pt and Au. Few of the transition elements are shown here; the vertical dashed lines represent average values of the electronegativities of the missing transition elements. (*Front "Chemistry", by Linus Pauling and Peter Pauling, W.H. Freeman and Company, San Francisco. Copyright © 1975.*)

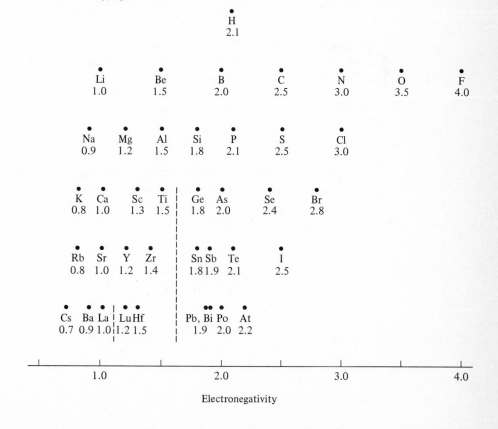

Electronegativity

Several features of Fig. 10-5 are worth remembering:

1. The values of the electronegativities of the first-row elements, Li through F, are rather widely and equally spaced, ranging from 1.0 for Li to 4.0 for F. Hydrogen (2.1) is near the middle of the table, slightly less electronegative than carbon (2.5).

2. If one considers the usual A columns of the periodic table, the diagram of electronegativities has a characteristic slant, the bottom part being displaced toward the left relative to the top. In other words, there is a decrease in electronegativity as one proceeds down a given A column. This effect is much more pronounced for the halogens (F, 4.0; Cl, 3.0; Br, 2.8; I, 2.4; At, 2.2) than for the alkali metals (Li, 1.0; Na, 0.9; K, 0.8; Rb, 0.8; Cs, 0.7).

3. Fluorine, the most electronegative element, is significantly more electronegative (by 0.5 units on this scale) than oxygen, which is the second most electronegative. Nitrogen and chlorine, which have about equal electronegativities, come next, about 0.5 units less than oxygen. At the other end of the scale the differences are much smaller. Cesium, the least electronegative element (other than the very rare Fr) is, for example, only 0.1 units less electronegative than rubidium and potassium.

4. The electronegativity of most metals is between 1.2 and 1.8, but the alkali metals lie significantly below this range. Gold and the "platinum metals", the elements in the last two rows of column VIII, have electronegativities between 2.2 and 2.4. Metalloids have electronegativities close to 2.0.

Other properties. We have already alluded to the way in which metallic character varies with position in the periodic table (Sec. 10-2) and shall have occasion in later chapters to examine various other properties in a similar fashion. For example, the relative effective sizes of ions, as well as of neutral atoms, both in bonding and in nonbonding situations, can be understood quite well in terms of their positions in the periodic table and their electronic structures. Thus, in the isoelectronic series O^{2-}, F^-, Na^+, Mg^{2+}, Al^{3+}, the effective radius of the ion decreases progressively and significantly with increasing atomic number because of the increased attraction by the nucleus without corresponding change in electron configuration. Another important type of correlation is that of the acid-base character of the oxides of elements with position in the periodic table: in general, the oxides of the elements with lowest electronegativity are strongly basic, those of elements with high electronegativity are acidic, and those of intermediate elements lie between and may even be amphoteric. There is a clear correlation with oxidation state as well, not entirely independent of that with electronegativity inasmuch as electronegativity varies somewhat with oxidation state.

Problems and Questions

10-1 **Predicted properties of a new element.** It has been predicted that a nucleus with 114 protons and 184 neutrons might be sufficiently stable to be isolated in appreciable quantities (with a half-life of 1000 years or more for its most rapid disintegration), if it could be made in some way. What would be the approximate atomic weight of this new element? What would be its probable outer-electron configuration (beyond the Rn core)? What element would it most closely resemble in properties? *Ans.* would resemble Pb

10-2 **Memorizing part of the periodic system.** (*a*) Reproduce from memory the first two short periods of the periodic system (extending from He through Ar) and the columns containing the inert gases, the alkali and alkaline-earth metals, the halogens, and the chalcogens (congeners of oxygen). (*b*) Do the same for the elements K to Kr.

10-3 **Inert gases.** Show the detailed electron configuration of all inert gases. Use formulas such as $1s^2 2s^2 2p^6$, with the orbitals approximately in sequence of increasing energy.

10-4 **Electron configurations of atoms.** Without consulting any of the tables in this chapter, give plausible electron configurations ($1s^2 2s^2 2p^6 \cdots$) of the following atoms (atomic numbers are given in parentheses): C (6), N (7), P (15), Sc (21), Cr (24), Ni (28), Cu (29), Zn (30), Ga (31), As (33), Zr (40), Pd (46), Te (52), and Au (79). Do the same for these ions: V^{3+} (23), Cr^{3+} (24), Fe^{2+} (26), Fe^{3+} (26), Co^{2+} (27), Co^{3+} (27), and Ni^{2+} (28).

10-5 **Excited state.** Define the term *excited state* as applied to an atom or molecule. Illustrate by giving the electron configuration for an excited state of a sodium atom.

10-6 **Electron configurations.** Give the electron configuration ($1s^2 2s^2 2p \cdots$) of (*a*) the ground state of Ga, I^-, Sr^+; (*b*) an excited state of Xe^+, Ca, O.

10-7 **Halogens.** What is the characteristic outer-shell configuration of the halogens? How is this correlated with their electron affinity?

10-8 **Hund's rule.** How many unpaired electrons are associated with the ground states of the atoms $_6C$, $_{24}Cr$, $_{64}Gd$, $_{67}Ho$, and $_{96}Cm$? Use the information in Table 10-3.
Ans. 2, 6, 8, 3, 8

10-9 **Ionization energy.** According to Table 10-5, the first ionization energy of helium is 2370 kJ mol^{-1}, the highest for any element. (*a*) Define *ionization energy* and discuss why that for He should be so high. (*b*) Which element would you expect to have the highest *second* ionization energy? Why? (*c*) Suppose that you wished to ionize some helium by shining electromagnetic radiation on it. What is the maximum wavelength you could use?
Ans. (*b*) Li; (*c*) 50.4 nm

10-10 **Ionization energies.** The energy needed to remove one electron from a gaseous potassium atom is only about two-thirds as much as that needed to remove one electron from a gaseous calcium atom, yet nearly three times as much energy is needed to remove one electron from K^+ as from Ca^+. What explanation can you give for this contrast? What do you expect to be the relation between the ionization energy of Ca^+ and that of neutral K?

10-11 **Arrangement of substances in order.** Without consulting any tables, arrange the following substances in order and explain your choice of order: (*a*) Mg^{2+}, Ar, Br^-, Ca^{2+} in order of increasing radius; (*b*) Na, Na^+, O, Ne in order of increasing ionization energy; (*c*) H, F, Al, O in order of increasing electronegativity.
Ans. (*a*) Mg^{2+}, Ca^{2+}, Ar, Br^-; (*b*) Na, O, Ne, Na^+; (*c*) Al, H, O, F

10-12 **Differences between Cl and S.** Both the electron affinity and the ionization energy of chlorine are higher than the corresponding quantities for sulfur. Explain in terms of the electronic structure of the atoms.

10-13 **Different electronic states.** (*a*) Give the complete electron configuration ($1s^2 2s^2 2p \cdots$) of Al in the ground state. (*b*) The wavelength of the radiation emitted when the outermost electron of Al falls from the 4*s* state to the ground state is about 395 nm. Calculate the energy separation (in joules) between these two states in the Al atom. (*c*) When the outermost electron in Al falls from the 3*d* state to the ground state, the radiation emitted has a wavelength of about 310 nm. Draw an energy diagram of the states and transitions discussed here and in (*b*). Calculate the separation (in joules) between the 3*d* and 4*s* states in Al. Indicate clearly which has higher energy.   *Ans.* (*b*) 5.03×10^{-19} J;   (*c*) 1.38×10^{-19} J

10-14 **Relative stabilities of electrons in different orbitals.** What experimental evidence does the periodic table provide that an electron in a 5*s* orbital is slightly more stable than an electron in a 4*d* orbital for the elements with 37 and 38 electrons?

"I am inclined to think, that when our views are sufficiently extended, to enable us to reason with precision concerning the proportions of elementary atoms, we shall find the arithmetical relation alone will not be sufficient to explain their mutual action, and that we shall be obliged to acquire a geometrical conception of their relative arrangement in all three dimensions of solid extension."

W. H. Wollaston,[1] 1808

"Before leaving the subject of the atom it is desirable to refer to the new idea which is revolutionizing chemistry, namely that the different atoms have different exchangeable values or valencies. An atom is the seat of attractive forces, and the valency is the number of centers of such force."

Anonymous essay,[1] 1869

11 Chemical Bonds: I. Experimental Facts and an Introduction to Bonding Theory

This chapter and the next one deal with the nature of the interactions between atoms that we characterize as *bonding*. Bonding is fundamental to all considerations of chemical structure and to all interpretations of chemical reactivity as well, inasmuch as reactions necessarily change the way atoms are combined. A brief introduction

[1]Quoted from W. G. Palmer, "A History of the Concept of Valency to 1930", pp. 27 and 89, Cambridge University Press, New York, 1965.

to chemical bonding was presented in Chap. 3 and it is assumed that the reader is familiar with that discussion. This chapter begins with an examination of some of the experimental evidence that leads us to believe that the usual concept of a chemical bond—a strong and specific attractive interaction between atoms—is a useful one. This discussion is followed by a consideration of the theoretical interpretation of the simplest bonds, the ionic bond on the one hand and the covalent bond, as exemplified in a simple diatomic molecule such as H_2, on the other. In the following chapter we examine bonding in more complex molecules and consider various theoretical interpretations of their observed properties.

11-1 Background

Although the *fact* that atoms are strongly attracted to one another in various combinations in different substances was self-evident from the earliest days of consideration of the atomic nature of matter, the interpretation of the *nature* of this attractive force developed only very slowly. Coulomb had firmly established experimentally the law of interaction of charged particles in the 1780s, only two decades before Dalton's atomic theory was published. It is thus not surprising that the earliest fruitful speculation about the nature of the attractive force between atoms was that it was essentially electrical in nature, resulting from the attraction of opposite charges. This view was developed first by Berzelius, a Swedish chemist, not long after Dalton's work had appeared, and it was very helpful in rationalizing the chemistry of many substances—the general class of compounds we now recognize as ionic. However, it could not explain the properties of many other substances, including most organic compounds, as Berzelius himself reluctantly recognized. It was only after more than a century had passed that it was possible to show, using the methods of quantum mechanics, that Berzelius' fundamental notion was correct.

As the concept of the combining power of atoms—valency—developed during the nineteenth century, the concept of the chemical bond developed simultaneously. The Russian chemist A. M. Butlerov first suggested in 1861 that a chemical formula should indicate clearly just how each atom is linked to other atoms in a molecule. He also suggested that all the properties of a compound were determined by its molecular structure and that it should be possible to deduce this formula by finding the different ways in which the molecule could be synthesized. Indeed, it was just through study of the chemical reactions by which molecules are formed, and of those in which molecules are fragmented in different ways, that all the detailed schemes of structural formulas involving bonds were gradually developed.

The use of symbols to represent bonds was a natural step. It provided at once a shorthand notation and a stimulus to thinking about the abstract concept of chemical bonds. The men who developed schemes of notation in the mid-nineteenth century—Couper, Crum Brown, Frankland, Kekulé, and others—were careful to point out that their early formulas had no geometric significance. However, a series of experimental and theoretical developments by Kekulé, Körner, Pasteur, van't Hoff, LeBel, Werner, and others gradually led to explicit inferences about the shapes of molecules and the relative orientations of the different bonds formed by various atoms. Most of these inferences have been completely verified by modern physical methods of structure analysis, developed during the last 50 years, which have revealed the precise shapes of molecules and the detailed three-dimensional arrangements of atoms in many different substances. Methods for studying the energies involved in chemical bonds have also been developed during the present century. In the next section we examine some of the experimental facts about the interactions of atoms in molecules, after which we turn to a more detailed consideration of theoretical interpretations of chemical bonding.

Before we proceed, however, a word of caution is needed. There is a tendency, among both beginning students and some experienced chemists, to regard chemical bonds in molecules as fixed entities with rather specific, well-defined properties. The fact that we have a conventional symbolic notation for bonds, draw schematic pictures of them involving orbital overlaps, present tables listing typical (equilibrium) bond distances and bond energies, and in other ways endow bonds with concrete properties should never obscure the fact that they are only useful abstractions for explaining the observed properties of certain substances. The concept of the chemical bond in nonionic molecules as a specific and rather strong attractive interaction between two atoms is limited. Later (Sec. 21-1) we shall encounter molecules whose properties are consistent with a current theory that involves the simultaneous mutual interaction of three or more atoms, rather than just two. There are also molecules for which recent experimental findings seem to show that the concept of fixed bonds between certain pairs of atoms is valid for short times but not for longer ones, so that the particular pairs of atoms involved in the bonding change with time.

This limitation in the power of an abstraction when we endow it with concrete properties is an example of a philosophical difficulty that A. N. Whitehead has termed the "fallacy of misplaced concreteness"—a trap that those working in science must always be wary of.

11-2 Experimental Information about Atomic Interaction in Molecules

Potential energy curves for interaction of pairs of atoms. Some typical energy curves for a covalent bond (H-H), an ionic bond (Na^+ Cl^-), and a van der Waals interaction (Ar $\cdots$ Ar) were shown and discussed in Fig. 3-13. Figure 11-1 is a schematic representation of such a curve for the H_2 molecule and illustrates the dissociation energy D_0 and the equilibrium internuclear separation r_e. In a diatomic molecule there is no ambiguity about the term *chemical bond*—there are only two atoms to be held together, and whatever attractive force there is operates to bind them. In molecules with more than two atoms, curves like that in Fig. 11-1 are often used to represent the interactions of pairs of atoms in bonds, e.g., the four C-Cl bonds in CCl_4, the two O-H bonds in water, or the C-C bond in ethane, H_3CCH_3. It is possible to measure internuclear separations, dissociation energies, and other properties of these polyatomic molecules. When this is done, the properties attributable to the O-H interactions in different molecules that contain this grouping are found to be very similar but not quite identical; in the same way, the properties

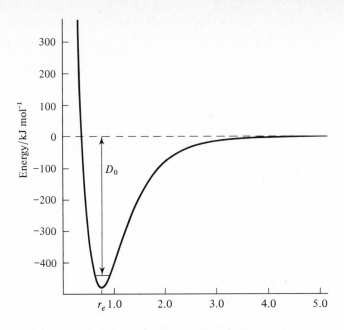

FIGURE 11-1   **Potential energy diagram for H_2.**   For diatomic molecules this energy is evaluated, by calculation or by elaborate analysis of spectroscopic data, for various values of the separation of the nuclei. It is measured relative to a zero point corresponding to infinite separation of the atoms. The dissociation energy of the molecule, D_0, is the energy needed to separate the atoms when they are initially in the lowest energy state. This state is not at the bottom of the potential well because, as discussed in Sec. 8-3, in accord with the uncertainty principle atoms vibrate relative to one another even at the lowest temperatures. This energy, called zero-point energy, cannot be removed.

The distance marked r_e represents the equilibrium internuclear separation for this molecule. Because the atoms are always in a state of vibration, their average separation may differ from r_e by a small amount.

of other bonds usually vary only by small amounts from one molecule to the next. Let us now examine some of these experimental data.

Bond energies.   The dissociation energy of a diatomic molecule—that is, the energy to dissociate a diatomic molecule from its lowest energy state into its component atoms in their lowest energy states—can be considered the energy of the chemical bond formed between the two atoms. Although the concept of the energy of a bond is quite natural for diatomic molecules, it cannot be extended to polyatomic molecules without some qualifications.

Consider the energy of the C-C bond in ethane. It might be taken to be the energy by which ethane is dissociated into two methyl radicals in the gas phase:

$$C_2H_6(g) \longrightarrow 2CH_3(g) \tag{11-1}$$

However, this procedure for determining the bond energy involves the assumption that the dissociation does not change the energies attributable to the C-H bonds, i.e., that the energies of the C-H bonds are the same in the methyl radical, CH_3, as they are in ethane. If these energies do not remain the same, then some of the energy change in (11-1) will be caused by changes in the C-H bonds.

A feeling for the variation of the energies of nominally identical bonds can be gotten by examining the energy changes in reactions such as

$$CH_4 (g) \longrightarrow CH_3 (g) + H (g) \qquad (11\text{-}2)$$

$$CH_3 (g) \longrightarrow CH_2 (g) + H (g) \qquad (11\text{-}3)$$

If there were no effect of molecular structure on the energy of a bond, then the energy required to remove successive H's from CH_4 would be constant. The data in Table 11-1 show that in fact these energies vary by as much as 26 percent from the mean value, 411 kJ mol^{-1}.

The variations indicated in Table 11-1 can be understood by considering the relative stabilities of the reactants and products, a principle emphasized in the discussion of Fig. 10-4. The dissociation energy will be smaller than average when the product is particularly stable or when the starting material is relatively unstable (or both). Conversely, it will be larger than average when opposite conditions hold.

Even though there is a spread in the values of the energy that can be associated with the C-H bond, it is sometimes useful in approximate calculations to assume that all C-H bonds have the same energy, irrespective of the molecule in which they are found. This energy, called the *bond energy*, is an average value derived from thermochemical and spectroscopic data. Consistent sets of bond energies can be derived for bonds between many other different pairs of atoms in a similar way. Some representative values are given in Table 11-2.

Bond energies can be used to predict approximate energy changes in many reactions, as demonstrated in Example 11-1.

EXAMPLE 11-1

Use of bond energies. Estimate the approximate energy changes accompanying the following reactions:

(a) $H_3CCH_3(g) + H_2(g) \longrightarrow 2CH_4(g)$

(b) $CH_4(g) + 2Cl_2(g) \longrightarrow CH_2Cl_2(g) + 2HCl(g)$

Solution. In reaction (a), one C-C bond is broken and one H-H bond is broken; two C-H bonds are formed. Thus the approximate energy released should be

$$2(\text{C-H}) - [(\text{C-C}) + (\text{H-H})] = [2(420) - (340 + 440)] \text{ kJ}$$

$$= \underline{60 \text{ kJ}} \text{ per mole of ethane reacted}$$

TABLE 11-1 **Representative bond dissociation energies/kJ mol^{-1}**

H—CH$_3$	423
H—CH$_2$	368
H—CH	519
H—C	335

TABLE 11-2 **Representative single-bond energies*/kJ mol^{-1}**

	H	C	N	O	F	Cl	Br	I	Si
H	440	420	390	460	560	430	370	300	300
C		340	290	350	440	330	280	240	290
N			160	180	270	200			
O				140	210	210	220	240	430
F					160	250	250	280	590
Cl						240	220	210	400
Br							190	180	290
I								150	210
Si									190

*Rounded to the nearest 10 kJ mol^{-1}. The bond energies given here are applicable only to reactions occurring in the gas phase. They can be used for liquids or solids when contributions from the energies of vaporization or sublimation are taken into account (Chap. 16).

The observed change corresponds to a release of 65 kJ per mole of ethane. A discrepancy of 10 or 20 kJ is not unusual when bond energies are used; nevertheless, the possibility of making estimates of energies of reaction within about this accuracy is valuable.

In reaction (b), two C-H bonds and two Cl-Cl bonds are broken and two C-Cl bonds and two H-Cl bonds are formed. Thus the energy change would be expected to be about

$$2[(C\text{-}Cl) + (H\text{-}Cl) - (C\text{-}H) - (Cl\text{-}Cl)] = 2(330 + 430 - 420 - 240)\,kJ$$

$$= 2(100)$$

$$= \underline{200\,kJ}\ \ \text{per mole of methane reacted}$$

The observed change corresponds to the release of 202 kJ per mole of methane. Thus for this reaction, bond energies give a remarkably good prediction of the energy change.

Some elements, most notably C, N, O, and their nearby congeners, can form double and triple bonds. In fact, as discussed in Chap. 12 (Secs. 12-1 and 12-3), bonds of character intermediate between single and double, and between double and triple, can also be considered to be present in many compounds. Bond energies for various kinds of multiple bonds have been estimated by the same methods used for single bonds. A few representative values are listed in Table 11-3. There is a

TABLE 11-3 **Representative single-, double-, and triple-bond energies*/kJ mol^{-1}**

Elements	Single bond	Double bond	Triple bond
O-O	140	400	
N-N	160	420	950
C-C	340	620	810
C-O	350	720	
C-N	290	620	890

*Rounded to the nearest 10 kJ mol^{-1}.

marked increase in bond energy with increasing multiplicity of the bond, which is in accord with the qualitative view that a multiple bond should be stronger than a single bond. The same view leads to the expectation that a multiple bond should be shorter than a single bond, a relationship that is confirmed experimentally. This is because the attractive interaction is stronger than for a single bond while the repulsive interaction is similar, leading to a potential curve (analogous to that in Fig. 11-1) with a deeper minimum that is displaced toward shorter distances.

Bond distances. As just implied, a characteristic bond distance can be associated with each kind of bond. Because, even at the lowest temperatures, atoms vibrate relative to one another at frequencies around 10^{13} per second, interatomic distances are continually changing in all molecules by 0.1 Å or more. Despite this fact, average or equilibrium distances between atomic nuclei in many molecules and ions, and in many unstable species as well, have been measured with considerable precision, often appreciably better than 0.01 Å (10^{-12} m). The principal methods used for such measurements are spectroscopy and the diffraction of X rays, neutrons, and electrons. Some average values for the lengths of different kinds of bonds are presented in Table 11-4.

The internuclear distance characteristic of a C-H bond varies over a range of only about 4 percent (about 0.04 Å) in different substances, and the variation is only around 1 percent if bonds in similar groups are compared (e.g., the C-H bonds in CH_3 groups in different molecules or the C-H bonds in benzene and molecules related to it). Comparable relative variations are found for other bonds that cannot have significant double- or triple-bond character (for example, O-H or C-halogen). On the other hand, some bonded interatomic distances, such as carbon-carbon and carbon-oxygen, vary by about 20 percent in different compounds. Most of this variation is attributable to multiple bond formation, and when this is taken into account, only small variations remain, comparable with those found for C-H bonds. Covalent bonding radii for different elements have been derived from tables such as Table 11-4. They make it possible to predict bond distances quite accurately for almost any molecule whose structural formula is known.

Bond angles. If the average positions of the atoms in a polyatomic molecule are known accurately, it is possible to calculate the angle subtended at an atom

TABLE 11-4   **Representative bond distances/Å**

N—N	1.45	C—N	1.48	C—C	1.54	C—O	1.43	O—O	1.45
N=N	1.25	C=N	1.28	C=C	1.34	C=O	1.21	O=O	1.21
N≡N	1.09	C≡N	1.16	C≡C	1.20				
H—H	0.74	C—H	1.09			N—H	1.01	O—H	0.97
F—F	1.42	C—F	1.38	H—F	0.92				
Cl—Cl	1.99	C—Cl	1.77	H—Cl	1.27				
Br—Br	2.28	C—Br	1.94	H—Br	1.41				
I—I	2.67	C—I	2.14	H—I	1.61				

TABLE 11-5   **Bond angles in some molecules and ions containing hydrogen**

CH_4	109.5°	OH_3^+	110°	H_2O	105°
NH_4^+	109.5°	NH_3	107°	H_2N^-	104°
BH_4^-	109.5°	PH_3	94°	H_2S	92°
SiH_4	109.5°	AsH_3	92°	H_2Se	91°
PH_4^+	109.5°				
AlH_4^-	109.5°				

by any two atoms bonded to it. For example, atoms X, Y, and Z may be arranged linearly or in a bent configuration. The angle of interest is the angle subtended at Y by X and Z. It is called the bond angle because it is normally assumed that a bond is coincident with the line between the nuclei, an assumption that is probably a very good approximation in most situations. In any event, the bond angle defined in this way is an experimentally accessible quantity whereas angles involving electron distributions in orbitals are not.

Experimentally observed values of bond angles range from around 60° (in three-membered rings) to 180°, but they tend to cluster around certain characteristic values. Many examples are given in Tables 11-5 to 11-7. The illustrations in Chap. 4 show typical bonded configurations. Angles near 90° are common at octahedrally coordinated atoms—such as Cu in the $CuCl_2$ structure (Fig. 4-3), Cd in the $CdCl_2$ structure (Fig. 4-5), or the central atom in many complex ions [for example, $PtCl_6^{2-}$ or $Al(H_2O)_6^{3+}$]—and in ions and molecules with other configurations as well. Bond angles near 109.5° are extremely common because this is the angle subtended at the center of a regular tetrahedron by any two of the corners. Many atoms have four bonded neighbors arranged either exactly or approximately tetrahedrally. Such arrangements are found in the methane molecule (CH_4), the diamond crystal (Fig. 4-6), and the $Zn(OH)_4^{2-}$ ion (Fig. 4-7), in all of which the angle is 109.5° (called the *tetrahedral angle*). There are many other molecules of similar structure in which the four substituents on the central atom are not identical and in which the angles depart by a few degrees from the ideal tetrahedral value. Bond angles near 120° are also common, usually (but not always) occurring whenever three atoms are bound to a central atom in a planar arrangement. Thus the bond angle in the layers of the boron nitride and the graphite structures (Fig. 4-4) is 120°, as is the angle

TABLE 11-6   **Comparison of bond angles at oxygen and sulfur atoms**

	X—O—Y		X—S—Y
Cl_2O	111°	Cl_2S	100°
H_2O	105°	H_2S	92°
CH_3OH	109°	CH_3SH	99°
$(CH_3)_2O$	112°	$(CH_3)_2S$	105°
$(SiH_3)_2O$	144°	$(SiH_3)_2S$	97°
O_3	117°	SO_2	120°

TABLE 11-7   **Bond angles in some XO_2 species**

	O—X—O		O—X—O
CO_2	180°	O_3	117°
NO_2^+	180°	SO_2	120°
NO_2	134°	ClO_2	116°
NO_2^-	115°	ClO_2^-	110°

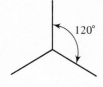

subtended at the carbon atom by any two of the oxygen atoms of the carbonate ion, CO_3^{2-}. Angles near 180°, giving rise to a linear or nearly linear bonded arrangement of three (or more) atoms, occur in a few situations, including some triatomic molecules such as CO_2 and $HgCl_2$, the simple organic molecule acetylene (HC≡CH), and the octahedral species just mentioned. It is also noteworthy that isoelectronic species show comparable angles. Thus CO_2 and NO_2^+ are both linear, and ozone (O_3) and nitrite ion (NO_2^-) are bent to about the same degree.

We shall see in Chap. 12 that a good empirical correlation of the configuration about a central atom can be made if bonded atoms and unshared pairs of electrons are taken into account.

Torsion angles about bonds. Information accumulated during the last few decades about the three-dimensional architecture of molecules has made it much easier to understand and visualize the ways in which molecules interact and react with one another, since molecular interactions depend intimately on the exact spatial arrangements of atoms. Thus chemists have had to learn to think about and visualize the three-dimensional implications of molecular formulas. In order to specify three-dimensional molecular geometry unambiguously, it is necessary to specify more than all the bond distances and bond angles. This is because any two points, corresponding to a bond distance, define only a one-dimensional object (a line), and any three points not on the same line, corresponding to a bond angle different from 180°, define only a two-dimensional object (a plane). One convenient and illuminating way of providing additional information for specifying the three-dimensional arrangement of the atoms in a molecule is to give the torsion angle about each bond (Fig. 11-2). The torsion angle about a bond X-Y can be defined by the positions of the two bonded atoms and of one bonded neighbor of each (W and Z in Fig. 11-2), that is, by the positions of four atoms W, X, Y, Z bonded in sequence. As explained in the legend of Fig. 11-2, the torsion angle represents the twist of the plane of one group of three consecutive atoms (X-Y-Z) relative to that of the other group of three (W-X-Y).

Rotation about a single bond is usually almost unhindered and so a wide range of torsion angles is possible. However, it has been found experimentally that certain torsion angles are preferred, varying to some extent with the nature of the bonded atoms X and Y. Rotation about *double* bonds is not possible at normal temperatures. Furthermore, all atoms directly bonded to two atoms joined by a double bond lie essentially in one plane, together with the double-bonded atoms (for example, H_2C=CHCl), and thus torsion angles for double bonds are always near 0 or 180°. For *triple* bonds, the directly bonded neighbors are collinear with the triple-bonded

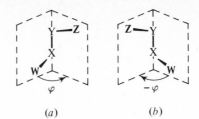

(a)   (b)

FIGURE 11-2   **Torsion angle.** A torsion angle φ can be defined for any sequence of four bonded atoms (W-X-Y-Z). It is the angle of twist, around the bond X-Y, of the plane of the three atoms X, Y, and Z relative to the plane of the three atoms W, X, and Y. The sense of rotation (clockwise or counterclockwise) is indicated by means of the sign of the torsion angle. Diagrams (a) and (b) show that two arrangements W-X-Y-Z that differ only in the sign of their torsion angles are mirror images of each other. Torsion angles are sometimes called conformational angles.

atoms (for example, HC≡CH and H—C≡N) and the torsion-angle concept does not apply.

Structures differing only in angles of rotation about single bonds are referred to as different *conformations*. For example, the observed torsion angle or conformational angle in peroxides such as hydrogen peroxide, H-O-O-H, is about 90°. It is not surprising that the preferred conformation for the corresponding compounds of the next higher congener of oxygen, sulfur (that is, compounds containing the disulfide linkage W-S-S-Z, where W and Z each represent some atom, to which other atoms may be joined), is very similar to that for the peroxides, the torsion angle being about 100°. The disulfide linkage plays an important role in the three-dimensional architecture of proteins. In all proteins containing this linkage whose structures have been studied in detail, the observed conformation about the S-S bond has been found to be near that in much simpler molecules.

A more complex example is the torsion angle about a single C-C bond, in which three other atoms are joined to each of the directly bonded carbon atoms, as in ethane. Two of the possible conformations for ethane are illustrated in Fig. 11-3. These two conformations, which represent the maximum and the minimum in the torsional potential energy, differ in energy by 12 kJ mol^{-1}, about five times the value of RT at room temperature. Conformations about C-C bonds in which the groups attached to each carbon atom are not all the same often have a larger difference in energy favoring the preferred conformation.

Conformational questions are of considerable importance in many areas of chemistry. They play an especially significant role in modern organic chemistry and biochemistry, but they are important in other areas also—for example, even in the chemistry of the elements themselves. The fact that sulfur normally exists as S_8 molecules, in a crown-shaped ring (Fig. 4-2b), can be explained readily as a consequence of the preferred S-S-S bond angle and S-S-S-S torsion angle (both about 100°). Rings of six sulfur atoms can be isolated for a short time but they are unstable, presumably because the torsion angle is nearer 60° than the preferred 100°.

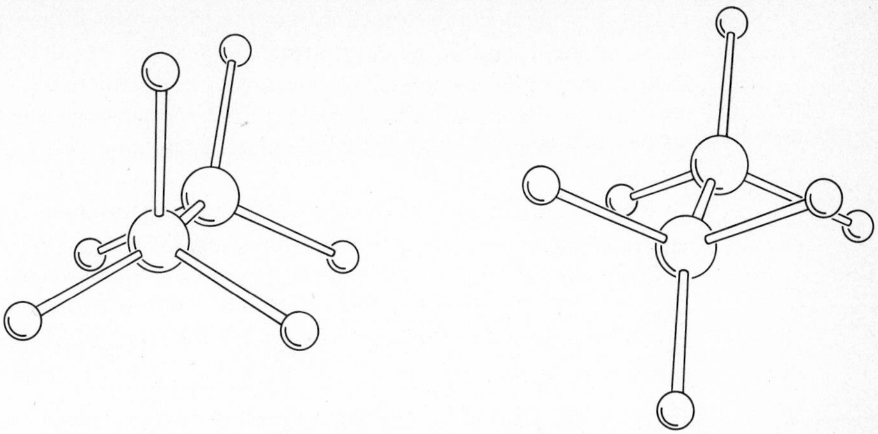

FIGURE 11-3   **Two conformations of ethane, H_3CCH_3.**   The two CH_3 groups in ethane may be rotated relative to each other through any angle. The figure shows two of the many possible relative orientations or conformations. The molecule on the left is said to be in the *eclipsed* conformation because, viewed along the C-C bond, the hydrogen atoms in one CH_3 group cover, or eclipse, those in the other. The molecule on the right is in the *staggered* conformation.

11-3   The Ionic Bond

The attraction between oppositely charged ions, which constitutes the ionic bond, is primarily the result of their electrostatic interaction [Equation (3-1)]. Repulsive forces caused by van der Waals interactions become appreciable as the ions approach, as indicated by the very steep portion on the short-distance side of the potential energy curves in Fig. 3-13. The fact that these curves are extremely steep implies that the model of an atom or ion as a hard (i.e., incompressible) sphere of definite radius is a good first approximation. This model has proved especially useful in consideration of ionic crystals.

Ionic crystals.   Many crystals can usefully be regarded as composed of ions packed together in a systematic and efficient way, with the only significant attractive forces between the ions arising from their electrostatic interaction, covalent forces being negligible. All the evidence for the existence of ions in solids cannot be reviewed here in detail, but the more important facts include:

1.   When many solids, including sodium chloride and other familiar salts, are melted (or vaporized), the resulting melt (or vapor) shows very high electrical conductivity, attributable to ions that are now highly mobile. Although the ions are in relatively fixed positions in a crystalline solid, they may migrate slowly, and this low mobility can also be observed.

2.   If it is assumed that a crystal of empirical formula MX is composed of ions M^+ and X^-, the *lattice energy* of the crystal is the energy released in the reaction

$$M^+(g) + X^-(g) \longrightarrow MX(s) \tag{11-4}$$

that is, it is the energy released during the formation of the crystal from its components in their gaseous states. Lattice energies can be determined indirectly from experimental measurements. They can also be calculated from a knowledge of the structure of the crystal and the energy of interaction of the ions as a function of their distance apart—essentially the electrostatic attraction and the van der Waals repulsion.

When comparisons are made between experimental and calculated lattice energies, the agreement is remarkably good for crystals believed to be essentially ionic in nature. For example, typical theoretical and experimental values are respectively, in kJ/gfw, 760 and 770 for NaCl, 670 and 670 for KCl, 610 and 570 for CsI. The agreement is significantly poorer for crystals in which the bonds between the ions presumed to be present are believed to have appreciable covalent character.

Ionic coordination geometry and ionic radii. The structures of ionic crystals of compositions described by the formulas AB or AB_2 are usually quite simple. Some typical structures are shown in Figs. 11-4 and 11-5. Electrostatic forces

FIGURE 11-4   **The sodium chloride structure.** The entire structure of a crystal can be built up by periodic repetition by translation in three dimensions of a small portion called the *unit cell*. One unit cell of the sodium chloride structure shown here is indicated by diagonal hatching; it has an edge length $a = 5.64$ Å at room temperature. The larger spheres represent Cl^-, the smaller ones Na^+. To show geometric relationships more clearly, the spheres have been reduced in scale and connected by lines that must not be mistaken for covalent bonds.

This is a face-centered cubic (FCC) structure, so called because the ions of each type occupy the corners and the centers of the faces of a cube (of side a). Each Cl^- has 6 Na^+ ions as closest neighbors, at distances $L = a/2$. Further close neighbors are 12 Cl^- ions at $L\sqrt{2}$, 8 Na^+ ions at $L\sqrt{3}$, and many more too numerous to list here. The environment of each Na^+ is similar; the only difference is that the roles of Cl^- and Na^+ are interchanged. Many substances have this structure. Among them are the halides of Li, Na, and K; CaO, MgO, NiO, PbS, AgF, AgCl, and AgBr. Some of them (e.g., PbS, AgCl, and AgBr) are considered to have substantial covalent character.

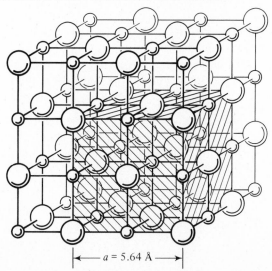

$\longmapsto a = 5.64$ Å $\longrightarrow$

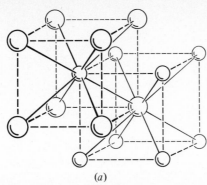
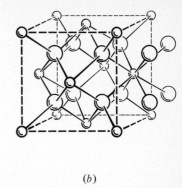

(a)   (b)

FIGURE 11-5   **(a) The cesium chloride structure; (b) the CaF₂ (fluorite) structure.**   In the cesium chloride structure each Cs^+ ion (smaller spheres) is surrounded by eight Cl^- ions (larger spheres) at the corners of a cube, and each Cl^- ion is surrounded in the same way by eight Cs^+ ions. Either of the two cubes shown in the figure can serve as the cubic unit cell of the structure. This structure is sometimes *erroneously* called body-centered cubic (BCC), but the ion in the center of either of the two cubes shown is of a different kind than the ions at the corners of the same cube. The structure is therefore *simple cubic,* an ion of a *given* kind occupying only the corners of a cube.

The calcium fluoride (fluorite) structure also has cubic symmetry; the dashed lines indicate the cubic unit cell of the crystal. The large spheres represent the anions (F^-), each being surrounded tetrahedrally by four cations (Ca^{2+}), indicated by the smaller spheres. Each Ca^{2+} is, in turn, surrounded by eight F^- ions at the corners of a cube.

Examples of crystals with the CsCl structure are the chlorides, bromides, and iodides of Cs and Tl. SrF_2, BaF_2, and CdF_2 are typical substances with the fluorite structure; Li_2O, Li_2S, and K_2Te have the so-called anti-fluorite structure, in which the roles of cations and anions are interchanged.

are nondirectional—that is, the electric field around an ion is spherically symmetric. Hence, to the extent that only electrostatic forces play a role, the characteristic arrangement of the ions of opposite charge immediately surrounding a given ion (an arrangement referred to as the coordination geometry of the ion) is determined by the relative sizes of the ions involved. The most common coordination geometries found in ionic crystals, those of coordination numbers 4, 6, and 8, are illustrated in Fig. 11-6, which also shows typical geometries of coordination numbers 2 and 3. The coordinated ions often occupy the corners of a regular polyhedron (a tetrahedron, an octahedron, or a cube, respectively), with the oppositely charged ion at its center.

The structures of many simple inorganic salts can be classified in a few structural types, of which the NaCl structure is a representative example. Different structural types correspond to different ways of combining the characteristic coordination polyhedra, in accord with the stoichiometry of the compound. The structural type is to some extent determined by the ratio of the sizes of the cation and the anion, as well as by the relative numbers of the different ions (the empirical formula of the compound).

The picture of ions as hard spheres with well-defined radii has proved very useful, particularly in interpreting the properties of ionic crystals. Within the range of

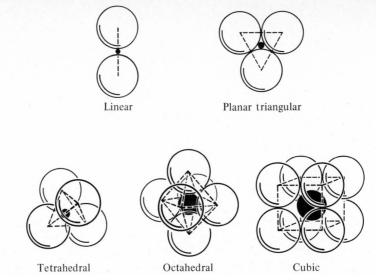

Linear   Planar triangular

Tetrahedral   Octahedral   Cubic

FIGURE 11-6   Five common coordination geometries. The names indicate the geometry of the coordination, that is, the geometric figure whose edges are represented by the dashed lines in each drawing and whose corners are occupied by the coordinated atoms (shown here by light spheres). The ion with which they are coordinated, shown dark, lies at the center of the geometric figure. The tetrahedron, octahedron, and cube, with 4, 6, and 8 corners, respectively, are three of the most common coordination polyhedra in ionic crystals. The respective coordination numbers are 4, 6, and 8. Linear and planar triangular coordination (coordination numbers 2 and 3) exist in isolated molecules and ions, as do the other coordination geometries shown.

applicability of this concept, the shortest distance between an anion and a cation (for example, Cl^- and Na^+) in a crystal is the sum of their ionic radii, but even if this distance is known it is not possible without some additional assumption to deduce the radius of either ion separately. However, if one can find a crystal in which the anions are in contact with one another, then the anion radius will be just half the distance between the centers of two anions in contact. It is possible to rationalize much of the data on ionic crystal structures by assuming that many crystals do contain anions in contact with each other. This makes it possible to derive anion radii and then to deduce cation radii from the shortest anion-to-cation distances in crystals in which the anions are not in contact.

The radii in Fig. 11-7 were obtained by refinements of this method, including judicious selection and averaging. Sets of ionic radii have proved of great value in crystal chemistry, in a variety of ways. They have provided the key to an understanding of the phenomenon of *isomorphism* (Greek: *isos,* equal; *morphe,* form) of ionic crystals, that is, the existence of different crystalline substances with very similar crystal shapes and chemical formulas. It has long been known that certain ions can often be substituted partially or completely in crystals for others of the same charge. However, this phenomenon long puzzled chemists because it seemed to bear no obvious relation to chemical similarity. For example, it is seldom possible to substitute K^+ for Na^+, or vice versa, in crystals, despite their close chemical similarity

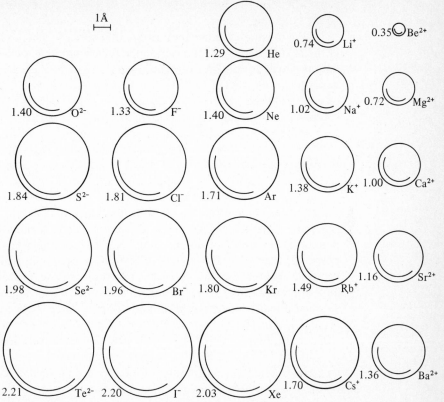

FIGURE 11-7 **Approximate sizes of some ions and isoelectronic inert-gas atoms.** The number below each ion or atom is its radius in angstroms. Note that the ions (and inert-gas atoms) in each horizontal row are isoelectronic with one another and that their radii decrease with increasing positive (or decreasing negative) charge, as expected. Each column contains a set of congeners, for which the radii increase as the atomic number increases. Whereas the radii of the spheres depicting the ions were obtained from crystal data, those for the inert-gas atoms were deduced from the detailed behavior of the gases, so that the two kinds of radii are not strictly comparable.

in many other ways. On the other hand, rubidium ion (Rb^+), ammonium ion (NH_4^+), and thallous ion (Tl^+), although chemically quite different, form many isomorphous crystals. These observations can be easily explained on the basis of the hard-sphere model for ions. Rb^+, NH_4^+, and Tl^+ have nearly identical radii, 1.49, 1.48, and 1.50 Å, respectively,[1] while the radii of Na^+ and K^+, 1.02 and 1.38 Å, respectively, are quite different from one another.

Another use of ionic radii is to provide an indication of the oxidation state of the ion concerned. The radii of different ions formed by a given element decrease markedly with increasing positive charge. For example, the radius of Tl^+, just cited, is 1.50 Å; that of Tl^{3+} is 0.88 Å. The radii of Cu^+ and Cu^{2+} are 0.96 and 0.73 Å, respectively. The effect is usually sufficiently pronounced, at least for lower oxidation

[1] The NH_4^+ ion is not spherical and the value quoted is an "effective" radius.

states, that the effective radius of an ion in a crystal can be a fairly reliable indicator of the oxidation state.

One of the most striking applications of ionic radii is in predicting the arrangement of the nearest-neighbor anions around a given cation in a crystal, that is, the coordination polyhedron of a given cation-anion pair. The predictions are based on the hard-sphere ion model and on the assumption that the distance between centers of oppositely charged ions will be as small as possible. It is also assumed that each cation will be surrounded by as many coordinated anions as possible, provided that each anion is in contact with the cation; this implies that the larger the relative size of the cation, the larger its coordination number. The important quantity is thus the ratio of the radii of the cation and the anion. The structures of many ionic crystals can be understood with the help of this model.

Ionic radii should not be taken too literally, despite their usefulness. No set of ionic radii is completely consistent in the sense that it can be used with unfailing reliability for the prediction of interatomic distances, the structures of ionic compounds, the existence of isomorphous crystals, the oxidation states of ions, and other properties that reflect effective ionic sizes. Ions are not, in reality, completely incompressible spheres. Furthermore, effective radii vary with coordination number and also with the number of unpaired electron spins when the ion has a partially filled subshell. Finally, the assumption that covalent forces between the atoms in "ionic" crystals are negligible is one that must eventually break down as one progresses from substances in which the bonds are almost entirely ionic (for example, CsF) to those in which the bonds are predominantly covalent (for example, ZnS) and a certain amount of deformation of ions from their ideal spherical shapes may also occur.

11-4   The Octet Rule

The valence shells of the elements in each of the two short eight-element periods contain four orbitals and thus can accommodate a total of eight electrons with paired spins—the orientation of four of the spins being opposed to that of the other four. G. N. Lewis postulated that there is a tendency of these atoms to have their valence shells either empty or else filled by a full complement of eight electrons. This is part of a rule known as the *octet rule*. There are two ways in which atoms may achieve a completely filled or a completely empty valence shell. The first is the outright *gain* or *loss* of electrons, that is, the formation of ions. A second and equally important way of attaining a full octet of valence electrons is the *sharing* of electrons, the formation of a covalent bond.

The octet rule is only infrequently violated for the elements of the first short period (Li through F), and then only by the atom's having *fewer* than eight electrons in its valence shell. A second part of the octet rule is, then, that the elements of the first short period are *never found to have more than eight valence electrons*. On the other hand, some of the elements in the second short period (Na through Cl) do have more than eight valence electrons in some of their compounds, although in many they have just an octet.

The reasons why the octet rule is strictly obeyed in the first short period, but may be broken occasionally in the second, become apparent when the energies of the orbitals involved are considered (Fig. 10-3). In the first short period only the $2s$ and $2p$ orbitals are available for use by valence electrons, because the energies of the $3s$ and $3p$ orbitals are so much higher as to be prohibitive. The situation is less stringent for the second short period, because the $3d$ orbitals are much closer in energy to the $3s$ and $3p$ orbitals than the $3s$ and $3p$ orbitals are to $2s$ and $2p$ orbitals. It is thus understandable that the elements Na through Cl may occasionally use $3d$ orbitals for purposes of sharing electrons, particularly if this is favored energetically by the formation of strong bonds.[1]

A corollary of the octet rule is that hydrogen and helium may never have more than two valence electrons. The reason is that the large energy gap between the $1s$ orbital and those in the next shell makes it unprofitable for H or He to use $2s$ or $2p$ orbitals.

The situation is more complex in the longer periods. Two general observations may be made. (1) In many compounds of elements in these periods the atoms contain—either by sharing or by outright acquisition—a complement of electrons that corresponds to the next higher or lower inert gas. This is particularly true for elements near the beginning or near the end of the period, such as the alkali and alkaline-earth metals, the halogens, and the chalcogens. (2) As the principal quantum number n increases, the level spacings narrow. Thus the use of orbitals outside the valence shell proper becomes less and less prohibitive. This explains, for example, why inert gases like Kr, Xe, and Rn are able to form compounds although no similar compounds have been prepared for He, Ne, and Ar.

11-5 The Covalent Bond: Lewis Model

As indicated in Chap. 3, the nature of chemical bonding in nonionic compounds began to be understood only after the discovery of the electron and the development of the model of the nuclear atom, with extranuclear electrons, by Rutherford and Bohr. The idea that covalent bonds involve shared pairs of electrons was first proposed by Lewis in 1916, but it was not until the development of quantum mechanics in the mid-1920s that a clear understanding of covalent bonding became possible. While the Lewis picture of bonding is qualitative and cannot account for the structures of some molecules, it is still in common use becase it is simple and, when used judiciously, is capable of predicting the structures and properties of many species. We shall use the Lewis model as an introduction to the detailed discussion of the covalent bond; the quantum-mechanical theory is developed in Sec. 11-6 and in Chap. 12.

Lewis formulas. The application of the octet rule to bonding is greatly helped by a simple scheme devised by Lewis for representing electronic formulas. In these

[1]The reader may wonder whether $4s$ electrons, which have about the same energy as $3d$ electrons, are similarly used; the answer is that they are not, because their spherically symmetric distribution does not favor the formation of strongly directed bonds, as we shall see in Chap. 12.

Lewis formulas, which were introduced in Sec. 3-2, shared pairs of electrons are represented by a line connecting the bonded atoms, unshared outer-shell electrons are represented by dots, and the nucleus and the inner-shell electrons are represented by the symbol for the element.

There is no necessary geometric significance to the relative positions of the atoms or the electron pairs in a Lewis formula; that is, the formula should in general be considered only schematic. However, when it is easily possible to display significant geometric features with such a formula, it is helpful to do so. Typical Lewis formulas (electron-dot formulas) are

$$H-\ddot{\underset{\cdot\cdot}{I}}: \qquad H-\ddot{\underset{|}{O}}: \qquad H-\ddot{\underset{|}{N}}: \qquad H-\underset{|}{\overset{|}{C}}-H \qquad (11\text{-}5)$$

Inspection shows that in each of these molecules the octet rule is satisfied; i.e., when the shared pairs are counted with *each* of the atoms that share them, the atoms achieve the electron configuration of the nearest inert gas. In each of the last three molecules the bond angle is not far from tetrahedral (Table 11-5) and an effort is sometimes made to display, in the formula, the bent shape of H_2O, the pyramidal shape of NH_3, and the tetrahedral shape of CH_4.

The Lewis formulas of charged molecules—that is, of ions containing several atoms—are no different from those of neutral molecules. It is convenient to place them within brackets so that the charge of the ion can properly be shown as an attribute of the whole group rather than of one of the constituent atoms. Examples of such ions are H_3O^+, BH_4^-, ClO^-, and SO_4^{2-}:

$$\left[H-\ddot{O}: \right]^+ \quad \left[H-\underset{|}{\overset{|}{B}}-H \right]^- \quad \left[:\ddot{\underset{\cdot\cdot}{Cl}}-\ddot{\underset{\cdot\cdot}{O}}: \right]^- \quad \left[:\ddot{\underset{\cdot\cdot}{O}}-\underset{|}{\overset{\cdot\cdot\overset{\cdot\cdot}{O}}{S}}-\ddot{\underset{\cdot\cdot}{O}}: \right]^{2-} \quad (11\text{-}6)$$

Isoelectronic species have similar Lewis formulas and, as mentioned earlier, similar geometric configurations. In the examples shown here, H_3O^+ is isoelectronic with NH_3, and BH_4^- is isoelectronic with CH_4 (and with NH_4^+). Both ClF and the isoelectronic ClO^- have Lewis formulas similar to those of any halogen molecule. In the sulfate ion, the four oxygen atoms are situated tetrahedrally around the sulfur atom, and the oxygens are all equivalent—i.e., there is nothing to distinguish any one of them from the others. It is difficult to represent this tetrahedral ion on a plane surface because of the numerous unshared electron pairs that must be portrayed.

EXAMPLE 11-2   **Lewis formulas for HSO_4^- and H_2SO_4.** Draw Lewis formulas for the hydrogen sulfate ion and sulfuric acid by starting with the Lewis formula for the sulfate ion in (11-6) and adding one or two protons to it.

Solution. It is immaterial to which of the four oxygen atoms each of the protons is added, except that they should be attached to different oxygen atoms. The electronic formula of SO_4^{2-} remains almost unchanged, since each proton carries no electrons with it. For each added proton one of the hitherto unshared pairs on a given oxygen atom becomes shared. The resulting formulas are

$$\left[\begin{array}{c} :\ddot{O}: \\ | \\ H-\ddot{O}-S-\ddot{O}: \\ | \\ :\ddot{O}: \end{array}\right]^{-} \qquad \begin{array}{c} :\ddot{O}: \\ | \\ :\ddot{O}-S-\ddot{O}: \\ | \quad \backslash H \\ :\ddot{O}: \\ | \\ H \end{array} \qquad (11\text{-}7)$$

In these species, the H-O-S bond angles are not far from tetrahedral, but it is awkward to represent them all as bent because of the crowding of the formula, so they are usually drawn as shown here. Note that in all these formulas, the octet rule is satisfied for every atom but hydrogen, which must achieve the He configuration for stability.

Lewis formulas for species containing multiple bonds are straightforward. Each shared pair of electrons is symbolized by a line between the atoms, as illustrated by the formulas for ethylene (C_2H_4), formaldehyde (H_2CO), acetylene (C_2H_2), nitrogen, and cyanide ion:

$$\begin{array}{cc} H & H \\ \backslash \quad / \\ C=C \\ / \quad \backslash \\ H & H \end{array} \qquad \begin{array}{c} H \\ \backslash \quad \cdot\cdot \\ C=\ddot{O}: \\ / \\ H \end{array} \qquad H-C\equiv C-H \qquad :N\equiv N: \qquad [:C\equiv N:]^{-} \qquad (11\text{-}8)$$

Again note that each atom has a completed valence shell (an octet for each atom except H) and also that the first two species above (ethylene and formaldehyde) are isoelectronic, as are the last three (acetylene, nitrogen, and cyanide ion).

EXAMPLE 11-3 **Lewis formula for N_2F_2.** Write a Lewis formula for dinitrogen difluoride, for which the sequence of atoms is FNNF.

Solution. Each F has 7 valence electrons and each N has 5, or 24 in all. If we start by joining the atoms in sequence by single bonds, leaving the other valence electrons of each atom unshared, we get

$$:\ddot{F}-\dot{N}-\dot{N}-\ddot{F}:$$

Each F has a completed octet, but each N has only seven electrons, one being unpaired on each N atom. The octet on each N can be completed if these two unpaired electrons become paired, forming a second N-N bond, leading to

$$:\ddot{F}-N=N-\ddot{F}:$$

All requirements are now satisfied. It is not evident from this formula that the molecule is actually bent at each N atom:

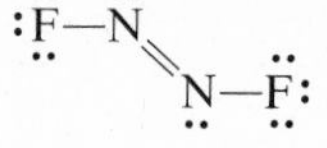

$$\ddot{:}\!\overset{..}{F}\!\!-\!\!\overset{..}{N}\!\!\diagdown\!\!\underset{\ddot{:}}{N}\!\!-\!\!\overset{..}{F}\!\!:$$

The geometric implications of the bonding are discussed in the following chapter.

As noted before, the octet rule is not entirely inviolable. Third-row elements (especially Si, P, S, and Cl) and those with higher atomic numbers, which have relatively low-lying *d* orbitals available, sometimes use more than four orbitals for bond formation and occupancy by unshared outer-electron pairs. The common orbital combinations used in such "expansions of the valence shell" are discussed in Chap. 12; only a few examples will be given here.

Phosphorus pentachloride (PCl_5) and sulfur hexafluoride (SF_6) both have structures that violate the octet rule as far as the P and S atoms are concerned. There are five bonds to the phosphorus atom in PCl_5, a molecule that has the shape of a *trigonal bipyramid*—that is, two triangular pyramids sharing an equilateral triangular face. A chlorine atom occupies each corner, with the phosphorus atom at the center of the polyhedron:

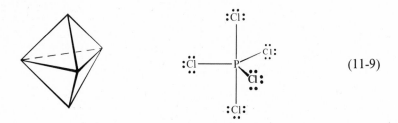

(11-9)

The total number of electrons displayed in this formula is 40 (20 pairs), corresponding to the seven valence electrons of each of five chlorine atoms and the five valence electrons of the phosphorus atom. The five P-Cl bonds in PCl_5 are not all equivalent; the three in the *equatorial* plane (the plane of the face shared by the two pyramids) are different from the two to the apexes of the trigonal bipyramid (called *axial* bonds).

The SF_6 molecule is octahedral, with the S atom at the center of the octahedron and an F atom at each of the six corners. All six bonds are equivalent in this molecule. Its Lewis formula is

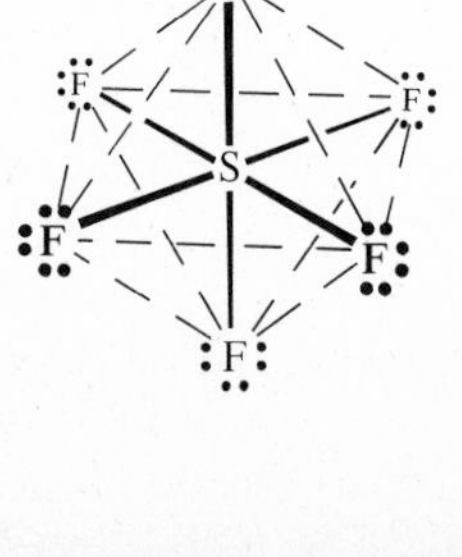

(11-10)

(The dashed lines are shown only to outline the octahedron whose six corners are occupied by the fluorine atoms.)

It is possible to draw satisfactory Lewis formulas for many compounds. There are also molecules and ions for which several possible formulations exist, each different, yet all about equally acceptable. As explained in the next chapter, when this is true only a combination of all satisfactory Lewis formulas provides an appropriate description of the molecule in valence bond terms.

In most of the molecules and ions for which Lewis formulas have been given so far, each of the atoms joined by a covalent bond has contributed one electron to the bond. However, there are some bonds for which both electrons may be regarded as supplied by only one of the atoms—for example, in the product of the reaction between ammonia and boron trifluoride (BF_3):

$$\begin{array}{c} H \\ | \\ H\!-\!N\!:\ +\quad \overset{\displaystyle :\!\overset{\cdot\cdot}{F}\!\cdot\quad \cdot\!\overset{\cdot\cdot}{F}\!:}{B} \\ / \\ H \end{array} \quad\begin{array}{c} :\!F\!: \\ | \end{array}\quad \longrightarrow \quad \begin{array}{c} H \\ \diagdown \\ H\!-\!N\!-\!B\!-\!\overset{\cdot\cdot}{F}\!: \\ \diagup \qquad | \\ H \qquad \overset{\cdot\cdot}{F}\!: \end{array}\qquad (11\text{-}11)$$

The product of this reaction is stable, whereas BF_3 is very reactive. Scrutiny of the Lewis formula given here for BF_3 shows that the B atom has only a sextet of electrons, not an octet, which helps to explain the reactivity of BF_3. In the product formed in (11-11), all atoms have completed valence shells.

The reaction between ammonia and a proton to form NH_4^+ is one in which the electrons in the fourth N-H bond are contributed by only one of the atoms:

$$H^+ +\ :\!\overset{\displaystyle H}{\underset{\displaystyle H}{N\!-\!H}} \quad \longrightarrow \quad \left[\ \overset{\displaystyle H\ \ H\ \ H}{\underset{\displaystyle H}{\diagdown\ |\ \diagup \atop N}}\ \right]^{+} \qquad (11\text{-}12)$$

Note, however, that once the NH_4^+ is formed, all four bonds in it are equivalent and indistinguishable.

Comparison of the reactions illustrated in (11-11) and (11-12) shows that there is a fundamental similarity. In each of them, the ammonia molecule donates its pair of electrons to another molecule or ion that needs additional electrons for greater stability. Since the reaction of ammonia with a proton (11-12) is a typical acid-base reaction, Lewis proposed that reactions like (11-11) also be regarded as acid-base reactions. In this general sense, a *Lewis acid* is any molecule or ion that will accept a pair of electrons and a *Lewis base* is any molecule or ion that will donate a pair of electrons. For example, H^+ and BF_3 are Lewis acids and OH^- and NH_3 are Lewis bases. Not all species with unshared pairs of electrons in the valence shell are good Lewis bases, although many are.

Rules for constructing Lewis formulas. A set of general rules can be given to serve as a guide in the construction of Lewis formulas; all the formulas in the preceding section will be found to satisfy these rules:

1.   The structural (or topological) formula of the molecule or ion concerned, that is, the way in which the atoms are connected to one another, must be known. One can sometimes make sensible guesses about which of two or more possible structural formulas might be more reasonable, but since isomers exist for many structures, it is generally essential to know which atoms are bonded to each other in the species of interest.

2.   The Lewis formula must show all the outer-shell (or valence) electrons of the atoms involved, appropriately increased or diminished to take the overall charge into account when considering an ion. For example, the formula

$$\left[:\overset{\displaystyle ..}{\underset{\displaystyle ..}{O}}-H \right]^{-}$$

shows a total of eight electrons, corresponding to six valence electrons for oxygen, another for hydrogen, and one additional electron for the single negative charge of the ion. Similarly, in the Lewis formula for sulfate ion [Formula (11-6)] 32 electrons are shown: six valence electrons for each of the four oxygen atoms, six for the sulfur atom, and two additional electrons corresponding to the charge -2.

3.   Each atom in the Lewis formula should have a completed valence shell (or possibly an expanded one for elements beyond Ne). This means two electrons for H, Li^+, and Be^{2+}, eight electrons for second-row elements, and eight (or sometimes ten or twelve) electrons for elements of higher atomic number. Expansion of the valence shell beyond twelve electrons (six pairs) is rare. Only occasionally does the valence shell contain fewer than eight shared and unshared electrons.

The partial ionic character of covalent bonds.   As we have indicated, most covalent bonds have some ionic character and most ionic bonds have some covalent character. Since electronegativity is defined (Sec. 10-4) as the tendency of a bonded atom to attract electrons from other atoms in the same molecule or ion, it is not surprising that the partial ionic character of a covalent bond should be related to the electronegativities of the bonded atoms. Table 11-8 shows the approximate

TABLE 11-8   **Approximate ionic character of covalent bonds**

$\lvert x_A - x_B \rvert$	Percent ionic character	$\lvert x_A - x_B \rvert$	Percent ionic character
0.2	1	1.8	55
0.4	4	2.0	63
0.6	9	2.2	70
0.8	15	2.4	76
1.0	22	2.6	82
1.2	30	2.8	86
1.4	39	3.0	89
1.6	47	3.2	92

degree of ionic character of a covalent bond as a function of the difference in electronegativity of the bonded atoms.[1] It can be seen that the larger the electronegativity difference, the greater the degree of ionic character. The halfway point (for a bond between atoms A and B) is at a value of about 1.7 for the electronegativity difference, $|x_A - x_B|$.

With a molecule such as HCl, one can draw both a covalent formula and an ionic formula:

$$H\text{—}\ddot{\underset{\cdot\cdot}{C}}l\colon \quad \text{and} \quad H^+ \left[\colon\ddot{\underset{\cdot\cdot}{C}}l\colon \right]^- \tag{11-13}$$

The electronegativity of Cl is about 3.0, that of H about 2.1 (Fig. 10-5). Thus, the difference in electronegativity is about 0.9 and, on the basis of Table 11-8, the bond is said to have about 18 percent ionic character. In agreement with these ideas, HCl is a polar molecule, the H atom being associated with a small residual positive charge and the Cl atom with a negative charge of equal magnitude. One of the most polar substances is LiF, which exists as a diatomic molecule only in the gaseous state at very high temperatures (the normal boiling point of LiF is 1949 K). The bond in LiF has around 90 percent ionic character, the electronegativity difference of Li and F being about 3.0. This gaseous "molecule" may thus reasonably be referred to as an ion pair.

Covalent formulas (Lewis formulas) are customarily drawn for molecular substances unless the degree of ionic character in the bonds is *extremely* high. For example, although gaseous LiF would not commonly be represented by a Lewis formula, the BF_3 molecule is usually so represented, as in (11-11), despite the fact that B-F bonds are estimated to have nearly two-thirds ionic character. The normal boiling point of BF_3 is $-100\,°C$ and the vapor contains individual BF_3 molecules, not dimers or more highly associated species.

Ionic contributions occur not only in bonds linking atoms of a diatomic molecule, but in polyatomic molecules as well. The rules are the same. Each bond may be considered separately and its ionic character judged by the electronegativity difference between the bonded atoms. The ionic character of a double bond or a triple bond is estimated from the difference in the electronegativities of the bonded atoms, just as if the bonds were two or three single bonds.

11-6   The Covalent Bond: Quantum Theory

General remarks about quantum-chemical theories of bonding. An accurate treatment of the binding together of atoms in molecules requires quantum mechanics. The systems involved may be represented by wave functions that are solutions of the Schrödinger wave equation appropriate to the molecule considered, just as for an atom. However, highly accurate solutions have been

---

[1] Although various quantitative relationships between electronegativity and ionic character have been proposed, they should not be taken too literally since neither property can be rigorously defined. These words of caution should be kept in mind when Table 11-8 is studied.

obtained to date only for the simplest diatomic ions and molecules, such as H_2^+ (the hydrogen molecule-ion), H_2, and HLi. For more complex species—even for such simple molecules as O_2 and H_2O—the exact solution of the Schrödinger equation, even if it were possible, would yield a wave function of such detail and complexity as to make it impossible even to record. What is of interest to the chemist is a method of constructing and improving simplified approximate wave functions that can be used for the calculation of energies, geometries, and other properties of molecular systems. Many different approaches to the problem of constructing suitable wave functions and thus obtaining approximate solutions to the Schrödinger equation for molecular systems have been proposed. We shall discuss qualitatively two of the simplest of them.

The justification of these quantum-chemical approximations is that, like any theory, they permit the explanation and categorization of a large variety of phenomena. Each of the approaches is based on wave-mechanical principles but includes additional postulates, different for each approach. These postulates are compatible with wave mechanics but are chosen so that they greatly simplify the process of obtaining solutions. In each of these theories, results obtained from application of wave mechanics to simple situations are assumed to be general and are applied to situations too complicated for rigorous treatment.

The most widespread of the current theories of molecular structure are the *valence bond* and the *molecular orbital* theories. An essential feature of both is that chemical bonding arises as a direct consequence of the overlap of orbitals on nearby atoms. The theories differ in their postulates and in details of treatment, and they are in part tailored to fit different types of molecules and ions. Thus in a given situation one or the other of them may be most easily applied and most revealing in its results. They would merge into one and the same definitive theory of molecular structure if it were possible sufficiently to refine the approximations used.

Molecular orbital treatment of the hydrogen molecule-ion and the hydrogen molecule. We start with a description of the basic features of molecular orbital theory and turn later to the valence bond method, even though the valence bond approach was developed first. The molecular orbital method was developed in its early form chiefly by Hund, Mulliken, Lennard-Jones, and Hückel. Its principal idea is to start with all the nuclei and inner shells of the atoms making up the molecule to be treated and formulate suitable approximate wave functions that extend over the entire molecular domain and are hence called *molecular orbitals*. These orbitals are then filled with the valence electrons of the original atoms, in accordance with the Pauli exclusion principle (Sec. 9-6) and (for the ground state of the molecule) in the order of increasing energy of the orbitals. In other words, an Aufbau principle is involved, similar to that used in the description of the electronic structure of atoms.

In the simplest version of the theory the molecular orbitals are formed by combining the atomic orbitals of the valence shells of the various atoms constituting the molecule. According to the principles of quantum chemistry, such combining of orbitals always yields the same number of resulting orbitals as there were individual orbitals at the start—there is, so to speak, an "orbital conservation principle".[1]

Consider a system of two hydrogen atoms. Each atom has a $1s$ orbital available in its valence shell, which has no radial nodes and which (like all s orbitals) is spherically symmetric (Figs. 9-6 and 9-9). Since we begin with two orbitals, there must be two ways to combine them so as to produce two molecular orbitals. By the rules of quantum chemistry the first combination is their sum, $\psi_+ = \psi_A + \psi_B$ (Fig. 11-8), and the second is their difference, $\psi_- = \psi_A - \psi_B$ (Fig. 11-9). When the atoms are far apart (Fig. 11-8a; even 5 Å is far apart for two hydrogen atoms), they hardly interact because neither of the two $1s$ wave functions has a value significantly different from zero near the position of the other atom. At smaller separations, however, the orbitals overlap noticeably—i.e., both wave functions have values significantly different from zero in the same region of space, called the region of overlap (Figs. 11-8b and 11-9a). The electron probability density ψ^2 is then significantly different from the sum of the probability densities of the two separate atoms. That is, if ψ_A and ψ_B are the orbitals of the separate atoms A and B, the square of their sum is $(\psi_+)^2 = (\psi_A + \psi_B)^2 = \psi_A{}^2 + \psi_B{}^2 + 2\psi_A\psi_B$. This function thus differs from $\psi_A{}^2 + \psi_B{}^2$, the sum of the probability distributions for the two separate atoms, when the product $2\psi_A\psi_B$ is significantly different from zero. This occurs when the wave functions of the two atoms overlap, that is, in a region in which both ψ_A and ψ_B are significantly different from zero. Similar considerations apply to the difference $\psi_A - \psi_B$, because $(\psi_-)^2 = (\psi_A - \psi_B)^2 = \psi_A{}^2 + \psi_B{}^2 - 2\psi_A\psi_B$.

Examination of Fig. 11-8c and d shows that when the orbitals (i.e., the wave functions) are added to each other there is enhancement of the electron density in the region between the nuclei and thus a relatively high probability of finding electrons in this region. This preponderance of negative charge between the (positively charged) nuclei tends to pull the nuclei together. At small separations, however, the mutual electrostatic repulsion between nuclei is sufficiently great to exceed the attraction caused by the electron density between them. At an intermediate distance there is a balance between attraction and repulsion and a stable configuration results.

Figure 11-9 shows the result of combining the atomic orbitals by taking their difference. At large distances (not shown), there is again no interaction, but as the atoms approach one another, the molecular wave function (the difference of the $1s$ functions of the two atoms) remains equal to zero halfway between the nuclei. Consequently the square of the wave function, the probability density, also falls to zero in this region. The electron density is somewhat enhanced on the side of each nucleus away from the other nucleus, but in contrast to the situation when the *sum* of the atomic wave functions is taken, there is no buildup of negative charge between the nuclei to offset the internuclear electrostatic repulsion.

The contrast between the effects of adding the atomic orbitals and of subtracting them (Figs. 11-8 and 11-9) is all-important to an understanding of bonding. In the first case, the molecular orbital that is formed is said to be a *bonding orbital;* that

---

[1]This principle, applied here to the combination of atomic orbitals on separate atoms, holds also for the combination of orbitals on the *same* atom, in the formation of what are termed *hybrid orbitals* (discussed in Chap. 12).

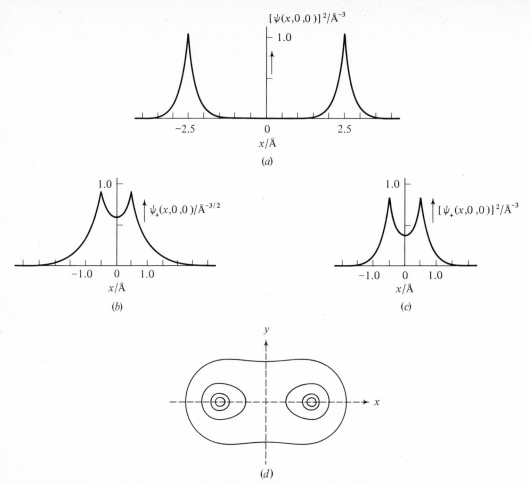

FIGURE 11-8   **Representations of ψ and ψ^2 for the sum of two hydrogen-atom 1s orbitals.**   (*a*) Cross section of the electron probability density, $[\psi(x,0,0)]^2$, for one electron and two hydrogen nuclei (protons) separated by 5.0 Å, along the line through the center of each proton (the line $y = 0$, $z = 0$). Interaction is essentially zero and diagrams of $(\psi_+)^2$ and $(\psi_-)^2$ are identical; i.e., the term $\pm\psi_A\psi_B$ is negligible.

(*b*) Cross section of the sum of two hydrogen 1s orbitals, $\psi_+ = \psi_A + \psi_B$, when the protons are 1.06 Å apart, the internuclear separation observed in H_2^+. The vertical scale and the units for ψ_+, $\text{Å}^{-3/2}$, are such that the square of ψ_+ [diagram (*c*)] has the units Å^{-3} and represents the probability density of one electron.

(*c*) Cross section of $[\psi_+(x,0,0)]^2$, the electron probability density, along the same line as in (*a*) and (*b*). The density is highest near the protons, but there is also a high probability of finding the electrons in the region between the protons.

(*d*) Contours of the electron probability density, $[\psi_+(x,y,0)]^2$, with a distance of 1.06 Å between the hydrogen nuclei. The view in (*c*) is a section of (*d*) along the x axis. The contours in (*d*) are at a constant increment of electron probability density of 0.2 Å^{-3}, the lowest contour being at 0.2 Å^{-3}. [The scale along x is expanded by a factor of about three relative to that in (*a*), (*b*), and (*c*).]

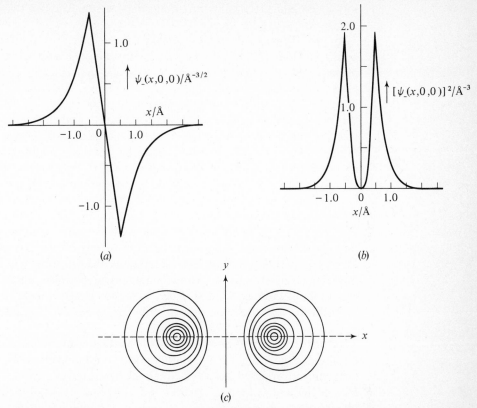

FIGURE 11-9 **Representations of ψ and ψ^2 for the difference of two hydrogen-atom 1s orbitals.**

(*a*) Cross section, along the line through the centers of the protons, of the difference of two hydrogen 1s orbitals, $\psi_- = \psi_A - \psi_B$, when the protons are 1.06 Å apart. The scale is the same as that for Fig. 11-8*b*.

(*b*) Cross section of the electron probability density, $[\psi_-(x,0,0)]^2$, along the same line as in (*a*). The density is highest near the protons, as it was for ψ_+^2, but it falls to zero in the region between them.

(*c*) Contours of the electron probability density, $[\psi_-(x,y,0)]^2$, with a distance of 1.06 Å between the protons. Note the enhancement of the density on the side of each proton away from the other proton. Diagram (*b*) is a cross section of (*c*) along the x axis. The contours are the same as in Fig. 11-8*d* except that the lowest contour is the nodal line $x = 0$ representing the nodal y, z plane, for which ψ_- and $(\psi_-)^2$ are zero.

is, when electrons occupy it, the atoms are held together in the stable combination that we refer to as a molecule. In the second case, the molecular orbital formed is referred to as an *antibonding orbital,* implying that if it is occupied by electrons, their density in the region between the nuclei is actually diminished relative to what it would be in the absence of any interaction, so that there is a repulsion of the nuclei instead of an attraction.

Let us consider in more detail what happens for the molecules H_2^+ and H_2. In H_2^+ there is only one valence electron available, which accordingly can occupy either the bonding or the antibonding molecular orbital. The two resulting energy curves are shown in Fig. 11-10.

In the H_2 molecule there are two valence electrons available, one from each hydrogen atom. If both electrons occupy the bonding orbital, they must, by the Pauli principle, have antiparallel spins. This is the molecular orbital description of the H_2 molecule in its ground state. It is also possible to place one of the two electrons in the bonding orbital and the other in the antibonding orbital, in which case the spins of the two electrons may be parallel or antiparallel. Detailed calculations show that for antiparallel spins there is still some attraction, so that this represents an excited state of the H_2 molecule. When the spins are parallel, there is repulsion at all distances and therefore no formation of a stable (but excited) molecular state. Finally, both electrons may be placed in the antibonding orbital, in which case their spins must be antiparallel. This again corresponds to repulsion at all internuclear distances and therefore does not represent a molecular state. The energy diagram of Fig. 11-10 applies only qualitatively to H_2 because, among other things, it does not take the electrostatic repulsion between the two electrons into account. However, elaborate quantum-mechanical calculations that include this and other possible interactions have yielded excellent agreement between observed and calculated properties of both $H_2{}^+$ and H_2, such as the internuclear distances (1.06 and 0.74 Å, respectively) and the bond energies (256 and 433 kJ mol^{-1}, respectively).

The valence bond method. This method is an extension of the original work on the H_2 molecule by W. Heitler and F. London in 1927, only a year after Schrödinger proposed his wave equation. It was developed in its early days chiefly by J. Slater and L. Pauling. Heitler and London considered the interaction between two *complete* hydrogen atoms, rather than the interactions between the individual nuclei and electrons. The resulting energy diagram is shown in Fig. 11-11. The two curves pertain to the interaction energies between two hydrogen atoms (including the electrostatic repulsion energy between the two electrons) rather than to the energies that result when molecular orbitals are filled with electrons. In the lower

FIGURE 11-10 **Energy of two hydrogen nuclei and one electron as a function of the internuclear distance.** The curves have been calculated by the molecular orbital method. The lower curve has a minimum that leads to the formation of the stable molecule-ion $H_2{}^+$. The upper curve represents repulsion between the hydrogen nuclei at all distances and no formation of a molecule.

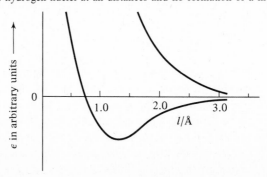

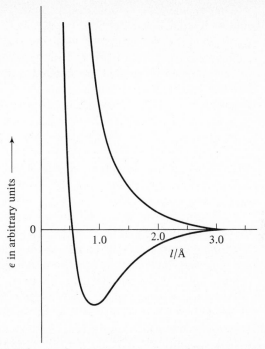

FIGURE 11-11   Energy for two hydrogen atoms as a function of their internuclear distance. The curves have been calculated by the valence bond method. The lower curve corresponds roughly to having both electrons in the bonding orbital of the molecular orbital theory. The upper curve is related to the state described by molecular orbital theory as having a bonding and an antibonding orbital, each occupied by one electron, the electrons having parallel spins. The other states possible according to the simple molecular orbital approach described in the text can also be obtained from refined valence bond theory.

curve the spins of the electrons of the two atoms are antiparallel; in the upper curve they are parallel.

An important feature of the valence bond approach results from the fact that for the two interacting atoms the situation in which a given electron is associated with the first H atom and the other electron with the second H atom cannot be distinguished from the situation in which the given electron is associated with the second H atom and the other electron with the first H atom. This is because electrons are indistinguishable. It is thus only meaningful to speak of electrons associated with a particular atom when the atoms do not interact because their separation is too large. At shorter distances, where the atoms begin to interact, the interchangeability of the electrons leads to a sizable energy term, sometimes called the exchange energy. It is stabilizing for electrons of antiparallel spin and destabilizing for electrons of parallel spin. While the idea of stabilization by electron interchange is very useful and illuminating, it must never be forgotten that this idea is part of the valence bond description of the interaction between atoms, rather than being an actual phenomenon. This is borne out by the fact that there is no equivalent concept in the molecular orbital theory (or other theories) of chemical bonding.

The valence bond method can be applied to other diatomic molecules in a fashion similar to that discussed here for H_2. Polyatomic molecules are also treated by considering the interactions of entire atoms. These interactions are described, in the simplest version of the theory, as arising from electron interchange between the atoms in each pair of atoms that are bonded in the molecule. The electrons are said to be *shared* by the two bonded atoms. In this regard the valence bond description corresponds closely to the conventional chemical picture, with bonding lines drawn between atoms regarded, on the basis of experimental evidence, as bonded. Such evidence includes, for example, close physical proximity that is maintained despite the strains of thermal motion or attacks by reactive molecular and ionic species. The sharing of two and three pairs of electrons describes the double and the triple bond, but sometimes only a single electron is shared (e.g., in the H_2^+ ion, which can be described by the valence bond method also).

In contrast, in the molecular orbital theory approximate orbitals are constructed that extend over the entire molecule. In its simplest version these are formed by combining atomic orbitals of the valence shells of the different atoms composing the molecule. For many molecules (of which H_2 is one), the molecular orbital and valence bond approaches lead to closely similar results. In fact, it is often possible to recombine truly molecular orbitals into orbitals that essentially extend across pairs of bonded atoms only, thus in a sense simulating the valence bond approach. For some molecules, however, including many for which it is difficult or impossible to write conventional Lewis electron-dot formulas, molecular orbital theory may provide additional insight. Molecular orbital theory also provides a better understanding of molecules in electronically excited states than does valence bond theory. Both approaches have their strengths and weaknesses, and neither can adequately replace the other. Molecular orbitals are discussed further in Chap. 12.

While quantum-mechanical theories enable us to calculate properties of molecules and provide important insights into bonding, the covalent bond should not be thought of as caused by a special quantum-mechanical force. The force acting on each nucleus in a molecule can in principle be calculated exactly by applying the laws of classical electrostatics to the electrons and the other nuclei in the molecule, according to a theorem discovered independently by H. Hellmann and R. P. Feynman. The forces responsible for covalent bonding are thus electrostatic in origin. This fact is chiefly of conceptual value; it is not of much help in the theoretical treatment of bonding in molecules, because to calculate the electrostatic forces due to the electrons, their distribution must be known, and this probability distribution can be calculated only by using quantum mechanics, which is usually prohibitively difficult.

Problems and Questions

11-1 Bond energies. Use bond energies to compute approximate values for the energy changes associated with the following gaseous reactions:

(a) $H_2 + Cl_2 \longrightarrow 2HCl$

(b) $3Cl_2 + N_2 \longrightarrow 2NCl_3$

(c) $C_2H_4 + H_2 \longrightarrow C_2H_6$
(d) $2H_2 + N_2 \longrightarrow N_2H_4$

> *Ans.* (a) 180 kJ released; (b) 470 kJ absorbed; (c) 120 kJ released; (d) 110 kJ absorbed

11-2 Bond energies. (a) Use bond energies to predict the approximate changes in energy that accompany the following reactions, where X is Cl, Br, or I:

$$CH_4 + X_2 \longrightarrow CH_3X + HX$$

(b) Would your prediction apply equally to hydrocarbons other than CH_4, such as C_2H_6 or C_8H_{18}?

11-3 Torsion angles. By inspection of Fig. 11-3, find the different torsion angles H-C-C-H for the hydrogen atoms bonded to the two carbon atoms in ethane (a) in the eclipsed conformation and (b) in the staggered conformation.

> *Ans.* (a) 0° and 120°; (b) 60° and 180°

11-4 Packing of unequal spheres. (a) The centers of three touching spheres (see Fig. 11-6) of radius a form the corners of an equilateral triangle (of edge 2a) while a fourth sphere of radius b, touching the other three, is centered in the middle of the triangle. What is the ratio b/a? (b) Repeat these calculations for a planar square with four spheres of radius a at the corners and one of radius b at the center. (c) Repeat for six spheres of radius a and one of radius b, with an octahedral configuration.

11-5 Tetrahedral angles. A set of four alternate corners A, B, C, and D of a cube (Fig. 11-12) define a regular tetrahedron. Make a sketch of the cube and the tetrahedron; let the tetrahedral edge (AB) be d, and let P be the center of the tetrahedron (and the cube). (a) By solid geometry, vector algebra, or any other means calculate the distance AP in terms of d. (b) Show that the angle APB is 109.5°. (c) Suppose four equal spheres of radius a are arranged at the corners of a tetrahedron. Find the radius b of the sphere located at the center of the tetrahedron that will just touch each of the four spheres at the corners; express the answer in terms of the ratio b/a, as in Prob. 11-4. *Ans.* (a) $d\sqrt{3}/(2\sqrt{2})$ (c) 0.225

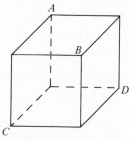

FIGURE 11-12

11-6 Radii of isoelectronic ions. What generalization can you make about the relation between the ionic radii of a pair of isoelectronic ions? Explain it with reference to some particular pair of isoelectronic ions of your own choosing.

11-7   Environment of the ions in CsCl.   The arrangement of the ions in a crystal of CsCl is shown in Fig. 11-5a. The edge of the cubic unit cell is 4.12 Å. (a) Consider a Cl^- ion and give the number of closest Cs^+ ions, their distance from the Cl^- ion, and the polyhedron these Cs^+ ions define. (b) Give the same kind of information for the Cl^- ions nearest the Cl^- ion considered. (c) Describe the environment of a Cs^+ ion in the same way.

Ans.   (a) $8Cs^+$, 3.57 Å, cube;   (b) $6Cl^-$, 4.12 Å, octahedron;

(c) $8Cl^-$, 3.57 Å, cube; $6Cs^+$, 4.12 Å, octahedron

11-8   Solid phosphorus pentachloride.   Although PCl_5 vapor has the structure shown in Fig. 11-13, the crystalline solid formed by condensation of the vapor is found to consist of a mixture of equal numbers of two ions, PCl_4^+ and PCl_6^-. Write Lewis formulas for these species and suggest geometries by analogy with known species.

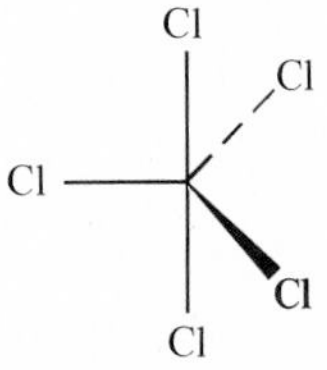

FIGURE 11-13

11-9   Environment of the ions in CaF₂.   The arrangement of the ions in a crystal of CaF_2 is shown in Fig. 11-5b, the edge of the unit cube being 5.46 Å. Describe (a) the arrangement of nearest F^- ions about each Ca^{2+} ion, as in Prob. 11-7, and (b) the arrangement of nearest Ca^{2+} and F^- ions about each F^- ion.

Ans.   (a) $8F^-$, 2.36 Å, cube;   (b) $4Ca^{2+}$, 2.36 Å, tetrahedron; $6F^-$, 2.73 Å,

octahedron

11-10   Ionic character of bonds.   Explain what is meant by the term *electronegativity*. Discuss briefly how it varies in the periodic table and how it may be used to assess qualitatively the ionic character of bonds.

11-11   Covalent and ionic compounds.   Diagram reasonable electronic structures for the following species. Use a line to represent a shared pair of electrons and indicate outer unshared electrons by dots. If the compound is predominantly ionic, indicate the ions by enclosing them in brackets, with the appropriate charge shown, and give the structures for the individual ions.

HBr, I_2, CCl_4, CH_2F_2, HCN, $SiBr_4$, Na_3PO_4, LiF, NCl_3, H_2Se,

NH_4BH_4, SCl_2, PH_3, Li_2S, $MgBr_2$

11-12   Unstable species.   Many molecules that are unstable under normal terrestrial conditions have been observed spectroscopically in comets, together with some familiar species as well. These include CN, C_2, OH, and CH, together with CH_4, NH_3, and CO. The pressure in the central portion of a comet has been estimated to be no higher than 10^{-10} torr, and it is presumably far lower in the tail of the comet. Suggest Lewis electronic structures for the various unstable species mentioned. What feature of these structures might explain why the species are unstable under normal conditions on earth? Why do they apparently persist in comets?

11-13 Ionic character of bonds. Find approximate values for the partial ionic character of the bonds in NH_3 and in NaCl.   *Ans.* about 19 and 67 percent

11-14 Lewis acid. Aluminum chloride, $AlCl_3$, is a strong Lewis acid. Explain what the term *Lewis acid* implies and how this property of aluminum chloride is interpreted in terms of its electronic structure.

11-15 Lewis acid. (*a*) The reaction of BF_3 with ammonia to form F_3BNH_3 is usually cited as the classic example of reaction of a Lewis acid (which is seeking an unshared electron pair) with a base. Is it reasonable to consider this also as an oxidation-reduction reaction? If so, why? If not, in what essential way does it differ? How does it differ from the reaction of a Brønsted acid with a base? (*b*) Contrast the electronic structures of the following two distinctly different compounds: NH_3BF_3 and NH_4BF_4.

"It is not unfair to say that in . . . practically the whole of theoretical chemistry, the form in which the mathematics is cast is suggested, almost inevitably, by experimental results. This is not surprising when we recognize how impossible is any exact solution of the wave equation for a molecule. Our approximations to an exact solution ought to reflect the ideas, intuitions, and conclusions of the experimental chemist."

C. A. Coulson,[1] 1952

"Consequently it is expected that of two orbitals in an atom the one that can overlap more with an orbital of another atom will form the stronger bond with that atom, and, moreover, the bond formed by a given orbital will tend to lie in that direction in which the orbital is concentrated."

L. Pauling,[2] 1939

12 Chemical Bonds: II. Theories of Bonding and Molecular Structure

This chapter is a continuation of the preceding one and deals in more detail with a number of topics alluded to there. The first section describes bonding in some molecules and ions in terms of resonance among different acceptable Lewis valence bond formulas. Next comes a discussion of the most common types of combinations of orbitals on a given atom (hybrid orbitals) that are used in explaining the geometry of molecules. An empirical scheme for correlating molecular shape with electronic structure is also presented. The chapter concludes with an introduction to molecular orbital theory.

[1]C. A. Coulson, "Valence", 2d ed., pp. 113–114, Oxford University Press, New York, 1961.

[2]Linus Pauling, "The Nature of the Chemical Bond", 3d ed., p. 108, Cornell University Press, Ithaca, N.Y., 1960. Copyright 1939 and 1940, 3d ed. © 1960, by Cornell University.

12-1   Resonance among Different Lewis Valence Bond Formulas

The basic concept.   There are some molecules and ions for which several acceptable Lewis formulas can be drawn, no one of which alone is adequate to describe the electronic structure or explain the observed properties. For example, there are three acceptable formulas for the nitrate ion:[1]

$$
\left\{
\left[
\begin{array}{c}
\overset{\cdot\cdot}{:}\overset{\cdot}{O} \\
\diagdown \\
N\!=\!\overset{\cdot\cdot}{\overset{\cdot\cdot}{O}} \\
\diagup \\
:\overset{\cdot\cdot}{O}:
\end{array}
\right]^{-}
,\
\left[
\begin{array}{c}
:\overset{\cdot}{O} \\
\diagdown \\
N\!-\!\overset{\cdot\cdot}{\overset{\cdot\cdot}{O}}: \\
\diagup \\
:\overset{\cdot\cdot}{O}:
\end{array}
\right]^{-}
,\
\left[
\begin{array}{c}
:\overset{\cdot}{O} \\
\diagdown \\
N\!-\!\overset{\cdot\cdot}{\overset{\cdot\cdot}{O}}: \\
\diagup \\
:\overset{\cdot\cdot}{O}:
\end{array}
\right]^{-}
\right\}
\qquad (12\text{-}1)
$$

(a) (b) (c)

Note that the formulas are indeed different; (b) and (c) should not be regarded simply as formula (a) turned 120° and 240°. The atomic nuclei (including inner shells) should be thought of as remaining fixed while the valence electrons are distributed in three different patterns. None of these formulas describes the nitrate ion satisfactorily. Each formula predicts two different kinds of nitrogen-oxygen bonds, but these bonds are known to be of equal length and equivalent in every way. The three oxygen atoms are thus equivalent to one another; they lie at the corners of an equilateral triangle, the center of which is occupied by the nitrogen atom.

No single Lewis formula, employing integral numbers of shared pairs and unshared pairs, can be drawn that *does* reflect these properties of the nitrate ion. This ion may instead be regarded as a blend or superposition of the three formulas shown. In the language of quantum chemistry, if ψ_a is the wave function describing a hypothetical nitrate ion for which the formula (12-1a) is a reasonable representation, and if ψ_b and ψ_c are wave functions of similar ions represented by the other two formulas given, then the nitrate ion is described by the wave function[2]

$$
\psi = \frac{1}{\sqrt{3}}(\psi_a + \psi_b + \psi_c)
\qquad (12\text{-}2)
$$

This relation can be expressed qualitatively by saying that the nitrate ion is a *resonance hybrid* of the three formulas (12-1), and it is often symbolized, as in (12-1), by separating the contributing formulas by commas and enclosing them within braces. The N-O bond is also said to have one-third double-bond character, because each of the N-O bonds is double in one of the three equally contributing formulas and single in the other two. A shorthand way of indicating that a bond is intermediate between single and double is to write the partial bond as a dashed line. Thus NO_3^{-}

---

[1]The implication of the braces, {}, will be explained shortly.

[2]The factor $1/\sqrt{3}$ is called a normalization factor and arises from the fact that the fractional contributions of the hypothetical formulas represented by ψ_a, ψ_b, and ψ_c are given by the squares of their coefficients in (12-2) rather than by the coefficients themselves (essentially because ψ^2 is more directly significant physically than ψ). The contributions of the three formulas (a) to (c) of (12-1) are, by symmetry, each $\tfrac{1}{3}$, of which $1/\sqrt{3}$ is the square root.

can be represented as

$$\left[\begin{array}{c} O \\ \diagdown \\ N{=}O \\ \diagup \\ O \end{array} \right]^{-} \tag{12-3}$$

A formula like (12-3) expresses the equivalence of the oxygen atoms, and of the three bonds, but it cannot be considered exactly comparable to a Lewis formula since it does not show unshared pairs (of which there is an average of $\frac{8}{3}$ on each oxygen atom).

Because the term *resonance* is used in physics in a sense that implies an oscillation between two states, the incorrect notion sometimes arises that the phenomenon here described as resonance implies an oscillation back and forth among the various contributing structures. The nitrate ion does *not* rapidly change structure so as to correspond at one instant to one of the possible formulas in (12-1) and at a later instant to another. Rather, it corresponds at any instant to a hybrid of all three formulas, as implied by (12-3). An analogy due to J. D. Roberts is helpful. A man who had come across a rhinoceros described the animal as something between a dragon and a unicorn. With this description he did not imply that the animal vacillated between being a dragon and a unicorn. Moreover he used in his description creatures that exist in the imagination only, just as formulas (*a*) to (*c*) of (12-1) relate to hypothetical species.

Thus resonance is not a phenomenon but rather a *description*. It serves to overcome a limitation of the description of molecules by Lewis formulas, which necessarily involve integral numbers of shared pairs of electrons. Whenever resonance is needed to describe the structure of a molecule, the molecule is found to have higher stability than would correspond to a single Lewis formula that might be used to describe it; in appropriate cases, it is found to have higher symmetry as well. Thus the nitrate ion itself is more stable than would be a hypothetical ion corresponding to any one of the three contributing formulas (12-1). The difference between the energy of one of these hypothetical ions and the actual energy of the nitrate ion is called the resonance energy. It is important to note that the apparently enhanced stability (the resonance energy) and other special properties are *not conferred* on the molecules *by resonance*. Rather, they are a consequence of the fact that *classical valence formulas were chosen as the starting point of the description*.

Benzene and related compounds. The most frequently cited example of a resonance hybrid is benzene, C_6H_6, for which two equivalent Lewis formulas can be drawn:

$$\tag{12-4}$$

(a) (b)

These valence bond formulas were first proposed by A. Kekulé and are often referred to as the Kekulé formulas for (or the Kekulé forms of) benzene. However, neither of them represents the properties of benzene satisfactorily. For example, compounds containing carbon-carbon double bonds, C=C, usually react rapidly with bromine, as illustrated by the reaction

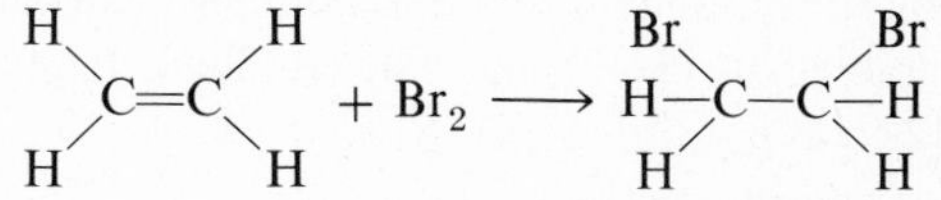

$$(12\text{-}5)$$

Similar reactions with benzene are extremely slow. Furthermore, it is known from spectroscopy and X-ray and electron diffraction that the carbon atoms of benzene form a *regular* planar hexagon, which would not be expected if (12-4a) or (12-4b) alone were appropriate. If benzene could be described correctly by one of the Kekulé formulas, isomerism of the type illustrated by formulas (a) and (b) of (12-6) would be observed:

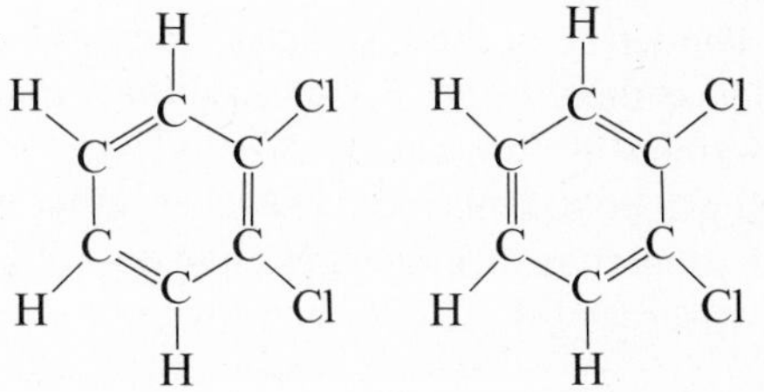

$$(12\text{-}6)$$

(a)   (b)

In (a) the two carbon atoms that carry the chlorine atoms are connected by a double bond, in (b) by a single bond, so that the formulas correspond to different compounds. No evidence for such isomerism has ever been found.

A shorthand notation is often used for benzene and related compounds. It shows only the six-membered ring skeleton of single and double bonds, as well as atoms *other than* hydrogen attached to ring carbons, as exemplified by

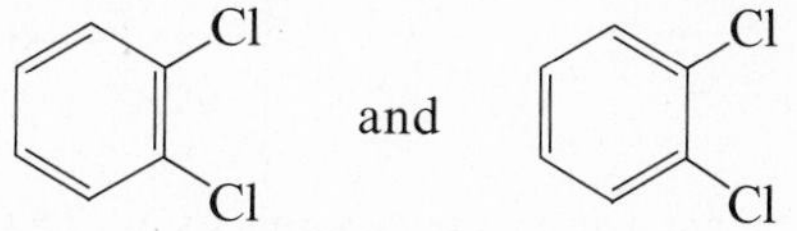

and

$$(12\text{-}7)$$

for the hypothetical isomers just discussed. In the same notation the resonance formulation for benzene is

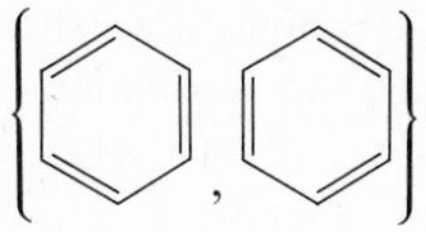

$$(12\text{-}8)$$

Alternative representations often used are

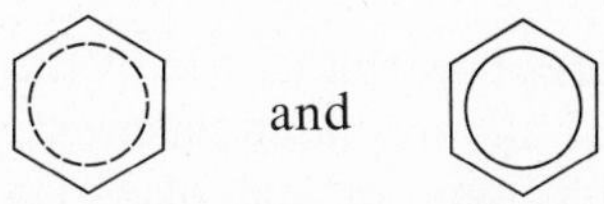

and

$$(12\text{-}9)$$

where the dashed or solid circles indicate, as before, additional partial bonding. In benzene each carbon-carbon bond has half double-bond character. As with the nitrate ion, the benzene molecule is significantly more stable than would be the hypothetical molecule with a structure like either one of the Kekulé forms of benzene.

Benzene is the simplest representative of a large number of organic compounds, termed *aromatic compounds*, many of which contain one or more six-membered rings like that in benzene. Examples of such compounds include the dichlorobenzene depicted in (12-7) and naphthalene, $C_{10}H_8$, for which the formula analogous to (12-9) is

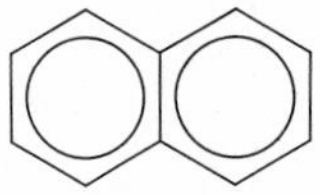

As before, the corners indicate carbon atoms and it is implied that a hydrogen atom is attached to each carbon that is linked to only two other carbons of which there are eight in all.

The virtue of the resonance description is that it extends the scheme of classical valence theory and permits relating the properties of molecules that cannot be described by the scheme of single Lewis formulas to the properties of molecules that can be so described. However, other ways have also been developed to describe the properties of molecules that do not correspond to single Lewis formulas. The most successful of them employs molecular orbitals, described in Sec. 12-3.

EXAMPLE 12-1   **Resonance formulas and variation in bond distances.** Write resonance formulas for nitric acid, HNO_3, and indicate how you would expect the different N-O bond lengths in this molecule to compare with each other and with that in NO_3^-. Recall that increasing double-bond character corresponds to a shorter distance.

Solution.   There are two equivalent resonance formulas for nitric acid:

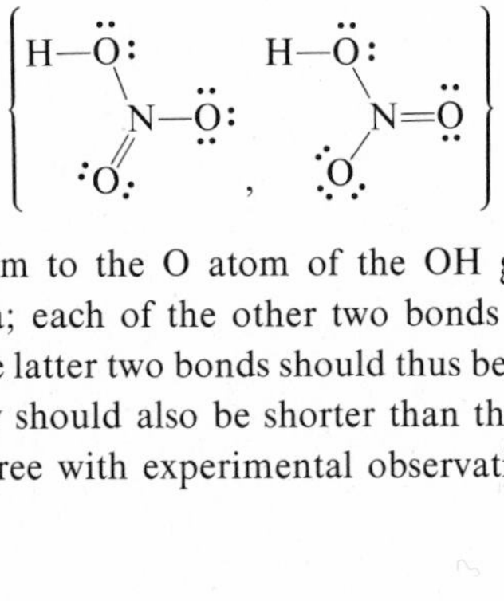

The bond from the N atom to the O atom of the OH group has no double-bond character in either formula; each of the other two bonds from N to O has one-half double-bond character. The latter two bonds should thus be equal in length and shorter than the N-OH bond; they should also be shorter than the N-O bonds in the nitrate ion. These expectations agree with experimental observations.

12-2   Hybridization

We consider now the nature of the atomic orbitals that are used to accommodate the shared and unshared electron pairs characteristic of Lewis formulas. Many of these same atomic orbitals also play an important role in the construction of molecular orbitals.

General remarks. In the valence bond picture of the H_2 molecule, the shared pair of electrons may be placed in the region of overlap between the two $1s$ orbitals of the two hydrogen atoms. The Li_2 molecule may be considered to be similar, except that $2s$ rather than $1s$ orbitals are used for it:

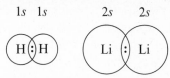

However, we run into trouble with this scheme when we consider most larger molecules,[1] for example, molecules of tetrahedral structure such as CH_4 (methane), SiH_4 (silane), and CCl_4 (carbon tetrachloride). As discussed in Chap. 11, the four peripheral atoms in each of these molecules are situated at the corners of a regular tetrahedron and are linked to the central C or Si atom by four covalent bonds that are equivalent to one another by symmetry. This equivalence of the bonds is at first sight difficult to understand, because the orbitals of the central atom available for bonding are the s and p orbitals in the valence shell (a $2s$ and three $2p$ orbitals for carbon; a $3s$ and three $3p$ orbitals for silicon), which are quite different in their directional properties and thus are by no means equivalent. How can these orbitals be used to form four equivalent bonds? The answer to this question lies in the description of the bonding by a formalism called *hybridization,* in which an s orbital and three p orbitals on the central atom are combined to create four new orbitals that are equivalent to each other and point toward the corners of a regular tetrahedron. Many other combinations of different orbitals on a given atom can also be formed, with other directional properties.

Hybridization of atomic orbitals and its application to bonding were first discovered and worked out independently by Pauling and Slater. The reason for the importance of hybridization is that suitable hybrid orbitals have highly directional character, leading to better overlap with orbitals of other atoms to which the atom of interest is bonded. In particular, the nondirectional character of s orbitals (which are spherically symmetric) makes them poorly suited for bonding unless they are combined in hybrids, and they are used by themselves only when no other orbitals of comparable energy are available, as in hydrogen.

Linear combinations of atomic orbitals. It will be recalled from the discussion in Sec. 9-3 that when several wave functions of a system lead to the same energy, the energy level is said to be degenerate. If there are k such independent wave functions, it is possible by linear combination of the k wave functions to form k linearly independent new wave functions that are *equally appropriate* to the energy level.

It is also possible to form linear combinations of orbitals of the same atom that are *not* degenerate, such as s and p orbitals. If the individual orbital energies are not greatly different, these linear combinations are called *hybrid orbitals*. The

---

[1]Even Li_2 is not as simple as just depicted, since there is also involvement of $2p$ orbitals, as will emerge shortly.

condition that the energies of the orbitals involved in making up the hybrid should not differ too much in energy means that, for example, no satisfactory hybrid would result from combining orbitals such as $2s$ and $3p$ although s and p orbitals of the same principal quantum number can be hybridized effectively.

We now turn to an examination of the geometric aspects of some important hybrid orbitals. Note that in each of the examples, the number of hybrid orbitals formed is just the same as the number of separate orbitals that we start with, as mentioned earlier (Sec. 11-6).

Hybridization of s and p orbitals.   Three important kinds of hybrids can be formed by combining s and p orbitals, as indicated schematically in Fig. 12-1. The principal quantum numbers of the s and p orbitals are usually not mentioned in designating the hybrids, but they must be the same for satisfactory hybridization. All the orbital graphs in Fig. 12-1 show only the *angular* dependence of the orbitals; they are polar graphs, with the distance from the origin of the graph in any particular direction representing the relative magnitude of the angular part of ψ in that direction.

One s and one p orbital can be hybridized to form what are known as *sp hybrids.* Two such hybrids can be formed, as shown in Fig. 12-1. They point in opposite directions from the center of the atom and thus are at 180° to each other. The two p orbitals not involved in the hybridization are directed at right angles to the sp orbitals. Note the directionality of Φ_1 and Φ_2: their magnitude is larger in the direction in which they point than are the magnitudes (in any direction) of the s and p orbitals of which Φ_1 and Φ_2 are composed. This feature is quite general for hybrid orbitals suitable to describe bonding.

A simple molecule that may be described by sp bonding is the hypothetical BeH_2 molecule,

$$H-Be-H$$

in which the bonds are formed by the overlap of the sp orbitals of the Be atom with the $1s$ orbitals of the hydrogen atoms, as portrayed schematically in Fig. 12-2. The expected geometry of the BeH_2 molecule is *linear,* in accord with the 180° angle between Φ_1 and Φ_2. Experimental verification of the structure of this molecule has not so far been possible, because only a polymeric form, represented as $(BeH_2)_n$, is known. A molecule for which sp hybrids are believed to be used is $HgCl_2$, which is linear; the two Hg-Cl bonds have appreciable covalent character. Presumably sp hybrids formed from the $6s$ and $6p$ orbitals of the Hg atom are involved in the bonding. An important application of sp hybridization is in one description of triple bonding (Fig. 12-12).

When one s and two p orbitals are hybridized in appropriate proportions, three equivalent hybrid orbitals result. These three hybrids, termed sp^2 hybrids, are shown schematically in Fig. 12-1, where they are labeled Φ_3, Φ_4, and Φ_5. The directions of maximum extension of the three hybrids are in the same plane and enclose angles of 120°. The plane is that characteristic of the two p orbitals; thus the remaining p orbital, not used in forming the hybrids, is perpendicular to this plane.

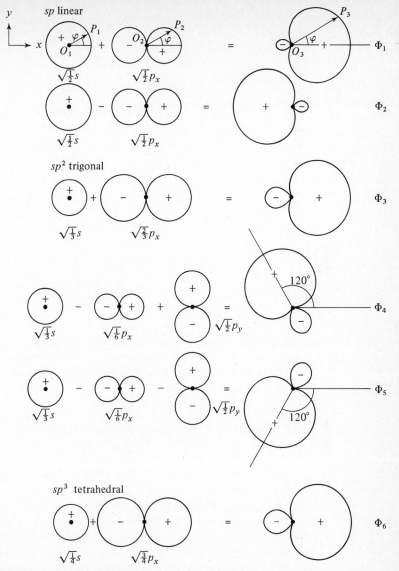

FIGURE 12-1 **Hybridization of s and p orbitals.** Each diagram represents a polar graph of the angle-dependent part of an orbital, at $\theta = 90°$, that is, in the x,y plane. For example, in the diagram labeled $\sqrt{\frac{1}{2}}s$ (top, left) the distance of each point P_1 from the origin O_1 is $\sqrt{\frac{1}{2}}|\Phi(\theta,\varphi)|$ for an s orbital, at any given single angle φ and at $\theta = 90°$. The next graph on the right is similar, referring to $\sqrt{\frac{1}{2}}p_x$. The plus and minus signs inside the graphs refer to the sign of Φ. The last graph on the top line depicts the combination $\Phi_1 = \sqrt{\frac{1}{2}}(s + p_x)$, so that $O_3P_3 = O_1P_1 + O_2P_2$. This combination is known as an sp hybrid. The second line of the diagram shows another sp hybrid, formed by subtraction: $\Phi_2 = \sqrt{\frac{1}{2}}(s - p_x)$. The shape of Φ_2 is the same as that of Φ_1. The two hybrids are therefore equivalent and are oriented at 180° to one another. The next three lines in the diagram show three sp^2 hybrids, which are equivalent and rotated 120° relative to each other. The last line depicts one of four sp^3 hybrids, all of which are equivalent and oriented at tetrahedral angles (109°28′) relative to each other. The diagrams cannot be readily compared with Figs. 9-10, 9-11, and 25-6, because the latter portray surfaces of constant ψ in ordinary three-dimensional space while the diagrams here represent the angle-dependent parts Φ of orbitals in terms of polar graphs.

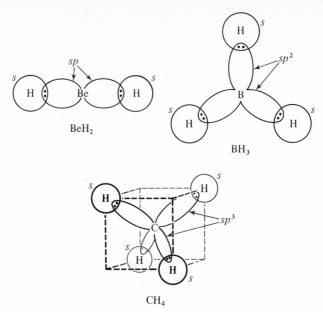

CH₄

FIGURE 12-2   **Bonding with** *sp, sp², and sp³* **hybridization.**   The *sp, sp²,* and *sp³* hybrids of the central atoms are shown in stylized form. BeH₂ is a hypothetical molecule that has not yet been observed. BH₃ has been observed, but its structure has not yet been proved. Nonetheless these two species serve as illustrative examples of *sp* and *sp²* hybridization of the orbitals on the central atom. CH₄ exemplifies *sp³* hybridization.

An example of a molecule with sp^2 hybridization is the unstable molecule BH_3,

$$\begin{array}{c} H \\ | \\ B \\ H \quad\; H \end{array}$$

which is observed in the mass spectrometer as a fragmentation product of B_2H_6 when the latter compound is bombarded with electrons of energy about 10 eV. BH_3 is expected (but not as yet known) to be planar, with 120° bond angles (Fig. 12-2). The methyl cation, $CH_3{}^+$, a short-lived reaction intermediate that is isoelectronic with BH_3, is expected to have this same *planar trigonal* configuration, a conjecture that has been verified experimentally. The planar trigonal configuration is also known to exist in trimethylboron, $B(CH_3)_3$, as well as in BCl_3 and other boron trihalides. Other instances of sp^2 hybridization are considered later, including its use in a description of double bonds (Fig. 12-12).

When one *s* and three *p* orbitals are appropriately combined, four equivalent hybrids directed toward the corners of a tetrahedron may be formed. They are termed *sp³ hybrids*. Figure 12-1 shows a polar graph for one of them, directed in this case[1] along the *x* axis and labeled Φ_6. The other three *sp³* hybrids, not shown in Fig.

---

[1]All three *p* orbitals are needed to form the set of four equivalent tetrahedral orbitals, although only p_x happens to be used in forming the one illustrated here.

12-1, make angles of 109°28′ with Φ_6 and with each other. This is how hybridization makes it possible to describe the bonding in CH_4, CCl_4, SiH_4, and other tetrahedral molecules; sp^3 orbitals on the central atom, together with suitable overlapping orbitals of the peripheral atoms, hold the shared electron pairs (Fig. 12-2).

It is also possible to combine s and p orbitals to form hybrids that are not equivalent to one another. For example, one s and three p orbitals can be hybridized in such a way that each resulting orbital does not have just $\frac{1}{4}s$ character and $\frac{3}{4}p$ character, which is the blend that makes the usual sp^3 hybrids exactly equivalent and tetrahedrally directed. If the proportions of s and p character differ in the four hybrid orbitals formed, the hybrids point toward the corners of a *distorted* tetrahedron rather than a regular one.

The promotion of electrons. Because the s and p states of the same principal quantum number are, in general, of different energy, the s states are completely filled before electrons occupy the p states. Thus, for example, the ground state of the Be atom is $1s^2 2s^2$ and that of the C atom is $1s^2 2s^2 2p^2$. To make the $2s$ orbital available for the sharing of electrons, one of the $2s$ electrons must be *promoted* to a $2p$ state, leading for C to the configuration $2s 2p^3$ (often referred to as the valence state). The source of the energy required for this promotion is the formation of covalent bonds, and the criterion for judging when promotion is likely is that the energy furnished by bond formation must be great enough to overcome the energy expended in electron promotion, so that an overall stabilization of the bonded group of atoms results. This criterion for promotion applies to the hybridization of any atomic orbitals. This energy give-and-take is illustrated in Fig. 12-3.

For CH_4, the promotion of a $2s$ electron of carbon to a $2p$ level requires about 400 kJ mol^{-1} whereas the energy liberated when four C-H bonds are formed is about

FIGURE 12-3   Relationship between promotion and bonding energies.

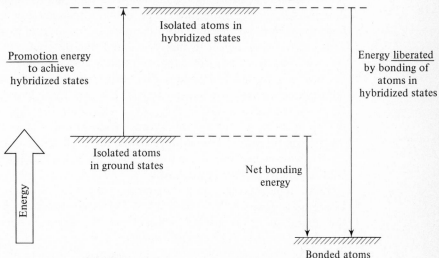

2100 kJ mol^{-1}. The net bonding energy, which is the difference between the energy expended for promotion and that gained in bonding, is about 1700 kJ mol^{-1}.

Atoms with unshared pairs.   There are many molecules and ions in which not all electron pairs of the valence-shell octet of an atom are shared. In spite of this it is often reasonable to describe the bonding orbitals by sp^3 hybridization, with the unshared pairs also occupying sp^3 orbitals. The oxygen atom of the water molecule may serve as an example. Its valence shell contains two unshared pairs of electrons, and two shared pairs that constitute the covalent bonds to the hydrogen atoms. In one extreme valence bond description the shared and unshared pairs may be considered to use four equivalent sp^3 hybrids; in the other extreme the unshared pairs occupy the $2s$ and one of the three $2p$ orbitals while the shared pairs use the remaining two $2p$ orbitals. The first of these situations would lead to an expected H-O-H angle equal to the tetrahedral angle of 109.5°, the second to an H-O-H angle of 90°, corresponding to the angle between two p orbitals. The bond angle in water is actually 104.5°, which suggests that the situation is closer to sp^3 hybridization, although other factors must also be considered, as mentioned below.

Another molecule containing an unshared pair on the central atom is NH_3. This molecule is pyramidal in shape, with the nitrogen atom at the apex of the pyramid and the three hydrogen atoms forming its base. Again we may consider that four equivalent sp^3 orbitals of the nitrogen atom are used, both for the shared pairs of electrons that bond the hydrogen atoms and for the unshared pair. An alternative is that the bonding pairs use the three $2p$ orbitals, with the unshared pair in the $2s$ orbital. The actual bond angle of 107.3° suggests hybridization close to equivalent sp^3 orbitals. The unshared pair then occupies an sp^3 orbital extending from the nitrogen atom in a direction away from the base of the pyramid.

How reasonable is it to expect sp^3 hybridization in these molecules? These hybrids provide greater bond strength than would pure p orbitals, since they are more highly directed, i.e., more concentrated in particular directions (Fig. 12-1), and also extend further out from the nucleus. Both these factors contribute to more effective overlap with bonded neighbors. On the other hand, the difference between the $2s$ and $2p$ states of oxygen is about 840 kJ mol^{-1} and the corresponding difference for nitrogen is about 630 kJ mol^{-1}. Hence a high price must be paid for promotion. Other important effects are the mutual repulsion that the two hydrogen atoms in H_2O or the three hydrogen atoms in NH_3 incur if they are too close to each other. There is also interaction between the hydrogen atoms and the unshared pairs (or pair) on the central atom. The balance of all these effects is hard to assess and only qualitative conclusions can be drawn, such as that the hybridization is somewhere between the two extremes described.

It is interesting that the bond angles in the hybrids of the congeners of N and O are quite close to 90° (Table 11-5). This may reflect in part the larger size of the central atoms involved, which increases the distances between hydrogen atoms and thus decreases their mutual repulsion, but whatever the cause, there is presumably little or no hybridization of s and p orbitals in these molecules. Tetrasubstituted congeners of carbon, as well as the corresponding ions in neighboring columns (for example, PH_4^+ and $GaBr_4^-$), are tetrahedral in shape and presumably use sp^3 orbitals on the central atom.

Hybridization involving d orbitals.

A number of different hybrids can be formed by combining d orbitals with s and p orbitals. Only two of the most important of these will be discussed. These are the hybrids sp^3d^2 and sp^3d, the first of which is more frequently encountered than the second.

Six equivalent sp^3d^2 *hybrids,* pointing toward the six corners of an octahedron, can be formed by linear combination of s, p_x, p_y, p_z, $d_{x^2-y^2}$, and d_{z^2} orbitals. The corners of the octahedron lie along the x, y, and z axes, with the atom assumed to be at the origin (Fig. 12-4). These hybrids are often called octahedral orbitals.

Five strongly directional orbitals, known as sp^3d *hybrids,* can be formed by linear combination of s, p_x, p_y, p_z, and d_{z^2} orbitals. These hybrids are *not* all equivalent. Two point in opposite directions along the z axis (the axial directions), and the three others, different from the first two, point toward the corners of an equilateral triangle in the x,y plane (equatorial plane) (Fig. 12-4). The five thus point toward the corners of a trigonal bipyramid [illustrated in formula (11-9)].

Both sp^3d^2 and sp^3d hybrids are especially suitable for bonding because of their high directionality. As a result the bonds that are formed are so strong that energy can be expended for the promotion required and even unshared (or "lone") pairs on the central atom occupy hybrid orbitals as well.

The d orbitals used in these hybrids may either have the same principal quantum number as the s and p orbitals, in which case they are known as *outer d orbitals* (for example, $4d$ orbitals hybridized with $4s$ and $4p$), or a principal quantum number one less, in which case they are called *inner d orbitals* (for example, $3d$ orbitals hybridized with $4s$ and $4p$). For the elements in columns IIIA to VIIA of the periodic table (the congeners of B, C, N, O, and F) and for the inert gases, the inner d orbitals

FIGURE 12-4   **Geometry of** sp^3d^2 **and** sp^3d **hybrids.**   The hybrid orbitals of the central atom are shown in stylized form. There are two different types of sp^3d orbitals—three equivalent by symmetry, in the x,y plane and at 120° to each other, and two, likewise equivalent by symmetry, pointing in opposite directions along the z axis. The six sp^3d^2 orbitals are equivalent by symmetry.

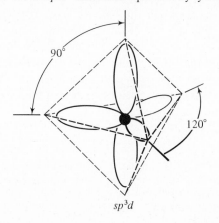

sp^3d

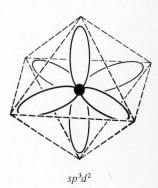

sp^3d^2

are either completely filled and therefore do not participate in hybridization or (when n is 3 or less) there are no inner d orbitals. Thus, for these elements only outer d orbitals can be involved in d orbital hybrids. Since these outer d orbitals are of higher energy than the ns and np orbitals, promotion energy is required. In the first short period, when n is less than 3, the promotion energy is prohibitive; the $3d$ orbitals would have to be hybridized with the $2s$ and $2p$ orbitals. This is why the valence shells of the atoms Li through F never contain more than eight unshared or shared electrons, in accord with the octet rule.[1]

The situation is different with the transition elements, for which at least some of the *inner d* orbitals are usually available for hybridization. Thus when some inner d orbitals are unfilled, all electrons in the $(n-1)d$, ns, and np levels of the atom may be counted as valence electrons. For example, the elements Ti, V, . . . , Ni have 4, 5, . . . , 10 valence electrons, respectively (see Table 10-3). Transition-metal compounds often exhibit geometries in accord with sp^3d^2, sp^3d, and other d orbital hybridization. However, this valence bond description has certain shortcomings, particularly in its failure to account for the spectra and reactivities of transition-metal complexes. A more satisfactory approach to bonding in transition-metal compounds is considered in Chap. 25.

Examples of structures involving outer d orbitals.   The valence shells of P and of S include the $3s$ and $3p$ orbitals and can be expanded by the inclusion of the $3d$ orbitals. In PCl_5 the total number of electrons to be accommodated in the valence shell of P is 10: five from the phosphorus atom and one from each of the five chlorine atoms. All 10 electrons are used to form the five covalent bonds in the molecule, so that there are no unshared pairs on the phosphorus atom. The required five bonding orbitals must therefore include one $3d$ orbital, and sp^3d hybrids are formed and used in the bonding. Phosphorus and its congeners form many trigonal bipyramidal molecules. Figure 12-5 illustrates the geometries of some representative molecules and ions in which outer d orbitals participate in sp^3d hybridization, including a number in which one or more unshared electron pairs occupy equatorial positions. The structures of some of these molecules were surprising when first discovered, such as the T-shaped arrangement in BrF_3 or the arrangement in SF_4.

In SF_6 there are 12 electrons in the valence shell of the sulfur atom, six from the S atom itself and one from each of the six fluorine atoms to which it is bonded. These 12 electrons are used to form the six covalent bonds in the molecule, so there are no unshared pairs on the sulfur atom. The sp^3d^2 hybridization in this molecule is also found in many other molecules and ions, a few of which are illustrated in Fig. 12-6. In a number of the examples one or two of the bonded atoms have been replaced by unshared pairs.

The concept of hybridization is an aid to the understanding of the geometric

---

[1] One might ask why the $(n+1)s$ orbital, which is of about the same energy as the nd orbital for most atoms, is not sometimes used in place of the outer nd orbital in formation of bonding hybrids. The answer is that the nondirectional character of the s orbital makes it an unsuitable substitute for a d orbital in the formation of highly directed hybrid orbitals.

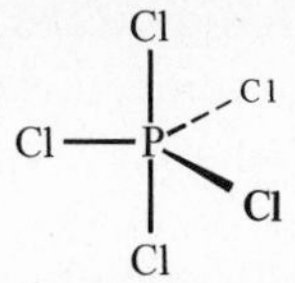

Trigonal bipyramid
PCl_5, PF_5, PF_3Cl_2, $SbCl_5$

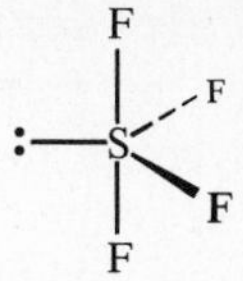

Trigonal bipyramid with missing equatorial corner
SF_4, SeF_4, $TeCl_4$, $[IO_2F_2]^-$, $Te(CH_3)_2Cl_2$

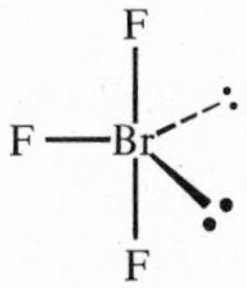

T-shaped
BrF_3, ClF_3, $(C_6H_5)ICl_2$

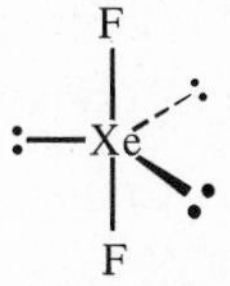

Linear
XeF_2, I_3^-, $[ICl_2]^-$, $[BrICl]^-$

FIGURE 12-5 **Representative structures based on trigonal bipyramidal (sp^3d) hybridization involving outer *d* orbitals.** One, two, or three of the shared electron pairs and the atoms bonded by them may be replaced by unshared pairs. The remaining bonded atoms still occupy some of the corners of a (sometimes slightly distorted) trigonal bipyramid around the central atom, always including the axial positions. The unshared pairs always occupy equatorial positions, although the maxima of electron density associated with them cannot easily be located. When the same kind of atom is bonded in both axial and equatorial positions, as in PCl_5, SF_4, and BrF_3, the equatorial bonds are slightly shorter and stronger. The promotion energy required for the use of outer *d* orbitals is furnished by the bonding energy. Note the inert-gas compound XeF_2.

configurations of many different molecules and ions. It must always be remembered, however, that hybridization itself is not a physical phenomenon, any more than is resonance. It is simply a useful way of describing the orbitals used in bonding in terms of particular sets of "atomic" orbitals (*s, p, d, . . .*) that are suitable for describing the properties of isolated atoms and ions but are far less suitable in descriptions of bonding.

FIGURE 12-6 **Representative structures based on octahedral (sp^3d^2) hybridization involving outer *d* orbitals.** One or two of the atoms or groups bonded to the central atom may be replaced by unshared electron pairs, the overall octahedral configuration (including, presumably, the unshared pairs) being at least approximately maintained. Note the inert-gas compounds $XeOF_4$ and XeF_4.

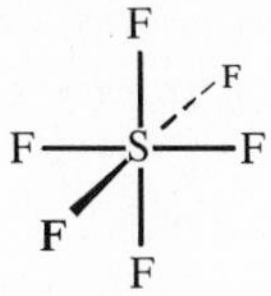

Octahedral
SF_6, SeF_6, TeF_6
PF_6^-, PCl_6^-, SbF_6^-
$GeCl_6^{2-}$, $SnCl_6^{2-}$, $SnBr_6^{2-}$
$PbCl_6^{2-}$, $ZrCl_6^{2-}$, $CdCl_6^{4-}$

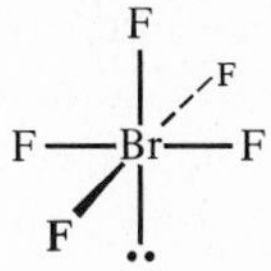

Square Pyramidal
BrF_5, IF_5, $[SbF_5]^{2-}$, $XeOF_4$

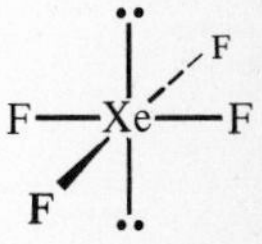

Square planar
XeF_4, ICl_4^-, BrF_4^-

Rules for predicting geometric configuration.   An empirical set of rules for the prediction of the geometric configuration of the bonded neighbors about a central atom can be summarized rather concisely. These rules can be rationalized and refined by considering that electron pairs in the valence shell of the central atom, whether unshared or in bonds, tend to remain as far apart as possible, and that repulsions between unshared pairs are somewhat greater than those between unshared pairs and bonded pairs, which are in turn somewhat greater than those between bonded pairs. An appreciation of hybridization also helps in understanding the rules, which are as follows:

1.   Formulate the Lewis structure(s) for the molecule or ion.

2.   Count the number of bonded atoms attached to the atom considered and the number of unshared electron pairs on that atom. (Do not include unshared pairs that occupy *inner d* orbitals.)

3.   Add the number of bonded neighbors to the number of unshared pairs. The geometric configuration about the central atom, including both the attached bonded atoms *and* the (presumed positions of the) unshared pairs, can then be predicted from Table 12-1. Since the positions of the atoms themselves, but not of the unshared pairs, are all that can be observed experimentally, the atomic arrangement in molecules with basically similar electronic structures may be described in quite different words (for example, CH_4, tetrahedral; NH_3, pyramidal; H_2O, bent), as in the right-hand column of Table 12-1.

4.   The following points are worth noting:

(*a*) It is the number of bonded neighbors, not the number of bonds, that is counted. Hence double or triple bonds, or possible resonance involving different numbers of bonds to the central atom, have no bearing on this correlation scheme.

(*b*) The configuration will not be completely symmetric unless all the atoms bonded to the central atom are the same, but it will usually *approximate* the regular figure—an equilateral triangle, a regular tetrahedron, and so on.

(*c*) Remember that any diatomic molecule is necessarily linear and any three atoms necessarily lie in the same plane. Thus, applying the term *linear* to a diatomic molecule or *planar* to a triatomic one is redundant.

(*d*) If the total of bonded neighbors plus unshared pairs is four and sp^3 orbitals are available, the configuration will be approximately tetrahedral. For some transition-metal complexes with one *d*, one *s*, and two *p* orbitals available, a square planar configuration about the central atom sometimes results, e.g., in some Ni(II) and Pt(II) complex ions. We do not consider these here, nor do we consider totals of bonded neighbors plus unshared pairs larger than six, although the scheme can be extended to cover these cases.

EXAMPLE 12-2   **Molecular geometry.**   Explain the square pyramidal molecular geometry of BrF_5 by the rules just given.

Solution.   The number of electrons in the valence shell of the Br atom is $7 + 5 = 12$; there are five shared pairs plus one unshared pair. The basic geometry is that of an

TABLE 12-1 Correlation of geometric configuration and electronic structure

No. of bonded neighbors plus unshared pairs	Possible suitable hybrids*	Expected configuration of bonded neighbors and *unshared* pairs		Observed examples of configurations of bonded neighbors
2	sp (linear)		Linear	CO_2, NO_2^+, NNO, OCS; $HgCl_2$, Hg_2Cl_2(Cl—Hg—Hg—Cl), $Ag(NH_3)_2^+$; H—C≡C—H, H—C≡N
3	sp^2 (trigonal planar)		Trigonal planar	BCl_3, NO_3^-, CO_3^{2-}, SO_3, GaI_3
			Bent	SO_2, NO_2^-
4	sp^3 (tetrahedral)		Tetrahedral	CH_4, $CHCl_3$; NH_4^+, PBr_4^+, SO_4^{2-}; $Ni(CO)_4$
			Trigonal pyramidal	NH_3, H_3O^+, PCl_3, SO_3^{2-}, XeO_3
			Bent	H_2O, SCl_2, $O(CH_3)_2$
5	sp^3d (trigonal bipyramidal†)		Trigonal bipyramidal	PCl_5, PF_5, PF_3Cl_2
			Trigonal bipyramidal with one corner unoccupied	SF_4, $TeCl_4$, $[IO_2F_2]^-$
			T-shaped	BrF_3, ClF_3
			Linear	I_3^-, ICl_2^-, XeF_2
6	sp^3d^2 (octahedral‡)		Octahedral	SF_6, $[Fe(CN)_6]^{3-}$, $[Fe(CN)_6]^{4-}$
			Square pyramidal	BrF_5, $[SbF_5]^{2-}$
			Square planar	$[ICl_4]^-$, XeF_4

*The orbitals in all but the trigonal bipyramidal set are equivalent by symmetry (provided the peripheral atoms are all of the same kind). In this set the three hybrids in the x,y plane and at 120° from each other are equivalent by symmetry, as are, separately, the two hybrids in the $\pm z$ direction.

†Unshared pairs always occupy equatorial postions. See Fig. 12-5 for additional examples.

‡See Fig. 12-6 for additional examples. When two unshared pairs are present they are opposite each other, leading to a square planar arrangement.

octahedron, one corner of which is occupied by the unshared pair and the other five by the five F atoms.

12-3   Molecular Orbitals

Introduction.   In principle, any quantum-chemical theory for the electronic structure of a molecule should describe orbitals that encompass the whole molecule (molecular orbitals), since electrons cannot be localized. In practice, the valence bond theory, with its localized electron pairs in bonding orbitals that result from the overlap of atomic orbitals on pairs of adjacent atoms, is a remarkably good approximation for many molecules and corresponds closely to the Lewis view of covalent bonding. The simple qualitative view of molecular orbital (MO) theory presented here is designed to give some notion of the way in which chemists attempt to construct orbitals characteristic of the molecule as a whole and also to give some insight into the usefulness of this approach. Molecular orbital theory turns out to be particularly helpful for just those molecules for which valence bond theory is not entirely adequate, especially those for which the concept of resonance must be invoked. It is illuminating for other molecules as well.

The usual procedure for formulating a molecular wave function (a molecular orbital) is by linear combination of atomic orbitals, one atomic orbital being chosen for each atom included. The atomic orbitals used are normally either the familiar hydrogenlike $s, p, d, \ldots$ or hybrids formed from them. Usually only atomic orbitals of valence shells are used, since the orbitals of the inner filled shells are so much more stable that they are little affected by interaction with other atoms and so the electrons in them are regarded as localized on specific atoms. The number of MOs formed is the same as the number of atomic orbitals taken initially. Each MO corresponds to a particular energy, usually different for different MOs, although sometimes several MOs correspond to the same energy and thus are said to be degenerate. Each orbital can hold at most two electrons, and if two electrons do occupy an orbital, their spins must be opposed, in accord with the Pauli principle. The MOs are filled with electrons in the order of increasing energy, just as atomic orbitals are. When a molecule is excited electronically, an electron is moved from a stable orbital (usually the highest occupied one) to a higher empty or half-filled orbital. We begin our discussion by considering MOs for some diatomic molecules and ions.

Diatomic molecules: general features.   The simplest diatomic molecule is H_2, the molecular orbitals for which are made up by combining the $1s$ orbital for each of the two H atoms, as discussed in Chap. 11 and illustrated in Figs. 11-8 and 11-9. It will be recalled that two molecular orbitals are formed, one corresponding to a greater stability and one to a lesser stability than for the separated atoms. The more stable, or *bonding*, orbital is formed by taking the sum of the two $1s$ orbitals and is thus designated ψ_+; the less stable, or *antibonding*, orbital corresponds to the difference of the two $1s$ orbitals and is designated ψ_-. The relationships between these orbitals are illustrated schematically in Fig. 12-7, which may be taken to apply

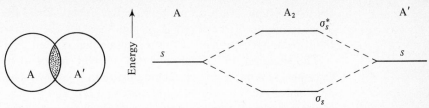

FIGURE 12-7   **Overlap of** s **orbitals (left) and the corresponding atomic and molecular energy levels.**   When s orbitals are added, the resulting molecular orbital ψ_+ is denoted σ_s (sigma s); the implications of the symbol σ are given in the text. Similarly, when s orbitals are subtracted, the antibonding orbital ψ_- formed is denoted by σ_s^*. The schematic energy-level diagram shows the energies of an electron in an orbital on each of the separated atoms and in the bonding orbital σ_s and the antibonding orbital σ_s^*. The s orbitals may be $1s$, $2s$, or higher s orbitals, as appropriate for the valence electrons involved. The representation of the orbitals in this and the next few figures is highly schematic.

more generally to the overlap of s orbitals of any pair of identical atoms (here designated A and A′) in a molecule A_2, rather than just to hydrogen. The difference between the energies of the two molecular orbitals and that of the isolated atomic orbitals depends upon the degree of orbital overlap and thus varies with both the nature of the orbitals and the identity of the atoms. These changes in energy have been calculated quantitatively for the overlap of many different kinds of atomic orbitals on different atoms.

Two cases of overlap of *directed* orbitals (such as p orbitals) must be considered: (1) when they are pointed along the molecular axis and (2) when they are pointed at right angles to the molecular axis. The overlap of two p orbitals directed toward each other along the internuclear axis (here designated the x axis) is depicted schematically in Fig. 12-8. At the left is shown the summation of the angle-dependent parts of the two overlapping p_x orbitals of atoms A and A′, corresponding to the formation of the bonding MO. The antibonding MO is formed by taking the difference of the orbitals (not shown), which is equivalent to reversing the signs of

FIGURE 12-8   **Overlap of** p_x **orbitals (left) and the corresponding atomic and molecular energy levels.**   The plus and minus signs near the p_x lobes represent, as always, the sign of the wave function and not electric charges. The combination shown here corresponds to the bonding orbital ψ_+; for ψ_- the signs on one orbital are reversed. Because overlap of the p_x orbitals is greater than for the s orbitals (Fig. 12-7), the energy separation of the bonding and antibonding MOs is greater than for the σ_s orbitals. The MOs formed by the combination of p orbitals directed along the internuclear axis are designated σ_p and σ_p^*.

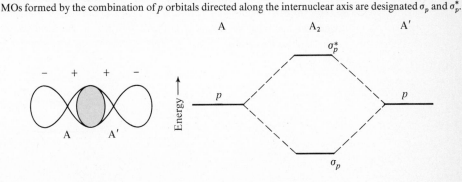

the lobes of one of them and adding. The energy levels are illustrated on the right. The separation of the levels is greater than for the overlap of s orbitals, because the directional character of the p orbitals makes their overlap more effective.

The situation for orbitals directed at right angles to the internuclear axis (for example, p_y or p_z) is shown schematically in Fig. 12-9. The overlap is smaller than in either of the two cases already discussed (s–s and p_x–p_x), and thus the separation of the energies of the corresponding MOs is also smaller.

As indicated in the legends to Figs. 12-7 to 12-9, the various MOs are designated by symbols that include the Greek letters σ and π, which parallel the Latin letters s and p. The rationale behind this nomenclature can be understood in terms of the symmetry of the orbitals with respect to the internuclear axis. Sigma orbitals, whether bonding or antibonding, have cylindrical symmetry about this axis—they remain unaltered by rotation around the line connecting A and A′, a property they share with s orbitals (which have spherical symmetry). Orbitals labeled π have a nodal plane, with the internuclear axis lying in this plane. The signs of these orbitals differ on opposite sides of this plane, a property that they share with p orbitals. The subscripts (s, p, x, and so on) of σ and π indicate the types of atomic orbitals used to construct the MOs; such subscripts are sometimes omitted when there is no ambiguity or when they are not of importance. The superscript * is used to denote an antibonding orbital.

Figure 12-10 represents a composite of the energy levels of the three cases discussed and illustrated in Figs. 12-7 to 12-9. One s level and three p levels for the two A atoms of an A_2 molecule are shown on the left and right. The levels corresponding to the MOs formed from these atomic orbitals are depicted in the center. The exact spacing and the relative order of the energy levels of a molecule can be determined only by experiment, but the general energy-level scheme shown in Fig. 12-10 is applicable to most diatomic molecules. The fact that the σ_p level is slightly higher in energy than the π_y and π_z levels in most molecules can be explained qualitatively. The electrons in the filled σ_s and σ_s^* orbitals are concentrated to a significant extent along the line through the two nuclei. The σ_p orbital also extends along this line and the electrons it contains are repelled by the electrons in the σ_s and σ_s^* orbitals. Electrons in π orbitals are concentrated in a different region of space.

FIGURE 12-9 **Overlap of p_y or p_z orbitals (left) and the corresponding atomic and molecular energy levels.** The overlap is smaller than for s orbitals or for p orbitals directed along the internuclear axis, and hence the separation of the MOs is smaller. The bonding orbitals are designated π_y and π_z, the antibonding orbitals π_y^* and π_z^*; they are sometimes abbreviated π_p and π_p^*. The symbol π is explained in the text.

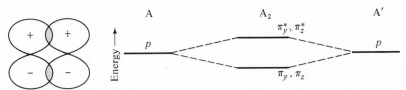

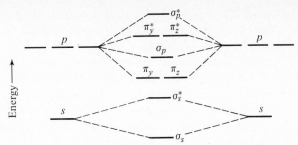

FIGURE 12-10 **Atomic and molecular energy levels involving s and p orbitals.** On the left and the right of the diagram are shown the energies of the s and the p orbitals of the atoms A and A′. In the center are the levels of the corresponding bonding and antibonding molecular orbitals. It is believed that in most A_2 molecules the σ_p level is slightly above the π_y and π_z levels, as indicated, but the positions of these levels may sometimes be reversed.

The different A_2 molecules. The electronic structures of various A_2 molecules can be discussed with the help of the energy level diagram of Fig. 12-10. The program is to construct molecular orbitals, find their relative energies, and fill the orbitals with the available valence electrons, starting at the lowest level. It is this last step that we now consider for different ions and molecules.

For *hydrogen and helium,* the only accessible atomic orbitals are $1s$ orbitals, so that only the σ_s and σ_s^* of Fig. 12-10 are relevant. We begin with H_2^+; the one electron available is placed in the σ_s orbital (see also the earlier discussion in Sec. 11-6). In H_2 the two valence electrons are both placed in the σ_s orbital, with antiparallel spins. The separation between bonding and antibonding levels should be somewhat different from that in H_2^+ because the internuclear distance is different and because there is energy of repulsion between the two electrons. The bond energy of H_2 is found to be about 1.7 times that of H_2^+ (Table 12-2).

In He_2^+ three electrons are available; two go into the σ_s orbital (as in H_2), while for the third only the relatively high-energy σ_s^* orbital is available. The occupancy of σ_s^* by one electron offsets in part the effect of the two electrons in the bonding orbital and thus has a destabilizing effect, which is why σ_s^* is called an antibonding orbital. The spacing of the levels and the energy of interaction of the electrons are different from those in H_2^+ and H_2, but we find the stability of He_2^+ to be comparable to that of H_2^+ (Table 12-2), which is not unexpected.

For the hypothetical molecule He_2, both σ_s and σ_s^* would be filled by electron

TABLE 12-2 **Simplest diatomic molecules and ions**

Species	Electron configuration	Bond distance/Å	Bond energy/kJ mol⁻¹
H_2^+	σ_s	1.06	256
H_2	σ_s^2	0.74	433
He_2^+	$\sigma_s^2\,\sigma_s^*$	1.08	244
He_2	$\sigma_s^2\,\sigma_s^{*2}$		

pairs. The bond energy would be essentially zero because the effect of bonding and antibonding electrons would balance. This is the explanation in MO theory for the nonexistence of the He_2 molecule.

We consider now the molecular orbital picture of A_2 molecules in which the element A ranges from Li through Ne. In all such A_2 molecules the $1s$ electrons of the separate atoms would completely fill the σ_s and σ_s^* orbitals formed by the atomic $1s$ orbitals, with zero total stabilization energy, as for He_2. It is thus considered that the $1s$ electrons of the separate atoms do not interact and that only the valence electrons need be considered. The same reasoning applies to all filled atomic shells.

The molecular orbitals constructed from the $2s$ and $2p$ orbitals of the atoms considered will be assumed to lead to the energy level system shown in Fig. 12-10, and it remains to fill the levels with the valence electrons available and to compare the results with the known experimental facts about these species. The configurations predicted on this basis are summarized in Table 12-3, where observed internuclear distances and bond energies are also shown. (A diatomic molecule A_2, with both atoms of the same atomic number, is called *homonuclear*; a molecule AB is called *heteronuclear*.)

At room temperature lithium is a metallic solid; the Li_2 molecule exists only in the vapor phase. The bonding in Li_2 is similar to that in H_2, but the bond energy of Li_2 is considerably smaller and the interatomic distance larger than in H_2. This is attributed to mutual repulsion of the $1s$ electrons of the two Li atoms.

In Be_2 the bonding by the electron pair in the σ_s orbital is nominally balanced by the pair in the σ_s^* orbitals, with a net result of zero bonding. The experimental fact is that at room temperature beryllium is a metallic solid. The normal boiling point of Be is 2970°C and the vapor is monatomic.

Boron is also a solid at room temperature, the boron atoms being linked into a three-dimensional network. The boiling point is 2550°C and the vapor contains B_2 molecules. Their electron configuration is shown in the diagram on the left of Fig. 12-11. The two electrons of highest energy are seen to be in the two degenerate levels π_y, π_z. Since two orbitals of equal energy are available, the spins of these

TABLE 12-3 **Homonuclear diatomic molecules for elements Li through Ne**

Species	σ_s	σ_s^*	π_y	π_z	σ_p	π_y^*	π_z^*	σ_p^*	Bond distance/Å	Bond energy/kJ mol^{-1}
Li_2	2								2.67	105
Be_2	2	2								
B_2	2	2	1	1					1.59	288
C_2	2	2	2	2					1.31	627
N_2^+	2	2	2	2	1				1.12	610
N_2	2	2	2	2	2				1.09	940
O_2^+	2	2	2	2	2	1			1.12	622
O_2	2	2	2	2	2	1	1		1.21	492
O_2^-	2	2	2	2	2	2	1		1.26	
F_2	2	2	2	2	2	2	2		1.42	150
Ne_2	2	2	2	2	2	2	2	2		

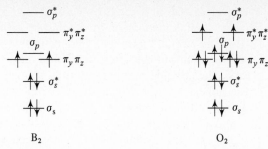

FIGURE 12-11   **The electron configuration of B_2 and O_2.** In both molecules the highest levels filled are doubly degenerate and contain two electrons. The spins of these electrons are parallel, analogous to the parallel spins encountered for degenerate atomic levels (Hund's rule).

electrons need not be opposed, and in analogy to the situation in atoms (Hund's rule) it is expected that parallel spins are energetically favored. This is found to be so: the B_2 molecule has two unpaired electrons.

The number of bonds in B_2 can be said to be one, because the electrons in the σ_s^* orbital largely offset the bonding effects of those in the σ_s orbital, so that only the electrons in the π_y and π_z orbitals contribute to the net bonding. In this sense the number of effective "bonds" corresponding to occupied molecular orbitals is half the difference between the number of electrons in bonding orbitals and the number of electrons in antibonding orbitals:

Nominal no. of "bonds" $= \frac{1}{2}$ (bonding electrons $-$ antibonding electrons) (12-10)

The room-temperature modifications of carbon are diamond and graphite, both of them solid. However, C_2 molecules exist at high temperatures in the vapor phase. This molecule has two bonds, as shown by applying (12-10) to the data in Table 12-3. The bond energy is 627 kJ mol^{-1} in C_2 and 288 kJ mol^{-1} in B_2, while the bond distance is 1.31 Å in C_2 and 1.59 Å in B_2. Thus all these experimental data are in accord with the MO picture.

In N_2 the bonding corresponds to a triple bond. It is interesting to compare the increase of bonding energy and the decrease of the internuclear distance in the sequence B_2, C_2, and N_2 with the increasing number of bonds. N_2 is diamagnetic[1] and thus has no unpaired electrons, as expected.

The fact that the O_2 molecule is observed to be paramagnetic, to a degree corresponding to two unpaired electrons, is inconsistent with the obvious Lewis formula that one can write for this species,

$$:\ddot{O}{=}\ddot{O}:$$

and so this formula was early recognized as inadequate. One of the first successes of molecular orbital theory was its explanation of the paramagnetism of oxygen.

---

[1]Most substances have only weak inherent magnetic properties. These show up when the substance is placed in a nonuniform magnetic field. If the substance is pulled into the field, it is said to be *paramagnetic;* if it is pushed away from the field, it is termed *diamagnetic*. Paramagnetism is related to the presence of unpaired electrons in a substance; in diamagnetic substances all the electrons are paired.

The level scheme in Fig. 12-10 leads to the prediction that the two highest-energy electrons occupy the degenerate π_y^* and π_z^* orbitals and thus remain unpaired (Fig. 12-11). The nominal number of bonds in the MO view of O_2 is two, as it is in the Lewis formula, but the interpretation is different.

In F_2 the nominal number of bonds is again one. In Ne_2 it is zero, corresponding to the nonexistence of Ne_2 and the chemical inertness of Ne.

Three ions of the type A_2^+ or A_2^- are also listed in Table 12-3. For two of these, N_2^+ and O_2^+, the relation (12-10) gives the number of bonds as 2.5. The bond energies and interatomic distances are seen to be similar in the two ions. It is particularly instructive to compare N_2 with N_2^+ and O_2 with O_2^+. When an electron is removed from N_2, the interatomic distance increases and the bond energy decreases, whereas removal of an electron from O_2 has just the opposite effects. The MO picture provides a ready explanation of this difference, since with N_2 the electron is removed from a bonding orbital, thus producing a destabilization, whereas with O_2 the electron is removed from an antibonding orbital, giving an increase in stability. The interatomic distance in O_2^- (its bond energy is not known) is also consistent with the MO view, for this distance is greater than that in O_2, in accord with the addition of an electron to the antibonding π_y^*, π_z^* system.

The present discussion can easily be extended to the higher periods of the periodic table. In considering the filling of molecular orbitals with electrons according to an Aufbau principle, one must always remember that this is purely a conceptual procedure, just as it is with atoms (Sec. 10-3). In the actual molecule, individual electrons cannot be identified or restricted to certain orbitals. Furthermore the energy levels are interdependent. The procedure is nonetheless very helpful as an accounting of filled and unfilled orbitals in the final molecule.

EXAMPLE 12-3   **MO treatment of Br_2 and S_2.**   Give the configuration of the valence electrons for the molecules Br_2 and S_2.

Solution.   In Br the valence shell consists of the $4s$ and $4p$ orbitals, the Ar shell and the $3d$ subshell being complete. Molecular orbitals can be constructed from the $4s$ and $4p$ orbitals with approximate energy levels as shown in Fig. 12-10. Filling these with the 14 valence electrons of the two bromine atoms, we obtain the configuration

$$\sigma_s^2(\sigma_s^*)^2(\pi_{y,z})^4\sigma_p^2(\pi_{y,z}^*)^4 \qquad (12\text{-}11)$$

for the ground state of Br_2. The number of electrons in bonding orbitals is 8 and that in antibonding orbitals is 6, so the number of bonds between the two Br atoms is 1. The molecules F_2, Cl_2, Br_2, I_2, and At_2 all have the same configuration of valence electrons.

In sulfur the valence shell consists of the $3s$ and the $3p$ orbitals, and again molecular orbitals with energies as shown on the right of Fig. 12-10 can be formed. The total of valence electrons for the two sulfur atoms is 12, and they fill the available orbitals as for O_2 (Fig. 12-11) with the π_y^* and π_z^* levels filled singly by electrons of parallel spin. The electron configuration is

$$\sigma_s^2(\sigma_s^*)^2(\pi_{y,z})^4\sigma_p^2\pi_y^*\pi_z^* \qquad (12\text{-}12)$$

The last two symbols, π_y^* and π_z^*, are written individually to signify that the orbitals are singly occupied. The electron configuration of S_2 parallels that for O_2, and S_2 is a paramagnetic molecule.

Molecules of type AB. Diatomic molecules in which the two atoms differ have, to a good approximation, the electron configurations of isoelectronic A_2 species. For example, CO has 10 valence electrons and is isoelectronic with N_2. Although the spacings of its molecular orbital energy levels differ from those of nitrogen because of the disparity in the electronegativities of carbon and oxygen, CO has the same electron configuration as nitrogen,

$$\sigma_s{}^2(\sigma_s^*)^2(\pi_{y,z})^4\sigma_p{}^2 \tag{12-13}$$

Similarly NO^+, which also has 10 valence electrons, has this same configuration, and NO^- is isoelectronic with and has the same electron configuration as O_2.

Bonding in C_2H_4 and C_2H_2. The bonding in ethylene (C_2H_4) and acetylene (C_2H_2) can be described in several ways, each consistent with the classical picture of a double bond in ethylene and a triple bond in acetylene. One important description involves the use of sp^2 hybrid orbitals for the carbon atoms in ethylene and sp hybrids for those in acetylene. We form sp^2 hybrids of the $2s$, $2p_x$, and $2p_y$ orbitals of each carbon atom and use them, with the $1s$ orbitals of the hydrogen atoms, to form a framework of five bonds that tie together the carbon and hydrogen atoms, using 10 of the 12 ($= 2 \times 4 + 4 \times 1$) valence electrons of the C_2H_4 molecule (Fig. 12-12). These bonds are called *sigma* (σ) *bonds* because they are cylindrically symmetric about each of the corresponding internuclear lines.

The $2p_z$ orbital on each carbon atom, which is not used in forming the sp^2 hybrid, extends perpendicular to the plane of the sp^2 hybrid. When the planes formed by the two CH_2 groups coincide—that is, when all six atoms of the molecule lie in the same plane, as they normally do in the unexcited ethylene molecule—these two p orbitals overlap to the maximum extent possible and form a π molecular orbital (Fig. 12-12). The final two valence electrons of the ethylene molecule occupy this orbital, and this electron pair constitutes a *pi* (π) *bond,* the second bond of the double bond in ethylene. The combination of the same two atomic p orbitals can also produce a π^* molecular orbital, and this orbital is occupied in one of the electronically excited states of ethylene, when an electron has been raised from the more stable π orbital. In the carbon-carbon double bond, the overlap of the p_z orbitals constituting the π bond is less effective than is that of the sp^2 hybrids composing the σ bond. This accounts for the fact that the energy of the carbon-carbon double bond (620 kJ mol^{-1}) is less than twice that for the C-C single bond (340 kJ mol^{-1}).

If the planes of the two CH_2 groups in ethylene are rotated relative to each other, the overlap of the p_z orbitals decreases (Fig. 12-12b, upper part). When the angle between the planes becomes 90°, the overlap is zero, which corresponds to breaking the π bond. Thus it is the π bond that keeps the molecule planar and keeps the two CH_2 groups from rotating around the σ bond that joins them. In fact, any six

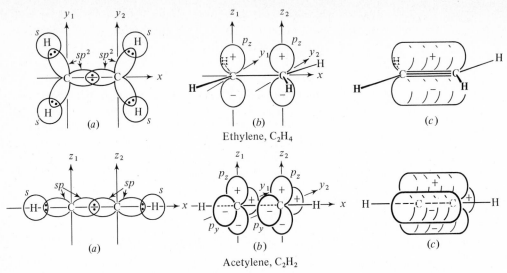

Ethylene, C_2H_4

Acetylene, C_2H_2

FIGURE 12-12   **Bonding in ethylene and acetylene.**   The two diagrams (*a*) are schematic cross sections through the *s* orbitals of the hydrogen atoms and the hybridized orbitals of the carbon atoms (sp^2 for ethylene and *sp* for acetylene), which are used by shared electron pairs to form frameworks of σ bonds. Diagrams (*b*) show the remaining *p* orbitals of the carbon atoms, which are joined as symbolized in diagrams (*c*) to be used by π electrons to form one π bond for ethylene and two for acetylene. The σ bonds are indicated by heavy lines (dashed if concealed) in diagrams (*b*) and (*c*).

atoms joined in the configuration

$$\begin{array}{cc} A & E \\ \diagdown & \diagup \\ B = D & \\ \diagup & \diagdown \\ C & F \end{array}$$ (12-14)

lie in the same plane, or very nearly so. The molecular rigidity resulting from the presence of the π bond makes possible the existence of geometrical isomers in double-bonded molecules when A differs from C and E differs from F (Sec. 26-1).

The triple bond can be described in a similar way. In acetylene (C_2H_2), for example, we begin with a framework of σ bonds that use the 1*s* orbitals of the hydrogen atoms and *sp* hybrids on the carbon atoms (Fig. 12-12). This leads to a linear H—C—C—H sequence. The remaining 2*p* orbitals of the carbon atoms, a pair of them on each atom, perpendicular to the H—C—C—H line, provide for two π bonds. A triple bond is thus pictured as the sum of a σ and two π bonds. This view not only explains the observed linear structure of any sequence of four atoms bonded thus:

$$A—B≡C—D$$ (12-15)

but also explains why quadruple bonds cannot be formed (at least not unless there is expansion of the valence shell), because there are no further overlapping orbitals by which atoms B and C could share an additional electron pair.

Delocalization of electrons. An area of chemistry in which the picture of σ and π bonds has been particularly fruitful is that of delocalization of electrons, illustrated here by the examples of the allyl cation, $(CH_2CHCH_2)^+$, and benzene, C_6H_6. The allyl cation is an important intermediate in organic reactions and its existence has been well established even though it is rather unstable and short-lived. It can be described by the valence bond resonance formulas

$$
\left\{
\left[
\begin{array}{c}
\mathrm{H} \qquad \mathrm{H} \\
\mathrm{C}{=}\mathrm{C} \\
\mathrm{H} \qquad \mathrm{C}{-}\mathrm{H} \\
\mathrm{H}
\end{array}
\right]^{+}
,
\left[
\begin{array}{c}
\mathrm{H} \qquad \mathrm{H} \\
\mathrm{C}{-}\mathrm{C} \\
\mathrm{H} \qquad \mathrm{C}{-}\mathrm{H} \\
\mathrm{H}
\end{array}
\right]^{+}
\right\}
\qquad (12\text{-}16)
$$

Its instability is related to the impossibility of writing Lewis formulas in which all three carbon atoms have a completed octet.

The alternative description of the allyl cation in terms of σ and π bonds is presented in the upper part of Fig. 12-13. The two π electrons are no longer restricted to forming a bond between two atoms as in C_2H_4; rather, they have available the p_z orbitals[1] (not as yet used) of all three carbon atoms. These three orbitals overlap to form an extended π *system,* a set of three molecular orbitals extending over all three atoms. The two π electrons that occupy the most stable of these π orbitals (the orbital illustrated in Fig. 12-13) are said to be *delocalized.*

The bonding of the atoms in benzene may be represented similarly, as in the lower part of Fig. 12-13, where six p_z orbitals on adjacent atoms form a set of six π orbitals that extend over all six carbon atoms. The six π electrons occupy the three lowest-energy orbitals of this set, the most stable of which is the one illustrated in Fig. 12-13. It is generally true that when the description of a molecule by the valence bond method requires a set of resonance formulas, the molecular orbital description will include delocalized π electrons. The term *resonance energy,* used to refer to the difference in stability between the actual molecule and the (less stable) individual valence bond formulas composing the resonance hybrid, is thus sometimes also called the delocalization energy.

Other molecules and ions for which we have earlier used the resonance description (for example, NO_3^-) can equally well be described in terms of a σ-bond framework and delocalized π electrons. The fact that delocalization leads to a lower energy, that is, to increased stability, can be understood qualitatively by noting that the electrons concerned have both kinetic and potential energy and that the lowering of the electronic energy that implies bonding refers to the total energy of the electrons. We stressed earlier (Chap. 11) that bonding occurs because negative charge is concentrated between the nuclei, that is, because the potential energy of the electrons is decreased. However, the kinetic energy of the electrons is also affected during the process of bonding, because in the separated atoms the electrons are confined in a smaller volume than in the final molecule.

Quantum theory shows that it is quite generally true that if an electron is confined

---

[1]We assume that the sp^2 hybrids were formed with p_x and p_y orbitals, so that the p_z orbitals are perpendicular to the plane of the hybrids.

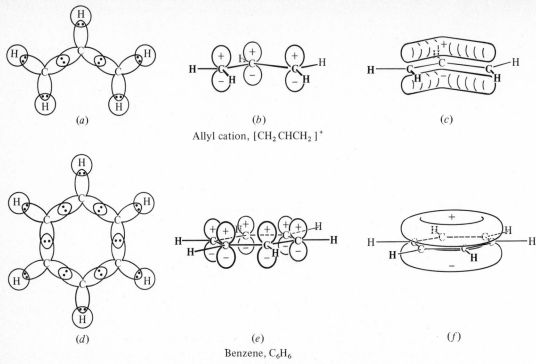

(a)   (b)   (c)

Allyl cation, [CH₂ CHCH₂]⁺

(d)   (e)   (f)

Benzene, C₆H₆

FIGURE 12-13   **An ion and a molecule with delocalized electrons.**   The σ-bond framework of the allyl cation requires 14 of the 16 ($= 3 \times 4 + 5 - 1$) valence electrons, as shown in (a). The unused p_z orbitals shown in (b) are occupied by the two remaining electrons and overlap to form a π orbital, depicted in a schematic and exaggerated way in (c).

In benzene, with a total of 30 ($= 6 \times 4 + 6$) valence electrons, the framework requires 24 σ electrons (d). The remaining 6 electrons are distributed over the remaining 6 unused p_z orbitals (e), which can be combined into two overlap regions, symbolized by the two doughnut-shaped arrangements (f).

The π orbitals depicted schematically in (b) and (c) and in (e) and (f) are the lowest-energy π orbitals for $C_3H_5^+$ and C_6H_6, respectively, and each is occupied by a pair of π electrons. For C_6H_6 the remaining four π electrons occupy the next two higher π orbitals, which have one node each and are degenerate (having the same energy).

Remember that the plus and minus signs in (b) and (e) represent only the sign of ψ and that both the upper and lower overlap regions shown in (c) and (f) correspond to positive signs for ψ^2 and thus (because the upper and lower regions are equivalent by symmetry in either diagram) to equally probable regions to be occupied by the π electrons.

in a large volume, its minimum kinetic energy is lower than when it is confined in a small volume. The more the electron is delocalized, the lower is its kinetic energy in the lowest possible state. This reduction in energy explains, at least in part, the stabilizing effect of the delocalization of the electrons over an entire molecular system.

Molecules with delocalized π electrons tend to absorb electromagnetic radiation more strongly and at longer wavelengths than do comparable molecules without such delocalization. This implies that the separation of the electronic energy levels has become smaller, so that less energy is required to excite an electron. One important class of molecules with delocalized electrons is that which can be described in valence bond terms by a sequence of alternating single and double bonds. Double bonds so arranged are said to be *conjugated;* examples of molecules with conjugated double bonds are 1, 3-butadiene, CH_2=CHCH=CH_2, and benzene. When the array of conjugated double bonds over which the delocalized electrons are able to range is large, there is often strong absorption in the visible region of the spectrum. In other words, the molecule is strongly colored. An example is the dye indigo, known since ancient times:

(12-17)

In conclusion, it should always be remembered that the theories and formalisms expounded in this and the preceding chapters are all *models*—extremely useful models of considerable predictive power, but models nonetheless. To what degree such models reflect features of the real world is a metaphysical question, as is true for any theory. The feeling of satisfaction conveyed by a "good" theory is largely tied to inherent aesthetic beauty, to compactness and economy of basic postulates, to logical structure, and to consistency and uniqueness of predictions.

Problems and Questions

12-1 **The resonance concept.** Explain why the concept of resonance was introduced into structural chemistry. Illustrate your explanation with its application to the structure of the nitrite ion, NO_2^-.

12-2 **Resonance and bond distances.** The carbonate ion, CO_3^{2-}, has a planar structure with three equal bond angles and three equal carbon-oxygen distances. Draw three equivalent valence bond structures for this ion. Draw two structures for the bicarbonate ion, $HOCO_2^-$, and predict which of the carbon-oxygen bonds in these two ions would be the shortest and which the longest. Would you expect the bonds in CO_2 to be similar in length to any of the carbon-oxygen bonds in these ions, or shorter, or longer? Explain.

12-3 **Paramagnetic molecules.** Molecules and ions in which the number of electrons with spins of one orientation is different from the number with spins of opposite orientation, so that the spins do not add to zero, are paramagnetic, and a permanent magnetic moment is associated with them. Which of the following molecules would you expect to be paramagnetic: NO_2, $NOCl$, N_2O_4, S_2, ClO_2, Cl_2O? *Ans.* three of the molecules are paramagnetic: NO_2, S_2, and ClO_2

12-4   **Lewis formulas.**   Draw Lewis formulas for the following sets of molecules or ions. Show important resonance formulas where appropriate.
(*a*) $Be(OH)_4^{2-}$, SO_2, H_3PO_2 (in which two H's, an O, and an OH group are linked to the phosphorus), NNO, PH_4^+
(*b*) SO_3, ClO_4^-, O_2^{2-}, NCS^-, H_2NNH_2, C_5H_5N (containing a six-membered ring with five C's, each attached to one H, and one N), ClO_2^-

12-5   **Electron configuration and molecular properties.**   (*a*) The B–F distance in BF_3 is 1.30 Å; that in BF_4^- is 1.40 Å. Explain. (*b*) NO is paramagnetic; explain. Would you expect NO^- to be paramagnetic?

12-6   **Hybrid orbitals.**   (*a*) Make a list of hybrid orbitals that include only *s* and *p* functions. Indicate their directional character. (*b*) Extend the list to include two hybrids that also include one or more *d* functions.

12-7   **Electron configuration and molecular shape.**   Give plausible electron configurations for the following ions and molecules. Indicate by an appropriate word or phrase the expected shape. (*a*) PF_3, PBr_5, HNO_3, AlH_4^-;   (*b*) ClO_3^-, PCl_6^-, NNN^-, D_2Te (D = deuterium).

12-8   **Geometry of molecules.**   Consider the molecules and ions listed below and describe the geometry of each by using one or two of the following terms: linear, bent, planar, tetrahedral, pyramidal.
(*a*) H_2S, SO_4^{2-}, BF_3, C_2H_4, NCS^-
(*b*) O_3, CS_2, ClO_2, $HNCS$, SbH_3
(*c*) SO_2, NCl_3, OCS, CH_2Cl_2, $PbCl_4$
(*d*) Cl_2CO, SO_3^{2-}, CO_3^{2-}, $ClNO$, HCN

12-9   **Dipole moment and molecular geometry.**   Consider the following molecules and ions and predict which of them have dipole moments (Fig. 3-9) and which do not: $CHCl_3$, CH_2F_2, $H_3C—O—CH_3$, $N(CH_3)_3$, $N(CH_3)_4^+$, $Cl_3Si—SiCl_3$, $Cl_3Si—O—SiCl_3$.

12-10   **Dipole moment and molecular geometry.**   On the basis of the structures of the molecules and the electronegativity differences between the bonded atoms, make qualitative comparisons of the dipole moments (see Fig. 3-9) of the following pairs of molecules by using the symbols > (greater than), = (equal to), or < (smaller than):

HF	HCl	F_2O	CO_2
CCl_4	CH_4	PH_3	NH_3
H_2S	H_2O	BF_3	NF_3

12-11   **Concepts of molecular structure theory.**   Explain what is meant by the following terms as used in modern discussions of atomic and molecular structure: electron affinity, hybrid atomic orbital, degenerate energy level, subshell, Pauli exclusion principle, outer-shell electron, antibonding molecular orbital, polar molecule.

12-12   **Shapes and signs associated with orbitals.**   (*a*) Give an example, including a sketch of the "shape" (indicating clearly what is being plotted), of a spherically symmetric orbital and of an orbital with one angular node (show where this node is and indicate what the term *angular node* implies). (*b*) Why is the sign of each component orbital wave function in different regions of space important in the formation of molecular orbitals and hybrid orbitals? Illustrate with a sketch.

12-13 Double bond and isomers. Draw three possible isomeric structures consistent with the formula C_2H_2FCl.

12-14 Molecular orbitals and diatomic molecules and ions. (*a*) Give the electron configurations in terms of filled molecular orbitals [such as $\sigma_s^2(\sigma_s^*)^2(\pi_{x,y})^4\sigma_p^2(\pi_{x,y}^*)^4(\sigma_p^*)^2$] for the following diatomic species: B_2, C_2, N_2, N_2^+, O_2^{2-}, O_2^-, O_2, and O_2^+. (*b*) The interatomic distances in the last four species are, in sequence, 1.49, 1.26, 1.21, and 1.12 Å. How are these values related to the bonding? (*c*) The interatomic distances in N_2 and N_2^+ are 1.09 and 1.12 Å; in Cl_2 and Cl_2^+ they are 1.99 and 1.89 Å. Explain why the values in each pair vary as they do. (*d*) Explain the existence of the species H_2^+ and He_2^+.

12-15 Molecular orbitals for NO. (*a*) Draw a molecular orbital diagram for NO as follows: on the left draw the $2s$ and $2p$ states for N; on the right those for O. The energies of the oxygen atomic states are somewhat below the corresponding nitrogen states, because O is more electronegative than N. Combine the states into σ, σ^*, π, and π^* molecular states. Label the states and show how they are filled with the valence electrons of N and O. (*b*) Predict the electron configurations of NO, NO^+, and NO^- in terms of occupancies of molecular orbitals. On the basis of these predictions, arrange the three species in order of decreasing bond energy and in order of decreasing bond distance. The known distances are NO, 1.15 Å; NO^+, 1.06 Å.

"When a system is in chemical equilibrium, a change in one of the parameters of the equilibrium produces a shift in such a direction that, were no other factors involved in this shift, it would lead to a change of opposite sign in the parameter considered."

Henri-Louis Le Châtelier, 1888

13 Introduction to Chemical Equilibrium

This chapter and the two that follow it deal with the quantitative aspects of a fundamental concept in chemistry, that of chemical equilibrium. In the present chapter we consider general features of systems at equilibrium, including the quantitative relationships embodied in the equilibrium constant. Typical ionic equilibria and equilibria involving gases are used as illustrations. Chapter 14 is concerned with equilibria in acid-base systems, and Chap. 15 deals with some other important equilibria involving aqueous solutions.

13-1 Background

The concept of dynamic equilibrium, involving a balance of two opposing processes that occur at equal rates, has been discussed in several different contexts in earlier chapters: the equilibrium between liquid and solid phases at the melting point of a crystalline solid (Sec. 6-1); the equilibrium between a gas and the corresponding condensed phases, leading to a definite vapor pressure of a liquid or a solid (Sec. 3-1); and the equilibrium between dissolved solute and excess solute in a saturated solution (Sec. 6-2). Furthermore, the entire discussion of the reactions of aqueous acids and bases in Chap. 7 can be profitably viewed in terms of chemical equilibria in which two (or more) bases compete for the available protons, as explained in Sec. 7-5 and in detail in Chap. 14.

Dynamic nature of equilibrium.   In a system at equilibrium, there is no perceptible change in the macroscopic properties of the system with time. However, it is the very essence of chemical equilibria that they are dynamic; there is a constant interchange of chemical species between different states of chemical combination or between different phases (or both). This is quite in contrast to certain static equilibria of interest in physics and engineering.[1] For example, the dynamic nature of the equilibrium between liquid and vapor in a closed flask can readily be demonstrated: if some "labeled" molecules—e.g., molecules containing some radioactive or otherwise easily identifiable isotope of one of the atoms present—are introduced into one phase, they are very quickly found to be uniformly distributed between the two phases present. Similarly, an equilibrium between different molecules is dynamic. For example, two molecules of NO_2, a red-brown gas, readily combine to form molecules of N_2O_4, a colorless gas, which in turn can dissociate to form NO_2. In a flask containing NO_2 and N_2O_4 in equilibrium with each other at a given temperature and pressure, the numbers of moles of NO_2 and of N_2O_4 in the flask remain constant, but there is a continuous and very rapid conversion of NO_2 into N_2O_4 and vice versa.

A fundamental characteristic of the equilibria that are of interest to chemists is that very large numbers of particles—atoms, molecules, or ions—are involved. In other words, we are dealing with macroscopic samples. The statement that the concentrations and partial pressures of the components of an equilibrium mixture remain constant with time does not mean that there are no fluctuations at all in these quantities, but rather that these fluctuations, measured relative to the actual concentrations or pressures, are imperceptible because the numbers of molecules involved are so large (see page 123).

An equilibrium of the type we are discussing may or may not involve any chemical reaction in the sense that different chemical species are formed, or even in the wider sense that chemical bonds are broken or formed. When dry ice sublimes, for example, no chemical bonds are broken; the van der Waals interactions among the CO_2 molecules merely become less important as the molecules move apart in the gaseous phase. On the other hand, when CO_2 is dissolved in water, there *is* chemical interaction. Most of the dissolved CO_2 becomes "hydrated" in a way that is not understood in detail, with a small fraction of it forming carbonic acid, H_2CO_3. In both examples, the sublimation of solid CO_2 and the dissolving of CO_2 in water, equilib-

---

[1]For example, the balanced arch of a bridge does not collapse, because it is in static equilibrium: nothing happens (excepting vibrations of the bridge). Similarly, a ball that comes to rest at the bottom of a hole is in static equilibrium, at a position corresponding to a minimum in its gravitational potential energy. These static equilibrium situations are quite different from the dynamic ones of interest in chemistry.

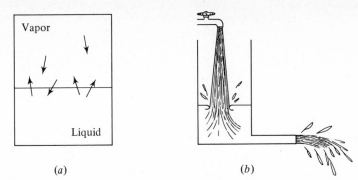

FIGURE 13-1   **(a) Equilibrium; (b) steady state.** In (a) liquid water and its vapor are at equilibrium because the rate of evaporation of water molecules is the same as their rate of condensation. In (b), a vessel receives water at a constant rate from a faucet. The water can leave the vessel through an opening at the bottom, at a rate that increases as the water level in the vessel rises. Unless the faucet is open too much or too little, the water level reaches a position where the rate of inflow just matches the rate of outflow; a steady state exists and the water level in the vessel remains constant.

rium situations may be established, characterized by constant partial pressures of $CO_2(g)$ (at a given temperature).

Contrast between equilibrium and steady state.   It is possible to have dynamic situations, called *steady states,* in which even though no equilibrium exists, there is no apparent change in the properties of the system with time. *Dynamic equilibrium requires processes that move in opposite directions,* one process being in essence the reverse of the other, as in Fig. 13-1a. In a steady state, the processes are not so opposed and a continuing supply of new material (or energy) is required. A macroscopic example of a steady state is illustrated in Fig. 13-1b. Although the level in the tank remains constant, the flow out of the tank is not the reverse of the flow from the faucet—water must be steadily supplied. A steady state may exist without supply to the system from outside; it might occur through the slow reaction of a substance that is present in a large and therefore essentially constant amount.

13-2   Qualitative Considerations

In the brief discussion of the equilibria involved in the sublimation of CO_2 and the chemical reaction of CO_2 with water in the previous section, we pointed out that in each equilibrium the partial pressure of CO_2 was constant at a given temperature. Both equilibria are sensitive to temperature. An equilibrium may also be affected by pressure: the bubbles that appear in a bottle of soda, beer, or champagne when the cap is removed are evidence that the equilibrium solubility of CO_2 in water decreases when the pressure is lowered.

A qualitative principle of great practical value in thinking about the effects on equilibria of changes in the conditions affecting the equilibria was discovered independently and almost simultaneously during the last century by H. L. Le Châtelier and F. Braun. For some reason, however, Braun has received little credit for his insight, and this useful generalization is commonly identified only with the name of Le Châtelier. Before we state and illustrate the principle, we must first digress briefly about the distinction between intensive and extensive properties.

Intensive and extensive properties. Many properties of a homogeneous phase may be classified as being either intensive or extensive. *Intensive* properties or quantities are those that are not affected by a division of the phase into parts—for example, temperature, pressure, density, and concentration. *Extensive* properties or quantities are those that are proportional to the amount of substance contained in the portion of the phase considered—for example, volume, mass, and energy. For heterogeneous systems an intensive property of the system is one that is defined for each component phase and has the same value for each. Examples are again temperature and pressure, but no longer density or concentration because these variables may be different for the different phases. Similarly, the term *extensive* may be applied to heterogeneous systems. For a heterogeneous system, the value of an extensive quantity, such as the volume or the mass, is the sum of the values for all component phases. Certain pairs of extensive and intensive properties are related. Important examples for present purposes are volume and pressure, and energy[1] and temperature.

Le Châtelier's principle. Le Châtelier's principle predicts qualitatively what happens to a system originally at equilibrium when an *extensive* quantity such as the volume or the energy is changed. The change affects one or more of the *intensive* properties of the system, such as the pressure or the temperature. The original change may cause a *shift in the equilibrium,* and the principle asserts that the shift will be in such a direction as to offset partially the effect of the change. The shift makes the changes in the related intensive quantities *less* than they would have been had the equilibrium not shifted, as illustrated in the following paragraphs.

Temperature dependence of the solubility. When a substance dissolves, energy is usually evolved or absorbed. If energy is evolved, the process is said to be *exothermic* (Greek: *exo,* out of); if energy is absorbed, the process is termed *endothermic* (Greek: *endo,* within). Normally, the energy that is evolved or absorbed causes either an increase or a decrease in the temperature of the solution itself. The only way to avoid this would be to carry out the dissolving of the solute very slowly in a constant-temperature bath.

It is found experimentally that an *increase in temperature* leads to an *increase in solubility* when the process of dissolving is *endothermic* and to a *decrease in solubility* when the process of dissolving is *exothermic*. To see how these facts are in accord with Le Châtelier's principle, consider a solution that is at equilibrium with undissolved solute at a particular temperature. Equilibrium implies that no additional solute dissolves, so that the solution is saturated at the given temperature. Suppose the solution is heated, increasing the energy of the system—a change in

---

[1]More exactly, the *internal* energy, defined in Chap. 16.

an extensive quantity. The related intensive quantity affected is the temperature. By Le Châtelier's principle the equilibrium must shift so that the temperature change is less than it would be if the equilibrium remained undisturbed. This happens as a result of a shift in the equilibrium—a change in the solubility—so that part of the energy introduced is used up, either because additional solute is dissolved or because some solute is precipitated, depending on whether the act of solution is endo- or exothermic.

Assume first that dissolving the solute is endothermic. The principle then predicts that the equilibrium shift is toward increased solubility, so that part of the energy introduced is used up as additional solute dissolves, because this lessens the temperature increase. In the end the solubility has increased and the temperature is higher (but not as high as it would have been had the solubility not been changed). Thus Le Châtelier's principle leads to a conclusion in accord with the experimental facts: if dissolving the solute is an endothermic process, the solubility increases with increasing temperature. Conversely, if dissolving the solute is exothermic, the solubility decreases upon introduction of energy, so that some of the energy introduced is used up by the precipitation of some of the dissolved solute. Finally, if no energy is evolved or absorbed when the solute dissolves, its solubility is not affected by a change in temperature.

The same result applies to any temperature-dependent equilibrium. As the temperature is increased, the products of any exothermic reaction will be less favored at equilibrium and the products of any endothermic reaction will be more favored. If a reaction is neither exo- nor endothermic, the equilibrium is not affected by a change in temperature.

Pressure dependence of the solubility. Consider a saturated solution at constant temperature. In the process of forming the solution, the total volume may increase or decrease, depending on the natures of the solvent and the solute; that is, the volume of the solution may be greater or less than (or, rarely, the same as) the sum of the volumes of its initially separated components. Le Châtelier's principle permits correlating the volume change with the pressure dependence of the solubility.

Suppose that the volume of the saturated solution is forcibly decreased by a certain amount (necessarily by applying pressure to the solution). Le Châtelier's principle says that if a volume change has any effect on the equilibrium at all, the equilibrium will shift in such a way that the increase in pressure accompanying the specified decrease in volume will be smaller than it would have been if there had been no shift in the equilibrium. Assume that the volume of the system decreases when additional solute is dissolved. An increase in solubility would then relieve some of the increase in pressure because it would permit the same total volume decrease at a lower pressure; that is, some of the volume decrease would result from the dissolving of additional solute. Thus an increase in pressure produces an increase in solubility. A parallel argument shows that if the volume of the system increases as solute dissolves, an increase in pressure will decrease the solubility. Finally, if dissolving the solute does not change the total volume of the system, then a change in pressure will have no effect on the solubility. These predictions are all in accord with experimental observations.

These same results apply to any equilibrium. If a shift in the equilibrium can result in a change in the volume of the system, then the shift of equilibrium with increasing pressure will be in such a direction that the volume decreases as a result of the shift.

Le Châtelier's principle[1] may be applied to many different equilibria of chemical interest, and we shall cite it again. The qualitative predictions from Le Châtelier's principle can be understood quantitatively in terms of equilibrium constants, considered in this chapter and the two that follow, and the change of equilibrium constants with temperature (Chap. 17).

Criteria for equilibrium. One very important feature of any chemical equilibrium is that, for a given set of conditions, the same equilibrium situation is achieved whether it is approached from one side or the other. For example, at a given temperature and total pressure, the same concentrations of NO_2 and N_2O_4 will be obtained at equilibrium whether one starts with pure NO_2, or an equivalent amount of pure N_2O_4, or any mixture of the two. Similarly, the pressure of a vapor in equilibrium with a liquid at a given temperature will be the same whether one starts with pure liquid and lets it evaporate until equilibrium is reached or starts with pure vapor and allows it to condense.

It is sometimes difficult in practice to distinguish a true equilibrium, in which the concentrations remain constant with time, from a nonequilibrium situation in which the concentrations also do not change perceptibly because the rate of any possible reaction is simply too small to observe. For example, at ordinary temperatures and pressures a mixture of gaseous hydrogen and oxygen in the mole ratio $2:1$ would be, at equilibrium, essentially completely converted into water, but such a mixture can be kept for many years apparently unchanged because the reaction is normally so slow.

To establish experimentally whether equilibrium does in fact exist, either of the following two criteria may be applied. (1) One or more of the variables (e.g., total pressure, temperature, or concentration of some component) that might affect a possibly existing equilibrium can be changed. If changes in the concentrations or partial pressures of the (other) substances involved then occur, and if these concentrations and partial pressures return to their initial values when the original values of the variables are restored, an equilibrium exists. (2) A possible equilibrium may be approached both from the side of the reactants and from the side of the products, in separate experiments carried out at the same temperature and pressure and with amounts of reactants in one experiment and of products in the other that are equivalent to one another. If the final concentrations and partial pressures are the same in each experiment, a true equilibrium exists.

Equilibrium and nonequilibrium situations may coexist in the same system. Thus,

---

[1]The principle is often stated in more general terms than those we have used, but it is more subtle than sometimes represented. If the distinction between extensive and intensive properties is not made, predictions may be made that are not in agreement with experience. Note, too, that the principle does not permit drawing direct conclusions about changes in intensive properties not related to the extensive quantity being changed.

hydrogen peroxide is less stable than its possible decomposition products, O_2 and H_2O,

$$2H_2O_2(l) \longrightarrow 2H_2O(l) + O_2(g) \tag{13-1}$$

but the rate of this reaction and therefore establishment of the equilibrium of (13-1) is very slow in the absence of catalysts. Another reaction, the dissociation into H^+ and HO_2^-,

$$H_2O_2(l) \rightleftharpoons H^+ + HO_2^-$$

reaches equilibrium rapidly[1] and can be examined by experiment in spite of the instability of the H_2O_2. [Incidentally, the H^+ ion is associated with at least one H_2O_2 molecule to form $H_3O_2^+$, which is probably solvated by further H_2O_2 molecules, in close analogy with the way in which H^+ ions in an aqueous medium are bound as H_3O^+, $H_9O_4^+$, and related species (Sec. 7-2).] As long as the rate of (13-1) is slow enough to be negligible, the H_2O_2 is said to be *metastable*, a term implying that a substance may persist (and thus *seem* stable) for a relatively long period, although it eventually changes (or may change) to a more stable form (see Sec. 3-1). Many commonly known substances are actually metastable. This is true, for example, of most organic compounds in the presence of oxygen, since they are readily combustible.

Position of equilibrium.   It is frequently convenient to refer to the *position of equilibrium* in a chemical system, even though it is difficult or impossible to define this term in such a general way as to apply to all possible equilibria. Often the position of equilibrium is a measure of the *relative* amounts of reactants and products present at equilibrium. For example, the position of equilibrium is said to be far to the left when the proportion of products (written on the right in a chemical equation) in the equilibrium mixture is small. Conversely, the position of equilibrium is said to be far to the right when the proportion of reactants remaining at equilibrium is small. In some cases, however, the position of equilibrium is not represented by the relative amounts of some of the reacting species. Thus, the position of equilibrium between a liquid and its vapor is characterized by the vapor pressure of the liquid and is completely independent of the amount of liquid present (as long as there is *some* liquid).

A common quantitative measure of the position of equilibrium is the fraction of some reactant that has been consumed.

EXAMPLE 13-1   **Calculation of concentrations from position of equilibrium.**   In a 0.10 F solution of potassium acetate, 8×10^{-3} percent of the original acetate ions are hydrolyzed. Calculate the concentrations of OAc^-, HOAc, and OH^- in this solution, neglecting the fact that H_2O can also act as a source of OH^-.

Solution.   The reaction involved is that of Equation (7-26):

$$H_2O + OAc^- \rightleftharpoons HOAc + OH^-$$

---

[1]The double arrow $\rightleftharpoons$ is the customary symbol for chemical equilibrium.

This is a typical acid-base reaction in which the bases OH^- and OAc^- are competing for the available protons. The data given imply that the acetate ion is a weaker base than OH^-, for it loses out in the competition: very few of the acetate ions combine with protons to form HOAc. Quantitatively, we are told that 8 acetate ions out of every 10^5 are converted into HOAc, forming an equal number of OH^- in the process. Since in the absence of any reaction there would be 0.10 mol acetate ion per liter, we must get $(0.10\ M)\ (8 \times 10^{-5}) = 8 \times 10^{-6}$ mol HOAc per liter of solution. The remainder of the OAc^- is unaltered. The reaction equation shows that equal quantities of the species HOAc and OH^- are produced, which makes $[OH^-]$ equal to $[HOAc]$. (A slightly more sophisticated approach is required to show that the amount of OH^- coming from H_2O is negligible here.) Thus $[HOAc] = [OH^-] = \underline{8 \times 10^{-6}\ M}$, and $[OAc^-] = 0.10 - 8 \times 10^{-6} = \underline{0.10\ M}$.

In many systems the position of equilibrium can be shifted by changing the concentrations. For example, the degree of dissociation of a weak acid can be altered by changing its formality. This is in marked contrast to the *equilibrium constant,* to be defined in Sec. 13-3, which is a constant (at a given temperature), independent of the concentrations of the components of the equilibrium mixture.

The two factors that affect chemical equilibrium. A very important distinction exists between the kinds of static equilibria exemplified by a ball resting in the bottom of a hollow and those of interest in chemistry. The position of equilibrium in such a static situation is simply the position of minimum potential energy, while in chemical systems the position of equilibrium depends on a balance between the tendency of the potential energy to be minimized and a second factor that is weighted increasingly as the temperature rises. Were it not for this additional factor, all spontaneous chemical processes would be accompanied by a release of energy, but it is well known that some are not—for example, the evaporation of a liquid into a vacuum or into any region not saturated with its vapor occurs spontaneously and with absorption of energy, as does the dissolving of many (although not all) solutes. This second factor, called entropy, affecting the position of equilibrium is discussed at length in Chap. 17.

Importance of equilibrium. All systems proceed spontaneously toward a position of equilibrium; that is, they tend to move toward equilibrium (although often only very slowly) in the absence of outside disturbances. They move away from a position of equilibrium only if they are perturbed from the outside, for example, by absorption of energy or by a sudden change of pressure. This is a generalization from many observations about the nature of our universe.

The phenomenon of chemical equilibrium is a very general one with applications to the entire range of chemistry. Consequently a firm grasp of the comparatively few general principles needed to understand equilibria can give one a broad command of many different aspects of chemistry, as illustrated throughout the remainder of this book. Many equilibria are established so slowly, however, that a large fraction of chemical problems involve nonequilibrium systems. For example, many of the most important chemical reactions in energy sources such as flames and in living organisms never reach equilibrium, and in many industrial processes the distribution of products depends on the relative rates of the reactions involved (as examined in Chap. 20) rather than on achieving equilibrium.

An understanding of nonequilibrium systems (which always tend toward equilibrium) is greatly helped by an appreciation of equilibrium systems. Hence the next few sections and chapters are important not only because of the significance of chemical equilibrium itself but also for the background they provide for situations in which equilibrium has not been achieved.

13-3   The Equilibrium Law

A simple quantitative relationship can be formulated among the concentrations of the reactants and products in any chemical system at equilibrium. For a reaction symbolized by the balanced equation

$$a\text{A} + b\text{B} + \cdots \rightleftharpoons d\text{D} + e\text{E} + \cdots \tag{13-2}$$

with capital letters denoting chemical species and lowercase letters signifying their coefficients in the balanced equation, we define the *reaction quotient*

$$Q \equiv \frac{[\text{D}]^d[\text{E}]^e \cdots}{[\text{A}]^a[\text{B}]^b \cdots} \tag{13-3}$$

Both experiment and theory show that at any particular temperature (and total pressure), Q as defined by (13-3) is a constant at equilibrium. The value of Q for a system at equilibrium is called the *equilibrium constant* for the reaction at the temperature (and pressure) in question. This constant is almost invariably symbolized by K, often with a subscript to indicate the reaction involved or the units used. What this all-important equilibrium law[1]

$$\frac{[\text{D}]^d[\text{E}]^e \cdots}{[\text{A}]^a[\text{B}]^b \cdots} = K \qquad \text{(equilibrium)} \tag{13-4}$$

implies is that, *if a system is at equilibrium,* the ratio of the product of the concentrations of the reaction products, each raised to a power equal to its coefficient in a balanced chemical equation (13-2) for the reaction, to the product of the concentrations of the reactants, each in turn raised to a power equal to its coefficient in the chemical equation, is always the same at a given temperature.[2] Although the concentrations of individual reactants or products may differ considerably from one

---

[1]The relationship given in (13-4) is often called, as it was by its nineteenth-century discoverers, C. M. Guldberg and P. Waage, *the law of mass action.* This usage of *mass* goes back to Berthollet (1801), but it is not at all illuminating and indeed is misleading in the modern context of this term, so we shall not use it.

[2]The "constant" K may also depend on the total pressure of the system, but this effect is negligible unless pressures of hundreds of atmospheres are involved (see also Sec. 13-5, item 8). In any event, K is independent of the pressure for reactions among ideal gases.

experiment to the next, they will always mutually adjust themselves by a shift in the position of equilibrium (provided that equilibrium is achieved) in such a way that (13-4) is satisfied.

In later chapters devoted to thermodynamics and reaction rates, we show why the particular quotient in (13-4) is a constant at a given temperature and pressure. For the moment we merely *assert* that (13-4) is true for any reaction (13-2) at equilibrium. The reaction quotient Q can, of course, be evaluated whether the reaction is at equilibrium or not, and its value is sometimes of interest even when equilibrium has not been reached. The quotient Q is *equal* to K when equilibrium has been established.

Conventions in formulating reaction quotients. There are several arbitrary rules or conventions followed in the formulation of reaction quotients that must be learned by anyone working with them:

I. The concentrations of the *products* of the reaction appear in the *numerator;* those of the reactants are in the *denominator.* This convention implies that if a reaction is reversed, the roles of the reactants and the products thus being interchanged, the new reaction quotient (call it Q') is the inverse of the initial one (Q); that is, $Q' = 1/Q$. Since at equilibrium $Q = K$, the equilibrium constant for the reverse process, K', is similarly related to K; that is, $K' = 1/K$.

II. Unless otherwise specified, solute concentrations are always expressed in molarities, symbolized by [ ]. (In some applications, mole fractions are used as concentration units instead, but this usage is then explicitly mentioned.)

III*a*. In a gaseous reaction mixture the concentrations of the gases in the reaction quotient are usually replaced by their partial pressures.

III*b*. In reactions of gases with substances in the solid or liquid states (including substances in solutions), the concentrations of the gases are also usually replaced by their partial pressures.

IV. Water that participates in a reaction occurring in a dilute aqueous solution is omitted in the formulation of the reaction quotient.

V. The rules given so far do not indicate how to deal with a *pure* solid or *pure* liquid that occurs as a reactant or a product, for the concept of concentration applies normally to solutions, not to pure phases. The convention for dealing with pure solids and pure liquids is a very simple one: they are *omitted* entirely from the reaction quotient.

Conventions I and II are self-explanatory, but the others require justification and further explanation. These five conventions establish a one-to-one relationship between a chemical equation and the reaction quotient associated with it (and thus also with the appropriate value of K).

Discussion of conventions III*a* and IV. For a gaseous reaction mixture, the gas concentrations appearing in Q are usually replaced by partial pressures (convention III*a*). This is permissible as long as the gases are at least nearly ideal, for if the gases participating in the reaction are specified by numbers $i = 1, 2, 3,$ ..., the molar concentration for gas i is $[i] \equiv n_i/V = P_i/(RT)$. In other words, the number of moles per liter of gas i is proportional to its partial pressure at a given temperature. The proportionality factor is constant at the given temperature and can be made part of the equilibrium constant. We examine this relationship further in Sec. 13-4 (where examples are given) and in some of the comments below. If the units of pressure are not explicitly stated, they should be assumed to be atmospheres.

Convention IV states that when water participates in a reaction in a dilute aqueous solution, the concentration of H_2O is omitted in formulating Q. This is always done when $[H_2O]$ in the solution does not deviate by more than 10 percent or so from the value it has for pure water. This value, the concentration of H_2O in pure water, is $(998 \text{ g liter}^{-1})/(18.0 \text{ g mol}^{-1})$ or about 55 mol liter^{-1} (55 M). Convention IV will be adhered to unless explicitly stated otherwise.

Thus in the very important autoprotolysis reaction of water (Sec. 7-2),

$$H_2O + H_2O \rightleftharpoons H_3O^+ + OH^- \tag{13-5}$$

the reaction quotient at equilibrium might be written as[1]

$$K_5' = \frac{[H_3O^+][OH^-]}{[H_2O]^2}$$

However, since $[H_2O]$ is always about 55 M in any dilute solution, this value is by convention combined with K_5', leading to

$$[H_3O^+][OH^-] = K_5'[H_2O]^2 \equiv K_5 \equiv K_w \tag{13-5a}$$

Finally, since $[H_3O^+]$ and $[H^+]$ mean the same thing (Sec. 7-2), we simply write

$$[H^+][OH^-] = K_w \tag{13-5b}$$

Thus, in water or any dilute aqueous solution, the product of the concentrations of H_3O^+ (or H^+) and OH^- is a constant at a given temperature. This constant, the *ion-product constant* of water, symbolized K_w, is equal to $1.008 \times 10^{-14} M^2$ at 25°C. Thus, in pure water, $[H^+] = [OH^-] = 1.00 \times 10^{-7} M$ at this temperature.

The following two important examples are further illustrations of convention IV:

$$NH_3 + H_2O \rightleftharpoons NH_4^+ + OH^- \tag{13-6}$$

$$\frac{[NH_4^+][OH^-]}{[NH_3]} = K_6 \tag{13-6a}$$

$$H_2O + HOAc \rightleftharpoons H_3O^+ + OAc^- \tag{13-7}$$

$$\frac{[H_3O^+][OAc^-]}{[HOAc]} \equiv \frac{[H^+][OAc^-]}{[HOAc]} = K_7 \tag{13-7a}$$

Again the concentrations of all hydrated proton species are usually represented by $[H^+]$—no attempt is made to distinguish among the various formulas.

Conventions III*b* and V are discussed later (Sec. 13-5).

---

[1]The subscript of K' used here refers to the equation number describing the pertinent chemical equilibrium.

General comments and examples. Many aspects of the chemical equilibrium law are difficult for those just learning to work with it, but the details that prove puzzling often vary from person to person. The following comments emphasize points worthy of special attention.

1. *Units of equilibrium constants:* The units of K often differ for different reactions. They always follow from and are identical with the units of the corresponding reaction quotient Q. Concentrations are always expressed in moles per liter (M) and partial pressures in atmospheres, unless explicitly stated otherwise. Thus the units of K_w [Equation (13-5a)] are (moles per liter)^2 or M^2, while those of K_6 and K_7 [Equations (13-6a) and (13-7a)] are moles per liter or M. The gas reaction

$$\mathrm{PCl_5}(g) \rightleftharpoons \mathrm{PCl_3}(g) + \mathrm{Cl_2}(g) \tag{13-8}$$

is associated with the equilibrium condition

$$\frac{P_{\mathrm{PCl_3}} P_{\mathrm{Cl_2}}}{P_{\mathrm{PCl_5}}} = K_8 \tag{13-8a}$$

and the units of K_8 are atmospheres (unless the partial pressures concerned are expressed in other units). The dissociation of gaseous HI into H_2 and gaseous I_2,

$$2\,\mathrm{HI}(g) \rightleftharpoons \mathrm{H_2}(g) + \mathrm{I_2}(g) \tag{13-9}$$

is governed at equilibrium by the equation

$$\frac{P_{\mathrm{H_2}} P_{\mathrm{I_2}}}{P_{\mathrm{HI}}^2} = K_9 \tag{13-9a}$$

with K_9 dimensionless because the units of the terms in the numerator cancel the units in the denominator.

Note that the numerical value of an equilibrium constant depends upon the units in which it is expressed, unless the sum of the exponents of the concentration terms is the same in the numerator as in the denominator of the reaction quotient and unless the same relation holds for the pressure terms. To illustrate, because there is one more pressure term in the numerator than in the denominator in (13-8a), K_8 has the dimensions of pressure and its numerical value depends upon the pressure units chosen. Thus, at 200°C, $K_8 = 2.2 \times 10^{-4}$ atm; if partial pressures are expressed in torr rather than atmospheres, however, the value of K_8 is $(2.2 \times 10^{-4}\text{ atm})$ $(760\text{ torr/atm}) = 0.17$ torr. On the other hand, K_9 is dimensionless and its value is thus independent of the units used.

2. *Equilibrium shift by addition or partial removal of a reacting species:* The law of chemical equilibrium embodied in (13-4) is a quantitative expression of the fact that when a reaction mixture is at equilibrium and an additional quantity of one of the reactants or products is supplied from the outside without significant change in volume, the equilibrium will shift so as to offset part of the increase in concentration of the added species.

Consider the dissociation of acetic acid, given by (13-7), with the corresponding equilibrium expression, (13-7a):

$$\frac{[H^+][OAc^-]}{[HOAc]} = K_7 \qquad (13\text{-}7a)$$

The equilibrium constant K_7 is frequently called the *dissociation constant* or *acid constant* of acetic acid.[1] If a strong acid like HCl is added to a solution containing acetic acid, acetate ions, and hydrogen ions at equilibrium, the added hydrogen ions increase [H$^+$] so that, at least momentarily, the quotient in (13-7a) no longer equals the dissociation constant of acetic acid. The equilibrium represented by (13-7) is, however, established extremely rapidly (in much less than a microsecond, 10^{-6} s, except perhaps for inhomogeneities arising from inadequate mixing). Some of the hydrogen ions react with acetate ions to form acetic acid, thereby reducing [OAc$^-$] and [H$^+$] and increasing [HOAc] until equilibrium again exists and (13-7a) is again satisfied. Thus, adding hydrogen ions to the original solution shifts the equilibrium of reaction (13-7) to the left; the addition of strong acid is said to repress the dissociation of acetic acid (or of any weak acid). Conversely, removal of H$^+$ from the original solution would cause the equilibrium to shift to the right.

3.   *All sources of a reacting species must be considered:* It is important to realize that the concentration or partial pressure terms in the reaction quotient of any reaction include contributions from *all available sources,* even though some of these sources may not be related to the reaction in question.

To illustrate, in the example just discussed the term [H$^+$] includes not only the hydrogen ions created by the dissociation of HOAc but also those coming from the added HCl. There is a further contribution from the dissociation of water, but it is completely negligible in this example and indeed is usually negligible in most of the applications of equilibria to weak acids and weak bases that we shall consider, other than pure water itself.

4.   *Equilibrium constant and degree of dissociation:* Beginners often confuse the equilibrium constant with the degree of dissociation. The two are quite distinct. The *equilibrium constant* is defined by (13-4); the *degree of dissociation* of a compound X is defined as the number of moles of X that have dissociated divided by the number of moles of X that would be present if none had dissociated.

Consider the dissociation of N_2O_4:

$$N_2O_4(g) \rightleftharpoons 2NO_2(g) \qquad (13\text{-}10)$$

The degree of dissociation, α, is the number of molecules (or moles) of N_2O_4 that have dissociated divided by the number of N_2O_4 molecules (or moles) that would have been present if no dissociation had occurred.

Dissociation of N_2O_4 increases the number of molecules in the gas; for each N_2O_4 molecule that dissociates, two NO_2 molecules appear. If pressure and temperature are kept constant, this means, by Avogadro's law, that dissociation increases the gas

---

[1] Acid constants are often symbolized as K_a; the corresponding constants for bases, a typical one of which is given above as K_6 [Equation 13-6a], are often symbolized as K_b. Acid and base equilibria are discussed in detail in Chap. 14.

volume. Since dissociation does not change the total mass of a given quantity of gas, it necessarily decreases the density (at a fixed P and T), which is consistent with the average molecular weight decreasing, as it must when dissociation occurs.

EXAMPLE 13-2

Dissociation and gas density. At 25°C and 1.000 atm, the degree of dissociation of N_2O_4 is 0.184. What is the density of the gas mixture?

Solution. If there were no dissociation, 1.000 gfw N_2O_4, which amounts to 92.0 g, would constitute 1.000 mol gas. Of this, 0.184 mol is dissociated, leaving (1.000 − 0.184) mol undissociated N_2O_4. By the dissociation equation (13-10) it is seen that each N_2O_4 molecule yields two molecules of NO_2. The dissociation of 0.184 mol N_2O_4 thus yields 2×0.184 mol NO_2. The total number of moles in the final gas mixture is (1.000 − 0.184) + 2 × 0.184 = 1.184. Thus

	Before dissociation	After dissociation
Moles N_2O_4	1.000	1.000 − 0.184
Moles NO_2	0.000	2 × 0.184
Total moles	1.000	1.184

The volume taken up by 1.184 mol gas at 25°C and 1.000 atm is $V = nRT/P = (1.184 \times 0.0821 \times 298/1.000)$ liters = 29.0 liters. The density is therefore 92.0 g/29.0 liters = 3.17 g liter^{-1}.

A fictional problem is discussed next as an illustration of the reasoning that is characteristic in dissociation problems.

EXAMPLE 13-3

Cracking walnuts. All the snowbound skier had was 10 dozen walnuts. He set about cracking them meticulously, separating them neatly into half shells and intact kernels. After a while the light failed and he decided to count the results of his labors—the uncracked nuts left, the half shells, and the kernels. He dared not remove his mittens in the bitter cold and therefore could not tell what his fingers were touching. All he knew was that in the end he had 194 pieces. What was the degree of dissociation of the walnuts?

Solution. If the kernels are symbolically denoted by K, the half shells by S, and the nuts therefore by KS_2, the cracking of nuts may be described by

$$KS_2 \longrightarrow K + 2S \qquad (13\text{-}11)$$
$$n_0 - x \qquad x \quad 2x$$

where $n_0 = 120$ is the original number of nuts and x is the number of cracked nuts. In the end there were x kernels, $2x$ half shells, and $(n_0 - x)$ whole nuts. This is the situation described by the symbols just beneath (13-11). The total number of pieces is therefore $n = x + 2x + (n_0 - x) = n_0 + 2x$, and this the skier found to be equal to 194: $n_0 + 2x = 120 + 2x = 194$. Thus, $x = \frac{74}{2} = 37$, so that 37 walnuts were cracked. To find the degree of dissociation we have to refer to the state in which all nuts were whole. Thus, $\alpha = x/n_0 = \frac{37}{120} = \underline{0.31}$.

5. *Implications of the magnitude of K:* In reactions for which K differs considerably from 1, equilibrium normally favors the side of the reactants when K is much smaller than 1 and the side of the products when K is much larger than 1. Intermediate situations are characterized by values of K of the order of unity.[1] In qualitative comparisons of different equilibria involving similar reactions—for example, acid dissociations—what matters is not so much the exact numerical values of the equilibrium constants but rather the powers of 10 that go with them. In other words, differences in equilibrium constants by a factor of 2 or 3 seldom lead to marked chemical effects, for which differences of at least a few powers of 10 are required.

6. *Driving a reaction in a desired direction:* If the equilibrium constant for a reaction is very small, so that the reactants are normally favored over the products, the reaction may still be driven to the right, as is implied by comment 2, either by removing one or more of the products as it is formed or by adding a large excess of one or more of the reactants. The first of these possibilities is usually easier to accomplish. For example, if one of the products is a gas, it may be removed by reducing the pressure or by increasing the temperature. Thus, it is possible to produce HCl by the reaction of nearly water-free H_2SO_4 with NaCl:

$$H_2SO_4 + NaCl \rightleftharpoons HCl + NaHSO_4$$

Because HCl is appreciably more volatile than any other component of this reaction mixture, it can readily be removed by heating and the reaction is thereby driven to the right.

Often a reaction appears to be driven (even though no external action is required) because a precipitate appears among the products and is effectively removed because of its small solubility. For example, H_2S is a very weak acid in aqueous solution, and the concentration of S^{2-} in a saturated solution of H_2S is less than $10^{-14}\ M$. Yet H_2S effectively dissociates completely in the presence of certain metal ions, such as Ag^+ or Hg^{2+}, that form very insoluble sulfides:

$$H_2S + 2Ag^+ \rightleftharpoons Ag_2S + 2H^+ \tag{13-12}$$

The right side of this equilibrium is favored because of the extreme insolubility of Ag_2S (Chap. 15). Similarly, one of the reaction products in an aqueous solution may be a slightly ionized substance, such as H_2O itself, which again causes the side of the products to be strongly favored. Thus the equilibrium position of the reaction

$$HCN + OH^- \rightleftharpoons CN^- + H_2O$$

---

[1]Unless K is dimensionless, its value depends on the units used, as stated in comment 1. This must be taken into account when considering the meaning of the numerical value of K. As discussed later, when the sum of the exponents in the numerator of the reaction quotient is different from that in the denominator, the position of equilibrium can be altered by changes in total concentration or pressure. Thus, although K_a for acetic acid is very small so that this acid is less than 1 percent dissociated in a $1\ F$ solution, it is more than half dissociated at a concentration of $10^{-5}\ F$ so that in this very dilute solution, the products are actually favored at equilibrium.

in aqueous solution is strongly in favor of the products, at least partly because H_2O is formed.

Another method of achieving almost complete conversion of reactants into products is recycling. For example, in the Haber process by which N_2 and H_2 are combined to form ammonia,

$$N_2(g) + 3H_2(g) \rightleftharpoons 2NH_3(g) \qquad (13\text{-}13)$$

the equilibrium constant is very small at the temperatures of 500°C and above at which this reaction is carried out industrially, and only a small fraction of the nitrogen and hydrogen is converted into ammonia at a given time. However, the ammonia formed is removed by being cooled until it liquefies. Not only does the equilibrium constant change in favor of NH_3 as the temperature is lowered, but the speed with which NH_3 decomposes into N_2 and H_2 at ordinary temperatures is essentially zero. The unreacted N_2 and H_2, still gaseous, are returned to the reaction chamber and combined with fresh supplies of these gases in a continuous process.

We return to additional general comments and examples in Sec. 13-5.

13-4   Equilibria Involving Gases

Our considerations in this section concern two types of reactions involving gases: homogeneous reactions (the only phase present being that containing the gases) and representative heterogeneous reactions between gases and solids. We assume that the reader is familiar with Chap. 5, especially Sec. 5-2.

Homogeneous gas-phase reactions.   Since all gases are miscible with one another, all reactions involving *only* gases are homogeneous reactions—that is, they occur in a single phase. Some such reactions come to equilibrium quickly, others slowly. We begin by considering examples of two common situations, the computation of an equilibrium constant from some experimental data and the calculation of the distribution of reactants and products from given equilibrium constants. Other examples of gas-phase equilibria are illustrated in the problems at the end of the chapter.

EXAMPLE 13-4   **Equilibrium constant from experimental data.**   At 25°C and 2.000 atm total pressure, N_2O_4 is 13.3 percent dissociated into NO_2. What is the value of the equilibrium constant at this temperature?

Solution.   Consider exactly 1 mol initially undissociated N_2O_4. After it comes to equilibrium with NO_2, 0.133 mol N_2O_4 has dissociated and twice this many moles of NO_2 have been formed, as indicated by the numbers beneath the symbols in (13-14):

$$N_2O_4 \rightleftharpoons 2NO_2 \qquad (13\text{-}14)$$
$$1.000 - 0.133 \qquad 2 \times 0.133$$

The resulting gas mixture contains a total of $0.867 + 0.266 = 1.133$ mol gas for each mole of N_2O_4 initially present. Since we know the total pressure and can calculate the mole fraction of each of the two components, we can calculate their partial pressures by using Dalton's law (Sec. 5-2). The mole fraction of N_2O_4 is $\frac{0.867}{1.133} = 0.765$ and that of NO_2 is $\frac{0.266}{1.133} = 0.235$. Since the total pressure is 2.000 atm, the partial pressures, each equal to the mole fraction of the corresponding component times the total pressure, are 1.530 and 0.470 atm, respectively. The equilibrium constant is

$$K_{14} = \frac{P^2_{NO_2}}{P_{N_2O_4}} = \frac{(0.470 \text{ atm})^2}{1.530 \text{ atm}} = \underline{0.144 \text{ atm}}$$

EXAMPLE 13-5   **Degree of dissociation of gaseous HI.**   The equilibrium constant (13-9a) for reaction (13-9) (restated below) is 0.0144 at 600 K. What is the degree of dissociation of gaseous HI at equilibrium at this temperature?

Solution.   For convenience, we assume that if none of the HI had dissociated there would be 1.000 mol of it present. Let the degree of dissociation be x. Then from our original sample of HI, x mol will have dissociated and $1.000 - x$ will remain undissociated. Since the chemical equation (13-9) indicates that one mole of H_2 and one mole of I_2 are formed for each *two* moles of HI dissociated, the amounts of H_2 and I_2 formed will each be $x/2$ mol:

$$2HI(g) \rightleftharpoons H_2(g) + I_2(g) \qquad (13\text{-}9)$$
$$1.000 - x \qquad x/2 \qquad x/2$$

The total number of moles is 1.000, regardless of the extent to which (13-9) has proceeded. Thus, the mole fraction of each component is, in this example, equal to the number of moles of that component. Hence, if the total pressure is P, the respective partial pressures are

$$P_{HI} = P(1.000 - x)$$
$$P_{H_2} = P_{I_2} = P(x/2)$$

and the reaction quotient, at equilibrium, given as 0.0144, is

$$\frac{P_{H_2}P_{I_2}}{P_{HI}{}^2} = \frac{P^2(x/2)^2}{P^2(1.000 - x)^2}$$

$$= \frac{x^2}{4(1.000 - x)^2} = 0.0144 \qquad (13\text{-}15)$$

The total pressure P cancels, so that the degree of dissociation x is independent of P. This is always true in a gas reaction when the number of gas molecules on each side of the chemical equation is the same. Equation (13-15) is readily solved by taking the square root of each side:

$$\frac{x}{2(1.000 - x)} = 0.120$$

or

$$x = 0.240 - 0.240x$$

and finally

$$x = \frac{0.240}{1.240} = \underline{0.194}$$

In the preceding example, the amounts of H_2 and I_2 present were equivalent by (13-9) because it was assumed that we started with pure HI and thus formed the decomposition products in equivalent quantities. However, this need not be so; we might be interested in the pressure of each constituent when we start with a mixture containing H_2 and I_2 in some proportion[1] other than 1:1, which is the proportion in which they would be present if originally only HI were present.

The following example illustrates this more complex situation for another reaction, the decomposition of CO_2 into CO and O_2. In this case, the products are *not* present in the proportion in which they are formed in the reaction.

EXAMPLE 13-6   **Equilibrium between CO_2, CO, and O_2.**   The equilibrium constant for the reaction

$$CO_2(g) \rightleftharpoons CO(g) + \tfrac{1}{2}O_2(g) \tag{13-16}$$

is 1.93×10^{-5} atm$^{1/2}$ at 1500 K. What are the equilibrium partial pressures of the constituents of a mixture that originally consists of 2.00 mol CO_2 and 1.00 mol CO and is heated to 1500 K at a total pressure of 1.00 atm?

Solution. Suppose that at equilibrium x mol CO_2 has dissociated by (13-16). This leads to x mol CO and $x/2$ mol O_2, leaving then $(2.00 - x)$ mol CO_2 mixed with $(1.00 + x)$ mol CO and $x/2$ mol O_2. The total number of moles of gas is $(2.00 - x) + (1.00 + x) + x/2 = 3.00 + x/2$, and each partial pressure is equal to the product of the total pressure of 1.00 atm and the respective mole fraction:

$$P_{CO_2} = \frac{2.00 - x}{3.00 + x/2} \text{ atm}$$

$$P_{CO} = \frac{1.00 + x}{3.00 + x/2} \text{ atm} \tag{13-17}$$

$$P_{O_2} = \frac{x/2}{3.00 + x/2} \text{ atm}$$

The equilibrium condition is

$$\frac{P_{CO}P_{O_2}^{1/2}}{P_{CO_2}} = K_{16} = 1.93 \times 10^{-5} \text{ atm}^{1/2}$$

Inserting the values for the pressures, we get

$$\frac{(1.00 + x)\,(x/2)^{1/2}\,\text{atm}^{1/2}}{(2.00 - x)\,(3.00 + x/2)^{1/2}} = 1.93 \times 10^{-5} \text{ atm}^{1/2}$$

This leads to a third-degree equation for x. In solving equations of this kind, it is important to realize that an accuracy of 5 to 10 percent is all that is needed, because of approximations inherent in the equilibrium law itself (discussed in Sec. 13-5). It is simple to get an approximate solution to the equation being considered by noting that the right-hand side is very small. This implies that the term $(x/2)^{1/2}$ on the left

---

[1]Under these more general circumstances, the definition of degree of dissociation might be ambiguous and usually some other means is used to describe the equilibrium position quantitatively.

must also be very small; there is no other way for the expression on the left to be small. (We know also from the way x is defined that it must be a positive number and smaller than 2.00, since it is the number of moles of CO_2 that decomposes and the original number of moles of CO_2 was only 2.00.) If we know that x is very small relative to 1, the quantities $1.00 + x$, $2.00 - x$, and $3.00 + x/2$ are, respectively, very near 1.00, 2.00, and 3.00. Thus as our first approximation we assume that we can neglect x in these three expressions, keeping it only in the term in which it stands by itself. We obtain in this way

$$\left(\frac{x}{2}\right)^{1/2} = 2(3)^{1/2}\,(1.93 \times 10^{-5})$$

or

$$\frac{x}{2} = 4(3)\,(1.93 \times 10^{-5})^2$$

$$x = 8.9 \times 10^{-9}$$

Checking the terms in which x was neglected, we see that x is indeed negligible with respect to 1.00, 2.00, and 3.00, so that our approximations are justified within the precision of the value of K and the other data of the problem. The equilibrium partial pressures are, then, by Equations (13-17),

$$P_{CO_2} = \underline{0.67 \text{ atm}}$$

$$P_{CO} = \underline{0.33 \text{ atm}}$$

$$P_{O_2} = \frac{x}{6} \text{ atm} = \underline{1.5 \times 10^{-9} \text{ atm}}$$

These results indicate that at this temperature and pressure, the decomposition of CO_2 into CO and O_2 is exceedingly small.

If the number of molecules on one side of an equation describing a gaseous equilibrium differs from the number of molecules on the other, a decrease in the volume of the system (a change corresponding to an increase in pressure) shifts the position of equilibrium toward the side with fewer molecules. This effect can be understood qualitatively in terms of Le Châtelier's principle.

Heterogeneous equilibria resulting from reactions of gases with solids.   Many heterogeneous equilibria in which gases play a role are interesting and important. We have already discussed in some detail vapor-pressure equilibria for both liquids and solids (Chaps. 3 and 6). Before giving illustrations that concern chemical equilibria we return to conventions III*b* and V for setting up the reaction quotient Q. According to convention III*b*, the concentrations of gases are replaced by their partial pressures even when the reaction mixture does not consist entirely of gases; by convention V, the "concentrations" of *pure* solids or liquids are deleted from Q. Two illustrative chemical equations and the related equilibrium expressions are

$$PCl_5(s) \rightleftharpoons PCl_3(g) + Cl_2(g) \tag{13-18}$$

$$P_{PCl_3}P_{Cl_2} = K_{18} \tag{13-18a}$$

$$2\text{HgO}(s) \rightleftharpoons 2\text{Hg}(l) + \text{O}_2(g) \qquad (13\text{-}19)$$

$$P_{\text{O}_2} = K_{19} \qquad (13\text{-}19a)$$

The symbols (s) and (l) imply that the substances concerned are *pure* solids and *pure* liquids, respectively, and also that these substances are *actually present* as pure solids or liquids under the equilibrium situations of interest. The symbol (g) implies purity only if just one gas is present, because all gases are miscible with one another.

How can we justify the rule that pure solids and pure liquids are omitted entirely from reaction quotients? The reason that solid PCl_5 need not be mentioned in (13-18a) is a direct consequence of the fact that, at whatever temperature (13-18a) is applicable, the vapor pressure of solid PCl_5 is a constant as long as the solid PCl_5 is pure. Let us call this vapor pressure $P^{\bullet}_{\text{PCl}_5}$. If the system is at equilibrium, the gas phase is saturated with PCl_5 vapor and we may rewrite (13-18) in an alternative form:

$$\text{PCl}_5(s) \rightleftharpoons \text{PCl}_5(g \text{ at } P^{\bullet}_{\text{PCl}_5}) \rightleftharpoons \text{PCl}_3(g) + \text{Cl}_2(g) \qquad (13\text{-}20)$$

In other words, since the vapor of PCl_5 (at a pressure $P^{\bullet}_{\text{PCl}_5}$) is in equilibrium with the solid, which is in turn in equilibrium with gaseous PCl_3 and Cl_2, the PCl_5 vapor is itself in equilibrium with the two decomposition products. We might then write for the right-hand equation in (13-20):

$$\frac{P_{\text{PCl}_3} P_{\text{Cl}_2}}{P^{\bullet}_{\text{PCl}_5}} = K_{20} \qquad (13\text{-}20a)$$

However, since $P^{\bullet}_{\text{PCl}_5}$ is a constant at this temperature, it is most convenient to combine it with K_{20}, thus leading to

$$P_{\text{PCl}_3} P_{\text{Cl}_2} = K_{20} P^{\bullet}_{\text{PCl}_5}$$

Above, in (13-18a), we have merely used the symbol K_{18} for this product $K_{20}P^{\bullet}_{\text{PCl}_5}$.

It is important to note the distinction between the equilibrium (13-18) and the corresponding one in which no solid PCl_5 is present, discussed in connection with (13-8):

$$\text{PCl}_5(g) \rightleftharpoons \text{PCl}_3(g) + \text{Cl}_2(g) \qquad (13\text{-}8)$$

for which we had

$$\frac{P_{\text{PCl}_3} P_{\text{Cl}_2}}{P_{\text{PCl}_5}} = K_8 \qquad (13\text{-}8a)$$

Because no solid PCl_5 is present, the pressure of gaseous PCl_5 can vary widely, from zero up to the vapor pressure, and hence must be included explicitly in the reaction quotient[1] for (13-8).

The justification for the omission of any pure liquid from the reaction quotient is based on identical reasoning. As long as the liquid is pure, its vapor pressure

---

[1] Equation (13-20a) is a special case of (13-8a), and K_8 and K_{20} are identical. On the other hand, the value and even the units of K_{18} and K_8 are different (atm^2 and atm, respectively, if partial pressures are expressed in atmospheres).

is constant at a given temperature. For example, in (13-19)—the equilibrium between solid mercuric oxide (HgO), liquid mercury, and gaseous oxygen—the only term that appears in the reaction quotient is the pressure of oxygen, which must be constant ($= K_{19}$) once equilibrium is established.

If, however, the mercury is diluted with some other material—e.g., by dissolving another metal in it to make a solution known as an amalgam—then the vapor pressure of the mercury will decrease in accord (at least approximately) with Raoult's law. Explicit account must then be taken in the reaction quotient of the fact that the liquid phase is no longer pure. Similarly, if any solid is present as a solid solution, its decrease in concentration must be included explicitly in formulating the reaction quotient. This is why it must be emphasized that only *pure* solids and liquids can be omitted from these quotients.

We now turn to a more detailed examination of some typical equilibria arising as a result of chemical reactions between gases and solids. Because intimate contact between different phases is difficult to establish, heterogeneous equilibria are often established slowly. For example, many common metals tend to react with some normal components of the earth's atmosphere, chiefly O_2, H_2O, and acidic oxides such as CO_2, NO_2, and SO_2. This corrosion, as it is called, is usually slow at ordinary temperatures, in part because some of the reactions involved are intrinsically slow and in part because a complex series of transformations usually takes place during corrosion. Many other heterogeneous gas-solid reactions also take time to reach equilibrium, but for some comparatively simple reactions equilibrium is achieved fairly rapidly. To illustrate the principles involved, we choose the reaction of CO_2 with a basic oxide to form a carbonate.

Carbon dioxide has been present in the earth's atmosphere at least since the dawn of life on this planet 2 billion or so years ago, and it plays a central role in almost all known forms of life today. It also figures prominently in many geological processes, because it is a ubiquitous acidic oxide, reacting with water to form the weak acid H_2CO_3 (carbonic acid). This acid, and CO_2 itself, react with bases to form carbonates, which are important constituents of many common rocks and minerals. The most abundant carbonate is $CaCO_3$, which occurs in an enormous variety of forms; it is produced by many marine organisms, being the major ingredient of coral, pearl, and the shells of most mollusks and crustaceans. The shells of dead marine animals accumulate in thick deposits on the floor of the sea and are slowly transformed by geological processes into such common rocks as limestone and marble.

When $CaCO_3$ is heated sufficiently, it decomposes to form calcium oxide (CaO) and CO_2:

$$CaCO_3(s) \rightleftharpoons CaO(s) + CO_2(g) \qquad (13\text{-}21)$$

The equilibrium condition for this reaction at any temperature is simply

$$P_{CO_2} = K_{21} \qquad (13\text{-}21a)$$

In other words, if at a particular temperature there is to be equilibrium between $CaCO_3(s)$ and $CaO(s)$, it can occur only at a specific pressure of CO_2 in this system, given by the value of K_{21} at that temperature. This equilibrium pressure of CO_2

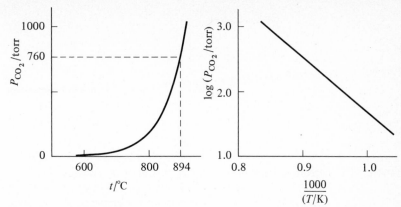

FIGURE 13-2 **Dissociation pressure of CaCO₃.** The partial pressure of carbon dioxide at equilibrium with both pure solid $CaCO_3$ and CaO is constant at a given temperature. The diagrams show this pressure as a function of the temperature. Note the similarity to vapor-pressure curves (see, for example, Fig. 3-4). In either case, a plot of log P versus $1/T$ is nearly a straight line.

is sometimes called the dissociation pressure of $CaCO_3$; it increases rapidly with increasing temperature (Fig. 13-2).

It is important to recognize that even though neither solid $CaCO_3$ nor solid CaO appears explicitly in the equilibrium expression (13-21a), both these substances must be present if the equilibrium (13-21) is to exist. Their absence from (13-21a) implies only that the position of equilibrium is independent of the amount of either solid *as long as some of each is present.* If only $CaO(s)$ and CO_2 are present, with no $CaCO_3$, the pressure of the CO_2 can have any value between zero and the equilibrium pressure (K_{21}) at the temperature of the system. If only $CaCO_3(s)$ and CO_2 are present, the pressure of the CO_2 must be greater than or equal to the equilibrium pressure. These conditions are analogous to those that govern vapor-liquid equilibria for pure substances. Recall (Sec. 3-1) that liquid and vapor coexist at a particular temperature only at a specific pressure, the vapor pressure. A single vapor phase exists at pressures below the vapor pressure, and a single liquid phase exists at higher pressures. The phase relations in the $CaCO_3$–CaO–CO_2 system are illustrated in Fig. 13-3, in which the pressure of CO_2 over a mixture of $CaO(s)$ and $CaCO_3(s)$ at 800°C is plotted as a function of the weight fraction y of $CaCO_3$. At this temperature, the dissociation pressure of $CaCO_3$ is 180 torr.

What can be said about the temperature at which some form of $CaCO_3(s)$ such as limestone[1] will decompose in air? To answer this question we need only the data of Fig. 13-2 and an estimate of the partial pressure of CO_2 in air. This quantity varies from place to place on the earth's surface. It depends on the density of plant life, of industrial activity, and of other consumers and producers of CO_2, but an average value for P_{CO_2} in the earth's atmosphere is about 0.2 torr. Thus log

[1] Actually, each different form of $CaCO_3(s)$ has its own dissociation pressure. The pressures vary according to the relative stabilities of the forms, being lowest for the most stable form, but this variation is very small for the different common forms, such as limestone, marble, and chalk.

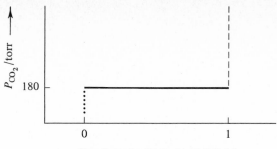

FIGURE 13-3   **Phase relations in the CaCO₃-CaO-CO₂ system at 800°C.** When y, the weight fraction of CaCO$_3$(s) in the solid phase, is between 0 and 1, both CaO(s) and CaCO$_3$(s) are present and P_{CO_2} must be equal to 180 torr at 800°C (solid line). If the pressure of the CO$_2$ were larger than 180 torr, the CaO(s) would react with the CO$_2$ until either all the CaO had been used up or P_{CO_2} had decreased to 180 torr. Likewise, if P_{CO_2} were smaller than 180 torr in the presence of both solids, some CaCO$_3$(s) would decompose, yielding CO$_2$, until either P_{CO_2} had increased to 180 torr or all the CaCO$_3$(s) had been used up.

When $y = 0$ and thus no CaCO$_3$(s) is present, P_{CO_2} (dotted line) may be smaller than 180 torr (and remain smaller) because no source of CO$_2$ exists to increase its pressure. Likewise, when $y = 1$ and no CaO(s) is present, P_{CO_2} (dashed line) may be larger than 180 torr (and remain larger) because the system contains no substance that can react with the excess CO$_2$ and thereby reduce its pressure. Only the horizontal line at 180 torr corresponds to an equilibrium situation, with all three substances present.

(P_{CO_2}/torr) is about -0.7. Extrapolation of the right-hand graph in Fig. 13-2 indicates that this pressure corresponds approximately to a value of 1.3 for 1000 $(T/\text{K})^{-1}$. Thus for atmospheric CO$_2$ there is equilibrium between CaCO$_3$(s) and CaO(s) when T is around $\frac{1000}{1.3}$ K, that is, somewhat below 800 K or around 500°C. This means that at any temperature below 500°C, the pressure of CO$_2$ in the atmosphere exceeds the dissociation pressure and CaCO$_3$(s) is stable whereas CaO(s) (quicklime) is unstable and reacts with CO$_2$ to form CaCO$_3$. The latter reaction is one of several important ones that occurs during the hardening of cement. Above 500°C, CaCO$_3$(s) is unstable relative to atmospheric CO$_2$ and decomposes into CaO(s) and CO$_2$.

With a substance such as CaCO$_3$(s) that can decompose to form a gaseous product, there is an important distinction between heating it in a closed system and in an open one such as the open air. In a closed system, the gas that is released accumulates and its concentration in the gas phase therefore rises, which causes an increase in its partial pressure (over and above that caused by any increase in temperature). Eventually the partial pressure of the accumulated gas reaches the dissociation pressure of the decomposing solid (if the sample of solid has not completely decomposed before this point) and equilibrium is established. On the other hand, in an open system, where the gas produced during the decomposition is swept away or diffuses away, the partial pressure of this gas does not increase significantly around the sample. Thus, if this pressure was initially below the dissociation pressure at the temperature of heating, it will remain so and the sample will eventually decompose completely.

Equilibria between other carbonates and the related oxides and CO$_2$ can be formulated in a fashion parallel to (13-21) and (13-21*a*). For those oxides appreciably more basic than CaO, such as Na_2O and the oxides of the other alkali metals, the dissociation pressure of the corresponding carbonate is much lower at a given temperature than it is for $CaCO_3$. Conversely, with carbonates such as Ag_2CO_3 and $PbCO_3$, formed by the reaction of less basic oxides with CO_2 (Ag_2O and PbO are appreciably less basic than CaO), the dissociation pressures are considerably higher at a given temperature and these substances are easier to decompose by heating. For many less basic oxides, the dissociation pressures are so high even at ordinary temperatures that the corresponding carbonates are not stable in the normal atmosphere and thus do not normally exist.

Another group of important heterogeneous equilibria involves substances capable of liberating oxygen when heated. Among these are the oxides of some of the less reactive metals such as mercury [Equation (13-19)] and silver and the peroxides of some of the alkaline-earth metals:

$$2BaO_2(s) \rightleftharpoons 2BaO(s) + O_2(g) \tag{13-22}$$

$$2Ag_2O(s) \rightleftharpoons 4Ag(s) + O_2(g) \tag{13-23}$$

The same principles apply to equilibria between water vapor and different hydrates of a substance (including the anhydrous form)—such as $CaSO_4$, $CaSO_4 \cdot \frac{1}{2}H_2O$, $CaSO_4 \cdot 2H_2O$—and also the saturated solution. Only *neighboring* forms (for example, $CaSO_4$ and $CaSO_4 \cdot \frac{1}{2}H_2O$) in this sequence of solid and liquid phases can coexist at equilibrium with water vapor, and for each pair of neighboring phases and a given temperature the equilibrium partial pressure of H_2O is fixed. If two phases that are not neighbors (for example, $CaSO_4$ and $CaSO_4 \cdot 2H_2O$) are present in a system, one will lose water or the other will gain water until only two neighboring phases exist, together with H_2O at the appropriate equilibrium pressure, or until only a single solid phase of composition intermediate between the original phases is present. When only one solid phase is present, the partial pressure of H_2O may vary within certain bounds.

13-5   Additional General Comments and Examples

We now return to important general features of the chemical equilibrium law, beginning with a further discussion of conventions V and III*b* for setting up the reaction quotient Q.

Conventions V and III*b*.   We start with convention V, according to which the "concentration" of any *pure* solid or liquid that is actually present at equilibrium is omitted from the quotient Q. Although we have discussed reactions of pure solids and liquids with gases (Sec. 13-4), we have not as yet considered reactions in aqueous solutions. We shall use the reaction described by the equation

$$AgCl(s) \rightleftharpoons Ag^+ + Cl^- \tag{13-24}$$

as an example. In this reaction solid AgCl dissolves to form the silver and chloride ions in solution. The resulting equilibrium might be written in the form[1]

$$AgCl(s) \rightleftharpoons AgCl(aq) \rightleftharpoons Ag^+ + Cl^- \qquad (13\text{-}25)$$

The concentration of molecular AgCl in the aqueous solution, though known experimentally to be extremely small, is constant as long as some solid AgCl is present and there is equilibrium. Thus we might write, considering the right-hand equilibrium in (13-25),

$$\frac{[Ag^+][Cl^-]}{[AgCl]} = K_{25} \qquad (13\text{-}25a)$$

but since [AgCl] is constant at a given temperature, its value may be combined with K_{25}, giving

$$[Ag^+][Cl^-] = K_{25}[AgCl] \equiv K_{24} \equiv K_{sp,AgCl} \qquad (13\text{-}24a)$$

The expression on the left side of this equation is often called the *ion product*. When it is evaluated with equilibrium concentrations, as here, its value is said to be equal to the *solubility-product* constant, symbolized K_{sp}, for the salt in question. There are many other situations in which *pure* solids or liquids participate in a reaction and are present at equilibrium. In each of these cases the quotient Q contains no term representing pure solid or liquid.

We return finally to convention III*b*, by which the concentration [*i*] of a gas that participates in a reaction is replaced by its partial pressure P_i. Again this convention also applies to reactions in aqueous solutions. The justification for this procedure lies in Henry's law (Sec. 6-4). For example, consider the following reaction occurring in a solution:

$$2Fe^{3+} + 2Cl^- \rightleftharpoons 2Fe^{2+} + Cl_2(g) \qquad (13\text{-}26)$$

where the letter *g* implies that a gas phase containing Cl_2 at a certain partial pressure P_{Cl_2} is in equilibrium with the solution. If the concentration of Cl_2 in the solution is not too high, the relationship

$$[Cl_2] = k_{Cl_2}P_{Cl_2} \qquad (13\text{-}27)$$

holds true, k_{Cl_2} being the appropriate proportionality constant of Henry's law. In terms of $[Cl_2]$, the equilibrium condition may be expressed as

$$\frac{[Fe^{2+}]^2[Cl_2]}{[Fe^{3+}]^2[Cl^-]^2} = K' \qquad (13\text{-}28)$$

Replacing $[Cl_2]$ by $k_{Cl_2}P_{Cl_2}$ and combining k_{Cl_2} with K', we obtain

$$\frac{[Fe^{2+}]^2 P_{Cl_2}}{[Fe^{3+}]^2[Cl^-]^2} = \frac{K'}{k_{Cl_2}} \equiv K_{26} \qquad (13\text{-}26a)$$

---

[1] The symbol AgCl(*aq*) emphasizes that the species in question is molecular AgCl, as it exists in aqueous solution, surrounded by H_2O molecules, as distinguished from solid silver chloride, represented as AgCL(*s*). Strictly speaking the symbols Ag^+ and Cl^- should be replaced by $Ag^+(aq)$ and $Cl^-(aq)$, because these species are hydrated also, but this is always assumed to be understood for *ions* in aqueous solutions.

On the left is the reaction quotient for reaction (13-26) set up according to convention III*b*; K_{26} is the pertinent equilibrium constant.

The equilibrium expression for the following reaction occurring in aqueous solution,

$$2I_2(s) + 2H_2O \rightleftharpoons O_2(g) + 4I^- + 4H^+ \qquad (13\text{-}29)$$

illustrates all five conventions on page 341:

$$P_{O_2}[I^-]^4[H^+]^4 = K_{29} \qquad (13\text{-}29a)$$

In setting up the quotient Q it is important to note what is on the left and what is on the right in the chemical equation and what the coefficients are (see, for example, comment 7 below). Moreover, symbols such as *s* and *g* must be used to specify phases completely whenever an ambiguity might exist. For example, the symbol $I_2(s)$ implies an excess of *solid* iodine at equilibrium and therefore the absence of any term referring to I_2 in the reaction quotient Q. If only I_2 vapor is present at equilibrium, the combination $I_2(g)$ must be used in the chemical equation to imply that the equilibrium quotient Q contains the term P_{I_2} (with the appropriate exponent). In contrast, no *g* is *needed* in combination with H_2, which is gaseous under all normal circumstances.

Further comments on the chemical equilibrium law. We present now some further general remarks about equilibria, continuing from the comments in Sec. 13-3.

7. *Combination of reactions:* When two or more reaction equations are combined in an overall equation, the corresponding equilibrium constants may be combined to give the equilibrium constant for the overall reaction by applying the following three rules:

(*a*) When two or more chemical equations are added, the equilibrium constant for the resulting equation is the product of the constants of the original equations.

(*b*) When a chemical equation is multiplied by a factor q, the original equilibrium constant K becomes K^q. The factor q may be an integer or a fraction.

(*c*) Subtracting one reaction from a second is the same as reversing the first one and adding them. We know already, as mentioned in convention I (Sec. 13-3), that reversing a reaction implies changing K to $1/K$. Thus if the equation for reaction 1, with equilibrium constant K_1, is subtracted from that for reaction 2, with constant K_2, the constant for the resulting equation is $K = K_2/K_1$.

These rules are a direct consequence of the way in which the reaction quotient (13-3) depends upon the coefficients of the chemical species in the related reaction equation. The following two examples illustrate the application of the rules.

EXAMPLE 13-7 **Changing the coefficients of a balanced equation.** The equilibrium constant for the reaction

$$N_2(g) + 3H_2(g) \rightleftharpoons 2NH_3(g) \qquad (13\text{-}30)$$

is $K_{30} = 3.3 \times 10^{-8}$ atm^{-2} at 1200 K. Suppose that the reaction is represented by the equation

$$\tfrac{1}{2}N_2(g) + \tfrac{3}{2}H_2(g) \rightleftharpoons NH_3(g) \qquad (13\text{-}31)$$

What is the value of the corresponding equilibrium constant K_{31}?

Solution. If we write the reaction quotients for the two reactions,

$$\frac{P_{NH_3}^2}{P_{N_2}P_{H_2}^3} = K_{30} \qquad \text{and} \qquad \frac{P_{NH_3}}{P_{N_2}^{1/2}P_{H_2}^{3/2}} = K_{31}$$

we see that K_{30} is just the square of K_{31}. This is in accord with rule (*b*) above: Equation (13-31) is obtained from (13-30) by multiplying through by $\tfrac{1}{2}$, and hence $K_{31} = K_{30}^{1/2} = \underline{1.8 \times 10^{-4} \text{ atm}^{-1}}$. (Note also the change in units.)

EXAMPLE 13-8   **Combination of reactions.**   The two reactions

$$16H^+ + 2MnO_4^- + 10Cl^- \rightleftharpoons 8H_2O + 2Mn^{2+} + 5Cl_2(g) \qquad (13\text{-}32)$$

and

$$Cl_2(g) + 2Fe^{2+} \rightleftharpoons 2Cl^- + 2Fe^{3+} \qquad (13\text{-}33)$$

have the respective equilibrium constants K_{32} and K_{33}. What is the equilibrium constant K_{34} for the reaction

$$8H^+ + MnO_4^- + 5Fe^{2+} \rightleftharpoons 4H_2O + Mn^{2+} + 5Fe^{3+} \qquad (13\text{-}34)$$

Solution. Equation (13-34) may be obtained by combining the two preceding equations as follows: $\tfrac{1}{2}$ (13-32) $+ \tfrac{5}{2}$ (13-33). The equilibrium constant for (13-34) is thus, by rules (*a*) and (*b*),

$$K_{34} = \sqrt{K_{32}K_{33}^{\,5}}$$

This result can be checked by writing out explicitly the expressions for the three quotients concerned.

A useful corollary of these rules is that in the treatment of an equilibrium situation, only those reactions need be considered that are *independent* of each other, that is, reactions with equations that are not combinations of others given. In other words, no reaction that can be represented as a combination of those explicitly considered needs to be included in examining the equilibrium situation, because the appropriate equilibrium conditions for it are automatically satisfied. Thus, if any two of the three reactions (13-32) to (13-34) are known to be at equilibrium, the third one necessarily is also at equilibrium.

8. *Concentrated solutions and high partial pressures:* Although we have stated that at a particular temperature the equilibrium constant *is* a constant (as its name implies), this is strictly true only in the limit of infinite dilution. In this sense the equilibrium relation is like the ideal-gas law—it is a very good approximation for low concentrations and becomes less exact at higher concentrations. The reason for the change in the value of K is the influence of molecular and ionic interactions on the equilibrium, which increase as the concentrations of the reacting species become higher.

In highly accurate work the reaction quotient must be formulated in terms of effective concentrations called *activities*. For equilibria involving gases, effective pressures called *fugacities* must be introduced. If these substitutions are made, the reaction quotient is truly constant at equilibrium (at a given temperature and total pressure). Tables of activities and fugacities of common species in different situations are available, as are formulas for their calculation.

Some numerical values will give an idea of the orders of magnitude involved. For gas reactions at pressures of 1 atm or less, use of partial pressures rather than fugacities leads to errors of no more than about 1 percent and thus such a correction is negligible for our purposes. For solutions of simple *ionic* substances such as NaCl, the difference between concentration and activity exceeds 5 percent at a formality of about 0.002; for *uncharged* species such as ethanol or acetone, the same order of discrepancy is reached only at concentrations of about 1 M. At intermediate concentrations typical activity corrections in aqueous solutions of ions are of the order of ± 20 percent; at high concentrations the ratio of the activity to the concentration (called the *activity coefficient*) may become very large.

9. *Adequacy of approximations:* It is seldom worthwhile to make highly precise calculations for equilibria in solution, and we shall only infrequently make them to a *precision* better than a few percent. Furthermore, the *accuracy* will often be even poorer.[1] There are three principal reasons for the low precision and accuracy of such calculations:

(*a*)   The use of concentrations in place of activities can lead to significant errors, as discussed in comment 8.

(*b*)   The values of the equilibrium constants used are usually precise to at best a few percent, and their accuracy is often considerably lower than their precision. For example, one of the most commonly used constants is that pertaining to the product of the concentrations of Ag^+ and Cl^- in a solution saturated with silver chloride, the *solubility product* of AgCl; yet the values for this constant quoted in various recent sources vary by almost a factor of 2, from $1.6 \times 10^{-10} M^2$ to $2.8 \times 10^{-10} M^2$ at 25°C.

(*c*)   Frequently a number of equilibria occur simultaneously but some of them are ignored in the calculations, either because they are believed to be of relatively minor importance and ignoring them simplifies the calculations appreciably or because one is unaware of them. Such ignored equilibria might often have had significant effects on the results of the calculations if they had been included.

Nevertheless, the relatively low precision and accuracy of even approximate equilibrium calculations are adequate for most purposes. The examples in Chaps. 14 and 15 show that the concentrations of species involved in equilibria in solution can vary over many powers of 10 (many *orders of magnitude*). Thus a calculation of these concentrations within 10 or 20 percent, or even within a factor of 2, almost always provides a quite meaningful picture of the actual equilibrium situation.

10. *Uniqueness of equilibrium concentrations:* An important and useful fact about

---

[1] Precision and accuracy are defined in Sec. 2-1.

chemical equilibria is that the concentrations of the different species involved, which satisfy the equilibrium equations and other conditions imposed upon the system, are *unique*. This means that there is only one set of concentration (and partial pressure) values that satisfies all the conditions that define a chemical system. If we wish to calculate these concentrations by solving the simultaneous equations describing the system, we may use any method that seems convenient, including guessing. The only criterion is that the values of the concentrations eventually arrived at satisfy all the equations concerned; if they do, they must necessarily be correct (provided that all the relevant equations have been included), regardless of how they were obtained.

It is for this reason that the intelligent use of approximations can be very helpful in analyzing equilibrium situations. If the initial approximations, made on the basis of an analysis of the problem and gradually accumulated experience, are reasonable, it is usually possible to simplify a complicated set of simultaneous equations so that they can be readily solved. Often the first answers so obtained will be sufficiently good—that is, the concentrations obtained will be consistent with all the equations, within the desired precision. If they are not, the *method of successive approximations* can be applied by substituting answers from the first approximation back into the equations and repeating the calculations to obtain a second approximation, and so on, until a set of values consistent within the desired precision is reached. We illustrate this procedure in examples in Chaps. 14 and 15. Often the first cycle is sufficient, as it was in Example 13-6.

11. *Effect of temperature:* The way in which equilibrium constants depend upon temperature is related to whether the corresponding reaction is exothermic or endothermic, as discussed in the consideration of Le Châtelier's principle (Sec. 13-2). The position of equilibrium of an exothermic reaction is shifted to the left, toward the reactants, when energy is added externally and the temperature is thus increased. Correspondingly, the position of equilibrium of an endothermic reaction is shifted to the right if energy is added. This means that in the first of these two cases the equilibrium constant decreases with increasing temperature and that in the second it increases. If energy is withdrawn from the system and the temperature is lowered, the effects are reversed. The equilibrium constant of a reaction that neither evolves nor absorbs energy is not affected by temperature.

For example, the reaction between nitrogen and oxygen to form nitric oxide,

$$N_2 + O_2 \rightleftharpoons 2NO \tag{13-35}$$

is endothermic and the equilibrium constant at 25°C is 4×10^{-31}, so that the left side is very heavily favored. An increase in temperature causes the equilibrium constant to become very much larger, however, and at 2000 K it is equal to 4×10^{-4}. It is possible, therefore, to produce NO by "burning" nitrogen at this temperature, and this is done commercially in an electric arc.[1] The yield of NO is only of the order of a few percent because the equilibrium constant still favors the left side of

---

[1] The nitrogen oxides found in the exhaust of automobiles and jet engines and involved in the production of smog are formed by a similar process. The air drawn into the cylinders of the typical auto engine is subjected to high temperatures during part of the combustion cycle.

(13-35), but the NO formed is removed, thus driving the reaction to the right. The NO must be quickly chilled to room temperature to prevent it from decomposing again to N_2 and O_2. At room temperature the rate at which equilibrium is established is so slow that the NO can be kept without decomposing. It is easily converted to NO_2 and then to nitric acid (Sec. 21-5).

12. *Names of constants:* Many different names are attached to equilibrium constants; usually they relate to the nature of the reaction involved, such as dissociation constants and association constants. Beginners are sometimes overwhelmed by these different names and get the notion that somehow the various constants differ in fundamental ways. In fact, they do not and the reaction quotient for any reaction can be formulated uniquely by following the five rules given earlier. The names associated with different constants are primarily a matter of convenience in identifying the nature of the reaction or whatever other feature may be of interest.

13. *A requirement—equilibrium must exist:* Finally, there is a rule that may seem most obvious but is occasionally overlooked by beginners (and by some experienced chemists as well): do not attempt to apply an equilibrium constant to a system unless all the substances appearing in the corresponding chemical equation are present and the system *is* at equilibrium. For example, the solubility product of AgCl, Equation (13-24a), applies *only* if some solid AgCl is in equilibrium with an aqueous solution containing Ag^+ and Cl^-, that is, only when the solution is saturated with AgCl. It is of course possible to have a solution containing Ag^+ and Cl^- that is not saturated with AgCl—for example, the ocean. In such a solution, the product of $[Ag^+]$ and $[Cl^-]$ will be less than $K_{sp,AgCl}$. Similar considerations apply to any equilibrium involving two or more phases—some of each phase must be present.

With a single-phase (homogeneous) equilibrium, the only question is whether the equilibrium is established with sufficient rapidity. If so, all species in the equilibrium equation will necessarily be present. We shall assume that the equilibria for which calculations are made in this chapter and the two that follow are established so rapidly that rate effects are unimportant.

Problems and Questions

13-1 **Reaction quotients and Le Châtelier's principle.** Consider the following reactions:
(I) $2C_2H_6(g) + 7O_2(g) \rightleftharpoons 4CO_2(g) + 6H_2O(g)$
(II) $COCl_2(g) \rightleftharpoons CO(g) + Cl_2(g)$
(III) $FeC_2O_4(s) \rightleftharpoons FeO(s) + CO(g) + CO_2(g)$
(*a*) Write the expression for the reaction quotient for each reaction. (*b*) Reaction II is endothermic. Suppose that an equilibrium mixture of the three gases is heated. How will the degree of dissociation of $COCl_2$ change? (*c*) Predict the effect on the amount of CO of a decrease in volume for reaction II and of a decrease in the amount of FeO(*s*) in reaction III.

Ans. (*b*) increase;   (*c*) decrease, no effect

13-2 **Reaction quotients.** Set up the reaction quotients Q that correspond to the following overall chemical equations (occurring in aqueous solution):

(a) $MnO_4^- + 5Cl^- + 8H^+ \rightleftharpoons Mn^{2+} + \frac{5}{2}Cl_2(g) + 4H_2O$
(b) $2MnO_4^- + 10Cl^- + 16H^+ \rightleftharpoons 2Mn^{2+} + 5Cl_2(g) + 8H_2O$
(c) $MnO_2(s) + 2Cl^- + 4H^+ \rightleftharpoons Mn^{2+} + Cl_2(g) + 2H_2O$
(d) $2MnO_4^- + 3Mn^{2+} + 2H_2O \rightleftharpoons 5MnO_2(s) + 4H^+$

Note any relationships between the different Q's.

13-3   Heterogeneous equilibrium. (a) Formulate the reaction quotient for the endothermic reaction

$$AgCl \cdot NH_3(s) \rightleftharpoons AgCl(s) + NH_3(g)$$

What is the result on P_{NH_3} at equilibrium if (b) additional $AgCl(s)$ is added or (c) additional $NH_3(g)$ is pumped into or out of the system, provided that neither of the two solid phases shown in the chemical equation is completely used up? (d) What is the effect on P_{NH_3} of lowering the temperature?

Ans.   (a) $P_{NH_3} = Q$;   (b) and (c) no effect;   (d) P_{NH_3} decreased

13-4   Calculation of an equilibrium constant. Sulfur trioxide, SO_3, can be formed by the oxidation of SO_2

$$2SO_2(g) + O_2(g) \rightleftharpoons 2SO_3(g)$$

When a mixture of SO_2 and O_2 is allowed to equilibrate at a high temperature, the following equilibrium pressures are observed:

$$P_{SO_2} = 0.340 \text{ atm} \qquad P_{O_2} = 0.250 \text{ atm} \qquad P_{SO_3} = 0.317 \text{ atm}$$

(a) Calculate the equilibrium constant for the reaction at this temperature. (b) What is the equilibrium constant for the reaction $SO_2(g) + \frac{1}{2}O_2(g) \rightleftharpoons SO_3(g)$?

13-5   Dissociation of a gas. Nitrosyl bromide, NOBr, is a gaseous compound that can be formed by the reaction

$$2NO + Br_2 \rightleftharpoons 2NOBr$$

At 25°C, the equilibrium constant for this reaction is 116 atm^{-1} when all substances are in the gaseous state. The reaction is exothermic. (a) Suppose that these compounds are introduced into a reaction chamber so that, if no reaction occurred, the partial pressures would be NOBr, 0.80 atm; NO, 0.40 atm; Br_2, 0.20 atm. Will any reaction take place? If so, will there be net formation or decomposition of NOBr? Explain. (b) Suppose that the temperature is raised to 100°C after the equilibrium is established at 25°C. Will there be net formation or decomposition of NOBr as a result of the temperature change? Explain. (c) Suppose that helium (which does not react with any of the ingredients of the mixture) is pumped into the flask when it is at equilibrium at 25°C until the total pressure in the flask is doubled. What will be the effect on the amount of NOBr present? Think carefully before answering.

Ans.   (a) yes, formation;   (b) decomposition

13-6   Gas reaction. At 1000 K the equilibrium constant of the exothermic reaction

$$H_2O(g) + CO(g) \rightleftharpoons CO_2(g) + H_2(g)$$

is 1.37. In which direction does the equilibrium position shift if (a) the pressure is raised at

1000 K and (*b*) the temperature is raised to 1100 K at constant pressure? (*c*) What is the composition of the equilibrium mixture at 1000 K if originally 1.00 mol H_2O and 1.00 mol CO are permitted to react and the total pressure is maintained at 1.00 atm? Would your answer be different if the total pressure were different?

13-7 **Gas equilibrium.** It is known from experiment that phosphorus pentachloride *vapor* dissociates as follows:

$$PCl_5(g) \rightleftharpoons PCl_3(g) + Cl_2(g)$$

When a flask that initially contained only pure PCl_5 is heated to 250°C, the PCl_5 is 69 percent dissociated at a total pressure of 2.00 atm. Calculate the equilibrium constant for the dissociation reaction at 250°C.   *Ans.* 1.81 atm

13-8 **Association of nitrogen dioxide.** One gram formula weight NO_2 occupies 15.7 liters at 20°C and 700 torr. (*a*) What is the number of moles of gas as computed from the ideal-gas equation? (*b*) Give a quantitative explanation for the result of (*a*) based on the reaction $2NO_2(g) \rightleftharpoons N_2O_4(g)$.

13-9 **Dissociation of sulfur trioxide.** At 627°C and 1 atm, SO_3 is partly dissociated into SO_2 and O_2: $SO_3(g) \rightleftharpoons SO_2(g) + \frac{1}{2}O_2(g)$. The density of the equilibrium mixture is 0.925 g liter^{-1}. What is the degree of dissociation of SO_3 under these circumstances?   *Ans.* 0.34

13-10 **Dissociation of O_2.** At 4000 K and 1.000 atm, O_2 is 53.3 percent dissociated to atomic O. What is the percent dissociation at 4000 K and 0.500 atm?

13-11 **Gas equilibrium constant.** To determine the equilibrium constant at 445°C of the reaction

$$H_2(g) + I_2(g) \rightleftharpoons 2HI(g)$$

a mixture of 0.915 g I_2 and 0.0197 g H_2 was heated to and held at 445°C until equilibrium was established. It was found that 0.0128 g H_2 was left unreacted. Compute the equilibrium constant K for this reaction, including appropriate units.   *Ans.* $K = 41$ (no units)

13-12 **Dissociation of a gas.** The gas X_2Z is stable at ordinary temperatures but dissociates partially to gaseous X_2 and gaseous Z when it is heated. Suppose a sample of X_2Z, initially at 1.00 atm and 27°C, is confined in a flask of fixed volume and is heated to 627°C. If the pressure is then found to be 4.20 atm, what is the degree of dissociation?

13-13 **Relation between equilibrium constants expressed in pressure units and in concentration units.** Although reaction quotients for reactions involving gases are usually formulated in terms of pressures, they are occasionally expressed in terms of molar concentrations. The respective equilibrium constants are often symbolized as K_p and K_c, the subscripts indicating whether pressure units or concentration units have been used. The relation between these two constants is particularly simple when the gases can be assumed to be ideal. Let Δn be the change in the number of molecules of gas during the reaction; that is, Δn is the number of molecules of gaseous products minus the number of molecules of gaseous reactants. Show then that $K_p = (RT)^{\Delta n}K_c$. (*Hint:* Consider the relation between the number of moles per liter of a gas and its pressure.)

13-14 **Dissociation of phosphorus vapor.** Between about 800 and 1200°C phosphorus vapor consists of a mixture of P_4 and P_2. At 1000°C and a pressure of 0.200 atm, the density of phosphorus vapor is found to be about 0.178 g liter^{-1}. What is the degree of dissociation of P_4 into P_2 under these conditions?

13-15 **Heterogeneous equilibrium.** Ammonium chloride is a somewhat volatile substance that dissociates completely in the vapor phase, forming gaseous ammonia and gaseous hydrogen chloride:

$$NH_4Cl(s) \rightleftharpoons NH_3(g) + HCl(g)$$

If solid ammonium chloride is heated in a previously evacuated container at 340°C, the total pressure is 1.00 atm when equilibrium is established. (*a*) Formulate the equilibrium constant for the reaction, and give its numerical value at 340°C. (*b*) From the fact that the reaction is endothermic as written, predict the effect on the total equilibrium pressure of lowering the temperature. (*c*) Suppose the partial pressure of ammonia at 340°C in equilibrium with solid NH_4Cl is 0.75 atm. What will be the pressure of HCl?

Ans. (*a*) 0.25 atm^2;   (*b*) decrease;   (*c*) 0.33 atm

13-16 **Thermal decomposition of barium peroxide.** A 20.0-g sample of BaO_2 is heated to 794°C in a closed evacuated vessel of volume 5.00 liters. How many grams of the peroxide are converted to BaO(*s*)? Neglect the volume of the solid.

$$2BaO_2(s) \rightleftharpoons 2BaO(s) + O_2(g) \qquad K = 0.50 \text{ atm}$$

13-17 **Dissociation of calcium carbonate.** One formula weight $CaCO_3(s)$ is heated to 850°C in a closed evacuated vessel of volume 10.0 liters. What fraction of the carbonate is converted to the oxide? Neglect the volume of the solids.

$$CaCO_3(s) \rightleftharpoons CO_2(g) + CaO(s) \qquad K(850°C) = 0.49 \text{ atm}$$

Ans. 5.3 percent

13-18 **Solid and three gases.** Solid ammonium carbonate decomposes according to the equation

$$(NH_4)_2CO_3(s) \rightleftharpoons 2NH_3(g) + CO_2(g) + H_2O(g)$$

At a certain elevated temperature the total pressure of the gases NH_3, CO_2, and H_2O generated by the decomposition of, and at equilibrium with, pure solid ammonium carbonate is 0.400 atm. Calculate the equilibrium constant for the reaction considered. What would happen to P_{NH_3} and P_{CO_2} if P_{H_2O} were adjusted by external means to be 0.200 atm without changing the relative amounts of $NH_3(g)$ and $CO_2(g)$ and with $(NH_4)_2CO_3(s)$ still being present?

13-19 **Calcium chloride as a drying agent.** Calcium chloride is an effective drying agent. It reacts with water vapor to form a series of hydrates, the simplest of which is the monohydrate:

$$CaCl_2(s) + H_2O(g) \rightleftharpoons CaCl_2 \cdot H_2O(s)$$

The equilibrium constant for this reaction as written is 25 torr^{-1} at 25°C. (*a*) Formulate the expression for the equilibrium constant of this reaction, and give the equilibrium pressure of water over a mixture of anhydrous calcium chloride and its monohydrate at 25°C. (*b*) The equilibrium pressure rises as the temperature increases; i.e., calcium chloride is a less effective drying agent at higher temperatures. Is the reaction above exothermic or endothermic? Explain.

(*c*) Suppose that a stream of air at 25°C and 1 atm, containing water vapor at a pressure of 0.025 atm, is passed over anhydrous $CaCl_2$ at a rate of 2.0 liters min^{-1}. If there is initially 1.00 mol $CaCl_2$ present, how long will it remain effective as a drying agent if the reaction above is assumed to be that which determines its effectiveness?

Ans. (*a*) 0.040 torr; (*b*) exothermic; (*c*) a little over 8 hours

13-20 **Relation between equilibrium constants.** Express the equilibrium constant for the reaction

$$NH_3 + HNO_2 \rightleftharpoons NH_4^+ + NO_2^-$$

in terms of the constants for the following reactions:

$HNO_2 \rightleftharpoons H^+ + NO_2^-$ $\qquad K_1 = 4.5 \times 10^{-4}$

$NH_3 + H_2O \rightleftharpoons NH_4^+ + OH^-$ $\qquad K_2 = 1.8 \times 10^{-5}$

$H_2O \rightleftharpoons H^+ + OH^-$ $\qquad K_w = 1.0 \times 10^{-14}$

13-21 **Combination of equilibria.** The equilibria

$$NH_3 + H_2O \rightleftharpoons NH_4^+ + OH^-$$
$$HOAc \rightleftharpoons H^+ + OAc^-$$

happen to have equilibrium constants of equal values, $1.8 \times 10^{-5}\ M$, while the ion product $[H^+][OH^-]$ has the value of $1.0 \times 10^{-14}\ M^2$, all at 25°C. What is the value, at 25°C, of the equilibrium constant for the reaction

$$NH_4^+ + OAc^- \rightleftharpoons NH_3 + HOAc$$

Ans. 3.1×10^{-5}

". . . Acids and bases are substances that are capable of splitting off or taking up hydrogen ions, respectively."

N. Brønsted, 1923

14 Ionic Equilibria in Aqueous Solution: Acids and Bases

14-1 Introduction

The early emergence of the concepts of acid, base, and salt, discussed in Chap. 7, constituted one of the first stages in the development of modern chemistry: the classification of compounds into types with characteristic properties and reactions. The implications of these concepts have evolved over the years, but an understanding of the effects of acids and bases is still of paramount importance for all those who deal with systems in which chemical reactions occur. Chemists frequently need to control pH precisely in order to ensure that a particular reaction mixture will yield the desired set of products rather than some quite different set. Those working in many other areas of science, technology, and medicine must understand acids and bases also—for example, engineers concerned with the properties of materials in various atmospheres and environments in which corrosion or other deterioration may occur; doctors and nurses who must recognize and if possible circumvent abnormalities in body fluids such as blood and urine, many of which develop from or are revealed by pH changes; geologists trying to understand the action of groundwater in the formation of limestone caves; biologists concerned with the regulation of metabolic processes in animals and plants. The list is almost endless.

Many of the properties of acids and bases have been described in detail in Chap. 7, and the reader should review that chapter carefully, especially Secs. 7-1 to 7-5, before studying the present chapter. A familiarity with Chap. 13 is also essential. The conventions for the formulation of reaction quotients (Sec. 13-3) and many of the general comments and examples are especially relevant.

14-2   The Strengths of Acids and Bases

Degree of dissociation.   As discussed in Sec. 7-2, what is usually spoken of as the dissociation of an acid or a base involves a proton transfer to or from H_2O rather than a simple dissociation alone. Nonetheless, the concept of degree of dissociation (Sec. 13-3) is commonly used for acids and bases. For example, acid HA may be said to be 1.4 percent dissociated in a 2.000 F aqueous solution; this means that 98.6 percent of the HA, or 1.972 mol liter^{-1}, remains undissociated while 1.4 percent, or 0.028 mol liter^{-1}, has formed H_3O^+ and A^-.

The degree of dissociation of an acid or base in a given solution can be estimated by performing approximate measurements of any of several properties of the solution and then comparing the results with those to be expected for various degrees of dissociation. For example, one might measure the pH of the solution, its electrical conductivity, or some colligative property such as its freezing point relative to that of the pure solvent (see Sec. 6-4). Solutions of a strong and a weak acid of comparable formality, or solutions of a strong and a weak base, will differ markedly with regard to each of these properties because essentially all molecules of strong electrolytes dissociate into ions whereas comparatively few molecules of weak electrolytes dissociate. As a result, the total concentration of solute species present is higher for strong electrolytes than for weak electrolytes in solutions of similar formality. Colligative properties depend on the total concentration of *all* solute species.

EXAMPLE 14-1   **Degree of dissociation.**   Estimate the degree of dissociation of the solute in each of the following solutions:

(*a*)   A 1.00 F solution of hydrofluoric acid, HF, which has a pH of 1.6.

(*b*)   A 0.010 F solution of sodium hydroxide in water, which melts at $-0.037°C$. The molal freezing-point-depression constant for water is $k_f = 1.86$ K kg mol^{-1} (Sec. 6-4).

Solution.   If you are uncertain about the precise significance of *degree of dissociation*, review comment 4 of Sec. 13-3 and Example 13-3.

(*a*)   The dissociation of HF in water is effected by transfer of a proton to a water molecule,

$$HF + H_2O \rightleftharpoons F^- + H_3O^+ \tag{14-1}$$

A pH of 1.6 implies that

$$[H^+] = 10^{-1.6} = 10^{0.4} \times 10^{-2.0} = 2.5 \times 10^{-2} \, M$$

Since one hydrogen ion is produced from each molecule of HF that dissociates, the number of moles of HF dissociated per liter must be 2.5×10^{-2}. The solution is 1.00 F, and thus there would be 1.00 mol HF liter^{-1} if none dissociated. Consequently the degree of dissociation is $2.5 \times 10^{-2}/1.00$, or 2.5 percent. HF is a weak acid in water.

(*b*)   The weight molality of the solution is

$$c_w = \frac{\Delta T}{k_f} = \frac{0.037}{1.86} = 0.020 \text{ mol kg}^{-1}$$

This means that there is 0.020 mol of solute particles in each kilogram of water, and since this aqueous solution is dilute, 1 liter weighs 1 kg (within the precision of the data). Thus a 0.010 F solution of NaOH contains about 0.020 mol of solute species per liter. The conclusion is that the NaOH has dissociated essentially completely, two ions being formed for every formula unit of the compound. (Measurements of colligative properties can give information about dissociation but *not* about whether the dissociation products are charged. Other means, such as tests of electrical conductivity, are needed for that. When we state here that two *ions* are formed from every formula unit of the compound, we are assuming that such other tests have been made.)

Dissociation constants.   Qualitative comparisons of acids and bases in terms of their degree of dissociation are of limited value because the degree of dissociation of a weak electrolyte varies with concentration, as discussed in Secs. 13-2 and 13-3 (comment 5). More precise statements about the relative strengths of various acids and bases can be made by comparison of the equilibrium constants for their dissociation, which are independent of concentration. As mentioned in Sec. 13-3, the dissociation constant for an acid in aqueous solution (the acid constant) is often symbolized by K_a and that for a base by K_b. For a monoprotic acid HA with only one acidic hydrogen atom, the reaction in water is

$$HA + H_2O \rightleftharpoons A^- + H_3O^+ \tag{14-2}$$

with
$$K_a = \frac{[H_3O^+][A^-]}{[HA]} = \frac{[H^+][A^-]}{[HA]} \tag{14-3}$$

For polyprotic acids, such as H_2SO_4 and H_3PO_4, the protons are furnished in successive steps, for each of which there is a corresponding equilibrium constant. These successive constants are usually distinguished by subscripts. For example, for phosphoric acid the successive equilibria are described by the following chemical equations, shown with the corresponding equilibrium constants:

$$H_3PO_4 + H_2O \rightleftharpoons H_2PO_4^- + H_3O^+ \tag{14-4}$$

$$K_1 = \frac{[H^+][H_2PO_4^-]}{[H_3PO_4]} = 7.1 \times 10^{-3}$$

$$H_2PO_4^- + H_2O \rightleftharpoons HPO_4^{2-} + H_3O^+$$

$$K_2 = \frac{[H^+][HPO_4^{2-}]}{[H_2PO_4^-]} = 6.2 \times 10^{-8} \tag{14-5}$$

$$HPO_4^{2-} + H_2O \rightleftharpoons PO_4^{3-} + H_3O^+$$

$$K_3 = \frac{[H^+][PO_4^{3-}]}{[HPO_4^{2-}]} = 4.4 \times 10^{-13} \tag{14-6}$$

These constants for phosphoric acid, K_1, K_2, and K_3, are usually referred to as the first, second, and third dissociation constants of phosphoric acid; however, K_2 and K_3 can equally well be regarded as the first dissociation constants of the acids $H_2PO_4^-$ and HPO_4^{2-}, respectively. The dihydrogenphosphate ion, $H_2PO_4^-$, behaves as a base in the equilibrium (14-4), accepting a proton, and as an acid in the equilibrium (14-5), donating a proton. Similarly, HPO_4^{2-} behaves as a base in (14-5) and as an acid in (14-6).

Any base B may react with water to form the conjugate acid BH^+ and hydroxide ions,

$$B + H_2O \rightleftharpoons BH^+ + OH^- \tag{14-7}$$

with the corresponding equilibrium constant, K_b, for B given by the expression

$$K_b = \frac{[BH^+][OH^-]}{[B]} \tag{14-8}$$

An extremely useful relation may be derived by considering the product of the acid constant for any acid HA, given by (14-3), and the base constant for the corresponding conjugate base A^-. By analogy with (14-7), the reaction of A^- as a base is

$$A^- + H_2O \rightleftharpoons HA + OH^- \tag{14-9}$$

with

$$K_{b,A^-} = \frac{[HA][OH^-]}{[A^-]} \tag{14-10}$$

The product of $K_{a,HA}$ and K_{b,A^-} is then

$$K_{a,HA}K_{b,A^-} = \frac{[H^+][A^-]}{[HA]} \frac{[HA][OH^-]}{[A^-]} = [H^+][OH^-] = K_w \tag{14-11}$$

which has the value 1.0×10^{-14} at 25°C. *Thus, the product of the acid constant of an acid and the base constant of its conjugate base is the ion product of water.* This completely general relationship for reactions in aqueous solutions makes it easy to calculate K_b of a base when K_a for the conjugate acid is known and vice versa. It implies that if an acid is strong—for example, if K_a is greater than 1—then the conjugate base must be very weak, with K_b less than 10^{-14}. In terms of the reactions involved, this implies that if HA is so strong an acid that it dissociates almost completely in water to form H_3O^+ and A^-, then A^- is so weak a base that it has essentially no tendency to take a proton away from the hydronium ion, H_3O^+.

Conversely, if a base is so strong that when dissolved in water it tends to be converted almost completely into the conjugate acid and hydroxide ions, by (14-7) or (14-9)—that is, if the base is much stronger than OH^- in a competition for protons —then the conjugate acid of this base must be a significantly weaker acid than H_2O.

EXAMPLE 14-2 K_a and K_b for conjugate pairs. (a) Estimate the acid constant for NH_4^+ in water, given that K_b is 1.8×10^{-5}. (b) Estimate the base constant for cyanide ion (CN^-) in water, given that K_a for HCN is 2×10^{-9}.

Solution. With (14-11) for any conjugate pair in water, we have

$$(a) \quad K_{a,NH_4^+} = \frac{K_w}{K_{b,NH_3}} = \frac{1.0 \times 10^{-14}}{1.8 \times 10^{-5}} = \underline{5.5 \times 10^{-10}}$$

and

$$(b) \quad K_{b,CN^-} = \frac{K_w}{K_{a,HCN}} = \frac{1.0 \times 10^{-14}}{2 \times 10^{-9}} = \underline{5 \times 10^{-6}}$$

Thus, ammonium ion is a very weak acid in water and cyanide ion is a moderately weak base (see Table 14-1), not very much weaker than ammonia.

The leveling effect of water on strong acids and bases. Since the equilibrium of the dissociation reaction for any strong acid lies, by definition, overwhelmingly to the right, there are essentially no undissociated HA molecules in the solution. All the HA that might be present is converted to H_3O^+ and the (very weak) conjugate base A^-. This is true, for example, for HI, HBr, HCl (but not HF), $HClO_4$, H_2SO_4 (but not HSO_4^-), and HNO_3. Aqueous solutions of any strong acid are thus essentially solutions of the same acid, H_3O^+, and different very weak bases, the anions of the corresponding acids. Any difference in the strengths of strong acids in water is thus masked by the fact that no acid appreciably stronger than H_3O^+ can exist in dilute aqueous solution in other than very small concentrations. Water is said to have a *leveling effect* on strong acids. To illustrate this effect, consider 0.10 F solutions of HNO_3 and HCl. The respective values of K_a have been estimated to be about 24 and about 10^5. This difference seems large, but it means little in terms of completeness of dissociation. In 0.10 F solutions, HNO_3 is about 99.6 percent dissociated and HCl is about 99.9999 percent dissociated; hence both solutions contain H_3O^+ at a concentration of essentially 0.10 M.

Differences between strong acids may become apparent when the acids are

TABLE 14-1   **Classification of the strengths of acids and their conjugate bases***

	Dissociation constants		
Strength of acid	Acid	Conjugate base	Strength of base
Strong	$K_a \geq 1$	$K_b \leq 10^{-14}$	Ineffectual
Moderately weak	$1 > K_a \geq 10^{-7}$	$10^{-14} < K_b < 10^{-7}$	Very weak
Very weak	$10^{-7} > K_a > 10^{-14}$	$10^{-7} \leq K_b < 1$	Moderately weak
Ineffectual	$K_a \leq 10^{-14}$	$K_b \geq 1$	Strong

*The categories used are based on suggestions by N. D. Cheronis.

dissolved in a nonaqueous solvent that can accept a proton and thereby form a stronger acid than H_3O^+—that is, a solvent that is a weaker base than water. For example, the acetic acid molecule HOAc can accept a second proton[1] and thus form the acid H_2OAc^+, which is a considerably stronger acid than H_3O^+. When HCl is dissolved in anhydrous acetic acid, the equilibrium of the reaction

$$HCl + HOAc \rightleftharpoons Cl^- + H_2OAc^+ \qquad (14\text{-}12)$$

is no longer completely in favor of the right side; there are substantial numbers of HCl molecules in this solution. This is true for some other strong acids as well. By comparisons in different solvents, it is possible to assess the relative strengths of acids that are almost equally strong in water. In this way it has been shown that $HClO_4 > HCl > H_2SO_4 > H_3O^+$ (see also Table 14-2).

Just as the strongest acid that can exist in significant quantity in an aqueous solution is H_3O^+, the strongest base that can be present in water in appreciable concentrations is the hydroxide ion, OH^-. Any base that is stronger than OH^- will take protons away from water molecules and form OH^- and the very weak conjugate acid of the base. Water thus has a leveling effect on strong bases, just as it does on strong acids. For example, the amide ion, NH_2^-, is a stronger base than OH^-. When the crystalline ionic compound sodium amide, $NaNH_2$, is dissolved in water, the amide ion reacts vigorously and almost completely to form its conjugate acid, NH_3, and OH^-:

$$NH_2^- + H_2O \rightleftharpoons NH_3 + OH^- \qquad (14\text{-}13)$$

Hydroxide ion is so much stronger a base than NH_3 that the NH_3 has little effect on the pH of the resulting solution. The pH is essentially the same as that of a solution of sodium hydroxide of the same formality.

The classification of acids and bases according to strength in just two categories, strong and weak, has some limitations, even for qualitative purposes. A modified qualitative classification that permits a consistent categorization of both members of a conjugate acid-base pair, depending upon the approximate values of the respective K_a and K_b, is given in Table 14-1. Strong acids and bases are considered to be those with K_a or K_b greater than or equal to 1. Weak acids and bases are divided into three categories, as indicated. The boundaries between categories are of course arbitrary, but the boundaries, and the names, have been chosen so as to emphasize the complementary nature of the strengths of the members of a conjugate pair.

Values of K_a and K_b for many of the more common acids and bases are included in Table 14-2 in order of decreasing acid strength (and in the alphabetized reference Table D-3). Few base constants are listed explicitly in Table 14-2, but any missing value of K_b can readily be derived from the corresponding value of K_a with the help of (14-11). Important moderately weak acids include the bisulfate ion, HSO_4^-, $K_a = 1.2 \times 10^{-2}$; phosphoric acid, H_3PO_4, $K_a = 7.1 \times 10^{-3}$; formic acid, HCOOH, $K_a = 1.7 \times 10^{-4}$; and acetic acid, HOAc, $K_a = 1.8 \times 10^{-5}$. The conjugate base of each of these moderately weak acids (respectively, SO_4^{2-}, $H_2PO_4^-$, $HCOO^-$, and

---

[1]The second proton is attached to the oxygen atom that is not already part of a hydroxyl group in the CH_3COOH molecule; H_2OAc^+ is thus $CH_3C(OH)_2^+$.

TABLE 14-2 Acid constants in water* at 25°C

(see also Table D-3, Appendix D)

Conjugate pair of acid and base		Acid constant	
$HClO_4$	ClO_4^-	$\sim 10^9$	
HI	I^-	$\sim 10^7$	
HBr	Br^-	$\sim 10^6$	
HCl	Cl^-	$\sim 10^5$	
H_2SO_4	HSO_4^-	$\sim 10^3$	
$HClO_3$	ClO_3^-	$\sim 10^3$	
H_3O^+	H_2O	55†	
HNO_3	NO_3^-	24	
H_2CrO_4	$HCrO_4^-$	1.2	
H_2SO_3	HSO_3^-	1.0×10^{-2}‡	
HSO_4^-	SO_4^{2-}	1.2×10^{-2}	
$HClO_2$	ClO_2^-	1.0×10^{-2}	
H_3PO_4	$H_2PO_4^-$	7.1×10^{-3}	
HF	F^-	6.7×10^{-4}	
HNO_2	NO_2^-	4.5×10^{-4}	
HCOOH	$HCOO^-$	1.7×10^{-4}	
HOAc	OAc^-	1.8×10^{-5}	$(K_b = 5.5 \times 10^{-10})$
H_2CO_3	HCO_3^-	4.4×10^{-7}§	
$HCrO_4^-$	CrO_4^{2-}	3.2×10^{-7}	
H_2S	HS^-	9.1×10^{-8}	
$H_2PO_4^-$	HPO_4^{2-}	6.2×10^{-8}	
HSO_3^-	SO_3^{2-}	5.0×10^{-8}	
HClO	ClO^-	1.1×10^{-8}	
HCN	CN^-	2.0×10^{-9}	$(K_b = 5.0 \times 10^{-6})$
H_3BO_3	$B(OH)_4^-$	6.4×10^{-10}	
NH_4^+	NH_3	5.5×10^{-10}	$(K_b = 1.8 \times 10^{-5})$
HCO_3^-	CO_3^{2-}	4.8×10^{-11}	
H_2O_2	HO_2^-	2.4×10^{-12}	
HPO_4^{2-}	PO_4^{3-}	4.4×10^{-13}	
HS^-	S^{2-}	1.2×10^{-15}	
H_2O	HO^-	1.8×10^{-16}¶	
NH_3	NH_2^-	$\sim 10^{-23}$	
HO^-	O^{2-}	$\sim 10^{-35}$	

*Values of equilibrium constants often differ widely in different reference sources, depending in part on whether they were extrapolated to zero concentration or not. Estimated values mainly by G. Schwarzenbach.

†By (14-2) and (14-3) with HA and A^- being, respectively, H_3O^+ and H_2O, $K_a = [H_3O^+][H_2O]/[H_3O^+] = [H_2O] = 55$.

‡Acid constant for which the term $[H_2SO_3]$ is the total concentration of both actual H_2SO_3 and dissolved (hydrated) SO_2 that has not reacted with H_2O to form H_2SO_3. The acid constant referred to just the species H_2SO_3 has the value 4.0×10^{-2}.

§Acid constant for which the term $[H_2CO_3]$ is the total concentration of both actual H_2CO_3 and dissolved (hydrated) CO_2 that has not reacted with H_2O to form H_2CO_3. The acid constant referred to just the species H_2CO_3 has the value 1.7×10^{-4}.

¶$K_a = [HO^-][H^+]/[H_2O]$. ($[H_2O]$ is retained in the denominator by analogy with other acid constants.)

OAc⁻) is a very weak base. Typical very weak acids include hydrocyanic acid, HCN, $K_a = 2.0 \times 10^{-9}$; boric acid, H_3BO_3, which is effectively monoprotic,[1] $K_a = 6.4 \times 10^{-10}$; ammonium ion, NH_4^+, $K_a = 5.5 \times 10^{-10}$; and bicarbonate ion, HCO_3^-, $K_a = 4.8 \times 10^{-11}$. The conjugate base of each of these very weak acids [respectively, CN^-, $B(OH)_4^-$, NH_3, and CO_3^{2-}] is a moderately weak base and thus is a stronger base than the conjugate form is an acid. It is an interesting coincidence that perhaps the most common weak acid and weak base, acetic acid and ammonia, have identical dissociation constants—that is, K_a for HOAc and K_b for NH_3 both equal 1.8×10^{-5}.

As Table 14-2 indicates, an enormous range of strengths of acids and bases has been measured, at least approximately. The values of K_a and K_b have been extended beyond the limits measurable in water by comparisons in other solvents.

Rules for estimating the strengths of inorganic oxygen acids. Two simple rules permit estimating the acid constants of an inorganic acid containing oxygen if the structure of the acid is known. In the form given here, as proposed by Pauling, these rules apply to acids with formula $(HO)_m XO_n H_p$ in the uncharged form, where X is the central atom and where the hydrogens of each OH group are acidic and the other hydrogens, which are attached directly to the central atom X, are not acidic. The strength of the acid is found to depend almost entirely on the number (n) of oxygen atoms to which no hydrogen atom is attached.

Rule 1. The first acid constant has approximately the value $K_1 \approx 10^{5n-7}$. For example, K_1 for sulfurous acid, H_2SO_3, which has the structure $(HO)_2SO$, with $n = 1$, is predicted to be about $10^{5 \times 1 - 7} = 10^{-2}$. It is actually 4×10^{-2}.

Rule 2. For polyprotic acids, the ratios of successive acid constants are approximately equal to 10^{-5}. Thus if there are three acidic hydrogens attached to different oxygen atoms, the three acid constants are in the approximate relationship $K_1 : K_2 : K_3 = 1 : 10^{-5} : 10^{-10}$.

As an example of application of both rules, consider the triprotic acid phosphoric acid, H_3PO_4 [or $(HO)_3PO$], for which $n = 1$. The constants are predicted to be $K_1 \approx 10^{-2}$, $K_2 \approx 10^{-7}$, and $K_3 \approx 10^{-12}$; the measured values are 7.1×10^{-3}, 6.2×10^{-8}, and 4.4×10^{-13}.

Phosphorous acid (H_3PO_3) has the structure

and is a diprotic rather than a triprotic acid. Its acid constants are 1.0×10^{-2} and

---

[1] H_3BO_3, or $B(OH)_3$, is unusual because it does not lose a proton when acting as an acid in water; it accepts a hydroxide ion instead:

$$B(OH)_3 + 2H_2O \rightleftharpoons B(OH)_4^- + H_3O^+$$

2.6×10^{-7}, close to the values predicted by the rules. Similarly, hypophosphorous acid (H_3PO_2) has the structure

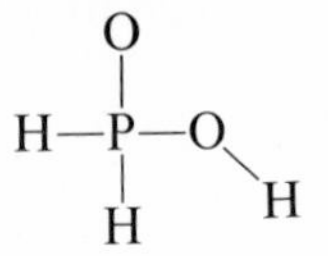

$$
\begin{array}{c}
O \\
\parallel \\
H-P-O \\
| \quad \diagdown \\
H \quad \quad H
\end{array}
$$

It is a monoprotic acid with $K_a = 1 \times 10^{-2}$, as expected from its structure and the rules.

Since acid constants can range over many powers of 10, any estimate that comes within one or two powers of 10 can be at least qualitatively very useful in indicating whether a given acid is likely to be strong, moderately weak, or very weak. The rules are good to about this degree.

No simple rules are available for estimation of the acid constants of inorganic acids other than oxygen acids with the structure discussed here.

14-3   Equilibrium Calculations: Dissociation and Hydrolysis

pH, pOH, and pK.   As discussed in Sec. 7-3, the use of the symbol pH for the negative base-10 logarithm of the concentration[1] of H^+ has been generalized, so that pOH means $- \log [OH^-]$, and the same symbol is used for equilibrium constants in similar logarithmic form, with $pK \equiv - \log K$.

Since

$$[H^+][OH^-] = K_w$$

it follows that

$$pH + pOH = pK_w$$

At 25°C, $pK_w = 14.00$. When the concentrations of H^+ and OH^- are equal, as they are in pure water, one speaks of a neutral medium, for which

$$[H^+] = [OH^-] = \sqrt{K_w}$$

and thus

$$pH = pOH = 0.5 \, pK_w \tag{14-14}$$

which is 7.00 at 25°C. Conversely, when the pH of any solution at 25°C is 7.00, it follows from (14-14) that $[H^+]$ and $[OH^-]$ are equal, so that the solution is neutral. Since K_w varies with temperature, it is only approximately true that a solution with pH = 7.00 is neutral at temperatures other than 25°C. For example, at 0°C, $K_w = 1.14 \times 10^{-15}$, so that $pK_w = 14.94$ and the pH of a neutral solution is 7.47. At 60°C, $K_w = 9.6 \times 10^{-14}$, corresponding to $pK_w = 13.02$, so that the pH of a

---

[1] In work with solutions more concentrated than those we shall normally be concerned with, and in work aiming at higher precision than we are interested in, it is necessary to use the stricter definition of pH as $-\log a_{H^+}$, where a_{H^+} is the activity or "effective concentration" of the hydrogen ion (Secs. 13-5 and 15-5). In highly concentrated solutions the differences between concentration and activity can be very large.

neutral solution is 6.51. At 37°C (body temperature), K_w is 2.6×10^{-14} and the pH of a neutral solution is 6.79.

As mentioned in Chap. 13, we shall not hesitate to use approximations in equilibrium calculations provided that no errors larger than 10 or 20 percent result. We shall therefore assume the pH of a neutral solution to be 7.0 and pK_w to be 14.0, at temperatures not far from 25°C. Acid and base constants also change with temperature, but we shall neglect these changes also. They are usually smaller than the change of K_w.

The pH scale in dilute solutions ranges from about 0 to 14. In more concentrated solutions, this range is exceeded at each end. On the acid side, the pH becomes negative for concentrated solutions. In concentrated H_2SO_4, the lowest existing pH has been estimated to be about -10; in concentrated solutions of NaOH and KOH, the pH is believed to reach 18 or 19. In such concentrated solutions, the more accurate definition of pH in terms of hydrogen-ion activity (rather than concentrations) is required.

Strong acids and bases. In an aqueous solution of a strong acid, essentially all molecules of the strong acid have transferred their protons to water molecules, thereby furnishing an equal quantity of hydrogen ions and of the base that is conjugate to the strong acid. The hydrogen-ion concentration and the concentration of conjugate base are thus equal to the formal concentration of the acid if we neglect the contribution to the hydrogen-ion concentration from the dissociation of water, which is almost always possible in such solutions, as discussed in the following paragraph. The concentration of the strong-acid species itself is essentially zero because it has completely dissociated. Thus, in a $1.0 \, F$ solution of HCl, for example, $[H^+] = [Cl^-] = 1.0 \, M$, so that the pH $= 0.0$, while $[HCl] \approx 0$.

The effect of the dissociation of water can be understood by noting that this reaction produces H^+ and OH^- in equal numbers. In a solution of a strong acid, water is the only source of hydroxide ions, so that $[OH^-]$ is just equal to the contribution to the total hydrogen-ion concentration that comes from the dissociation of H_2O. In an acid solution, $[OH^-]$ is necessarily less than $10^{-7} \, M$ and the contribution to $[H^+]$ from the dissociation of water is thus less than $10^{-7} \, M$. For example, suppose that calculation of $[H^+]$ with complete neglect of the dissociation of water gives $[H^+] = 10^{-6} \, M$, or pH $= 6$. This implies that $[OH^-] = 10^{-8} \, M$, so that the contribution to $[H^+]$ from the dissociation of water is $10^{-8} \, M$, only 1 percent of the value obtained when this contribution is neglected. Any error made by ignoring the contribution from the dissociation of H_2O is thus indeed negligible for our purposes. This source of H^+ is always negligible unless the pH is close to 7. Thus for strong acids the dissociation of water can be ignored unless the concentration of the acid is below $10^{-6} \, F$, which is so low a concentration that it has little importance in practical situations: it corresponds to about one drop of $1 \, F$ acid added to more than 40 liters of water.

In a solution of a strong base the situation is similar to that for a strong acid. For example, in $0.1 \, F$ NaOH, $[OH^-] = [Na^+] = 0.1 \, M$, so that pOH $= 1$ and pH $= 14 - 1 = 13$, while $[NaOH] \approx 0$. The contribution to $[OH^-]$ from the dissociation of water is negligible. In either an acidic or a basic solution, the effect of the dissociation of water can be ignored unless the pH is within less than one unit of 7.

Weak acids. A weak monoprotic acid is not completely dissociated in an aqueous solution; that is, proton transfer from the acid to H_2O is incomplete. The interrelation between the degree of dissociation and the acid constant K_a is illustrated in the following example.

EXAMPLE 14-3   **Calculation of K_a from degree of dissociation.** A weak acid HA is 2 percent dissociated in a 1.00 F solution. What is the dissociation constant of HA?

Solution. The reaction of interest is

$$H_2O + HA \rightleftharpoons H_3O^+ + A^-$$

Since the solution is 1.00 F in HA, the concentration of HA that would be present *if there were no dissociation* is 1.00 M. However, 2 percent of the HA present has dissociated, so its actual concentration is 98 percent of 1.00 M (that is, 0.98 M) and by its dissociation it forms H_3O^+ and A^-, each at a concentration of 0.02 M. Summarizing:

	[HA]	$[H_3O^+]$	$[A^-]$
If no dissociation	1.00	~0	0
At equilibrium, after dissociation	0.98	0.02	0.02

Thus, when we evaluate the equilibrium constant by inserting the *equilibrium* concentrations into the reaction quotient, we get

$$K_a = \frac{[H^+][A^-]}{[HA]} = \frac{(0.02)(0.02)}{0.98} M = \underline{4 \times 10^{-4} \ M}$$

The degree of dissociation is, as given, 2 percent; that is,

$$\text{degree of dissociation} = \frac{[A^-]}{[HA] + [A^-]} = \frac{0.02}{0.02 + 0.98}$$

$$= \tfrac{0.02}{1.00} = \underline{0.02}$$

Note that the denominator in the expression for the degree of dissociation of a weak electrolyte in solution is always its formality, since this is just the number of moles per liter that would be present if none had dissociated.

To calculate the pH of a solution of a weak monoprotic acid, we proceed as in Example 14-4.

EXAMPLE 14-4   **The pH of dilute acetic acid.** What is the pH of 0.20 F acetic acid?

Solution. We start with 0.20 mol HOAc molecules $liter^{-1}$ and assume that at equilibrium x mol liter^{-1} have lost their protons, leaving $(0.20 - x)$ mol HOAc liter^{-1}. The concentrations of the different species are then as shown on the line below the reaction equation:

$$HOAc \rightleftharpoons H^+ + OAc^-$$
$$0.20 - x \qquad x \qquad x$$

This neglects the contribution made to $[H^+]$ by the dissociation of water, which causes an error of less than 1 percent if $[H^+]$ turns out to be larger than $10^{-6}\,M$ and $[OH^-]$ smaller than $10^{-8}\,M$, as discussed earlier. Insertion of the concentrations in terms of x into the equilibrium expression results in the equation

$$\frac{[H^+][OAc^-]}{[HOAc]} = \frac{x^2}{0.20 - x} = K_a = 1.8 \times 10^{-5} \qquad (14\text{-}15)$$

This is a quadratic equation, and such an equation can always be solved by an exact method, but that is not necessary here. The use of judicious approximations is frequently faster and simpler; one need only check the answer obtained to be sure that it is consistent with the assumptions made in the initial approximations.

Inspection of (14-15) shows that the right side is very small, 1.8×10^{-5}. The left side can only be small if x is small relative to 0.20, for if x is comparable to 0.20, the x^2 term in the numerator cannot be very small and the term $0.20 - x$ in the denominator might itself be small, which would only make the whole expression on the left larger.

Thus we suspect that a good approximation is that x is small compared to 0.20, that is, that

$$0.20 - x \approx 0.20 \qquad (14\text{-}16)$$

within the desired precision. Substitution of the approximation (14-16) into the denominator of the quadratic expression in (14-15) gives

$$x^2 \approx 0.20 \times 1.8 \times 10^{-5} = 3.6 \times 10^{-6}$$

or
$$x = [H^+] \approx 1.9 \times 10^{-3}\,M$$

and
$$pH \approx 3.0 - \log 1.9 = \underline{2.7}$$

Since x is less than 1 percent of 0.20, our approximation (14-16) was indeed justified. Furthermore, because the pH is so low, we know also that the contribution to $[H^+]$ from the dissociation of water, measured by the $[OH^-]$ in the solution, is completely negligible.

As emphasized in Sec. 14-2, and especially in the discussion accompanying Example 14-2 and Table 14-1, the conjugate form of a moderately weak base is itself a very weak acid. Thus, for example, salts containing ammonium ion can have an acidic reaction by the equilibrium $NH_4^+ + H_2O \rightleftharpoons NH_3 + H_3O^+$, provided that the other ion present does not have an even stronger opposing effect. Salts formed from other weak bases behave similarly.

EXAMPLE 14-5   **pH of a solution of a salt of a weak base with a strong acid.** Ethylamine, $C_2H_5NH_2$, is a weak base very similar to ammonia. It reacts with acids to form salts containing the ethylammonium ion, $C_2H_5NH_3^+$, and the conjugate base of the acid. For example, ethylamine reacts with HCl to form ethylammonium chloride, $C_2H_5NH_3Cl$, a strong electrolyte that dissociates completely in solution into $C_2H_5NH_3^+$ and Cl^-. Since Cl^- is the conjugate base of a strong acid, HCl, it shows no significant tendency to react with water. The ethylammonium ion is, however, a very weak acid, just as is NH_4^+. If K_b for ethylamine is 4.7×10^{-4}, what is the pH of a 0.10 F solution of ethylammonium chloride?

Solution. The solution is acidic because of the reaction

$$C_2H_5NH_3^+ + H_2O \rightleftharpoons C_2H_5NH_2 + H_3O^+ \tag{14-17}$$

and the acid constant for $C_2H_5NH_3^+$ can be derived from K_b for its conjugate base by the usual relation analogous to (14-11):

$$K_{a,BH^+} K_{b,B} = K_w = 1.0 \times 10^{-14} \tag{14-18}$$

Thus, here

$$K_a = \frac{1.0 \times 10^{-14}}{4.7 \times 10^{-4}} = 2.1 \times 10^{-11} \tag{14-19}$$

If we symbolize the ethylammonium ion by BH^+ and ethylamine by B and let the unknown $[H^+]$ be x, we have

$$[B] = [H^+] = x \quad \text{and} \quad [BH^+] = 0.10 - x$$

and

$$K_a = \frac{[B][H^+]}{[BH^+]} = \frac{x^2}{0.10 - x} = 2.1 \times 10^{-11} \tag{14-20}$$

It is quite evident from the smallness of K_a that the value of x must itself be very small and hence that the approximation $0.10 - x \approx 0.10$ is a plausible one to try.

Thus we get

$$x^2 \approx 0.10 \, (2.1 \times 10^{-11})$$

$$[H^+] = x \approx 1.5 \times 10^{-6} \, M$$

Checking the validity of our approximation, we find it completely justified. Furthermore, $[H^+]$ is slightly greater than $10^{-6} \, M$ so that ignoring the contribution from the dissociation of water is also reasonable since it causes an error of less than 1 percent. Finally, the pH of the solution is

$$pH = 6.0 - \log 1.5 = \underline{5.8}$$

We now consider an example in which the approximation $[H^+] \ll c$, where c is the formality of the acid, is not valid.

EXAMPLE 14-6   **pH of a solution of $HClO_2$.** What is the pH of a 0.10 F solution of chlorous acid, $HClO_2$, for which K_a is 1.0×10^{-2}?

Solution. Proceeding as in Examples 14-4 and 14-5, we obtain the equation for the hydrogen-ion concentration x:

$$\frac{x^2}{0.10 - x} = 1.0 \times 10^{-2} \tag{14-21}$$

Since K_a is not very small, we suspect that x is not negligible with respect to 0.10 and thus that our earlier approximation, $c - x \approx c$, is no longer valid. However, even if this suspicion is correct, we can solve (14-21) either by the method of successive approximations or exactly with the formula for the solution of a quadratic equation. We illustrate both approaches.

In applying the *method of successive approximations,* we proceed as in Examples 14-4 and 14-5, using the approximation $0.10 - x \approx 0.10$ on the left side of (14-21) even though we believe that this may lead to a significant error. The result is

$$x \approx (0.10 \times 1.0 \times 10^{-2})^{1/2} = 0.032 \tag{14-22}$$

which confirms our suspicion that the approximation $0.10 - x \approx 0.10$ is not a good one. However, we can consider (14-22) to be a good *first approximation* to the correct value of x and substitute it for x in the expression $0.10 - x$ in (14-21). This leads to

$$x^2 \approx (0.10 - 0.032) \, 1.0 \times 10^{-2}$$

and hence

$$x \approx (0.068 \times 10^{-2})^{1/2} = 0.026 \tag{14-23}$$

This is a *second approximation* to x. Repeating the procedure, we obtain $x^2 \approx (0.10 - 0.026)1.0 \times 10^{-2}$ and

$$x \approx (0.074 \times 10^{-2})^{1/2} = 0.027$$

This is within 4 percent of the preceding value and therefore as close to the correct value as we may expect, unless activities are taken into account. We therefore conclude that $[H^+] = [ClO_2^-] = 0.027 \, M$ and $[HClO_2] = 0.10 - 0.027 = 0.07 \, M$. As a check we note that if these concentrations are inserted into the reaction quotient the correct value of K_a is obtained:

$$\frac{[H^+][ClO_2^-]}{[HClO_2]} = \frac{(2.7 \times 10^{-2})^2}{0.07} = \frac{7.3 \times 10^{-4}}{7 \times 10^{-2}} = 1.0 \times 10^{-2}$$

One characteristic of the method of successive approximations is illustrated by the application here: the successive approximate values often straddle the correct answer, being alternatively on one side of it or the other, greater or smaller. If convergence is slow, that is, if successive values differ markedly, it can sometimes be accelerated by averaging two successive values, which will normally give an approximation better than either value. However, when convergence is rapid, as it was in this example, no such averaging is needed.

We might also have solved (14-21) by using the familiar formula for the exact solution of a quadratic expression,

$$ax^2 + bx + c = 0 \tag{14-24}$$

which is

$$x = \frac{-b \pm \sqrt{b^2 - 4ac}}{2a} \tag{14-25}$$

with $a = 1$, $b = 0.010$, and $c = -1.0 \times 10^{-3}$ in this example. This approach leads to

$$x = \frac{-0.010 \pm (1.0 \times 10^{-4} + 4.0 \times 10^{-3})^{1/2}}{2}$$

$$= \frac{-0.010 \pm 0.064}{2}$$

Only the plus sign gives a physically meaningful value for x, the hydrogen-ion concentration, which cannot be negative. This leads to

$$[H^+] = x = \tfrac{0.054}{2} = 0.027 \ M$$

which is the same result obtained in the third approximation above. The pH of the solution is $2.00 - \log 2.7 = \underline{1.57}$, with the final digit uncertain.

Which method, successive approximation or exact solution, should be used to solve problems like this one? The techniques lead to the same answer—the choice is a matter of personal taste.

Weak bases. Calculations for weak bases are in all respects analogous to those for weak acids. We shall consider as examples both an uncharged weak base, whose reaction may be represented by

$$B + H_2O \rightleftharpoons HB^+ + OH^- \tag{14-26}$$

and the salt of a weak acid, such as sodium acetate, which acts as a very weak base in water because acetate ion is the conjugate base of a moderately weak acid. This reaction can be represented by

$$A^- + H_2O \rightleftharpoons HA + OH^- \tag{14-27}$$

which differs from (14-26) only in the charges on the basic and the acidic species.

EXAMPLE 14-7 **The pH of a solution of ammonia.** What is the pH of 0.50 F ammonia, for which $K_b = 1.8 \times 10^{-5}$?

Solution. We start with 0.50 mol NH_3 liter^{-1} and assume that at equilibrium x mol liter^{-1} has reacted by the equation given below. The final concentrations are as indicated:

$$NH_3 + H_2O = NH_4^+ + OH^-$$
$$0.50 - x \qquad\qquad x \qquad\quad x$$

The contribution of OH^- from the dissociation of H_2O has been neglected. The equilibrium condition yields

$$\frac{[NH_4^+][OH^-]}{[NH_3]} = \frac{x^2}{0.50 - x} = K_b = 1.8 \times 10^{-5}$$

If we assume that x is small compared to 0.50, then $0.50 - x \approx 0.50$ and we find $x^2 \approx 0.50 \times 1.8 \times 10^{-5}$ and $x = [OH^-] \approx 3.0 \times 10^{-3}\ M$. All assumptions are seen to be justified. Therefore, $pOH = 3.00 - \log 3.0 = 2.5$ and $pH = 14.0 - pOH = \underline{11.5}$.

EXAMPLE 14-8

Estimation of K_a for an acid from the pH of a solution of the sodium salt of the acid. The pH of a $0.10\ F$ solution of the sodium salt of a certain acid HX is 9.5. What is K_a for the acid?

Solution. The reaction involved is (14-27). Since the pH is 9.5, we know that $pOH = 14 - pH = 4.5$, so that $[OH^-] = 10^{-4.5} \approx 3 \times 10^{-5}\ M$. This must also be the concentration of HX, inasmuch as HX and OH^- are formed in equal numbers in reaction (14-27). Consequently the concentration of unaltered X^- is $0.10 - 3 \times 10^{-5} = 0.10$. Substitution of these equilibrium concentrations in the reaction quotient for the reaction of X^- as a base gives

$$K_{b,X^-} = \frac{[OH^-][HX]}{[X^-]} = \frac{(10^{-4.5})^2}{0.10} = 1.0 \times 10^{-8}$$

Hence

$$K_{a,HX} = \frac{K_w}{K_{b,X^-}} = \frac{1.0 \times 10^{-14}}{1.0 \times 10^{-8}} = \underline{1.0 \times 10^{-6}}$$

Note that in this example we have ignored the effect of Na^+ in the solution. The fact that NaOH is a very strong base in water implies that the sodium ion has essentially no tendency to act as an acid, even though the hydrated species, $Na(H_2O)_n^+$, could in principle lose a proton. Thus, the position of equilibrium in the potential reaction

$$Na(H_2O)_n^+ + H_2O \rightleftharpoons NaOH(H_2O)_{n-1} + H_3O^+ \qquad (14\text{-}28)$$

lies almost completely to the left, with no detectable amount of undissociated NaOH present. The species $Na(H_2O)_n^+$ also has no tendency to accept a proton. Similar arguments support our neglect of the effect of Cl^- in the example dealing with the pH of ethylammonium chloride (Example 14-5).

Hydrolysis of salts. Salts are ionic compounds that often show little covalent character; they are usually strong electrolytes. However, the anion A^- of a strong electrolyte may be conjugate to a weak acid HA. In this case some of the A^- in an aqueous solution of the electrolyte reacts with H_2O to form HA and OH^-. Similarly, if the cations HB^+ of such a salt are conjugate to a weak base, some of them transfer their protons to water, forming the base B and H_3O^+. These are, of course, the very reactions illustrated in Examples 14-5 and 14-8. The term *hydrolysis* has long been applied to reactions of this general type, as emphasized in Sec. 7-5. We avoided using the term *hydrolysis* in presenting Examples 14-5 and 14-8 in order to stress that, in fact, hydrolytic reactions are in no way different from other acid-base reactions and may be handled identically. It is important to know what the term *hydrolysis* implies, since it is still commonly used, but it is equally important to recognize that it is not some new and distinct phenomenon for which new principles and techniques must be learned.

No hydrolysis occurs in solutions of salts of strong acids with strong bases (for example, NaCl, KNO_3, $LiClO_4$), since neither the anion nor the cation tends to react with water. Thus such solutions are neutral, with pH = 7.*

Aqueous solutions of salts formed from weak bases and strong acids (for example, NH_4NO_3) are acidic because of reactions analogous to (14-17) in Example 14-5. Similarly, solutions formed from strong bases and weak acids (for example, NaOAc) are alkaline because of reactions analogous to (14-27). If both the acid and the base from which the salt is formed are weak, hydrolysis of each ion occurs but the reaction that produces the weaker conjugate form predominates. Thus if a salt BH^+A^- is formed from a weak acid HA and a weak base B, solutions of this salt will be basic if K_a for HA is smaller than K_b for B and will be acidic if the converse is true. In the unusual event that K_a for HA is equal to K_b for B, solutions of the salt will be neutral, with pH = 7, even though extensive hydrolysis of each ion occurs. Such is the case for ammonium acetate, NH_4OAc (see Prob. 14-28 at the end of the chapter).

14-4   Buffer Solutions

Many chemical reactions are sensitive to changes in pH; that is, the yield of products and even the nature of the products may be altered appreciably if the pH changes significantly during the course of the reaction. This is especially true of biochemical reactions important to the proper metabolism and functioning of animals and plants, but it is by no means limited to them. Since many chemical processes either produce or consume protons or other acids, large pH changes normally accompany such reactions unless there is a mechanism for preventing such changes. All organisms are equipped with naturally occurring chemical systems that minimize the effects of the addition or removal of acid, and many similar systems have been devised for use with reactions that do not occur in organisms. Any chemical system that can provide a resistance to pH change is termed a *buffer*, in analogy with the common use of this word to denote anything that can absorb the effect of a shock or otherwise minimize the effects of some drastic change.

It is characteristic of chemical buffer systems that *they contain substantial amounts of both a moderately weak (or very weak) acid and its correspondingly very weak (or moderately weak) conjugate base.* Typical components of buffer systems include HOAc and OAc^-, NH_4^+ and NH_3, $H_2PO_4^-$ and HPO_4^{2-}, and HCO_3^- and CO_3^{2-}.

Buffer solutions resist pH changes because added H^+ or OH^- does not remain as

---

*This statement may not be true if either the anion or the cation is derived from a polyfunctional acid or base and is capable of further reaction. In a solution of $NaHSO_4$, for example, the hydrated Na^+ does not furnish protons and HSO_4^- does not accept protons readily, but HSO_4^- is itself a moderately weak acid and dissociates further to make the solution acidic.

such in the solution but is removed by reaction with one of the forms of the conjugate acid-base pair that constitutes the buffer system. For example, if H^+ is added to an acetic acid–acetate buffer, most of it reacts with some of the acetate, converting it to acetic acid; conversely, if OH^- is added to the same buffer, most of it is removed by reaction with acetic acid to form acetate ion. There is always *some* pH change when acid or base is added to a buffer, but the change is far less than it would be for a solution of the same pH that did not contain the ingredients of a buffer. This is illustrated shortly in Example 14-9. Before examining specific problems, however, we consider qualitatively the reactions in a buffer.

Equilibria of acids and bases in aqueous solutions are established so quickly that for practical purposes we may assume that acid-base systems, including buffers, are at equilibrium as soon as the components are uniformly mixed. Thus if we have an aqueous solution containing an acid HX, with dissociation constant K_a, we may always assume that

$$\frac{[H^+][X^-]}{[HX]} = K_a$$

or
$$\frac{[X^-]}{[HX]} = \frac{K_a}{[H^+]} \tag{14-29}$$

Hence, if we know the pH of any aqueous solution and the acid constants of any acids present, we also know the ratio of the molar concentrations of the basic and acidic forms of each conjugate pair present. Conversely, if we know the ratio of $[X^-]$ to $[HX]$, we know $K_a/[H^+]$ and can, for example, calculate K_a for an acid if the pH is known or calculate the pH if K_a is known. These relations hold for *any* solution of an acid or a base. What makes them particularly relevant and useful for buffer solutions is that the ratio of the molar concentrations of the members of a conjugate acid-base pair in a buffer [i.e., the left-hand side of (14-29)] can almost always be assumed to be the same as the ratio of their formalities in the solution. If the respective formalities of conjugate acid and base are c_a and c_b, then

$$[H^+] = K_a \frac{[HX]}{[X^-]} \approx K_a \frac{c_a}{c_b} \tag{14-30}$$

as long as *both* members of the conjugate pair are present in significant concentration, as must be true in a buffer.

The reason that the substitution of formalities for molarities is almost always justified for a buffer is easy to understand qualitatively. Imagine a solution containing both acetic acid and sodium acetate, each 0.10 *F*. Their reactions in water are

$$HOAc + H_2O \rightleftharpoons OAc^- + H_3O^+ \qquad K_a = 1.8 \times 10^{-5} \tag{14-31}$$
$$OAc^- + H_2O \rightleftharpoons HOAc + OH^- \qquad K_b = 5.5 \times 10^{-10} \tag{14-32}$$

From the earlier discussion and examples (see especially Example 14-4), we know that a weak acid with $K_a \approx 10^{-5}$ in a 0.1 *F* solution is only about 1 percent dissociated even in the absence of a large added quantity of its conjugate base. Thus in such a solution the molarity of the acid is for our purposes the same as its formality. If an appreciable quantity of the conjugate base (here acetate ion) is added to the solution, it suppresses the dissociation indicated by (14-31) for reasons under-standable in terms of the equilibrium law or of Le Châtelier's principle; hence the approximation molarity $\simeq$ formality becomes even better. In the same way, the hydrolysis of acetate ion represented by (14-32) is less than 0.01 percent in a 0.1 F acetate solution even in the absence of added acetic acid; in a buffer containing added HOAc an even smaller fraction of the OAc$^-$ reacts. Here again the approxi-mation that the molarity of the acetate is the same as its formality is an excellent one. These same general arguments apply to any buffer for which the formalities of X$^-$ and HX are each appreciably greater than the concentration of OH$^-$ and H$^+$, which is true of most buffers.

EXAMPLE 14-9   **Comparison of an acidic buffer with a solution of a strong acid at the same pH.** Suppose that you have a large volume of each of the following solutions:

1.   Water brought to pH = 5.0 by addition of strong acid (e.g., to 10 liters of water, add 0.10 ml of 1.0 F HCl).

2.   A buffer solution prepared by adding 1.0 gfw HX ($K_a = 1.0 \times 10^{-5}$) and 1.0 gfw NaX to several liters of water and diluting to 10 liters. Thus this buffer is 0.10 F in HX and 0.10 F in X$^-$, and

$$[H^+] = K_a \frac{[HA]}{[A^-]} \simeq 1.0 \times 10^{-5} \left(\tfrac{0.10}{0.10}\right) = 1.0 \times 10^{-5} \, M$$

Compare the effects of adding to separate 1.0-liter portions of each of these solutions

(*a*)   50 ml of 1.0 F HCl (that is, 0.05 mol H$^+$)

(*b*)   50 ml of 1.0 F NaOH (that is, 0.05 mol OH$^-$)

Solution.
1.   *Unbuffered solution:* At the start the strong-acid solution at pH = 5.0 contains 1.0×10^{-5} mol H$^+$ liter^{-1} and no other acid.

(*a*)   Addition of the 0.05 mol H$^+$ in 50 ml of 1.0 F HCl to the 1.0 liter of solution gives a total of 0.05 mol H$^+$ in 1.05 liters, so that $[H^+] = 5 \times 10^{-2} \, M$ and pH = 2.0 − 0.7 = 1.3, a change of 3.7 pH units.

(*b*)   Addition of the 0.05 mol OH$^-$ in 50 ml of 1.0 F NaOH neutralizes the minute quantity of H$^+$ initially present (10^{-5} mol) and since there is no other acid available leaves an excess of essentially all the OH$^-$. Hence $[OH^-] = \tfrac{0.05}{1.05} = 5 \times 10^{-2} \, M$, pOH = 1.3, and pH = 12.7, a change of 7.7 units.

2.   *Buffered solution:* Addition of H$^+$ to the buffer results in the conversion of X$^-$ into HX because X$^-$ is a sufficiently strong base (HX a sufficiently weak acid) that no large concentrations of both H$^+$ and X$^-$ can coexist in water. Similarly, addition of OH$^-$ to the buffer causes conversion of HX to X$^-$ because no appreciable concen-trations of both HX and OH$^-$ can coexist. As long as large quantities of both X$^-$ and HX are present in the buffer, almost all the added H$^+$ or OH$^-$ is used up in this way.

(*a*)   When 0.05 mol H^+ is added, it changes 0.05 mol X^- into HX, so we have

$$\text{Mol HX in solution} \approx 0.10 + 0.05 = 0.15$$
$$\text{Mol } X^- \text{ in solution} \approx 0.10 - 0.05 = 0.05$$

The total volume has increased slightly, by 50 ml (0.05 liter). This has no significant effect on the resulting pH, however, since it is the *ratio* of the concentrations of X^- and HX that is important [see (14-29)] and in the calculation of this ratio from the corresponding numbers of moles of each member of the conjugate pair, the volume occurs in both the numerator and the denominator. Thus

$$\frac{[X^-]}{[HX]} \approx \frac{\frac{0.05}{1.05}}{\frac{0.15}{1.05}} = \frac{0.05}{0.15} = \frac{1}{3} \tag{14-33}$$

and by (14-29),

$$\frac{K_a}{[H^+]} = \frac{[X^-]}{[HX]} \approx \frac{1}{3} \tag{14-34}$$

or
$$[H^+] = 3K_a$$

and
$$pH = pK_a - \log 3 = 5.0 - 0.5 = 4.5$$

The addition of acid has indeed lowered the pH, as is reasonable, but it has lowered it only 0.5 pH units instead of 3.7, as it did for the unbuffered solution.

(*b*)   The addition of 0.05 mol OH^- to a portion of the original buffer at pH = 5.0 may be analyzed in a parallel fashion. It results in the conversion of 0.05 mol HX into 0.05 mol X^-. Thus, after the addition we have

$$\text{Mol } X^- \text{ in solution} \approx 0.10 + 0.05 = 0.15$$
$$\text{Mol HX in solution} \approx 0.10 - 0.05 = 0.05$$

and the ratio of the molar concentration of X^- to that of HX is 3. Hence,

$$\frac{K_a}{[H^+]} = \frac{[X^-]}{[HX]} \approx 3 \tag{14-35}$$

Consequently pH = pK + log 3 = 5.0 + 0.5 = 5.5. The addition of 0.05 mol OH^- has, as expected, made the solution more basic, but the *change* is only 0.5 pH units as contrasted with the 7.7-unit change for the unbuffered solution.

This example shows that even a relatively dilute buffer resists pH change considerably more than does a solution of a strong acid at the same pH.

The two fundamental characteristics of a buffer are its pH and its capacity to resist pH change. A solution can be buffered at almost any pH in the range from about 3 to about 11 by appropriate choice of the conjugate pair to be used and of the ratio of their concentrations. The pair selected must be one with a pK_a near the desired pH, preferably within 0.5 units. The ratio of the concentrations of the two forms of the conjugate pair is chosen in accord with (14-30) to give the desired pH.

The capacity of a buffer to resist pH change depends solely on the amounts of the two conjugate forms present in the solution. The larger these amounts, the smaller is the change in their ratio and hence in the ratio of their concentrations, for a given addition of acid or base, and by (14-30) the smaller is the change in $[H^+]$ and thus in pH.

EXAMPLE 14-10

Selection of components of a buffer. Suppose that you need to prepare a buffer at pH = 9.0. Select several different acid-base pairs from Table 14-2 that would be appropriate and describe the composition of one such buffer.

Solution. From (14-30) we see that, since we want the concentrations of both X^- and HX to be appreciable, the most appropriate acid to choose must have pK within about 0.5 units of 9.0; that is, K_a should fall in the range between 3×10^{-9} and 3×10^{-10}. Two acids in Table 14-2 do fall in this range, H_3BO_3 ($K_a = 6.4 \times 10^{-10}$) and NH_4^+ ($K_a = 5.5 \times 10^{-10}$); HCN ($K_a = 2 \times 10^{-9}$) is just outside it and could be used.

If an ammonia–ammonium ion buffer were used, the ratio of the concentrations of the conjugate forms should be

$$\frac{[NH_3]}{[NH_4^+]} = \frac{K_a}{[H^+]} = \frac{5.5 \times 10^{-10}}{1.0 \times 10^{-9}} = \frac{5}{9}$$

Any solution in which the ratio of the concentration of ammonium ion to ammonia was $9:5$ would suffice—for example, $0.18\ F$ ammonium chloride and $0.10\ F$ ammonia, or $0.9\ F$ ammonium nitrate and $0.5\ F$ ammonia. The latter would have a higher capacity for a given volume.

Boric acid and one of its salts, or even HCN and NaCN, could also be used, with different relative proportions, chosen in accord with (14-30).

The following additional points about buffers are worth remembering:

1. As long as the concentrations of the conjugate acid and base are much greater than $[OH^-]$ or $[H^+]$, the pH of the buffer can be calculated from the ratios of the numbers of gram formula weights of acid and base present in the solution, without need to calculate the concentrations explicitly. Thus, if we let V be the volume of the buffer solution, we can multiply *both* the numerator and the denominator of the ratio of formalities c_a/c_b by V without affecting the value of this ratio:

$$\frac{K_a}{[H^+]} = \frac{[X^-]}{[HX]} \approx \frac{c_b}{c_a} = \frac{Vc_b}{Vc_a} = \frac{\text{gfw } X^- \text{ present}}{\text{gfw HX present}} \qquad (14\text{-}36)$$

2. The pH of a buffer is to a first approximation independent of its volume, at least for volume changes that are not excessive. Thus dilution will not markedly change the pH of a buffer unless the volume changes by several orders of magnitude.

3. A buffer that may be used for absorption of *either* acid or base is most effective when $c_a = c_b$, which implies pH = pK_a. Under these conditions the maximum amount of either H^+ or OH^- can be absorbed for a given change in the ratio $[X^-]/[HX]$ and thus for a given pH change.

14-5 Titration of Acids and Bases

Titration has been mentioned briefly in Sec. 7-3. It is a method for determining the amount or the concentration of a substance in solution by quantitative reaction with a solution of some second substance, called the titrant, that is gradually added to the first. The solutions used are usually liquid but may be gaseous; sometimes the titrant is generated in the first solution rather than being added to it externally. The titrant is added gradually until some signal from the reaction mixture indicates that the amount of reagent added is stoichiometrically equivalent (in the sense used in Sec. 7-3) to the amount of the first substance initially present. The signal may be a change in color of some added compound (called an *indicator*) so chosen that it changes color at or near the desired point in the reaction. Alternatively, the signal might be the appearance of a precipitate, an electrical indication of some kind, or any other event that occurs at a point in the reaction sufficiently well defined to be suitable for quantitative purposes.

Our present concern is with the titration in aqueous solutions of acids with bases, or vice versa, but some of the results are more widely applicable, both to acid-base reactions in nonaqueous media and to other kinds of reactions as well—e.g., oxidation-reduction reactions or the formation of a slightly soluble precipitate, as in the reaction of a solution of a silver salt with a soluble chloride.

During the titration of an acid with a base, or a base with an acid, there is a continual change of the acid into its conjugate base, or vice versa, and there is a simultaneous change in the pH of the solution. Before analyzing the general problem of the variation of pH as a function of the amount of base (or acid) added during a titration, we consider the variation in the proportions of the two members of a conjugate acid-base pair at different pH values.

Distribution of species for a monoprotic acid. It is useful to write expressions for the fraction of the total concentration of a given conjugate acid-base species in solution that is present in each separate form, HA and A^-. This is readily done with the help of the equilibrium expression for the acid-base pair in solution in the form

$$\frac{[A^-]}{[HA]} = \frac{K_a}{[H^+]} = 10^{pH}10^{-pK_a} = 10^{pH-pK_a} \tag{14-37}$$

We then have, for the fraction in the form HA,

$$\frac{[HA]}{[HA] + [A^-]} = \frac{[HA]}{[HA]\,(1 + [A^-]/[HA])} = \frac{1}{1 + [A^-]/[HA]} = \frac{1}{1 + 10^{pH-pK_a}} \tag{14-38}$$

and similarly, for the fraction in the form A^-,

$$\frac{[A^-]}{[HA] + [A^-]} = \frac{1}{[HA]/[A^-] + 1} = \frac{1}{1 + 10^{pK_a-pH}} \tag{14-39}$$

The sum of the fractions (14-38) and (14-39) is unity, and these fractions can thus be represented by a single curve as shown in Fig. 14-1. Comparison of (14-38) and (14-39) shows that the two expressions are symmetric about the point $pH = pK_a$, or $pH - pK_a = 0$, at which point each fraction is 0.5 because the concentrations of A^- and HA are equal. The fractions (14-38) and (14-39) depend only on the difference between the pH and the pK_a for the acid of interest. When the fractions are expressed as a function of $pH - pK_a$, as here, the expressions are completely general; that is, they apply to any monoprotic acid. The corresponding curves for different acids are therefore all identical when plotted as a function of $pH - pK_a$. When the fractions are expressed as a function of pH, rather than $pH - pK_a$, each curve is centered at the pK_a of the corresponding acid. A set of such curves is shown in Fig. 14-2. Curves of the general shape illustrated in Figs. 14-1 and 14-2 are said to be sigmoidal because of their resemblance to the letter S (Greek: *sigma*, S).

Titration curves: the change in pH during titration. A knowledge of the pH at different points during a titration is valuable because it makes possible the choice of an appropriate indicator for signaling the conclusion of the titration, provides insight into the precision with which the titration can be done, and helps to explain a simple method for measuring K_a (or K_b) for a given weak acid–weak base conjugate pair. In this section, we illustrate the calculation of the pH at representative points during a titration and outline the shapes of typical titration curves, i.e., graphs of pH against the fraction of added reagent needed to reach equivalence.

For simplicity we assume in all the following calculations that there is no volume change during titration. The titrant may, for example, be assumed to be generated directly in the titration vessel with no volume change, and this is in fact sometimes possible. However, even when titrant is added from a buret, the volume change

FIGURE 14-1   **Fractions of HA and of A^- as a function of $pH - pK_a$.** The curve represents the fraction of HA + A^- that is HA if the left-hand scale is used and the fraction that is A^- if the right-hand scale is used. The curve is symmetric about its midpoint, which corresponds to $pH = pK_a$ and to equal concentrations of HA and A^-, so that each fraction is 0.5. The change from about 99 percent HA to about 1 percent HA takes place within a pH range of four units, and the largest part of this change, from 90 to 10 percent, occurs within about two pH units.

To obtain the fractions of HA and A^- at a given pH, a vertical line is drawn with the pH value as the abscissa. The portion of the line that is in the field labeled HA represents the fraction that is HA; the portion in the field labeled A^- represents the fraction that is A^-. Thus, the dashed line indicates a distribution of 76 percent HA and 24 percent A^- at a $pH - pK_a$ of -0.50.

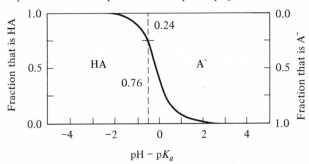

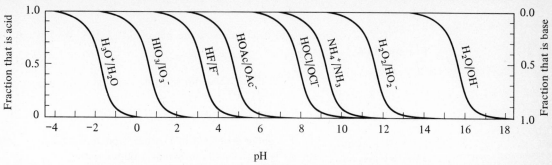

FIGURE 14-2 **Distribution of species for several conjugate acids and bases.** The figure represents a superposition of distribution curves for several separate conjugate acid-base pairs. The curves are of the kind shown in Fig. 14-1, each moved to the proper pK_a value. The abscissa here is the actual pH, while in Fig. 14-1 it is the pH relative to pK_a. The curves at the extreme left and extreme right apply to water itself, that is, to the pairs H_3O^+/H_2O and H_2O/OH^-. This and similar figures apply also to mixtures of different conjugate pairs that might be present together at any specified pH. Since the different pairs are independent, each curve shows the distribution between acidic and basic forms for the given pair.

during titration is almost always sufficiently small that the resulting dilution does not change the characteristic features of the titration curves. At most it only shifts the pH by a few tenths of a unit over part of the curve but does not change the shape of the curve significantly.

First, suppose that we titrate a monoprotic acid HA with a base. The acid may be weak or strong; the base is assumed to be strong, because this is an experimental advantage, as will be seen. During the course of the titration, the added base changes HA into A^- (unless the acid is so strong that all HA is dissociated to begin with) and H_3O^+ changes into H_2O. We assume that the original solution contains only HA and A^-; the total of HA + A^- remains constant and represents the equivalents of acid present initially. For convenience we define the degree of titration τ as

$$\tau = \frac{\text{equivalents of base added}}{\text{equivalents of acid present initially}} \tag{14-40}$$

At the beginning of the titration, τ is 0. At the point at which τ is exactly 1, the amount of base added is equivalent to the total of HA + A^-. This is called the equivalence point, and the titration should be stopped as close to this value as possible. We will calculate the pH at the start (before any base is added), at the equivalence point, and at several other points and plot the pH as a function of τ (Fig. 14-3).

EXAMPLE 14-11 **Titration curve for a weak acid and a strong base.** Consider the titration of a 0.10 F solution of a weak acid HA, $K_a = 1.0 \times 10^{-5}$, with a strong base. Assume no volume changes and calculate the pH at the following values of τ: 0.00, 0.09, 0.50, 0.91, 0.99, 1.00, 1.01, 1.10.

Solution.

$\tau = 0.00$. This corresponds to calculation of the pH of $0.10\ F$ HA. We have, as in Example 14-4,

$$\frac{[H^+]^2}{0.10 - [H^+]} = 1.0 \times 10^{-5}$$

which leads to $[H^+] = 1.0 \times 10^{-3}\ M$ and pH = $\underline{3.0}$.

$\tau = 0.09$.   At this value of τ, about 9 percent of HA has been converted to A^- and about 91 percent remains as HA. The solution is thus a buffer, with

$$\frac{K_a}{[H^+]} = \frac{[A^-]}{[HA]} \approx \frac{9}{91} = 0.10$$

whence $\log [H^+] = \log 10 + \log K_a$,   or $-\log [H^+] = -1.0 - \log K_a$.   Thus pH $\approx -1.0 + pK_a = \underline{4.0}$.

$\tau = 0.50$.   Here $[A^-] \approx [HA]$, so pH $\approx pK_a = \underline{5.0}$.

$\tau = 0.91$.   At this point in the titration,

$$\frac{[A^-]}{[HA]} \approx \frac{91}{9} = 10$$

so $[H^+] \approx 0.10\ K_a$ and pH $\approx 1.0 + pK_a = \underline{6.0}$.

$\tau = 0.99$.   Here it is approximately true that

$$\frac{[A^-]}{[HA]} = \frac{99}{1} = 10^2$$

so pH $\approx pK_a + 2.0 = \underline{7.0}$.

$\tau = 1.00$.   This is the equivalence point, and a little reflection about the nature of the species present at this juncture simplifies the calculation of the pH. In the titration of $0.10\ F$ HA with NaOH, the solution at the equivalence point is indistinguishable from a solution of NaA at an equivalent concentration, $0.10\ F$, if we ignore volume changes. In such a solution the very weak base A^- reacts with water to form equal numbers of OH^- ions and HA molecules:

$$A^- + H_2O \rightleftharpoons HA + OH^- \tag{14-41}$$

$$0.10 - x \qquad\qquad x \qquad x$$

If the equilibrium represented by (14-41) is approached from the opposite direction by mixing OH^- and HA in equivalent amounts, the position of equilibrium is necessarily identical. Thus the reverse of (14-41), the titration reaction, does not go quite to completion even at the equivalence point—a little OH^- and HA remain in the solution—precisely because A^- is so weak a base that it hydrolyzes perceptibly in water.

Calculation of the pH at the equivalence point is now simply a matter of calculating the pH of a solution of a weak base A^- at the appropriate formality and with $K_b = K_w/K_a = 1.0 \times 10^{-14}/1.0 \times 10^{-5} = 1.0 \times 10^{-9}$. The procedure is identical to that illustrated in Example 14-7. From (14-41) we have $[OH^-] = [HA] = x$ and $[A^-] = 0.10 - x$. Hence,

$$\frac{x^2}{0.10 - x} = 1.0 \times 10^{-9}$$

which leads to $x^2 \approx 1.0 \times 10^{-10}$ and $x \approx 1.0 \times 10^{-5}$. Thus pOH ≈ 5.0 and pH $\approx \underline{9.0}$.

$\tau = 1.01$.   The solution now contains $0.10\ F\ A^-$ and a 1 percent excess of OH^-, which means that $[OH^-] \approx 1.0 \times 10^{-3}$. Thus pOH ≈ 3.0 and pH $\approx \underline{11.0}$.

$\tau = 1.10$.   Here there is a 10 percent excess of OH^- in a $0.10\ F$ solution of A^-. Thus $[OH^-] \approx 1.0 \times 10^{-2}$, pOH ≈ 2.0, and pH $\approx \underline{12.0}$.

These results are plotted in Fig. 14-3. Titration curves for acids of different strengths, as well as for bases, can be derived in a similar manner. A family of curves that is quite generally applicable to the titration of $0.10\ F$ solutions of weak acids with strong bases, or weak bases with strong acids, is presented in Fig. 14-4. The precision with which the equivalence point can be found is directly related to the magnitude of the change of pH (with volume of titrant added) in the vicinity of

FIGURE 14-3   **Titration curve for a weak acid ($K_a = 10^{-5}$) with a strong base.**   The initial concentration of the acid is $0.10\ F$; volume changes during titration are ignored. The eight points calculated in Example 14-11 lie on this curve. The titration starts at $\tau = 0$, with no base added as yet. As base is added, τ increases. At $\tau = 1$ the equivalence point has been reached, and values of τ larger than 1 correspond to an excess of base.

The rate of pH change is greatest in the region of the equivalence point; it changes by four units between $\tau = 0.99$ and $\tau = 1.01$. The broad, rather flat region of the curve between $\tau \approx 0.1$ and $\tau \approx 0.9$ is called the buffer region because in this region of composition, characteristic of buffer solutions, the pH changes only slowly as strong base is added. The pH at the midpoint of this region ($\tau = 0.5$) is equal to pK_a of the acid, just as in Figs. 14-1 and 14-2 (where pH is plotted horizontally instead of vertically). The present curve is closely related to those in Figs. 14-1 and 14-2.

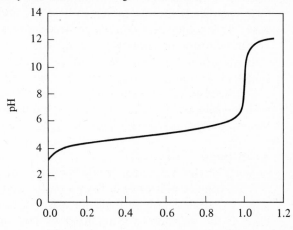

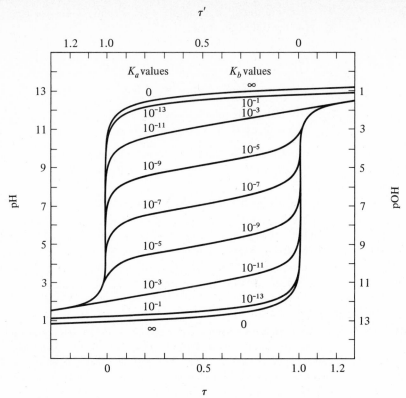

FIGURE 14-4   **Family of titration curves for 0.10 F solutions.**   The curves apply to the titration of an acid with dissociation constant K_a with strong base (bottom scale, τ) as well as to the titration of a base with dissociation constant $K_b = K_w/K_a$ with strong acid (top scale, τ').

the equivalence point ($\tau = 1.0$). Figure 14-4 shows that if K_a is significantly smaller than 10^{-7}, the substance can be titrated more precisely by first converting the acid form completely into its conjugate base (by adding a known quantity of a strong base). This conjugate form, which necessarily has K_b greater than 10^{-7}, can then be titrated quite precisely with strong acid.

It is worth noting that when one speaks of "neutralizing" an acid with a base, it usually means adding a quantity of the base just equivalent to that of the acid. The pH of the resulting solution will *not* be neutral—that is, the pH will *not* be 7—unless both the acid and the base are strong (or unless they are both weak and it happens that K_a for the acid is just equal to K_b for the base).

Acid-base indicators.   An acid-base indicator is a substance that has the property of changing color within a certain pH range when acid or base is added. It is a substance that gains or loses a proton and changes color when doing so. This implies that an acid-base indicator exists in two forms, an acid and its conjugate base. If the indicator is symbolized by HIn and its conjugate base by In$^-$, the equilibrium between the two forms,

$$HIn + H_2O \rightleftharpoons In^- + H_3O^+ \tag{14-42}$$

is governed by the equilibrium expression

$$\frac{[H^+][In^-]}{[HIn]} = K_i \tag{14-43}$$

with K_i called the indicator constant.

A good example of an indicator is methyl orange, the indicator constant of which is $K_i = 3 \times 10^{-4}$ ($pK_i = 3.5$). It is intensely red in the acid form and intensely yellow in the basic form; HIn is red, In^- yellow. Equation (14-43) may be rearranged,

$$\frac{[In^-]}{[HIn]} = \frac{[yellow]}{[red]} = \frac{K_i}{[H^+]} \tag{14-44}$$

to give the concentration ratio of yellow ions to red molecules as a function of $[H^+]$.

To show how the color of this indicator changes with pH, the concentration ratio of yellow to red species will be considered at three pH values, $pK_i + 1$, $pK_i - 1$, and pK_i. (1) When $pH = pK_i + 1$ or $[H^+] = K_i/10$, it follows from (14-44) that $[yellow]/[red] = 10$. Because the color intensities of HIn and In^- are comparable, the tenfold numerical superiority of yellow ions makes the solution yellow. For more basic solutions the ratio of yellow ions to red molecules exceeds 10. (2) When $pH = pK_i - 1$ or $[H^+] = 10K_i$, it follows that $[yellow]/[red] = \frac{1}{10}$, so that the solution is red. In more acid solutions the red molecules are favored even more. (3) At $pH = pK_i$ or $[H^+] = K_i$, the concentrations of yellow and red species are equal, $[yellow] = [red]$; the indicator is at a transition color of orange, the exact hue depending on the color intensities of the yellow and the red species. Thus as the pH decreases, methyl orange undergoes a transition from yellow through orange to red and we expect this to occur in a pH interval of no more than about two units, centered approximately at pK_i. The actual visually observed transition occurs in the pH interval 3.1 to 4.4, while $pK_i = 3.5$ (Table 14-3), in reasonable agreement with the semiquantitative considerations just given.

For some indicators, one of the two forms is colorless and the foregoing analysis

TABLE 14-3 **Some common acid-base indicators and their transition ranges**

Indicator	Color on acidic side	Color on basic side	Approximate transition range, pH units	pK_i
Bromphenol blue	Yellow	Lavender	3.0–4.6	3.8
Methyl orange	Red	Yellow	3.1–4.4	3.5
Bromcresol green	Yellow	Blue	3.8–5.4	4.7
Methyl red	Red	Yellow	4.2–6.1	5.0
Bromthymol blue	Yellow	Blue	6.0–7.6	7.1
Phenol red	Yellow	Red	6.8–8.1	7.8
Thymol blue	Yellow	Blue	8.0–9.6	8.9
Phenolphthalein	Colorless	Red	8.0–9.3	9.3
Thymolphthalein	Colorless	Blue	9.4–10.5	9.7

has to be modified. For example, the acid form of phenolphthalein ($pK_i = 9.3$) is colorless; the basic form is intensely red. Inspection of an equation analogous to (14-44) shows that there is a tenfold increase in red In^- ions for an increase of the pH by one unit. Experimentally it is found that phenolphthalein changes from colorless to red over the approximate pH interval 8.0 to 9.3 (Table 14-3).

To be useful in an aqueous medium, an indicator must have pK_i within the range about 2 to 12. Furthermore, the concentration of the indicator in the solution being titrated must be very small; otherwise the indicator will react with a significant amount of the acid or base being used and will interfere with the accuracy of the titration. Thus the color of at least one of the two forms of the indicator must be intense, so that the color is visible even when the indicator is highly diluted.

14-6   Polyprotic Acids

A number of common and important acids have more than one dissociable proton. These include carbonic acid, H_2CO_3, phosphoric acid, H_3PO_4, sulfuric acid, H_2SO_4, and oxalic acid, $(COOH)_2$. Some highly charged cations—for example, $Al(H_2O)_6^{3+}$ and $Fe(H_2O)_6^{3+}$—also behave as polyprotic acids in water. The successive reactions and corresponding equilibrium expressions for the three-stage dissociation of phosphoric acid were given in (14-4) to (14-6). The acid constants for the successive dissociation steps of a number of polyprotic acids are listed in Tables 14-2 and D-3.

No new principles need be learned to understand the equilibria and reactions of polyprotic acids in water. The equilibria seem more complicated and formidable than for a monoprotic acid because the successive stages of dissociation lead to more species (for example, H_3A, H_2A^-, HA^{2-}, A^{3-}) and, of course, all equilibria must be satisfied simultaneously. In fact, however, the consideration of solutions of many polyprotic acids can be greatly simplified because at most only two of the successive species can be present in appreciable concentration in any solution. Dealing with these is then not much more difficult than analyzing a situation involving a monoprotic acid.

We discuss first the distribution of species in a diprotic acid as a function of pH and then consider pH calculations.

Distribution of species as a function of pH.   As an example we choose solutions containing the species H_2CO_3, HCO_3^-, and CO_3^{2-}. The simplest approach is to express the concentration of each form in terms of any one of them—we shall choose HCO_3^-—and then find the fraction of each by dividing by their sum. We then have[1]

$$[H_2CO_3] = \frac{[HCO_3^-][H^+]}{K_1} \qquad (14\text{-}45)$$

[1] The designation $[H_2CO_3]$ here and in Fig. 14-5 refers to the total concentration of the species of formulas CO_2 and H_2CO_3, both presumably hydrated. Most of it is actually CO_2. See also footnote § in Table 14-2.

$$[CO_3^{2-}] = \frac{[HCO_3^-]K_2}{[H^+]} \qquad (14\text{-}46)$$

with the values 4.4×10^{-7} and 4.8×10^{-11} for K_1 and K_2. If the total carbonate in all different forms is denoted by c,

$$c = [H_2CO_3] + [HCO_3^-] + [CO_3^{2-}] \qquad (14\text{-}47)$$

it follows that

$$c = [HCO_3^-]\left(\frac{[H^+]}{K_1} + 1 + \frac{K_2}{[H^+]}\right) \qquad (14\text{-}48)$$

The relative proportions of the species in solution are thus

$$[H_2CO_3]:[HCO_3^-]:[CO_3^{2-}] = \frac{[H^+]}{K_1}:1:\frac{K_2}{[H^+]} \qquad (14\text{-}49)$$

Consideration of the right-hand side of (14-49), which of course applies to any diprotic acid, shows why only two of the forms can be present in significant concentration whenever K_1 and K_2 differ by 10^3 or more, as they do for most common diprotic acids. Imagine, for example, that $pH = pK_1$, so that $[H^+]/K_1 = 1$. The solution then contains equal concentrations of H_2CO_3 and HCO_3^-. Since K_2 is smaller than K_1 by a factor of 10^4, $K_2/[H^+] = 10^{-4}$ and the relative proportions in (14-49) are $1:1:10^{-4}$. The fraction of CO_3^{2-}, $[CO_3^{2-}]/c$, is negligible and $[H_2CO_3]/c$ and $[HCO_3^-]/c$, the fractions of H_2CO_3 and HCO_3^-, are both 0.5. If the pH is still lower (the solution more acidic), the proportions favor H_2CO_3 even more and the proportion of CO_3^{2-} is even smaller. On the other hand, when $[H^+] = K_2$ the proportions are $10^{-4}:1:1$. That is, the concentration of H_2CO_3 is negligible, and it is even more so at higher pH. When $pH = \frac{1}{2}(pK_1 + pK_2) \approx pK_1 + 2 \approx pK_2 - 2$, the proportions are about $10^{-2}:1:10^{-2}$; the solution contains about 98 percent of bicarbonate ion and only 1 percent each of H_2CO_3 and CO_3^{2-}. At no pH are more than two of the forms present in appreciable concentrations, because K_1 and K_2 differ so greatly. This conclusion is illustrated in a graphic representation of the fractional distribution of species as a function of pH (Fig. 14-5). These fractions are calculated from (14-45), (14-46), and (14-48). The sum of the fractions at any pH is, of course, 1.

A similar analysis can be applied to a typical triprotic acid such as H_3PO_4. Since the successive K_a values for this acid differ by about 10^5, it is again true that only two (neighboring) species can be present in appreciable concentrations at the same time.

pH calculations.   It turns out to be relatively straightforward to calculate the pH of any solution containing species related to a polyprotic acid with dissociation constants that differ by 10^3 or more. We illustrate this for an acid H_2A.

1.   A solution containing appreciable quantities only of H_2A itself is almost entirely governed by the reaction $H_2A \rightleftharpoons HA^- + H^+$, with the equilibrium determined then by K_1, unless the acid is exceptionally strong. Even for $0.1\ F$ sulfuric acid, for

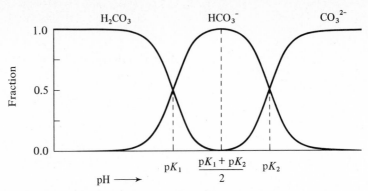

FIGURE 14-5   **Fractions of H_2CO_3, HCO_3^-, and CO_3^{2-} as a function of pH.** Each species is represented by a distinct curve, as indicated, in contrast to the graphs for monoprotic acids in Figs. 14-1 and 14-2. The sigmoidal curve on the left represents the fraction of H_2CO_3 and that on the right the fraction of CO_3^{2-}. The central bell-shaped curve represents the fraction of HCO_3^-, which rises to nearly 1.0 in the pH range about 8 to 9 and diminishes at higher and at lower pH.

The crossing points correspond respectively to pK_1 (on the left) when $[H_2CO_3] = [HCO_3^-]$ and to pK_2 (on the right) when $[HCO_3^-] = [CO_3^{2-}]$; the maximum of $[HCO_3^-]$ is at $(pK_1 + pK_2)/2$.

The graph shows that HCO_3^- and CO_3^{2-} cannot coexist with $[H^+]$ at appreciable concentrations; even if $[H^+]$ is as low as 10^{-4} (pH = 4), essentially all carbonate exists as H_2CO_3. Similarly, HCO_3^- and H_2CO_3 cannot coexist with OH^- in significant concentrations; when $[OH^-]$ is 10^{-2} or higher, essentially all carbonate species are in the form of CO_3^{2-}.

which K_1 is about 10^3, the contribution of the further dissociation of HSO_4^- is only about 10 percent; for acids with smaller values of K_1 (or for more concentrated sulfuric acid solutions), the contribution of the second dissociation is negligible. The calculation can then be handled exactly as for a monoprotic acid.

2.   A solution containing only the salt Na_2A (or any other salt of A^{2-} with a nonhydrolyzing cation) has a pH dominated by the reaction of A^{2-} as a base with water:

$$A^{2-} + H_2O \rightleftharpoons HA^- + OH^- \tag{14-50}$$

with $K_b = K_w/K_2$. Further reaction of HA^- with water contributes negligibly.

3.   Any solution that contains appreciable concentrations of *two* of the neighboring forms (H_2A and HA^-, or HA^- and A^{2-}) will be a buffer if both the acidic form and its conjugate base are weak. Formalities of the acid and base may be substituted for molarities in pH calculations with equilibrium expressions involving K_1 and K_2—provided, as before, that these formalities are much greater than both $[H^+]$ and $[OH^-]$.

4.   Finally, the only remaining possibility is a solution prepared from or corresponding to a salt of HA^-, for example, $NaHA$. The calculation for such a solution is somewhat lengthy, but the result is simple: $[H^+] \approx \sqrt{K_1 K_2}$, or pH $\approx$

0.5 ($pK_1 + pK_2$), provided that $[HA^-] \gg K_1$ and $K_2 [HA^-] \gg K_w$. This result does not depend on the formality of the solution.

This analysis of solutions prepared from a diprotic acid H_2A can be extended to solutions of acids with three or more protons, provided that the consecutive acid constants differ by three or more powers of 10. For example, the pH of a NaH_2PO_4 solution is 0.5 ($pK_1 + pK_2$), where K_1 and K_2 are the first two acid constants of H_3PO_4, and that of a Na_2HPO_4 solution of $0.5(pK_2 + pK_3)$. These pH values are also those of the two equivalence points in a titration of H_3PO_4 or of PO_4^{3-}.

Buffer solutions involving polyprotic acids and their salts are common, particularly in biological systems. The following example illustrates the preparation of a phosphate buffer and the calculation of its pH.

EXAMPLE 14-12   **pH of a phosphate solution.** What is the pH of a solution made by dissolving and mixing quantities of reagents that correspond to the following simultaneous formal concentrations of the resulting solution: 0.30 F NaOH, 0.25 F Na_2HPO_4, and 0.25 F H_3PO_4?

Solution. Remembering that at most two neighboring members of the series OH^-, PO_4^{3-}, HPO_4^{2-}, $H_2PO_4^-$, H_3PO_4, H^+ can coexist in significant concentrations, we begin by mentally reacting members at extreme ends of this sequence. Thus we use up as much of the OH^- as is needed to convert all H_3PO_4 to $H_2PO_4^-$. The resulting concentrations are $[H_3PO_4] \approx 0$, $[OH^-] = 0.05$, and $[H_2PO_4^-] = 0.25$. ($[Na^+]$ remains at $0.30 + 2(0.25) = 0.80$ throughout.) Next the remaining OH^- is imagined to react with the $H_2PO_4^-$, the result being $[OH^-] \approx 0$, $[H_2PO_4^-] = 0.20$, and $[HPO_4^{2-}] = 0.05 + 0.25 = 0.30$. These, then, are the major components. They lead to the hydrogen-ion concentration $[H^+] = K_2(\frac{0.20}{0.30}) = 0.67 \times 6.2 \times 10^{-8} = 4.1 \times 10^{-8}$ and pH = 7.4. The other concentrations in the solution are $[OH^-] = 1.0 \times 10^{-14}/4.1 \times 10^{-8} = 2.4 \times 10^{-7}$; $[H_3PO_4] = [H^+][H_2PO_4^-]/K_1 = 0.20 \times 4.1 \times 10^{-8}/7.1 \times 10^{-3} = 1.2 \times 10^{-6}$; and $[PO_4^{3-}] = K_3 [HPO_4^{2-}]/[H^+] = 0.30 \times 4.4 \times 10^{-13}/4.1 \times 10^{-8} = 3.2 \times 10^{-6}$. All these concentrations are seen to be small compared to those of $H_2PO_4^-$ and HPO_4^{2-}.

Hydrated species. Hydrated metal ions (Sec. 7-5) sometimes act as acids in water, and their conjugate bases—metal ions associated with one or more hydroxide ions, as well as with some water molecules perhaps—often show basic properties. The higher the charge and the smaller the size of the ion, the greater its acid strength when hydrated. For example, $Al(H_2O)_6^{3+}$, with $K_1 \approx 10^{-5}$, is about as strong an acid as acetic acid. This is primarily a consequence of the high electric field of the cation, which polarizes the attached water molecules sufficiently to make dissociation of protons from them far easier than from ordinary water molecules. The hydrated Al^{3+} ion is, in fact, a polyprotic acid, losing one proton from each water molecule successively to form a series of products:[1]

---

[1]As mentioned in Chap. 7, the water molecules are frequently omitted from the formulas for hydrated species; thus $Al(H_2O)_6^{3+}$ is abbreviated as Al^{3+} and $Al(OH)_4(H_2O)_2^-$ is written as $Al(OH)_4^-$. In the present context of conjugate acid-base pairs, the water molecules are of course crucial, being the source of the protons.

$$
\begin{array}{cc}
\textit{Acid} & \textit{Base} \\
[Al(H_2O)_6]^{3+} & \rightleftharpoons [AlOH(H_2O)_5]^{2+} + H^+ \\
[AlOH(H_2O)_5]^{2+} & \rightleftharpoons [Al(OH)_2(H_2O)_4]^+ + H^+ \\
[Al(OH)_2(H_2O)_4]^+ & \rightleftharpoons Al(OH)_3(H_2O)_3 + H^+ \\
Al(OH)_3(H_2O)_3 & \rightleftharpoons [Al(OH)_4(H_2O)_2]^- + H^+
\end{array}
\tag{14-51}
$$

Note that $Al(OH)_3(H_2O)_3$ is the only uncharged species in the series. It is rather insoluble, as is often true of uncharged coordination species containing OH groups. When acid is added to $Al(OH)_3(H_2O)_3$, the aluminum hydroxide dissolves with the formation of $Al(OH)_2(H_2O)_4^+$ and the other positively charged species in (14-51); one, two, or all three of the OH groups in the neutral species are converted to H_2O. It also dissolves when base is added, forming $Al(OH)_4(H_2O)_2^-$ and perhaps other negatively charged species. This is the amphoteric behavior of aluminum hydroxide mentioned in Sec. 7-4.

Many other metal ions form series of hydrates and hydroxides similar to those of aluminum. Some of the uncharged hydroxides are amphoteric, as are those of Zn(II), Sb(III), and Sn(II), while others are not. For example, the species $Ca(OH)_2(H_2O)_4$ is not amphoteric and forms a series with $[CaOH(H_2O)_5]^+$ and $[Ca(H_2O)_6]^{2+}$, for which the usual symbols are $Ca(OH)_2$, $[CaOH]^+$, and Ca^{2+}. Again the hydrated $Ca(OH)_2$ is only slightly soluble in water. It readily dissolves upon addition of acid, but not of base, because species like $[Ca(OH)_3(H_2O)_3]^-$ do not appear to form. The tendency of $Ca(H_2O)_6^{2+}$ to lose protons is so much weaker than that of $Al(H_2O)_6^{3+}$ that Ca^{2+} shows essentially no acidic reaction in water; in other words, $CaOH(H_2O)_5^+$ is a relatively strong base, far stronger than $AlOH(H_2O)_5^{2+}$.

Many oxygen acids and hydroxides tend to form large aggregates by what is sometimes called a condensation reaction in which water is eliminated and oxygen bridges are formed between different groups. An example was given in Sec. 7-6, Equation (7-36). Weak acids such as silicic acid, $Si(OH)_4$, and boric acid, $B(OH)_3$, have a strong tendency to form stable condensation products. Most silicate minerals, which constitute a large portion of the earth's crust—oxygen and silicon are the most abundant elements (by weight) on the earth's surface—represent condensation products of this kind. They are discussed further in Chap. 22.

Strong acids and bases tend to form less stable condensation products. For example, while the very strong acid $HClO_4$ may undergo the reaction

$$2HClO_4 \rightleftharpoons Cl_2O_7 + H_2O$$

the resulting chlorine heptoxide, Cl_2O_7, is highly unstable and explosive. The same is true of the very strong acid $HMnO_4$, which yields the explosive Mn_2O_7. Sulfuric acid forms pyrosulfuric acid, $H_2S_2O_7$, when strongly heated. Although this condensation compound is not explosive, it is unstable in aqueous solutions. Phosphoric acid, H_3PO_4, is a weaker acid. It forms pyrophosphoric acid, $H_4P_2O_7$, and higher condensation products such as $H_5P_3O_{10}$ and eventually $(HPO_3)_n$ when heated to $215°C$ and higher.

The tendency to form aggregates in solution is especially noteworthy for certain transition metals. The elements of column VIB—Cr and especially Mo and W—tend to form soluble polynuclear aggregates with specific compositions and structures.

On the other hand, $Fe(H_2O)_6^{3+}$, which is a stronger acid than $Al(H_2O)_6^{3+}$, with $pK_1 = 3.1$ and $pK_2 = 3.3$, forms a mixture of polynuclear aggregates in aqueous solution when the pH is above about 3. With increasing pH these aggregates grow so large that the solution becomes viscous and a gelatinous precipitate forms, with no regular structure or composition. This precipitate is often referred to as $Fe(OH)_3$, ferric hydroxide or hydrated ferric oxide.

Problems and Questions

14-1 **Degree of dissociation of a base.** Suppose that a 0.40 F solution of a weak base has the same hydroxide-ion concentration as a 0.020 F solution of sodium hydroxide. What is the degree of dissociation of the weak base in its solution? Explain. *Ans.* 0.050

14-2 **Acid constants.** (*a*) The acid HX is 4.5 percent dissociated in a 0.10 F solution. What is its acid constant? (*b*) The degree of dissociation of the acid HY in a 0.50 F solution is 0.15. What is its acid constant?

14-3 **Very weak acid.** The pH of a 0.10 F solution of the very weak acid HA is 5.0. (*a*) What is the acid constant for HA? (*b*) What is the degree of dissociation of HA in its 0.10 F solution? (*c*) What would be the formality of a solution of HA in which the degree of dissociation was twice that in a 0.10 F solution?
Ans. (*a*) $1.0 \times 10^{-9} M$; (*b*) 0.010; (*c*) 0.025 F

14-4 **Weak base.** A certain weak base B has a base constant of $6 \times 10^{-4} M$, corresponding to the reaction $B + H_2O \rightleftharpoons BH^+ + OH^-$. Calculate the pH of a 0.20 F aqueous solution of B.

14-5 **Base constant.** A 0.10 F solution of a certain base has a pH of 11.50. What is the base constant K_b of this base? *Ans.* $1.0 \times 10^{-4} M$

14-6 **Concentration and pH.** A certain solution of acetic acid ($K_a = 1.8 \times 10^{-5} M$) has a pH of 3.30. What is its formality?

14-7 **Relative strength of bases.** Arrange the following bases in the order of their tendency to combine with protons, putting that with the greatest proton affinity first: NH_3, Cl^-, OH^-, acetate ion, H_2O. *Ans.* OH^-, NH_3, OAc^-, H_2O, Cl^-

14-8 **Acid constants and structure.** Predict the successive dissociation constants of H_3AsO_4 and H_2SO_4. Compare them with the experimental values in Tables D-3 and 14-2.

14-9 **Dissociation of acetic acid.** Although acetic acid is normally regarded as a weak acid, it is about 34 percent dissociated in a $10^{-4} F$ solution at 25°C. It is less than 1 percent dissociated in 1 F solution. Discuss this variation in degree of dissociation with dilution in terms of Le Châtelier's principle, and explain how it is consistent with the supposed constancy of equilibrium constants.

14-10 **Ammonia solution.** A 10.0-ml sample of aqueous ammonia weighs 9.60 g and has a pH of 12.0. What is the percentage by weight of ammonia in this solution? ($K_b = 1.8 \times 10^{-5} M$)

14-11 **Acid constant of a weak acid.** Suppose that a 0.10 F aqueous solution of a monoprotic acid HX has just 11 times the conductivity of a 0.0010 F aqueous solution of HX. What is the approximate dissociation constant of HX? (*Hint:* In thinking about this problem, consider what the ratio of the conductivities would be if HX were a strong acid and if HX were extremely weak, as limiting cases.)   *Ans.* 4.5×10^{-5} M

14-12   Properties of an unknown acid.   A certain compound C has a formula weight of 120. A 0.100 F solution of C in water is acidic, conducts electricity moderately well, and freezes at $-0.25°C$. When 20.0 ml of the solution is titrated with base, 16.0 ml 0.250 N NaOH is required to reach an end point with phenolphthalein. Calculate the equivalent weight of C in acid-base reactions and its degree of dissociation in 0.100 F aqueous solution, assuming that one molecule of C dissociates to form only two ions in this solution. Why may this assumption be questionable?

14-13   Ammonium ion.   Discuss the justification for this statement: "Although one does not normally regard NH_4^+ as an acid, it is actually only slightly weaker as an acid than hydrocyanic acid, HCN, in aqueous solution."

14-14   Hydrolysis.   State clearly what information you would need in order to predict whether a given salt BX would give a neutral solution, an acidic solution, or a basic solution when dissolved in water.

14-15   Solution of NaCN.   Calculate the concentrations of Na^+, CN^-, HCN, H^+, and OH^- in a 0.20 F solution of sodium cyanide. Calculate also the degree of hydrolysis of CN^- in this solution. (K_a for HCN = 2×10^{-9} M.)   *Ans.* 0.20 M, 0.20 M, 0.001 M, 1×10^{-11} M, 0.001 M; 0.005

14-16   Comparison of three solutions.   Three flasks, labeled A, B, and C, contained aqueous solutions of the same pH. It was known that one of the solutions was 1.0×10^{-3} F in nitric acid, one was 6×10^{-3} F in formic acid, and one was 4×10^{-2} F in the salt formed by the weak organic base aniline with hydrochloric acid ($C_6H_5NH_3Cl$). (Formic acid is monoprotic.) (*a*) Describe a procedure for identifying the solutions. (*b*) Compare qualitatively (on the basis of the preceding information) the strengths of nitric and formic acids with each other and with the acid strength of the anilinium ion, $C_6H_5NH_3^+$. (*c*) Show how the information given may be used to derive values for K_a for formic acid and K_b for aniline. Derive these values.

14-17   Solution of Novocain.   Novocain, the commonly used local anaesthetic, is a weak base with $K_b = 7 \times 10^{-6}$ M. (*a*) If one had a 0.020 F solution of Novocain in water, what would be the approximate concentration of OH^- and the pH? (*b*) Suppose that you wanted to determine the concentration of Novocain in a solution that is about 0.020 F by titration with 0.020 N HCl. Calculate the expected pH at the equivalence point and select an appropriate indicator from Table 14-3.   *Ans.* (*a*) 4×10^{-4} M, 10.6;   (*b*) 5.4, methyl red

14-18   Solution of potassium benzoate.   Benzoic acid is a weak acid with K_a about 6×10^{-5} M. Explain with the aid of Le Châtelier's principle the effects of the following changes in conditions on the degree of hydrolysis of potassium benzoate in a dilute solution. Assume that activity effects are negligible. (*a*) Addition of KNO_3 to the solution; ignore any volume change. (*b*) Addition of NaOH to the solution; ignore any volume change. (*c*) Dilution of the solution with water.

14-19   Solution of sodium pivalate.   A 0.100 F solution of pivalic acid has a pH of 3.0. What would be the pH of 0.10 F sodium pivalate?   *Ans.* 9.0

14-20   Acetic acid–acetate buffer.   What volume of 1.0 F KOH must be added to 10 ml of 1.0 F HOAc to give a pH of 5.0? (K_a for HOAc = 1.8×10^{-5} M.)

14-21 Dimethylamine and its solutions. Dimethylamine, $(CH_3)_2NH$, is very soluble in water, forming a weakly basic solution similar to that formed by ammonia. (*a*) Write a reaction, analogous to that of ammonia with water, that can account for the basicity of a solution of dimethylamine in water. If the OH^- concentration of a 0.50 *F* solution of this compound is about 5×10^{-3} *M*, what is K_b for dimethylamine? (*b*) Suppose that 40.0 ml of a 0.40 *N* solution of dimethylamine in water is neutralized by 32.0 ml of a solution of a strong acid. What can you conclude about the normality and the formality of the acid solution? (*c*) [Assume answer from part (*a*).] What will be the pH of a 1.0 *F* solution of dimethylamine that has been 90 percent neutralized with a strong acid?

Ans. (*a*) 5×10^{-5} *M*; (*b*) 0.50 *N*; 0.50/*nF*, with *n* an integer; (*c*) 8.7

14-22 Formic acid and its solutions. The dissociation constant of formic acid, which is monoprotic, is 1.7×10^{-4} *M*. (*a*) What is the approximate degree of dissociation of formic acid in a 0.10 *F* solution? (*b*) What is the approximate pH of 0.10 *F* potassium formate? (*c*) What is the approximate pH of a solution prepared by adding 10 ml of 1.0 *N* HCl to 30 ml of 1.0 *F* sodium formate?

14-23 Buffer capacity: buffers of different concentration. Suppose you have a buffer at pH 5.0 that is 1.00 *F* in HX and 1.00 *F* in X$^-$, where HX is a weak acid with $K_a = 1.0 \times 10^{-5}$ *M*. Calculate the effects of adding 50 ml of 1.0 *F* HCl to a 1-liter portion of this buffer and 50 ml of 1.0 *F* NaOH to a separate 1-liter portion. Compare the results with those in Example 14-9 for the similar buffer with 0.1 *F* concentrations.

Ans. pH changes by -0.04 units, $+0.04$ units

14-24 Preparation of a buffer. Suppose that you want to prepare a buffer solution with pH = 4.0 and have available 2 liters of 1.0 *F* NaOH and 2 liters of 1.0 *F* phenylacetic acid, for which K_a is 5×10^{-5} *M*. Describe and explain how you could prepare at least 2 liters of the desired buffer, using only the solutions given. By how much would the pH of 1 liter of your buffer change if $\frac{1}{3}$ liter of 1.0 *N* KOH were added to it?

14-25 Ammonia-ammonium ion buffer. (*a*) Describe how you would prepare 1 liter of a buffer at pH = 9.0, using 1.0 mol NH_3 and as much of any strong acid as needed. (*b*) A 40-ml sample of a 0.10 *F* solution of nitric acid is added to 20 ml of 0.30 *F* aqueous ammonia. What is the pH of the resulting solution?

Ans. (*a*) dissolve 1.0 mol NH_3 and 0.64 equivalent strong acid in enough water to make total volume 1 liter; (*b*) 8.95

14-26 Arsenate buffer. Arsenic acid, H_3AsO_4, is very similar to phosphoric acid. It is a triprotic acid with $K_1 \approx 6 \times 10^{-3}$ *M*, $K_2 \approx 2 \times 10^{-7}$ *M*, and $K_3 \approx 3 \times 10^{-12}$ *M*. What is the ratio of $HAsO_4{}^{2-}$ to $H_2AsO_4{}^-$ in a solution at pH = 7.0? What volume of 0.10 *N* NaOH should be added to 100 ml of 0.10 *F* H_3AsO_4 to give a solution with pH = 7.0?

14-27 Preparation of a buffer. Imagine that you want to do physiological experiments at a pH of 6.0 and the organism with which you are working is sensitive to most available materials other than a certain weak acid, H_2Z, and its sodium salts. K_1 and K_2 for H_2Z are 3×10^{-2} *M* and 5×10^{-7} *M*. You have available 1.0 *F* aqueous H_2Z and 1.0 *N* NaOH. How much of the NaOH solution should be added to 1.0 liter of the acid solution to give a buffer at pH = 6.0?

Ans. $\frac{4}{3}$ liters

14-28 Ammonium acetate solution. Show that NH_4OAc is about 0.6 percent hydrolyzed in aqueous solution, that to a good approximation this degree of hydrolysis is independent of the concentration of the salt, and that the pH of the solution is close to 7.0. (*Hint:* Since OAc^- is a much stronger base than water, and NH_4^+ a much stronger acid than water, the only important reaction in the solution is the direct reaction $NH_4^+ + OAc^- \rightleftharpoons NH_3 + HOAc$.)

14-29 Exact neutralization. (*a*) How many milliliters of 0.1000 F NaOH must be added to 100.0 ml of 0.1000 F HClO to reach pH = 7.0? $K_a = 1.1 \times 10^{-8} M$. (*b*) Repeat the calculation for 0.1000 F solutions of acids with $K_a = 1.0 \times 10^{-6} M$ and $1.0 \times 10^{-4} M$.

Ans. (*a*) 10 ml; (*b*) 91 ml, 99.9 ml

14-30 Degree of dissociation. For many weak-acid solutions, it is sometimes said to be a good approximation that the degree of dissociation α is related to the acid constant K_a and the formality of the acid, c, by $\alpha = \sqrt{K/c}$. Show that in fact this relation is only valid (within 5 percent) if $K/c \leq 10^{-2}$, and that if K/c is greater than about 10^2, then $\alpha \approx 1 - c/K$.

14-31 Titration curve for pyridine. Pyridine is a very weak base with $K_b = 2 \times 10^{-9} M$. Like ammonia, it forms salts with strong acids. (*a*) Calculate the pH of a 0.20 F aqueous solution of pyridine. (*b*) Calculate the pH of a 0.10 F solution of the salt formed by the reaction of pyridine with HCl. (*c*) Using the results of (*a*) and (*b*), as well as calculations for at least two other points, sketch the titration curve for the titration of 0.20 F pyridine with 0.20 N HCl. (*d*) Give the approximate pK of an indicator appropriate for this titration. Suggest why in practice the titration would not be easy to perform very precisely. *Ans.* (*a*) 9.3; (*b*) 3.15

"If we think of the metal atom as the center of the whole system, then we can most simply place six molecules connected with it at the corners of an octahedron."

A. Werner, 1893

15 Ionic Equilibria in Aqueous Solutions: Ionic Solids and Complex Ions

15-1 Introduction

In this chapter we continue the discussion of important ionic equilibria in aqueous solutions begun in Chap. 14. The emphasis on ionic aqueous solutions in these chapters is a direct consequence of our environment. Many natural processes on this planet's surface, as well as many of the reactions that have been discovered and applied in the laboratory and in industry, involve ionic aqueous solutions.

We begin with a consideration of the equilibria between ionic solids and their solutions—equilibria that are important in the chemistry of both the ocean and "fresh" water, which (except for rain) invariably contains dissolved salts picked up as the water passes over rocks and clays. Ionic solubility equilibria also underlie several practical methods of analysis, a few of which are illustrated in examples below.

We continue then with a consideration of complex ions and their equilibria. Complex ions are of enormous significance in many areas of chemistry and bio-chemistry, from the fixing of photographic film and the solvent power of aqua regia for gold and platinum to the prevention of the yellowing of the leaves of certain plants grown in iron-poor soils and the proper functioning of thousands of biochem-ical catalysts (enzymes) in the cells of every organism.

In a later section we examine a number of heterogeneous equilibria in which several of the kinds of reactions considered individually in Chap. 14 and in this chapter take place simultaneously. The chapter ends with a brief discussion of activity effects, important in all but quite dilute solutions.

15-2 Equilibria between Ionic Solids and Aqueous Solutions: The Solubility-Product Principle

In formulating the reaction quotient for the equilibrium between a pure ionic solid and its aqueous solution, the solid is conventionally omitted (Sec. 13-3, convention V, and especially Sec. 13-5). The corresponding equilibrium constant is referred to as K_{sp}, the solubility product or the solubility-product constant for the ionic com-pound being considered.

Although the dissolving of *any* ionic compound may be described in terms of a solubility equilibrium and a corresponding solubility-product constant, the usual application of solubility products is to substances of low solubility. This is true in part because these applications are often the most interesting ones and in part because substances of high solubility lead to concentrations at which activities differ greatly from concentrations, so that there are large deviations from the solubility-product relationship formulated with concentrations (see Sec. 13-5, comment 8).

The form of the solubility-product relation depends on the coefficients in the chemical equation that describes the dissolving of the substance. This is illustrated in the following example.

EXAMPLE 15-1   **The form of the ion product.** Formulate the ion products for calcium phos-phate, $Ca_3(PO_4)_2$, and magnesium ammonium phosphate, $MgNH_4PO_4 \cdot 6H_2O$.

Solution. The dissolving of $Ca_3(PO_4)_2$ is described by

$$Ca_3(PO_4)_2(s) \rightleftharpoons 3Ca^{2+} + 2PO_4^{3-} \tag{15-1}$$

The ion product is therefore given by the solubility-product relationship

$$[Ca^{2+}]^3 \, [PO_4^{3-}]^2 \leq K_{sp} \text{ for } Ca_3(PO_4)_2 \tag{15-2}$$

The "smaller than" sign in (15-2) applies in the absence of the solid; the "equal" sign applies in the presence of the solid at equilibrium. For the second salt, the corre-sponding relationships are

$$MgNH_4PO_4 \cdot 6H_2O \rightleftharpoons Mg^{2+} + NH_4^+ + PO_4^{3-} + 6H_2O \tag{15-3}$$

and $\qquad\qquad$ $\underline{[\text{Mg}^{2+}][\text{NH}_4^+][\text{PO}_4^{3-}]} \leq K_{sp}$ for $\text{MgNH}_4\text{PO}_4 \cdot 6\text{H}_2\text{O}$ $\qquad$ (15-4)

Note that $[\text{H}_2\text{O}]$ does not appear in the ion product (15-4).

The value of K_{sp} for a given substance must be determined experimentally. This may be done by measuring the solubility, as illustrated in the following example, unless the solubility is so small that its measurement presents difficulties. Other methods are available that do not suffer from this limitation; one of them is discussed in Chap. 19, which deals with electrochemistry.

EXAMPLE 15-2

K_{sp} **from solubility measurement.** In reference works, solubilities are frequently expressed in units of grams per 100 ml. For Ag_2CrO_4 the solubility in water at 25°C is 0.0027 g per 100 ml. Estimate K_{sp} for this substance.[1]

Solution. The formula weight of Ag_2CrO_4 is 332. Thus the formal solubility is

$$\left(\frac{0.0027}{332}\text{ gfw}\right)\left(\frac{1}{100 \text{ ml}}\right)\left(\frac{10 \times 100 \text{ ml}}{\text{liter}}\right) = 8.1 \times 10^{-5}\ F$$

This means that in the saturated solution the concentration of CrO_4^{2-} is $8.1 \times 10^{-5}\ M$ and the concentration of Ag^+ is, from the stoichiometry of the salt, twice as great, $1.6 \times 10^{-4}\ M$. $\qquad$ Therefore, $\qquad$ $K_{sp} = [\text{Ag}^+]^2[\text{CrO}_4^{2-}] = (1.6 \times 10^{-4})^2(8 \times 10^{-5}) = 2.1 \times 10^{-12}\ M^3$.

Solutions in pure water. In the foregoing example we assumed that the ionic substance was dissolving in pure water rather than in water containing other ions, and we also ignored any possible hydrolysis. The effects of hydrolysis can be estimated with the methods of Chap. 14, but to simplify the presentation we shall neglect them. In most cases the degree of hydrolysis is sufficiently small that its effects are not significant. They can occasionally be appreciable, however, especially because the concentrations of ions produced in saturated solutions of slightly soluble compounds are very low and the degree of hydrolysis increases with dilution (by Le Châtelier's principle).

The next example illustrates how the solubility of a slightly soluble substance in water can be calculated from its solubility-product constant.

EXAMPLE 15-3

Solubility of lanthanum iodate, $\text{La(IO}_3)_3$. Estimate the solubility at 25°C of $\text{La(IO}_3)_3$, for which $K_{sp} = 6.2 \times 10^{-12}$. Give the answer in gram formula weights per liter and in grams per 100 ml.

Solution. Let the solubility be y gfw liter^{-1}, so that $[\text{La}^{3+}] = y$ and $[\text{IO}_3^-] = 3y$, because each formula unit of the salt yields one La^{3+} and three IO_3^- ions. [We neglect

---

[1]Note that activity effects are liable to be particularly large for multiply charged ions such as CrO_4^{2-}. Moreover the neglect of the concentration of dissolved species such as $\text{Ag}_2\text{CrO}_4(aq)$, which exists largely as closely associated and aquated groups of ions, is by no means necessarily justified. Both effects increase the actual solubility from that calculated by the straightforward application of the solubility-product principle, sometimes by a factor of 2 or even more. Nevertheless, the results of simple calculations of the type presented are adequate in all but the most exacting considerations.

the concentration of any undissociated $La(IO_3)_3$.] Therefore

$$[La^{3+}][IO_3^-]^3 = y(3y)^3 = 27y^4 = 6.2 \times 10^{-12}$$

whence $y^4 = \frac{6.2}{27} \times 10^{-12} = 0.23 \times 10^{-12}$ and $y = 7.0 \times 10^{-4}$ gfw   liter^{-1},   or $y = 7.0 \times 10^{-4}$ F. To calculate the solubility in grams per 100 ml, we need to know the formula weight of $La(IO_3)_3$, which is 664. Hence for 100 ml the solubility is

$$\left(\frac{7.0 \times 10^{-4} \times 664 \text{ g}}{\text{liter}}\right)\left(\frac{\text{liter}}{10 \times 100 \text{ ml}}\right) = \frac{0.046 \text{ g}}{100 \text{ ml}}$$

It is instructive to note that each 3 in the term $(3y)^3$ in Example 15-3 has its origin in the coefficient of IO_3^- in the formula of $La(IO_3)_3$ and thus in the equation for its dissolution, $La(IO_3)_3(s) \rightleftharpoons La^{3+} + 3IO_3^-$. In the solubility product, the *exponent* 3 arises from the rules for formulating Q's; the *coefficient* 3 arises because we have chosen to express the concentration of IO_3^- in terms of the formal solubility of the salt and there are three IO_3^- in each formula unit.

A common error made by many beginners studying problems like Example 15-3 is to imagine, through inattention to definitions, that the concentration of iodate has been *tripled* as well as raised to the third power. It has not been; $[IO_3^-]$ has been multiplied by no factor. The concentration of IO_3^- is three times the formal solubility, and thus, since there is no other source of either IO_3^- or La^{3+}, $[IO_3^-] = 3[La^{3+}]$. If we chose to represent the iodate concentration by z, as we might, then $[IO_3^-] = z$, $[La^{3+}] = z/3$, and $K_{sp} = (z/3) z^3 = z^4/3$. This would not imply that we had multiplied the La^{3+} concentration by one-third, but only that we had defined the formal solubility as $z/3$. Remember: *no ion concentration in a solubility-product expression, or in any other reaction quotient, is ever multiplied by any factor.*

Two other caveats are worth noting in order to avoid common pitfalls. First, remember that the solubility-product principle can be applied only to saturated solutions, i.e., only when equilibrium exists (see Sec. 13-5, comment 13). Second, distinguish carefully between the *solubility product,* which is an equilibrium constant, and the *solubility,* which is a measure of the concentration of some species in a saturated solution in equilibrium with another phase, in this case a solid phase.

The magnitudes of the solubilities of two substances need not parallel the magnitudes of their solubility products. For example, the formal solubility of $La(IO_3)_3$ (7×10^{-4} F) is somewhat greater than that of $SrSO_4$ (5.3×10^{-4} F), despite the fact that the numerical value of K_{sp} for the second compound (2.8×10^{-7}) is greater by nearly 10^5 than that of the first (6.2×10^{-12}). The point is that comparison of the values of the two solubility products is meaningless because they have different units, those of K_{sp} for $SrSO_4$ being M^2 and those of K_{sp} for $La(IO_3)_3$ being M^4.

Table D-2 lists K_{sp} values for many common substances. The units are not given in the table but follow at once from the stoichiometry of the compounds and the convention that all concentrations are molarities.

Solubility of an ionic substance in a solution containing one of its ions: the common-ion effect. The solubility of a sparingly soluble ionic substance is sharply decreased by the presence of another ionic substance when the two have an ion in common. The following example demonstrates this *common-ion effect.*

EXAMPLE 15-4   **Solubility of SrSO$_4$ in 0.10 *F* Na$_2$SO$_4$.** What is the formal solubility of SrSO$_4$ ($K_{sp} = 2.8 \times 10^{-7}$) in 0.10 *F* Na$_2SO_4$?

Solution. Because Na$_2$SO$_4$ is a strong electrolyte, the sulfate concentration is 0.10 *M* before any SrSO$_4$ dissolves. Letting the unknown solubility be z, we have [Sr^{2+}] = z and [SO$_4^{2-}$] = 0.10 + z. When these expressions for the concentrations are substituted into the solubility-product expression $K_{sp} = $ [Sr^{2+}][SO$_4^{2-}$] = 2.8 × 10^{-7}, we obtain $z(0.10 + z) = 2.8 \times 10^{-7}$, a quadratic equation in z. Instead of applying the usual formalism for solving quadratic equations, we note that z is probably very small compared to 0.10, so that z can be neglected in the combination 0.10 + z. This gives $z(0.10) = 2.8 \times 10^{-7}$, or $z = 2.8 \times 10^{-6}$. This result is indeed small relative to 0.10, so the approximation is justified and the solubility is 2.8 × 10^{-6} *F*.

The solubility of SrSO$_4$ in pure water is $\sqrt{K_{sp}} = 5.3 \times 10^{-4}$ *F*, so the presence of 0.10 *F* Na$_2$SO$_4$ (or of sulfate ion at this concentration from any other source) has decreased the solubility nearly 200-fold. Similarly, the presence of any soluble strontium salt, such as SrCl$_2$ or Sr(NO$_3$)$_2$, would provide Sr^{2+} ions and hence decrease the solubility of SrSO$_4$. Figure 15-1 shows how the solubility of SrSO$_4$ in solutions of Na$_2$SO$_4$ depends on the concentration of Na$_2$SO$_4$. The same figure applies to the solubility of SrSO$_4$ in solutions of SrCl$_2$.

The common-ion effect is often important in quantitative analysis. Suppose, for example, that the amount of SO$_4^{2-}$ in a solution is to be determined by precipitating

FIGURE 15-1   **The common-ion effect.** The solubility x of SrSO$_4$ in a solution of Na$_2$SO$_4$ is plotted as a function of the concentration c of Na$_2$SO$_4$. Since [Sr^{2+}] = x and [SO$_4^{2-}$] = $c + x$, we obtain $x(c + x) = K_{sp} = 2.8 \times 10^{-7}$. This equation is quadratic in x and linear in c. For plotting x against c it is therefore easiest to assume values for x and solve for c. Note the decrease in solubility caused by the increase in the concentration of the SO$_4^{2-}$ ion, the ion common to both SrSO$_4$ and Na$_2$SO$_4$. There is a similar decrease of the solubility when SrSO$_4$ is dissolved in a SrCl$_2$ solution, the common ion now being Sr^{2+}. The solubility of SrSO$_4$ is measured by [SO$_4^{2-}$] in this new situation so that now $x = $ [SO$_4^{2-}$] while [Sr^{2+}] = $c + x$. Thus again, $x(c + x) = 2.8 \times 10^{-7}$.

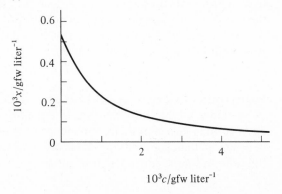

the sulfate as $BaSO_4$ ($K_{sp} = 1.1 \times 10^{-10}$) and weighing the dried precipitate. It is necessary that essentially all the SO_4^{2-} be removed from the solution. This can be accomplished by using an excess of barium ions in precipitating the sulfate, so that the concentration of SO_4^{2-} remaining is appreciably smaller than it would be if Ba^{2+} and SO_4^{2-} were present in equal numbers.

EXAMPLE 15-5   **Precipitation of Fe(OH)$_2$.** The solubility of ferrous hydroxide, $Fe(OH)_2$, in water is about 0.6 mg liter^{-1}. Calculate its K_{sp} and then calculate the minimum pH needed to precipitate ferrous ion so completely that no more than 1.0 μg (microgram, or 10^{-6} g) of the ion remains in solution per liter.

Solution.   The formula weight of $Fe(OH)_2$ is 90. Thus the formal solubility is (0.6 mg liter^{-1}) (1 g/1000 mg) [1 gfw $Fe(OH)_2$/90 g $Fe(OH)_2$] $= 7 \times 10^{-6}$ F, and $[Fe^{2+}] = 7 \times 10^{-6}$ M and $[OH^-] = 1.4 \times 10^{-5}$ M. The value of K_{sp} is then $[Fe^{2+}][OH^-]^2 = (7 \times 10^{-6})(1.4 \times 10^{-5})^2 = \underline{1.4 \times 10^{-15}}$.

The maximum concentration of Fe^{2+} that can remain is 1.0×10^{-6} g liter^{-1}, or $(1.0 \times 10^{-6}$ g liter^{-1}) (1 mol Fe^{2+}/56 g) $= 1.8 \times 10^{-8}$ mol liter^{-1} Fe^{2+}. Thus the minimum concentration of OH^- can be deduced from the solubility-product principle. When the solution is saturated, $[OH^-]^2 = (1.4 \times 10^{-15})/[Fe^{2+}]$. Hence, if $[Fe^{2+}] \leq 1.8 \times 10^{-8}$ M, $[OH^-]^2 \geq (1.4 \times 10^{-15})/(1.8 \times 10^{-8}) = 8 \times 10^{-8}$,   or $[OH^-]^2 \geq 8 \times 10^{-8}$ M. Thus $[OH^-] \geq 2.8 \times 10^{-4}$ and pOH ≤ 3.6, so pH $\geq \underline{10.4}$.

Precipitation titrations.   The formation of a precipitate of a sparingly soluble salt is sometimes useful as an end-point indicator in titrations in volumetric quantitative analysis. The principles governing such titrations are analogous to those of acid-base titrations, discussed in Sec. 14-5; the end point of the titration, signaled by some characteristic change, should be as close as possible to the equivalence point, at which the quantity of titrant added is equivalent to the quantity of substance being titrated.

For example, in the Mohr titration of Cl^- with a solution of $AgNO_3$ of known concentration (called a standard solution), a small amount of potassium chromate, K_2CrO_4, is added to the chloride solution and acts as an indicator by the formation of the slightly soluble red Ag_2CrO_4. The K_{sp} values of $AgCl$ and Ag_2CrO_4 are such that the titration produces $AgCl$ as long as significant amounts of Cl^- remain in the solution. As further titrant is added the reddish silver chromate is formed, which can be recognized because the color of the precipitating solid changes from a white to a reddish hue (see Problem 15-13).

Solutions containing several sparingly soluble salts.   The solubilities of sparingly soluble salts that are in equilibrium with the same solution but that have no ions in common are independent of each other. If the salts do have ions in common, their solubility-product equations must be satisfied simultaneously. The following example illustrates this point.

EXAMPLE 15-6   **Simultaneous solubilities of PbSO$_4$ and SrSO$_4$.** The solubility-product constants of PbSO$_4$ and SrSO$_4$ are, respectively, 1.7×10^{-8} and 2.8×10^{-7}. What are the values of [SO$_4{}^{2-}$], [Pb^{2+}], and [Sr^{2+}] in a solution at equilibrium with both substances?

Solution. Since the dissolving of each formula unit of PbSO$_4$ gives one Pb^{2+} ion and one SO$_4{}^{2-}$ ion, the concentration of sulfate ions originating from this source is given by [Pb^{2+}]. A similar argument applies to the SO$_4{}^{2-}$ ions arising from the dissolving of SrSO$_4$, so that the total concentration of SO$_4{}^{2-}$ is

$$[SO_4{}^{2-}] = [Pb^{2+}] + [Sr^{2+}] \tag{15-5}$$

The solubility-product relationships yield the equations $[Pb^{2+}] = 1.7 \times 10^{-8}/[SO_4{}^{2-}]$ and $[Sr^{2+}] = 2.8 \times 10^{-7}/[SO_4{}^{2-}]$. When these are inserted into (15-5), we get

$$[SO_4{}^{2-}] = \frac{1.7 \times 10^{-8} + 2.8 \times 10^{-7}}{[SO_4{}^{2-}]} \tag{15-6}$$

and thus

$$[SO_4{}^{2-}]^2 = (0.17 + 2.8) \times 10^{-7} = 3.0 \times 10^{-7}$$

or
$$[SO_4{}^{2-}] = \underline{5.5 \times 10^{-4}\ M}$$

From this result we obtain

$$[Pb^{2+}] = \frac{1.7 \times 10^{-8}}{5.5 \times 10^{-4}} = \underline{3.1 \times 10^{-5}\ M}$$

and
$$[Sr^{2+}] = \frac{2.8 \times 10^{-7}}{5.5 \times 10^{-4}} = \underline{5.1 \times 10^{-4}\ M}$$

As a check, we note that, within their precision, the three concentrations found satisfy (15-5).

Limitations of solubility-product calculations. As noted earlier, concentrations arrived at in solubility-product calculations sometimes deviate appreciably from those found experimentally. In part, this low accuracy arises from failure to correct for activity effects; however, the chief factor is usually that other equilibria simultaneously involving the ions of interest are disregarded. We have mentioned hydrolysis as one possible complication; others include the formation of undissociated neutral molecules and of complex ions. These latter effects are well illustrated by solutions of AgCl in the presence of excess Ag$^+$ and especially in the presence of excess Cl$^-$. First, when AgCl(s) dissolves, it yields not only Ag$^+$ and Cl$^-$ ions, but also a very small but not always negligible concentration of undissociated AgCl *molecules* in solution. A second effect, the formation of the complex ion AgCl$_2{}^-$ and of others containing even more chloride, enters when [Cl$^-$] begins to exceed about $10^{-3}\ M$. This effect substantially increases the solubility of AgCl(s). Thus, when silver ions are precipitated for purposes of analysis by adding a reagent containing Cl$^-$ ions, the solubility of the resulting AgCl(s) is decreased in the presence of a small excess of Cl$^-$ but is increased substantially by a large excess.

15-3 Equilibria Involving Complexes

Background. Some of the properties and structural features of representative complex ions have been described briefly and illustrated in earlier chapters, especially in Secs. 4-5 and 7-6. The modern picture of the geometry, bonding, and reactions of complexes is a direct outgrowth of the research of an Alsatian-Swiss chemist, Alfred Werner, who worked during the late years of the nineteenth century and the early years of this one. Although many of what are now called coordination compounds (Sec. 7-6), a term originated by Werner, were known before his day, few systematic studies of their properties and reactions had been made and ideas concerning their structure were almost completely fallacious. The reasons for the comparatively late development of a correct structural view of these relatively simple inorganic compounds are of some interest. They provide insight into the ways in which fashions in one area of science inhibit progress in other areas, even those quite closely related, a sad testimony to the limitations of the imaginations of all but a few rare individuals in any period.

Around the middle of the nineteenth century organic chemistry had come to dominate the chemical scene, attracting many of the best minds with a bent for chemistry and, in fact, for science generally. By the mid-1870s, the structural theory of organic chemistry was well established and even some three-dimensional aspects of structure were understood. As complicated a molecule as that of the blue dyestuff indigo (page 329), which had been used for thousands of years, had been synthesized in the laboratory. During the last quarter of the century the structures of a great many physiologically important organic compounds were determined or were under vigorous investigation, including sugars, amino acids (the building blocks of proteins), fragments of nucleic acids, and chlorophyll.

In contrast, relatively little progress was made toward a real understanding of the structures of any but the simplest inorganic compounds. The very success of the structural principles applicable to organic compounds led to attempts to apply these principles to the inorganic field. However, they were of limited value for compounds of elements other than carbon, because the chief bonding characteristic of carbon atoms, their tendency to join together to form chains and rings, is rare in the chemistry of other elements. Modification of the prevalent structural ideas of the time, as well as the creation of new ones, was a necessary prerequisite to the development of an understanding of inorganic complexes. Werner almost single-handedly provided these new insights, on the basis of years of careful and ingenious experimental work.

When Werner began his work, *incorrect* structural formulas such as

$$
\begin{array}{ccc}
\text{Cl} & & \text{NH}_3\text{—Cl} \\
& \diagdown \text{Pt} \diagup & \\
\text{Cl—NH}_3 & & \text{NH}_3\text{—NH}_3\text{—NH}_3\text{—Cl}
\end{array}
\tag{15-7}
$$

were common. This formula, strongly influenced by the concept of chains of bonded atoms, common in organic chemistry, represented the compound with composition $PtCl_4(NH_3)_5$ or $PtCl_4 \cdot 5NH_3$, one of a series of compounds of platinum with chlorine and ammonia (Table 15-1). It was known that the addition of $AgNO_3$ to solutions

TABLE 15-1 **Fraction of chloride reacting with Ag⁺ and conductivity for seven platinum complexes**

Compound	Fraction of chloride reacting with Ag⁺	Modern formula	Number of conducting species	Relative electrical conductivity of solution*
$PtCl_4 \cdot 6NH_3$	$\frac{4}{4}$	$Pt(NH_3)_6^{4+}$, $4Cl^-$	5	523
$PtCl_4 \cdot 5NH_3$	$\frac{3}{4}$	$Pt(NH_3)_5Cl^{3+}$, $3Cl^-$	4	404
$PtCl_4 \cdot 4NH_3$	$\frac{2}{4}$	$Pt(NH_3)_4Cl_2^{2+}$, $2Cl^-$	3	229
$PtCl_4 \cdot 3NH_3$	$\frac{1}{4}$	$Pt(NH_3)_3Cl_3^{+}$, Cl^-	2	97
$PtCl_4 \cdot 2NH_3$	$\frac{0}{4}$	$Pt(NH_3)_2Cl_4$	0	~0
$PtCl_4 \cdot NH_3 \cdot KCl$	$\frac{0}{5}$	K^+, $Pt(NH_3)Cl_5^{-}$	2	109
$PtCl_4 \cdot 2KCl$	$\frac{0}{6}$	$2K^+$, $PtCl_6^{2-}$	3	256

*Arbitrary scale. Conductivity measured for solutions of same formality.

of compounds such as those in this series does not necessarily cause the precipitation as AgCl of all the chlorine they contain. For example, only three-quarters of the chlorine of the second compound in the table, that represented in (15-7), reacts with Ag⁺. The different behavior of the Cl atoms toward Ag⁺ was explained by the hypothesis that the Cl atom attached directly to the Pt atom does not react with Ag⁺ to form AgCl, whereas the Cl atoms attached to the NH₃ groups do. As more experimental facts were established—not all of the evidence in Table 15-1 and indeed not all of the compounds were known before Werner's day—it became clear that formulas such as (15-7), attributing (in modern terminology) just four covalent bonds to the platinum atom in these compounds, could not be correct.

Werner used many different methods, including syntheses of series of compounds such as that in Table 15-1 and a variety of physical measurements, to establish the modern picture of the structure of complexes. The modern formulas listed in Table 15-1, which are essentially those he proposed, show that in each compound all the NH₃ molecules are attached directly to the Pt, together with enough chlorine atoms to make the total number of ligands six. The remaining chlorine atoms, if any, are present as Cl⁻. The oxidation state of Pt in each compound is +IV. The first four complexes listed are positively charged and are thus cationic; the fifth is electrically neutral, and the last two are anionic. The measurements of electrical conductivity listed in the final column of the table were made by Werner. They provide a rough indication of the number of conducting, and therefore charged, species in each compound and are in good accord with the formulas Werner suggested.

Werner also realized the importance of three-dimensional structural features and showed by clever experiments and reasoning that the configuration of the ligands in complexes with ligancy 6, such as those in Table 15-1, is octahedral. These arguments, too lengthy and intricate to summarize here, were based on the existence and properties of different isomeric complexes: the number of different possible structures corresponding to a given formula and the symmetry of these structures as revealed by their influence on plane polarized light, a phenomenon referred to as *optical activity,* discussed briefly in Sec. 26-4.

Final unambiguous confirmation of many of Werner's deductions had to await the development of newer experimental aproaches. Direct structure determination by analysis of the X-ray diffraction patterns of single crystals has been perhaps the most fruitful technique, but other methods, such as infra-red, Raman, and nuclear magnetic resonance spectroscopy have been valuable in the study of both the structures and the reactions of complexes.

We turn now to a consideration of equilibria involving complexes. In recent decades, careful studies of such equilibria, and the evaluation of equilibrium constants involving different complex species, have provided much information about coordination compounds.

The stepwise attachment of the ligands.   Complexes may be formed by the reaction of ligands with metal ions in solution. Usually the ligands are attached successively, and for each concentration of ligands and metal ions there are corresponding equilibrium concentrations of species that contain different numbers of ligands. For example, the reaction between silver ion and thiosulfate ion, important in the fixing of a photographic image on film, can be described by the following equations and the complex *formation* constants related to them:

$$Ag^+ + S_2O_3^{2-} \rightleftharpoons AgS_2O_3^- \qquad K_1 = 6.6 \times 10^8 \qquad (15\text{-}8)$$

$$AgS_2O_3^- + S_2O_3^{2-} \rightleftharpoons Ag(S_2O_3)_2^{3-} \qquad K_2 = 4.4 \times 10^4 \qquad (15\text{-}9)$$

$$Ag(S_2O_3)_2^{3-} + S_2O_3^{2-} \rightleftharpoons Ag(S_2O_3)_3^{5-} \qquad K_3 = 4.9 \qquad (15\text{-}10)$$

The constant for the overall reaction,

$$Ag^+ + 3S_2O_3^{2-} \rightleftharpoons Ag(S_2O_3)_3^{5-} \qquad (15\text{-}11)$$

is the product of the individual constants, since (15-11) is the sum of the preceding three chemical equations:

$$K_{tot} = K_1K_2K_3 = 1.4 \times 10^{14} \qquad (15\text{-}12)$$

In reactions of this kind the ligands do not react with a bare metal ion but rather with one that is hydrated, as emphasized in Sec. 7-6. In other words, the ligands replace H_2O in an aquo complex. However, because the concentration of water is constant in dilute solutions it can be omitted from the equilibrium expressions (and the chemical equations). This is usually done, partly for simplicity and partly because the number of water molecules is frequently unknown.

When consulting tables of equilibrium constants for complex ions, it is essential to note carefully just how the constants are defined. They may be given for *each step separately,* as in (15-8) to (15-10), or the *overall* constant may be given, as in (15-12). Furthermore, quite frequently, especially in textbooks, *dissociation* constants are quoted rather than *formation* constants. If stepwise dissociation constants are considered for a complex with several ligands, it must be remembered that the dissociation of the first ligand is the reverse of the addition of the last one. For example, silver ion forms two stable complexes with ammonia, $AgNH_3^+$ and $Ag(NH_3)_2^+$. The equations for the formation of these complexes, with the corresponding constants subscripted f to identify them as formation constants, are

$$Ag^+ + NH_3 \rightleftharpoons AgNH_3^+ \qquad K_{1f} = 1.6 \times 10^3 \qquad (15\text{-}13)$$

$$AgNH_3^+ + NH_3 \rightleftharpoons Ag(NH_3)_2^+ \qquad K_{2f} = 6.8 \times 10^3 \qquad (15\text{-}14)$$

Dissociation of one ammonia molecule from $Ag(NH_3)_2{}^+$ is then the reverse of (15-14).

$$Ag(NH_3)_2^+ \rightleftharpoons AgNH_3^+ + NH_3$$

and thus its equilibrium constant is $K_{1d} = 1/K_{2f} = 1.5 \times 10^{-4}$, the subscript $1d$ implying dissociation of the first ligand.

It is perhaps easier to think in terms of dissociation constants because this is how we have discussed acids and bases (Chap. 14). A small dissociation constant implies a very stable complex, just as a small acid dissociation constant implies a very stable acid with little tendency to transfer its proton to the base water. Table 15-2 gives values of both overall ($K_{\text{tot},d}$) and stepwise dissociation constants for a few representative complexes.

Distribution of complex species as a function of ligand concentration. The relative proportions of the different complex species at different ligand concentrations in a particular metal-ion–ligand system may be described by distribution diagrams similar to Fig. 14-5, used to illustrate the stepwise dissociation of a polyprotic acid. Figures 15-2 and 15-3 illustrate typical situations. As abscissas we have used the negative logarithms of the ligand concentrations, denoted pL by analogy with pH:

$$pL \equiv -\log [L] \qquad (15\text{-}15)$$

Sometimes complex formation (or dissociation) occurs in widely separated steps (Fig. 15-2), but the more usual situation is that the steps are not well separated because the constants for successive steps do not differ greatly (Fig. 15-3).

EXAMPLE 15-7   **Distribution of species in the silver-ammonia system.** Calculate the approximate fractions of the different complex species in the Ag^+–NH_3 system when the equilibrium concentration of NH_3 is (a) 10^{-2} M; (b) 10^{-3} M; (c) 10^{-4} M; (d) 10^{-5} M. The relevant equilibrium constants are given in Table 15-2.

Solution. It is convenient to express the concentrations of all three species containing Ag(I) in terms of one of them. We choose Ag^+. The relevant equilibrium expressions are

$$\frac{[Ag^+][NH_3]}{[AgNH_3^+]} = K_{2d} = 6 \times 10^{-4} \qquad (15\text{-}16)$$

and

$$\frac{[Ag^+][NH_3]^2}{[Ag(NH_3)_2^+]} = K_{1d}K_{2d} = K_{\text{tot},d} = 9 \times 10^{-8} \qquad (15\text{-}17)$$

which lead to

$$[AgNH_3^+] = [Ag^+]([NH_3] \times 1.7 \times 10^3) \qquad (15\text{-}18)$$

$$[Ag(NH_3)_2^+] = [Ag^+]([NH_3]^2 \times 1.1 \times 10^7) \qquad (15\text{-}19)$$

TABLE 15-2 **Dissociation constants for representative complexes***

Complex	$K_{tot,d}$	K_{1d}	K_{2d}	K_{3d}	K_{4d}
$Ag(NH_3)_2^+$	9×10^{-8}	1.5×10^{-4}	6×10^{-4}		
$Cu(NH_3)_4^{2+}$	2×10^{-13}	7×10^{-3}	1.3×10^{-3}	3×10^{-4}	7×10^{-5}
$Ni(NH_3)_4^{2+}$	1.1×10^{-8}	7×10^{-2}	1.9×10^{-2}	6×10^{-3}	1.6×10^{-3}
$Zn(NH_3)_4^{2+}$	9×10^{-10}	9×10^{-3}	4×10^{-3}	5×10^{-3}	5×10^{-3}
$Co(NH_3)_6^{2+}$	8×10^{-6}	4	6×10^{-1}	1.8×10^{-1}	9×10^{-2}
$Co(NH_3)_6^{3+}$	7×10^{-36}	4×10^{-5}	9×10^{-6}	2.5×10^{-6}	8×10^{-7}
$Zn(OH)_4^{2-}$†	3×10^{-16}	5×10^{-2}	6×10^{-5}	1.4×10^{-6}	7×10^{-5}
$Fe(SCN)_4^-$	4×10^{-4}	1.3	2.5	1.0×10^{-2}	1.1×10^{-2}
$Ag(S_2O_3)_3^{5-}$	7×10^{-15}	2.0×10^{-1}	2.3×10^{-5}	1.5×10^{-9}	
$Ca(OH)^+$	3×10^{-2}				
$Sr(OH)^+$	1.5×10^{-1}				
$Ba(OH)^+$	2.3×10^{-1}				

* The subscript d denotes dissociation, to avoid confusion with equilibrium constants for the formation reactions. K_{tot} is the product of the individual stepwise K's; for the two cobalt complexes only the first four of these stepwise K's are listed here.

† The constants K_{2d} and K_{3d} for $Zn(OH)_4^{2-}$ refer to equilibria involving the dissolved neutral species $Zn(OH)_2$. Solid zinc hydroxide, $Zn(OH)_2(s)$, is almost insoluble ($K_{sp} = 7 \times 10^{-18}$); the concentration of undissociated $Zn(OH)_2$ in equilibrium with the solid is about 10^{-7} M.

FIGURE 15-2 **Distribution curves for the Ag^+-$S_2O_3^{2-}$ system.** These distribution curves resemble those for the carbonate system (Fig. 14-5), except for the number of species, and are to be interpreted in the same way. Each curve is labeled with the formula of the ion to which it corresponds. The abscissa is the negative logarithm of the $S_2O_3^{2-}$ concentration, $pS_2O_3 \equiv -\log [S_2O_3^{2-}]$. The fraction of the total Ag(I) present [Ag(I) represents silver in oxidation state (+I)] in the form of each ion at any particular value of the thiosulfate concentration can be found by drawing a vertical line at the corresponding value of pS_2O_3 and reading the appropriate fraction from each curve. Because the successive stepwise dissociation constants differ from each other by about four powers of 10, no more than two forms of the complex are ever present simultaneously in significant concentrations and there are ranges of $[S_2O_3^{2-}]$ in which practically all Ag(I) is present in a single species. Thus at thiosulfate concentrations below about 10^{-10} M (pS_2O_3 above 10), almost all the Ag(I) is present as free (hydrated) Ag^+; between pS_2O_3 about 5.5 and 7.5, it is chiefly $AgS_2O_3^-$; between pS_2O_3 about 2 and 3.5, it is mostly $Ag(S_2O_3)_2^{3-}$; and at a concentration of thiosulfate above about 2 M (pS_2O_3 below -0.3), more than 90 percent is $Ag(S_2O_3)_3^{5-}$. The pK values for the successive dissociation steps may be obtained from these curves by noting that $pK = pS_2O_3$ at the point at which the corresponding curves cross, since at this point the concentrations of successive forms are equal.

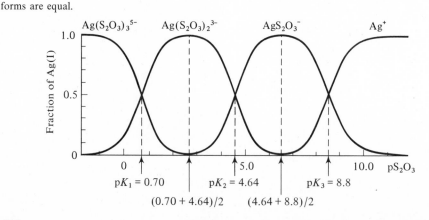

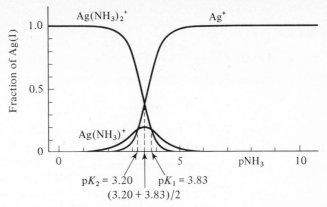

FIGURE 15-3 **Distribution curves for the Ag⁺-NH₃ system.** Each curve is labeled with the formula of the ion to which it corresponds. The abscissa is $pNH_3 \equiv -\log[NH_3]$. The interpretation of this diagram is similar to that of Fig. 15-2. The second dissociation constant for $Ag(NH_3)_2^+$ is about four times as large as the first ($K_{1d} = 1.5 \times 10^{-4}$, $K_{2d} = 6 \times 10^{-4}$). Consequently the field for $AgNH_3^+$ is very narrow; this field is centered at $pNH_3 = (pK_1 + pK_2)/2 = 3.5$. At no time does more than about 20 percent of the Ag(I) exist as $AgNH_3^+$, and this species is present in significant quantities only between pNH_3 about 2.5 and 4.5, that is, when $[NH_3]$ is between about 3×10^{-3} and 3×10^{-5} M. When $[NH_3]$ is below about 3×10^{-5} M, Ag(I) is present chiefly as Ag^+; when it is above about 3×10^{-3} M, essentially all Ag(I) is present as the diammine complex, $Ag(NH_3)_2^+$.

This diagram is applicable only when the concentration of OH^- is sufficiently low to prevent precipitation of AgOH. The maximum allowable pH depends on the concentration of Ag^+ and can be calculated from the solubility product of AgOH, $K_{sp} = 2 \times 10^{-8}$. When $[NH_3]$ is above about 10^{-3} M, there is no chance of precipitation of AgOH unless a high concentration of strong base is present also, since so small a fraction of the Ag(I) is present as Ag^+.

The proportions of the three forms can then be expressed, in terms of the ammonia concentration, as

$$[Ag(NH_3)_2^+]:[AgNH_3^+]:[Ag^+] = ([NH_3]^2 \times 1.1 \times 10^7):([NH_3] \times 1.7 \times 10^3):1$$

(15-20)

For convenience, we shall abbreviate the diammine complex as Di and the monoammine complex as Mon. To get the proportions at different ammonia concentrations, we merely substitute the appropriate value of $[NH_3]$ in (15-20).

(*a*)   When $[NH_3] = 10^{-2}$ M,

$$Di:Mon:Ag^+ = 1.1 \times 10^3:1.7 \times 10^1:1$$

There is a great preponderance of the diammine form; for every 1100 or so diammine ions, there are about 17 monoammine and 1 free Ag^+. The fraction of free Ag^+ is thus about 0.001 [that is, $1/(1100 + 17 + 1)$]; that of $AgNH_3^+$ is about 0.015 [that is, $17/(1100 + 17 + 1)$]; and that of $Ag(NH_3)_2^+$ is, by difference, about 0.984.

(*b*)   When $[NH_3] = 10^{-3}$ M,

$$Di:Mon:Ag^+ = 1.1 \times 10^1:1.7:1 = 11:1.7:1$$

In other words, for every 13.7 silver-containing species, 11 are Di, 1.7 are Mon, and 1 is Ag^+. The respective fractions are thus about 0.80, 0.12, and 0.07 (which add up only to 0.99 because of cumulative rounding-off errors). The diammine form still predominates at this concentration, but all the ions are present in significant quantity.

(c)   When $[NH_3] = 10^{-4}\ M$,

$$Di:Mon:Ag^+ = 1.1 \times 10^{-1}:1.7 \times 10^{-1}:1 = \underline{0.11:0.17:1}$$

For every 1.28 silver-containing species, 0.11 is Di, 0.17 is Mon, and the remainder is Ag^+. The respective fractions are about 0.09, 0.13, and 0.78. Now free Ag^+ is the predominant species; significant amounts of all three ions are still present.

(d)   When $[NH_3] = 10^{-5}\ M$,

$$Di:Mon:Ag^+ = \underline{1.1 \times 10^{-3}:1.7 \times 10^{-2}:1}$$

and the respective fractions are thus about 0.001, 0.017, and 0.982. The fractions of the diammine and even the monoammine are now almost negligible.

Complexometric titrations.   The question may be raised as to whether the formation of stable complexes ML_i permits the titration of solutions of M with titrant containing a standard solution of the ligand L. For many ligands the answer is no, because if several different complexes are formed (as they usually are), the individual equilibrium constants for the successive stages frequently differ by so little that the titration curve shows no well-developed step or steps.

An ingenious way to remedy this situation is to link several ligands together, giving a multidentate complexing group. Such a group can form a chelate (Sec. 7-6) if the different potential ligands have the proper geometry so that all can coordinate with the metal ion simultaneously. One of the most widely used molecules of this kind is ethylenediaminetetraacetic acid (abbreviated EDTA), one form of which has the formula

$$\begin{array}{ll} HOOC-H_2C & CH_2-COOH \\ \qquad\qquad N-CH_2-CH_2-N \\ HOOC-H_2C & CH_2-COOH \end{array} \qquad (15\text{-}21)$$

EDTA has four acidic hydrogen atoms, and we shall abbreviate its formula as H_4Y. When it forms chelate complexes with metal ions, the complexing agent is actually $EDTA^{4-}$ or Y^{4-}. Both nitrogen atoms of this anion and one oxygen atom of each of its four $-CH_2COO^-$ (acetate) groups can coordinate with a single metal atom. The EDTA anion is thus a sexidentate (Latin: *sex*, six) complexing agent. It forms stable complexes with many metal ions, coordinating the central ion octahedrally. The structure of one such complex, as found by X-ray crystallography, is illustrated in Fig. 15-4.

The formation constants for complexes between Y^{4-} and metal ions are usually very large. The complex ZnY^{2-} is typical:

$$Zn^{2+} + Y^{4-} \rightleftharpoons ZnY^{2-} \qquad K_f = 3.2 \times 10^{16} \qquad (15\text{-}22)$$

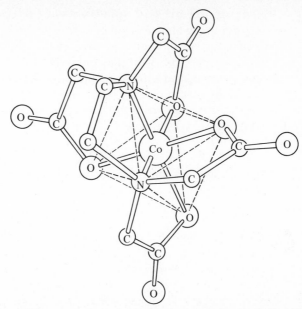

FIGURE 15-4 Structure of the EDTA complex of Co(III) in crystals of Rb[CoEDTA] · 2H$_2$O. The structure of this complex was determined by X-ray crystallography. The molecular geometry of EDTA^{4-}—bond lengths and angles and the flexibility allowed by rotation about single bonds—is such that each of six of the atoms of the anion, two nitrogens and four oxygens, can be simultaneously at one of the six different corners of an octahedron about the central metal atom and at a distance from that atom appropriate for bonding. The dashed lines are an aid to visualizing the octahedral surroundings of the Co atom.

Because this is a single-step process with no intermediate forms, and because K_f is so large, it is possible to titrate Zn^{2+} with EDTA. Many other metal ions can equally well be titrated. The metal-ion concentration changes abruptly in the region of the equivalence point, so that a sharp end point can be observed with a suitable indicator.

15-4 Simultaneous Competing Heterogeneous Equilibria

We have considered several different situations in which a number of equilibria are operative simultaneously and compete with each other for at least one of the ions or molecules present. These include all acid-base equilibria, in which two bases (one of them water) compete for a proton, and especially polyprotic acid equilibria, in which at least three (and often more) bases are present. They also include the competition for the ligand among the various complexes present in a multistage complex-ion equilibrium. Example 15-6 concerns a heterogeneous competing equilibrium in which Pb^{2+} and Sr^{2+} are simultaneously in equilibrium with their corresponding sulfates and thus compete for SO$_4{}^{2-}$. In this section we examine a number of other heterogeneous competing equilibria of some importance, chiefly by means of illustrative examples.

Effect of pH on the solubility of salts of weak acids and bases.
There are many almost insoluble salts of weak acids, and it is a perfectly general
rule that the solubility of these salts increases with increasing acidity of a solution
(decreasing pH), provided that the weak acid is itself soluble. The examples that
follow are concerned with the quantitative aspects of this phenomenon; qualitatively
it is easy to understand. Imagine a salt MX formed by the metal ion M^+ and the
anion X^-, with X^- being the conjugate base of a weak acid HX. In a saturated
solution of MX(*s*) we have

$$MX(s) \rightleftharpoons M^+ + X^- \qquad (15\text{-}23)$$

with the concentrations of the ions satisfying the solubility-product expression for
MX(*s*),

$$[M^+][X^-] = K_{sp} \qquad (15\text{-}24)$$

If the pH is so high that no appreciable quantity of X^- picks up a proton in solution,
the concentrations of M^+ and X^- will be equal, and each will equal $\sqrt{K_{sp}}$, which
is then a measure of the solubility of MX. As the pH is lowered, however, a significant
number of the X^- ions will pick up protons and be converted into HX, because
HX is a weak acid. When $pH = pK_a$, $[X^-] = [HX]$ and thus 50 percent of the
X^- will have been changed to HX; when $pH = (pK_a - 1)$, 91 percent will have been
converted. As the acidity increases, more and more of the X^- that goes into solution
is transformed into HX. Consequently it is no longer true that $[M^+] = [X^-]$. However,
it is true[1] that $[M^+] = [X^-] + [HX]$, so that $[M^+] > [X^-]$, the disparity becoming
larger as the pH decreases. Thus, to satisfy (15-24), more MX dissolves than in a
solution in which essentially all the X^- remains as X^-—the solubility, measured by
[M] or by $([X^-] + [HX])$, becomes greater.

The argument is in principle the same regardless of the charges on M and
X or of their proportions in the salt; for example, the salt might be $(M^{2+})_3(X^{3-})_2$.
It applies to salts of many different soluble weak acids—phosphates, carbonates,
acetates, even sulfates at high enough acidity, sulfides, and others. It applies as well
to the solubility of hydroxides and oxides, which may in this context be regarded
as salts of the very weak acid H_2O.

EXAMPLE 15-8   **Solubility of silver acetate at pH = 3.0.**   What is the solubility of AgOAc in
a solution buffered at pH = 3.0? K_{sp} for AgOAc = 2.5×10^{-3} and K_a for HOAc =
1.8×10^{-5}.

Solution.   As AgOAc dissolves, acetate ions are produced, some of which combine
with H^+ to form HOAc. This permits more AgOAc to dissolve until the concentrations
of Ag^+ and OAc^- reach the values permitted by the solubility-product constant. If
we let the solubility of AgOAc in this solution be S gfw liter^{-1}, then

---

[1]This condition, sometimes misleadingly termed "mass balance", expresses the fact that atoms are
conserved. M^+ and X^- are present in the salt in 1:1 proportion, and thus enter the solution in that
ratio. All the M^+ remains as M^+; although some of the X^- becomes protonated, forming HX, the sum
of the moles of X^- and HX in the solution must necessarily equal the total number of moles of X^-,
and thus of M^+, that entered the solution.

$$[Ag^+] = S = [OAc^-] + [HOAc]$$

since the total concentration of acetate in all forms must equal the silver-ion concentration inasmuch as Ag^+ and OAc^- are present in $1:1$ proportion in AgOAc. However, we know that

$$[HOAc] = [OAc^-]\frac{[H^+]}{K_a} \tag{15-25}$$

so that

$$S = [OAc^-] + [HOAc] = [OAc^-]\left(1 + \frac{[H^+]}{K_a}\right) \tag{15-26}$$

Multiplying by $[Ag^+]$ gives

$$[Ag^+]\,S = [Ag^+][OAc^-]\left(1 + \frac{[H^+]}{K_a}\right) \tag{15-27}$$

However, we know that $[Ag^+] = S$, so the left side is just S^2, and if undissolved AgOAc is present at equilibrium, the right-hand side is a function of pH only:

$$S^2 = K_{sp}\left(1 + \frac{[H^+]}{K_a}\right) \tag{15-28}$$

In this particular example, with $[H^+] = 1.0 \times 10^{-3}$ and $K_a = 1.8 \times 10^{-5}$,

$$S^2 = K_{sp}\left(1 + \frac{1.0 \times 10^{-3}}{1.8 \times 10^{-5}}\right) = K_{sp}\,(1 + 56)$$

and

$$S = (57 K_{sp})^{1/2} = 7.5\,K_{sp}^{\,1/2} = 7.5 \times 5.0 \times 10^{-2}$$
$$= \underline{0.38\ F}$$

At high pH, where essentially no HOAc is formed, the solubility of AgOAc is $K_{sp}^{\,1/2} = 5.0 \times 10^{-2}\ F$. Thus, it has been increased $7\frac{1}{2}$-fold by the increased acidity.

Equation (15-28) is perfectly general for a salt M^+X^-. When the pH is well above pK_a, then $[H^+]/K_a$ is very small and $S \approx K_{sp}^{\,1/2}$. In solutions sufficiently acidic that $[H^+]/K_a$ is large, the solubility is increased by approximately the factor $([H^+]/K_a)^{1/2}$; that is, $S \approx K_{sp}^{\,1/2}\,([H^+]/K_a)^{1/2}$.

EXAMPLE 15-9   **Prevention of precipitation of $Mg(OH)_2$.** How many moles of NH_4Cl must be added to 100 ml of $0.10\ F\ NH_3$ so that, when the latter is added to 100 ml of $0.020\ F$ $MgCl_2$, no precipitate of $Mg(OH)_2$ will form? K_{sp} for $Mg(OH)_2$ is 1.8×10^{-11} and K_b for NH_3 is 1.8×10^{-5}. No complexes with NH_3 are formed by Mg^{2+}.

Solution. Addition of NH_4Cl, which is a weak acid, reduces the OH^- concentration in the NH_3 solution and creates a buffer. We must calculate the maximum concentration of OH^- that can be present without the formation of $Mg(OH)_2$ and then the minimum concentration of NH_4^+ needed in a solution with $[NH_3] = 0.10\ M$ to produce at most this concentration of OH^-.

Since the $MgCl_2$ solution is $0.020\ F$ and is to be diluted twofold by addition of the buffer, the final $[Mg^{2+}] = 0.010$. Thus, to prevent precipitation,

$$[Mg^{2+}][OH^-]^2 < 1.8 \times 10^{-11}$$

or
$$[OH^-]^2 < \frac{1.8 \times 10^{-11}}{0.010} = 1.8 \times 10^{-9}$$

so that $[OH^-] = \sqrt{18} \times 10^{-5} = 4.2 \times 10^{-5}$ is the upper limit of permissible concentrations.

The ammonia buffer is governed by the relationship

$$\frac{[NH_4^+]}{[NH_3]} = \frac{K_b}{[OH^-]} = \frac{1.8 \times 10^{-5}}{[OH^-]}$$

Replacement of $[OH^-]$ by its upper permissible limit just calculated yields a lower limit for the ratio $[NH_4^+]/[NH_3]$, because $[OH^-]$ is in the denominator.

$$\frac{1.8 \times 10^{-5}}{4.2 \times 10^{-5}} = 0.42 \le \frac{[NH_4^+]}{[NH_3]}$$

Hence the concentration of ammonium ion must be above 0.42 times that of NH_3. In 100 ml of 0.10 F NH_3, this means about 0.05 F NH_4^+, or about $\underline{5 \times 10^{-3}\ mol}$ of NH_4Cl should be added to the 100-ml sample.

A few salts of common weak bases are not very soluble. Their solubility increases with increasing pH for the same reason that the solubility of the salt of a weak acid increases with decreasing pH, and they can be treated in an exactly analogous fashion.

Effects of complexing on solubility of ionic solids. The formation of complexes incorporating at least one of the ions present in a solid can increase the solubility of the solid by decreasing the concentration of this ion in the solution in equilibrium with the solid. Two distinct cases can be distinguished. (1) The complexing ligand is also present in the solid phase, e.g., the dissolving of AgCl in excess Cl^- because of the formation of $AgCl_2^-$ and $AgCl_3^{2-}$, mentioned earlier. (2) The complexing ligand is not present in the solid. An example is the dissolving of AgCl(s) by a solution containing NH_3. The formation of $Ag(NH_3)_2^+$ decreases $[Ag^+]$ to such an extent that the solubility-product expression for AgCl(s) can be satisfied only by the dissolving of more AgCl to increase $[Cl^-]$. This is exactly the same effect that a significant concentration of protons has on the anion of a weak acid, and the analysis of a problem involving the complexing of a cation is thus just parallel to that of a problem about the effect of acid on the solubility of the salt of a weak acid, e.g., Example 15-8.

EXAMPLE 15-10   **Solubility of AgCl(s) when $[NH_3] = 1.0\ M$.** What is the solubility of AgCl(s) in a solution in which the final (equilibrium) concentration of NH_3 is 1.0 M? Show that within the accuracy of such a calculation (at best about 10 percent) the answer would be nearly the same if the initial (not final) concentration of NH_3 had been specified as 1.0 M. K_{sp} for AgCl(s) $= 1.8 \times 10^{-10}$; $K_{1d} = 1.5 \times 10^{-4}$ and $K_{2d} = 6 \times 10^{-4}$ for $Ag(NH_3)_2{}^+$.

Solution. Let the solubility be S gfw liter^{-1}. We know that

$$S = [Cl^-] = [\text{total Ag(I)}] = [Ag^+] + [AgNH_3{}^+] + [Ag(NH_3)_2{}^+]$$

and since the ammonia concentration is so high, essentially all the Ag(I) is in the form of $Ag(NH_3)_2{}^+$ (Fig. 15-3), so that

$$S = [Cl^-] \approx [Ag(NH_3)_2{}^+] \tag{15-29}$$

We then express $[Ag(NH_3)_2{}^+]$ in terms of $[Ag^+]$, since it is $[Ag^+]$ that occurs in K_{sp}. To do this, we use $K_{tot,d} = K_{1d}K_{2d} = 9 \times 10^{-8}$:

$$K_{tot,d} = \frac{[Ag^+][NH_3]^2}{[Ag(NH_3)_2{}^+]} = 9 \times 10^{-8}$$

or, with $[NH_3] = 1.0\ M$,

$$[Ag(NH_3)_2{}^+] = \frac{[Ag^+][NH_3]^2}{9 \times 10^{-8}} = 1.1 \times 10^7\ [Ag^+] \tag{15-30}$$

Finally, multiplying (15-30) by $[Cl^-]$ and noting (15-29),

$$[Ag(NH_3)_2{}^+][Cl^-] = S^2 = 1.1 \times 10^7\ [Ag^+][Cl^-]$$

or $\qquad S^2 = 1.1 \times 10^7 \times K_{sp} = (1.1 \times 10^7)(1.8 \times 10^{-10}) = 2.0 \times 10^{-3}$

Thus $S = \underline{4.5 \times 10^{-2}\ F}$, or a little more than $\underline{6\ g\ liter^{-1}}$.

With an *initial* ammonia concentration of 1.0 M, the equilibrium concentration of NH_3 would be somewhat smaller because two NH_3 molecules are removed for each $Ag(NH_3)_2{}^+$ formed. As a first approximation, we assume that the solubility is about the same as that just calculated, $4.5 \times 10^{-2}\ F$, and thus the concentration of NH_3 removed by complexing is $2 \times 4.5 \times 10^{-2} = 0.09\ M$. The equilibrium concentration of NH_3 is then about 0.91 M. This implies

$$S = [Ag(NH_3)_2{}^+] = 1.1 \times 10^7 (0.91)^2 [Ag^+]$$

and substituting as above we find

$$S^2 = 1.1 \times 10^7 K_{sp} (0.91)^2$$

Thus the solubility is about 0.91 times what it was when the final (rather than initial) NH_3 concentration was 1.0 M, or about $0.91\ (4.5 \times 10^{-2}) = \underline{4.1 \times 10^{-2}\ F}$. This second approximation then implies an equilibrium value of $[NH_3]$ of $1.0 - 2(0.041) = 0.92\ M$, which in turn implies that the solubility is reduced by about 8 percent, rather than the 9 percent calculated in the first approximation. The successive approximations have converged adequately.

The foregoing examples illustrate methods of dealing with simultaneous competing equilibria in some heterogeneous systems. An important point to remember here, as in Chap. 14, is that consideration of the relative magnitudes of the equilibrium constants involved frequently permits making simplifying assumptions that eliminate many of the apparent complexities. The treatment of simultaneous equilibria can be complicated when only the total amounts of substances added are known. It is simpler when the final concentration of any of the species involved is known, because this makes it possible to find the concentrations of all other species that are related to the first by equilibrium expressions. The situation is analogous to that with acids and bases: knowledge of the final pH greatly simplifies calculations of concentrations, which are more difficult to find when only the total formalities of all acids and bases present (and their respective dissociation constants) are known.

15-5 Activity

As pointed out at the start of this chapter and in Sec. 13-5, comment 8, the use of concentrations in equilibrium constant expressions is an approximation valid for dilute solutions: it is exact only in the limit of infinite dilution. In more concentrated solutions, the values of equilibrium expressions written in terms of concentrations are no longer independent of concentrations; they deviate markedly from those obtained in dilute solutions. These deviations are in large part attributable to the interactions between solute molecules. It is to be expected, then, that the deviations become more significant as the likelihood for an encounter between solute species increases, that is, as the concentration increases. Deviations are much more pronounced for ions than for neutral molecules because of the strong electrostatic forces between charged particles.

However, all these deviations of the values of equilibrium expressions from constancy can be removed if *activities,* or "effective concentrations", are substituted for the actual concentrations of the reacting species. The activity of a given species in a particular solution can be uniquely defined; it can be measured in a variety of ways and the same value is found to be applicable to different phenomena. Thus, the activities of HCl (and H^+ and Cl^- ions) in an aqueous solution can be determined from the freezing-point depression; the same values of the activities will be found to explain quantitatively the vapor pressure of HCl above the solution, the pH of the solution, and other effects that for a dilute solution depend only on the concentration of the solute.

Figure 15-5 shows how the ratio of the activity to the concentration varies with concentration for some typical solutes in aqueous solution. Although activities are usually determined experimentally, they can also be calculated in some situations that are simple enough for detailed theoretical analysis.

At low pressures, expressions for equilibria involving gases can be written in terms of partial pressures, but as the pressure is increased, deviations from the low-pressure values of the equilibrium constants are observed. These deviations are attributable to molecular interactions and can be taken into account by the use of "effective pressures", called *fugacities.*

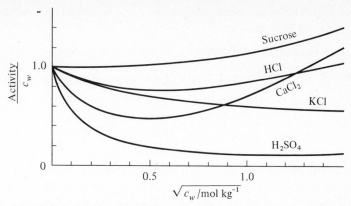

FIGURE 15-5 **The ratio of activity to concentration for several solutes.** The ratio of activity to concentration is often plotted against the *square root* of the weight molality, c_w, for theoretical reasons. This ratio is equal to unity at infinite dilution and, for ionic substances, decreases as the concentration is increased, because of the mutual attraction of oppositely charged ions such as Na^+ and Cl^-. For substances with doubly and triply charged ions, such as $BaCl_2$, Na_2SO_4, and $Al_2(SO_4)_3$, the deviations from unity are larger than for singly charged ions. The deviations are smallest for uncharged molecules, as shown by the curve for sucrose.

Problems and Questions

15-1 **Form of the solubility product.** Let S equal the solubility in gram formula weights per liter for each of the following salts. Express the solubility product of each salt in terms of S: (*a*) AgBr; (*b*) PbF_2; (*c*) Ag_2CrO_4; (*d*) Ag_3AsO_4; (*e*) $BaSO_4$.
$\qquad$ *Ans.* (*a*) S^2; (*b*) $4S^3$; (*c*) $4S^3$; (*d*) $27S^4$; (*e*) S^2

15-2 **Solubilities from K_{sp} values.** Calculate the solubility in water of each of the following substances from its K_{sp} value. Express the solubility in gram formula weights per liter and in grams per 100 ml.
(*a*) $BaCrO_4$ $\quad K_{sp}, 3 \times 10^{-10} M^2$ $\qquad$ (*d*) $PbSO_4$ $\quad K_{sp}, 1.7 \times 10^{-8} M^2$
(*b*) Ag_2SO_4 $\quad K_{sp}, 1.6 \times 10^{-5} M^3$ $\qquad$ (*e*) $Fe(OH)_3$ $\quad K_{sp}, 10^{-36} M^4$
(*c*) $Mg(OH)_2$ $\quad K_{sp}, 1.8 \times 10^{-11} M^3$

15-3 **K_{sp} from solubility data.** The solubilities of the following ionic compounds are given in grams per 100 ml of saturated solution in water. Calculate the solubility-product constant for each substance.
(*a*) SrF_2 $\quad 1.1 \times 10^{-2}$ $\qquad$ (*d*) Ag_3PO_4 $\quad 6.7 \times 10^{-4}$
(*b*) Ag_2S $\quad 9 \times 10^{-16}$ $\qquad$ (*e*) $PbCrO_4$ $\quad 4.5 \times 10^{-6}$
(*c*) CuI $\quad 4 \times 10^{-5}$
$\qquad$ *Ans.* (*a*) $3 \times 10^{-9} M^3$; (*b*) $2 \times 10^{-49} M^3$; (*c*) $4 \times 10^{-12} M^2$;
$\qquad\qquad$ (*d*) $1.8 \times 10^{-18} M^4$; (*e*) $2 \times 10^{-14} M^2$

15-4 **Solubility of silver carbonate.** The solubility product of silver carbonate is $5 \times 10^{-12} M^3$ at 25°C. (*a*) Calculate the approximate formal solubility of silver carbonate in water and its solubility in milligrams per 100 ml. (*b*) Calculate the formal solubility of silver carbonate in 0.1 *F* $AgNO_3$ and in 0.1 *F* Na_2CO_3.

15-5   **Solutions of lead fluoride.** The solubility product of PbF_2 is about 4×10^{-8} M^3 at 25°C. (*a*) Calculate the solubility of PbF_2 in water in milligrams per 100 ml. (*b*) Calculate the formal solubility of PbF_2 in 0.20 *F* $Pb(NO_3)_2$. (*c*) The solubility of PbF_2 in nitric acid is appreciably greater than its solubility in pure water, but its solubility in aqueous ammonia is no greater than in water. Discuss each of these observations.

> *Ans.*   (*a*) 5×10^1 mg per 100 ml;   (*b*) 2×10^{-4} gfw liter^{-1};
> (*c*)   *Hint:* Note that HF is not a strong acid.

15-6   **Test for chloride ion.** Suppose that a 100-ml sample of a solution is to be tested for chloride ion by addition of 1.0 ml of 0.20 *F* $AgNO_3$. What is the minimum number of grams of Cl$^-$ that must be present in order for some AgCl to be formed? (K_{sp} for AgCl is 1.8×10^{-10} M^2.)

15-7   **Insoluble carbonates.** The solubility of $CaCO_3$ in water is about 7 mg liter^{-1}. Show how one can calculate the solubility product of $BaCO_3$ from this information and from the fact that when sodium carbonate solution is added slowly to a solution containing equimolar concentrations of Ca^{2+} and Ba^{2+}, no $CaCO_3$ is formed until about 90 percent of the Ba^{2+} has been precipitated as $BaCO_3$. *Ans.* K_{sp} for $BaCO_3 = 5 \times 10^{-10}$ M^2

15-8   **Precipitation of $PbBr_2$.** The solubility product of $PbBr_2$ is 7×10^{-5} M^3. If 10 ml of a 0.02 *F* NaBr solution is added to 10 ml of 0.2 *F* $Pb(NO_3)_2$ solution, will any solid $PbBr_2$ be present at equilibrium?

15-9   **Solubility of carbonates in acids.** It is sometimes asserted that carbonates are soluble in strong acids because a gas is formed that escapes (CO_2). Suppose that CO_2 were extremely soluble in water (as, for example, ammonia is) so that it did not leave the site of the reaction, but that otherwise its chemistry was unchanged. Would calcium carbonate be soluble in strong acids? Explain. *Ans.* yes

15-10   **Solubility of CaF_2 in an acidic solution.** The solubility of CaF_2 at pH = 1.0 is about 5.4×10^{-3} *F*. Use this information, together with the solubility product of CaF_2 ($K_{sp} = 3.4 \times 10^{-11}$ M^3), to estimate the dissociation constant of HF.

15-11   **Strontium carbonate and strontium fluoride.** The solubility-product constant for $SrCO_3$ is 1.1×10^{-10} M^2 and that for SrF_2 is 2.9×10^{-9} M^3. Calculate the formal solubility of strontium carbonate in a solution in which the fluoride-ion concentration is held constant at 0.10 *M* by calculating the equilibrium concentration of CO_3^{2-} in such a solution. Is $[Sr^{2+}]$ equal to $[CO_3^{2-}]$ at equilibrium? If not, why not? *Ans.* 3.8×10^{-4} *F*

15-12   **Separation of Fe(III) and Ni(II).** The solubility products of $Fe(OH)_3$ and $Ni(OH)_2$ are about 10^{-36} M^4 and 6×10^{-18} M^3, respectively. Find the approximate pH range suitable for the separation of Fe^{3+} and Ni^{2+} by precipitation of $Fe(OH)_3$ from a solution initially 0.01 *M* in each ion, as follows: (*a*) calculate the lowest pH at which all but 0.1 percent of the Fe^{3+} will be precipitated as $Fe(OH)_3$; (*b*) calculate the highest pH possible without precipitation of $Ni(OH)_2$.

15-13   **Simultaneous precipitation of AgCl and Ag_2CrO_4.** The solubility products of AgCl and Ag_2CrO_4 are $1.8 \times 10^{-10}\ M^2$ and $2.0 \times 10^{-12}\ M^3$, respectively. Suppose that a very dilute $AgNO_3$ solution is added dropwise to a solution that is $0.0010\ M$ in Cl^- and $0.010\ M$ in CrO_4^{2-}. Ignore volume changes. Will $AgCl$ or Ag_2CrO_4 precipitate first? Approximately what fraction of the anion first precipitated will have been removed when the second one starts to precipitate? Comment on the answer in relation to the Mohr titration of a solution containing $[Cl^-] = 0.0010\ M$. *Ans.* $AgCl$ precipitates first; a little less than 99 percent

15-14 **Solubility of CdC_2O_4 in water and in an ammonia solution.** Cadmium oxalate, CdC_2O_4, is a relatively insoluble salt with $K_{sp} = 4 \times 10^{-8}\ M^2$. The element cadmium lies directly below zinc in the periodic table, and like Zn^{2+}, Cd^{2+} forms a tetrammine complex, $Cd(NH_3)_4^{2+}$, for which $K_{tot,d} = 2 \times 10^{-7}\ M^4$. (a) Calculate the solubility of CdC_2O_4 in water, in both gram formula weights per liter and grams per 100 ml. Ignore possible hydrolysis and other competing reactions. (b) Calculate the formal solubility of CdC_2O_4 in a solution in which the concentration of NH_3 after equilibrium has been established is $1.0\ M$. Assume that the only complex formed in significant quantity at this ammonia concentration is $Cd(NH_3)_4^{2+}$.

15-15 **Solubility of AgBr in a thiosulfate solution.** Calculate the solubility of AgBr ($K_{sp} = 5.2 \times 10^{-13}\ M^2$) in a solution in which sufficient sodium thiosulfate has been dissolved that, at equilibrium, the concentration of $S_2O_3^{2-}$ is $0.2\ M$. Use constants for the silver-thiosulfate system from Table 15-2. {*Hint:* Note that the equilibrium value of $[S_2O_3^{2-}]$ is just equal to K_{1d} for $Ag(S_2O_3)_3^{5-}$, so that the concentrations of the trithiosulfato and dithiosulfato complexes are equal and, as is evident from Fig. 15-2, that of the monothiosulfato complex is negligible.} Use the resulting solubility of AgBr in this solution and the known distribution of the complex species, to calculate what the *initial* concentration of $Na_2S_2O_3$ must have been before any AgBr dissolved. *Ans.* about $1.1\ F$; about $2.95\ F$

15-16 **Distribution of complex species.** With the help of Fig. 15-2, evaluate within 5 percent the fraction of the total Ag(I) present in a solution as Ag^+ and in the form of each of the possible complexes with thiosulfate when the concentration of thiosulfate at equilibrium is (a) $10^{-1.0}\ M$; (b) $10^{-5.0}\ M$; (c) $10^{-9.0}\ M$. Show that your answers are consistent with the data in Table 15-2.

15-17 **Distribution of complex species.** With the help of Fig. 15-3, evaluate the approximate percentage of the total Ag(I) present in a solution as Ag^+ and in the form of the monoammine and diammine complexes when the concentration of ammonia at equilibrium is (a) $10^{-2.5}\ M$; (b) $10^{-3.5}\ M$; (c) $10^{-4.5}\ M$. Show that your answers are consistent with the results of Example 15-7. *Ans.* (a) about 1, 5, 94; (b) about 41, 18, 41; (c) about 93, 5, 2

"A theory is the more impressive the greater the simplicity of its premises, the more varied the subjects it ties together, and the wider the domain of its applicability. Thus the deep impression that classical thermodynamics made upon me. It is the only physical theory of general content about which I have the conviction that it will never be overturned within the bounds of applicability of its fundamental concepts (for the special attention of inveterate sceptics)."

Albert Einstein, 1946[1]

16 The First Law of Thermodynamics: Energy and Enthalpy

The chief topic of this chapter and the next one is the relationship between the energy and the temperature of a system that contains a large number of atoms or molecules. Our approach is macroscopic, through the postulates of thermodynamics. We begin with a brief exposition of the nature and the purpose of thermodynamics and continue with a development of the first law of thermodynamics (the principle of conservation of energy), starting with basic concepts of work, internal energy, and

[1] P. A. Schlipp (ed.), "Albert Einstein, Philosopher-Scientist", Tudor Publishing Company, New York, 1951.

heat. Some chemical applications follow. A review of the discussion of temperature and temperature scales given in Sec. 1-3 is recommended.

16-1 The Nature of Thermodynamics

Thermodynamics is the science of heat and temperature and, in particular, of the laws governing the conversion of heat into mechanical, electrical, or other macroscopic forms of energy. It is a central branch of science with important applications in chemistry and other physical sciences, in biology, and in engineering.

Thermodynamics is a macroscopic theory, concerning quantities such as pressure, temperature, or volume. It is both the strength and the weakness of thermodynamics that the relationships based upon it are completely independent of any microscopic explanation of physical phenomena. The *strength* is that thermodynamic relationships are not affected by the changes in microscopic explanations that continue to occur as the theories of atomic and molecular interactions and structure are modified and improved. On the contrary, the conclusions of atomic and molecular theories *must not* contradict those of thermodynamics, so that thermodynamics can be used as a guide or as a touchstone in the development of microscopic theories. Much of thermodynamics was, in fact, developed at a time when many scientists did not believe in atoms and molecules, long before detailed atomic theories were available.

The *weakness* of thermodynamics is that it does not provide the deep insight into chemical and physical phenomena that is afforded by microscopic models and theories. Although thermodynamics is a completely self-contained macroscopic theory, it is nevertheless possible to find a microscopic interpretation of it in what is called statistical mechanics, which provides considerable insight and is of great value for a full understanding of thermodynamics. On the other hand, to understand statistical mechanics a great deal of mathematics is required, whereas the more simple aspects of calculus suffice for thermodynamics.

Several postulates, referred to as *laws,* are basic to thermodynamics. Although these laws are consistent with the results of all known experiments, this great mass of observational material serves only as support, not as proof. The "laws" of thermodynamics are postulates, or axioms, as are all "laws" of nature.

An important characteristic of thermodynamics is that it permits the derivation of *relationships* among various laws, even though those laws themselves are not a consequence of thermodynamics. For example, if Raoult's law of vapor-pressure lowering is assumed, other colligative properties of a solution, such as the osmotic pressure law, can be derived. Neither Raoult's law nor the osmotic pressure law, however, is an individual consequence of thermodynamics. One or the other must be accepted on the basis of experiment or derived from a detailed molecular theory of solutions.

As we shall see in the following chapter, it is possible to give thermodynamic criteria for the direction in which a chemical reaction is likely to proceed. However, it is only in this sense that thermodynamics is concerned with time—in the sense of the direction of the flow of time, or what has been called "time's arrow". It is not possible to establish by thermodynamics how rapidly a reaction proceeds toward equilibrium. In favorable cases this may be established by detailed molecular theories, but in general each reaction must be investigated experimentally (Chap. 20). Thermodynamics might thus more properly be called thermostatics, but the term *thermodynamics,* coined in the early days of steam-engine theory, is somehow more enticing and is firmly entrenched.

16-2   Some Basic Concepts and Definitions

System and surroundings.   As a prelude to thermodynamic reasoning, the physical universe is divided conceptually into the *system,* the part under particular scrutiny, and the *surroundings.* The system may be complex and contain all kinds of substances, machinery, and equipment, or it may be very simple and consist only of a homogeneous piece of matter, such as a certain volume of air. What is included in the system is a matter of convenience that depends on the question being considered. The boundaries of the system must be well defined; they may be real, such as the walls of a box, or they may be purely imaginary. Physical boundaries may be part of the system and their properties may be important, or they may not; they may be movable or fixed. A system is *open* if a flow of *matter* across the boundaries is possible; otherwise it is *closed.* We shall normally be concerned with closed systems.

Internal energy.   The concepts of energy and the conservation of energy are easily visualized in mechanical systems. We expend energy when we lift a weight and we get energy back when we allow the weight to fall. If there are no frictional or other losses, the energy expended and the energy regained are equal.

We recognize that energy can also be stored in matter and that there are material sources of energy (e.g., a charged storage battery and an explosive before detonation); in these cases, however, the energy relationships are not easily visualized. A charged and an uncharged battery look alike, and a sample of explosives may be difficult to distinguish from a very stable compound.

However, it is possible by experimental means to establish a relation between mechanical energy and energy stored in matter. The principle of the experiment is simple: perform mechanical work on a substance and observe the change produced in the properties of the substance. If the experiment is repeated for varying amounts of work, it is possible to establish a one-to-one relationship between the amount of work performed and the extent of the change in the substance. The experimental conditions must be carefully controlled, however, so that the only changes induced in the substance come from the mechanical work.

If a beaker of water at 10°C is placed in a room at 20°C, the properties of the water will change—its temperature will rise to 20°C and there will be an attendant increase in its vapor pressure and a decrease in its density. Other physical properties, such as the viscosity and the electrical conductivity, will also change. The rate at which these changes occur can be decreased by insulating the beaker—for example, by placing it inside a jacket made of foam plastic. Addition of more layers of insulation will slow the rate of change still further until it is imperceptible. It is possible to imagine a system so well insulated that its properties are unaffected by the temperature of its surroundings. Such a system is said to be thermally insulated or enclosed by an *adiabatic* (Greek: *a*, not; *diabatos*, passable) wall. If careful attention is paid to the design of apparatus, it is possible to perform mechanical work on a system that is surrounded by an adiabatic wall.[1] This is the experimental procedure that leads to a unique relationship between mechanical work and energy stored in matter.

Consider the experiment shown in Fig. 16-1. A mixture of 1 kg of ice and 1 kg of liquid water at 0°C enclosed by adiabatic walls can be stirred by a paddle powered through a pulley system by a falling lead weight. (For simplicity we assume that the pulley system is frictionless.) When the mixture is stirred, ice melts. Suppose that the stirring is continued only until the last trace of ice has melted; the temperature of the water has not yet risen above 0°C. If the weight was initially at height h_i and has dropped to h_f, an amount of work $w = mg(h_i - h_f)$ has been performed, where m is the mass of the lead and g is the gravitational acceleration. Where has this energy gone? All observations on systems such as this are consistent with the concept of the conservation of energy; the change in the properties of the system (that is, the melting of the ice) has been caused by the absorption of energy of an

---

[1] Although a truly adiabatic wall is an idealization, it is possible to carry out an experiment in such a way that the effect of a temperature difference between the system and the surroundings is negligible.

FIGURE 16-1 **Melting ice by performing mechanical work.** The descending weight rotates a paddle in a beaker containing water and ice. The beaker is thermally insulated. The transfer of energy to the water by friction causes ice to melt.

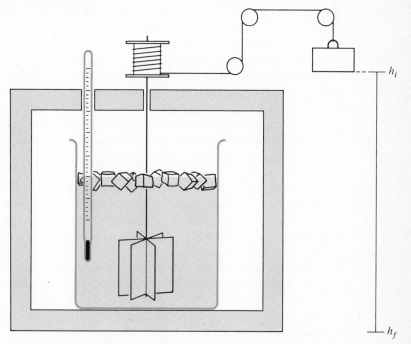

amount w. If the initial energy of the system is designated by U_i and the energy after stirring by U_f, then the transfer of energy from the weight to the system can be described by the equation

$$\Delta U = U_f - U_i = w \qquad \text{(adiabatic)} \qquad (16\text{-}1)$$

where the condition of thermal insulation is noted by the word *adiabatic*.

If we replace the paddle-and-pulley arrangement with a resistor of value **R**, a similar experiment can be performed using electrical energy. When a current **I** is passed through the resistance **R** for a time t, an amount of electrical energy $w_{\text{elec}} =$ **I**^{2}**R**t (see Study Guide) is dissipated. If we start with the same amount of ice and measure the electrical work required to melt it, we find that w_{elec} is exactly the same as the mechanical work, as long as the experiment is performed under adiabatic conditions. Also, if some portion of the ice is melted by mechanical work and electrical work is used to melt the rest, the sum of the adiabatic work steps is equal to ΔU.

Experience shows that these observations are not unique to the ice-water mixture described. The amount of energy required to bring a thermally insulated system from one set of conditions to another is well defined and reproducible. It is therefore possible to use (16-1) to define the energy that can be contained in matter, the internal energy U. More precisely, (16-1) is a definition of the *change* in internal energy, ΔU, and not of the internal energy itself. The *absolute internal energy* of a system remains undefined in thermodynamics, and an arbitrary constant may be added to any set of U values. Since we deal only with differences of internal energy, such a constant always drops out. Microscopically, the internal energy of a substance is the entire energy of its atoms or molecules, kinetic and potential. For thermodynamic considerations, however, analysis of the many interactions on the molecular level is not needed.

States and state functions.   In the preceding discussion of the transfer of energy to a thermally insulated system, we were able to calculate the work performed by a falling weight by specifying the change in height, $h_i - h_f$. Only the initial and final heights need be specified; the position of the weight in any intermediate stage is irrelevant. This principle follows from the conservation of energy. If the work obtainable from a weight were dependent on the intermediate steps between the initial and final positions, energy could be produced in a cyclic process in which a weight is shifted between two positions along different paths (Fig. 16-2).

The condition of a system is called its *state*[1] and is described by specifying the values of all the macroscopic properties that are needed to define it unambiguously, so that the system could be duplicated precisely from this information. Different arrangements of molecules and their velocities that lead to the same macroscopic properties are thermodynamically indistinguishable. Description on the molecular level is therefore neither pertinent nor necessary to define the state of a system.

A considerable body of experimental evidence shows that under adiabatic condi-

---

[1]The uses of the word *state* in thermodynamics and in quantum mechanics (as in "the ground state of an atom") are unrelated.

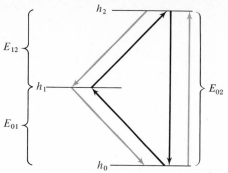

FIGURE 16-2 **Energy changes in a cyclic process.** A weight is lifted from h_0 to h_1 and then from h_1 to h_2. It is then lowered from h_2 back to h_0. The changes in the gravitational potential energy of the weight in the steps are E_{01}, E_{12}, and $-E_{02}$, respectively. If the potential energy of the weight did not depend only on its height, the sum of E_{01} and E_{12} would not necessarily equal E_{02} and it would be possible to create energy. Suppose $E_{01} + E_{12}$ were less than E_{02}; then each time the weight were brought through one complete cycle $h_0 \longrightarrow h_1 \longrightarrow h_2 \longrightarrow h_0$ an amount of energy $E_{02} - (E_{01} + E_{12})$ would be produced. If $E_{01} + E_{12}$ were greater than E_{02}, moving the weight through the cycle $h_0 \longrightarrow h_2 \longrightarrow h_1 \longrightarrow h_0$ would produce energy. Experience shows that the creation of energy by such cyclic processes is not possible; hence $E_{01} + E_{12} = E_{02}$.

tions the work required to produce a given change of state in a system, including a change in its internal energy, is independent of the path. A basic tenet of thermodynamics is a generalization: the change in internal energy of a system as the result of *any* process, adiabatic or not, can depend only on the initial and final conditions of the system and not on any intermediate steps. This statement is one form of the first law of thermodynamics. The first law cannot be derived from first principles; however, its validity is supported by the success of the thermodynamic structure that can be built upon it. Alternative statements of the first law are given in the following section.

Properties that depend only on the state of the system and not on its history (i.e., how it was prepared) are called *state variables* or *state functions*. Typical state variables for a homogeneous system include the volume, temperature, pressure, density, and internal energy. It will be seen shortly that work is *not* a state function.

State variables are not necessarily independent of each other. For example, the volume and density of a specified quantity of gas are determined by the pressure and temperature. The relationship among temperature, pressure, and volume is called the equation of state (Sec. 5-1) and depends, of course, on the nature of the substance. The number of variables that need to be specified to define the state of any particular system depends in part on the nature of the system and the focus of one's interest. If unpaired electrons are present, for example, the strength of any magnetic field present must be specified for a complete description; with a two-phase system containing a solid and a gas, it may be necessary to specify the surface area in contact with the gas if adsorption is important. However, we shall normally ignore such specialized effects.

When the state of a system is altered, the *change* of any *state function* depends, by definition, only on the initial and final states. It must therefore be *independent of the path* along which the change in state has taken place. This important fact is sometimes overlooked in involved situations. If, as is often done, the initial state is designated by 1 and the final state by 2, the change ΔX of a state function X is defined by

$$\Delta X = X_2 - X_1 \tag{16-2}$$

Note that the definition of a Δ quantity is always analogous to that given in (16-2); that is, the initial value is subtracted from the final value. In other words, the Δ quantity is what must be added to the initial value to reach the final value, $X_2 = X_1 + \Delta X$.

An extremely useful aspect of the path independence of a state function is that we can always mentally invent a path (Fig. 16-3) to replace the actual path. Whatever tortuous path might describe the actual physical way in which the change took place, whatever the fluctuations in pressure and temperature, whatever the temporary chemical changes, we can select the new path specifically so that the calculation of the change in the state function of interest is particularly simple. The calculation can be performed even if it is impossible to define the path along which the *actual* change in state takes place, as when there is turbulent internal motion along the way. Such difficulties do not affect the existence of state functions and (16-2) remains valid.

If states 1 and 2 are identical, one speaks of a *cyclic process* or a *cycle,* which thus leads from a given state through a sequence of changes back to the given state.

FIGURE 16-3   **Changes of a state function.** Instead of calculating ΔX along the actual path it may be simpler to calculate it along a fictitious path, e.g., as the sum of two changes, the first of which is at constant pressure and the second at constant temperature. Alternatively, if X is known explicitly as a function of the variables describing the initial and final states (for example, T and P), it may be simplest to calculate directly $X(T_2,P_2) - X(T_1,P_1) = \Delta X$, without elaborations concerning the path. The actual path may contain turbulent and even violent portions, for which T or P may not even be defined, so that it may be impossible to describe the actual path in terms of T and P alone. Nevertheless, for any specified initial and final states the quantity ΔX is defined.

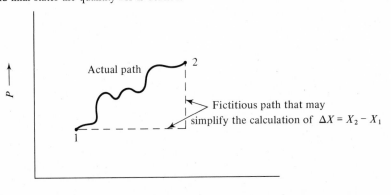

For any cycle and any state function X,

$$\Delta X = 0 \tag{16-3}$$

Heat.   Consider a mixture of 1 kg of ice and 1 kg of water at 0°C, in surroundings at a higher temperature. This ice-water mixture is identical with the system in the experiment illustrated in Fig. 16-1, in its initial state. If the system is *not* thermally insulated, that is, not enclosed by adiabatic walls, the ice will begin to melt. When the 1 kg of ice has just melted and the temperature of the water is still 0°C, the system is in exactly the same state it was in when the ice was melted by the performance of either mechanical or electrical work. Since internal energy is a state function, the system has gained an amount of energy ΔU exactly equal to that transferred to it in the mechanical and electrical experiments. In this case, however, no work has been performed.

The source of the energy is the surroundings. Whenever an uninsulated system is placed in surroundings that are at higher temperature, energy is transferred from the surroundings to the system. If the temperature of the system is higher than that of the surroundings, energy is transferred in the other direction—the system loses energy to the surroundings. Only when the temperatures of the system and surroundings are identical is there no energy transfer.

Energy that is transferred as the result of a temperature difference is called *heat*. When the internal energy of a system is changed by heat transfer alone, that is, when there is no other mode of energy transfer such as mechanical or electrical work, the conservation of energy can be expressed by the equation

$$\Delta U = q \qquad \text{when } w = 0 \tag{16-4}$$

where q is the symbol for the energy gained by the system by heat transfer.[1]

It is important to note that heat is defined only in terms of macroscopic processes involving energy *transfer*. It is as meaningless to say a system contains heat as to say it contains work. In this context a system may only contain *energy*. To give an example from mechanics, a tightly coiled spring contains potential energy that may be used to perform work, but the spring does not therefore contain work. A hot body may transfer heat to a colder body, but neither body can properly be said to contain heat.

16-3   The First Law of Thermodynamics

Until now, we have been careful to choose the conditions of the experiments described so that only one form of energy transfer can take place between the system and the surroundings. In the more general situation the internal energy of a system can be changed by the transfer of both work and heat; that is,

$$\Delta U = q + w \tag{16-5}$$

Equation (16-5) is another statement of the first law of thermodynamics, and (16-1)

---

[1] Energy that is transferred by radiation is considered as heat because a temperature difference is required.

and (16-4) can be considered special cases of the first law for $q = 0$ and $w = 0$, respectively. Another specialized statement of the first law is

$$\Delta U = 0 \qquad \text{when } w = 0, q = 0 \qquad (16\text{-}6)$$

There can be no change in the internal energy of an *isolated* system, a system that cannot exchange energy either as work or as heat. All the statements of the first law are generalizations of the principle of conservation of energy.

We have used the convention that work is positive when it is performed by the surroundings on the system. Work performed by the system on the surroundings is therefore negative. We have used the same sign convention for heat. Heat that is absorbed by the system, that is, heat transferred from the surroundings to the system, is by convention positive and heat transferred from the system to the surroundings negative.

Note that the symbol Δ has not been used with either w or q, because they are not state functions. It is only the sum of q and w, ΔU, that represents a change in a state function [(16-5)]. As we have shown for the melting-ice experiments, the same change in the internal energy can be produced by a process in which $w = 0$ or one in which $q = 0$. This change can also be accomplished along an infinity of other paths for which $q + w = \Delta U$ with $q \neq 0$, $w \neq 0$.

P,V **work.**   An important type of work is *pressure-volume work* (or *P,V* work for short). This is work performed when a system expands or contracts against an external pressure P_{surr}.

As a special case, consider the system to be a substance contained in a cylinder of cross section A with a movable (frictionless) piston (Fig. 16-4). The piston is to be moved a distance dl against an external pressure P_{surr}. By definition, pressure is force divided by area. The work[1] performed by the system is $F \, dl = -Dw$, and since $F = P_{surr} A$ is the force exerted by the piston on the system, $Dw = -P_{surr} A \, dl$. The important minus sign arises from the fact that for a positive dl (that is, an expansion) the system performs work on the surroundings, which by our convention

---

[1]So that state functions and nonstate functions can be kept clearly distinct, we use Dw rather than dw to denote an infinitesimal quantity of work and reserve the symbol d for state functions (as in the "infinitesimal" dl; see Study Guide).

FIGURE 16-4   *P,V* **work.**   When the gas expands and moves the piston of cross-sectional area A through the distance dl, it performs work $F \, dl$, where $F = AP_{surr}$. The volume increase of the gas is $dV = A \, dl$ and the work is thus $P_{surr} A \, dl = P_{surr} \, dV$. This is equal to $-Dw$, because work is performed *on* the surroundings when dV is positive; positive work is, by convention, assumed to be performed *by* the surroundings on the system. The distance dl is assumed to be "infinitesimally" small, so any possible change in P_{surr} caused by the expansion can be neglected.

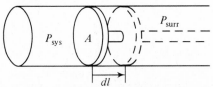

represents negative work. In general, work of expansion is negative and work of contraction, performed by the surroundings on the system, is positive. Since $A\,dl$ is the volume increase dV of the system,

$$Dw = -P_{surr}\,dV \qquad (16\text{-}7)$$

A common situation is one in which P_{surr} is kept constant during a change of volume, so that

$$w = -P_{surr}(V_2 - V_1) = -P\,\Delta V \qquad (16\text{-}8)$$

where V_1 and V_2 are the initial and final volumes. The subscript *surr* has been dropped in the latter half of (16-8) because the context should imply that it is P_{surr} that is meant. The subscripts *sys* (Fig. 16-4) and *surr* are often omitted. In their absence it is always important to reflect whether a given P indicates the pressure in the system or an external pressure exerted by the surroundings on the system.

EXAMPLE 16-1   **Change in internal energy of a gas on expansion.** A 1-liter sample of gas is thermally insulated. The gas is allowed to expand to 10.0 liters against a constant pressure of 1 atm. The expansion is slow enough that only P,V work is performed and any work expended to overcome friction or to accelerate the piston is negligible. Calculate (*a*) the work performed by the gas, (*b*) the change in its internal energy, and (*c*) the change in the internal energy of the surroundings.

Solution. (*a*) Since the pressure in the surroundings is constant, the work performed by the gas, $-w$, is $P(V_2 - V_1) = 1.00\,(10.0 - 1.0)$ liter atm. Thus $-w = (9.0$ liter atm$)(101$ J liter^{-1} atm$^{-1}) = \underline{9.1 \times 10^2\,\text{J}}$. (See Table A-4 for the conversion from liter atmospheres to joules.) (*b*) The change in internal energy of the gas is $\Delta U = q + w = \underline{-9.1 \times 10^2\,\text{J}}$ since $q = 0$. (*c*) The internal energy of the surroundings is increased by $\underline{9.1 \times 10^2\,\text{J}}$.

More generally, for a finite change in volume the external pressure P_{surr} in (16-7) need not remain constant during the change in volume. If the work is to be calculated, however, P_{surr} must be known for each point of the path along which the system expands. The external pressure may be given as a function of the temperatures T and volumes V that the system assumes along the path, $P = P(V,T)$. Thus for a finite change in volume,[1]

$$w = -\int_{V_1}^{V_2} P_{surr}(V,T)\,dV \qquad (16\text{-}9)$$

The limits of the integral denote the initial and final states of the system.

As suggested by their general form, the expressions just given for P,V work, Dw and w, are not restricted to the simple cylinder-and-piston arrangement for which they were derived. They apply to any P,V work. In other words, a change in the volume of any system of completely irregular shape by dV against an external pressure P_{surr} is accompanied by the work $Dw = -P_{surr}\,dV$.

[1] The integral sign $\int$ implies in essence just a sum of "infinitesimals". More details are given in the Study Guide; a deeper understanding requires a knowledge of elementary calculus.

Reversible and irreversible paths. Changes of state may be brought about along reversible or irreversible paths. A *reversible path* is one that may be followed in all its details in either direction. At any point along the path the direction of the change may be reversed by a small alteration in a variable such as the temperature or pressure.

Consider, for example, a gas-filled cylinder equipped with a frictionless piston that exerts pressure on the gas. If this external pressure equals the internal pressure of the gas, there is equilibrium—the piston does not move. By a slight decrease of the piston pressure, the gas can be made to expand, and by a slight increase, to contract. Suppose the gas is expanded by slowly decreasing the piston pressure so that at all times this pressure is very slightly smaller than the pressure the gas exerts on the piston. Figure 16-5 shows the pressure P_{sys} the gas exerts on the piston and the pressure P_{surr} the piston exerts on the gas, as the gas expands from state 1 to state 2. The smaller the pressure difference between piston and gas, the closer the lower curve lies to the upper curve and the slower the rate at which the gas expands. In the idealized case where the two curves coincide, expansion takes an infinite time because there is no pressure difference left to act on the piston. In fact, in this limiting case, expansion can be changed into compression by an *infinitesimal* increase of the outside pressure; similarly, compression can be changed into expansion by an infinitesimal decrease of the outside pressure. Only under these circumstances is the expansion of the gas reversible. Reversible expansion of a gas is thus an idealized concept. Although any actual expansion is necessarily irreversible, one can in practice carry out expansion or compression of a gas, and other changes of state, in such a way as to approach reversible conditions very closely without the need to prolong the experiment unduly.

Reversible processes are important in many thermodynamic considerations, in which so-called thought experiments are imagined rather than actually being carried out. Note that irreversibility does not imply that it is impossible to force a process

FIGURE 16-5   **Pressure as a function of volume during the expansion of a gas.** The gas expands only if there is a net force outward on the piston, requiring that the outside pressure be slightly below the inside pressure as long as expansion is occurring. This makes the expansion irreversible. The expansion is reversible only in the limit that the dashed curve coincides with the solid curve.

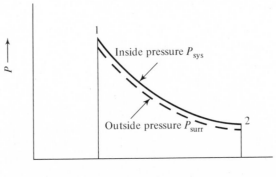

in the reverse direction but rather that *such a reversal cannot be achieved merely by changing a variable by an infinitesimal amount.* We return to a further consideration of reversibility and irreversibility in Chap. 17.

We calculate now the work accompanying the *isothermal reversible* expansion of an ideal gas. The term *isothermal* means that the temperature remains constant.[1] Because the expansion is reversible, the external pressure P_{surr} is to be successively reduced so that it always balances the internal pressure P_{sys}. By the ideal-gas equation,

$$P_{surr} = P_{sys} = \frac{nRT}{V} \tag{16-10}$$

and thus (see Study Guide)

$$w = -\int_{V_1}^{V_2} P_{surr}\, dV = -nRT \int_{V_1}^{V_2} \frac{dV}{V} = -nRT \ln \frac{V_2}{V_1} \tag{16-11}$$

Note that w is negative when $V_2 > V_1$ (expansion) and positive when $V_2 < V_1$ (compression), because the logarithm of a number between 0 and 1 is negative. Figure 16-6 is a pressure-volume diagram of this process; the shaded area represents the work performed by the system and is therefore equal to $-w$. Energy has to be supplied to the system to keep the temperature of the expanding gas constant.

[1] Beginners sometimes confuse the terms *isothermal* and *adiabatic,* which refer to quite distinct conditions. Consider an exothermic reaction. If it is carried out in an insulated container (adiabatic conditions), the energy given off during the reaction raises the temperature of the system because no heat can escape from the container ($q = 0$). This change is adiabatic but not isothermal. On the other hand, the reaction might be allowed to proceed slowly in a vessel immersed in a large constant-temperature bath (a thermostat). The heat of reaction will then be absorbed by the bath without significant change in the temperature of the reaction mixture. This process is isothermal but not adiabatic ($q < 0$). (The changes in state are not the same in the two cases because the final states differ.)

FIGURE 16-6   **Reversible expansion of an ideal gas.**   When a gas is expanded the work performed is $\int_{V_1}^{V_2} P\, dV$, which is equal to $-w$ because the work is performed on the surroundings. The integral is equal to the shaded area. For the reversible expansion of an ideal gas at a constant temperature T, the external pressure must be equal to the ideal-gas pressure, $P = nRT/V$.

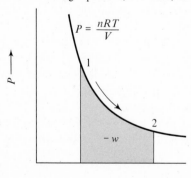

EXAMPLE 16-2   **Isothermal reversible expansion of an ideal gas.**   What is the work performed on the surroundings by the reversible expansion of 1.00 liter of an ideal gas at a constant temperature of 300 K, from 10.0 to 1.00 atm?

Solution.   The final volume of the gas is, by Boyle's law, 10.0 liters. By (16-11) the work performed on the surroundings is $-w = nRT \ln (V_2/V_1)$. Here the unknown number of moles of gas, n, is related to P_1 and V_1, the initial pressure and volume of the gas, by $n = P_1V_1/RT$. Hence

$$-w = P_1V_1 \ln \frac{V_2}{V_1} = (10.0 \text{ atm})(1.00 \text{ liter})(\ln \tfrac{10.0}{1.00})$$

$$= (10.0)(\ln 10.0) \text{ liter atm}$$

The value of $\ln 10.0$ can be obtained directly from tables of natural logarithms or from the relationship $\ln x = 2.303 \log x$, where log is the base-10 logarithm. Hence $\ln 10.0 = 2.303$ and

$$-w = 10.0 \times 2.303 \text{ liter atm} = 23.03 \text{ liter atm}$$

$$= 23.03 \text{ liter atm} \times 101.3 \text{ J/liter atm}$$

$$= \underline{23.3 \times 10^2 \text{ J}}$$

An alternative, slightly more elaborate approach is to calculate n explicitly:

$$n = \frac{10.0 \text{ liter atm}}{(0.0821 \text{ liter atm mol}^{-1} \text{ K}^{-1})(300 \text{ K})}$$

$$= 0.406 \text{ mol}$$

Hence $$-w = nRT \ln 10.0$$

$$= (0.406 \text{ mol}) (8.31 \text{ J mol}^{-1} \text{ K}^{-1})(300 \text{ K})(2.30)$$

$$= \underline{23.3 \times 10^2 \text{ J}}$$

Comparison with the result of Example 16-1, the irreversible expansion of 1.00 liter of gas against a constant pressure of 1.00 atm, shows that the work performed by the gas in the reversible expansion is $\frac{23.3}{9.1} = 2.6$ times larger.

Heat capacity.   The heat required to raise the temperature of a system by one kelvin is called the *heat capacity* C of the system. If the system is one mole of a substance, one speaks of the *molar heat capacity*. The heat capacity for one gram of a substance is called the *specific heat,* so that the molar heat capacity is the specific heat multiplied by the molecular weight. (For substances such as NaCl that do not consist of molecules, the specific heat multiplied by the formula weight is usually referred to as the molar heat capacity.) Molar quantities are indicated by a tilde, as for the molar volume $\tilde{V}$; hence the molar heat capacity is given the symbol $\tilde{C}$.

The energy required to change the temperature of a substance at constant volume from T to $T + \Delta T$ by heat transfer alone is given by the product of the heat capacity at constant volume, C_V, and the temperature change ΔT:

$$q_V = C_V \Delta T = n\tilde{C}_V \Delta T \tag{16-12}$$

where the subscript V has been used to specify the constant-volume path along which the temperature change occurs. If only P, V work is considered, as is often useful, $w = 0$ for a constant-volume process and hence

$$\Delta U = w + q = 0 + q_V = C_V \Delta T$$

Another common mode of heating is at constant pressure, and the corresponding heat capacity is denoted by C_P. Thus, if by heat transfer the temperature of a substance is to be changed from T_1 to T_2 at constant pressure, the amount of heat involved is

$$q_P = C_P \Delta T = n\widetilde{C}_P \Delta T \tag{16-13}$$

The constant-volume and constant-pressure heat capacities are usually not identical because a substance expands when heated at constant pressure and so some energy is also used in P, V work.[1] For *ideal* gases the difference between $\widetilde{C}_P$ and $\widetilde{C}_V$ is

$$\widetilde{C}_P - \widetilde{C}_V = R$$

The heat capacity of a substance is temperature-dependent. For our purposes, however, the change of $\widetilde{C}_P$ and $\widetilde{C}_V$ with temperature is usually sufficiently small to be neglected. For example, it takes 75.2 J to raise the temperature of 1 mol of water at 1 atm from 25 to 26°C; that is, at this temperature $\widetilde{C}_P = 75.2$ J mol^{-1} K^{-1}. (Note that the temperature intervals of 1°C and 1 K are equal by definition.) If the temperature change is from 95 to 96°C, 75.8 J are required; thus at this higher temperature $\widetilde{C}_P = 75.8$ J mol^{-1} K^{-1}, different by less than 1 percent. Heat capacities are always positive; thus if ΔT is positive, heat must be supplied to the system. If ΔT is negative, heat is liberated from the system.

EXAMPLE 16-3   **Temperature change and heat transfer.** Find the heat transferred to or from the surroundings when the temperatures of the following systems are changed: (*a*) a copper penny at 1.0 atm from 20.0 to 25.0°C; (*b*) 150 g water at 1.0 atm from 25.0 to 10.0°C; (*c*) 0.500 mol methanol at 1.0 atm from 15.0 to 25.0°C; and (*d*) the same system and temperature change as in (*c*) but at constant volume. At constant pressure the heat capacity of the penny is $C_P = 1.21$ J K^{-1} and the specific heat of water is 4.18 J g^{-1} K^{-1}. The pertinent molar heat capacities of methanol are $\widetilde{C}_P = 81.6$ J mol^{-1} K^{-1} and $\widetilde{C}_V = 64.0$ J mol^{-1} K^{-1}. These heat capacities remain constant within the precision of the values given and in the temperature ranges specified.

Solution. Applying (16-12) and (16-13) as needed, we have:

(*a*) $q_P = (1.21 \text{ J K}^{-1})(5.0 \text{ K}) = \underline{6.0 \text{ J}}$

(*b*) $q_P = (150 \text{ g})(4.18 \text{ J g}^{-1}\text{K}^{-1})(-15.0 \text{ K}) = \underline{-9.4 \text{ kJ}}$

(*c*) $q_P = (0.500 \text{ mol})(81.6 \text{ J mol}^{-1}\text{K}^{-1})(10.0 \text{ K}) = \underline{408 \text{ J}}$

(*d*) $q_V = (0.500 \text{ mol})(64.0 \text{ J mol}^{-1}\text{K}^{-1})(10.0 \text{ K}) = \underline{320 \text{ J}}$

In (*b*) heat is transferred to the surroundings (that is, q is negative); in the other three cases heat is transferred to the system (q is positive).

[1]Moreover, a volume increase is usually accompanied by a change in internal energy, an important exception being ideal gases.

16-4   Energy Changes in Chemical Reactions

The application of thermodynamics to chemical reactions is a straightforward extension of the principles that have already been discussed. The system consists of a quantity of some substance or mixture of substances that, in the initial state, is termed the reactants. In the final state these substances have been converted into new substances called the products. The change in internal energy as a result of a chemical reaction

$$a\text{A} + b\text{B} + \cdots \longrightarrow c\text{C} + d\text{D} + \cdots$$

is then given by

$$\Delta U = U_f - U_i = \Sigma U_{\text{prod}}(P_f, T_f, \ldots) - \Sigma U_{\text{react}}(P_i, T_i, \ldots) \tag{16-14}$$

where the quantity ΣU_{prod} is the sum of the internal energies of all products. If c mol of product C and d mol of product D are formed, then

$$\Sigma U_{\text{prod}} = c\tilde{U}_{\text{C}} + d\tilde{U}_{\text{D}}$$

ΣU_{react} is similarly defined for all reactants. The number of moles used in the sums must be those in the balanced equation if ΔU is to be meaningful. Note that it is necessary to specify the initial and final conditions, such as pressure and temperature, in order to characterize the initial and final states.

For reactions that take place at constant volume there is no P,V work. Therefore when $w = 0$ (absence of frictional, electrical, and all other non-P,V work), the heat q_V that is transferred to the system as a result of the reaction at constant volume equals the increase in internal energy:

$$q_V + w = q_V = \Delta U = \Sigma U_{\text{prod}} - \Sigma U_{\text{react}} \tag{16-15}$$

The quantity called the *heat of reaction* is usually defined as the heat transferred *to the surroundings*. Thus it is $-q$. For an exothermic reaction at constant volume, heat is liberated by the reaction; $-q_V = -\Delta U$ is positive and ΔU is negative. For an endothermic reaction, heat is absorbed and ΔU is positive.

Chemical processes are usually run in open vessels at atmospheric pressure, so that they take place at constant pressure rather than constant volume. Hence heats of reaction at constant pressure rather than at constant volume are of primary interest. In fact, unspecified "heats of reaction" imply constant pressure. Such heats of reaction are considered at the end of this section.

Calorimetry.   How do we measure energies of reaction and the energies that are associated with phase changes? There are no heat meters that can somehow probe the system and give a direct reading of an energy change. Heats of reaction are determined by calorimetric (Latin: *calor,* heat) measurements in which (in spite of the derivation of the name) the quantities directly measured are usually temperature and electrical work.

In a typical determination of the heat of a reaction, a vessel containing the separated reactive substances is submerged in a large quantity of water held in a thermally insulated container. The temperature of the water is monitored and when it reaches a constant value T_i, the reaction is initiated without affecting the adiabatic system. This can be done for example by a mechanical linkage that opens a valve and allows the reactants to mix. After the reaction is complete, the temperature is again monitored until it reaches a constant value T_f.

Since the conditions are such that there is essentially no heat exchange with the surroundings, any temperature change of the system caused by the reaction is related to the heat of reaction, q_{react}, by the equation

$$q_{react} = C(T_f - T_i) = C\,\Delta T \tag{16-16}$$

where C is the heat capacity of the system, comprising water, vessel, and reaction products. This heat capacity can be determined by performing a measurement in which for a time t a known current $\mathbf{I}$ is passed through a heater of resistance $\mathbf{R}$ located in the vessel. If the temperature change produced by the electrical work is $\Delta T'$, C is given by

$$C = \frac{\mathbf{I}^2 \mathbf{R} t}{\Delta T'} = \frac{w_{elec}}{\Delta T'} \tag{16-17}$$

The assumption that the heat capacity of the system is independent of the temperature is an approximation that will not lead to a significant error if ΔT is small. For precise work the temperature dependence of C must be taken into account.

Combining (16-16) and (16-17), we have

$$q_{react} = w_{elec}\,\frac{\Delta T}{\Delta T'}$$

Thus q_{react} is determined by measuring, under adiabatic conditions, electrical work and changes in temperature. Various other methods are also used for measuring heats of reaction, but all measurements of q are indirect.

Enthalpy. When a reaction does not take place at constant volume, P,V work must be taken into account. Although for some processes the P,V work transferred may be of major interest, as it is for heat engines, it may also be just incidental, and this is the usual situation in chemistry. Chemical reactions are usually performed in vessels open to the atmosphere, so that the pressure during a reaction remains essentially constant. If gases are evolved or used up, or if the volume of the system changes in any other way during the reaction, the resulting P,V work done on the system is, by (16-8), $-P\,\Delta V$. Frequently the work transferred to or from the surroundings is of no particular interest but must nevertheless be taken into account.

By introduction of a new state function, the enthalpy, this $-P\,\Delta V$ term can be taken into account automatically. For the present we assume that the only possible work that can be done is P,V work; this is a common situation in chemical systems, although there are occasions when electrical work and sometimes even gravitational, magnetic, and other kinds of work are also performed. We return to these more general but less common cases shortly.

The *enthalpy H* (Greek: *enthalpein*, to warm) is defined by the equation

$$H = U + PV \tag{16-18}$$

Because U, P, and V are functions of state, H is also. For a system undergoing a change of state along a constant-pressure path so that $\Delta P = 0$, with no work other than P,V work, we have[1]

$$\Delta H = \Delta U + \Delta(PV) = \Delta U + P\,\Delta V \qquad (16\text{-}19)$$

and also

$$w = -P\,\Delta V \qquad (16\text{-}20)$$

From the first law, $\Delta U = q + w$, we obtain, for a constant-pressure path with only P,V work,

$$\Delta U = q_P - P\,\Delta V \qquad (16\text{-}21)$$

which leads, with (16-19), to

$$q_P = \Delta U + P\,\Delta V = \Delta H \qquad (16\text{-}22)$$

Thus the heat absorbed during a change in state at constant pressure[2] when only P,V work is done is equal to the change in a state function, the enthalpy. Constant-pressure processes are endothermic when ΔH is positive and exothermic when ΔH is negative.

Equations (16-15) and (16-22) show that when no non-P,V work is done, the heat absorbed during a change in state at constant volume or at constant pressure is equal to the change in a state function, the internal energy or enthalpy, respectively. These are two important and common special situations; more generally, the heat effect accompanying a given change in state cannot be equated with a change in a state function.

For the more general case that the non-P,V work (the *net work, w'*) is no longer zero, we have

$$w = w' \qquad \text{(constant-volume process)}$$

and

$$w = w' - P\,\Delta V \qquad \text{(constant-pressure process)}$$

which lead to

$$q_V = \Delta U - w' \qquad (16\text{-}23a)$$

and

$$q_P = \Delta H - w' \qquad (16\text{-}23b)$$

While the enthalpy is an extremely useful thermodynamic function, it does not occupy as central a position as the internal energy U. For example, for an isolated system $\Delta U = 0$ [by (16-6)] whereas $\Delta H = \Delta U + \Delta(PV) = \Delta(PV)$ is zero only if there is no change in the product PV. Thus while U is conserved for an isolated system, H need not be. Similarly, ΔH_{sys} need not be equal to $-\Delta H_{\text{surr}}$ whereas $\Delta U_{\text{sys}} = -\Delta U_{\text{surr}}$. In fact, ΔH_{surr} need not even be defined.

---

[1] Note that while $\Delta(PV) = P\,\Delta V$ when $P = \text{const}$, in general the expanded form of $\Delta(PV)$ is more complicated. By definition $\Delta(PV) = P_2 V_2 - P_1 V_1 = (P_1 + \Delta P)(V_1 + \Delta V) - P_1 V_1$; expanded, $\Delta(PV) = V_1\,\Delta P + P_1\,\Delta V + \Delta P\,\Delta V$. The last term must not be overlooked, because ΔP and ΔV are not necessarily very small quantities, although the equation $d(PV) = P\,dV + V\,dP$ may tempt one to do so.

[2] We have implicitly assumed in deriving (16-22) that the pressure in the *system*, which is what P stands for in (16-18) and (16-19), is the same as the pressure in the *surroundings*, which is the meaning of P in (16-20). This is commonly true for constant-pressure processes.

When a solid substance melts, the increase in its enthalpy is called the *heat of fusion* or *enthalpy of fusion*. Similarly the increases in enthalpy that accompany the vaporization of a liquid or of a solid are called the *heats* or *enthalpies of vaporization* or *sublimation*. Constant pressure is, of course, implied. Table 16-1 gives some examples of enthalpy changes for phase transitions as well as some accompanying chemical reactions. Reversing a reaction implies a change in sign of the value of ΔH. The usual SI units of ΔH are *kilo*joules, but many compilations are in terms of kilocalories; ΔH refers, as it always will in this book, to as many moles of reactants and products as are indicated by the coefficients of the overall chemical equation given.

Confusion sometimes arises because of differing sign conventions. For historical reasons, the terms heat of fusion, evaporation, and sublimation stand for the enthalpy that is *absorbed* during the melting, evaporation, or sublimation of a substance; the heat of reaction is the enthalpy *liberated* by a chemical reaction. This is why it is best *not* to speak of *heats* of reaction, but rather of the ΔH, the *enthalpy change*, or just the *enthalpy of a reaction*. The enthalpy change is always defined in the same way, $\Delta H = \Sigma H_{prod} - \Sigma H_{react}$.

Sometimes the heat of reaction $-q_P$ is indicated in combination with the chemical equation, as shown by the following example:

$$H_2(g) + \tfrac{1}{2}O_2(g) \longrightarrow H_2O(l) + 285.8 \text{ kJ} \qquad (16\text{-}24)$$

This symbolism is intended to imply that 285.8 kJ of energy in the form of heat is liberated—is a "product" of the reaction. However, the plus sign can be confusing; it is important to realize that the 285.8 kJ represents $-\Delta H$ of the reaction.

Enthalpy and heat capacity.

The heat capacity at constant pressure, C_P, is closely related to H. To increase the temperature of any substance from T to $T + \Delta T$ at constant pressure (with $P_{sys} = P_{surr}$) by a process involving only P,V

TABLE 16-1   Enthalpy changes for some phase changes and chemical reactions*

Phase change or reaction	$\Delta H/kJ$	$t/°C$
(1) $H_2O(s) \longrightarrow H_2O(l)$	6.02	0
(2) $H_2O(l) \longrightarrow H_2O(g)$	44.0	25
	40.6	100
(3) $Br_2(l) \longrightarrow Br_2(g)$	31.0	59
(4) $I_2(s) \longrightarrow I_2(g)$	62.3	25
(5) $C(graphite) \longrightarrow C(g)$	716.7	25
(6) $C(graphite) + O_2(g) \longrightarrow CO_2(g)$	-393.5	25
(7) $CO(g) + \tfrac{1}{2}O_2(g) \longrightarrow CO_2(g)$	-283.0	25
(8) $H_2(g) + \tfrac{1}{2}O_2(g) \longrightarrow H_2O(l)$	-285.8	25

*The values refer to as many moles of reactants and products as are shown by the coefficients of the chemical equations, the initial states being reactants (or initial phase) only and the final states products (or final phase) only. The pressure is 1 atm.

work, a quantity of heat

$$q_P = \Delta H = C_P \Delta T \tag{16-25}$$

must be transferred to the system. In this equation it is assumed that C_P does not depend on temperature; hence the equation is strictly valid only for small values of ΔT.

EXAMPLE 16-4

Mixing of snow and water. In a thermos bottle supplied with a heat-insulating stopper, 100.0 ml of liquid water at 25°C and 10.0 g of snow at -10.0°C are mixed under adiabatic conditions. The pressure is 1.00 atm. What is the final temperature? The enthalpy of fusion of ice is 334 J g^{-1} and the constant-pressure specific heats (assumed to be temperature-independent) are 2.09 J g^{-1} K^{-1} for ice and 4.18 J g^{-1} K^{-1} for liquid water.

Solution. The enthalpy change of the contents of the thermos bottle is zero for the prescribed adiabatic, constant-pressure process. Consideration of the relative amounts and temperatures of the snow and the water at the start indicates that the final temperature will not be below 0°C. As the temperature of the liquid water decreases, its loss of enthalpy is balanced by the enthalpy increase of the snow being warmed to 0°C and the snow being melted. If there is sufficient snow to cool all the original liquid to 0°C, the final temperature will be 0°C; otherwise it will be somewhat higher, with all the snow melted. We begin by assuming the latter condition; we note below how to check whether this assumption is correct. Let the final temperature be t_f and replace the unit K^{-1} by the unit °C^{-1} (the degrees being equal in magnitude). We then have (100.0 g)(4.18 J g^{-1} °C^{-1})($t_f - 25.0$°C) + (10.0 g)(2.09 J g^{-1} °C^{-1})(10.0°C) + (10.0 g)(334 J g^{-1}) + (10.0 g)(4.18 J g^{-1} °C^{-1})$t_f = 0$. The final temperature is

$$t_f = \frac{6900 \text{ J}}{460 \text{ J °C}^{-1}} = \underline{15.0\text{°C}}$$

and our assumption that the final temperature was above 0°C was justified. If, however, the amount of snow at low temperature had been sufficient to cool all the water to 0°C without melting all the snow, our equations would have yielded a negative value for t_f, contrary to our assumption. An alternative relation could then be used to calculate the fraction of the snow that would have to melt in order to cool all the liquid to 0°C.

Combining enthalpies of reaction. Since enthalpy is a state function, the enthalpy increase for any change of state is independent of the path. This has important implications for the enthalpy changes of chemical reactions: when the chemical equations describing such reactions are added or otherwise combined into an equation describing a new reaction, the corresponding ΔH values, combined in the same way, yield the ΔH for the new reaction. This is because the ΔH does not depend on the actual course of the reaction. This important result is called *Hess'* *law.* Its application is illustrated by the following three examples.

EXAMPLE 16-5 **Combustion of graphite to carbon monoxide.** Find the enthalpy increase that accompanies the combustion of graphite to carbon monoxide at 25°C by considering the reactions in Table 16-1.

Solution. The equation of the reaction,

$$C \text{ (graphite)} + \tfrac{1}{2}O_2(g) \longrightarrow CO(g)$$

may be obtained by subtracting reaction (7) of Table 16-1 from reaction (6). The enthalpy change for the new reaction may therefore be obtained by subtracting ΔH of (7) from that of (6): $\Delta H = [-393.5 - (-283.0)] \text{ kJ} = -110.5 \text{ kJ}$. It is possible to combine graphite and oxygen to form CO not only directly, or as a sequence of the reactions represented by (6) and (7), but along many other paths involving any number of steps. For all these paths the total of the ΔH values is -110.5 kJ.

Many reaction enthalpies have to be measured indirectly by following a roundabout path leading from the reactants to the products or from the products to the reactants. An experimental procedure that is often followed is to burn separately, in oxygen, all reactants and products that are combustible, to measure the corresponding individual *enthalpies of combustion,* and to combine the results.

EXAMPLE 16-6 **Use of enthalpies of combustion.** Calculate the enthalpy ΔH for the reaction

$$2C \text{ (graphite)} + 2H_2(g) + H_2O(l) \longrightarrow C_2H_5OH(l) \qquad (16\text{-}26)$$

from the measured enthalpies or heats of combustion of ethanol, graphite, and hydrogen at 25°C:

$$C_2H_5OH(l) + 3O_2(g) \longrightarrow 2CO_2(g) + 3H_2O(l) \qquad (16\text{-}27)$$
$$\Delta H_{27} = -1366.7 \text{ kJ}$$

$$C \text{ (graphite)} + O_2(g) \longrightarrow CO_2(g) \qquad (16\text{-}28)$$
$$\Delta H_{28} = -393.5 \text{ kJ}$$

$$H_2(g) + \tfrac{1}{2}O_2(g) \longrightarrow H_2O(l) \qquad (16\text{-}29)$$
$$\Delta H_{29} = -285.8 \text{ kJ}$$

Solution. The chemical equation (16-26) can be obtained by combining the other equations in the way indicated symbolically by $2[(16\text{-}28) + (16\text{-}29)] - (16\text{-}27)$. Thus

$$\Delta H_{26} = 2(\Delta H_{28} + \Delta H_{29}) - \Delta H_{27} = \underline{8.1 \text{ kJ}}.$$

EXAMPLE 16-7 **Reaction enthalpy and specification of initial and final states.** Calculate ΔH for a reaction like (16-26) but with $H_2O(l)$ on the left replaced by $H_2O(g)$. Use data from Table 16-1 and Example 16-6.

Solution. The chemical equation of the desired reaction

$$2C \text{ (graphite)} + 2H_2(g) + H_2O(g) \longrightarrow C_2H_5OH(l)$$

results by adding the chemical equation

$$H_2O(g) \longrightarrow H_2O(l)$$

with $\Delta H = -44.0$ kJ to Equation (16-26), for which ΔH is 8.1 kJ. The desired enthalpy change is therefore the sum of the values quoted: $\Delta H = (-44.0 + 8.1)$ kJ $= \underline{-35.9 \text{ kJ}}$. Note the difference of the result from that in Example 16-6, and therefore the importance of carefully specifying initial and final states.

Standard enthalpies.   Within the framework of thermodynamics, only relative values of $H = U + PV$ are defined because only relative values for U are defined. It is, however, possible to assign zero enthalpy to suitable reference states *by convention*. This is of great value because it permits the assignment and tabulation of a standard enthalpy for each substance. These standard enthalpies are useful in many computations, as will be seen. The conventions chosen fulfill the experimental requirement that a reference state be realized easily and reproducibly.

These conventions may be separated into two parts, the definition of standard states for all substances and the definition of reference states on which tabulated enthalpy values are based.

1.   A substance is in its *standard state* when under a pressure of 1 atm and at a standard temperature, usually chosen to be 25°C or 298 K (more accurately 298.15 K, but the difference of 0.15 K is normally insignificant and will be neglected). For dissolved molecular or ionic species a concentration (or more accurately, an "effective" concentration or activity) of 1 mol liter^{-1} is implied. Quantities referring to standard states are designated by a superscript $\ominus$ (such as $X^{\ominus}$ for a quantity X);[1] if the standard state is chosen to be at a temperature T that differs from 298 K, as is sometimes useful, this is indicated by showing the value of T in parentheses, $X^{\ominus}(T)$. Thus $X^{\ominus}$ specifies a standard state of 1 atm, 1 M for dissolved species, and 298 K, whereas $X^{\ominus}(1000)$ refers to the same pressure and concentration but at 1000 K. In other words, the temperature of a standard state is flexible and may need special mention, whereas pressure and concentration are fixed.

The standard states discussed here must not be confused with the standard conditions customarily abbreviated by STP (standard temperature and pressure); STP conditions are often used when specifying quantities of a gas and imply 1 atm and 0°C (Sec. 5-2).

2.   The *reference states* for tabulated enthalpies are the *chemical elements in their most stable forms and in their standard states*. Under these conditions *zero enthalpy* is assigned to each element. To all pure substances are then assigned *standard molar enthalpies of formation* $\widetilde{H}_f^{\ominus}$ that represent the enthalpy change, positive or negative, when 1 mol of the substance is formed at the standard pressure and temperature from the elements in their most stable forms. In some texts this quantity is also given, among others, the symbol $\Delta\widetilde{H}_f^{\ominus}$, $\Delta\widetilde{H}_f^{\circ}$, or $\Delta\widetilde{H}f^{\circ}$. Often we shall call it more briefly the *standard molar enthalpy*.

---

[1]The plimsoll $\ominus$ is a load-line mark on merchant vessels to prevent their overloading; it is named after Samuel Plimsoll, whose efforts resulted in its legal adoption by the English Parliament in 1876.

For the chemical elements in their most stable states, $\widetilde{H}_f^{\ominus}$ is thus by definition zero. At a temperature T other than 298 K the reference state for the enthalpy of formation $\widetilde{H}_f^{\ominus}(T)$ of a compound is again the elements at 1 atm in their most stable form, but now at the temperature T. Thus $\widetilde{H}_f^{\ominus}(T)$ for the elements in their most stable form at T is zero. Most of our discussion refers to a temperature of 298 K.

The conventions about zero enthalpy are just as arbitrary as the choice of the zero level for measurement of terrestrial altitudes.

EXAMPLE 16-8   **Standard molar enthalpy.** What is the standard enthalpy (of formation) $\widetilde{H}_f^{\ominus}$ of liquid water?

Solution. For the reaction

$$H_2(g) + \tfrac{1}{2}O_2(g) \longrightarrow H_2O(l) \qquad (16\text{-}30)$$

$$\Delta H^{\ominus} = H_f^{\ominus}\{H_2O(l)\} - H_f^{\ominus}\{H_2(g)\} - \tfrac{1}{2}\widetilde{H}_f^{\ominus}\{O_2(g)\}$$

The enthalpy change at standard conditions is -285.8 kJ (Table 16-1). The standard molar enthalpy $\widetilde{H}_f^{\ominus}$ of liquid water is thus $\underline{-285.8 \text{ kJ mol}^{-1}}$ since $H_2(g)$ and $O_2(g)$ are elements in their standard states.

Some substances or modifications of substances are not stable at standard conditions—for example, water vapor at 25°C and 1 atm. In such cases standard enthalpies may be assigned by extrapolation. For elements that are not in their most stable forms at standard conditions, $\widetilde{H}_f^{\ominus}$ is the enthalpy change when these forms are produced from the most stable modification. For example, standard enthalpies of free atoms are equal to the enthalpy increase accompanying the formation of 1 mol of atoms from the element in its most stable state, all at 25°C and 1 atm. The dissolving of a substance is accompanied by an enthalpy change called the *enthalpy of solution*, ΔH_{soln}. The heat liberated upon dissolving the substance is $-\Delta H_{\text{soln}}$. Its value may depend on the solute concentration achieved, so that there may also be enthalpy changes when a solution is diluted.

Table D-4 contains the standard molar enthalpies $\widetilde{H}_f^{\ominus}$ for a number of elements, compounds, atoms, and aqueous ions, some in forms that are not those most stable at standard conditions.

When the values of $\widetilde{H}_f^{\ominus}$ for the reactants and products of a reaction are known, it is possible to compute the standard enthalpy change $\Delta H^{\ominus}$ for the reaction from the relation

$$\Delta H^{\ominus} = \Sigma H_f^{\ominus}{}_{\text{prod}} - \Sigma H_f^{\ominus}{}_{\text{react}}$$

This extremely useful procedure is illustrated in the following examples, with $\widetilde{H}_f^{\ominus}$ values shown in kilojoules per mole beneath the chemical symbols.

EXAMPLE 16-9   **Combination of standard molar enthalpies.** Water gas, an equimolar mixture of CO and H_2, is made from carbon and water by the reaction

$$H_2O(g) + C \text{ (graphite)} \longrightarrow CO(g) + H_2(g) \qquad (16\text{-}31)$$

$$\begin{array}{cccc} -241.8 & 0 & -110.5 & 0 \end{array}$$

What is ΔH° for this reaction?

Solution. The change in standard enthalpies is

$$\Delta H^\circ = (-110.5 + 241.8)\,\text{kJ} = \underline{131.3\,\text{kJ}}$$

The reaction is actually carried out at around 600°C by passing steam over hot coal. The ΔH° just given applies to 25°C and shows the reaction to be endothermic at that temperature. It is also endothermic at 600°C, and to maintain this temperature while making water gas the steam is turned off every few minutes and replaced by a brief blast of hot air.

The enthalpy change when forming molecules from atoms in the gas phase is frequently of interest.

EXAMPLE 16-10   **Combination of atoms into molecules.** What would be the enthalpy change if benzene vapor were made from carbon and hydrogen atoms?

Solution.

$$6C(g) + 6H(g) \longrightarrow C_6H_6(g) \qquad (16\text{-}32)$$
$$716.7 \quad\ \ 217.9 \qquad\quad 82.9$$

The enthalpy change under standard conditions is

$$\Delta H^\circ = (82.9 - 6 \times 716.7 - 6 \times 217.9)\,\text{kJ} = \underline{-5525\,\text{kJ}}$$

The energy and enthalpy of melting are nearly equal, but for vaporization there is a significant difference between ΔU and ΔH.

EXAMPLE 16-11   **Energy of vaporization and of melting.** What is $\Delta\widetilde{U}$ for the vaporization of water at 25°C? What is $\Delta\widetilde{U}$ for the melting of ice at 0°C? Use the values given in Table 16-1.

Solution. The value of $\Delta\widetilde{H}$ is 44.0 kJ mol^{-1}. (Note the temperature dependence of $\Delta\widetilde{H}$, the value at 100°C being 40.6 kJ mol^{-1}.) For any constant-pressure process $\Delta\widetilde{U} = \Delta\widetilde{H} - P\,\Delta\widetilde{V}$. Assuming ideal-gas behavior of the water vapor and neglecting the volume of the liquid water compared to that of the vapor, we get

$$-P\,\Delta\widetilde{V} \approx -P\widetilde{V}_g \approx -P\frac{RT}{P} = -RT$$

Thus $\Delta\widetilde{U} \approx (44.0 - 8.31 \times 298 \times 10^{-3})\,\text{kJ mol}^{-1} = (44.0 - 2.48)\,\text{kJ mol}^{-1} = \underline{41.5\,\text{kJ}}$ $\underline{\text{mol}^{-1}}$.

In contrast, when ice is melted at 0°C and 1 atm, $P\,\Delta\widetilde{V}$ is so small as to be negligible. The density of ice at 0°C is 0.915 g ml^{-1} and that of liquid water is 0.99987 g ml^{-1}, so that for the melting of 1 mol of ice, $\Delta\widetilde{V} = 18.0\ (\frac{1}{1.000} - \frac{1}{0.915})$ ml mol$^{-1} \approx 18.0$ $(1.000 - 1.093)$ ml mol$^{-1} = -1.67$ ml mol^{-1}. Thus $P\,\Delta\widetilde{V} = (-1.67 \times 10^{-3}$ liter atm mol$^{-1})$ [1 J/(0.00987 liter atm)] $= -0.169$ J mol$^{-1} = -1.7 \times 10^{-4}$ kJ mol^{-1} and $\Delta\widetilde{U}$ is almost exactly equal to $\Delta\widetilde{H}$. Hence, by Table 16-1, $\Delta\widetilde{U} \approx \Delta\widetilde{H} = \underline{6.02\,\text{kJ mol}^{-1}}$.

Problems and Questions

16-1 **Concepts and definitions.** Distinguish clearly: (*a*) heat and molecular kinetic energy; (*b*) an isothermal process and an adiabatic process; (*c*) intensive and extensive properties; (*d*) a change in state that occurs adiabatically and a change in state that occurs in an isolated system.

16-2 **Heat capacities.** How much heat is required to change the temperature of 100 ml benzene from 20 to 50°C (*a*) at constant pressure, (*b*) at constant volume? $\tilde{C}_P = 133$ J mol^{-1} K^{-1}; $\tilde{C}_V = 92$ J mol^{-1} K^{-1}; MW = 78; density = 0.88 g ml^{-1} at 20°C.

16-3 **Iced drink.** Ice cubes at 0.0°C are used to cool 250 ml water from 30.0 to 0.0°C. How much of the ice melts, assuming no heat transfer between the water and its container or other parts of its surroundings? The enthalpy of fusion of ice is 6.02 kJ mol^{-1}, $\tilde{C}_P$ of liquid water is 75.3 J mol^{-1} K^{-1}, and the density of liquid water is 1.000 g ml^{-1}.   *Ans.*   94 g

16-4 **Cooling a gas at constant pressure.** A 2.00-mol sample of an ideal gas initially at 350 K and 1.50 atm was cooled at constant pressure until its volume was 35.0 liters. During this process the gas lost 1256 J as heat. Assume the gas to be the system and calculate (*a*) the initial volume, (*b*) the final temperature, (*c*) the work done, (*d*) the change in internal energy, (*e*) the change in enthalpy, and (*f*) $\tilde{C}_P$ for the gas.

16-5 **Isothermal expansion of a gas.** Exactly 3 mol of an ideal gas is expanded isothermally at 300 K so that the volume is increased by a factor of exactly 4 and work against a constant outside pressure of 2.00 atm is performed. Find *w*, *q*, ΔU, and ΔH. The initial pressure is 8.00 atm.
                Ans. w = -5.61 kJ = $-q$; $\Delta U = \Delta H = 0$

16-6 **Mixing of gases.** A box with walls of a thermally nonconducting material is divided into two compartments by a partition of the same material. One compartment contains 0.40 mol He at 20°C and 1.00 atm; the other contains 0.60 mol N_2 at 100°C and 2.00 atm. The partition is removed so that the two gases mix. What is the final temperature and pressure? The molar heat capacities of He and of N_2 at constant volume are 12.5 and 20.7 J mol^{-1} K^{-1}, respectively. The gases may be assumed to be ideal gases and to behave as such on mixing. (*Hint:* Imagine the gases first to come to the final temperature while at their original volumes; then allow them to mix.)

16-7 **Changing ice to steam.** Suppose that 1.00 mol ice at -20°C is heated at a constant pressure of 1.00 atm until it has been transformed into 1.00 mol steam at 127°C. Calculate *q*, *w*, ΔH, and ΔU for this process. Use relevant data from Table 16-1 as needed. The densities of ice and water are 0.92 and 1.00 g ml^{-1} and may be assumed to be independent of temperature. The heat capacities of ice, water, and steam at 1 atm are, respectively, 38, 75, and 36 J mol^{-1} K^{-1} and may be assumed to be independent of temperature. Assume that steam behaves like an ideal gas.   *Ans.*   55.9 kJ, -3.3 kJ, 55.9 kJ, 52.6 kJ

16-8 **Implications of the first law.** Indicate whether each of the following statements is true or false as it stands. If the statement is not true, indicate in what way it is false and whether it could be made into a true statement by a slight change in wording. If the statement is true but unnecessarily restricted, indicate what qualifying words or phrases can be omitted.

(*a*) The work done by the system on the surroundings during a change in state is never greater than the decrease in the energy of the system.

(*b*)   The enthalpy of a system cannot change during an adiabatic process.

(*c*)   When a system undergoes a given isothermal change in state, its change in enthalpy does not depend upon the process involved.

(*d*)   When a change in state occurs, the increase in the enthalpy of the system must equal the decrease in the enthalpy of the surroundings.

(*e*)   The equation $\Delta U = q + w$ is applicable to any macroscopic process, provided no electrical work is done by the system on the surroundings.

(*f*)   No change in state occurring within an isolated system can cause a change in its energy or its enthalpy.

(*g*)   For any constant-pressure process, the increase in enthalpy equals the heat absorbed whether or not electrical work is done during the process.

(*h*)   A reversible process is one in which the amount of energy lost by the system is just sufficient to restore the system to its initial state.

(*i*)   When an imperfect gas expands into a vacuum, it does work because the molecules of the gas have been separated from one another against an attractive (van der Waals) force.

16-9   **Reversible expansion of a gas.**   Exactly 5 mol of an ideal gas is expanded reversibly from 50.0 to 150.0 liters at a constant temperature of 150°C. Calculate w in liter atmospheres and in joules.   *Ans.* -191 liter atm; -19.3 kJ

16-10   **Enthalpy of formation.**   (*a*) The enthalpy of combustion of benzoic acid, C_6H_5COOH, at 25°C is -3227 kJ mol^{-1}. What is the enthalpy of formation? Use values of standard enthalpies of formation listed in Appendix D as needed. (*b*) The enthalpy of combustion of succinic acid, $HOOCCH_2CH_2COOH$, is -1494 kJ mol^{-1}. What is its enthalpy of formation?

16-11   **Hess' law of heat summation.**   Given the enthalpy changes at a certain temperature for reactions (*a*) and (*b*), calculate that for (*c*). Is (*c*) endothermic or exothermic at this temperature?

(*a*) $3H_2(g) + N_2(g) \longrightarrow 2NH_3(g)$ $\Delta H = -92.4$ kJ
(*b*) $2H_2(g) + O_2(g) \longrightarrow 2H_2O(g)$ $\Delta H = -483.7$ kJ
(*c*) $4NH_3(g) + 3O_2(g) \longrightarrow 2N_2(g) + 6H_2O(g)$

Ans. -1266 kJ; exothermic

16-12   **Enthalpy of reaction.**   (*a*) With the help of data in Table D-4, calculate the heat evolved or absorbed (say which) when the following reaction is carried out at 25°C at constant pressure:

$$CCl_4(l) + H_2(g) \longrightarrow CHCl_3(l) + HCl(g)$$

(*b*) Calculate also the enthalpy of vaporization of CCl_4 at 25°C from data in Table D-4, and determine the enthalpy of the above reaction if gaseous CCl_4 instead of liquid CCl_4 were used as the reactant.

16-13   **Combustion of nitromethane.**   Nitromethane, CH_3NO_2, is a good fuel. It is a liquid at ordinary temperatures. When the liquid is burned, the reaction involved is chiefly

$$2CH_3NO_2(l) + \tfrac{3}{2}O_2(g) \longrightarrow 2CO_2(g) + N_2(g) + 3H_2O(g)$$

The standard enthalpy of formation of liquid nitromethane at 25°C is -112 kJ mol^{-1}; other relevant values may be found in Appendix D. (*a*) Calculate the enthalpy change in the burning of 1 mol liquid nitromethane to form gaseous products at 25°C. State explicitly whether the reaction is endothermic or exothermic. (*b*) Would more or less heat be evolved if *gaseous* nitromethane were burned under the same conditions? Indicate what additional information (if any) you would need to calculate the exact amount of heat, and show just how you would use this information.   *Ans.*   (*a*) -644 kJ mol^{-1};   (*b*)   more

16-14   **Energy and enthalpy of combustion.**   The combustion of 1 mol gaseous acetylene, C_2H_2, to liquid water and gaseous CO_2 yields 1303 kJ at constant volume and at 20°C. What is the enthalpy of combustion of acetylene under the same conditions except that the pressure is kept constant at 1 atm? Consider all gases involved in the reaction to be perfect and neglect the volume of the liquid water produced.

16-15   **Combustion of isobutene.**   When 1 mol isobutene, a gas with formula C_4H_8, is burned at 25°C and 1 atm to form CO_2 and gaseous water, the enthalpy change is -2528 kJ. (*a*) Calculate, with the aid of any information needed from Table D-4, Appendix D, the standard enthalpy of formation of isobutene. (*b*) Suppose that 0.50 mol isobutene is burned adiabatically at constant pressure in the presence of an excess of oxygen, with 5.0 mol oxygen left at the end of the reaction. The heat capacity of the reaction vessel is 700 J K^{-1} and pertinent molar heat capacities (in joules per kelvin per mole) are $CO_2(g)$, 37; $H_2O(g)$, 34; $O_2(g)$, 29. What is the approximate final temperature of this system (including the reaction vessel)?
                                   Ans.   (*a*) -13 kJ mol^{-1};   (*b*)   about 1300°C

"Suppose we think of . . . a body of a higher and one of a lower temperature . . . ; through interaction between the bodies this state is changed; according to the second law this change must always occur so that the total entropy of all bodies increases; according to our present interpretation this means nothing else than that the probability of the entire state of these bodies is always increasing; the system of bodies always proceeds from a less probable to a more probable state."

L. Boltzmann, 1877

"The second law of thermodynamics has the same degree of truth as the statement that if you throw a tumblerful of water into the sea, you cannot get the same tumblerful of water out again."

J. C. Maxwell, 1870

17 The Second Law of Thermodynamics: Entropy and Free Energy

17-1 Introduction

It is a central fact of experience that there is a natural direction for all kinds of changes in the physical universe. At ordinary pressures, ice always melts at temperatures above 0°C; water at such temperatures and pressures never freezes. Similarly, if a piece of ice is dropped into some boiling water, the net result is the formation of water at some intermediate temperature. The reverse never happens: a sample of water at, say, 40°C never spontaneously turns into a mixture of ice and boiling water. To give still another example, a gas at a given temperature expands spontaneously to fill all the volume accessible to it; it never shrinks in volume spontaneously at the same temperature. For any of the changes just described, the process can occur in either direction in accord with the first law of thermodynamics; yet in fact each happens in only one direction.

Everyone is aware that there is a natural direction for most processes. Suppose that a movie of an egg being broken is run backward. Everyone watching it immediately recognizes that what he seems to see is impossible, that time has apparently been reversed by the trick of reversing the sequence in which the individual frames are projected. When we speak of a natural direction of reactions, we are implicitly assuming a direction for time. Processes that occur in this "natural direction" are referred to as *spontaneous*. The fact that a process is spontaneous implies nothing about its rate. Many spontaneous processes proceed at an imperceptibly slow rate for reasons that we consider in Chap. 20.

What factors determine the natural direction of a process? Consider a stone held at some distance above the floor. When the stone is released it falls. In the process of falling its potential energy is converted to kinetic energy and the kinetic energy is in turn dissipated when the stone collides with the floor. When it has come to rest on the floor it has reached a position of minimum gravitational energy. The decrease in gravitational energy governs the direction of this process. The inverse process, the rise of the stone from the floor, requires an investment of energy and never occurs spontaneously.

Minimization of energy cannot be the only factor that determines the direction of spontaneous changes, however, because many spontaneous chemical processes are accompanied by an absorption, rather than a release, of energy. For example, the evaporation of a liquid into a vacuum or into any region not saturated with its vapor occurs spontaneously and sometimes rapidly. Unless there is a simultaneous absorption of energy from the surroundings, this evaporation will be accompanied by a lowering of the average kinetic energy of the molecules of the remaining liquid as well as of the vapor (and thus a lowering of the temperature), in order to supply the energy that is needed by the escaping molecules to overcome the forces that hold the condensed phase together. This increase in the potential energy of the escaping molecules occurs also if energy is supplied to the system from the outside to keep the temperature constant. Thus during spontaneous evaporation there is an *increase* of the potential energy, not the decrease that we associate with equilibria in mechanics. It is as if a stone suddenly jumped up into the air from the ground—or is it? Clearly some additional factor must be at work.

What is this factor? It is related to what might be called the intrinsic probability of various arrangements of the molecules concerned, a factor that plays a role whenever the number of molecules is sufficiently large that we may speak of a macroscopic system. Suppose that there are *no* forces between the molecules; then there will be no reason for the existence of a condensed phase and there will be a uniform probability of finding the molecules anywhere in the container, as with any gas (unless the container is so tall that there is appreciable gravitational settling of the molecules). On the other hand, if there are extremely strong forces between the molecules, they will remain together in the small volume characteristic of the solid or liquid phase, with almost none escaping into the surrounding space. There is a wide range of possible intermediate situations.

The probability that any particular distribution of the molecules will occur can be evaluated from statistical *and* energetic considerations. For example, a distribution in which a condensed phase exists is extremely unlikely unless the molecules attract one another with energies large compared to the average molecular kinetic energy at the temperature of the system. In other words, the probability that most of the molecules of a gas would occupy only a very small fraction of the available volume without some strong energetic impetus to do so is exceptionally small, so small that we regard it as "impossible". Correspondingly, the probability is high that molecules initially confined in a very small fraction of the total volume available will spread throughout all the available volume unless the energies of attraction of the molecules for one another are large relative to their average kinetic energies. Thus with nonpolar substances of low polarizability, such as helium and hydrogen, it is impossible to form a condensed phase unless the temperature is very low, for at most temperatures the average kinetic energy of the molecules is great enough to overcome the energy of intermolecular attraction. Whatever the magnitude of the intermolecular forces, the probability of escape of molecules from the condensed phase into the vapor increases with increasing temperature, and thus the vapor pressure increases with increasing temperature. The escape of molecules from the condensed phase into the vapor phase corresponds to an increase in the potential energy of these molecules.

The principle that emerges is that the position of equilibrium (in the case discussed, the vapor pressure at any particular temperature) depends upon the balance between the tendency for the potential energy to be minimized (corresponding to condensation) and the tendency of the molecules to fill the available space uniformly.[1] In this balance, the second factor is weighted increasingly with increasing temperature, because the kinetic energy of the molecules must be sufficient to overcome the intermolecular attraction.

It develops that the position of equilibrium in any chemical system can be analyzed in similar terms, as the result of a balance between the tendency of the potential energy to be minimized and a second factor that is weighted increasingly as the temperature rises. Although we have described the second factor here qualitatively in microscopic terms—that is, by considering distributions of molecules—it is possible to arrive at it by purely macroscopic reasoning in terms of thermodynamics. This factor may be expressed as the product of the temperature of the system and the change in a function of the system called its *entropy*. It is also possible

---

[1]There are many more ways of distributing a given number of molecules over a large volume than over a small one and so, *if* these different distributions are equally favorable energetically, it is much more likely that the molecules will be spread over the large volume. However, in the presence of relatively strong intermolecular attractions, distributions in which the molecules are concentrated in a small volume *are* favored energetically. The molecules may then remain concentrated, i.e., in a liquid or solid phase.

to find a function of any chemical system that plays for chemical processes the same role that potential energy does for ordinary mechanics of macroscopic objects.[1] In other words, this function always decreases in any spontaneous process and is at a minimum when equilibrium is achieved. It is termed the *free energy* (Sec. 17-4).

Reversibility and irreversibility. The concept of reversibility is essential to an understanding of entropy and its applications. We considered in Sec. 16-3 the reversible expansion of an ideal gas and emphasized that a reversible change is one whose direction can be reversed at any point along its path by an infinitesimal change in the conditions. Table 17-1 gives some examples of various kinds of reversible processes and parallel irreversible processes.

17-2 Entropy

We begin this section with an *assertion* about the properties of the state function known as entropy, symbolized by S. A definition of entropy follows shortly.

For any process that can actually occur, the accompanying entropy change always satisfies this condition:

$$\Delta S_{tot} = (\Delta S_{sys} + \Delta S_{surr}) \geq 0 \qquad (17\text{-}1)$$

Here ΔS_{tot} is the entropy change of the entire "universe", composed of that portion

---

[1] An object such as a stone is, of course, composed of enormous numbers of molecules, but when the stone moves, its molecules move essentially in concert and its motion can be idealized as that of a single macroscopic unit.

TABLE 17-1   **Examples of reversible and irreversible processes**

Reversible process	*Parallel irreversible process*
Expansion of a gas when $P_{sys} \approx P_{surr}$. A slight increase in P_{surr} will cause the gas to contract.	Expansion of a gas into a vacuum. No slight change in pressure can stop the expansion.
Heating of a system immersed in a large bath, with T_{sys} almost equal to T_{bath}. A slight decrease in T_{bath} will cause heat to flow out of the system into the bath.	Heating of a system by immersing it in a bath at a temperature much higher than T_{sys}. The direction of heat flow, from the bath into the system, cannot be altered by small changes in T_{bath}.
Ice in water at 0°C. A slight heat flow into the ice-water mixture will melt some ice; a slight withdrawal of heat from the mixture will freeze some water.	Ice in water at 20°C. The ice will melt even if the water is cooled several degrees (to any temperature above 0°C).
A chemical reaction at equilibrium, e.g., $$H_2O + HOAc \rightleftharpoons H_3O^+ + OAc^-$$ A slight increase in the concentration of HOAc will produce more ions; a slight increase in the concentration of either ion will produce more HOAc.	A chemical reaction far from equilibrium, e.g., a piece of zinc dropped into acid, $$Zn + 2H^+ \longrightarrow Zn^{2+} + H_2$$ No small (or even large) increase in the concentration of Zn^{2+} or in the pressure of H_2 can reverse the direction of the reaction.

identified as the system and the remainder characterized as surroundings. The statement (17-1) provides a criterion for finding the natural direction of any process. That direction must be such that the total entropy increases; we show shortly that the condition $\Delta S_{tot} = 0$ corresponds to equilibrium conditions or a change carried out reversibly, so that there is no natural direction. For an irreversible change, that is, a system moving toward the equilibrium condition (although not necessarily reaching it), the total entropy always increases.

The total entropy can never decrease. This does not mean that the entropy of *either* the system *or* the surroundings cannot decrease during some process, but only that if either of these entropies does decrease, the other must increase at least as much so that the sum of the entropy changes is either zero or positive.

The second law of thermodynamics.   Equation (17-1) is one of a number of ways of expressing the second law of thermodynamics. The implications of the second law and its applications are far from being as intuitively evident as those of the first law. However, intuitively reasonable consequences of the second law can be demonstrated by specific examples once a way of evaluating entropy changes has been defined. We turn now to this question.

Entropy changes for isothermal processes.   Consider first a special case, that of an isothermal process, for which we *postulate* that the change in the state function S is given by

$$\Delta S = \frac{q_{rev}}{T} \tag{17-2}$$

that is, the change in entropy is equal to the heat transferred to the system along a reversible path divided by the temperature T of the system. While there is an important restriction on the path, that of reversibility, the path is otherwise arbitrary provided that it leads from the given initial state to the desired final state. The restriction to reversibility applies only to (17-2), whereas ΔS itself is entirely independent of the path relating the states considered, because S is a state function. In effect (17-2) provides a way to *calculate* ΔS for a specified change of state, and the result applies whether or not the change is *actually* effected along the reversible path used in the computation, along any other reversible path, or along *any irreversible path*.

Application to changes of state.   Let us apply (17-2) to a reversible change of state, such as the vaporization of a liquid at its boiling temperature T_b, that is, at the temperature at which the outside pressure on the liquid is equal to the vapor pressure of the liquid (Sec. 3-1). Pressure and temperature stay constant for this process. We know from Sec. 16-4 and in particular Equation (16-22) that as long as only P,V work is done, $q_{rev} = \Delta H_{vap}$ at T_b (for the pressure considered). Therefore

$$\Delta S_{vap} = \frac{\Delta H_{vap}}{T_b} \tag{17-3}$$

that is, the entropy of vaporization at the boiling temperature at the pressure considered is the enthalpy of vaporization divided by the (absolute) boiling temperature.

EXAMPLE 17-1

Molar entropy of vaporization of water. What is the entropy of vaporization per mole of water at 100°C and 1 atm? Also find $\widetilde{q}$ and estimate $\widetilde{w}$ and $\Delta\widetilde{U}$. Neglect the volume of the liquid and assume that the vapor is an ideal gas. The molar enthalpy of vaporization under the conditions stated, for which there is equilibrium between liquid water and water vapor, is 40.6 kJ mol^{-1}.

Solution. By (17-3) we have $\Delta\widetilde{S}_{\text{vap}} = 4.06 \times 10^4$ J mol^{-1}/373 K $=$ 109 J mol^{-1} K^{-1}; by (16-22) with $w' = 0$, $\widetilde{q} = \Delta\widetilde{H} =$ 40.6 kJ mol^{-1}. The work along the path described is, by (16-8),

$$\widetilde{w} = -P\,\Delta\widetilde{V} = -P(\widetilde{V}_{\text{vap}} - \widetilde{V}_{\text{liq}}) \approx -P\widetilde{V}_{\text{vap}} \tag{17-4}$$

(Note that $\widetilde{w}$ is negative; work is done by the system on the surroundings.) Here $P = 1$ atm and $\Delta\widetilde{V} \approx \widetilde{V}_{\text{vap}} \approx RT/P$. Thus $\widetilde{w} \approx -RT = -(8.31 \times 373)$ J mol^{-1} $=$ -3.10 kJ mol^{-1}, and $\Delta\widetilde{U} = \widetilde{q} + \widetilde{w} \approx (40.6 - 3.1)$ kJ mol^{-1} $=$ 37.5 kJ mol^{-1}.

EXAMPLE 17-2

Entropy change for an irreversible process. Exactly 1 mol liquid water at 1 atm and 100°C is injected into an evacuated container of fixed volume, where it vaporizes. The temperature is maintained at 100°C and the container volume is such that it is precisely filled by 1 mol water vapor at 100°C and 1 atm. Calculate $\Delta\widetilde{U}$, $\Delta\widetilde{H}$, $\Delta\widetilde{S}$, $\widetilde{w}$, and $\widetilde{q}$ for the water and the process described, using any relevant information from the preceding example.

Solution. Entropy is a function of state, as are energy and enthalpy, and the change in state sustained by the water is the same as in the preceding example. Thus, $\Delta\widetilde{S} = \Delta\widetilde{S}_{\text{vap}} =$ 109 J mol^{-1} K^{-1}, $\Delta\widetilde{U} \approx$ 37.5 kJ mol^{-1}, and $\Delta\widetilde{H} =$ 40.6 kJ mol^{-1}. No work is performed on the surroundings by the water, so that $\widetilde{w} = 0$ and $\widetilde{q} = \Delta\widetilde{U} \approx$ 37.5 kJ mol^{-1}.

The relation between these two examples illustrates a principle fundamental to the evaluation of the entropy change for an irreversible process. A reversible path is sought connecting the two states considered, and the entropy change is then found for this path; for an isothermal process, ΔS is the ratio q_{rev}/T. The value of ΔS evaluated for the reversible path is equal to ΔS for the specified change in state, irrespective of the actual path, because S is a state function.

Equations (17-1) and (17-2) permit us to make another important statement about isothermal processes, whether they are reversible or not. In any process the heat q (positive, negative, or zero) is defined as being transferred from the surroundings to the system. In the isothermal process considered we can always arrange for a suitable constant-temperature bath in the surroundings that makes the heat transfer reversible as far as the surroundings are concerned. The heat transferred reversibly to the surroundings is then $-q$, and therefore

$$\Delta S_{surr} = \frac{-q}{T} \tag{17-5}$$

This relationship, together with (17-1), permits us to draw conclusions about ΔS_{sys} even if we know no details about the process occurring. Substitution of (17-5) into (17-1) gives

$$\Delta S_{sys} + \Delta S_{surr} = \Delta S_{sys} - \frac{q}{T} \geq 0$$

or if we drop the subscript (because absence of a subscript implies in itself that the quantity refers to the system),

$$\Delta S \geq \frac{q}{T} \tag{17-6}$$

The "larger than" symbol applies to irreversible isothermal processes in the system; the "equal" symbol refers to reversible isothermal processes, for which (17-6) and (17-2) are identical. The relationship (17-6) is an important thermodynamic criterion that makes it possible to decide whether an isothermal process is reversible (and thus at equilibrium) or irreversible (and therefore permitted to proceed spontaneously in the direction considered). Example 17-3 illustrates how the inequality portion of (17-6) works.

EXAMPLE 17-3   **Irreversible vaporization of water.** Apply (17-6) to the irreversible vaporization described in Example 17-2, using $\Delta \tilde{S}$ and $\tilde{q}$ obtained there.

Solution. From Example 17-2, we note that $\Delta \tilde{S} = \Delta \tilde{S}_{vap} = 109$ J mol^{-1} K^{-1} while $\tilde{q}/T \approx (3.75 \times 10^4/373)$ J mol^{-1} K^{-1} = $\underline{101}$ J mol^{-1} K^{-1} $< \Delta \tilde{S}$. We see that the inequality portion of (17-6) applies, in agreement with the known irreversibility of the process described.

Equation (17-6) can also be combined with (17-2)

$$\Delta S = \frac{q_{rev}}{T} \geq \frac{q}{T}$$

to obtain

$$q_{rev} \geq q \tag{17-7}$$

Thus, for a given isothermal change of state the heat transferred to the system is largest along a reversible path. According to the first law, $\Delta U = q + w$ and ΔU is independent of path. Thus,

$$\Delta U = q_{rev} + w_{rev} = q + w \qquad \text{for any path}$$

Combination of this with (17-7) gives $w_{rev} \leq w$, or

$$-w_{rev} \geq -w \tag{17-8}$$

Since w is the work done on the system by the surroundings, $-w$ is the work delivered to the surroundings by the system. Equation (17-8) then says that the work delivered to the surroundings during an isothermal change is largest for a reversible path.

General statement of the second law. For isothermal changes of state, ΔS has been defined by (17-2) to be equal to q_{rev}/T. This definition must be amplified when the temperature is not constant during the process, and the same applies to the more general statement (17-6) that $\Delta S \geq q/T$ for isothermal processes. Processes that are not isothermal may be considered to take place in infinitesimal steps, each of which is essentially at a constant temperature. If such a step is reversible and involves the transfer of the heat Dq_{rev}, then we postulate in analogy to (17-2) that

$$dS = \frac{Dq_{rev}}{T} \tag{17-9a}$$

which constitutes a definition of dS for an infinitesimal change of state. For an infinitesimal step along an irreversible path,

$$dS > \frac{Dq_{irrev}}{T} \tag{17-9b}$$

where dS is defined by (17-9a), which implies that to find dS for a given infinitesimal change in state we must find a reversible path by which the desired change of state can be effected.

For a finite change in state, we must take the sum over all the infinitesimal steps that constitute the path; more precisely, we take the integral over Dq/T for the path considered. This leads to the following two related statements that express the second law of thermodynamics:

1. For any reversible change of state of a system, the integral

$$\int_{1}^{2} \frac{Dq_{rev}}{T} = S_2 - S_1 = \Delta S \tag{17-10a}$$

is independent of the path and defines the change ΔS of a function of state S called the entropy. The temperature T is that of the system.

While the present statement demands reversibility for the processes in the system, no such restriction is required for the processes in the surroundings because all we are concerned with is the system. We may also replace the surroundings by any other surroundings without affecting the validity of statement 1, as long as this replacement does not change in any way the processes considered *in the system*. The surroundings may, for example, be at a different temperature than the system, making the transfer of heat into or out of the system irreversible; the subscript *rev* in Dq_{rev} refers only to the reversibility of the processes in the system. We can also replace the surroundings by a thermostat at the temperature of the system, if we wish to make the transfer of heat between system and surroundings reversible.

It is quite remarkable that although $\int_{1}^{2} Dq$ is path-dependent (even if we restrict ourselves to reversible paths) it is possible to find a function that is independent of the path along any of the reversible paths existing between states 1 and 2 simply by dividing each Dq_{rev} by the temperature T.

2.   Along an irreversible path in the direction actually traversed, the integral over Dq/T is path-dependent and is always smaller than the same integral over any reversible path relating the same initial and final states:

$$\int_1^2 \frac{Dq_{irrev}}{T} < \int_1^2 \frac{Dq_{rev}}{T} \qquad (17\text{-}10b)$$

The second integral in (17-10b) is, by (17-10a), the change of the entropy, ΔS, so that

$$\int_1^2 \frac{Dq_{irrev}}{T} < \Delta S = \int_1^2 dS \qquad \text{or} \qquad \frac{Dq_{irrev}}{T} < dS \qquad (17\text{-}10c)$$

for any irreversible or spontaneous process in the direction in which it may proceed. The inequalities (17-10c) thus apply to all *actual* or natural processes; we might therefore change the subscripts from *irrev* to *act*. If the path of an irreversible process could be traversed in the opposite direction by infinitesimal changes in conditions, which by the definition of irreversibility is forbidden, the signs of the values of all Dq_{act} involved would be changed, the inequalities in (17-10c) would change into their opposites, and the first integral would be larger than ΔS.

The procedure for finding ΔS for an irreversible process depends on the fact that S is a function of state. All details of the actual process except the initial and final states can be ignored because the details of the path are irrelevant to the value of ΔS. The actual path is replaced by a reversible path that connects the two states and that is specifically invented for the convenient evaluation of the integral in (17-10a). Strictly, this reversible path can exist only in thought because any actual path is irreversible. However, reversibility can be approximated sufficiently well in many actual situations that reliable experimental values of ΔS can be obtained. By this we mean that all effects of irreversibility are small compared to the main effect considered or measured, small enough to remain within the limits of the errors of observation. Such "practical reversibility" is achieved by appropriate adjustment of the conditions affecting the process, e.g., the pressure or the temperature of the surroundings, and by proceeding sufficiently slowly. The change in entropy, evaluated along the reversible path, is independent of the path for the desired change in state. The only general statement that can be made about the first integral in (17-10b) along the actual path is that it is more negative than ΔS.

Entropy change during heating.   For nonisothermal changes in state, we evaluate the entropy change with the help of (17-10a). For example, the change in temperature of a system that is being heated is related to the amount of heat absorbed, q, and the heat capacity C by

$$q = C \Delta T \qquad (17\text{-}11)$$

a relation given in Sec. 16-3 [Equations (16-12) and (16-13)]. To evaluate the entropy change with (17-10a), we rewrite (17-11) for an infinitesimal, and hence reversible, change:

$$Dq_{\text{rev}} = C\,dT \tag{17-12}$$

Substitution of (17-12) into (17-10a) leads to

$$\Delta S = \int_1^2 \frac{Dq_{\text{rev}}}{T} = \int_1^2 \frac{C\,dT}{T} \tag{17-13}$$

Because dT/T is just equal to $d \ln T$ (see Study Guide), $C\,dT/T$ may also be written $C\,d \ln T$. Thus if we assume C to be independent of temperature,

$$\Delta S = \int_{T_1}^{T_2} C\,d \ln T = C \ln \frac{T_2}{T_1} \tag{17-14}$$

If the change in state is carried out at constant pressure or at constant volume, the C to be used in (17-14) is C_P or C_V.

EXAMPLE 17-4   **Mixing portions of water of different temperatures.** What is the entropy increase when 100 g $H_2O(l)$ at 0°C is mixed adiabatically with 100 g $H_2O(l)$ at 100°C?

Solution. This is an example of an irreversible adiabatic process. Assuming a constant specific-heat capacity C_P of 4.18 J g^{-1} K^{-1}, we find the final temperature to be 50°C, because for this final temperature the enthalpy gain of the water being heated is just compensated by the enthalpy loss of the water being cooled. To reach the same final state reversibly, we heat the cold water from 273 to 323 K by placing it in contact with a temperature bath that is kept adjusted to be warmer than the water by only an infinitesimal amount. The hot water is cooled from 373 to 323 K in a similar way. The total entropy change for this system is

$$\Delta S = \int_{273}^{323} C_P\,d \ln T + \int_{373}^{323} C_P\,d \ln T$$

$$= C_P \ln \tfrac{323}{273} + C_P \ln \tfrac{323}{373}$$

$$= C_P \ln \frac{(323)^2}{273 \times 373}$$

where C_P is the heat capacity of 100 g water, or 418 J K^{-1}. In numbers,

$$\Delta S = (418 \times 2.30 \log 1.025)\ \text{J K}^{-1} = \underline{10.3\ \text{J K}^{-1}}$$

Entropy change in adiabatic and in isolated systems. For any process in an adiabatic system, Dq remains zero by definition. Therefore, for any process in such a system,

$$\int_1^2 \frac{Dq}{T} = 0 \tag{17-15}$$

Any adiabatic path

If the process is reversible, the integral represents ΔS, which is therefore zero for a reversible adiabatic process,

$$\Delta S_{adiab,rev} = 0 \qquad (17\text{-}16a)$$

For an irreversible adiabatic process, the integral, while still zero, no longer represents ΔS, but we deduce from (17-10c) that for the change in state now under consideration,

$$\Delta S_{adiab,irrev} > 0 \qquad (17\text{-}16b)$$

Comparison with (17-16a) shows that the present change in state cannot be achieved by a reversible adiabatic process. All statements in this paragraph apply to an isolated system also, because such a system is necessarily adiabatic.

Equations (17-16a) and (17-16b) state that the *entropy of an adiabatic system or of an isolated system can never decrease*. It remains constant for reversible processes and it increases for irreversible ones, as illustrated in Example 17-4.

All processes that actually occur are irreversible and thus increase the entropy of an adiabatic system or of an isolated system. Once the entropy has attained a value beyond which it cannot increase, all processes must cease. *Equilibrium in an adiabatic or isolated system is characterized by the maximum possible value of the entropy.* The tendency toward increase in entropy represents an important driving force for chemical and physical processes. If the direction of an irreversible adiabatic process were to be reversed, the entropy would decrease. This is prohibited; the second law tells in which direction irreversible processes may proceed. It gives a direction, an arrow, to time.

Isothermal expansion of an ideal gas.   Consider the entropy change accompanying the isothermal expansion of an ideal gas, a process treated in Sec. 16-3, Example 16-2. For an ideal gas U is a function of T only, and so $\Delta U = 0$ for any isothermal change. Hence $w = -q$ for such changes. For a reversible expansion of n moles of gas from the volume V_1 to the volume V_2, we found by (16-11) that

$$w_{rev} = -nRT \ln \frac{V_2}{V_1} = -q_{rev}$$

Combining the second portion of this equation with (17-2), we get

$$\Delta S = \frac{q_{rev}}{T} = nR \ln \frac{V_2}{V_1} \qquad (17\text{-}17)$$

This equation holds for *any* isothermal expansion of an ideal gas, reversible or not, because S is a function of state.

The enthalpy of an ideal gas also depends only on T because

$$H = U + PV = U + nRT$$

EXAMPLE 17-5   **Change in state for an ideal gas.**   The pressure on 1.00 mol helium is reduced from 1.00 to 0.100 atm at 50°C. Assume the gas to be ideal. What are ΔU, ΔH, and ΔS?

Solution. For an ideal gas U and H depend only on the temperature, and therefore $\Delta U = \Delta H = 0$. By (17-17), $\Delta S = nR \ln (V_2/V_1) = nR \ln (P_1/P_2) = 1 \times R \times \ln \frac{1.00}{0.100} = (8.31 \times 2.30)$ J K^{-1} = $\underline{19.1 \text{ J K}^{-1}}$.

Throughout the preceding chapter and this one, we have emphasized the difference between state functions, such as U, H, and S, and other quantities characteristic of changes in state, such as heat and work. To underscore this essential distinction, we now examine a number of different isothermal expansions of an ideal gas from the same initial state to the same final state by different paths, including the reversible process characterized by (16-11) just mentioned, and two irreversible processes. The first of the latter is the expansion against a constant pressure P_2, chosen so that $P_2 V_2 = P_1 V_1 = nRT$, where P_1 is the initial pressure and the expansion goes from V_1 to V_2 as earlier. At a constant external pressure P_2, $w = -P_2 \Delta V = -P_2(V_2 - V_1)$, so that

$$q = -w = P_2 V_2 - P_2 V_1 = P_2 V_2 \left(1 - \frac{V_1}{V_2} \right) = nRT \left(1 - \frac{V_1}{V_2} \right)$$

The other irreversible isothermal expansion that we examine here occurs into an initially evacuated chamber of fixed volume; for this process $q = -w = 0$. In all these expansions, ΔS is given by (17-17). For comparison we consider three specific parallel examples, supposing that $n = 1$, $T = 300$ K, and $V_2 = 2V_1$. We calculate q, w, q/T, ΔS, and ΔU and enter the results in Table 17-2. For the reversible expansion, $q_{rev} = -w_{rev} = RT \ln 2 = (8.31 \times 300 \times 0.693)$ J = 1.73 kJ; for the expansion against constant pressure, $q = -w = RT/2 = (8.31 \times 300/2)$ J = 1.25 kJ. For ΔS we find $R \ln 2 = (8.31 \times 0.693)$ J K^{-1} = 5.8 J K^{-1}, which is also equal to the ratio q/T for the reversible path. For the two irreversible paths specified, q/T has the respective values of $\frac{1.25}{0.300} = 4.2$ J K^{-1} and zero. Comparison of the values in Table 17-2 shows that (17-6) to (17-8) are all satisfied.

Important properties of entropy. Among the properties of entropy the following deserve particular emphasis. (1) Entropy is a function of state, but it is not conserved in an isolated system, in the way energy is. It cannot be destroyed, so that in an isolated system the entropy can only stay constant or increase. (2) Only entropy *differences* are defined by the second law of thermodynamics. (3) The entropy difference between two states can be defined thermodynamically only if it is possible

TABLE 17-2 **Comparison of thermodynamic quantities** The quantities refer to the isothermal expansion of 1 mol ideal gas to twice its volume at 300 K.

	w	q	q/T	ΔS	ΔU
Reversible expansion	− 1.73 kJ	1.73 kJ	5.8 J K^{-1}	5.8 J K^{-1}	0
Expansion against constant pressure	− 1.25 kJ	1.25 kJ	4.2 J K^{-1}	5.8 J K^{-1}	0
Expansion into evacuated chamber	0	0	0	5.8 J K^{-1}	0

to invent a reversible path that connects the two states, at least in thought. (4) The temperature scale denoted by T is the only one for which the infinitesimal dS of a function of state S is obtained when Dq_{rev} is divided by T, except that T might be multiplied by any constant factor. The quantity T may therefore be defined entirely on thermodynamic grounds, without recourse to the behavior of ideal gases. This is the real basis of its absolute nature, and it is called the *thermodynamic temperature scale* for this reason. There is, however, no compelling reason to choose a particular size of the degree, and this size is the only aspect of the Kelvin or the Rankine scale (Sec. 1-3) that is not absolute. (5) Entropy is an extensive quantity. When the total amounts of substances undergoing a specified change of state are multiplied by a given factor, the amounts of heat transferred during the change, and therefore the total entropy change, are multiplied by the same factor. Similarly, the entropies of two independent systems are additive, just as are their energies and enthalpies. (6) The SI units used for entropy are joules per kelvin (J K^{-1}); many compilations utilize the units calories per kelvin (cal K^{-1}), sometimes called "entropy units" (eu).

17-3   The Statistical Interpretation of Entropy

Microstates and macrostates.   Thermodynamics permits (although it is independent of) interpretations on a molecular basis. These microscopic interpretations are often valuable in imparting an intuitive feeling for entropy and allowing the development of plausible arguments for its properties.

A state in which all details that can be known about atoms and molecules are specified is called a microscopic state or *microstate*. Usually, a huge number of possible microstates corresponds to a given macroscopic state or *macrostate*. They are indistinguishable at the macroscopic level and are called *realizations* of the macrostate. To illustrate, there are many different microstates in which the velocities of individual molecules differ but for which the *average* kinetic energy is unchanged, so that all these microstates correspond to the same macroscopic quantity, called temperature. The number of microscopic realizations of a macroscopic state of fixed energy will be denoted by Ω (omega).

Suppose that an isolated system is in a certain macroscopic state 1 with Ω_1 microscopic realizations and that it is possible for it to go to a macrostate 2 with Ω_2 microscopic realizations. If Ω_2 is larger than Ω_1, the system tends to assume state 2 under the randomizing influence of temperature motion. The state of a system tends to move in the direction of increasing numbers of realizations.

As an illustration, consider a box containing many red and blue spheres of equal size and mass. At the beginning, all red spheres are placed in layers at the bottom of the box and all blue spheres are put on top. The box, which is not full, is then shaken vigorously. At the end, the distribution of the colored spheres is random. The many different microscopic realizations of the *initial* state are obtained by interchanging all the red spheres among themselves and, separately, all the blue spheres among themselves. The microstates that correspond to the *final* macrostate are obtained by interchanging *all* spheres with each other, regardless of their color. The final state has many more microscopic realizations than the initial state. There are parallels between this example and the dissolving of a substance. Initially, one has pure solvent and pure solute, but temperature motion eventually causes the solute molecules to be distributed evenly throughout the solvent, just as continued shaking of the box results eventually in a random distribution of the colored spheres. It is extremely unlikely that the solution will ever become unmixed into pure solvent and solute, just as it is unlikely that shaking the box will ever substantially decrease the randomness of the distribution of the colored spheres.

In another example, the first state is the initial state of the expansion of a gas, with all gas at volume V_1. State 2 corresponds to the complete expansion of the gas to the final volume V_2, which is larger than V_1. The number of microscopic realizations is proportional to the volume so that this number is larger for the final state than for the initial state.

Both entropy and Ω increase during spontaneous processes in isolated systems and assume their maximum values when equilibrium has been reached. The same behavior is shared by any function that increases steadily (monotonically) with Ω, and the question is whether it is possible to equate any such function with S. Boltzmann demonstrated that the natural logarithm of Ω is just such a function. He showed that for an isolated system the entropy must be proportional to $\ln \Omega$:

$$S = k \ln \Omega \qquad (17\text{-}18)$$

The proportionality factor is the same Boltzmann constant that we encountered in Chap. 5, the gas constant R divided by Avogadro's number N_A: $k = R/N_A = 1.381 \times 10^{-23}\,\text{J K}^{-1} = 1.381 \times 10^{-16}\,\text{erg K}^{-1}$. Note that (17-18) refers not to an entropy *difference* but to absolute entropy. The change in entropy on going from a state with Ω_1 realizations to a state with Ω_2 realizations is given by

$$\Delta S = k \ln \frac{\Omega_2}{\Omega_1} \qquad (17\text{-}19)$$

The evaluation of Ω for a given situation usually requires quantum mechanics and provides an important method for obtaining entropy values for many substances. However, Boltzmann's relationship (17-18) can be used in a qualitative way to understand the nature of entropy. As the examples discussed earlier have shown, Ω (and therefore the entropy) are measures of molecular disorder, and an increase in entropy parallels an increase in molecular disorder. The second law is a matter of probabilities. Under the influence of temperature motion a system tends to assume the state of maximum probability. The arrow of time is in the direction of spreading the total energy of a system over as many energy levels as are accessible.

It is of interest to find the order of magnitude of Ω corresponding to a representative entropy value, such as $50\,\text{J K}^{-1}$. Since $k = 1.38 \times 10^{-23}\,\text{J K}^{-1}$, we have

$$50 = k \ln \Omega = 1.38 \times 10^{-23} \times 2.30 \log \Omega$$
$$\log \Omega = 1.58 \times 10^{24}$$
$$\Omega = 10^{1.58 \times 10^{24}}$$

It is hard enough to imagine the magnitude of Avogadro's number N_A, but the value just given, with a number of digits of the order of N_A itself, strains the imagination incomprehensibly.

The statistical interpretation of the second law makes it the law of the probable rather than the law of the certain. There are always minute local fluctuations in density, pressure, and temperature. However, improbabilities of the order of 10^{-N_A} are so exceedingly small that the statistical interpretation of the second law makes it the law of the overwhelmingly probable, and macroscopic violations of the second law are unthinkable. Furthermore, the keystone of science is reproducible behavior. Even if some "impossible" phenomenon were observed once, no one would believe it because it would not be reproducible and would be ruled out as a "fluke". Examples of such essentially impossible violations are these: (1) The velocities of all molecules of the air in a flask become directed the same way by accident so that a vacuum is spontaneously created on one side of the flask. (2) A sufficient number of molecular velocities in a macroscopic body become directed upward by accident so that the body suddenly moves upward. (3) The velocity distribution changes so that one part of an isolated body becomes warmer, the other cooler.

Return to phase transformations. Equation (17-3) is an example of a more general expression for a phase transformation,

$$\Delta S_{tr} = \frac{\Delta H_{tr}}{T_{tr}} \tag{17-20}$$

relating the entropy of transformation ΔS_{tr} to the enthalpy of transformation ΔH_{tr} and the transformation temperature T_{tr}, all quantities being associated with phase equilibrium at the temperature and pressure considered. An example of such a transformation might be from one solid phase to another, as from orthorhombic to monoclinic sulfur or from graphite to diamond. Other common phase transformations are fusion (melting) and vaporization.

EXAMPLE 17-6   **The entropy of melting of bromine.** Bromine melts at $-7°C$ and the molar enthalpy of melting is 10.8 kJ mol^{-1}. What is the molar entropy of melting?

Solution. The melting temperature in kelvins is $(273 - 7)$ K $= 266$ K, so that $\Delta \widetilde{S}_{melt}$ $= \Delta \widetilde{H}_{melt}/T_{melt} = (10.8 \times 10^3$ J mol$^{-1})/(266$ K$) = \underline{40.6$ J mol^{-1} K$^{-1}}$.

The entropy always increases when a solid is melted or sublimed or when a liquid is vaporized, because ΔH_{tr} is always positive in these cases.[1] This is readily understood in microscopic terms because the state of disorder increases in each case.

Caution must, however, be used in arguments that relate entropy increase to loss of order. An example of an apparent contradiction concerns a supercooled melt that suddenly crystallizes. This is a spontaneous process, so that the entropy must increase. Yet how can it, since crystallization of the liquid results in a more orderly state?

---

[1] The only known exception concerns the ^{3}He isotope, for which ΔH_{melt} is negative, e.g., at 28.9 atm and the corresponding melting temperature of 0.318 K. Under the conditions stated, solid ^{3}He has a larger entropy than liquid ^{3}He, a fact that can be understood in terms of quantum mechanics.

The answer is that if the system is adiabatic, the heat of fusion liberated increases the temperature (and thus the thermal motion) sufficiently to cause a net increase in disorder. If the system is kept at constant temperature, the heat liberated increases the entropy of the surroundings sufficiently to cause a net gain of the *total* entropy of system and surroundings.

Table 17-3 contains a number of entropies of fusion and of vaporization. It will be noted that the entropy of fusion is generally smaller than the entropy of vaporization. This indicates that the increase in randomness is smaller when going from the solid to the liquid than when going from the liquid to the vapor. Note also the orders of magnitudes of these molar entropy changes, from about 8 to some 100 J mol^{-1} K^{-1}. It is *joules* per kelvin that are involved, whereas the corresponding molar enthalpy changes are best expressed in *kilo*joules.

Inspection of Table 17-3 shows that the entropies of vaporization for many liquids are approximately the same, although there is no similar regularity shown by the entropies of fusion. The approximate constancy of $\Delta \widetilde{S}_{vap}$ is called *Trouton's rule,* according to which

$$\frac{\Delta \widetilde{H}_{vap}}{T_b} = \Delta \widetilde{S}_{vap} \approx 88 \text{ J mol}^{-1} \text{ K}^{-1} \tag{17-21}$$

where T_b is the boiling temperature at 1 atm and the approximate value for $\Delta \widetilde{S}_{vap}$ is an average over many more substances than are listed in the table. The rule holds fairly well for many liquids, but there are exceptions.

17-4 Equilibrium and Spontaneous Processes

We began this chapter with a qualitative discussion of the two factors that determine the direction of spontaneous change—energy and entropy—and showed, with (17-16*a*) and (17-16*b*), that for an adiabatic process or an isolated system the criterion for the spontaneous direction of change is $\Delta S > 0$ and the criterion of equilibrium is $\Delta S = 0$. It would be useful to have criteria applicable more generally, not just

TABLE 17-3 **Molar entropies of fusion and vaporization***

	$\Delta \widetilde{S}_{fus}$/J mol^{-1} K^{-1}	$\Delta \widetilde{S}_{vap}$/J mol^{-1} K^{-1}
N_2	11.4	72.0
O_2	8.2	75.7
Cl_2	37.2	85.3
CS_2	27.3	83.8
CH_4	10.3	74.5
CH_3Cl	43.9	88.8
CCl_4	9.7	85.8
C_6H_6	35.3	87.1
NH_3	28.9	97.5
Hg	10.0	92.0

*At mp and bp, respectively.

for adiabatic paths and isolated systems. Such criteria were established by the American physical chemist J. W. Gibbs in terms of a function of the system that he called the free energy.

The Gibbs free energy.   We begin with a definition of the Gibbs free energy or Gibbs function G:

$$G = H - TS \qquad (17\text{-}22)$$

It is a function of state, because H, T, and S are, and it has the dimensions of an energy, as do H and TS. The usefulness of G can be established by considering its behavior for a finite change in state at constant pressure and temperature. For such a change

$$\Delta G = \Delta H - T \Delta S \qquad (17\text{-}23)$$

because at constant temperature $\Delta(TS) = T_2 S_2 - T_1 S_1 = T(S_2 - S_1) = T \Delta S$. We know, by (17-6), that at constant temperature $\Delta S \geq q/T$. Furthermore, at constant pressure with no work other than P,V work, $q_P = \Delta H$ by (16-23b) (Sec. 16-4). Hence, under these circumstances

$$T \Delta S \geq q_P = \Delta H$$

or $$-T \Delta S \leq -\Delta H$$

Addition of ΔH to each side and substitution of (17-23) gives

$$\Delta H - T \Delta S = \Delta G \leq 0 \qquad (17\text{-}24)$$

when $P = $ constant, $T = $ constant, and no work is done other than P,V work. Many chemical processes occur under just these conditions.

Conditions for equilibrium and spontaneous processes.   Three cases can be distinguished for chemical processes that occur *at constant temperature and pressure with only P,V work:*

1.   A chemical change, or any other process, is at equilibrium when

$$\Delta G = 0 \qquad (17\text{-}25a)$$

as the process is made to proceed in either direction.

2.   An actual process must occur in the direction for which

$$\Delta G < 0 \qquad (17\text{-}25b)$$

3.   A process for which

$$\Delta G > 0 \qquad (17\text{-}25c)$$

is thermodynamically forbidden.

For chemical processes at constant T and P and with only P,V work, conditions (17-25a) and (17-25b) imply that equilibrium is characterized by a minimum of G.

The situation is analogous to that of an object at rest in a gravitational field. The object is in (stable) equilibrium when its potential energy is a minimum.

For changes at constant P and T when there is some net work w' other than P,V work, we have, by (16-23b),

$$q_P = \Delta H - w'$$

and hence

$$T \Delta S \geq q_P = \Delta H - w'$$

or

$$-T \Delta S \leq -\Delta H + w'$$

Addition of ΔH to each side gives

$$\Delta G = \Delta H - T \Delta S \leq w' \qquad \text{when } P = \text{const}, \ T = \text{const} \qquad (17\text{-}26a)$$

To see one of the implications of (17-26a), we write, from $w' \geq \Delta G$,

$$-w' \leq -\Delta G \qquad (17\text{-}26b)$$

On the left side, $-w'$ is the net work done by the system on the surroundings; on the right side, $-\Delta G$ is the loss of the free energy of the system. The amount of work obtainable depends on the path followed; ΔG, of course, does not. The relationship (17-26b) thus states that $-\Delta G$ is the maximum net work that the system is capable of performing at constant temperature and pressure. This fact is the origin of the name *free energy*—that part of the energy of the system that is "freely convertible" into net work at constant T and P. This maximum net work can be obtained only under reversible or equilibrium conditions, that is, in the limit of an infinitesimally slow, idealized process. The free energy is, of course, altered by the same amount for a given change of state, no matter what work has actually been performed. We return to (17-26a) when we consider the thermodynamics of chemical reactions capable of doing electrical work (Sec. 18-3).

The importance of ΔH **and** ΔS. The condition (17-25b) for the spontaneous direction of a reaction at constant temperature and pressure with only P,V work can be written in the form

$$\Delta G = \Delta H - T \Delta S < 0 \qquad (17\text{-}27)$$

A spontaneous process is therefore favored not only by a decrease of H (negative ΔH) but also by an increase of S (positive ΔS). The influence of the entropy change becomes more important as the absolute temperature increases; the temperature is a weighting factor for the entropy change relative to the enthalpy change.

Four basically different situations can be envisioned, differing in the signs of ΔH and ΔS for the process under consideration:

1. As long as $\Delta H < 0$ and $\Delta S > 0$ at all T, then ΔG is necessarily negative *at any temperature*. (If the sign of ΔH or ΔS changes as T changes, then ΔG will not be negative at all temperatures and the situation falls under 2 or 3 below.)

2. When $\Delta H < 0$ and $\Delta S < 0$, the process may occur *only* when $|T \Delta S| < |\Delta H|$. The process must thus be sufficiently exothermic to overcome the handicap of the entropy decrease. Processes that fall in this category include condensation of a gas or freezing of a liquid. At sufficiently large T, the term $T\Delta S$ will overcome the ΔH term and the process can no longer proceed. If ΔH and ΔS are known at some temperature, then the temperature T' above which the process becomes prohibited by thermodynamics is, by (17-27), given approximately by $T' = \Delta H/\Delta S$. This relationship is only approximate, however, because both ΔH and ΔS depend to some extent on the temperature.

3. When $\Delta H > 0$ and $\Delta S > 0$, the process tends to occur *only if $T\Delta S > \Delta H$*. Here the advantage of the entropy increase must overcome the handicap of the endothermic nature of the process, and for this T must be sufficiently large. Examples include melting and evaporation. The heat required by the process comes, of course, from the surroundings—that is, from the constant-temperature bath needed to maintain the temperature at a constant level. When T is too small the reaction is prohibited.

4. When $\Delta H > 0$ and $\Delta S < 0$, the process is *prohibited thermodynamically at any temperature,* unless a sign change of ΔH or ΔS with temperature occurs in such a way that the sign of ΔG changes also.

Table 17-4 summarizes these four situations and the following examples of chemical reactions furnish illustrations. The numbers on the right of the chemical equations are, in sequence, ΔH, $T\Delta S$, and ΔG, all in kilojoules at 25°C and 1 atm.

Case 1

$$2H_2O_2(g) \xrightarrow{?} 2H_2O(g) + O_2(g) \qquad -211; +39; -250$$

There is a strong tendency of the reaction to proceed to the right. At no temperature is this tendency expected to be reversed.

Case 2

$$3H_2(g) + N_2(g) \xrightarrow{?} 2NH_3(g) \qquad -92; -59; -33$$

Again, this reaction tends to go to the right. As the temperature is increased, however, this tendency becomes less because the $T\Delta S$ term becomes increasingly negative. At sufficiently elevated temperatures the reaction tends to proceed to the left when ΔG becomes positive.

TABLE 17-4   **Influence of enthalpy and entropy changes on chemical reactions at constant P and T with only P,V work**

Case	ΔH	ΔS	Reaction may proceed
1	<0	>0	At any T
2	<0	<0	When T sufficiently low
3	>0	>0	When T sufficiently high
4	>0	<0	Not at all

Case 3

$$N_2O_4(g) \xrightarrow{?} 2NO_2(g) \qquad +58; \ +53; \ +5$$

At 25°C, ΔG is positive, and the reaction cannot proceed to the right when both gases are at 1 atm. As the temperature is raised, however, $T\Delta S$ becomes more positive and eventually outweighs ΔH, so that ΔG becomes negative and the reaction may proceed to the right.

Case 4

$$N_2(g) + 2O_2(g) \xrightarrow{?} 2NO_2(g) \qquad +68; \ -36; \ +104$$

Since $\Delta H > 0$ and $\Delta S < 0$, there is no temperature at which the reaction is expected to proceed to the right. There is instead a strong tendency toward the left.

These examples illustrate again the importance of rates (see Chap. 20). In the absence of catalysts, H_2O_2 does not decompose appreciably at room temperature and the rate of formation of NH_3 from H_2 and N_2 is zero for all practical purposes. Although NO_2 is unstable at room temperature (reaction of case 4 going to the left), the rate of decomposition is immeasurably small. If the rate of reaction is small or zero and the reaction is permitted by thermodynamics, there is no reason why a diligent search for a suitable catalyst that can accelerate the reaction should not be made. However, the presence of catalysts cannot affect ΔG, because G is a function of state, dependent on the nature of the reactants and products but not on the path that relates them. Thus if $\Delta G > 0$, there is no point in searching for a catalyst for the corresponding reaction.

For a reaction taking place at constant T and P with P,V work only, the *condition for equilibrium* is that $\Delta G = \Delta H - T\Delta S = 0$ or that

$$\Delta H = T\Delta S \qquad (17\text{-}28)$$

The two terms ΔH and $T\Delta S$ are thus exactly equal *at equilibrium*. If they are both positive, the tendency toward greater randomness, measured by ΔS and weighted by multiplication by T, is just sufficient to compensate for the endothermic nature of the reaction. If they are both negative, the exothermic character of the reaction just compensates for the decrease in randomness, again weighted by T.

At equilibrium there is no change in free energy G when one phase is transformed into another or when reactants are transformed into products, provided that equilibrium concentrations and pressures of all reaction participants are maintained.

EXAMPLE 17-7   **Free-energy change during a phase transformation.** Consider the phase transformation

$$H_2O(l) \longrightarrow H_2O(g)$$

at 25°C and 1 atm. Calculate ΔG from the values of ΔH and ΔS ($\Delta H = 44.0$ kJ, $\Delta S = 118.7$ J K^{-1}) and discuss the result.

Solution. We find $T \Delta S = 35.4 \, \text{kJ}$ and, therefore, $\Delta G = \Delta H - T \Delta S = 8.6 \, \text{kJ}$. At 25°C, $T \Delta S$ is smaller than ΔH, and ΔG is positive. For the reverse process, condensation, ΔG is therefore negative, so that water vapor at 1 atm and 25°C is unstable and condenses.

A more general question than that discussed in the example is this: At what temperature would equilibrium between vapor at 1 atm and liquid water be achieved? At equilibrium, ΔH and $T \Delta S$ must be equal, and since ΔH is larger than $T \Delta S$ at 298 K, the equilibrium temperature must be higher. The equilibrium temperature T_{eq} may be estimated to be

$$T_{eq} \approx \frac{\Delta H(298)}{\Delta S(298)} = 371 \, \text{K} \tag{17-29}$$

on the assumption that ΔH and ΔS do not vary with temperature. The actual equilibrium temperature is, of course, 100°C or 373 K. Figure 17-1 shows the actual behavior of ΔH and $T \Delta S$ for the vaporization of water at 1 atm, as a function of temperature. It is seen that ΔH slowly decreases with temperature. For the approximate relationship (17-29), a constant ΔH was assumed. The reasonable agreement between the estimated and actual equilibrium temperatures (371 and 373 K) is due in part to a cancellation of errors: ΔS slowly decreases with temperature also, and the curve $T \Delta S$ rises less steeply than it would if ΔS were constant.

Standard free energies. Because the free energy is a function of state, values of ΔG for different chemical reactions may be combined in the same way that the chemical equations for the reactions considered may be combined into the equations for new reactions. All this is analogous to the way in which ΔH values can be combined (Sec. 16-4).

FIGURE 17-1 ΔH **and** $T \Delta S$ **for the phase transformation** $H_2O(l) \longrightarrow H_2O(g, 1 \, \text{atm})$. At the point where the two curves cross (373 K) there is equilibrium, since $\Delta H = T \Delta S$ implies that $\Delta G = 0$. To the left of this intersection the difference between the two curves is $+\Delta G$ because ΔH is larger than $T \Delta S$; on the right the difference is $-\Delta G$ because ΔH is now smaller than $T \Delta S$. Thus, ΔG for the reaction $H_2O(l) \longrightarrow H_2O(g, 1 \, \text{atm})$ is positive below 373 K and negative above. Below 373 K, liquid water is more stable than water vapor at 1 atm; above 373 K, vapor at 1 atm is the more stable; and at 373 K there is equilibrium.

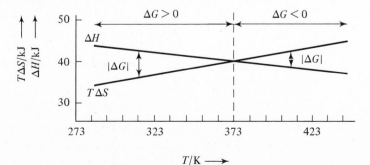

EXAMPLE 17-8 | **Combination of ΔG values.** The overall equations of two chemical reactions, (1) and (2), are given below. Under each is the value of ΔG at 298 K for transforming as many moles of reactants into products as are indicated by the coefficients in the equations, with ions at concentrations of 1 mol liter^{-1} and gases at partial pressures of 1 atm. What is ΔG for reaction (3), which may be obtained by suitably combining reactions (1) and (2)?

$$(1) \quad 16H^+ + 2MnO_4^- + 10Cl^- \longrightarrow 5Cl_2(g) + 2Mn^{2+} + 8H_2O(l)$$

$$\Delta G_1 = -142.0 \text{ kJ}$$

$$(2) \quad Cl_2(g) + 2Fe^{2+} \longrightarrow 2Fe^{3+} + 2Cl^-$$

$$\Delta G_2 = -113.6 \text{ kJ}$$

$$(3) \quad 8H^+ + MnO_4^- + 5Fe^{2+} \longrightarrow 5Fe^{3+} + Mn^{2+} + 4H_2O(l)$$

$$\Delta G_3 = ?$$

Solution. The third equation can be obtained by adding one-half of the first equation and five-halves of the second equation. Therefore, $\Delta G_3 = \frac{1}{2}\Delta G_1 + \frac{5}{2}\Delta G_2 = [\frac{1}{2}(-142.0) + \frac{5}{2}(-113.6)]$ kJ = $-\underline{355.0 \text{ kJ.}}$

Since only differences in free energy are defined by thermodynamics, it is possible, just as in the case of enthalpies, to agree on certain reference states that are assigned zero free energy of formation *by convention*. These are exactly the same reference states that were assigned zero standard enthalpies of formation—the elements at standard conditions in their most stable forms (see Sec. 16-4). The *standard molar free energies of formation* $\widetilde{G}_f^\ominus$ of compounds, and of elements in other than their stable modifications, are then their free energies of formation at standard conditions from the elements in their most stable forms. For substances in aqueous solution, standard conditions imply unit concentration (or more precisely, activity) and the standard free energies of formation of ionic species are defined by the convention that the standard free energy of H^+ is zero. Table D-4 contains a representative list of such standard free energies of formation or, more simply, *standard free energies,* as they will usually be called. In many texts these quantities are given the symbols $\Delta \widetilde{G}_f^\ominus$, $\Delta \widetilde{G}_f^\circ$, or $\Delta \widetilde{G} f^\circ$. Unless otherwise indicated, standard free energies refer to 25°C or 298 K. Occasionally another temperature T is more useful and molar standard free energies at T are indicated by the symbol $\widetilde{G}_f^\ominus(T)$. Thus $\widetilde{G}_f^\ominus(500)$ is the free energy of formation of the substance in question from the elements at standard conditions and 500 K. Again the convention is to take as reference states the elements in their most stable forms at 1 atm and the temperature T, to which are thus assigned zero standard free energies of formation.

The standard free-energy change for any reaction can be found from the standard free energies of all substances involved by the relationship

$$\Delta G^\ominus = \Sigma G_{f, \text{ prod}}^\ominus - \Sigma G_{f, \text{ react}}^\ominus \tag{17-30}$$

This quantity is meaningful only if the explicit overall balanced equation of the reaction it applies to is given. It refers to the complete conversion into products of as many moles of reactants as are indicated by the coefficients in the chemical equation, all at standard conditions. A temperature of 298 K is implied here also unless another temperature is specifically mentioned.

EXAMPLE 17-9 ΔG° **for burning NH$_3$ to NO.**   What is the standard free energy for burning ammonia in oxygen to NO?

$$4NH_3(g) + 5O_2(g) \longrightarrow 4NO(g) + 6H_2O(l)$$
$$\quad -16.7 \qquad\quad 0 \qquad\qquad 86.7 \qquad -237.2$$

This is an important industrial reaction, a step in the manufacture of nitric acid from synthetic ammonia. The standard molar free energies of formation of the substances involved (Table D-4) are shown under the chemical symbols.

Solution.

$$\Delta G^{\circ} = (4 \times 86.7 - 6 \times 237.2 + 4 \times 16.7)\, kJ = \underline{-1010\ kJ}$$

This value refers, of course, to 298 K or 25°C and to the numbers of moles specified in the reaction equation given.

Standard free energies of formation of substances may be obtained from experimental values of ΔH and ΔS [(17-23)]. The enthalpies and entropies may in turn be derived from *thermal measurements alone*—calorimetric studies of heats of transition (17-20) and by application of (17-13) and (17-14). The Boltzmann relation (17-18) provides another important approach for deriving S values. Other experimental approaches are the measurement of equilibrium constants (Sec. 17-5) and of electrical cell potentials, discussed in Secs. 18-3 and 19-1.

17-5   Chemical Equilibria

The chemical potential for mixtures of substances.   Many systems of chemical interest involve mixtures, and the free energy of a mixture is not simply the sum of the free energies of the pure components. The difference is associated with the disorder introduced when the components are mixed and with the interactions between them. It is possible to associate each component i of a mixture with a quantity μ_i, called the chemical potential of i, in such a way that the free energy of the mixture can be expressed as

$$G = \Sigma n_i \mu_i \qquad\qquad (17\text{-}31)$$

where n_i is the number of moles of component i in the mixture. The chemical potential of each component, μ_i, is a function not only of the temperature and pressure but also of the concentration or partial pressure of that component (and often of the other components also). For pure substances μ_i is identical with the molar free energy. The dependence of chemical potentials on concentrations and partial pressures cannot be derived from thermodynamics alone. It requires experimental data or sometimes results from the statistical theory of the behavior of large numbers of molecules. Among the most important cases are the following.

In a *mixture of perfect gases* labeled by the subscripts $i = 1, 2, \ldots$, the chemical potential of the species i is given by the expression

$$\mu_i = \widetilde{G}^{\ominus}_{f,i} + RT \ln P_i \tag{17-32}$$

where P_i is the partial pressure in atmospheres of the species i and where $\widetilde{G}^{\ominus}_{f,i}$ is the standard molar free energy of formation of the pure gas i.

For *dilute solutions* the chemical potential of the solute species i is, to a good approximation,

$$\mu_{i,s} = \widetilde{G}^{\ominus}_{f,i} + RT \ln c_i \tag{17-33}$$

where c_i is the concentration in moles per liter of the species i—also denoted by $[i]$ in reaction quotients—and $\widetilde{G}^{\ominus}_{f,i}$ is the standard molar free energy of formation of the species i in solution. The subscript s is used to differentiate the chemical potential of the species i in a solution from that in a gas mixture as shown in (17-32); $\mu_{i,s}$ depends on T and the total pressure P.

Expressions (17-32) and (17-33) are very similar. In each there is a logarithmic term, and both μ_i and $\mu_{i,s}$ are independent of the nature and the concentrations of the other components of the mixture. For other than dilute solutions, expression (17-33) remains valid only if concentrations are replaced by activities (Sec. 15-5):

$$\mu_{i,s} = \widetilde{G}^{\ominus}_{f,i} + RT \ln a_i \tag{17-34}$$

Similarly, for a mixture of *real* gases, (17-32) is valid only if the partial pressures P_i are replaced by fugacities.

EXAMPLE 17-10   **Chemical potentials for the components of a mixture.** What are the chemical potentials at 298 K for the species in a three-phase system composed of AgCl(s); H$_2$O(l) containing Ag$^+$, H$^+$, and Cl$^-$; and HCl(g)?

Solution. Using the values for $\widetilde{G}^{\ominus}_f$ in Appendix D, we find:

Solid phase: The chemical potential of AgCl(s) is equal to the molar free energy because AgCl is a pure substance:

$$\mu_{\text{AgCl}} = \widetilde{G}^{\ominus}_{f,\text{AgCl}} = \underline{-109.7 \text{ kJ mol}^{-1}}$$

Liquid phase: The solution is dilute; hence it can be assumed that the chemical potential of H$_2$O is approximately constant and equal to that of pure water:

$$\mu_{\text{H}_2\text{O}} \approx \text{const} = \widetilde{G}^{\ominus}_{f,\text{H}_2\text{O}} = \underline{-237.2 \text{ kJ mol}^{-1}}$$

$$\mu_{\text{Ag}^+} = \underline{77.1 \text{ kJ mol}^{-1} + RT \ln [\text{Ag}^+]}$$

$$\mu_{\text{H}^+} = \underline{RT \ln [\text{H}^+]}$$

$$\mu_{\text{Cl}^-} = \underline{-131.2 \text{ kJ mol}^{-1} + RT \ln [\text{Cl}^-]}$$

Vapor phase:

$$\mu_{\text{HCl}} = \underline{-95.3 \text{ kJ mol}^{-1} + RT \ln P_{\text{HCl}}}$$

The general thermodynamic equilibrium expression. An important application of chemical potentials is to chemical equilibrium. The balanced equation for a chemical reaction can be put in the form

$$a\text{A} + b\text{B} + \cdots \longrightarrow d\text{D} + e\text{E} + \cdots \qquad (17\text{-}35)$$

Suppose that we transform a moles of reactant A, b moles of reactant B, and perhaps other reactants, into d moles of product D, e moles of product E, and perhaps other products. Assume that the entire transformation is done at constant temperature and pressure and at constant concentrations. The free energy of the reactants is

$$G(\text{reactants}) = a\mu_\text{A} + b\mu_\text{B} + \cdots$$

and that of the products is

$$G(\text{products}) = d\mu_\text{D} + e\mu_\text{E} + \cdots$$

The change in free energy resulting from the reaction (17-35) is the difference between the last two expressions,

$$\Delta G = (d\mu_\text{D} + e\mu_\text{E} + \cdots) - (a\mu_\text{A} + b\mu_\text{B} + \cdots) \qquad (17\text{-}36)$$

This quantity must be zero if there is chemical equilibrium at constant temperature and pressure:[1]

$$(d\mu_\text{D} + e\mu_\text{E} + \cdots) - (a\mu_\text{A} + b\mu_\text{B} + \cdots) = 0 \qquad (17\text{-}37)$$

This is as far as we can go with general thermodynamic considerations. Any further steps require knowledge of specific expressions for the chemical potentials, such as (17-32) to (17-34).

Gas reactions. To illustrate some implications of the general condition for chemical equilibrium just stated, consider the reaction between hydrogen and nitrogen to produce ammonia:

$$3\text{H}_2(g) + \text{N}_2(g) \rightleftharpoons 2\text{NH}_3(g) \qquad (17\text{-}38)$$

When the appropriate expressions of the type (17-32) for the chemical potentials of the three gases are inserted into (17-36), we get

$$\Delta G = 2\widetilde{G}^{\ominus}_{\text{NH}_3} - 3\widetilde{G}^{\ominus}_{\text{H}_2} - \widetilde{G}^{\ominus}_{\text{N}_2} + RT(2 \ln P_{\text{NH}_3} - 3 \ln P_{\text{H}_2} - \ln P_{\text{N}_2}) \quad (17\text{-}39)$$

The combination of $\widetilde{G}^{\ominus}$ terms in (17-39) is just the standard free energy of the reaction, $\Delta G^{\ominus}$:

$$\Delta G^{\ominus} = 2\widetilde{G}^{\ominus}_{\text{NH}_3} - 3\widetilde{G}^{\ominus}_{\text{H}_2} - \widetilde{G}^{\ominus}_{\text{N}_2} \qquad (17\text{-}40)$$

The term in (17-39) that includes the factor RT can be rearranged as follows:

$$RT(2 \ln P_{\text{NH}_3} - 3 \ln P_{\text{H}_2} - \ln P_{\text{N}_2}) = RT(\ln P^2_{\text{NH}_3} + \ln P_{\text{H}_2}^{-3} + \ln P_{\text{N}_2}^{-1})$$

---

[1]At equilibrium the relationship (17-37) can be shown to hold regardless of whether T and P are kept constant, or T and V, or any other two variables that specify the state of the system.

$$= RT \ln \frac{P_{NH_3}^2}{P_{H_2}{}^3 P_{N_2}} = RT \ln Q \qquad (17\text{-}41)$$

where Q is the reaction quotient for the chemical equation (17-38). Substituting from (17-40) and (17-41) into (17-39), we obtain

$$\Delta G = \Delta G^{\ominus} + RT \ln Q \qquad (17\text{-}42)$$

for the free-energy change when (17-38) is carried out at the particular partial pressures specified in the quotient Q. This free-energy change applies for the number of moles of reactants being changed into products that is indicated by the overall chemical equation. The pressures of both reactants and products are assumed to remain constant during the reaction.

If the ΔG indicated by (17-42) is negative, there is a tendency for the reactants of the corresponding equation to change into products under the conditions specified; if ΔG is positive, the reverse is true. In other words, the reaction tends to proceed in such a direction that the overall free energy decreases.

When the partial pressures in (17-42) are equilibrium pressures, $\Delta G = 0$ so that

$$RT \ln Q_{eq} = -\Delta G^{\ominus} \qquad (17\text{-}43)$$

Since $\Delta G^{\ominus}$ is a constant at constant temperature, the left side in (17-43) must also be constant under these circumstances. Hence at equilibrium

$$Q_{eq} = K(T) \qquad (17\text{-}44)$$

and
$$-RT \ln K(T) = \Delta G^{\ominus}(T) \qquad (17\text{-}45)$$

Equation (17-44) is a shorthand statement of the equilibrium law discussed in Chap. 13; Equation (17-45) gives the relationship between the equilibrium constant K at a given temperature and the *standard* free energy of the reaction at that temperature. Although these equations were derived in terms of a specific example, they are of completely general validity for reactions among ideal gases. If the values of $\widetilde{G}_f^{\ominus}$ in Appendix D are used to evaluate $\Delta G^{\ominus}$, the result applies to 298 K. For other temperatures more extensive tabulations are available. Since the standard states used for molar free energies of formation are at 1 atm, all partial pressures in Q must be expressed in atmospheres. Any other pressure unit, such as the torr, could be used provided that standard states were redefined so as to pertain to that unit (e.g., to $P = 1$ torr), with corresponding changes in the tabulated values of $\widetilde{G}_f^{\ominus}$ and $\widetilde{H}_f^{\ominus}$.

EXAMPLE 17-11   **Equilibrium constant for the formation of ammonia.** What are the values for $\Delta G^{\ominus}$ and K at 298 K for the formation of ammonia by reaction (17-38)?

Solution. From Table D-4, $\Delta G^{\ominus} = 2\widetilde{G}_{f,NH_3}^{\ominus} - 3\widetilde{G}_{f,H_2}^{\ominus} - \widetilde{G}_{f,N_2}^{\ominus} = [2(-16.7) - 3 \times 0 - 0]\,\text{kJ} = -33.4\,\text{kJ}$. To obtain $\log K$ we divide by $-2.30 RT$, using (17-45) and the relation $\log K = (\ln K)/2.30$. It is important that the units of RT employed in this division be the same as those of $\Delta G^{\ominus}$. We use $R = 8.31\,\text{J K}^{-1}$ and convert $\Delta G^{\ominus}$ from kilojoules to joules. Thus, $\log K = 33.4 \times 10^3/(8.31 \times 298 \times 2.30) = 5.86$; $K = 7.2 \times 10^5\,\text{atm}^{-2}$.

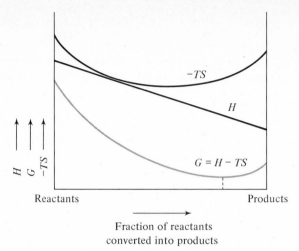

FIGURE 17-2   **The free energy of a typical reaction mixture.** The graph shows H, $-TS$, and G for situations in which different fractions of reactants have been converted into products, from reactants only (left edge) to products only (right edge). As the reaction proceeds from reactants to products, the concentrations or partial pressures of all participants change in such a way that the free energy of the system is lowered. Equilibrium exists when G is at its minimum value. If the reaction mixture initially contained only products, the reaction would proceed toward the left until the same minimum value of G had been reached. Note the importance of the term $-TS$ for the existence of a minimum in G.

It is important to recognize the essential distinction between $\Delta G^\ominus$ and ΔG. The former quantity refers to a change in state in which the reactants in their standard states are transformed into products in their standard states. In general this will not be a situation in which these participants in the reaction are in equilibrium with one another, except in the unlikely event that the equilibrium constant happens to be equal to 1. It may at first seem somewhat paradoxical that this standard free-energy change, which usually refers to a nonequilibrium situation, is what determines the magnitude of the equilibrium constant, by (17-45). However, as our derivation shows, this is true just because in an equilibrium situation ΔG itself, which applies to the change in state at the concentrations or pressures actually existing, is necessarily zero.

While ΔG in (17-42) refers to the change in free energy incurred when reactants are changed into the products at the prevailing partial pressures that must be kept constant, it is instructive to consider the total free energy of the reaction mixture for changing compositions and partial pressures that correspond to different degrees of conversion of reactants into products, from no reaction (no products formed) to complete conversion (no reactants remaining). A typical case is shown in Fig. 17-2. As the pressures or concentrations of the reactants decrease, so does the free energy associated with them, while the free energy associated with the products begins to rise as their pressures or concentrations increase. The total free energy decreases at first, reaches a minimum value, and then increases again. The minimum corresponds to the equilibrium composition of the reaction mixture.

Equilibrium for general solution reactions. Most of the results and discussion of the preceding section apply to any kind of chemical equilibrium. Consider, for example, the following reaction in a dilute solution:

$$4H^+ + MnO_2(s) + 2Cl^- \rightleftharpoons Mn^{2+} + Cl_2(g) + 2H_2O$$

Aside from the ions, this reaction involves H_2O, as well as a gas and a solid that are both at equilibrium with the solution. The chemical potentials of the ions depend on their concentrations (or activities) as shown by (17-33) or (17-34). For H_2O it can be assumed that the chemical potential is approximately constant and equal to that of pure water:

$$\mu_{H_2O} \approx \text{const} \approx \widetilde{G}^{\ominus}_{f,H_2O}$$

The chemical potential of the dissolved Cl_2 is equal, at equilibrium, to that of the Cl_2 in the gas phase, so that

$$\mu_{Cl_2} = \widetilde{G}^{\ominus}_{f,Cl_2} + RT \ln P_{Cl_2}$$

Finally, because the solution is saturated with MnO_2, whose solubility is in fact extremely low, the chemical potential of the MnO_2 in the solution is equal at equilibrium to that of pure solid MnO_2, $\widetilde{G}^{\ominus}_{f,MnO_2}$:

$$\mu_{MnO_2} = \widetilde{G}^{\ominus}_{f,MnO_2}$$

Combination of the chemical potentials for the different species gives

$$\Delta G = \mu_{Mn^{2+}} + \mu_{Cl_2} + 2\mu_{H_2O} - 4\mu_{H^+} - \mu_{MnO_2} - 2\mu_{Cl^-}$$

$$= \underbrace{\widetilde{G}^{\ominus}_{f,Mn^{2+}} + \widetilde{G}^{\ominus}_{f,Cl_2} + 2\widetilde{G}^{\ominus}_{f,H_2O} - 4\widetilde{G}^{\ominus}_{f,H^+} - \widetilde{G}^{\ominus}_{f,MnO_2} - 2\widetilde{G}^{\ominus}_{f,Cl^-}}_{\Delta G^{\ominus}}$$

$$\underbrace{+ RT \left(\ln [Mn^{2+}] + \ln P_{Cl_2} - 4 \ln [H^+] - 2 \ln [Cl^-] \right)}_{RT \ln Q}$$

This is equal to $\Delta G^{\ominus} + RT \ln Q$ [see (17-42)], with

$$Q = \frac{P_{Cl_2}[Mn^{2+}]}{[H^+]^4 [Cl^-]^2}$$

Again, the equilibrium condition $\Delta G = 0$ leads to the equilibrium law, $Q = K(T)$, and to $\Delta G^{\ominus}(T) = -RT \ln K(T)$. Note that Q has the form specified by the conventions for reaction quotients (Sec. 13-3). Similar reasoning applies to any other chemical equilibrium.

EXAMPLE 17-12   **Equilibrium constant from standard free energies.** What are $\Delta G^{\ominus}$ and K at 298 K for the oxidation of Cl^- by $MnO_2(s)$ in the reaction just considered?

$$4H^+ + MnO_2(s) + 2Cl^- \rightleftharpoons Mn^{2+} + Cl_2(g) + 2H_2O$$

$4H^+$	$MnO_2(s)$	$2Cl^-$	Mn^{2+}	$Cl_2(g)$	$2H_2O$
0	−464.8	−131.2	−227.6	0	−237.2

The free energies $\tilde{G}_f^\ominus$ of the species involved as found in Appendix D are listed beneath the chemical symbols.

Solution. $\Delta G^\ominus = -227.6 - 2 \times 237.2 + 464.8 + 2 \times 131.2 = \underline{25.2 \text{ kJ}};$ $\log K = -25.2 \times 10^3/(8.31 \times 298 \times 2.30) = -4.42;$ $K = \underline{3.8 \times 10^{-5} \text{ atm mol}^{-5} \text{ liter}^5}.$

The important relationship (17-45) between K and $\Delta G^\ominus$ shows how values of $\Delta G^\ominus$ may be determined experimentally by measuring equilibrium constants. Experimental determination of K usually offers no problems for equilibria that are readily established and not strongly one-sided. Many standard free-energy values have been determined by this valuable method. On the other hand, many equilibria are either not readily established or are so one-sided that it is not a simple matter to measure the equilibrium constant. In such cases K may be calculated, provided the standard free energies of all participating species are known. Equilibrium constants can therefore be determined by thermal measurements alone (resulting in ΔH and ΔS values), so that it is possible to make quantitative statements about chemical reactions without ever performing them.

For example, it is possible to determine ΔH for the reaction of ethylene with water to form ethanol,

$$C_2H_4(g) + H_2O(g) \longrightarrow C_2H_5OH(g)$$

by determining the heats of combustion of C_2H_4 and C_2H_5OH, both at any convenient temperature T. The entropies at T of all three substances, and thus ΔS, can be determined from measurements of the molar heat capacities $\tilde{C}_P$ and the enthalpies of all pertinent phase transformations of the three substances. Adjustments to standard conditions yield $\Delta H^\ominus(T)$, $\Delta S^\ominus(T)$, and thus $\Delta G^\ominus(T) = \Delta H^\ominus(T) - T \Delta S^\ominus(T)$, from which $K(T)$ can be calculated. The reaction considered forms the basis for the large-scale industrial production of ethanol from ethylene.

Furthermore, a listing of equilibrium constants of reactions can be replaced by a listing of standard free energies $\tilde{G}_f^\ominus$ of the substances involved, with far fewer entries. Equilibrium constants can be calculated from such a free-energy table for all the innumerable possible combinations of the substances listed as reactants and products in potential reactions.

Temperature dependence of the equilibrium constant.   The fundamental relationship between an equilibrium constant K and the standard free-energy change $\Delta G^\ominus$ associated with changing reactants into products under standard conditions is, by (17-45),

$$-RT \ln K = \Delta G^\ominus = \Delta H^\ominus - T \Delta S^\ominus \qquad (17\text{-}46)$$

Division by $-RT$ yields

$$\ln K = \frac{\Delta S^\ominus}{R} - \frac{\Delta H^\ominus}{RT} \qquad (17\text{-}47)$$

Thus the temperature dependence of $\ln K$ has the form

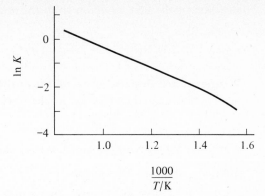

FIGURE 17-3 **Relation between ln K and $1/T$ for the reaction $CO_2(g) + H_2(g) \rightleftharpoons CO(g) + H_2O(g)$.** The slope of the curve is equal to $-\Delta H^\oplus/R$. The value of $\Delta H^\oplus$ (35 kJ at 1000 K) is seen to change slightly with temperature.

$$\ln K = 2.303 \log K = \frac{-\Delta H^\oplus}{RT} + \text{const} \qquad (17\text{-}48)$$

and a graph of ln K against $1/T$ is expected to yield a straight line with the slope $-\Delta H^\oplus/R$ (Fig. 17-3) provided that $\Delta H^\oplus$ and $\Delta S^\oplus$ are constant in the temperature interval considered.[1] It is therefore possible to determine enthalpies of reaction without any calorimetric experiments by measuring the equilibrium constant at several temperatures and plotting the measurements as in Fig. 17-3. This indirect method of finding enthalpy changes is usually simpler than a direct calorimetric method. The calorimetric method of measuring $\Delta H^\oplus$ requires that the reaction go to completion or that a correction be applied if it does not.

Equation (17-48) implies exponential dependence of K on $\Delta H^\oplus/RT$,

$$K = \text{const} \times \exp \frac{-\Delta H^\oplus}{RT}$$

a temperature-dependence characteristic of many phenomena.

Equation (17-47) can also be used to derive a convenient relation between K and $\Delta H^\oplus$. Suppose we consider two temperatures T_1 and T_2 that are sufficiently close to each other that $\Delta S^\oplus$ and $\Delta H^\oplus$ can be considered to remain constant as T_1 is changed into T_2. If K_1 and K_2 are the equilibrium constants at these two temperatures, we have

$$\ln K_1 = \frac{\Delta S^\oplus}{R} - \frac{\Delta H^\oplus}{RT_1} \qquad (17\text{-}49)$$

and

$$\ln K_2 = \frac{\Delta S^\oplus}{R} - \frac{\Delta H^\oplus}{RT_2} \qquad (17\text{-}50)$$

[1] A more detailed derivation shows that when $\Delta S^\oplus$ and $\Delta H^\oplus$ are not constant, the slope of a graph of ln K against $1/T$ still is equal to $-\Delta H^\oplus/R$, but now at the temperature corresponding to the point at which the slope is taken.

Subtraction of the first equation from the second yields

$$\ln K_2 - \ln K_1 = \frac{\Delta H^\ominus}{R}\left(\frac{1}{T_1} - \frac{1}{T_2}\right) = \frac{\Delta H^\ominus(T_2 - T_1)}{RT_1T_2} \tag{17-51}$$

EXAMPLE 17-13   **Temperature dependence of an equilibrium constant.**   At 25°C the value of the equilibrium constant for the reaction

$$\tfrac{1}{2}I_2(g) + \tfrac{1}{2}Br_2(g) \rightleftharpoons IBr(g)$$

is 20.5 and $\Delta H^\ominus$ is -5.29 kJ mol^{-1}. What is the value of K at 100°C?

Solution. From (17-51) we have

$$\ln \frac{K_{100}}{K_{25}} = \frac{(-5.29 \times 10^3 \text{ J mol}^{-1})(75 \text{ K})}{(8.31 \text{ J mol}^{-1}\text{ K}^{-1})(298 \text{ K})(373 \text{ K})}$$

$$= -0.430 = 2.303 \log \frac{K_{100}}{K_{25}}$$

Hence

$$\frac{K_{100}}{K_{25}} = 10^{-0.187} = 0.651$$

$$K_{100} = \underline{13.3}$$

Problems and Questions

17-1   **Entropy of vaporization.**   The normal boiling point of methanol is 65.0°C and the enthalpy of vaporization is 35.3 kJ mol^{-1}. What is its entropy of vaporization?

Ans.   104 J K^{-1} mol^{-1}

17-2   **Entropy.**   Is it possible to decrease the entropy of a fluid (*a*) by an adiabatic expansion? (*b*) By an adiabatic compression? Explain.

17-3   **Spontaneous processes.**   If any of the following is impossible, explain why. If not, give one example of each: (*a*) a spontaneous process in which the entropy of the system decreases; (*b*) a spontaneous process at constant T and P in which the free energy increases when only P,V work is done; (*c*) a spontaneous process that is endothermic.

17-4   **Entropy changes.**   Indicate for each of the following pairs of processes which process is accompanied by the larger entropy increase. Explain your reasoning.

(*a*)   (i) Solid water at 0°C changes to water vapor at 0°C; (ii) liquid water at 0°C changes to water vapor at 0°C.

(*b*)   (i) A liquid is vaporized to gas at a pressure lower than the vapor pressure at a given temperature; (ii) a liquid is vaporized to gas at a pressure equal to the vapor pressure at a given temperature.

(*c*)   (i) A gas at 30°C is heated reversibly to 70°C; (ii) the same gas at 30°C is heated irreversibly to 80°C.

17-5 **Entropy, a state function.** (*a*) One mole of an ideal gas at 300 K and 1 atm is compressed reversibly and isothermally to half its volume. (*b*) Another mole of the gas at the same initial conditions is heated reversibly to 600 K at constant volume and then cooled reversibly to 300 K at constant pressure. Assure yourself that the final state is the same in each case and calculate ΔS for the two paths, assuming a constant $\widetilde{C}_P$ of 20.9 J mol^{-1} K^{-1}.

Ans. -5.76 J K^{-1}

17-6 **Heating a gas.** Consider an ideal gas for which, in the temperature range of interest, $\widetilde{C}_P$ is constant and equal to 29.3 J mol^{-1} K^{-1}. (*a*) The temperature of 1.00 mol of the gas, contained in a cylinder with a piston exerting a constant pressure of 1.00 atm, is raised gradually and reversibly from 250 to 500 K. Calculate q, ΔV, w, ΔU, ΔH, and ΔS for this sample of gas. (*b*) Suppose that the same change in state described in (*a*) had occurred irreversibly (e.g., by plunging the cylinder into a large "thermostat" at 500 K). Explain how ΔV, ΔU, ΔH, and ΔS for the gas would differ (if at all) from the values obtained in (*a*).

17-7 **Melting ice.** Find $\Delta \widetilde{S}$ for the change in state $H_2O(s, 0°C) \longrightarrow H_2O(l, 25°C)$. The molar enthalpy of fusion of ice at 0°C is 6.02 kJ mol^{-1}; $\widetilde{C}_P$ for $H_2O(l)$ is 75.3 J mol^{-1} K^{-1} and is to be assumed constant.

Ans. 28.7 J K^{-1} mol^{-1}

17-8 **Enthalpy of vaporization.** Benzene boils at 80°C. Estimate $\Delta \widetilde{H}_{vap}$ from Trouton's rule.

17-9 **Crystallization of supercooled water.** The molar enthalpy of fusion of ice at 0°C is 6.02 kJ mol^{-1}; the molar heat capacity of supercooled water is 75.3 J mol^{-1} K^{-1}. (*a*) One mole of supercooled water at $-10°C$ is induced to crystallize in a heat-insulated vessel. The result is a mixture of ice and water at 0°C. What fraction of this mixture is ice? (*b*) What is ΔS for the system?

Ans. $\frac{1}{8}$; 0.05 J K^{-1}

17-10 **Chemistry of molecules.** The strongest chemical bond known is that in carbon monoxide, CO, with a bond enthalpy of 1.05×10^3 kJ mol^{-1}. Furthermore, the entropy increase in a gaseous dissociation of the kind $AB \rightleftharpoons A + B$ is about 110 J mol^{-1} K^{-1}. These factors establish a temperature beyond which there is essentially no chemistry of molecules. Show why this is so and find the temperature.

17-11 **Two solid phases.** A certain substance consists of two modifications A and B; $\Delta G^{\ominus}$ for the transition from A to B is positive. The two modifications produce the same vapor. Which has the higher vapor pressure? Which is the more soluble in a solvent common to both?

Ans. B, B

17-12 **Vaporization of ethanol.** The enthalpy of vaporization of ethanol is 38.7 kJ mol^{-1} at its boiling point, 78°C. Calculate q, w, ΔU, ΔS, and ΔG for vaporizing 1 mol ethanol reversibly at 78°C and 1.00 atm. Neglect the volume of liquid ethanol and assume that the vapor is an ideal gas.

17-13 **Iron oxides.** From the values in Table D-4 calculate $\Delta H^{\ominus}$ and $\Delta G^{\ominus}$ for the reaction $3Fe_2O_3(s) \longrightarrow 2Fe_3O_4(s) + \frac{1}{2}O_2(g)$ at 25°C. Which of the two oxides is more stable at 25°C and $P_{O_2} = 1$ atm?

Ans. 232 kJ; 195 kJ; Fe_2O_3 more stable

17-14 **Entropy of vaporization.** Calculate the entropy of vaporization of ethanol at 1 atm and 50°C from the molar enthalpy of vaporization at the normal boiling point, 78°C, which is 38.7 kJ, and from the constant-pressure heat capacities of liquid and gaseous ethanol, 111 and

65 J K^{-1} mol^{-1}, respectively. Is this change in state spontaneous? Explain how your answer to this question is consistent with the sign of ΔS and the known boiling point.

17-15 Equilibrium constants. Find $\Delta G^{\ominus}$ and the equilibrium constant K for the following reactions by referring to the values of the standard free energies of formation listed in Table D-4:

(a) $4I^-(aq) + O_2(g) + 4H^+ \rightleftharpoons 2H_2O(l) + 2I_2(g)$
(b) $CO(g) + 2H_2(g) \rightleftharpoons CH_3OH(l)$
(c) $3H_2(g) + SO_2(g) \rightleftharpoons H_2S(g) + 2H_2O(l)$
(d) $Ca(s) + CO_2(g) \rightleftharpoons CaO(s) + CO(g)$

Ans. value of K:(a) 1.2×10^{40};   (b) 1.2×10^5;   (c) 1.8×10^{36};   (d) 6.4×10^{60}

17-16 Solubility products. Find K_{sp} for the following salts by using the values of the standard free energies of formation given below and those listed in Table D-4: (a) AgCl(s), $\tilde{G}_f^{\ominus} = -110$ kJ mol^{-1};   (b) AgBr(s), $\tilde{G}_f^{\ominus} = -96$ kJ mol^{-1};   (c) AgI(s), $\tilde{G}_f^{\ominus} = -66$ kJ mol$^-$

17-17 Free energy of atoms. What is the partial pressure of atomic chlorine that would be at equilibrium with $Cl_2(g)$ at 25°C and 1 atm?   *Ans.* 3×10^{-19} atm

17-18 Equilibrium pressure of CO$_2$. From $\tilde{G}_f^{\ominus}$ for $CO_2(g)$ and $H_2CO_3(aq)$ find the partial pressure of CO_2 at equilibrium with a 1 F solution of H_2CO_3.

17-19 Gas equilibrium. At 3500 K the equilibrium constant for the reaction $CO_2(g) + H_2(g) \rightleftharpoons CO(g) + H_2O(g)$ is 8.28. What is $\Delta G^{\ominus}(3500)$ for this reaction? What is ΔG at 3500 K for transforming 1 mol CO_2 and 1 mol H_2, both held at 0.1 atm, to 1 mol CO and 1 mol H_2O, both held at 2 atm? In which direction would this last reaction run spontaneously?
Ans. -61.5 kJ; 112.8 kJ; to left

17-20 Free energy from equilibrium measurements. At 35°C and a total pressure of 1 atm, $N_2O_4(g)$ is 27.2 percent dissociated into $NO_2(g)$. What is $\Delta G^{\ominus}$ at 35°C for the reaction $N_2O_4(g) \rightleftharpoons 2NO_2(g)$?

17-21 Enthalpy of neutralization and K_w. At 24°C, $K_w = 1.0 \times 10^{-14}$. Find the enthalpy of the reaction

$$H_2O(l) \rightleftharpoons H^+(aq) + OH^-(aq)$$

from the values in Table D-4. Assume that this enthalpy of neutralization is temperature-independent and use it to estimate K_w at normal body temperature (about 37°C) and at 100°C.   *Ans.* 55.8 kJ, 2.6×10^{-14}, 1.0×10^{-12}

17-22 Entropy and enthalpy of formation. The free energy of formation $\tilde{G}_f^{\ominus}$ for CO is -182.3 kJ mol^{-1} at 800 K and -200.5 kJ mol^{-1} at 1000 K. What are the entropy and the enthalpy of formation, $\tilde{S}_f^{\ominus}$ and $\tilde{H}_f^{\ominus}$, in this temperature range, assuming both to be constant?

17-23 Vapor pressure of benzene. (a) From the values in Table D-4 find the enthalpy and the free energy of vaporization of benzene, C_6H_6, at 25°C. (b) What is the vapor pressure of benzene at 25°C? (c) Assume that the entropy and enthalpy of vaporization are constant and estimate the normal boiling point of benzene.   *Ans.* (a) $\Delta H_{vap}^{\ominus} = 33.9$ kJ, $\Delta G_{vap}^{\ominus} = 5.2$ kJ;   (b) 0.122 atm;   (c) 79°C

17-24 $\Delta G^{\ominus}$ from equilibrium data. At 1200 K and in the presence of solid carbon, an equilibrium mixture of CO and CO_2 ("producer gas") contains 98.3 mol percent CO and 1.69 mol percent CO_2 with the total pressure at 1 atm. What are P_{CO} and P_{CO_2}, what is the equilibrium constant, and what is ΔG° associated with the reaction

$$CO_2(g) + C(\text{graphite}) \rightleftharpoons 2CO(g)$$

at 1200 K?

"In the third series of these Researches, after proving the identity of electricities derived from different sources, and showing, by actual measurement, the extraordinary quantity of electricity evolved by a very feeble voltaic arrangement, I announced a law, derived from experiment, which seemed to me of the utmost importance to the science of electricity in general, and that branch of it denominated electrochemistry in particular. The law was expressed thus: The chemical power of a current of electricity is in direct proportion to the absolute quantity of electricity which passes."

Michael Faraday, 1834

18 Electrochemistry I

18-1 Introduction

The field of electrochemistry deals with redox reactions, reactions in which electrons are exchanged between two species. It encompasses a wide range of practical processes and their interpretation in terms of thermodynamics, kinetics, and other general chemical and physical principles. Some spontaneous redox reactions are valuable as sources of electrical energy—for example, the reactions in a lead storage battery (the kind of battery in most automobiles) or a flashlight battery or an efficient fuel cell like those used in spacecraft. Redox reactions are also responsible for the corrosion and consequent malfunction of all manner of man-made objects, from ships and bridges to "tin" cans and microscopic circuits in electronic equipment. The chemistry of corrosion is closely related to the chemistry of batteries.

How can redox reactions be harnessed to perform electrical work? This must be done by keeping the reacting species apart and allowing them to exchange electrons through an external circuit. To illustrate, consider the reaction of metallic zinc, a bright silvery metal, with a solution of a salt of Cu^{2+}. If a piece of zinc is dropped into a solution of cupric sulfate, there is an immediate reaction. A reddish brown deposit of metallic copper forms on the surface of the zinc, the solution becomes warm, and the intensity of the blue color of the solution (due to hydrated Cu^{2+}) is gradually reduced. Zinc atoms in the metal transfer electrons to cupric ions. The zinc is oxidized to Zn^{2+} and the Cu^{2+} is reduced to Cu:

$$Zn(s) + Cu^{2+} \longrightarrow Zn^{2+} + Cu(s) \tag{18-1}$$

When reaction (18-1) proceeds in the manner described, it is useless as a source of electrical energy. However, if the reductant (Zn) and the oxidant (Cu^{2+}) are separated and a path for electron flow between them is provided, the reaction can serve as a source of electrical energy, a chemical battery. Later we consider practicable ways in which this can be managed.

Many substances that are not stable in the normal atmosphere on the earth's surface can be produced by using electrochemical methods to force appropriate redox reactions to go in a direction opposite to their natural or spontaneous direction. These techniques are used to produce the most reactive elements, including those with broad applications in modern consumer and industrial technology, such as aluminum and magnesium, chlorine and fluorine.

We begin the discussion of electrochemistry with a summary of its historical background and some of its basic concepts, particularly Faraday's law of electrolysis.

Historical background. The first recognition of a relation between electrical and chemical phenomena dates to the end of the eighteenth century. Luigi Galvani, a lecturer in anatomy at the University of Bologna, observed in 1786 that a frog's leg muscle contracted if it was in contact with any of various metals, provided that the metal was in turn in contact with a second metal that touched a nerve of the frog. Galvani attributed this effect to "animal electricity", but his countryman, the physicist Alessandro Volta, showed that it was due to the contact of the two dissimilar metals. This led Volta in 1799 to the invention of the electric battery. His famous "pile" was a sequence of disks of silver and zinc in contact with one another, each pair separated from the next by a sheet of heavy paper moistened with salt water. When the disk of zinc at one end of the pile was connected by a wire to the disk of silver at the other end, a current was generated.

Electrochemical methods were first applied to the isolation of strongly electropositive elements by Sir Humphry Davy, who discovered potassium and sodium in 1807 by electrolyzing melts of their carbonates. Later he produced four of the alkaline-earth metals in a similar way. The success of these experiments confirmed in Davy's mind the idea that chemical and electrical attraction were essentially the same phenomenon, a view developed into a system of electrochemical dualism by Berzelius (Sec. 11-1).

The key quantitative discovery in electrochemistry was made by the self-educated English physicist and chemist Michael Faraday. His studies of electrolysis led him in 1833 to the discovery of a relation between the quantity of electricity passed through a solution and the equivalent weight of the substance deposited. He thought, however, that the ions that carried the electricity in the solution were created by the current. It was more than half a century before Arrhenius recognized that ions exist in conducting solutions whether or not there is actual passage of current.

18-2 Electrodeposition and Faraday's Law

Half-reactions and electrochemical cells. Electrons carry electric current in metals, but electrons do not exist by themselves in solutions in water or other ionizing solvents, or in molten electrolytes. Electrical conduction in these substances depends on the movement of ions (Sec. 7-1). Consequently when an electric current passes from a metallic conductor through an electrode into or out of an electrolyte solution, an electron-transfer reaction must occur—an electron must leave the electrode and combine with some species in solution or conversely.

At an electrode where electrons are delivered to the electrolyte there is reduction of an oxidized species (ox 1) to some reduced form (red 1):[1]

$$\text{ox } 1 + \mathbf{n}_1 e^- \longrightarrow \text{red } 1 \tag{18-2}$$

At an electrode where electrons are received from the electrolyte there is oxidation of a reduced species (red 2) to some oxidized form (ox 2),

$$\text{red } 2 \longrightarrow \mathbf{n}_2 e^- + \text{ox } 2 \tag{18-3}$$

The solution and the external circuit can remain electrically neutral only if the number of electrons delivered by an electrode to the electrolyte in a given time is matched by the number removed at another electrode. The electrical potentials produced by even a slight departure from electrical neutrality are so enormous[2] that for all practical purposes exact neutrality is maintained at all times on a macroscopic basis. Thus the half-reactions at the two electrodes are mutually adjusted so that the number of electrons is the same in each. The corresponding equations can be suitably adjusted if (18-2) is multiplied by $\mathbf{n}_2$ and (18-3) by $\mathbf{n}_1$, so that when $\mathbf{n}_1\mathbf{n}_2$ electrons are delivered to the electrolyte, the same number is also withdrawn from it. If the equations are then added, the electrons cancel and only an equation relating two redox couples remains:

$$\mathbf{n}_2 \text{ ox } 1 + \mathbf{n}_1 \text{ red } 2 = \mathbf{n}_2 \text{ red } 1 + \mathbf{n}_1 \text{ ox } 2 \tag{18-4}$$

This approach is indeed one way of balancing redox equations (Appendix C).

The flow of electric current through an electrolyte is thus tied to the progress of the electrode reactions. Reduction and oxidation are intimately coupled, even though they occur in separate places. The combination of electrodes and mobile ions in a solution or a molten electrolyte is the essence of what is called an *electrochemical cell*. When the reaction of such a cell occurs in the spontaneous direction, one speaks of a *galvanic* or *voltaic* cell, or sometimes a *battery*, although originally this term was reserved for a sequence of galvanic cells joined together.

[1] See footnote, p. 190.

[2] A free charge of about 10^{-17} mol electrons causes a potential of 1 V at a distance of 1 cm.

In some cells the reaction is driven by an external potential applied to the electrodes. The reaction then goes in a direction opposite to that in which it would go spontaneously and is termed *electrolysis;* the cell is called an *electrolytic* cell (Fig. 18-1). Chemists find it useful to label electrical terminals by the type of electrode reaction associated with them. If the reaction is a reduction, the terminal is called a *cathode.* If the reaction is an oxidation, the terminal is called an *anode.* The electrons therefore always enter an electrochemical cell at the cathode and leave at the anode. Signs are sometimes associated with the electrodes of a cell, but they are a frequent source of confusion and we shall not consider them here, except to

FIGURE 18-1   **Electrolysis of a $Cu(NO_3)_2$ solution.**   At the cathode there is reduction and metallic Cu is deposited:

$$Cu^{2+} + 2e^- \longrightarrow Cu(s)$$

At the anode there is oxidation, but NO_3^- is oxidized much less readily than H_2O, so H_2O is oxidized to oxygen:

$$2H_2O \longrightarrow O_2(g) + 4H^+ + 4e^-$$

with hydrogen ions as by-product. Since the number of electrons entering must be the same as the number leaving, two copper atoms are deposited for each O_2 molecule produced. The current is carried through the electrolyte by the migration of Cu^{2+} and NO_3^- ions. The solution stays electrically neutral at all times. For every Cu^{2+} that leaves the solution, two H^+ enter it; the migration of NO_3^- from the cathode toward the anode maintains electrical neutrality throughout the solution.

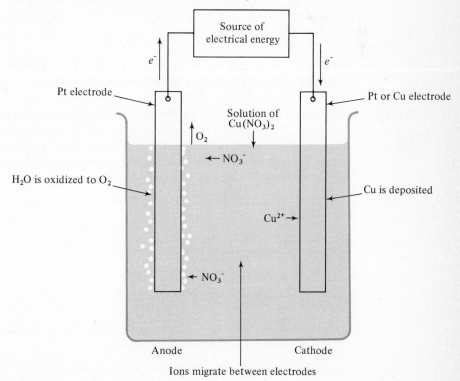

point out that on commercial batteries, such as lead storage batteries and dry cells, the terminal labeled $(+)$ is the cathode when the battery is acting as a source of electricity. In a storage battery, this same terminal is the anode when the battery is being recharged.[1] More details of electrochemical cells and batteries are considered in Secs. 18-4 and 18-5.

Faraday's law. In any electrochemical process the number of electrons leaving through the anode equals the number of electrons entering through the cathode, since the net charge on the electrolyte remains zero. This number of electrons also governs the stoichiometry of the electrode reactions. The charge on one mole of electrons, called the faraday ($\mathcal{F}$) and expressed in coulombs (Appendix A), is

$$\mathcal{F} = 96,485 \text{ C}$$

This value is usually rounded to 96,500 C, correct to 0.02 percent. If **n** is the number of *electrons* participating in the reaction of *one molecule* or ion, then the reaction of *one mole* of substance involves **n** *faradays*. A consequence of this relationship is that the weights of different substances formed or used up in electrode reactions are proportional to their redox equivalent weights as defined by these reactions. This is Faraday's law of electrolysis.

The magnitude of an electric current is defined as the amount of electric charge flowing through a conductor per unit time. Since a current of one ampere (A) is defined as a flow of one coulomb per second, $1 \text{ A} = 1 \text{ C s}^{-1}$ and conversely $1 \text{ C} = 1 \text{ A s}$. More generally,[2] if a current **I** flows for a time t, the total charge **Q** involved is

$$\mathbf{Q} = \mathbf{I}t \qquad (18\text{-}5)$$

and if t is in seconds and **I** in amperes, **Q** is in coulombs. The application of Faraday's law is illustrated in the following example.

EXAMPLE 18-1   **Two electrolytic cells in series.** An electric current first passes through a cell in which 506 mg silver is deposited from an $AgNO_3$ solution. It continues through a cell in which copper is plated out from a $CuSO_4$ solution. The duration of current flow is 1 hour. What is the average current in amperes and how many grams of copper are deposited?

Solution. The half-reaction for the first cell is the reduction of Ag^+,

$$Ag^+ + e^- \longrightarrow Ag(s)$$

while that for the second is the reduction of Cu^{2+},

$$Cu^{2+} + 2e^- \longrightarrow Cu(s)$$

For every faraday or 96,500 C passing through the first cell, 1 mol Ag atoms or 107.9 g

---

[1]The phrase "charging a battery" must be understood in the sense of supplying with energy rather than the sense of building up a positive or negative charge.

[2]Electrical quantities are denoted by boldface capitals throughout this text.

Ag is plated out. The 506 mg actually deposited corresponds to $\frac{0.506}{107.9} \times 96{,}500$ C = 453 C. Spread over 3600 s, this corresponds to an average current of $\mathbf{I} = \mathbf{Q}/t =$ 453 A s/3600 s = 0.126 A.

The cells are connected so that for every faraday passing through the first cell a faraday must pass through the second cell. Let x be the amount of Cu deposited. Since the reduction of each Cu^{2+} requires *two* electrons, there is a ratio of 1:2 between moles of Cu and Ag deposited,

$$\frac{x}{63.5} = \frac{1}{2} \times \frac{0.506}{107.9}$$

and $x = (506 \times 63.5$ mg$)/(2 \times 107.9) =$ 149 mg Cu.

No equation for a half-reaction is intended to imply anything more than the stoichiometry of the electrode reaction. Nothing is implied about the detailed sequence of changes actually occurring—the "mechanism" of the electrode reaction. The situation is the same as for chemical equations in general, as discussed in Chap. 20.

A shorthand notation for cells. Electrochemical cells, and especially galvanic cells, are often described by a shorthand notation illustrated by the following example. Consider a cell with a zinc electrode dipping into a 1 *F* zinc sulfate solution and a copper electrode surrounded by a 0.1 *F* copper sulfate solution. Such a cell is illustrated in Fig. 18-2*a*. The two solutions must be in electrical contact to allow the passage of current between them, yet the solutions must not be allowed to mix, for this would lead to the *direct* reaction of Cu^{2+} with the Zn electrode and there would be no need for electrons to flow through the external circuit. The solutions can be separated, as shown, by a thin layer of unglazed porcelain or another porous membrane that is permeable to ions but slows the mixing process. Under such conditions, however, potential differences called *liquid-junction potentials* arise at the interface between the solutions. These potentials, which complicate the analysis of cells, can be minimized by the use of a salt bridge, illustrated in Fig. 18-2*b*.

A *salt bridge* is an inverted U-tube containing a KCl solution given a jellylike consistency by addition of a substance such as agar. Like a porous membrane, a salt bridge allows the passage of ions and prevents mixing of the solutions. The reasons why its use practically eliminates liquid-junction potentials will not be examined here, but they depend on the fact that K^+ and Cl^- move with approximately equal speed in an electric field.

The shorthand notation for the cell in Fig. 18-2*b* is

$$Zn|Zn^{2+}(1\ M)||Cu^{2+}(0.1\ M)|Cu \qquad (18\text{-}6)$$

The single vertical lines indicate phase boundaries across which there are potential differences; the double line signifies that there is no significant liquid-junction potential between the two solutions. Note that the anions present in the solutions $(SO_4{}^{2-})$ need not be mentioned explicitly, although such ions must always be present to maintain electrical neutrality.

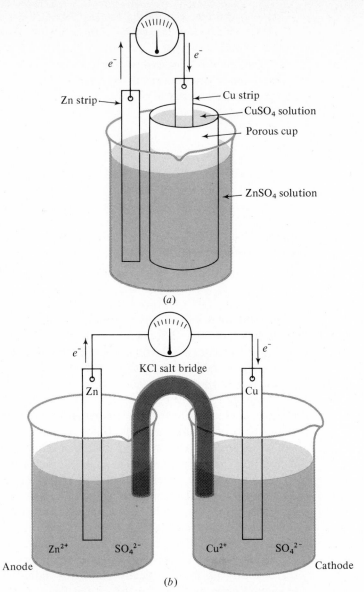

FIGURE 18-2 **Typical galvanic or voltaic cells.** This cell is composed of an anode of zinc in a solution containing Zn^{2+} ions and a cathode of copper in a solution containing Cu^{2+} ions. In (*a*) the two solutions are separated by a porous membrane, which inhibits mixing. In (*b*) a KCl salt bridge is used for this purpose.

There exists *by convention* a one-to-one relationship between the direction of the sequence of symbols in a shorthand description of a cell, such as (18-6), and the direction in which the corresponding overall cell reaction is written. By this convention, the cell is thought of as operating in such a way that *electrons leave* the cell by the electrode *on the left*[1] and enter the cell by the electrode on the right, regardless of whether this is the direction in which the electrons would flow spontaneously in such a cell or whether an external potential would be needed to drive the electrons in this direction. This implies that, for purposes of discussion, the electrode reaction at the right is a reduction, making the right-hand electrode of this conventional description the cathode and the electrode on the left the anode.

In the example given, the electrons that would enter on the right would reduce Cu^{2+} to Cu by the half-reaction

$$Cu^{2+} + 2e^- \longrightarrow Cu(s) \tag{18-7}$$

The electrons leaving on the left would result from the oxidation of Zn to Zn^{2+} by the half-reaction

$$Zn(s) \longrightarrow Zn^{2+} + 2e^- \tag{18-8}$$

The overall cell reaction would be the combination of (18-7) and (18-8),

$$Zn(s) + Cu^{2+} \longrightarrow Zn^{2+} + Cu(s) \tag{18-9}$$

The Cu electrode would be the cathode and the Zn electrode the anode.

If it is to be implied that the cell reaction is run in the opposite direction,

$$Cu(s) + Zn^{2+} \longrightarrow Cu^{2+} + Zn(s) \tag{18-10}$$

then the Zn electrode would be the cathode and the Cu electrode the anode. The conventional cell description would then have to be inverted also,

$$Cu|Cu^{2+}(0.1\ M)||Zn^{2+}(1\ M)|Zn$$

because electrons are always assumed to leave on the left.

Changes accompanying the cell reaction: the function of the salt bridge. When the cell described by (18-6) and illustrated in Fig. 18-2*b* produces an electric current, the reaction (18-9) occurs, zinc being oxidized to Zn^{2+} and Cu^{2+} being reduced to metallic copper. Thus the concentration of Zn^{2+} ions in the left-hand (anode) compartment of the cell gradually increases, while that of Cu^{2+} in the cathode compartment decreases. Simultaneously, in order to maintain electrical neutrality, Cl^- ions from the salt bridge move into the anode compartment to balance the increase in positive charge caused by the increase in $[Zn^{2+}]$, and K^+ ions move from the salt bridge into the cathode compartment to balance the decrease in $[Cu^{2+}]$.* The flow of ions between the two compartments exactly parallels the external flow of electrons through the wire. Were there no provision for these ion migrations, the reaction would stop very quickly as unneutralized charges accumulated around the electrodes and built up a potential that offset the electrode potential difference.

---

[1]The alliteration "electrons *leave* on the *left*" is helpful in remembering this convention.

*Small quantities of Cu^{2+}, Zn^{2+}, and SO_4^{2-} ions also diffuse through the salt bridge, but these quantities are normally negligible.

As the cell reaction proceeds, the concentrations of the ions involved in the electrode reactions change significantly and the potentials of the electrodes simultaneously change, for reasons that we consider shortly. The difference in the potentials of the two electrodes, which is the voltage that can be supplied by this battery, invariably decreases as the spontaneous cell reaction proceeds, approaching zero as a limit. The battery runs down. This is a perfectly general phenomenon, characteristic of all galvanic cells; some can be recharged by forcing a current to pass through them in the opposite direction, but this is not always practicable.

Electrodes of different kinds.   In the example just given, the electrodes are simply pieces of metal. More complex electrodes also exist. The hydrogen electrode is a typical gas electrode. Hydrogen gas is bubbled over a platinum electrode that has been coated with finely divided platinum (called platinum black) and that is surrounded by a solution containing hydrogen ions. The half-reaction at this electrode is

$$2H^+ + 2e^- \longrightarrow H_2(g) \qquad (18\text{-}11)$$

The platinum acts as catalyst for the reaction between the H^+ ions and H_2 molecules and acquires a potential characteristic of this reaction. Other gas electrodes may be constructed similarly. Examples include the oxygen electrode, with the half-reaction

$$4H^+ + O_2(g) + 4e^- \longrightarrow 2H_2O$$

and the chlorine electrode, with the half-reaction

$$Cl_2(g) + 2e^- \longrightarrow 2Cl^-$$

Electrodes may be characterized by the redox couple they represent, such as Cu^{2+}/Cu, Zn^{2+}/Zn, H^+/H_2, O_2/H_2O, and Cl_2/Cl^- for the electrodes mentioned so far. We always list the oxidized form first, followed by the reduced form.

Some electrodes represent redox couples in which both oxidized and reduced forms are ionic species in the solution and the electron transfer occurs at an inert electrode. An example is a solution containing Fe^{2+} and Fe^{3+} ions and a gold or platinum electrode. The potential corresponds to the Fe^{3+}/Fe^{2+} couple, with the half-reaction

$$Fe^{3+} + e^- \longrightarrow Fe^{2+}$$

In another type of electrode, the concentration of the cations associated with the electrode metal is kept fixed through the solubility-product principle by the presence of a salt of low solubility. Such electrodes are used primarily as reference electrodes for measurements made with a negligibly small flow of current. A silver surface coated with solid silver chloride, written AgCl/Ag, is representative. Since AgCl(s) is present, the solution is saturated with silver chloride, so that at all times $[Ag^+][Cl^-] = K_{sp}$, where K_{sp} is the solubility-product constant. If any Ag^+ ions are reduced to Ag by the gain of electrons, they are replenished by the dissolving of AgCl(s). The dissolving of AgCl(s) also produces Cl^- ions, and the concurrent reactions may be described by the equations

$$Ag^+ + e^- \longrightarrow Ag(s)$$
$$AgCl(s) \longrightarrow Ag^+ + Cl^-$$

Conversely, if any Ag(s) is oxidized to Ag^+ by loss of electrons, AgCl(s) is precipitated to keep the product $[Ag^+][Cl^-]$ equal to K_{sp}. The half-reaction of this electrode is most simply described as the sum of the foregoing reactions:

$$AgCl(s) + e^- \longrightarrow Ag(s) + Cl^- \qquad (18\text{-}12)$$

EXAMPLE 18-2   **Cathode, anode, and overall cell reactions.**   What are the cathode, anode, and overall reactions of the cell

$$Ag(s), AgCl(s)|Cl^-(0.5\ M)||H^+(0.1\ M)|H_2(0.5\ atm), Pt \qquad (18\text{-}13)$$

Solution.   Since for purposes of writing cell reactions electrons are always thought of as leaving on the left, the electrode at the left functions as anode. The anode reaction is

$$\underline{Ag(s) + Cl^- \longrightarrow AgCl(s) + e^-} \qquad (18\text{-}14)$$

The cathode consists of a hydrogen gas electrode with H_2 at 0.5 atm and an H^+ concentration of 0.1 M. The cathode reaction is

$$\underline{2H^+ + 2e^- \longrightarrow H_2(g)} \qquad (18\text{-}15)$$

The overall cell reaction is that combination of (18-14) and (18-15) in which the electrons just cancel, that is, $2 \times (18\text{-}14) + (18\text{-}15)$, or

$$\underline{2Ag(s) + 2Cl^- + 2H^+ \longrightarrow 2AgCl(s) + H_2(g)} \qquad (18\text{-}16)$$

18-3   Free Energy and Cell EMF

Cell emf.   When a quantity Q of electricity is passed reversibly through an electrochemical cell, work is performed on the system or on the surroundings, depending on the direction in which the electricity is passed. If Φ_1 and Φ_2 are the respective potentials of the two cell electrodes, relative to an arbitrary point of reference, the quantity of work equals[1] $|Q(\Phi_2 - \Phi_2)| = |Q\,\Delta\Phi|$, where we ignore the sign for the moment. When a cell is operated reversibly, this electrical work is the absolute value of the net work done, $|w'| = |Q\,\Delta\Phi|$, where w' is the total work minus the P,V work (Sec.16-4).

When an electrochemical cell is galvanic $\Phi_{cathode}$ must be more positive than Φ_{anode} in order to drive the electrons in the external circuit from the anode to the cathode. Moreover, the cell does electrical work on its surroundings, and this work is by definition negative ($w' < 0$). When the cell is electrolytic, $\Phi_{cathode} < \Phi_{anode}$;

---

[1] This is basically the same expression for electrical work as $w_{elec} = I^2Rt$ on page 430 and in Equation (16-17), because the potential drop $\Delta\Phi$ across a resistance R through which a current I is flowing is $\Delta\Phi = IR$. By (18-5), $Q = It$, so that $|w_{elec}| = |(It)(IR)| = |Q\Delta\Phi|$. See also the Study Guide.

the surroundings must perform electrical work on the cell to make the cell reaction proceed and $w' > 0$. For convenience in associating the correct sign with this electrical work we introduce the quantity $\mathbf{E}$, called the *cell electromotive force* or *cell emf*. It is defined to be the potential of the cathode minus the potential of the anode:

$$\mathbf{E} = \mathbf{\Phi}_{cathode} - \mathbf{\Phi}_{anode}$$

With this definition of the cell emf $\mathbf{E}$, the relation

$$w' = -\mathbf{E}Q \qquad (18\text{-}17)$$

holds *regardless of whether the cell reaction proceeds in the galvanic or the electrolytic direction.* Since the labeling of electrodes as anode and cathode depends on the direction in which the current passes through the cell, the sign of $\mathbf{E}$ likewise depends on this direction. In terms of our shorthand description of a cell, $\mathbf{E}$ is the potential of the right side minus the potential of the left side.

In a cell in which $\mathbf{n}$ mol electrons is transferred from one electrode to the other, $Q = \mathbf{n}\mathfrak{F}$, where $\mathfrak{F}$ is the faraday, the charge on a mole of electrons. Thus we have, from (18-17),

$$w' = -\mathbf{n}\mathfrak{F}\mathbf{E} \qquad (18\text{-}18)$$

The net work obtainable under reversible conditions, w', which is also the maximum net work obtainable, is just equal at constant temperature and pressure to the change in free energy for the process under consideration [Equation (17-26*a*) with the equality sign].

We have, finally,

$$w' = \Delta G = -\mathbf{n}\mathfrak{F}\mathbf{E} \qquad (18\text{-}19)$$

where ΔG is the difference between the free energy of the products and that of the reactants, at the prevailing concentrations. Equation (18-19) may be solved for $\mathbf{E}$:

$$\mathbf{E} = \frac{-\Delta G}{\mathbf{n}\mathfrak{F}} \qquad (18\text{-}20)$$

A special case of this relationship is that of a cell with all reactants and products at standard conditions, in which case the emf is called the *standard emf* $\mathbf{E}^{\ominus}$ and is related to $\Delta G^{\ominus}$ of the cell reaction by

$$\mathbf{E}^{\ominus} = \frac{-\Delta G^{\ominus}}{\mathbf{n}\mathfrak{F}} \qquad (18\text{-}21)$$

Note that the right sides of both (18-20) and (18-21) contain the ratio of two extensive quantities, ΔG and $\mathbf{n}$. This is in agreement with the fact that $\mathbf{E}$ and $\mathbf{E}^{\ominus}$ are *intensive* quantities. The voltage of a cell is independent of the size of the cell.

It will be recalled that the criterion for a reaction proceeding spontaneously in the direction written is that ΔG is negative. By (18-20) this corresponds to a positive cell emf $\mathbf{E}$. Thus, as stated earlier, *when $\mathbf{E}$ for an electrochemical cell as written is positive, the cell is galvanic and the cell reaction proceeds spontaneously as written.* When a current flows, the potential available at the electrodes is smaller than $\mathbf{E}$

because some energy is required to overcome the internal resistance of the cell. The potential available may be diminished further by phenomena that will be discussed later.

Suppose an opposing external voltage, equal in magnitude to **E**, is applied to the cell. The cell reaction is thus kept exactly in balance and the cell is just on the point of being run as an *electrolytic cell,* with the electrode reactions reversed and the roles of anode and cathode interchanged. The voltage required to bring the cell to this point is called the *equilibrium decomposition potential.*

To make electrolysis proceed at a finite rate, additional voltage is needed to overcome the electrical resistance represented by the cell. This voltage is equal to **IR**, where **I** is the current and **R** the resistance. Sometimes additional voltage, called *overvoltage* (Sec. 19-2), is required because of irreversible phenomena that occur when a gas is evolved or a solid substance is deposited at an electrode.

When **E** *for an electrochemical cell as written is negative, the cell is electrolytic,* because the spontaneous direction of the cell reaction is opposite to the way the cell is written. The equilibrium decomposition voltage is $|\mathbf{E}|$, and to run the cell in the direction written requires additional voltage for overcoming resistance and possible overvoltage.

The Nernst equation. To derive an expression for the emf **E** of an electrochemical cell, we recall the expression for ΔG in terms of $\Delta G^{\ominus}$ and the reaction quotient Q developed in Sec. 17-5, especially Equation (17-42):

$$\Delta G = \Delta G^{\ominus} + RT \ln Q \qquad (18\text{-}22)$$

Insertion of (18-22) into (18-20) gives

$$\mathbf{E} = -\frac{\Delta G^{\ominus} + RT \ln Q}{n\mathfrak{F}}$$

or, by use of (18-21),

$$\mathbf{E} = \mathbf{E}^{\ominus} - \frac{RT}{n\mathfrak{F}} \ln Q \qquad (18\text{-}23)$$

This is the *Nernst equation.* It expresses the emf of an electrochemical cell in terms of the standard emf, the temperature T, the number **n** of electrons involved in the overall cell reaction, and the concentrations and partial pressures of the chemical species involved. The conventions that apply to Q in connection with the equilibrium law apply here also.

It is convenient to replace the natural logarithm in the Nernst equation by the base-10 logarithm and to combine the conversion factor between these two logarithms with the value of $RT/\mathfrak{F}$ at 25°C, the temperature commonly assumed in calculations with cell potentials. First, we change the faraday to new units. When one coulomb (C) of charge moves through a potential of one volt (V), the work performed, $w_{\text{elec}} = \mathbf{Q}\Delta\mathbf{\Phi}$, is one joule (J). That is, 1 C V = 1 J (Appendix A), and equivalently, 1 C = 1 J V^{-1}. Hence,

$$1 \text{ faraday} = 96{,}485 \text{ C} = 96{,}485 \text{ J V}^{-1} \qquad (18\text{-}24)$$

Next, $RT = 8.314 \times 298.15 = 2479$ J mol^{-1} and $\ln x = 2.303 \log x$, so that at 298 K,

$$\frac{RT}{\mathbf{nF}} \ln Q = \left(\frac{2479 \times 2.303}{\mathbf{n} \times 96,485} \log Q\right) \text{volts} = \left(\frac{0.0592}{\mathbf{n}} \log Q\right) \text{volts}$$

Thus, at 25°C,

$$\mathbf{E} = \mathbf{E}^{\ominus} - \left(\frac{0.0592}{\mathbf{n}} \log Q\right) \text{volts} \qquad (18\text{-}25a)$$

$$\mathbf{E} \approx \mathbf{E}^{\ominus} - \left(\frac{0.06}{\mathbf{n}} \log Q\right) \text{volts} \qquad (18\text{-}25b)$$

In applications a temperature of 298 K will always be assumed unless another temperature is explicitly indicated.

EXAMPLE 18-3   **The emf of a cell.** Write the Nernst equation for a cell like that shown in Fig. 18-2, with the overall reaction $Zn(s) + Cu^{2+} \longrightarrow Zn^{2+} + Cu(s)$.

Solution. Since $\mathbf{n} = 2$,

$$\mathbf{E} = \mathbf{E}^{\ominus} - \tfrac{0.0592}{2} \log \frac{[Zn^{2+}]}{[Cu^{2+}]} \qquad (18\text{-}26a)$$

$$\mathbf{E} \approx \mathbf{E}^{\ominus} - 0.03 \log \frac{[Zn^{2+}]}{[Cu^{2+}]} \qquad (18\text{-}26b)$$

where $\mathbf{E}^{\ominus}$ is the cell emf at standard conditions. Note that when $[Zn^{2+}]$ and $[Cu^{2+}]$ are both 1 *M*, the logarithmic term in (18-26) vanishes, so that $\mathbf{E} = \mathbf{E}^{\ominus}$, as it should, since these concentrations correspond to standard conditions. In fact, the logarithmic term vanishes whenever $[Zn^{2+}] = [Cu^{2+}]$. Furthermore, when the cell is *written* in the opposite direction, $\mathbf{E}^{\ominus}$ changes sign and the logarithmic term is $\log ([Cu^{2+}]/[Zn^{2+}]) = -\log ([Zn^{2+}]/[Cu^{2+}])$. Since both terms change sign, $\mathbf{E}$ retains its magnitude but changes sign, as anticipated. Experimental measurements show that $\mathbf{E}^{\ominus}$ for this cell is about 1.1 V.

EXAMPLE 18-4   **The emf of another cell.** What is the Nernst equation for a cell described by (18-13), with the cell reaction of (18-16)?

Solution. Since silver and silver chloride are solids, no terms corresponding to them appear in the equation. Again we have $\mathbf{n} = 2$ in the reaction as formulated in (18-16). The value of $\mathbf{n}$ can always be checked by considering the half-reactions combined, in this case (18-15) and twice (18-14). Thus

$$\mathbf{E} = \mathbf{E}^{\ominus} - \tfrac{0.0592}{2} \log \frac{P_{H_2}}{[H^+]^2[Cl^-]^2} \qquad (18\text{-}27)$$

where again for most purposes the coefficient in front of the logarithm can be set equal to $0.06/\mathbf{n} = 0.03$. At standard conditions, $P_{H_2} = 1$ atm and $[H^+] = [Cl^-] = 1$ *M*, so the logarithmic term vanishes and $\mathbf{E} = \mathbf{E}^{\ominus}$ at standard conditions, as it should. From measurements it has been determined that $\mathbf{E}^{\ominus}$ for this cell is about -0.22 V.

18-4   Electrode Potentials

Electrode potentials.   Each electrode of an electrochemical cell can be assigned an electrode potential such that the observed cell emf is the difference between these values for the two electrodes. The zero point of the scale of electrode potentials is quite arbitrary, since only differences between the values can be measured experimentally. This arbitrary reference point is conventionally chosen to be the standard hydrogen electrode, a hydrogen electrode in which the H_2 gas is at 1 atm pressure and the H^+-ion concentration is 1 M:

$$E^{\ominus}_{H^+/H_2} = 0.000 \tag{18-28}$$

Consider, for example, a Zn electrode immersed in a Zn^{2+} solution. The half-reaction is

$$Zn^{2+} + 2e^- \longrightarrow Zn(s)$$

The electrode potential for this electrode is *defined* as the emf of the cell

$$Pt, H_2(1 \text{ atm})|H^+(1\ M)||Zn^{2+}|Zn(s) \tag{18-29}$$

for which the cell reaction is

$$H_2(g) + Zn^{2+} \longrightarrow 2H^+ + Zn(s)$$

By the Nernst equation this emf is

$$\mathbf{E} = \mathbf{E}^{\ominus} - \frac{RT}{n\mathcal{F}} \ln Q = \mathbf{E}^{\ominus} - \frac{0.0592}{2} \log \frac{[H^+]^2}{P_{H_2}[Zn^{2+}]} \tag{18-30}$$

where $\mathbf{E}^{\ominus}$ is the emf of this cell when $[Zn^{2+}] = [H^+] = 1\ M$ and $P_{H_2} = 1$ atm. Since this emf is the difference between that of the standard hydrogen electrode, assumed to be 0.000 because of (18-28), and that of the Zn^{2+}/Zn half-cell at standard conditions ($[Zn^{2+}] = 1\ M$), it is often given an identifying subscript—$\mathbf{E}^{\ominus}_{Zn^{2+}/Zn}$—and is called the *standard* electrode potential of the Zn^{2+}/Zn half-cell. Furthermore, $Q = 1/[Zn^{2+}]$ in (18-30) because the reference electrode is always the *standard* hydrogen electrode, so $[H^+]$ is by definition 1 M and P_{H_2} is equal to 1 atm. Equation (18-30) thus has the form

$$\mathbf{E} \approx \mathbf{E}^{\ominus}_{Zn^{2+}/Zn} - 0.03 \log \frac{1}{[Zn^{2+}]} \tag{18-31}$$

Comparison with the Zn^{2+}/Zn half-reaction shows that (18-31) is formally identical with the Nernst equation for this half-reaction, provided that the electrons appearing in the half-reaction are ignored when forming Q. This result is general.

By convention, we *always* write the half-reaction as a *reduction* in formulating the analog of the Nernst equation, as in (18-31). Standard conditions for half-cells pertain to the species that appear in the half-reaction. They are unit molarity for dissolved species and partial pressures of 1 atm for gases. When the solubility of a substance is small, standard conditions require an excess of the solid form (or liquid—for example, Hg) of the substance to be present to assure saturation. For example, the standard conditions for the Ag^+/Ag couple are $[Ag^+] = 1$, presence of $Ag(s)$; for the $AgCl/Ag$ couple, $[Cl^-] = 1$, presence of $AgCl(s)$ and $Ag(s)$.

Values of standard electrode potentials.

Values of standard electrode potentials can be obtained from measurement of voltages of appropriate cells and also from thermodynamic data. A representative set of standard electrode potentials is given in Tables 18-1 and D-5. The choice of the standard hydrogen electrode as the reference for measurement of electrode potentials is analogous to the choice of mean sea level as the zero point for measurement of heights on the earth's surface; it is merely a convenience. Any other electrode—AgCl/Ag, for example—might have been used as a reference instead. The difference between two heights on the earth need not be measured directly if each has been established relative to mean sea level; their difference may be obtained directly by subtraction, taking into account the sign of each if one is above and the other below sea level. The same is true for electrode potentials (Fig. 18-3).

It is only by convention that we assign a negative sign to the altitudes of points that are on the same side of mean sea level as the center of the earth (points such as the low point of the floor of Death Valley) and a positive sign to the altitudes of mountain peaks and other points "outside" mean sea level. The opposite convention could just as well be used. Similarly, the convention[1] for the signs of electrode potentials is arbitrary. A positive sign is given to the electrode emfs of redox couples for which the oxidized form is a better oxidizing agent—that is, gains electrons more

[1]The convention given here was adopted internationally in 1953. Actually, the opposite convention, with the half-reactions written as oxidations so the electrons are shown on the right side, was widely used in the United States for many years and is still found in some books. The convention used should always be checked when data are taken from an unfamiliar source.

TABLE 18-1
Standard electrode potentials

Reaction	E°/volts	Reaction	E°/volts
$F_2(g) + 2e^- = 2F^-$	2.87	$Hg_2Cl_2(s) + 2e^- = 2Hg(l) + 2Cl^-$	0.268
$H_2O_2 + 2H^+ + 2e^- = 2H_2O$	1.77	$AgCl(s) + e^- = Ag(s) + Cl^-$	0.222
$MnO_4^- + 8H^+ + 5e^- = Mn^{2+} + 4H_2O$	1.51	$Cu^{2+} + e^- = Cu^+$	0.153
$PbO_2(s) + 4H^+ + 2e^- = Pb^{2+} + 2H_2O$	1.47	$S(s) + 2H^+ + 2e^- = H_2S$	0.14
$Cl_2(g) + 2e^- = 2Cl^-$	1.359	$HSO_4^- + 3H^+ + 2e^- = SO_2(g) + 2H_2O$	0.14
$Cr_2O_7^{2-} + 14H^+ + 6e^- = 2Cr^{3+} + 7H_2O$	1.33	$S_4O_6^{2-} + 2e^- = 2S_2O_3^{2-}$	0.09
$O_2(g) + 4H^+ + 4e^- = 2H_2O$	1.229	$2H^+ + 2e^- = H_2(g)$	0.000
$2IO_3^- + 12H^+ + 10e^- = I_2(s) + 6H_2O$	1.19	$Pb^{2+} + 2e^- = Pb(s)$	-0.126
$Br_2(l) + 2e^- = 2Br^-$	1.065	$Ni^{2+} + 2e^- = Ni(s)$	-0.23
$Cu^{2+} + I^- + e^- = CuI(s)$	0.85	$Fe^{2+} + 2e^- = Fe(s)$	-0.440
$Ag^+ + e^- = Ag(s)$	0.799	$Zn^{2+} + 2e^- = Zn(s)$	-0.763
$Fe^{3+} + e^- = Fe^{2+}$	0.771	$Mg^{2+} + 2e^- = Mg(s)$	-2.37
$O_2(g) + 2H^+ + 2e^- = H_2O_2$	0.69	$Na^+ + e^- = Na(s)$	-2.698
$I_3^- + 2e^- = 3I^-$	0.535	$K^+ + e^- = K(s)$	-2.925
$I_2(s) + 2e^- = 2I^-$	0.534	$Li^+ + e^- = Li(s)$	-3.03
$Cu^{2+} + 2e^- = Cu(s)$	0.337		

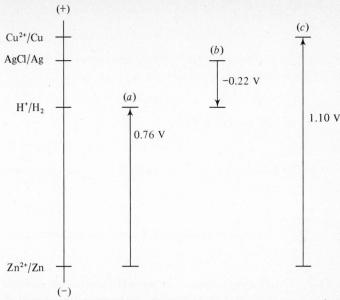

FIGURE 18-3   **Relative values of a few standard electrode potentials.**   The vertical line at the left represents a linear voltage scale; the short horizontal lines on it are spaced in proportion to the relative values of the standard emfs of the indicated half-cells. Only differences of electrode potentials can be measured experimentally. The differences in standard electrode potentials for three pairs of half-cells are given beside the arrows labeled (*a*), (*b*), and (*c*). These are the pairs of half-cells combined in the cells described in (18-29), (18-13), and (18-6), respectively. The emf values given are those that would be measured for such cells under standard conditions. Each arrow in the figure points from the electrode *assumed* to be the anode to that *assumed* to be the cathode according to the convention corresponding to the written descriptions in (18-29), (18-13), and (18-6).

The (+) and (−) at the ends of the scale are the signs conventionally associated with electrode potentials in these regions of the scale. When the standard hydrogen electrode is assigned the reference value $E^\circ = 0.000$ V, the standard value for the Zn^{2+}/Zn electrode is then -0.76 V, that for AgCl/Ag is $+0.22$ V, and that for Cu^{2+}/Cu is $+0.34$ V, that is, $(1.10 - 0.76)$ V.

easily, is more easily reduced—than is H^+ in the standard hydrogen electrode. Thus couples containing strong oxidizing agents, such as F_2/F^-, Cl_2/Cl^-, and MnO_4^-/Mn^{2+}, have highly positive standard emfs. Couples containing the best reducing agents, including strongly electropositive metals like aluminum, the alkaline-earth metals, and the alkali metals, have quite negative standard emfs.

EXAMPLE 18-5   **Electrode potential.**   What is the electrode potential at 25°C for an AgCl/Ag electrode?

Solution.   The half-reaction, written conventionally as a reduction, is

$$AgCl(s) + e^- \longrightarrow Ag(s) + Cl^-$$

and $E^\circ_{AgCl/Ag}$ is found in Table 18-1 to have the value of 0.222 V. Therefore, with

n = 1,

$$\mathrm{E} = (0.222 - 0.059 \log [\mathrm{Cl}^-]) \text{ volts} \tag{18-32}$$

It is instructive to set up the cell *defining* the electrode potential desired in the example. It is

$$\text{Pt, H}_2 \text{ (1 atm)|H}^+ \text{ (1 } M\text{)||Cl}^-\text{|AgCl, Ag}$$

with the cell reaction

$$\text{H}_2(g) + 2\text{AgCl}(s) \longrightarrow 2\text{Ag}(s) + 2\text{H}^+ + 2\text{Cl}^-$$

Inserting $P_{\mathrm{H}_2} = 1$ atm and $[\mathrm{H}^+] = 1$ M into the Nernst equation for *this* cell reaction, we obtain

$$\mathrm{E} = \mathrm{E}^\ominus - \tfrac{0.059}{2} \log [\mathrm{Cl}^-]^2$$

This reduces to (18-32) because the $\tfrac{1}{2}$ in front of the logarithm cancels the exponent in $[\mathrm{Cl}^-]^2$.

EXAMPLE 18-6 **pH dependence of an electrode potential.** What is the pH dependence of the emf of the hydrogen electrode at a constant gas pressure, $P_{\mathrm{H}_2} = 1$ atm?

Solution. Since $\mathrm{E}^\ominus = 0.000$ by definition and the half-reaction is $2\text{H}^+ + 2e^- \longrightarrow \text{H}_2(g)$,

$$\mathrm{E} = -\left(\tfrac{0.0592}{2} \log \frac{P_{\mathrm{H}_2}}{[\mathrm{H}^+]^2}\right) \text{volts}$$

When $P_{\mathrm{H}_2} = 1$ atm,

$$\mathrm{E} = -(0.0592 \text{ pH}) \text{ volts} \tag{18-33}$$

because pH = $-\log [\mathrm{H}^+]$. The electrode potential is seen to be directly proportional to the pH, and a hydrogen electrode may thus be used to measure the pH of the solution that surrounds it. Indeed, any other half-cell representing a half-reaction that includes H^+ as a reactant or product may be used to measure the pH of a solution. The concentrations of all other species involved in the half-reaction must, of course, be kept constant. Electrodes of this kind have been used to measure pH in the past, but all (including the hydrogen electrode) suffer from the disadvantage that, besides responding to the pH, their potential is changed by the presence of redox couples. Thus, they may be used to measure the pH only when there is no interference from oxidizing or reducing agents. In the modern pH meter this disadvantage is overcome by the use of a glass electrode, which responds only to the pH (see Sec. 19-4).

EXAMPLE 18-7 **Dependence of electrode potential on concentrations and amounts of reactants.** Consider the cathode reaction in the lead storage battery (at 25°C):

$$\text{PbO}_2(s) + 4\text{H}^+ + 2e^- \longrightarrow \text{Pb}^{2+} + 2\text{H}_2\text{O}$$

for which $\mathrm{E}^\ominus = 1.47$ V. By how much will the electrode potential increase or decrease if (*a*) the concentration of H^+ increases by a factor of 10; (*b*) the concentration of Pb^{2+} increases by a factor of 100; (*c*) the amount of $PbO_2(s)$ is doubled?

Solution.   The electrode reaction given is written as a reduction in accord with the convention for the electrode potential. A formal Nernst equation parallel to (18-31) can be written

$$\mathbf{E}_{PbO_2/Pb^{2+}} = \mathbf{E^{\circ}}_{PbO_2/Pb^{2+}} - \frac{0.06}{2} \log \frac{[Pb^{2+}]}{[H^+]^4}$$

$$= (1.47 - 0.03 \log [Pb^{2+}] + 0.12 \log [H^+]) \text{ volts}$$

Thus, (*a*) if $[H^+]$ increases by a factor of 10, $\log [H^+]$ is larger by one unit; the electrode potential will therefore increase by 0.12 V. (*b*) If $[Pb^{2+}]$ increases by a factor of 100, $\log [Pb^{2+}]$ increases by two units, so the electrode potential changes by -0.03×2 V; that is, it decreases by 0.06 V. (*c*) Since PbO_2 is a solid, the amount of it present has no effect on the emf (as long as *some* is present).

Electrode potentials are by definition independent of the direction in which the electrode reaction is *written;* that is, they do not change sign when the reaction is reversed. This is because the electrode potential is a physical quantity, the potential the electrode assumes relative to the standard hydrogen electrode when the half-cell being considered is combined with a standard hydrogen half-cell. Nevertheless many chemists find it convenient in calculations to change the sign associated with the potential of the electrode when the reaction is written as an oxidation.

Combination of half-cells to form electrochemical cells.

When two half-cells are combined to form an electrochemical cell, it is convenient to specify the cell in the shorthand notation exemplified in (18-6), without regard to whether or not the cell reaction for the cell as written would be spontaneous, that is, whether or not the cell is galvanic. In other words, the cell is written down in a certain way and this implies, by the convention discussed earlier, that the electrode on the left is *considered* to be the anode and the electrode on the right is *considered* to be the cathode. Oxidation is thus *assumed* to occur at the left electrode, reduction at the right.

The emf of the cell as it is assumed to be operating is

$$\mathbf{E}_{cell} = \mathbf{E}_{reducing \ electrode, \ right} - \mathbf{E}_{oxidizing \ electrode, \ left} = \mathbf{E}_{cathode} - \mathbf{E}_{anode} \quad (18\text{-}34)$$

If the value of $\mathbf{E}_{cell}$ so calculated is positive, the assumed cell reaction is spontaneous and thus the cell as written is galvanic, since a positive $\mathbf{E}_{cell}$ implies a negative free-energy change for the corresponding cell reaction. If on the other hand $\mathbf{E}_{cell}$ as calculated is negative, then the reversed cell, with anode and cathode interchanged, would give a spontaneous reaction. Equation (18-34) is equally applicable to the reversed cell, and since the electrodes have been switched, the value of $\mathbf{E}_{cell}$ for the *reversed* cell is now positive, corresponding to the spontaneous cell reaction for the reversed cell.

When $\mathbf{E}_{cell}$ is negative for a cell written in a particular way, the cell reaction is spontaneous from right to left rather than in the usual direction, left to right. The cell reaction can be made to go from left to right *only* if sufficient external energy is supplied. The external voltage must be sufficient not only to overcome $|\mathbf{E}_{cell}|$, the emf corresponding to the tendency of the cell to run in the spontaneous (galvanic) direction, but also to overcome the internal resistance of the cell and possible overvoltages (Sec. 19-2).

The value and sign of $\mathbf{E}_{cell}$ depend not only on the $\mathbf{E}^{\ominus}$ values of the half-cells but also on the concentrations of the species participating in the cell reaction in the way indicated by the Nernst equation. Consider the two half-reactions and $\mathbf{E}^{\ominus}$ values

$$Ag^+ + e^- \longrightarrow Ag(s) \qquad \mathbf{E}^{\ominus} = 0.80 \text{ V} \qquad (18\text{-}35)$$
$$Zn^{2+} + 2e^- \longrightarrow Zn(s) \qquad \mathbf{E}^{\ominus} = -0.76 \text{ V} \qquad (18\text{-}36)$$

At 25°C the electrode potentials are

$$\mathbf{E}_{Ag^+/Ag} = \left(0.80 - 0.06 \log \frac{1}{[Ag^+]}\right) \text{volts} \qquad (18\text{-}37)$$

$$\mathbf{E}_{Zn^{2+}/Zn} = \left(-0.76 - \frac{0.06}{2} \log \frac{1}{[Zn^{2+}]}\right) \text{volts} \qquad (18\text{-}38)$$

To obtain the cell reaction the half-reactions have to be multiplied by suitable coefficients to make the electrons cancel. Let the anode reaction arbitrarily be the second of the half-reactions, so the Zn is oxidized to Zn^{2+}. A proper combination is $2 \times (18\text{-}35) - 1 \times (18\text{-}36)$:

$$2Ag^+ + Zn(s) \longrightarrow 2Ag(s) + Zn^{2+} \qquad (18\text{-}39)$$

the cell reaction corresponding to an interchange of two electrons. The cell emf is $\mathbf{E}_{cell} = \mathbf{E}_{cathode} - \mathbf{E}_{anode} = \mathbf{E}_{Ag^+/Ag} - \mathbf{E}_{Zn^{2+}/Zn}$. Before (18-37) and (18-38) are combined, however, it is convenient to make the coefficients of the logarithms equal, with the exponents of the concentrations changed correspondingly. A convenient common factor is $\frac{0.06}{2} = 0.03$; the logarithmic terms in (18-37) and (18-38) then become, respectively, $-0.03 \log (1/[Ag^+]^2)$ and $-0.03 \log (1/[Zn^{2+}])$. Combining everything, we get

$$\mathbf{E}_{cell} = 0.80 - (-0.76) - 0.03 \left(\log \frac{1}{[Ag^+]^2} - \log \frac{1}{[Zn^{2+}]}\right)$$

$$= \left(1.56 - 0.03 \log \frac{[Zn^{2+}]}{[Ag^+]^2}\right) \text{volts} \qquad (18\text{-}40)$$

The same result could have been obtained by applying the Nernst equation directly to the whole cell, noting that the standard cell emf is the difference of the standard electrode potentials, $\mathbf{E}^{\ominus}_{cell} = \mathbf{E}^{\ominus}_{cathode} - \mathbf{E}^{\ominus}_{anode}$. Indeed, (18-40) is identical with

$$\mathbf{E}_{cell} = \mathbf{E}^{\ominus}_{Ag^+/Ag} - \mathbf{E}^{\ominus}_{Zn^{2+}/Zn} - \frac{0.06}{2} \log Q$$

where Q is the reaction quotient of the cell reaction (18-39) and the number of electrons interchanged is $\mathbf{n} = 2$.

What changes occur in **E** and in the concentrations of the ions in the cell as the cell reaction proceeds? Assume first that the concentrations of Ag^+ and Zn^{2+} are such that $\mathbf{E}_{cell}$ as calculated from (18-40) is positive. The cell is galvanic and, if current is permitted to flow, the reaction as written in (18-39) proceeds from left to right. As a consequence, the concentration of Zn^{2+} increases and that of Ag^+ decreases, so Q increases and the value of $\mathbf{E}_{cell}$ decreases as the reaction proceeds. Similarly if $\mathbf{E}_{cell}$ is negative initially, which means that for the concentrations present the reaction in (18-39) would tend to go spontaneously from right to left, the reaction may be driven from left to right by an external voltage; the resulting concentration changes would make Q larger and $\mathbf{E}_{cell}$ would become still more negative. Under these circumstances the cell is electrolytic, and the external voltage required increases as the electrolysis proceeds. In contrast, the voltage available from a galvanic cell, that is, a cell in which the reaction is proceeding spontaneously, decreases toward zero as the reaction goes on. All this makes good physical sense.

EXAMPLE 18-8   **Number of electrons exchanged.** What value of **n** must be used in the Nernst equation for each of the following reactions:

(*a*) $2Fe^{3+} + Pb \longrightarrow 2Fe^{2+} + Pb^{2+}$

(*b*) $Ag(s) + Cu^{2+} + 2Cl^- \longrightarrow AgCl(s) + CuCl(s)$

(*c*) $Pt(s) + 6Cl^- + 4H^+ \longrightarrow 2H_2 + PtCl_6{}^{2-}$

(*d*) $6H^+ + IO_3{}^- + 5I^- \longrightarrow 3I_2(s) + 3H_2O$

Solution. Since the number of electrons gained must be the same as that lost, the value of **n** can be found by considering either the substance oxidized or that reduced. As a check, we shall consider both.

(*a*)   Each of two ferric ions gains one electron, so **n = 2**; this is consistent with the single Pb atom's losing two electrons to form Pb^{2+}.

(*b*) $Ag(s)$ combines with one Cl^- to form uncharged $AgCl(s)$, so one electron is lost in this half-reaction. The same value of **n**, 1, is derived from the fact that Cu^{2+} and one Cl^- form uncharged $CuCl(s)$, a process that must be accompanied by the gain of one electron to maintain charge balance.

(*c*)   Each of four protons gains an electron as $4H^+$ forms $2H_2$, so **n = 4.** The single Pt atom combines with $6Cl^-$ to form $PtCl_6{}^{2-}$ with a loss of four units of negative charge, again consistent with **n** = 4.

(*d*)   Each of five I^- loses an electron in going to I_2, which implies **n = 5.** Simultaneously, the I in $IO_3{}^-$, which is in the +V oxidation state, is also changed into iodine in the zero oxidation state, corresponding to a gain of five electrons by each $IO_3{}^-$, in agreement with the value **n** = 5.

EXAMPLE 18-9   **Cell emf.** What is the emf at 25°C of the cell

$$Zn|Zn^{2+}(1\ M)||Cu^{2+}(0.1\ M)|Cu$$

Solution. By Table 18-1 we note that $E^{\ominus}_{Zn^{2+}/Zn} = -0.76$ V and $E^{\ominus}_{Cu^{2+}/Cu} = 0.34$ V. Furthermore, the Zn electrode is the anode for the cell as written, and the cell reaction is

$$Zn(s) + Cu^{2+} \longrightarrow Zn^{2+} + Cu(s)$$

Inserting $[Zn^{2+}]/[Cu^{2+}] = 10$ into the Nernst equation yields

$$E = (0.34 + 0.76 - \tfrac{0.06}{2} \log 10) \text{ volts}$$

$$= \underline{1.07 \text{ V}}$$

Since **E** is positive, the cell is galvanic.

EXAMPLE 18-10   **Cell reaction and emf.** Calculate the cell emf for the reaction

$$5Cl^- + MnO_4^- + 8H^+ \longrightarrow \tfrac{5}{2}Cl_2(g) + Mn^{2+} + 4H_2O$$

at 25°C, when $P_{Cl_2} = 1$ atm, $[H^+] = 1.0$ M, and the concentrations of all other ionic species are 0.1 M. The standard electrode potentials are $E^{\ominus}_{Cl_2/Cl^-} = 1.36$ V and $E^{\ominus}_{MnO_4^-/Mn^{2+}} = 1.51$ V.

Solution. For the cell reaction **n** = 5, so that

$$E = \left(1.51 - 1.36 - \tfrac{0.06}{5} \log \frac{P_{Cl_2}^{5/2}[Mn^{2+}]}{[Cl^-]^5[MnO_4^-][H^+]^8}\right) \text{volts}$$

The logarithmic term is equal to $-\tfrac{0.06}{5} \log (0.1)^{-5} = -0.06$, and the final result is $E = \underline{0.09 \text{ V}}$. Since **E** is positive, the reaction is spontaneous and MnO_4^- is capable of oxidizing Cl^- under the conditions stated. However, the smallness of the value of **E** indicates that the concentrations are not far from equilibrium concentrations, as we shall see in Chap. 19.

18-5   **Technological Applications**

Some common batteries. In principle any redox reaction can be used to produce an electric current, but in commercial practice a number of requirements must be satisfied. These requirements include low cost of materials and manufacture (except in very specialized applications such as space exploration or heart-pacer batteries, for which cost is secondary), high energy per unit volume and unit weight, ruggedness, portability, long shelf life, and relatively constant potential during discharge even at high discharge rates. The total energy a battery is capable of delivering is measured in watthours and is sometimes called the capacity of the cell.

In the common *flashlight battery* or *dry cell* (Fig. 18-4a) the electrolyte is a saturated solution of ammonium chloride, also containing zinc chloride. The electrolyte is made into a paste with an excess of solid ammonium chloride and diatomaceous earth or another inert filler. This cell is not really dry, for water is an essential component of the electrolyte.

The cathode of the dry cell is a graphite rod in the center, surrounded by a

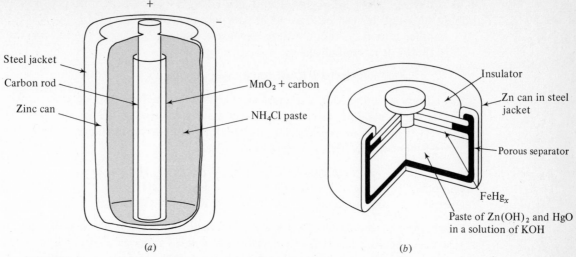

+

−

Steel jacket

Carbon rod

Zinc can

MnO_2 + carbon

NH_4Cl paste

Insulator

Zn can in steel jacket

Porous separator

$FeHg_x$

Paste of $Zn(OH)_2$ and HgO in a solution of KOH

(a)

(b)

FIGURE 18-4   **Typical cells.**   (a) Dry cell; (b) mercury cell.

powdered mixture of manganese dioxide and graphite. The MnO_2 serves as the oxidizing agent, accepting electrons that enter the cell at the cathode; if the MnO_2 were not present, the electrons would be accepted instead by the water of the electrolyte and gaseous H_2 would be formed. This would make it impossible to seal the battery and would also produce a film of gas around the cathode, impeding further reaction and therefore making it impossible to draw more current from the battery. Most present-day batteries contain oxidizing agents such as MnO_2, called depolarizers, to prevent evolution of H_2 at the cathode. The cathode reaction is complex, but is often represented as

$$2MnO_2 + 2NH_4^+ + 2e^- \longrightarrow 2MnO(OH) + 2NH_3$$

The anode of a dry cell is a zinc cylinder that acts as a container for the electrolyte and is essentially the casing of the battery, except for a heavy paper, plastic, or steel shield wrapped around it. The zinc surface is amalgamated to take advantage of the high overvoltage of H_2 on mercury (Sec. 19-2). The electrolyte is quite acidic because of the ammonium chloride present, and ordinary zinc would react quite rapidly with this solution to form H_2 and Zn^{2+}. However, the high overvoltage resulting from the amalgamation of the zinc reduces the rate of this reaction enormously, and the lifetime of the battery when not in use (the shelf life) is greatly increased. The direct reaction does go on slowly nevertheless, consuming the zinc by corrosion rather than by the current-producing reaction, and dry cells kept for several years may eventually leak when the zinc shell has been eaten through, unless they have an outer jacket.

The voltage produced by a dry cell, about 1.5 V, drops slightly during use because of the accumulation of Zn^{2+} ions near the zinc anode. The cell recovers partially when not used, as Zn^{2+} ions diffuse into the interior of the electrolyte where they can react with NH_3 to form $Zn(NH_3)_4^{2+}$ and other ammine complexes. In the *alkaline dry cell* ammonium chloride is replaced by potassium hydroxide. It is more expensive than the usual dry cell but has 50 percent more capacity.

Another important though more expensive modern battery is the *mercury cell* (Fig. 18-4*b*). It is symbolized by

$$Zn(amalgam)|8\ F\ KOH, \text{sat'd. with } ZnO|HgO(s) + C(s)|Hg(l)$$

The battery is contained in a stainless steel cylinder, with the Zn anode at the center. The strongly alkaline electrolyte is saturated with zinc oxide in the form of $Zn(OH)_4^{2-}$ ions. Powdered carbon provides for adequate conductance in the region around the cathode. The electrode reactions are

Anode: $\quad Zn(amalgam) + 2OH^- \longrightarrow ZnO(s) + H_2O + 2e^-$
Cathode: $\quad HgO(s) + H_2O + 2e^- \longrightarrow Hg(l) + 2OH^-$
Overall: $\quad Zn(amalgam) + HgO(s) \longrightarrow ZnO(s) + Hg(l)$

This battery has an excellent voltage characteristic, maintaining a potential of about 1.34 V during its entire life. The reason the voltage remains constant is that the overall cell reaction involves only solid substances; the electrolyte acts solely as an intermediary. The battery was developed during World War II when a self-contained source of sustained voltage, despite current drain, was badly needed. Numerous applications for it have since been found in the electronics industry and in such devices as hearing aids, electric wristwatches, and light meters.

Of great importance are *storage batteries,* which usually consist of three or six identical, rechargeable cells joined together in sequence. One of these is the lead storage battery, one cell of which can be represented as

$$Pb(s), PbSO_4(s)|H_2SO_4(35\% \text{ soln})|PbSO_4(s), PbO_2(s)$$

and has a potential of about 2.0 V. The electrodes are grids of lead alloy, the interstices being filled with spongy lead at the anode and lead dioxide at the cathode. The reactions that occur as the cell produces current are

Anode [terminal stamped $(-)$]:
$$Pb(s) + SO_4^{2-} \longrightarrow PbSO_4(s) + 2e^- \tag{18-41}$$

Cathode [terminal stamped $(+)$]:
$$PbO_2(s) + SO_4^{2-} + 4H^+ + 2e^- \longrightarrow PbSO_4(s) + 2H_2O \tag{18-42}$$

Overall:
$$Pb(s) + PbO_2(s) + 2H_2SO_4 \longrightarrow 2PbSO_4(s) + 2H_2O \tag{18-43}$$

These reactions produce $PbSO_4$ at each electrode; this salt is only slightly soluble and remains on the plates. Sulfate ions react not only at the cathode (with PbO_2) but also at the anode (with Pb), in spite of their negative charge. Some of the sulfuric acid is used up and water is produced as the spontaneous reaction (18-43) goes on, so that the sulfuric acid electrolyte is gradually diluted as the cell reaction proceeds to the right. It is therefore possible to determine the state of the battery by measuring the density of the liquid in the battery.

During the recharging of a lead storage battery, reaction (18-43) is reversed by the application of an external potential. This can be done with no difficulty when the $PbSO_4$ is freshly precipitated, but aging slowly changes the lead sulfate on the plates to a relatively unreactive form. A battery that remains in a discharged state for weeks can be recharged only partially, if at all.

Another important storage cell is the *nickel-cadmium cell,*

$$Cd(s),\ CdO(s)|KOH\ (20\%\ soln)|Ni(OH)_3(s),\ Ni(OH)_2(s),\ Ni(s)$$

with the overall discharge reaction

$$Cd(s) + 2Ni(OH)_3(s) \longrightarrow CdO(s) + 2Ni(OH)_2(s) + H_2O(l)$$

This cell is more expensive than the lead storage cell but it is more rugged, lighter, has a longer life, and may remain uncharged after use with no damage. It is used in cordless electric toothbrushes, electric razors, pocket calculators, and other small portable appliances. The potential of this cell is about 1.35 V.

Fuel cells. The batteries discussed so far are closed systems with fixed amounts of oxidant and reductant, although in storage cells the oxidant and reductant can be regenerated by recharging from an external energy source. Batteries can also be made in which reactants are continuously fed to the electrodes, and such batteries, known as *fuel cells,* are of increasing importance. They are analogous in some ways to ordinary furnaces fed with coal, oil, or gas and to internal-combustion engines that burn gasoline or oil, each of which produces energy on demand from a continuously fed fuel supply.

Fuel cells, however, have an advantage in efficiency over heat engines that operate between two temperatures and convert some of the heat generated into work. Consider the production of electrical energy in a steam plant. Coal or oil is burned and the resulting energy is used to produce steam that drives a turbine coupled to an electric generator. If the fuel is coal, the reaction is

$$C(s) + O_2(g) \longrightarrow CO_2(g) \tag{18-44}$$

for which $\Delta H^\ominus$ and $\Delta G^\ominus$ are essentially the same, -393 kJ. This is not a very efficient way to produce electrical energy, for at the temperature used in modern steam plants, at best about 60 percent of the enthalpy released can be converted to work (a limitation imposed by the second law of thermodynamics). In practice, the overall efficiency of this conversion is only 35 to 40 percent. The situation would be vastly improved if it were possible to invent an electrochemical cell having (18-44) as its overall reaction and operating as a fuel cell so that it could supply energy continuously. Fuel cells are operated isothermally, not as heat engines, and the maximum energy theoretically available for conversion into electrical energy is ΔG, not some fraction of ΔH. Although efforts to design a practical fuel cell based on (18-44) are being made, they have not yet proved successful.

A number of practicable fuel cells are, however, currently in use. Those best developed are based on the hydrogen-oxygen reaction

$$2H_2(g) + O_2(g) \longrightarrow 2H_2O(g)$$

for which $\Delta H^\ominus = -485$ kJ and $\Delta G^\ominus = -456$ kJ at 25°C. The automated source of auxiliary power for the command and service modules of the Apollo space vehicles was a hydrogen-oxygen fuel cell of a type originally designed by F. T. Bacon more than two decades ago. The electrodes in this cell are of porous nickel and the cell runs at about 200°C and at gas pressures below 1 atm. The electrolyte is 80 percent KOH; hydrogen is fed to the anode, oxygen to the cathode.

Another version of the hydrogen-oxygen fuel cell has a solid electrolyte in the form of membranes capable of interchanging ions. This cell was used in the earlier Gemini space program. In both this cell and that described above, the water formed in the cell reaction is used for drinking, a particularly desirable feature in a manned spaceship, where mass and storage space are at a premium and a life-support system must be provided. These cells represent the lightest available methods for packaging energy because the atomic weights of hydrogen and oxygen are so low and their reaction is so exothermic.

Of great potential value are biological fuel cells, in which bacterial oxidations of low-cost abundant fuels, including sewage, simultaneously provide electric power and improve the environment, rather than polluting it as some power sources do. While it has been possible to "convert" sewage into electrical energy on a small scale, practical large-scale application is not yet a reality.

Problems and Questions

18-1 **Deposition of zinc.** A current of 0.52 A flowing for 88 min will deposit what weight of zinc?   *Ans.* 0.93 g

18-2 **Fuel cell.** A current of 7.50 A is produced in a fuel cell when CO is oxidized to CO_2 at the anode while oxygen is reduced at the cathode. How many grams of CO and of O_2 does the cell use up per hour?

18-3 **Electroplating of copper.** During the electroplating of copper from a $CuSO_4$ solution, an ammeter shows that a steady current of 0.50 A is passing through the cell. (*a*) How many electrons pass into or out of the cell per minute? (*b*) What is the rate of deposition of copper in grams per hour?   *Ans.* (*a*) 1.9×10^{20};   (*b*) 0.59 g hour^{-1}

18-4 **Electrolysis of magnesium iodide.** Hydrogen and iodine are formed at the electrodes when aqueous magnesium iodide is electrolyzed. In a particular experiment starting with 500 ml of 0.200 *F* magnesium iodide, the volume of hydrogen collected was exactly 1.00 liter when measured over water at 25°C and a pressure of 746 torr. The vapor pressure of water is 24 torr at 25°C. (*a*) How many faradays of electricity had been used? (*b*) If the current flowing was constant at 5.0 A, how long did the electrolysis take? (*c*) What were the final concentrations of I$^-$, Mg^{2+}, and OH$^-$ in the solution? (Assume that the volume does not change and that no reactions occur among these ions.)

18-5 **Galvanic cell.** A galvanic cell is made up of a 3.6-g aluminum rod in an essentially infinite volume of 1.00 *F* aluminum nitrate and a 6.4-g copper bar in an equally large volume of 1.00 *F* cupric nitrate. The electrodes are connected externally by a wire, and the solutions are joined through a porous connector to complete the circuit. A steady current of 2.0 A flows through the external circuit. Because the volume of the solutions is so large, concentration changes may be ignored. (*a*) State what changes occur at each electrode and why; then write the overall chemical reaction that takes place. (*b*) What maximum voltage will this cell supply? (*c*) What is the direction of flow of electrons in the external circuit? (*d*) What is the maximum time for which this cell can supply current? (*e*) Would the voltage of the cell be increased or decreased by replacing the 1.00 *F* cupric nitrate solution by 0.0100 *F* cupric nitrate? Show your reasoning.

Ans. (*a*) Al dissolves at anode, Cu plates out at cathode; (*b*) 2.00 V; (*c*) from Al to Cu; (*d*) 5.4 hours; (*e*) decreased

18-6 Electrolysis of KCl solution. In an electrolysis of a KCl solution, hydrogen is evolved at the cathode and a mixture of chlorine and oxygen at the anode, the oxygen accounting for 6.0 percent of the total current. (*a*) What is the molar ratio of O_2 to Cl_2 at the anode? (*b*) What is the ratio of the volumes of gas produced at the anode and cathode?

18-7 Galvanic cell. A battery is made with a cadmium electrode in a solution of 0.100 *F* $CdSO_4$ and a Zn electrode in a solution of 1.00 *F* $ZnSO_4$. The two solutions are kept apart by a porous partition. (*a*) What is the emf of this cell? (*b*) Which electrode is the anode? (See Table D-5.)

Ans. (*a*) 0.331 V; (*b*) Zn anode

18-8 Electrochemical cells. Consider the following cells and state for each the cell reaction and the number of electrons involved:

(*a*) Pt, $Cl_2(g)|Cl^-||Fe^{2+}$, $Fe^{3+}|Pt$
(*b*) $Hg(l)$, $Hg_2Cl_2(s)|HCl(aq)|H_2(g)$, Pt
(*c*) Pb, $PbSO_4(s)|H_2SO_4(aq)|PbSO_4(s)$, $PbO_2(s)$, Pb

18-9 Lead storage cell. A lead storage cell is discharged, which causes the sulfuric acid electrolyte to change from a concentration of 34.6 wt percent (density = 1.261 g ml^{-1} at 60°F) to one of 27.0 wt percent. The original volume of electrolyte is 1.000 liter. How many faradays have left the anode of the battery? Note that water is produced by the cell reaction as H_2SO_4 is used up.

Ans. 1.25 𝔉

18-10 Cell emf. What is **E** for the reaction $Pb(s) + 2H^+ \longrightarrow Pb^{2+} + H_2(g)$ when the concentrations are $[H^+] = 1.00 \times 10^{-2}$ *M*, $[Pb^{2+}] = 0.100$ *M*, and $P_{H_2} = 1$ atm?

18-11 Peroxydisulfuric acid. This acid, $H_2S_2O_8$, can be prepared by electrolytic oxidation of H_2SO_4: $2H_2SO_4 \longrightarrow H_2S_2O_8 + 2H^+ + 2e^-$. Hydrogen and oxygen are by-products. In such an electrolysis 9.72 liters H_2 and 2.35 liters O_2 were generated at STP. How many moles of $H_2S_2O_8$ were produced? How long did a current of 1.00 A have to flow to achieve this result? (*Hint:* Consider the reactions at the anode and the cathode separately, noting that the number of faradays involved must be the same at each. There are two separate reactions at the anode and one at the cathode.)

Ans. 0.224 mol; 23.2 hours

". . . If we knew the energy and entropy of oxygen and hydrogen at the temperature and pressure at which they are disengaged in an electrolytic cell, and also the energy and entropy of the acidulated water from which they are set free . . . , we could at once determine the electromotive force for a reversible cell. This would be a limit below which the electromotive force required in an actual cell used electrolytically could not fall, and above which the electromotive force of any such cell used to produce a current . . . could not reach."

J. W. Gibbs, 1887

19 Electrochemistry II

The discussion of electrochemistry begun in Chap. 18 continues here. First, the connection between cell emf and chemical equilibrium is derived by consideration of the relationship between free-energy changes and cell emfs. Methods for obtaining equilibrium constants from electrochemical data are explained. A section on electrolysis follows, including the relation to equilibrium processes and the nonequilibrium complications of concentration polarization and overvoltage. The remainder of the chapter is devoted to applications, such as corrosion, disproportionation, and ion-selective electrodes.

19-1 Cell EMF and Equilibrium

As a spontaneous cell reaction proceeds, the cell emf drops toward zero—the battery runs down. This fact of experience is consistent with the Nernst equation, as illustrated in the discussion of the cell corresponding to reaction (18-39) (Sec. 18-4). When E_{cell} finally becomes zero, there is equilibrium between the two redox couples present, at their final concentrations. Of course it may occasionally happen that equilibrium exists at the initial concentrations chosen when the cell is prepared; if so, the initial value of $\mathbf{E}_{cell}$ is zero and the cell is of no use as a battery. The fact that $\mathbf{E}_{cell} = 0$ at equilibrium is reasonable in light of the relation $\Delta G = -\mathbf{n\mathfrak{F}E}$ at constant temperature and pressure, for we know that under these conditions the criterion of equilibrium is $\Delta G = 0$.

The equilibrium eventually reached when a spontaneous cell reaction has been allowed to proceed is the *same* as that which would have been reached had the reactants been mixed together directly. When the reaction is harnessed in a galvanic cell, some of the energy released in the reaction is available as electrical energy as a consequence of the flow of electrons through the external circuit. On the other hand, when the reactants are mixed directly, the energy released is not utilized—it is dissipated, chiefly as thermal energy.

Equilibrium constants for redox reactions can be calculated from the standard electrode potentials of the half-cells that are coupled together in a reaction. No cell need be considered specifically; in fact, it does not matter whether a cell could actually be constructed with the two redox couples considered. As discussed in Chap. 17, an equilibrium constant can be calculated from standard free energies regardless of whether the corresponding reaction has ever been carried out or could be carried out in practice. This must also be true of standard electrode potentials because they are proportional to standard free energies per mole of electrons exchanged.

If we write the Nernst equation (18-23) for a cell for which $\mathbf{E}_{cell}$ is zero, replacing Q by K because the reactants are at equilibrium with each other under these conditions, we obtain an equation for K:

$$\mathbf{E}_{cell} = 0 = \mathbf{E}^{\ominus}_{cathode} - \mathbf{E}^{\ominus}_{anode} - \frac{RT}{\mathbf{n\mathfrak{F}}} \ln K$$

or

$$RT \ln K = \mathbf{n\mathfrak{F}}(\mathbf{E}^{\ominus}_{cathode} - \mathbf{E}^{\ominus}_{anode}) = \mathbf{n\mathfrak{F}E}^{\ominus}_{cell} \tag{19-1}$$

and at 25°C, with $\mathbf{E}^{\ominus}$ in volts,

$$\log K = \frac{\mathbf{n E}^{\ominus}_{cell}}{0.0592} \tag{19-2a}$$

or

$$K = 10^{\mathbf{n E}^{\ominus}{}_{cell}/0.0592} \tag{19-2b}$$

Table 19-1 shows the relationships between the characteristic features of $\Delta G^{\ominus}$, $\mathbf{E}^{\ominus}_{cell}$, K, and the direction in which a cell reaction *under standard conditions* occurs spontaneously. No special significance attaches to the second case ($\Delta G^{\ominus} = 0$), because to

TABLE 19-1 **Correlation among $\Delta G^{\ominus}$, $\mathbf{E}^{\ominus}_{cell}$, K, and the behavior of the cell reaction**

$\Delta G^{\ominus}$	$\mathbf{E}^{\ominus}_{cell}$	K	*Cell reaction under* standard conditions *is:*
< 0	> 0	> 1	Spontaneous toward right
0	0	1	At equilibrium
> 0	< 0	< 1	Spontaneous toward left

have the reactants and products at standard conditions and at equilibrium at the same time is a most unlikely situation.

The relationship between cell emf and ΔG shows the feasibility of obtaining free-energy changes by electrical measurements. Values of ΔG may also be deduced by measurement of appropriate thermal properties and of equilibrium constants, as discussed in Chap. 17.

EXAMPLE 19-1   **Equilibrium constants and standard electrode potentials.**   What is the equilibrium constant at 25°C for the cell reaction

$$Zn(s) + Cu^{2+} \longrightarrow Zn^{2+} + Cu(s)$$

The standard electrode potentials are 0.34 V for Cu^{2+}/Cu and -0.76 V for Zn^{2+}/Zn.

Solution.   Since the Zn^{2+}/Zn couple is run as an oxidation in the cell as written, its emf is to be subtracted from that of the other couple and $\mathbf{E}^{\ominus} = 0.34 - (-0.76) = 1.10$ V. The number of electrons interchanged is two. Thus, by (19-2), log $K = 2 \times 1.10/0.0592 = 37.2$ and $K = 10^{37.2} = \underline{1.6 \times 10^{37}}$.

Actually, the answer might better be written as 2×10^{37} because the last digit in the figure 37.2 is uncertain, and this digit represents the entire mantissa of the logarithm. Hence the antilogarithm of that mantissa is only approximate. (Since the antilog of 0.1 is 1.3, an uncertainty of 0.1 in a logarithm to the base 10 corresponds to an uncertainty of a factor of 1.3, or about 30 percent, in the antilogarithm.)

EXAMPLE 19-2   **Equilibrium constant and standard electrode potentials: the iodate-iodide reaction.**   What is the equilibrium constant at 25°C for the reaction

$$6H^+ + IO_3^- + 5I^- \longrightarrow 3I_2(s) + 3H_2O$$

In this reaction, iodide is oxidized to iodine by iodate, and the iodate is itself reduced to iodine. The standard electrode potentials are 1.19 V for the IO_3^-/I_2 couple and 0.534 V for the I_2/I^- couple.

Solution.   In this reaction the direction in which the I_2/I^- couple proceeds is that of an oxidation, so its emf is to be subtracted from that of the IO_3^-/I_2 couple. Thus, $\mathbf{E}^{\ominus} = (1.19 - 0.534)$ V $= 0.66$ V. The number of electrons interchanged is five (Example 18-8d). Thus, by (19-2), log $K = 5 \times 0.66/0.0592 = 56$ and $K = \underline{10^{56}}$. Note that in this example even the power of 10 is somewhat uncertain because of the limited precision of the $\mathbf{E}^{\ominus}$ value for the IO_3^-/I_2 couple.

EXAMPLE 19-3   **Solubility product and standard electrode potentials.**   Consider the following two half-reactions and standard electrode potentials at 25°C:

$$Ag^+ + e^- \longrightarrow Ag(s) \qquad \mathbf{E}^{\ominus} = 0.799 \text{ V} \qquad (19\text{-}3)$$
$$AgCl(s) + e^- \longrightarrow Ag(s) + Cl^- \qquad \mathbf{E}^{\ominus} = 0.222 \text{ V} \qquad (19\text{-}4)$$

The reaction in each system may be considered to be the reduction of Ag(I) to Ag(0). Why then is there a difference in the standard emfs? Calculate the solubility product of AgCl(s) from the data given.

Solution.   Recalling what is meant by standard conditions, we see that in (19-3), $[Ag^+] = 1\ M$, whereas in (19-4), $[Cl^-] = 1\ M$ *in the presence of solid* AgCl. The solubility of silver chloride is, however, so small that when $[Cl^-]$ in a solution is $1\ M$, $[Ag^+]$ in that same solution must be very small. Thus we understand why the standard emfs are so different for (19-3) and (19-4): the concentration of Ag^+ in equilibrium with the silver electrode is very different in the two solutions, $1\ M$ in (19-3) and very small in (19-4). From the Nernst equation for the Ag^+/Ag electrode,

$$\mathbf{E}_{Ag^+/Ag} = \left(0.799 - 0.0592 \log \frac{1}{[Ag^+]}\right) \text{volts}$$
$$= (0.799 + 0.0592 \log [Ag^+]) \text{ volts} \tag{19-5}$$

We see that $\mathbf{E}_{Ag^+/Ag}$ decreases as $[Ag^+]$ decreases. Thus the much lower standard emf associated with (19-4) is qualitatively reasonable.

Quantitatively, in a solution saturated with AgCl and thus at equilibrium we know that

$$[Ag^+] = \frac{K_{sp,AgCl}}{[Cl^-]}$$

and when $[Cl^-] = 1\ M$, as in (19-4), then $[Ag^+] = K_{sp,AgCl}$. Applying (19-5) and taking into account that the potential of the silver electrode has been reduced to 0.222 V in the solution saturated with AgCl when $[Cl^-] = 1\ M$, we have

$$0.222 = 0.799 + 0.0592 \log K_{sp,AgCl}$$

$$\log K_{sp,AgCl} = \frac{0.222 - 0.799}{0.0592} = -9.75$$

$$K_{sp,AgCl} = 10^{-9.75} = 10^{0.25} \times 10^{-10} = \underline{1.8 \times 10^{-10}}$$

This same result may be derived in an alternative but equivalent way by considering the cell for which (19-3) is the anode reaction and (19-4) is the cathode reaction:

$$Ag|Ag^+(1\ M)\|Cl^-(1\ M)|AgCl(s), Ag \tag{19-6}$$

For this cell the (standard) emf is $0.222 - 0.799 = -0.577$ V, $\mathbf{n} = 1$, and the overall reaction is (19-4) − (19-3):

$$AgCl(s) \longrightarrow Ag^+ + Cl^- \tag{19-7}$$

This is just the reaction for which we want the equilibrium constant, which is $K_{sp,AgCl}$. Applying (19-2), we have

$$\log K = \tfrac{-0.577}{0.0592} = -9.75$$

as before.

It is worth noting that reaction (19-7) is not a redox reaction. Nonetheless its equilibrium constant can be calculated from electrochemical information. The reason is that the potential of the silver electrode depends on the concentration of Ag^+ in the surrounding solution. A cell with the same voltage as that of (19-6) can be prepared, without using AgCl(s) and $[Cl^-] = 1\ M$, by having two silver electrodes and two solutions of a silver salt (for example, $AgNO_3$) at concentrations in the ratio $1/1.8 \times 10^{-10}$:

$$Ag|Ag^+(1\ M)\|Ag^+(1.8 \times 10^{-10}\ M)|Ag \tag{19-8}$$

Cells such as that represented in (19-8) are called concentration cells. The spontaneous reaction is such as to make the two concentrations equal, which implies a reduction of Ag^+ in the 1 M solution and an oxidation of Ag to Ag^+ at the electrode in the very dilute solution. This is opposite to the reaction implied by the way the cell (19-8) is written, which is consistent with the fact that the emf of the cell in (19-8), like that in (19-6), is negative.

Free-energy changes and half-reactions.   The free-energy change for the reaction that defines $E^\ominus$ for the Ag^+/Ag half-cell

$$H_2(g) + 2Ag^+ \longrightarrow 2H^+ + 2Ag(s)$$

is given by

$$\begin{aligned}
\Delta G^\ominus &= 2\widetilde{G}^\ominus_{H^+} + 2\widetilde{G}^\ominus_{Ag} - \widetilde{G}^\ominus_{H_2} - 2\widetilde{G}^\ominus_{Ag^+}\\
&= 2(\widetilde{G}^\ominus_{Ag} - \widetilde{G}^\ominus_{Ag^+}) - (\widetilde{G}^\ominus_{H_2} - 2\widetilde{G}^\ominus_{H^+})
\end{aligned} \tag{19-9}$$

This experimentally measurable free-energy change (19-9) can be assigned to the Ag^+/Ag half-reaction by assigning the arbitrary value of zero to the free-energy change of the hydrogen half-reaction, $\widetilde{G}^\ominus_{H_2} - 2\widetilde{G}^\ominus_{H^+}$. This is consistent with the convention that the electrode potential of the standard hydrogen electrode is zero. According to these conventions, then, the standard free-energy change for any half-reaction *written as a reduction* may be derived from the fundamental relationship

$$\Delta G^\ominus = -n\mathfrak{F}E^\ominus \tag{19-10}$$

If the half-reaction were written as an oxidation, the sign of the free-energy change would have to be reversed because this quantity depends on the direction of the reaction. Confusion can be avoided by consistently writing half-reactions as reductions when relating them to electrode potentials. Because of (19-9), $E^\ominus$ is a measure of the tendency of a half-reaction to proceed to the right or left *under standard conditions*. The more positive $E^\ominus$ is, the more the products of the half-reaction are favored; the more negative $E^\ominus$ is, the more the reactants are favored—all relative to the reactants and products of the standard hydrogen electrode.

EXAMPLE 19-4   **Electrode potential and free energy.**   Find the standard electrode potential at 25°C for the reaction

$$14H^+ + Cr_2O_7^{2-} + 6e^- \longrightarrow 2Cr^{3+} + 7H_2O$$
$$\quad 0 \qquad -1319 \qquad\qquad\quad -215 \quad\ -237$$

from the free energies of formation of the species involved, which are listed (in kilojoules) beneath the chemical symbols.

Solution.   The half-reaction is already written as a reduction, so the calculated standard free-energy change will be related to $E^\ominus$ for this half-reaction by (19-10). We calculate

$\Delta G^{\ominus} = (-2 \times 215 - 7 \times 237 + 1319)$ kJ $= -770$ kJ and therefore, with $\mathscr{F} = 96.5$ kJ V^{-1},

$$E^{\ominus} = \frac{770}{6 \times 96.5} \text{ V} = \underline{1.33 \text{ V}}$$

Combining half-cells into new half-cells. It is possible to combine half-reactions in several ways. We have already considered two: combining two different half-cells so that electrons cancel and a balanced redox equation results, and combining the same half-reactions to give a cell in which the species are present at different concentrations on the two sides. A third possibility involves two different half-reactions with a species in common that changes oxidation state in different ways in the two different reactions. The reactions may then be combined so as to eliminate this species, rather than eliminating electrons. The result is a new half-reaction, with a corresponding standard electrode potential, which we shall consider in a moment. This procedure is often of considerable value because it makes it possible to derive information about half-reactions for which data are not immediately available or perhaps have not even been measured.

Consider, for example, the half-reactions (19-11) and (19-12) below, which involve the couples MnO_4^-/Mn^{2+} and MnO_2/Mn^{2+}. They have the ion Mn^{2+} in common. This ion may be eliminated by subtracting (19-11) from (19-12) to form the half-reaction (19-13) for the couple MnO_4^-/MnO_2, written in the customary way with electrons on the left:

$$8H^+ + MnO_4^- + 5e^- \longrightarrow 4H_2O + Mn^{2+} \qquad E_{11}^{\ominus} = 1.51 \text{ V} \qquad (19\text{-}11)$$
$$4H^+ + MnO_2(s) + 2e^- \longrightarrow 2H_2O + Mn^{2+} \qquad E_{12}^{\ominus} = 1.23 \text{ V} \qquad (19\text{-}12)$$
$$4H^+ + MnO_4^- + 3e^- \longrightarrow MnO_2(s) + 2H_2O \qquad E_{13}^{\ominus} = ? \qquad (19\text{-}13)$$

The question is this: How are the $E^{\ominus}$ values of (19-11) and (19-12) to be combined to furnish $E^{\ominus}$ for reaction (19-13)? The answer is given readily when it is remembered that standard free energies of reactions combine in the same way as do the reaction equations. The values for $\Delta G^{\ominus}$ are $\Delta G_{11}^{\ominus} = -5\mathscr{F}E_{11}^{\ominus}$; $\Delta G_{12}^{\ominus} = -2\mathscr{F}E_{12}^{\ominus}$; $\Delta G_{13}^{\ominus} = \Delta G_{11}^{\ominus} - \Delta G_{12}^{\ominus} = -5\mathscr{F}E_{11}^{\ominus} + 2\mathscr{F}E_{12}^{\ominus}$. But we also know that $\Delta G_{13}^{\ominus} = -3\mathscr{F}E_{13}^{\ominus}$. Therefore,

$$E_{13}^{\ominus} = \frac{5E_{11}^{\ominus} - 2E_{12}^{\ominus}}{3} = 1.70 \text{ V}$$

In other words, standard emf values $E_i^{\ominus}$, which are intensive quantities, are not (necessarily) additive. The quantities that may be added or subtracted are the products $n_i E_i^{\ominus}$, which contain as factors the numbers n_i of electrons involved in the half-reactions. The quantities $n_i E_i^{\ominus}$ are extensive and proportional to the standard free-energy changes, $\Delta G_i^{\ominus}$. Values of $E_i^{\ominus}$ are additive only when the numbers of electrons in the different half-reactions are the same. When we combine half-reactions in such a way as to eliminate electrons and give a balanced overall redox reaction, we necessarily first multiply through by coefficients so chosen that the number of electrons *is* the same in each half-reaction, which is why $E^{\ominus}$ values for half-cells can be directly combined to get the overall $E^{\ominus}$ for the corresponding cell.

19-2   Electrolysis

Electrolytic methods are used in the manufacture of many of the more reactive elements, as discussed in Secs. 22-1 and 23-1. They are also used in the refining of some impure metals (notably copper), in the synthesis of various organic and inorganic materials, and in electroplating, which has extensive applications in precise analysis and in the creation of durable and handsome finishes on all kinds of objects. We illustrate some of the general principles of electrolytic methods with discussions of the electrolytic refining of copper and the electrolysis of water.

Electrolytic refining of copper.   This is a particularly illustrative application of electrochemistry (Fig. 19-1). The impure copper, formed into slabs directly from the crude metal as it is obtained from its ores, is made the anode in an electrolytic cell and a sheet of pure copper is used as the cathode. The electrolyte is a dilute solution of a salt of Cu^{2+}. As electricity passes through the cell, copper is oxidized to Cu^{2+} at the impure copper anode and Cu^{2+} is reduced to metallic copper at the pure copper cathode, which is usually coated with a thin layer of graphite so that the fresh copper deposit can be easily removed. Metals more noble than copper—that is, metals that are harder to oxidize, such as silver, gold, and platinum—will not be oxidized to their cations in the presence of Cu. If a cation of any such noble metal were formed, or were introduced, it would immediately be reduced again by the copper, with formation of Cu^{2+} and the corresponding metal. Hence such metals fall to the bottom of the cell as the anode dissolves and they can be recovered from the sludge that accumulates there. Metals that are more easily oxidized than copper do indeed dissolve as the anode dissolves, but they are not plated out at the cathode. No metal more easily oxidized than copper can be plated from a solution containing a significant concentration of Cu^{2+}; if it were formed, it would at once reduce Cu^{2+}

FIGURE 19-1   **Electrolytic refining of copper.**

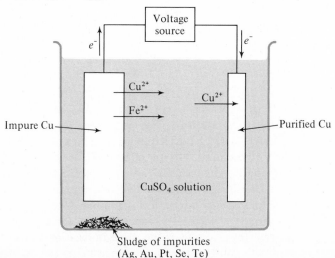

CuSO$_4$ solution

Impure Cu

Purified Cu

Sludge of impurities
(Ag, Au, Pt, Se, Te)

to Cu and itself be oxidized to its cation. Electrolytically refined copper is typically about 99.95 percent pure.

Electrolysis of water. Suppose we wish to electrolyze distilled water that is either at pH 7 or has been acidified by the addition of $HClO_4$ or made alkaline by the addition of NaOH. At the cathode H(+I) is reduced to H(0) and liberated as $H_2(g)$; at the anode O(−II) is oxidized to O(0) and liberated as $O_2(g)$. The cathode reaction may be written in either of two ways, shown below to be equivalent:

$$2H^+ + 2e^- \longrightarrow H_2(g) \qquad \mathbf{E}_{14}^{\ominus} = 0.000 \text{ V} \qquad (19\text{-}14)$$
$$2H_2O + 2e^- \longrightarrow H_2(g) + 2OH^- \qquad \mathbf{E}_{15}^{\ominus} = -0.829 \text{ V} \qquad (19\text{-}15)$$

Each of these electrode reactions involves the reduction of H(I) to H(0). They differ only in standard states—(19-14) is written for an acidic solution, $[H^+] = 1$ *M*, while (19-15) is written for a basic solution, $[OH^-] = 1$ *M*. It is customary to use (19-14) for electrolysis of acidic solutions and the equivalent (19-15) for neutral and alkaline solutions. It must be remembered, however, that any half-reaction equation describes only the overall change at the electrode and implies nothing about the particular species that may be involved, perhaps transiently, in the actual electrode reaction.

To show the equivalence of (19-14) and (19-15) we note that, for specific concentrations of all species involved, an actual electrode must have a unique potential, no matter which set of pertinent reactants and products are chosen to describe the electrode reaction. The Nernst equations associated with the two half-reactions are, at 25°C,

$$\mathbf{E} = 0.000 - \frac{0.0592}{2} \log \frac{P_{H_2}}{[H^+]^2} \qquad (19\text{-}16)$$

for (19-14) and

$$\mathbf{E} = -0.829 - \frac{0.0592}{2} \log (P_{H_2}[OH^-]^2) \qquad (19\text{-}17)$$

for (19-15). We set the two right sides equal to each other:

$$-\frac{0.0592}{2} \log P_{H_2} + \frac{0.0592}{2} \log [H^+]^2$$
$$= -0.829 - \frac{0.0592}{2} \log P_{H_2} - \frac{0.0592}{2} \log [OH^-]^2 \qquad (19\text{-}18)$$

The terms in P_{H_2} cancel, and the exponents of $[H^+]$ and $[OH^-]$ cancel the $\frac{1}{2}$ in the factors $\frac{0.0592}{2}$. Rearrangement and combination of the logarithmic terms then yield

$$\log ([H^+][OH^-]) = -\frac{0.829}{0.0592} = -14.00 \qquad (19\text{-}19)$$

This relation is identical with the relation $[H^+][OH^-] = 1.00 \times 10^{-14} = K_w$, the ion product of water. Thus the fact that (19-14) and (19-15) are equivalent to one another implies the relation $[H^+][OH^-] = K_w$. At a temperature other than 25°C, $\mathbf{E}_{14}^{\ominus}$ is still 0.000 V, by definition, while the value of $\mathbf{E}_{15}^{\ominus}$ is such that equivalence of the Nernst equations for (19-14) and (19-15) is in accordance with the value of K_w for that temperature.

There are also two equivalent ways of writing the electrode reaction at the anode. We write these two equations as reductions:

$$O_2(g) + 4H^+ + 4e^- \longrightarrow 2H_2O \qquad \mathbf{E}^\ominus = 1.229 \text{ V} \qquad (19\text{-}20)$$
$$O_2(g) + 2H_2O + 4e^- \longrightarrow 4OH^- \qquad \mathbf{E}^\ominus = 0.400 \text{ V} \qquad (19\text{-}21)$$

As oxidations, these reactions in fact proceed from the right to the left. Note that the standard emfs for these two reactions differ by 0.829 V, just as do (19-14) and (19-15); the reason is the same.

Arbitrarily choosing (19-14) and (19-20) to describe the cathode and anode reactions, we find for the electrode potentials of the two electrodes, at 25°C,

$$\mathbf{E}_{\text{cathode}} = \left(0.000 - \frac{0.0592}{2}\log \frac{P_{H_2}}{[H^+]^2}\right) \text{volts}$$

$$= (0.000 - \frac{0.0592}{2}\log P_{H_2} + 0.0592 \log [H^+]) \text{ volts} \qquad (19\text{-}22)$$

and

$$\mathbf{E}_{\text{anode}} = \left(1.229 - \frac{0.0592}{4}\log \frac{1}{P_{O_2}[H^+]^4}\right) \text{volts}$$

$$= (1.229 + \frac{0.0592}{4}\log P_{O_2} + 0.0592 \log [H^+]) \text{ volts} \qquad (19\text{-}23)$$

The two electrode potentials depend on the pH, as shown by the terms in $\log [H^+]$. For pressures of 1 atm for both H_2 and O_2, this pH dependence is given by the equations

$$\mathbf{E}_{\text{cathode}} = (-0.0592 \times \text{pH}) \text{ volts} \qquad (19\text{-}24)$$

$$\mathbf{E}_{\text{anode}} = (1.229 - 0.0592 \times \text{pH}) \text{ volts} \qquad (19\text{-}25)$$

The cell potential is the difference between (19-22) and (19-23):

$$\mathbf{E}_{\text{cell}} = \mathbf{E}_{\text{cathode}} - \mathbf{E}_{\text{anode}}$$

$$= [-1.229 - \frac{0.0592}{4}\log (P_{H_2}{}^2 P_{O_2})] \text{ volts} \qquad (19\text{-}26)$$

or at pressures of 1 atm,

$$\mathbf{E}_{\text{cell}} = \mathbf{E}_{\text{cathode}} - \mathbf{E}_{\text{anode}} = -1.229 \text{ V} \qquad (19\text{-}27)$$

It is independent of the pH, as would be expected from the overall electrolysis reaction $[2 \times (19\text{-}14) - (19\text{-}20)]$,

$$2H_2O \longrightarrow 2H_2(g) + O_2(g) \qquad (19\text{-}28)$$

The minimum voltage needed to decompose water by electrolysis, the *equilibrium decomposition potential*, is $-\mathbf{E}_{\text{cell}}$ or $+1.229$ V. This can be understood by considering that the electrolysis produces H_2 at 1 atm at one electrode and O_2 at 1 atm at the other. These elements tend to combine spontaneously by the reverse of (19-28) and would produce a voltage of 1.229 V; hence at least this opposing voltage is needed to prevent the combination of H_2 and O_2 and drive the reaction in the nonspontaneous direction.

For the actual operation of the electrolytic cell a larger external voltage is needed. One reason is that the resistance of the cell must be overcome. This requires an excess voltage $\mathbf{IR}$ for a current $\mathbf{I}$ and a cell resistance $\mathbf{R}$. One speaks of the $\mathbf{IR}$

drop across the cell. Additional voltage may be required because of the irreversible nature of electrode processes, as discussed in the next two subsections.

Suppose the solution being electrolyzed has been made alkaline with NaOH. Is there a possibility that Na^+ might be reduced at the cathode? The answer is no, because any metallic Na plated out would immediately react with water to form $H_2(g)$ and Na^+. For similar reasons, in a solution made acidic with $HClO_4$, the ClO_4^- is not affected in the electrolysis of dilute solutions and the product formed at the anode is O_2, as in the electrolysis of most dilute aqueous solutions.

Concentration polarization. Part of the voltage needed, in addition to the equilibrium decomposition potential and the **IR** drop, is attributable to concentration changes in the neighborhood of the electrodes. Thus, during the electrolysis of water, the volume around the cathode becomes partially depleted of H^+ ions and the pH rises. Near the anode, the hydrogen-ion concentration increases because (19-20) is run from the right to the left, and the pH falls. The result is that, by (19-24), $E_{cathode}$ becomes more negative and, by (19-25), E_{anode} more positive, so that $|E_{cell}|$ becomes larger than 1.229 V. This effect is called *concentration polarization* and is a common electrochemical phenomenon. It can be eliminated or at least minimized in electrolysis by vigorous stirring of the electrolyte.

Overvoltage. Another effect may increase the voltage required for electrolysis. Even in a stirred solution, hydrogen is evolved at a more negative emf than the equilibrium value given by (19-24). The difference between the two values is called the cathode overvoltage. Similarly, oxygen is evolved at a more positive emf than is indicated by (19-25) because of what is called anode overvoltage. Overvoltages make electrolysis more difficult by increasing the voltage needed above that required at equilibrium. They arise from rate effects, such as the slowness of electron transfer at electrodes.

Overvoltages for the deposition of metals are generally small (of the order of a few millivolts). They are often large for the evolution of gases, however, varying with the nature of the gas and the electrode material and increasing with increasing current density at the electrode. For many gases overvoltages are smallest on platinum electrodes, but even on this material oxygen shows an overvoltage of about 0.5 V. On many other metallic electrodes, overvoltages for gases range up to more than 1 V at current densities as low as 0.01 A cm^{-2}.

Selection of electrode reactions. Suppose a solution contains a number of different cations and anions, each of which can undergo one or more electrode reactions. What happens when the solution is electrolyzed? At the *cathode,* where reduction occurs, the most likely half-reaction is the one with the strongest tendency for reduction to take place, as evidenced by the *most positive* (or *least negative*) *value of* **E** at prevailing concentrations, gas pressures, electrode materials, and other relevant conditions. Overvoltage must be included in calculations of the relative values of different possible electrode potentials. The most likely *anode* reaction is that with the largest tendency for oxidation to occur and thus that with the *least positive* (or *most negative*) *value of* **E**, again with overvoltage taken into account.

The reactions chosen according to these principles correspond to the pair of possible electrode reactions that requires application of the smallest external voltage.

19-3 Corrosion

Corrosion is the deterioration of a substance as a result of reactions involving the environment. The term is applied almost exclusively to the deterioration of metals, on which the corrosive attack is invariably an oxidation. Electrochemical phenomena, including overvoltage, usually play a major role. Corrosion is manifested by such familiar and seemingly inevitable phenomena as the rusting of iron, the tarnishing of silver, or the development of a green patina on copper, brass, or bronze. The damage it causes amounts to billions of dollars each year—and would be far greater if strenuous preventive efforts were not made.

The complexities of corrosion phenomena are enormous, in part because the reactions are invariably heterogeneous, with cracks, defects, and impurities in the metal often playing a key role. Our discussion will be limited to a consideration of some of the factors that affect rates of corrosion, an examination of some typical reactions (those occurring as iron rusts), and mention of some common methods used to prevent or minimize corrosion.

Many factors accelerate corrosion. The presence of both oxygen and water appears to be necessary for most corrosion processes; both these substances are generally abundant on the earth's surface, except in extremely arid regions, and indeed corrosion is usually far slower in deserts than in more normal climes. It is markedly accelerated near bodies of salt water or in other places where salts can contaminate metal surfaces (e.g., on automobiles driven over roads on which salts have been used to melt ice or prevent dust from blowing). Acids also enhance rates of corrosion, and they are more ubiquitous in the atmosphere than is commonly realized. Carbon dioxide, which dissolves to form the weak acid H_2CO_3, is always present; other volatile acidic oxides, such as NO_2 and SO_2, are also present, especially near industrial areas. Elevated temperatures speed up corrosion, just as they do most other reactions. It is no surprise that corrosion is extremely rapid in tropical jungles, where moisture, elevated temperatures, and acidic products from decaying vegetation act in concert.

Many of the basic features of the process of rust formation apply to other corrosion processes as well. Rust is a hydrated form of iron(III) oxide, $Fe_2O_3 \cdot xH_2O$, with x approximately equal to 3. Iron does not rust in a dry atmosphere or in oxygen-free water. The reactions involved in rust formation are complex and not completely understood, but in the main the steps are thought to be the following. Metallic iron is first oxidized to the dipositive state,

$$Fe(s) \longrightarrow Fe^{2+} + 2e^- \qquad (19\text{-}29)$$

The resulting electrons may reduce H^+ to H_2, but this normally occurs only in an acidic medium and is not usually involved in rusting. Instead, the electrons reduce atmospheric oxygen in a reaction that is facilitated by the slightly acidifying presence of dissolved CO_2,

$$\tfrac{1}{2}O_2(g) + 2H^+ + 2e^- \longrightarrow H_2O \tag{19-30}$$

At the same time, atmospheric oxygen may also combine with the Fe^{2+} ions and H_2O to form rust,

$$\tfrac{1}{2}O_2(g) + 2Fe^{2+} + (2 + x)\,H_2O \longrightarrow Fe_2O_3 \cdot xH_2O + 4H^{+,} \tag{19-31}$$

This reaction is itself capable of yielding sufficient H^+ ions to keep reaction (19-30) going.

Reactions (19-29) and (19-30) need not occur at the same position. A galvanic cell, in which (19-29) provides the anode reaction and (19-30) that at the cathode, may be set up on the metal surface. The cells formed in this way may be on a microscopic scale, but anodic and cathodic sites may also be separated by macroscopic distances. The electric circuit must be completed by the migration of ions, which explains why the presence of salt water or other good electrolytes accelerates corrosion. Portions of the metal that are under mechanical stress often act as anodic corrosion sites, leading to stress-corrosion cracks that can very markedly weaken the metal. Reaction (19-31) may occur at still another site, because the Fe^{2+} ions and H^+ ions involved may migrate by diffusion. Pitting and rust formation may thus occur in different locations (Fig. 19-2).

FIGURE 19-2   **Electrochemistry of rusting.** In a common situation ferrous ions are formed at an anodic corrosion site that is protected from oxygen and then diffuse to a location accessible to the atmosphere, where rust is formed. The cathodic site may be at yet another location. Thus the shanks of bolts that fasten iron plates, even though protected from air, may be thinned by corrosion while rust forms on the neighboring plate surface and in other locations. While the corrosion of many reactive metals such as Al and Zn is minimized by a protective oxide layer that adheres tightly to the underlying metal, the formation of rust provides only minor and temporary protection to the underlying iron, because rust flakes off.

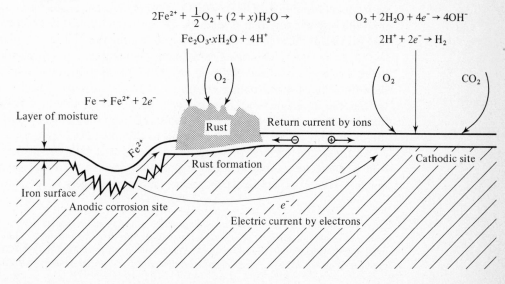

Inhibition of corrosion. Corrosion can be curbed in a number of ways. The most obvious method is to cover the metal surface with a coating impervious to air and moisture. Because corrosion may begin in even microscopic cracks, this coating must itself be completely intact. It might be a layer of paint, plastic, or some similar material, a coating of a strongly adherent oxide, or a layer of some metal that is itself resistant to corrosion.

An interesting phenomenon is *passivation,* in which a metal surface is made inactive by covering it with a very thin oxide layer. Iron can, for example, be passivated by treatment with concentrated nitric acid or dichromate. Once in this state it does not react with acid or reduce cupric ions. Passivation is easily destroyed, however, and if the surface is scratched anywhere, the protection breaks down—not only at the location of the scratch but in a steadily increasing area with a fast-moving boundary.

Some metals that are inherently reactive in the presence of water and oxygen, including aluminum, magnesium, titanium, and various stainless steels, become passive spontaneously. They acquire a tough and adherent oxide film in air, and this film protects them effectively from serious corrosion damage, except over long periods of time or in unusual atmospheres. Although normally the oxide film placed on iron by passivation offers little corrosion protection by itself, it can help considerably if it is covered appropriately. Paints that contain oxidizing substances such as "red lead" (Pb_3O_4) or potassium dichromate ($K_2Cr_2O_7$) are especially effective on iron because they make the surface passive. They are regularly used on structural steel exposed to severe weathering—for example, on ships and bridges.

Protective films of other metals, such as Zn or Sn, are also used on iron because these metals form self-protective oxide coats. Zinc plating (galvanizing) has the advantage that zinc is more easily oxidized than iron, $E^{\ominus}_{Zn^{2+}/Zn}$ being -0.76 V while $E^{\ominus}_{Fe^{2+}/Fe}$ is -0.44 V. If a scratch exposes the iron, it is still the zinc coating that is attacked. Since the product of that attack, zinc oxide, is somewhat basic, it reacts with H_2O and CO_2 in the air to form a basic zinc carbonate with typical formula $Zn(OH)_2 \cdot ZnCO_3$. This material has a sufficiently larger volume than the original zinc that it tends to fill the scratch, thereby covering the exposed iron (as well as the zinc) at least in part and healing the wound. Tin plating, although less expensive than galvanizing, is not nearly as reliable. Tin is more noble[1] than iron, being less easily oxidized ($E^{\ominus}_{Sn^{2+}/Sn} = -0.14$ V), so that iron corrodes more easily than tin. Thus if a tin coating on iron is damaged, it is the iron that corrodes first. Indeed, the iron sometimes then corrodes more rapidly than it would have without the tin coating, in part because a galvanic cell is set up, the iron being oxidized and the oxide film on the tin being reduced to the metal.

An important method of corrosion prevention is cathodic protection, in which iron is made to assume a negative potential relative to its surroundings so that reaction (19-29) is less likely to occur. This can be done by placing pieces of Zn

---

[1] Actually, under the anaerobic conditions *within* a tinned container, tin appears to be more easily oxidized than iron in the presence of mildly acidic materials such as grapefruit juice. This altered potential relationship, which is one reason for the usefulness of tinned cans, has been attributed in part to the presence of an alloy, $FeSn_2$, at the interface between the tin and the underlying steel.

or Mg as sacrificial electrodes near the object to be protected (pipeline, ship hull) and connecting them to it electrically by direct contact or by heavy wires. The Mg and Zn act as anodes and are consumed while the iron remains uncorroded. A carbon electrode with an external voltage source may also be used. Zinc plating is, in effect, a form of cathodic protection.

If the sacrificial Zn or Mg electrodes were replaced by electrodes of Cu or Sn, the iron would become the sacrificial electrode; instead of being protected, the iron would be made more vulnerable. This is what happens, for example, when iron pipes are connected to copper pipes or brass fittings, and it explains the resulting corrosion problem. Impurities of noble metals in the iron or a damaged coating of tin act in a similar manner.

Many alloys—compounds or mixtures of two or more metals—are much less readily corroded than the separated components. Stainless steel is the best-known example. A number of different alloys of iron go by this generic name; a typical one is 18-8 stainless, which contains 18 percent chromium and 8 percent nickel. Even this material corrodes fairly rapidly under some circumstances—for example, when ground to a fine, thin cutting edge, exposed to hot water and considerable stress, and then left wet in air (the normal state of a used razor blade). A steel with lower chromium content is now used for such blades, and research continues on ways to minimize its corrosion still further.

19-4   Other Applications of Electrochemistry

Disproportionation and the use of Latimer diagrams.   When a substance can exist in three or more oxidation states, there is always the possibility that a species representing an intermediate level of oxidation may be unstable. Part of it could be oxidized to a higher level by the remainder, which would thereby be reduced to a lower oxidation state. Such a reaction is called a *disproportionation*. It occurs, for example, with H_2O_2, which contains oxygen in the —I oxidation state and decomposes spontaneously to molecular oxygen (oxidation state 0) and water, containing O(—II). An understanding of the relationship between different oxidation states and the possible disproportionation reactions involving them is facilitated by a kind of diagram popularized by the American chemist W. M. Latimer, which shows the electrode potentials relating the different levels.

Consider, for example, the half-reactions and $\mathbf{E}^{\ominus}$ values relating the three oxidation states of oxygen mentioned above:

$$O_2 + 2H^+ + 2e^- \longrightarrow H_2O_2 \qquad \mathbf{E}^{\ominus} = 0.69 \text{ V} \tag{19-32}$$

$$H_2O_2 + 2H^+ + 2e^- \longrightarrow 2H_2O \qquad \mathbf{E}^{\ominus} = 1.77 \text{ V} \tag{19-33}$$

The Latimer diagram is a device for summarizing these data:

$$O_2 \xrightarrow{\;0.69\;} H_2O_2 \xrightarrow{\;1.77\;} H_2O \tag{19-34}$$

By convention, the species of highest oxidation state is shown on the far left.

How can the electrode potentials be used to show that H_2O_2 does indeed have a tendency to disproportionate into oxygen and water? To do this, we combine the half-cells corresponding to (19-32) and (19-33) into a cell, with reaction

$$2H_2O_2 \longrightarrow O_2 + 2H_2O \tag{19-35}$$

In combining the half-reactions to form (19-35), reaction (19-33) is the reduction (cathode) reaction and reaction (19-32) is the oxidation (anode) reaction; therefore the overall standard emf corresponding to (19-35) is

$$\mathbf{E}^{\ominus}_{cell} = \mathbf{E}^{\ominus}_{cathode} - \mathbf{E}^{\ominus}_{anode} = 1.77 - 0.69 = 1.08 \text{ V}$$

The fact that $\mathbf{E}^{\ominus}_{cell}$ is positive demonstrates that (19-35) is spontaneous in the indicated direction. The equilibrium constant is, in fact, very large, about $10^{2 \times 1.08/0.06} = 10^{36}$.

In general terms, we may denote the successive oxidation states as *A, B,* and *C, A* being the highest and *C* the lowest, so that the Latimer diagram is

$$A \xrightarrow{\mathbf{E}_{AB}} B \xrightarrow{\mathbf{E}_{BC}} C \tag{19-36}$$

Disproportionation of *B* into *A* and *C* occurs if $\mathbf{E}_{cell} = \mathbf{E}_{BC} - \mathbf{E}_{AB}$ is positive, that is, if $\mathbf{E}_{BC} > \mathbf{E}_{AB}$. Conversely, if $\mathbf{E}_{BC} < \mathbf{E}_{AB}$, no significant disproportionation can take place.

This result can be visualized and remembered by studying the $\mathbf{E}$ values shown in the Latimer diagram (19-36) and recalling that $\mathbf{E}$ is a measure of the tendency of a reaction to go to the right. We see that when $\mathbf{E}_{BC}$ is larger than $\mathbf{E}_{AB}$ the reaction leading from *B* to *C* is more strongly favored than that leading from *A* to *B*, so that the disproportionation would tend to occur. Conversely, if $\mathbf{E}_{BC} < \mathbf{E}_{AB}$ the reaction from *A* to *B* is more strongly favored than that from *B* to *C* and there can be no significant disproportionation.

EXAMPLE 19-5   **Disproportionation of Cu^+ and of Fe^{2+}.**   Consider the Latimer diagrams of the oxidation states of copper and iron:

$$Cu^{2+} \xrightarrow{0.153} Cu^+ \xrightarrow{0.521} Cu$$

$$Fe^{3+} \xrightarrow{0.771} Fe^{2+} \xrightarrow{-0.440} Fe$$

Predict whether Cu^+ disproportionates into Cu^{2+} and Cu and whether Fe^{2+} disproportionates into Fe^{3+} and Fe.

Solution.   For the copper system we have

$$\mathbf{E}_{cell} = 0.521 - 0.153 = 0.368 \text{ V}$$

and for the iron system

$$\mathbf{E}_{cell} = -0.440 - 0.771 = -1.211 \text{ V}$$

Reasoning along the lines just stated therefore leads to the conclusion that Cu^+ ion is expected to (and indeed does) disproportionate. Ferrous ion is stable, however, and cannot disproportionate significantly.

Ion-selective electrodes.   Many electrodes for which the emf is indicative of the concentrations of specific ions have been developed in recent years. They are of particular value in analytical applications—for example, in analysis of solutions containing small amounts of many different substances, such as clinical specimens or seawater samples. Such electrodes are also especially adaptable to systems for automatic control of chemical processes on a large scale.

Probably the oldest of such electrodes is the glass electrode (Fig. 19-3*a*), which is the sensing probe of a pH meter. This device is based on the experimental finding that an emf develops across a very thin glass membrane separating two solutions of different pH and that this emf depends on the difference of the pH values. For some purposes the emf developed across the glass membrane could be determined sufficiently precisely by inserting two wires, one into the bulb and the other into the solution outside, and measuring the difference in emf between them. However, the stability and reproducibility of the emf developed across the interface between the two solutions is far greater if the system contains instead an AgCl/Ag electrode within the bulb, the connection to it being by a liquid junction through a porous wall. For maximum precision, the second electrode in the pH-measuring system is usually a standard calomel reference electrode in which liquid mercury (connected with the outside by a Pt wire) is in contact with solid calomel (Hg_2Cl_2) and a solution

FIGURE 19-3   **Ion-selective electrodes.**   (*a*) Glass electrode; (*b*) lead-selective electrode.

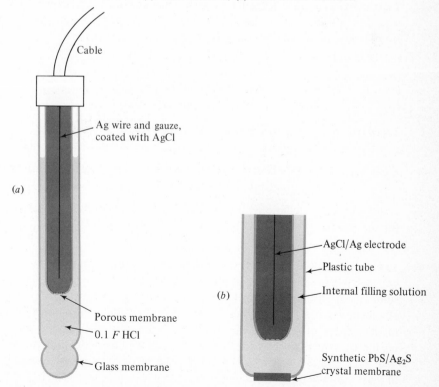

Cable

Ag wire and gauze,
coated with AgCl

(*a*)

AgCl/Ag electrode

Plastic tube

(*b*)

Internal filling solution

Porous membrane

0.1 *F* HCl

Glass membrane

Synthetic PbS/Ag$_2$S
crystal membrane

containing chloride ions. The half-reaction is

$$Hg_2Cl_2(s) + 2e^- \longrightarrow 2Hg(l) + 2Cl^- \tag{19-37}$$

Electrical contact between the calomel electrode and the solution whose pH is to be measured is established through porous membranes and a KCl solution. Because the AgCl/Ag and calomel electrodes develop constant and highly reproducible emf values, the total emf between the leads to the meter depends only on the pH of the solution.

Electrodes with selective sensitivity for more than 20 other ions have also been developed and more are in prospect. The ions include those derived from the alkali and alkaline-earth elements, as well as Ag^+, Pb^{2+}, NO_3^-, ClO_4^-, and S^{2-}, among others. The lead-selective electrode (Fig. 19-3b) is representative. It is sensitive at the parts-per-billion level. Potential applications include the detection of traces of Pb^{2+} in urine and other bodily fluids in the diagnosis of lead poisoning, as well as the analysis of atmospheric samples (after they have been dissolved and concentrated) in studies of air pollution.

Problems and Questions

19-1 **Equilibrium position of redox reactions.** Predict whether the following reactions would tend to occur to a significant extent in aqueous solution. Assume that the initial concentrations are about $1\ F$. Explain the basis of your predictions.

(a) $Cl_2 + H_2O_2 \longrightarrow O_2 + 2H^+ + 2Cl^-$
(b) $Sn + Cd^{2+} \longrightarrow Cd + Sn^{2+}$
(c) $2I^- + Sn^{4+} \longrightarrow I_2(s) + Sn^{2+}$
(d) $2Fe^{2+} + Br_2(l) \longrightarrow 2Fe^{3+} + 2Br^-$

19-2 **Equilibrium constants from electrode potentials.** Calculate the equilibrium constants for the reactions given in Prob. 19-1. [Assume that (c) takes place in $1\ F$ HCl.]

19-3 **Cell emf and equilibrium constant.** The measured emf of the cell

$$Pb(s)|Pb^{2+}(10^{-2}\ M)||VO^{2+}(10^{-1}\ M),\ V^{3+}(10^{-5}\ M),\ H^+(10^{-1}\ M)|Pt(s)$$

is $+0.67$ V. (a) Calculate $E^{\ominus}$ for the VO^{2+}/V^{3+} couple and (b) calculate the equilibrium constant for the reaction $Pb(s) + 2VO^{2+} + 4H^+ \rightleftharpoons Pb^{2+} + 2V^{3+} + 2H_2O$.

Ans. (a) 0.37 V;   (b) $5 \times 10^{16}\ M^{-3}$

19-4 **Predicting reactions from redox couples.** An imaginative manufacturer, who had seen what happened when a zinc rod was placed in a copper solution, decided to sell $1\ M$ solutions of chromic nitrate, $Cr(NO_3)_3$, for use in "instant chromium plating" of iron objects. His idea was that the iron need only be dipped for a short time in the solution. He shipped his first batch of solutions in aluminum containers; the process did not work. He blamed it on the containers and switched to glass. (a) Was he right that the Al containers might have had some effect? Explain. (b) What are the prospects for his success if his solution is shipped in glass containers?

19-5 **Predicting reactions from redox couples.** (*a*) An excess of powdered nickel is added to a 2 *M* solution of silver nitrate. Write an equation for any reaction that will occur, and calculate the equilibrium concentrations of the metal ions present. (*b*) Give the standard free-energy change for the reaction in (*a*).

Ans. (*a*) $[Ag^+] = 4 \times 10^{-18}$ *M*, $[Ni^{2+}] = 1$ *M*; (*b*) -199 kJ per mole of nickel

19-6 **Complex formation constant.** Use the following data

$$Al^{3+} + 3e^- \rightleftharpoons Al(s) \qquad\qquad E^\circ = -1.66 \text{ V}$$
$$Al(OH)_4^- + 3e^- \rightleftharpoons Al(s) + 4OH^- \qquad E^\circ = -2.35 \text{ V}$$

to calculate the equilibrium constant for the reaction $Al^{3+} + 4OH^- \rightleftharpoons Al(OH)_4^-$. (*Hint:* Refer to Example 19-3.)

19-7 **Solubility-product constant.** The solubility product of silver iodide is frequently given as 1×10^{-16}. Calculate an approximate value for this quantity from the following data, and comment on any difference from the value given.

$$Ag^+ + e^- \rightleftharpoons Ag \qquad\qquad E^\circ = +0.80 \text{ V}$$
$$AgI(s) + e^- \rightleftharpoons Ag + I^- \qquad E^\circ = -0.15 \text{ V}$$

Ans. 9×10^{-17}

19-8 **Free energy from cell voltage.** The cell $Ag(s)$, $AgCl(s)|HCl(0.1 \ M)|Cl_2(1 \text{ atm})$, $Pt(s)$ is galvanic and has the voltage 1.137 V. (*a*) What is the cell reaction? (*b*) What are the values of ΔG and ΔG° for this reaction?

19-9 **Cell reaction, cell emf, and free-energy change.** This problem illustrates that the cell emf depends on the number of electrons interchanged as well as on the overall cell reaction, while ΔG depends on the latter alone. Consider three cells, in which all ionic species indicated are at standard concentrations: (*a*) $Tl|Tl^+||Tl^{3+}$, $Tl^+|Pt$; (*b*) $Tl|Tl^{3+}||Tl^{3+}$, $Tl^+|Pt$; (*c*) $Tl|Tl^+||Tl^{3+}|Tl$. For each cell, find the cell reaction, the number of electrons involved, the cell emf, and ΔG°. The standard electrode potentials for the three pertinent half-cells are as follows (in volts): Tl^+/Tl, -0.34; Tl^{3+}/Tl, 0.74; Tl^{3+}/Tl^+, 1.28.

Ans. in all three cases $2Tl + Tl^{3+} \longrightarrow 3Tl^+$, $\Delta G^\circ = -313$ kJ.
For (*a*), (*b*), (*c*) in sequence, **n** = 2, 6, 3; **E**$^\circ$ = 1.62 V, 0.54 V, 1.08 V

19-10 **Free energy of formation of ions.** From the standard electrode potentials given below, calculate the standard free energy of formation of the ions Cu^{2+}, Fe^{2+}, and Fe^{3+}, all at 25°C:

$$Cu^{2+}/Cu, \ 0.337 \text{ V} \qquad Fe^{2+}/Fe, \ -0.440 \text{ V} \qquad Fe^{3+}/Fe^{2+}, \ 0.771 \text{ V}$$

19-11 **Standard electrode potential from nonelectrical measurements.** (*a*) Find the standard free energy of formation of $Mg(OH)_2$ at 25°C from that of $H_2O(l)$ and the following free energies of reaction at 25°C and under standard conditions: $Mg(s) + \frac{1}{2}O_2(g) \longrightarrow MgO(s)$, -569.6 kJ and $MgO(s) + H_2O(l) \longrightarrow Mg(OH)_2(s)$, -27.0 kJ. (*b*) Calculate the standard free energy of $Mg^{2+}(aq)$ from that of $OH^-(aq)$ and from K_{sp} for $Mg(OH)_2(s)$, 1.8×10^{-11}. (*c*) Calculate **E**$^\circ$ for the half-cell reaction $Mg^{2+} + 2e^- \rightleftharpoons Mg(s)$.

Ans. (*a*) $Mg(OH)_2(s)$, -833.8 kJ; (*b*) $Mg^{2+}(aq)$, -457.9 kJ mol^{-1}; (*c*) -2.37 V

19-12 **Behavior of electrochemical cell.** An electrochemical cell consists of a silver electrode dipping into 0.10 *F* $AgNO_3$ and a nickel electrode dipping into 0.10 *F* $Ni(NO_3)_2$:

$$Ni^{2+} + 2e^- \longrightarrow Ni(s) \qquad E^\circ = -0.23 \text{ V}$$
$$Ag^+ + e^- \longrightarrow Ag(s) \qquad E^\circ = +0.80 \text{ V}$$

(*a*) Calculate the initial voltage of this cell. (*b*) Write the overall cell reaction that occurs as this cell generates electric current, indicate the reaction that occurs at the cathode and that at the anode, and state in which direction electrons flow in the external circuit (from Ag to Ni or from Ni to Ag). Explain. (*c*) If a steady current of 0.50 A is supplied by this cell for a period of 3.0 hours, what weight of nickel will be *deposited* or *dissolved*? (*d*) Calculate the equilibrium constant for the overall reaction given in (*b*). (*e*) If the volumes of Ni^{2+} and Ag^+ solutions are equal and the reaction is allowed to proceed until equilibrium is reached, what will be the approximate final concentrations of Ni^{2+} and Ag^+?

19-13 **Combination of electrode potentials.** Calculate $E^{\ominus}$ for the half-reaction

$$Co^{3+} + 3e^- \rightleftharpoons Co(s)$$

from the following standard half-reaction potentials:

$$Co^{2+} + 2e^- \rightleftharpoons Co(s) \qquad E^{\ominus} = -0.28 \text{ V}$$
$$Co^{3+} + e^- \rightleftharpoons Co^{2+} \qquad E^{\ominus} = +1.82 \text{ V}$$

Ans. 0.42 V

19-14 **Combination of electrode potentials.** Calculate $E^{\ominus}$ for the half-cell $MnO_4^- + 8H^+ + 7e^- \rightleftharpoons Mn(s) + 4H_2O$ from the standard electrode potentials for two half-cells:

$$Mn^{2+} + 2e^- \rightleftharpoons Mn(s) \qquad E^{\ominus} = -1.190 \text{ V}$$
$$MnO_4^- + 8H^+ + 5e^- \rightleftharpoons Mn^{2+} + 4H_2O \qquad E^{\ominus} = 1.51 \text{ V}$$

(Note that electrode potentials are not necessarily accessible to direct measurement.)

19-15 **Cleaning of silverware.** In a procedure for removing Ag_2S tarnish from silverware the silver is placed in a warm Na_2CO_3 solution in an aluminum pan. Use the data given in Appendix D and below to explain this procedure. $E^{\ominus}_{Ag_2S/Ag} = -0.71 \text{ V}$; $E^{\ominus}_{Ag^+/Ag} = 0.80 \text{ V}$.

Ans. $Ag_2S \longrightarrow Ag + S^{2-}$; $Al \longrightarrow Al^{3+}$

19-16 **Refining of copper.** In the electrolytic method of refining copper, the impure copper is made the anode of an electrolytic cell and the purified metal deposits on the cathode. This works because some impurities in the anode never dissolve while others never get plated out again. Which among the following metals belong in the first category and may be recovered from the sludge that accumulates at the bottom of the cell? Which belong in the second category? Possible impurities: Pt, Pb, Fe, Ag, Zn, Au, Ni.

19-17 **Effects of concentration polarization.** The running of a galvanic cell represented by the symbols $Zn(s)|ZnSO_4(1 \text{ } F)||CuSO_4(1 \text{ } F)|Cu(s)$ causes the depletion of Cu^{2+} ions in the surroundings of the cathode to a concentration of $1.0 \times 10^{-4} \text{ } M$. It also causes an increase of $[Zn^{2+}]$ to 1.50 M near the anode (and the appropriate changes in the SO_4^{2-} concentration to maintain an electrically neutral solution). (*a*) What is the original voltage of the cell exclusive of **IR** drops? (*b*) What is the voltage after the polarization has taken place?

Ans. (*a*) 1.100 V; (*b*) 0.976 V

19-18 **pH dependence of electrode potentials.** The pH dependence of the electrode potentials for the reactions (19-14) and (19-20) is exactly the same. Why is this necessarily so?

19-19 **Latimer diagram.** The Latimer diagram of the species MnO_4^-, MnO_4^{2-}, and $MnO_2(s)$ at $[H^+] = 1.00 \text{ } M$ is

$$MnO_4^- \xrightarrow{\;0.56\;} MnO_4^{2-} \xrightarrow{\;2.26\;} MnO_2(s)$$

(a) Does a solution of MnO_4^{2-} tend to disproportionate into MnO_4^- and $MnO_2(s)$ (at least if all soluble species involved are $1\ M$)?

$$3MnO_4^{2-} + 4H^+ \longrightarrow 2MnO_4^- + MnO_2(s) + 2H_2O$$

(b) What is $E^\ominus_{MnO_4^-/MnO_2(s)}$?   *Ans.* (a) yes;   (b) 1.69 V

19-20 Latimer diagram. The Latimer diagram below pertains to the oxidation states IV, V, and VI of uranium in acid solution. Is the oxidation state V stable?

$$UO_2^{2+} \xrightarrow{\;0.05\;} UO_2^+ \xrightarrow{\;0.62\;} U^{4+}$$

"The rate of chemical reactions is a very complicated subject. This statement is to be interpreted as a challenge to enthusiastic and vigorous chemists; it is not to be interpreted as a sign of defeat."

H. S. Johnston,[1] 1966

20 Chemical Kinetics: The Rates and Mechanisms of Chemical Reactions

20-1 Introduction

The conceptual fabric of modern chemistry is woven principally from the threads of structure and reactivity. Each of these threads is itself composed of many strands. Structure, in the broad sense, implies not only the geometric arrangements of atoms in molecules and in larger aggregates but also the electron configurations responsible for bonding, and the various associated energy levels as well.

What is meant by chemical reactivity? The term is used in two fundamentally distinct senses: to refer to the rapidity with which a mixture of substances tends to react under a given set of circumstances (the kinetic criterion of reactivity) and to refer to the relative proportions of reactants and products present when equilibrium is achieved for a given reaction mixture (the thermodynamic criterion). Our

[1]H. S. Johnston, "Gas Phase Reaction Rate Theory", The Ronald Press Company, New York, 1966.

present concern is with the first of these, with what is called *reaction kinetics*—that is, measures of the rates at which chemical reactions proceed and especially the factors that affect these rates and their interpretation in terms of specific events and structures at the molecular level. The effort to correlate structure and reactivity is a major goal of chemistry and is thus a recurring theme in certain later chapters.

Relative to the mature science of thermodynamics and the firmly grounded, if still maturing, field of structural chemistry, kinetics is in an early stage of development. Most of our knowledge of kinetics is empirical, that is, derivable only from experimentation rather than from some fundamental and well-tested theory, for the theoretical foundations of kinetics are still relatively rudimentary. The first two sections of this chapter constitute an introduction to some of the terminology and fundamental concepts of kinetics and to the two general measures of rates of reaction, the differential (or instantaneous) rate expression and the corresponding integrated rate expression. These sections are followed by a discussion of homogeneous reactions in the gas phase, which illustrates some of the basic concepts of kinetics, and an introduction to studies of reactions in liquid solutions, with emphasis primarily on the role of the solvent and thus on the differences from reactions in gaseous systems. The chapter concludes with a brief discussion of catalysis.

20-2 Some Terminology and Basic Concepts of Kinetics

Kinetics, like other fields of science, has its own specific terminology, the implications of which must be recognized by any student of the subject. The phrases and ideas introduced here are fundamental to modern chemical kinetics. They can be appreciated only in the context of examples, such as those in the sections that follow.

Mechanisms and elementary processes. The goal of chemical kinetics is to obtain an exact description of the sequence of processes that occurs on the molecular level as reacting species are converted into products. This microscopic reaction sequence is called the *mechanism* of the reaction. Each of the individual encounters between atomic or molecular species in the mechanistic sequence is termed an *elementary process,* a phrase implying that the chemical equation representing this particular step in the overall reaction includes the exact atomic or molecular species reacting with one another at this stage. The sum of the equations that represent the successive elementary processes making up the entire reaction is necessarily the chemical equation for the overall reaction. However, the sequence of elementary processes involved in any particular reaction can *never* be deduced from the overall stoichiometric equation. In other words, there is *no necessary relation between the stoichiometric equation for a reaction and the actual mechanism of that reaction,* other than that the sum of the equations representing the individual steps in the mechanism must be the overall equation.

To illustrate, the gas-phase reaction of molecular hydrogen with molecular bromine to form hydrogen bromide is represented by the stoichiometric equation

$$H_2(g) + Br_2(g) \longrightarrow 2HBr(g) \tag{20-1}$$

A naive interpretation of (20-1) as an equation representing an elementary process would imply that this reaction occurred by the direct interaction of one hydrogen molecule with one bromine molecule to form, in a single step, two molecules of hydrogen bromide. Experimental studies have shown, however, that in fact this direct reaction does not occur to any significant degree. Rather, as we discuss later, H_2 and Br_2 react by a series of elementary processes involving first the dissociation of a bromine molecule by collision with some other molecule, subsequent reaction of a resulting bromine atom with a hydrogen molecule to produce hydrogen bromide and a hydrogen atom, and a number of further steps. The important point is that the complex mechanism by which this reaction proceeds could not be deduced from the stoichiometric equation. Indeed, it cannot even be deduced from the most advanced theories currently available, but only from the results of carefully planned experiments.

Rate-determining step.   When a reaction occurs in a series of successive steps, with some products of one step participating as reactants in a subsequent step, it often happens that one step is much slower than any of the others. The overall rate of the reaction is then determined by the rate of this slowest step, which is said to be *rate-determining*.

A reaction that occurs in a series of successive steps may be compared with an assembly line. In a typical automobile assembly line, for example, parts are added or modified at successive stages. Each step is necessary before the next one can be performed and thus each step is essential to the completion of the car. Nothing can come off the line at a rate greater than that of the slowest step in the line. Suppose there is a bottleneck at the stage at which the body of the car is painted, either because this step is intrinsically slower or because of a breakdown of equipment. The rates of all subsequent steps (and thus the rate at which cars come off the line) are limited by the rate at which the cars can be painted.

The synthesis or alteration of a molecule by a series of successive reactions is in many ways analogous to a manufacturing assembly line, but there is one important difference. As we shall see shortly, the rates of chemical processes usually depend on the concentrations of some of the reactants and consequently the rates vary as these concentrations change during the course of a reaction. Furthermore the relative rates of different steps can also change with other changes in reaction conditions, e.g., in the temperature, the pressure, or the solvent. Thus a step that is rate-determining under one set of conditions may not be under others.

Nature of the barrier to reaction.   The range of speeds observed for chemical reactions is enormous. Some reactions are immeasurably slow; others are essentially complete in times as short as 10^{-10} s. Perhaps the most fundamental question that chemical kineticists attempt to answer is, "What is the limiting factor in the rate of reaction?", or put in another way, "What is the barrier to reaction?" If, for example, every collision between reactants leads to reaction, the rate is limited by the rate at which collisions between reactants occur. With such reactions, which are not very common, the barrier is simply the rate at which the reactants diffuse together; in the gaseous or liquid state the reaction is normally a very fast one on the ordinary human time scale.

More commonly, the rate of reaction is limited by the fact that only a very small fraction of the reacting molecules have enough energy, even when they do collide with one another, to bring about the necessary electronic and atomic rearrangement that characterizes a chemical change. In other words, a certain minimum energy, or "threshold energy", may be needed for a colliding pair to react. The barrier is then an energetic one.

With some reactions, especially those of large polyatomic molecules, it frequently happens that the potential reactants are not properly oriented relative to one another when they do come together, even though their mutual energy may exceed the threshold. Because of this inappropriate mutual orientation, atoms that would be linked together if a product molecule were formed never come sufficiently close for the reaction to take place. In this case the barrier to reaction is said to be configurational or orientational.

20-3 The Rates of Chemical Reactions

The speed of a chemical reaction may be defined as the rate at which one of the reactants disappears or, alternatively, the rate at which one of the products is formed. Normally the rate is expressed in terms of the change with time in the concentration of a reactant or product. Consider the hypothetical reaction

$$A + 3B \longrightarrow 2D \tag{20-2}$$

Its rate could be determined by studying the concentration of one of the species involved as a function of the time. Figure 20-1 represents a possible variation of the concentration of A with time. If the concentration of A at time t is $[A]_t$ and it has become $[A]_t + \Delta[A]$ at $t + \Delta t$, the rate is given by (see Fig. 20-1)

$$\text{Rate} = -\frac{[A]_t + \Delta[A] - [A]_t}{t + \Delta t - t} = -\frac{\Delta[A]}{\Delta t} \tag{20-3}$$

From the shape of the curve in Fig. 20-1 it is clear that the rate of the reaction changes with time. There is increasingly less change in [A] per unit time as the time increases—the rate decreases continually.

Because the reaction rate usually changes with time, the definition given in (20-3) is not adequate; the value of the rate at a given time depends on the size of the time increment Δt. A more precise definition of the rate is obtained by taking the limit of (20-3) as Δt approaches zero:

$$\text{Rate} = \lim_{\Delta t \to 0} -\frac{[A]_t + \Delta[A] - [A]_t}{t + \Delta t - t} = -\frac{d[A]}{dt} \tag{20-4}$$

which, in calculus, is the negative of the derivative of [A] with respect to time, the negative of the slope at a given point of a graph of [A] plotted against t. The rate

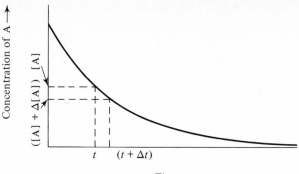

FIGURE 20-1   **The change in concentration of a reactant A with time** t. The concentration of the species A participating in a chemical reaction decreases with time. For this reaction the concentration of A does not vary linearly with time; the rate of the reaction changes as time passes. Consequently the rate at any time t is defined as the limit approached by the ratio $-\Delta[A]/\Delta t$ as Δt becomes smaller and smaller. (The minus sign is used because it is customary to define rates as positive quantities and $\Delta[A]$ is negative if A is a reactant.) In other words, the rate of reaction at any time t is the negative of the slope of the curve at that instant, equal to $-d[A]/dt$.

can be expressed in terms of the change in concentration of any reactant or product. To avoid ambiguity when the numbers of molecules of the various reactants and products differ, it is customary to take the stoichiometric coefficients into account. Thus for reaction (20-2), the rate $-d[A]/dt$ can be expressed equally well as $-\frac{1}{3}d[B]/dt$ or $+\frac{1}{2}d[D]/dt$. These alternative expressions take into account the fact that, by (20-2), three molecules of B disappear for every molecule of A reacting and two molecules of D are formed for each A that is removed. Hence,

$$\text{Rate of reaction} = -\frac{d[A]}{dt} = -\frac{1}{3}\frac{d[B]}{dt} = \frac{1}{2}\frac{d[D]}{dt} \tag{20-5}$$

The units of the rate are usually $M\,\text{s}^{-1} = \text{mol liter}^{-1}\,\text{s}^{-1}$, since concentrations are usually expressed as molarities and time as seconds. Occasionally other units are used. The quantity of reactant or product may be expressed in units that are proportional to concentration and can be conveniently measured, e.g., the pressure of a gas, units of electrical conductivity, or some optical property.

The rate expressions (20-5) imply that the concentrations of A, B, and D change only because of the reaction considered, that is, reaction (20-2) occurring in the indicated direction. This is not always true, since other reactions involving these species may occur simultaneously. Frequently there is a back reaction in which the products (here D molecules) react to form the reactants again, the reverse of (20-2). In such a situation, we define the net rate as the difference between the rates of the forward and reverse reactions,

$$\text{Net rate} = \text{forward rate} - \text{reverse rate} \tag{20-6}$$

At the very beginning of a reaction, when no products are yet present in any appreciable concentration, the reverse rate is negligible so that the net rate is equal to the forward rate. As the forward reaction proceeds, the products accumulate and, unless the position of equilibrium lies very far toward the product side, the rate of the reverse reaction may need to be taken into account. At equilibrium, the rates of the forward and reverse reactions are equal and the net rate is zero.

Sometimes the products of one reaction are themselves depleted by further reaction to form other substances. If this were true of reaction (20-2), the rate of accumulation of D could not be used as a measure of the rate of reaction. For the present, we shall ignore such complications and assume, except where specifically noted otherwise, either that the only significant rate is that of the reaction under discussion or that the effects of different reactions can be separated from one another.

The form of the rate law. The instantaneous rate of a reaction most commonly depends on a number of variables, including the concentrations of at least some of the reactants, the temperature, and the nature of the solvent (if any). The first problem in any kinetic study is to find, by appropriate experiments, the form of the functional dependence of the rate on the concentrations of the reactants, which is called the *rate law*. Consider a reaction with stoichiometry such that n_A molecules of A react with n_B molecules of B (and perhaps with some other reactants as well) to form certain products:

$$n_A A + n_B B + \cdots \longrightarrow \text{products} \tag{20-7}$$

If there are no interfering reactions that consume or produce A or B and if all the variables that might affect the rate of the reaction except the concentrations of the reactants are held constant, then it is often found (by experiment) that the rate law may be expressed as

$$\text{Rate} = -\frac{1}{n_A}\frac{d[A]}{dt} = -\frac{1}{n_B}\frac{d[B]}{dt} = k[A]^a[B]^b \cdots \tag{20-8}$$

Here the exponents a and b are usually (but not always) integers, with *no necessary relation to the stoichiometric coefficients* n_A and n_B in the equation[1] for the reaction (20-7). The proportionality constant k is called the *rate constant;* it depends not only on the nature of the reaction, but normally also on the temperature, the nature of the solvent, and other possible variables.

Since the *rate* of any reaction has the dimensions concentration time^{-1}, the dimensions of the rate constant k for that reaction depend on the exponents of the concentration terms in the rate law. In particular, if we let p be the sum of the exponents of the concentration terms in the experimental rate law (20-8),

$$p = a + b + \cdots \tag{20-9}$$

then the dimensions of k are concentration^{1-p} time^{-1}.

Order of a reaction. The experimental quantity p defined by (20-9) is called the *order of the reaction;* the individual exponents $a, b, \ldots$ in (20-8) are said to be the orders of the reaction with respect to the individual reactants A, B, ...

---

[1] For some reactions the rate law contains concentration terms of species that do not appear in the stoichiometric overall reaction equation.

For example, if in a reaction with two reactants A and B it is found experimentally that $a = 1$ and $b = 2$, the reaction is said to be third order overall and to be first order in A and second order in B.

There are many reactions for which the rate law cannot be expressed in the form (20-8), that is, as a product of concentration terms each raised to some power. For example, the rate may be expressible only as the sum of several terms or even as a ratio of sums. The term *order of reaction* has no meaning for a reaction with such a rate law. Furthermore, such complicated rate expressions invariably imply that the mechanism of the reaction includes a number of steps, no one of which is rate-determining.

EXAMPLE 20-1   **Concentration dependence of a reaction rate.**   The rate of oxidation of bromide ions by bromate in an acidic aqueous solution,

$$6H^+ + BrO_3^- + 5Br^- \longrightarrow 3Br_2 + 3H_2O$$

has been found experimentally to be first order in bromide, first order in bromate, and second order in hydrogen ion; that is,

$$-\frac{d[BrO_3^-]}{dt} = \frac{1}{3}\frac{d[Br_2]}{dt} = k[Br^-][BrO_3^-][H^+]^2$$

What happens to the rate if, in separate experiments, (*a*) $[BrO_3^-]$ is doubled; (*b*) the pH is increased by 1.0 units; (*c*) the solution is diluted to twice its volume, with the pH kept constant by use of a buffer?

Solution.   (*a*) Since the rate is proportional to $[BrO_3^-]$, doubling this concentration doubles the rate. (*b*) Increase of the pH by 1.0 means that $[H^+]$ is decreased by a factor of 10. The rate is therefore decreased by a factor of 100 since the rate is proportional to $[H^+]^2$. (*c*) Both $[Br^-]$ and $[BrO_3^-]$ decrease by a factor of 2, so that the rate decreases by a factor of 4 at constant $[H^+]$.

EXAMPLE 20-2   **Determining a rate law.**   Hypophosphite ion, $H_2PO_2^-$, decomposes in alkaline solution into hydrogen phosphite ion, HPO_3^{2-}, and hydrogen according to the stoichiometric equation

$$H_2PO_2^- + OH^- \longrightarrow HPO_3^{2-} + H_2 \tag{20-10}$$

The rate of disappearance of hypophosphite ion at 100°C has been found to depend on the concentrations of the reactants as follows:

$[H_2PO_2^-]/M$	$[OH^-]/M$	$\dfrac{-d[H_2PO_2^-]}{dt}$ / M min^{-1}
0.10	1.0	3.2×10^{-5}
0.50	1.0	1.6×10^{-4}
0.50	4.0	2.5×10^{-3}

(*a*)   Deduce the order of the reaction with respect to each reactant, and write the rate expression.

(*b*) Calculate the rate constant for the reaction at this temperature.

(*c*) How many moles of HPO_3^{2-} are being formed each minute in a solution of volume 0.50 liter at the moment at which $[OH^-] = [H_2PO_2^-] = 1.0\ M$?

Solution. (*a*) The first two lines of the tabulated data indicate that when $[OH^-]$ is kept constant, a fivefold increase in the concentration of $H_2PO_2^-$ leads to a fivefold increase in the rate. Thus, the rate is proportional to $[H_2PO_2^-]$, that is, the reaction is first order in $H_2PO_2^-$. Comparison of the last two lines of the tabulated data indicates that at constant $[H_2PO_2^-]$ a fourfold increase in the OH^- concentration leads to an approximately sixteenfold increase in the rate. Consequently the reaction is second order in OH^-. Thus, the rate expression is

$$-\frac{d[H_2PO_2^-]}{dt} = k\,[H_2PO_2^-][OH^-]^2 \qquad (20\text{-}11)$$

(*b*) Substitution of any of the tabulated data—perhaps those of the first line are most convenient—into (20-11) leads to $k = 3.2 \times 10^{-4}\ M^{-2}\ min^{-1}$.

(*c*) Substitution of the 1.0 *M* concentrations and the value of k from (*b*) into (20-11) gives the rate of disappearance of $H_2PO_2^-$ under these conditions as 3.2×10^{-4} (mol liter^{-1} min^{-1}). Since the volume is 0.50 liter and since the stoichiometric equation (20-10) indicates that one HPO_3^{2-} ion is formed for each $H_2PO_2^-$ that disappears, there must be 1.6×10^{-4} mol HPO_3^{2-} ions being formed per minute.

Integrated rate expressions. It is frequently inconvenient, and sometimes even impossible, to measure the instantaneous rate of a reaction directly with much precision. This is especially true if the reaction is so fast that the concentrations of the reactants, and hence the rate, change appreciably in the time it takes to make measurements. A common practice in experimental studies of kinetics is to measure the concentration of some reactant (or product) at intervals of time so chosen that a smooth curve can be drawn representing the variation of concentration with time during the course of most of the reaction—for example, a curve like that in Fig. 20-1. If the rate at any instant is desired, it can be found by measuring the slope of the curve at that particular time, as indicated in the legend of Fig. 20-1.

Curves like that in Fig. 20-1 represent what are sometimes called *integrated rate expressions,* the name implying that they can be derived from the instantaneous (or differential) rate laws, such as (20-11), by the technique of integration. Suppose, for example, that the concentration of a certain substance X is altered by a first-order reaction. This implies that

$$-\frac{d[X]}{dt} = k[X] \qquad (20\text{-}12)$$

or
$$-\frac{d[X]}{[X]} = k\,dt \qquad (20\text{-}13)$$

Let the concentration of X at time t_0 be $[X]_0$ and that at a later time t be $[X]$. To integrate (20-13) between these limits, we need to find

$$-\int_{[X]_0}^{[X]} \frac{d[X]}{[X]} = \int_{t_0}^{t} k \, dt \qquad (20\text{-}14)$$

Since the integral of dx/x is $\ln x$, the natural logarithm of x (see Study Guide), we have, for first-order reactions,

$$-(\ln [X] - \ln [X]_0) = k(t - t_0)$$

It is customary to take the time of the initial measurement as the zero of time so that $t_0 = 0$, whence we get

$$-\ln \frac{[X]}{[X]_0} = kt$$

$$\frac{[X]}{[X]_0} = e^{-kt} \qquad (20\text{-}15)$$

or

Thus, the concentration of X at any time t is related to its concentration at $t = 0$, $[X]_0$, by

$$[X] = [X]_0 \, e^{-kt} \qquad (20\text{-}16)$$

This negative exponential dependence on time is uniquely characteristic of first-order reactions. An enormous variety of chemical and physical processes follow first-order kinetics; one of the most familiar of these is radioactive decay.

Equation (20-15) implies that the *half-life of a first-order reaction,* defined as the time required for the concentration of reactant to fall to half its initial value, is independent of the concentration. If we let $\tau_{1/2}$ represent the half-life, we have, by (20-15),

$$-\ln \frac{0.5[X]_0}{[X]_0} = -\ln 0.5 = k\tau_{1/2}$$

Since $\ln 0.5 = -\ln 2 = -0.693$,

$$-(-0.693) = k\tau_{1/2}$$

or

$$\tau_{1/2} = \frac{0.693}{k} \qquad (20\text{-}17)$$

Thus it is easy to calculate the half-life of a substance that decomposes by a first-order reaction if the rate constant is known or to calculate the rate constant if the half-life is known. The length of time required for reaction of any other particular fraction, e.g., nine-tenths, of the original sample is also independent of the initial concentration, as implied by (20-15).

EXAMPLE 20-3   **First-order kinetics.** Paraldehyde, P, is formed from three molecules of acetaldehyde, A.

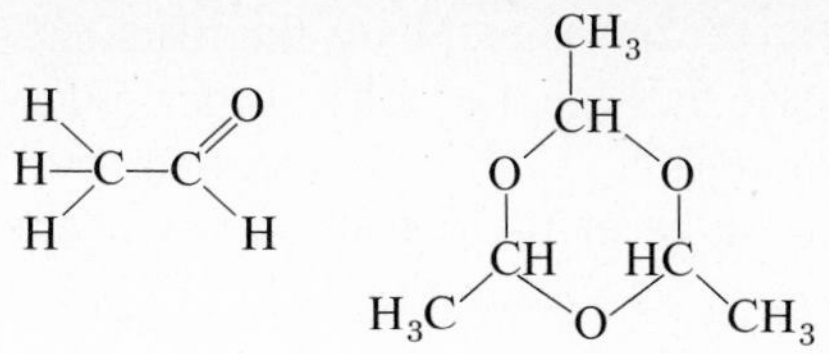

At temperatures around 500 K, paraldehyde vapor decomposes by a first-order reaction to acetaldehyde, P $\longrightarrow$ 3A. At 519 K the rate constant for this reaction is $3.05 \times 10^{-4}\,s^{-1}$. What is the half-life of a sample of paraldehyde under these conditions? What is its "tenth-life" (the time required for nine-tenths of the sample to decompose, leaving just one-tenth of the original material)?

Solution. The half-life, by (20-17), is

$$\tau_{1/2} = \frac{0.693}{3.05 \times 10^{-4}}\,s = \underline{2.27 \times 10^3\,s}$$

or about 38 min. From (20-15) we know that the time needed for the concentration to fall to one-tenth of its initial value is given by

$$-\ln 0.10 = \ln 10 = k\tau_{0.1}$$

or

$$\tau_{0.1} = \frac{\ln 10}{k} = \frac{2.303}{3.05 \times 10^{-4}}\,s = \underline{7.55 \times 10^3\,s}$$

(The ratio of the "tenth-life" to the half-life must always be the same, $\frac{2.303}{0.693} = 3.32$.)

It is possible to integrate rate expressions that are appreciably more complicated than that for a first-order reaction, (20-13), and integrated forms of the most commonly encountered rate laws have been worked out. Many of them can be derived in a straightforward fashion. We shall give the integrated expression for only one other, that for a reaction that is second order in a single reactant, for which the rate law is thus

$$-\frac{d[X]}{dt} = k[X]^2 \tag{20-18}$$

The integrated relation between [X] and the time is

$$\frac{1}{[X]} - \frac{1}{[X]_0} = kt \tag{20-19}$$

where the time is measured from a zero point at which the concentration was $[X]_0$. The time dependence of the concentration of a single reactant[1] that is consumed by a second-order process, Equation (20-19), is quite different from that for a reactant consumed by a first-order process, Equation (20-16). This difference is easily

---

[1]Equation (20-19) also applies when two different substances react by a second-order process if, and only if, their initial concentrations are equal.

observed experimentally if the reaction is followed for a period significantly longer than the half-life. For a first-order process, the half-life is independent of the initial concentration, as already stressed, but for a second-order reaction of the kind discussed here, the half-life varies inversely with the initial concentration. This can be shown by substituting $[X] = [X]_0/2$ (and $t = \tau_{1/2}$) into (20-19):

$$\frac{1}{[X]_0/2} - \frac{1}{[X]_0} = k\tau_{1/2}$$

$$\frac{2}{[X]_0} - \frac{1}{[X]_0} = k\tau_{1/2}$$

or

$$\tau_{1/2} = \frac{1}{k[X]_0} \tag{20-20}$$

Thus, in marked contrast to a first-order reaction, the half-life increases steadily as the initial concentration decreases.

The distinction between the rate laws expressed in (20-19) and (20-16), or (20-15), is often most apparent if the experimental data are graphed. For the first-order reaction, a plot of ln $[X]$ against the time is a straight line; for the second-order reaction leading to (20-19), a plot of $1/[X]$ against time is a straight line. If the fraction of the reactant consumed is small, the data for a given reaction may appear to give a straight line when plotted either way, but if the elapsed time is a half-life or more (and the data are precise), only one of the lines can be straight. Of course, if the reaction is neither first order nor second order in X, neither line will be straight.

Experimental measurement of rates.   Because of the marked effect of temperature on the rates of most reactions it is essential to keep the temperature of the reaction mixture constant during kinetic measurements. Reaction vessels are kept in a constant-temperature bath or similar thermostatic device.

Ideally, the progress of the reaction is measured without disturbing the system. This can be done for many reactions by observing continuously, or at appropriate intervals, some physical change that is related to the progress of the reaction. For example, in a gas reaction the number of product molecules may differ from the number of reactant molecules, as in the reaction

$$2N_2O_5(g) \longrightarrow 4NO_2(g) + O_2(g)$$

The pressure change at constant volume and temperature can therefore be used to follow the reaction. With reactions in solution, there are also sometimes small volume changes that can be measured, or a gas may be evolved and its volume measured. Other useful physical changes include the appearance or disappearance of some species that absorbs electromagnetic radiation in an accessible region of the spectrum, or the formation or removal of ions whose presence can be monitored by measurements of conductivity or electrical potential. A special case is a changing pH that can be observed with a pH meter. With appropriate instrumentation, many of these methods can be adapted for studying extremely fast reactions with half-lives of microseconds or less.

If no suitable method can be devised for following the reaction by some continuous observation of its properties without disturbing it, samples can be withdrawn at intervals and analyzed for a particular reactant or product. The reaction must be stopped or *quenched* in each sample at the time it is taken from the reaction mixture. This may be done by lowering the temperature of the sample appreciably, a method especially valuable for reactions studied at elevated temperatures. A reaction can also be quenched by lowering the concentration of the reactants sufficiently. The sample can be rapidly diluted with solvent, or an appropriate reagent can be added. For example, an acid-catalyzed reaction can be quenched by neutralizing the acid catalyst. An alternative to sampling the mixture at intervals is to prepare a number of identical mixtures and to stop the reaction in each at different times.

20-4 Homogeneous Reactions in the Gas Phase

The kinetic[1] theory of gases provides a sound foundation for interpreting the results of experiments involving gaseous molecules. There is, on the other hand, no manageable quantitative theory of liquids or solutions to serve as a guide in attempts to understand experiments in these media. Consequently, in order to show how some of the simplest experimental facts of chemical kinetics can be interpreted theoretically, we begin with a discussion of reactions in the gaseous phase.

Concentration dependence of the rate of molecular collisions.

Collisions between molecules[2] play an essential role in most chemical reactions. For example, at some stage in a reaction in which a new bond is formed, there must be close contact of the species containing the atoms that are eventually joined by the new bond. Clearly the rate of such a reaction cannot exceed the rate at which the reacting molecules come together (although it might be much less than this rate of collision if only a small fraction of the collisions lead to reaction). We now consider how the rate of molecular collisions depends on the concentrations of the reacting species.

Imagine a box containing three gases, E, F, and G, moving at speeds characteristic of a temperature T. Let the numbers of molecules of E, F, and G per liter be n_E, n_F, and n_G, respectively. Consider first the collisions that involve only one molecule of E and one molecule of F. We denote the number of such collisions per second in each liter by coll (E,F). Since any molecule of E can collide with any molecule of F, there are $n_E n_F$ distinct possible pairs of colliding molecules of this type per liter and the frequency of such collisions per liter at any temperature is thus proportional to $n_E n_F$:

$$\text{coll (E,F)} = c_1 n_E n_F \qquad (20\text{-}21)$$

---

[1] The adjective *kinetic* (Greek: *kinetikos,* moving, pertaining to motion) is used in the phrase *kinetic theory* to imply that the theory is one in which the motions of molecules, and the energies associated with these motions, play a dominant role. It is only indirectly related to the term *kinetics* as applied to the study and interpretation of the rates of chemical reactions.

[2] We are again using the term *molecules* in a general sense to include not only molecules but other possible reacting species: atoms, ions, and radicals.

Here c_1 is a constant that depends on the temperature and the masses and sizes of E and F. As the temperature increases, c_1 increases because the increased molecular speeds result in a higher collision rate. The numbers of molecules per liter are related to the respective molar concentrations by Avogadro's number, N_A; for example, $n_E = [E]N_A$. Therefore,

$$\text{coll (E,F)} = c_2[E][F] \tag{20-22}$$

where $c_2 = c_1 N_A{}^2$. The rate of collisions per unit volume at a given temperature when different molecules are involved is therefore proportional to the product of the concentrations of the colliding species.

For collisions between two molecules of the same type, such as two E molecules, the collision rate is proportional to $n_E (n_E - 1)/2 \approx n_E{}^2/2$. Without the factor $\frac{1}{2}$ each collision would be counted twice. Thus

$$\text{coll (E,E)} \approx c_3[E]^2 \tag{20-23}$$

where $c_3 = c_2/2$ for the special situation that the masses, sizes, and molar concentrations of E and F are the same.

Similar arguments can be used to show that the rates of collisions in which more than two molecules come together are also proportional to the products of the concentrations of the colliding species. For example, the rate of collision per unit volume of one molecule of E, two molecules of F, and one molecule of G is given by

$$\text{coll (E,F,F,G)} = c_4[E][F]^2[G] \tag{20-24}$$

Collisions that involve two molecules, whether of the same or of different kinds, are called *bimolecular;* those among three molecules are said to be *termolecular.* The intrinsic probability of collisions involving more than two molecules simultaneously, which is reflected in the magnitude of the proportionality constants in equations such as (20-24), is far smaller than that of bimolecular collisions because the likelihood that three or more molecules will be in the same place at the same time is very low, especially for gases and for solutes in dilute solutions. Nonetheless, termolecular collisions are important in some reactions, because even though they are rarer, such collisions are for these reactions far more fruitful than are bimolecular collisions. This brings us to a consideration of the fraction of collisions that lead to reaction. For simplicity, we examine this question for bimolecular processes.

Collision rates for bimolecular reactions. The rate at which bimolecular collisions occur in a gas can be calculated with fair precision from the kinetic theory of gases. In essence this amounts to evaluating the constants c_2 and c_3 in (20-22) and (20-23). These constants can then be compared with the experimentally observed rate constants for various second-order gas-phase reactions, for which the rate laws are (by definition) of the form

$$\text{Rate} = k[A][B] \quad \text{or} \quad \text{rate} = k'[A]^2 \tag{20-25}$$

just parallel to (20-22) and (20-23), respectively.

An "order-of-magnitude" calculation leads to the result that at room temperature

k(or k') is of the order of 10^{11} liter mol^{-1} s^{-1}. Suppose that two gases A and B are mixed, each at about 0.5 atm pressure, which corresponds to concentrations of about 0.02 mol liter^{-1}. If $k \approx 10^{11}$ liter mol^{-1} s^{-1}, there would be of the order of $10^{11}(0.02)^2 = 4 \times 10^7$ mol of collisions each second between the molecules in a 1-liter sample. Since in this volume there is only 0.02 mol of each reactant to be used up, any reaction in which every collision of A with B yielded products would be complete in a time of the order of 0.02 mol/4×10^7 mol s^{-1} or in less than 10^{-9} s.

Thus gas reactions for which the rate-controlling step involves a bimolecular collision, called *bimolecular reactions,*[1] would be over almost instantaneously if every collision were fruitful in forming products. In fact, very few reactions occur this rapidly. Most known bimolecular rate constants are smaller by many powers of 10 than the value 10^{11} M^{-1} s^{-1}. In other words, most of the collisions do not result in reaction; the colliding molecules just bounce off each other and go their separate ways, unchanged except in their direction of motion and, usually, their energy. How can we explain the fact that so small a fraction of the collisions are effective in producing chemical change? We shall see shortly that the explanation is in part tied to the Maxwell-Boltzmann distribution of kinetic energies (Sec. 5-5).

Temperature dependence of reaction rates: activation energy. The rates of most chemical reactions increase markedly as the temperature is increased. There are exceptions, but they are uncommon. For a typical bimolecular reaction, an increase of 10 K at temperatures not far from room temperature will increase the reaction rate by a factor of 2 to 3; that is, the rate will increase by 100 to 200 percent. What is the reason for this marked temperature effect? If the volume is constant, the only temperature-dependent term in the collision rate is the molecular velocity. This velocity is proportional to $T^{1/2}$ and enters to the first power in the collision rate. For a 10 K rise in temperature near 300 K (about a 3 percent increase in T or a 1.5 percent increase in $T^{1/2}$), the increase in the collision rate will therefore be only about 1.5 percent, an effect completely negligible in comparison with the usual increase in the reaction rate of 100 percent or more. Thus the temperature dependence of the reaction rate cannot be attributed to the increased collision rate.

Arrhenius observed that for many reactions in gases and in other media the rate constant was related to the absolute temperature exponentially:

$$k = Ae^{-E_a/RT} \tag{20-26}$$

The constant A is called the frequency factor; E_a, which has the dimensions of energy, is called the *activation energy*. For most reactions E_a is a positive quantity; that is, k increases with increasing T (decreasing $1/T$).

Equation (20-26) can be arranged to

$$\ln k = \ln A - \frac{E_a}{R}\frac{1}{T} \tag{20-27}$$

---

[1] Note carefully the distinction between the terms *bimolecular* and *second order*. A reaction scheme involving a bimolecular step does not necessarily lead to an overall rate equation that is second order. Molecularity (unimolecular, bimolecular, termolecular, etc.) refers only to the number of reacting entities in an elementary process, not to the order in a rate expression.

If A and E_a are independent of temperature, a graph of $\ln k$ against $1/T$ should be a straight line with slope $-E_a/R$ and intercept $\ln A$. Such graphs do give remarkably straight lines for many reactions over a significant range of temperature; one is shown in Fig. 20-2. Values of activation energies can thus be derived readily from experimental data. If (20-27) is assumed to hold, then only two experimental points, that is, values of k at two precisely measured temperatures, are needed to find E_a and A. If we let the temperatures be T_2 and T_1 and let the corresponding rate constants be k_2 and k_1, we have, from (20-27),

$$\ln k_2 - \ln k_1 = -\frac{E_a}{R}\left(\frac{1}{T_2} - \frac{1}{T_1}\right)$$

or
$$\ln \frac{k_2}{k_1} = 2.303 \log \frac{k_2}{k_1} = \frac{E_a}{R}\frac{T_2 - T_1}{T_1 T_2} \tag{20-28}$$

This equation confirms our earlier remark that E_a is positive if k increases with temperature; if T_2 is greater than T_1, then E_a must have the same sign as $\log (k_2/k_1)$ and so must be positive if k_2 is greater than k_1.

EXAMPLE 20-4   **Activation energy.** The rate constant of a certain reaction increases by a factor of 18 when the temperature is increased from 20 to 40°C. What is the activation energy for this reaction?

FIGURE 20-2   **Log k as a function of $1/T$ for the thermal decomposition of HI.** The reaction whose rate constant was measured at different temperatures is

$$2HI \rightarrow H_2 + I_2$$

The experimental data have been plotted in a form to test their fit to (20-27). Since $\log k = (\ln k)/2.303$, the slope of the curve is $-E_a/(2.303R)$. The value of E_a found from the slope is 184 kJ mol^{-1}. Extrapolation of the line to $1/T = 0$ gives an experimental value of A [Equations (20-27) and (20-26)] between 10^{10} and 10^{11} M^{-1} s^{-1}. The data shown here are from the classic work of M. Bodenstein, a pioneer of experimental chemical kinetics.

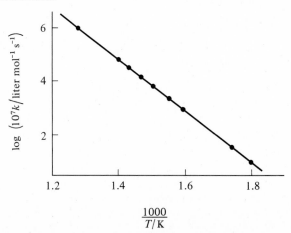

Solution. We know that $k_2/k_1 = 18$ when $T_2 = 313$ K and $T_1 = 293$ K. Equation (20-28) then gives

$$2.303 \log 18 = \frac{E_a}{R} \frac{20}{313 \times 293} \text{K}^{-1}$$

$$E_a = [2.3\,(1.26)(4.6 \times 10^3)R] \text{K} = \underline{1.1 \times 10^2 \text{ kJ mol}^{-1}}$$

Note that the precision with which E_a can be determined experimentally with the help of (20-28) depends directly on the precision with which the difference in temperature, $T_2 - T_1$, can be measured, as well as on the precision of the ratio of the rate constants. The activation energy of 110 kJ mol^{-1} found in this example is quite typical, although it is somewhat higher than that corresponding to the rule cited earlier, "a factor of about 2 to 3 in k for a 10 K rise in temperature near room temperature". In this case there is an increase of a factor of more than 4 for each 10 K.

Collision theory and the Arrhenius equation. We now consider the observed temperature variation of the rate constant in terms of the Maxwell-Boltzmann energy distribution. Arrhenius interpreted (20-26) to imply that molecules could not react unless they had sufficient energy to reach an "activated state" with energy E_a above the initial state. This idea is illustrated in Fig. 20-3.

The abscissa in this schematic diagram is called the *reaction coordinate*. For a simple reaction it is some measure of the interatomic distance or another parameter that changes smoothly in the progression from reactants to products. At the left the system is in the form of the reactants; at the right the system has been transformed into products. Arrhenius suggested that in order for the reaction to proceed from

FIGURE 20-3 **Activation energy.** A reaction proceeds in a forward direction from some well-defined initial state (reactants), through an intermediate activated state, to a well-defined final state (products). For the reverse reaction, the roles of reactants and products are interchanged. The activation energy for either reaction, which appears in an expression like (20-26) for the rate constant k, is the energy needed to go from the initial state for the reaction to the intermediate activated state.

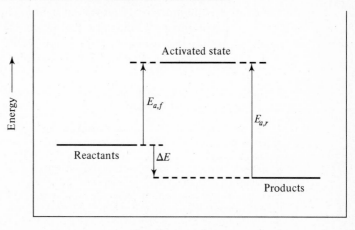

left to right the system has to move through some intermediate state, the activated state. Thus, for the forward reaction, the activation energy is $E_{a,f}$, the difference in energy between the reactants and the activated state. Unless the reactants have energy equal to or greater than $E_{a,f}$, they cannot progress through the activated state to products.

For the reverse reaction, the activation energy is $E_{a,r}$, the difference in energy between the activated state and the products. The overall energy change during the forward reaction is given by

$$\Delta E = E_{a,f} - E_{a,r}$$

To relate the Arrhenius concept of activation to kinetic theory, it is necessary to consider the distribution of kinetic energy among the molecules in a gas or liquid. There is a continual exchange of kinetic energy among the constantly colliding molecules. At thermal equilibrium the energies are distributed at any time according to curves very much like those for the distribution of velocities in Fig. 5-10. The fraction of molecules with energies far in excess of the average—essentially the tail to the right in the curves in Fig. 5-10—increases very rapidly with increasing temperature, even though the average energy increases only in proportion to T. If the activation energy lies well above the average kinetic energy, only highly energetic molecules can reach the activated state. The rapid increase with temperature of the fraction of highly energetic molecules therefore corresponds to a large increase in the number of molecules with sufficient energy to react, which explains the large increase with temperature of most reaction-rate constants.

A careful analysis has been made of the frequency of molecular collisions for which the molar kinetic energy of the colliding pair along the line between their centers exceeds a specified minimum value E_c, the *collisional activation energy*. The energy is specified in this way because only the energy due to the relative motion of the molecules along this line is available on collision to be converted into potential energy and thus to raise the molecules to some activated state.[1] The result turns out to be a very simple one: the fraction of collisions for which the kinetic energy along the line of centers exceeds E_c is given by the Boltzmann factor, $e^{-E_c/RT}$. Consequently, if Z is the number of molecular collisions per unit volume per unit time, $Ze^{-E_c/RT}$ is the number of collisions for which the collisional activation energy is exceeded. Hence this is also the number of collisions per unit volume per unit time that lead to reaction, and this number should equal the rate of reaction per

---

[1] The following analogy given by W. Moore, "Physical Chemistry", 3d ed., p. 278, Prentice-Hall, Inc., Englewood Cliffs, N.J., 1962, is illuminating:

The process called *activation* consists essentially in the transfer of translational kinetic energy into energy stored in internal degrees of freedom, especially vibrational degrees of freedom. The mere fact that a molecule is moving rapidly, i.e., has a high translational kinetic energy, does not make it unstable. In order to cause reaction, the energy must get into the chemical bonds, where high-amplitude vibrations will lead to ruptures, decompositions, and rearrangements. The transfer of energy from translation to vibration can occur only in collisions with other molecules or with the wall. The situation is like that of two rapidly moving automobiles; their kinetic energies will not wreck them unless they happen to collide and the kinetic energy of the whole is transformed into internal energy of the parts.

unit volume. We can then write

$$k = Ze^{-E_c/RT} \qquad (20\text{-}29)$$

a result almost identical in form with the empirical Arrhenius law (20-26). The difference is that while the constant A in the Arrhenius equation is assumed to be independent of temperature, Z depends on the temperature because it is proportional to the molecular velocity and thus to $T^{1/2}$. We can substitute $B = Z/T^{1/2}$, with B independent of temperature, which leads to

$$k = Ze^{-E_c/RT} = BT^{1/2}e^{-E_c/RT} \qquad (20\text{-}30)$$

It is only in the most precise work that it is necessary to distinguish the collisional activation energy of (20-30), E_c, from the Arrhenius activation energy of (20-26), E_a. It can be shown[1] that $E_a = E_c + RT/2$, and since $RT/2$ is only 1.3 kJ mol^{-1} at room temperature, the difference between E_a and E_c is usually negligible.

Orientational or steric[2] effects in reaction rates. The activation energy for a particular reaction must be found experimentally from a plot of $\ln k$ against $1/T$, as described earlier. Once this has been done, it is possible to test the validity of the collision theory expression for the rate constant, Equation (20-29). An approximate value of the collision frequency Z can be calculated for molecules of known diameter at any temperature, and thus a value of k at that temperature can be calculated from (20-29) for any bimolecular reaction of known activation energy. When these calculated k's are compared with the experimental k's, it is usually found that the experimental values are too small, sometimes only slightly, sometimes by many powers of 10. These discrepancies can be explained, at least in part, by the assumption that there is an orientational or steric requirement for reaction as well as an energetic one. Even though the potentially reacting molecules collide with sufficient energy, if the atoms to be joined by a bond in the new species that might be formed are too far apart at the time of collision, that particular collision cannot result in reaction. Other collisions with similar energy in which the mutual orientation of the reactants is different and the reacting atoms are appropriately positioned can yield product molecules.

A *steric factor,* usually designated p, may be introduced as a correction factor into (20-29):

$$k_{\text{exp}} = pk_{\text{calc}} = pZe^{-E_c/RT} \qquad (20\text{-}31)$$

Values of p of 0.1 or more are found for a few bimolecular reactions between small gaseous molecules (such as H_2 and I_2), but with larger molecules p may be as small as 10^{-3} or even 10^{-5}. There is no a priori way to predict values of p at all precisely, and although part of the observed variation in them can reasonably be attributed to the postulated orientational effects, p values for some systems are too small to be so explained.

---

[1]This follows by comparison of $d\,(\ln k)/dT$ as calculated from (20-26) and from (20-30).

[2]The adjective *steric* (Greek: *stereos,* solid) is used in chemistry to refer to an effect arising as a result of the three-dimensional arrangement of the atoms in a molecule.

Reactions of hydrogen with halogens.   As examples of homogeneous gas-phase reactions for which detailed mechanisms have been deduced, we consider the formation of hydrogen halides. The reaction of H_2 with gaseous Br_2 to form HBr,

$$H_2 + Br_2 \longrightarrow 2HBr \qquad (20\text{-}32)$$

was one of the earliest reactions for which a complicated rate law was established.

Bodenstein had previously made extensive studies of the analogous reaction of H_2 and I_2 and found its rate law to be a simple second-order one, first order in each reactant. In contrast, he found in 1906 that reaction (20-32) had the surprising rate law

$$\frac{d[HBr]}{dt} = \frac{k[H_2][Br_2]^{1/2}}{1 + k'[HBr]/[Br_2]} \qquad (20\text{-}33)$$

where k and k' are constants. This rate law implies that in the initial stages of the reaction, when [HBr] is so small that the denominator in the right side of (20-33) reduces to approximately 1, the reaction is of three-halves order, first order in H_2 and one-half order in Br_2. Furthermore, as the product of the reaction, HBr, accumulates, the ratio [HBr]/[Br$_2$] increases and so the denominator of the right side of (20-33) increases and the rate falls. Thus HBr is said to be an inhibitor of the reaction.

It took more than a decade before anyone could devise a mechanism to account for the peculiar rate law (20-33), and then three people did it independently and almost simultaneously. They proposed a chain reaction involving bromine atoms and hydrogen atoms as *intermediates,* transient species that exist for times of the order of 10^{-13} s—short-lived by ordinary chemical standards. The term *chain reaction* implies a sequence of steps in some of which there is regeneration of reactants as well as formation of products. A single step to initiate the chain may eventually result in the formation of a large number of product molecules. The proposed steps were

Chain initiation:	$Br_2 \longrightarrow 2Br$	(20-34a)
Chain propagation:	$Br + H_2 \longrightarrow HBr + H$	(20-34b)
	$H + Br_2 \longrightarrow HBr + Br$	(20-34c)
Chain inhibition:	$H + HBr \longrightarrow H_2 + Br$	(20-34d)
Chain termination:	$2Br \longrightarrow Br_2$	(20-34e)

The chain is initiated by bromine atoms formed by the thermal or photochemical dissociation of molecular bromine, reaction (20-34a). In the chain-propagating steps, (20-34b) and (20-34c), two molecules of HBr are formed and a bromine atom is regenerated so that these steps can be repeated as long as bromine atoms (and the supply of Br_2 and H_2) remain. The inhibiting effect of HBr is accounted for by reaction (20-34d): H atoms produced in (20-34b) are consumed in (20-34d), as well as in (20-34c), forming H_2 and using up HBr. Why was reaction (20-34d) proposed rather than the alternative possible inhibiting reaction, $Br + HBr \longrightarrow Br_2 + H$? Because, according to the rate law (20-33), the inhibition depends on the ratio [HBr]/[Br$_2$], implying that HBr and Br_2 compete with each other for some atom.

Although by (20-34c) bromine molecules would be expected to compete effectively with HBr for H atoms, they would not compete for Br atoms since there is no driving force for a reaction of Br_2 with Br.

We shall not attempt to derive (20-33) from a combination of the rate laws for the elementary processes represented by Equations (20-34). It is usually necessary in this and similar situations to make what is called a *steady-state assumption,* which implies that once the reaction is established, the concentrations of very reactive intermediates do not change with time because they are produced at the same rate as they are consumed. In the present system this implies that when the steady state is established, $d[H]/dt = 0$ and $d[Br]/dt = 0$. With these assumptions, a rate law of exactly the form (20-34) can be derived, in which k and k' represent combinations of rate constants for the elementary processes.

Does this agreement between the derived and experimental rate laws prove that the mechanism represented by (20-34) is correct? It does not. As we have stressed before, it is impossible in science to prove that some model is *the* correct one. All that can be said is that it is consistent with the evidence that has been considered. Other models, not tested, might also be consistent with that evidence. Furthermore, new evidence might be discovered that is inconsistent with the proposed mechanism, and then either the mechanism must be revised or the supposed evidence proved to be spurious or irrelevant.

Consider, for example, the reaction of gaseous iodine with hydrogen. This reaction is kinetically very simple. It was long believed that its second-order rate law,

$$\frac{d[HI]}{dt} = k[H_2][I_2] \tag{20-35}$$

implied the bimolecular mechanism $H_2 + I_2 \longrightarrow 2HI$. However, it has recently been shown by J. H. Sullivan that very little if any HI is produced by direct bimolecular collisions of H_2 with I_2. Rather, two distinct mechanisms, which occur simultaneously and thus compete with each other, are needed to explain all the known experimental facts. One of these has a termolecular rate-determining step,

$$H_2 + 2I \longrightarrow 2HI \tag{20-36}$$

The iodine atoms normally exist in equilibrium with molecular iodine,

$$I_2 \rightleftharpoons 2I \tag{20-37}$$

so that $[I]^2$ in the rate law for the elementary process represented in (20-36) can be replaced by $K[I_2]$, where K is the equilibrium constant for (20-37). Thus the rate law for (20-36) agrees with that observed experimentally:

$$\frac{d[HI]}{dt} = k'[H_2][I]^2 = k'[H_2]K[I_2] = k[H_2][I_2] \tag{20-38}$$

The other mechanism consists of a chain reaction much like that of the hydrogen-bromine reaction, but differing in one step so that the net result is the simple rate law rather than a complex one.

The reaction of hydrogen with chlorine is similar to that with bromine, although it is more difficult to study experimentally. The contrast between the kinetics of these stoichiometrically identical reactions of hydrogen with the three common halogens is instructive—and should be sobering to anyone contemplating easy generalizations about kinetics.

The transition from reactants to products.   So far we have focused attention on three states in kinetics: the reactants, the products, and the activated state between them. However, the transition from reactants to products involves a succession of many states, from which these three are singled out because they represent, respectively, a position of relative stability at the start, a stable position at the end, and a maximum in potential energy during the transition between the other two states.

Consider what happens as molecules collide and react, for example, in a simple bimolecular reaction

$$XY + Z \longrightarrow X + YZ \qquad (20\text{-}39)$$

When a collision of Z and XY occurs with sufficient kinetic energy and appropriate orientation for reaction, some of the mutual kinetic energy is transferred into vibrational energy of the X-Y bond. The amplitude of these vibrations is thereby greatly increased. As a result the average X-Y bond distance is lengthened and the overlap of the bonding orbitals of X and Y is greatly diminished. Simultaneously, if atom Z is in an appropriate position, it can bond weakly with Y. As the X-Y interaction weakens, the Y-Z interaction can grow stronger. These two effects will eventually be comparable in strength, and we can represent the state at that stage as

$$X \text{------} Y \text{------} Z \qquad (20\text{-}40)$$

with the long lines implying greatly stretched and therefore weak bonds. The species (20-40) may pass smoothly over into the products, X plus YZ, or it may dissociate back into the reactants, XY and Z. In any event, it is itself unstable and will have a very short lifetime. It presumably corresponds to the maximum, or very close to the maximum, in the potential energy during the course of the reaction. Thus (20-40) represents a possible structure for the activated state, or what is often called the transition state, for this reaction. Such specific arrangements of atoms in the activated state are called *activated complexes*. They are in many ways like ordinary molecules with specific geometry, bond force constants, and other properties, but differ in that they are invariably extremely reactive, decomposing almost instantly.

An ideal theory of kinetics would permit one to calculate for any set of reactants and products the most likely path by which they could be interconverted and also the rate at which they would travel over this path. The various possible paths for any reaction can be thought of as laid out on a multidimensional potential-energy surface. In two dimensions this is analogous to the variation of altitude over the earth's surface. Altitude can be thought of as proportional to gravitational potential energy, and position on the earth's surface can be specified by two variables, e.g., longitude and latitude. In this analogy, valleys represent positions of relative stability (reactants and products) and the mountains between the valleys represent the barriers that must be surmounted in passing from one stable state to another. If reactants in valley *A* (Fig. 20-4) are to form products, they must cross the mountains to valley *B*. There are many paths that can be followed between the two valleys. The path requiring the minimum expenditure of energy is that through the lowest accessible pass, which represents the activated complex. This is the most probable path for the reaction. Some molecules may get to valley *B* by higher energy paths and some will wander off to other valleys; i.e., other products will be formed by side reactions.

The calculation of multidimensional potential-energy surfaces adequate for prediction of the likely course and activation energy for a reaction is far too complex

FIGURE 20-4 (*a*) **Analog to a potential energy surface;** (*b*) **symbolic representation of different reaction paths.** In the symbolic picture in (*a*) the varying altitudes of mountains and valleys are a measure of varying gravitational potential energy. This energy varies over the earth's surface and thus is a function of two variables, such as latitude and longitude. The surface might also represent the potential energy of a reaction for which the potential energy depends on just two variables. Reactants are symbolized by the "start" line in one valley, products by the "finish" line in another valley; the wanderings of people symbolize the progress of individual reactions as reactant molecules change into product molecules.

In (*b*) the numbered curves represent the paths with the same numbers in (*a*). The three actual paths are, of course, quite different; placing them on the same two-dimensional graph is only schematic.

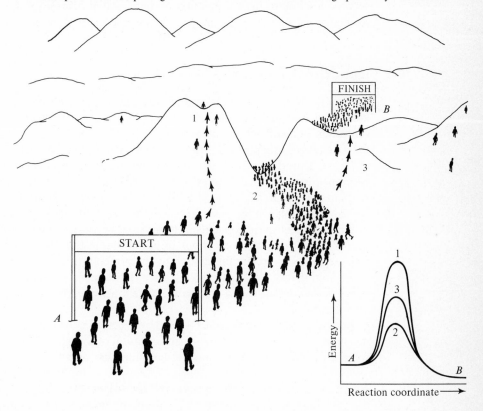

(*a*)   (*b*)

for any but the simplest reactions; even then it can be done only approximately. What is needed is the potential energy for all possible interatomic distances of the interacting atoms. As discussed earlier (especially in Chap. 12), precise calculation of molecular properties is laborious and uncertain even for very simple chemical systems, and present-day quantum-chemical methods are not adequate for complex molecules. Furthermore, when such calculations are done at all, they are done for specific structural arrangements of atoms corresponding to the normal states, or particular excited states, of molecules of interest. To do the calculations for all possible combinations of parameters in order to define a potential energy surface completely is beyond our current capabilities. Even if such a function could be somehow approximated, its representation would pose insurmountable problems even for as simple a reaction as that of a triatomic molecule with a diatomic one. Such a reaction would have a five-atom transition state, with 10 independent parameters (for example, 10 interatomic distances), and the potential energy would need to be specified for every reasonable combination of these 10 parameters.

Since detailed calculations are impractical, one must make plausible guesses about possible structures for the transition state in a reaction. The observed rate law provides evidence about the mechanism and thus about the nature of the reactants that come together in the transition state. The principles of structural chemistry then serve as a guide in the formulation of an activated complex that can reasonably be formed from the reactants and can decompose readily into the products.

20-5   Reactions in Liquid Solutions

Most of the qualitative and quantitative concepts developed through analysis of the kinetics of gaseous reactions are also applicable to reactions in solutions. The concentration dependence of reaction rates can be expressed by laws similar to those for gaseous reactions, and frequently these laws are of the particularly simple form of (20-8), a product of concentration terms each raised to some power. The effects of temperature on reaction rates are very frequently expressible in terms of the simple Arrhenius relation, Equation (20-26), and the concepts of activation energy and the activated complex have been extended to reactions in solution. To be sure, the interpretation of these concepts in terms of a molecular mechanism is usually more subtle and uncertain for reactions in liquid media than for those in the gas phase because of the complications caused by the presence of the solvent. It can play many roles, from that of a comparatively passive bystander to that of an active participant in the reaction, and it can change the reaction from what it would have been with no solvent present.

Before we consider various specific effects of solvents on reaction rates and mechanisms, we can gain a better appreciation of the general effect of even an inert solvent by examining the nature of collisions between reactants in a solution.

Encounters between reactants in solution.   If two molecules collide in the gas phase and do not react, they normally move apart again at once and there is little likelihood that this same pair will soon collide again. In contrast, when two potentially reacting solute molecules A and B diffuse together in a solution, they cannot normally move apart again quickly because they are surrounded closely by other molecules, chiefly of solvent. The reactants are, so to speak, in a cage of solvent. It is not a rigid or permanent cage. Sooner or later, after much jostling and agitation, an opening appears in some direction and, if they have not already reacted with each other, either A or B may escape from the cage and move away from its potential reacting partner.

Thus, the *cage effect* causes reactants to remain together far longer in solution than in the gaseous state, long enough for them to undergo tens or hundreds of collisions with each other before they drift apart. The term *encounter* is often used to distinguish this situation, in which potential reactants may remain together long enough to undergo many successive collisions, from the fleeting and nonrecurring collisions characteristic of gases.

For reactants with a high probability of reacting on each collision, the cage effect virtually ensures reaction during each encounter. Hence the rate of such a reaction is limited only by the rate at which the reactants can diffuse together, and the reaction is said to be *diffusion-controlled*. For aqueous solutions, this rate is of the order of $10^{10} \, M^{-1} \, s^{-1}$ for unit concentrations, corresponding to a bimolecular rate constant of about $10^{10} \, M^{-1} \, s^{-1}$—a little smaller for large or massive reactants (which move slowly) and a little larger for reactions between small ions of opposite charge (which are mutually attracted).

This rate constant represents an upper limit for bimolecular reactions in aqueous solutions, for no reaction can proceed faster than the reactants can come together. Various ionic reactions, including many proton-transfer processes, have rate constants of this order of magnitude, as do the recombination reactions of many free radicals in nonpolar solvents. These reactions typically have activation energies in the range 12 to 20 kJ mol^{-1}, which corresponds to the activation energy for diffusion in most liquids. Thus, the evidence is strong that these reactions are indeed diffusion-controlled.

The cage effect can also play a role in solution reactions that are not diffusion-controlled. For example, for those reactions in which the mutual orientation of the reactants is critical, the relatively long duration of encounters in solution may increase the probability of reaction. Because the reacting partners may collide in many different orientations as they are buffeted by their neighbors, they are more likely to achieve the proper orientation for reaction before they drift apart than if they underwent only a single collision.

Specific solvent effects. Because of the overriding importance of water as a solvent, we focus attention on it in most of the following discussion. However, each of the effects mentioned can be illustrated with many other solvents as well. Only the details would differ.

The most general effects of water on the rate and even the course of a reaction are a consequence of its highly polar nature and resulting high dielectric constant. When ions are separated by water, their energy of interaction and the force between them are greatly diminished (see Study Guide).[1] Furthermore, most ions, especially cations, are hydrated in aqueous solution, as are some neutral polar molecules. These are, of course, just the factors that make water so good a solvent for many ionic and polar compounds; they can markedly affect reaction rates in various ways.

For example, when a hydrated ion or molecule reacts to form an activated complex, it may lose some of its hydration shell. Since the energy of hydration is often large, the energy of activation for this path is increased over what it would have been if the hydration shell had been undisturbed. Conversely, molecules that are originally not highly hydrated may react by way of an ionic transition state that is stabilized by hydration. The activation energy for this reaction will then be much lower in water than in some less polar medium, and the rate will be correspondingly much greater.

One of the best examples of the last-mentioned effect is the hydrolysis of tertiary butyl (*t*-butyl) chloride, $(CH_3)_3CCl$, a reaction frequently used as a laboratory demonstration of first-order kinetics:

$$(CH_3)_3CCl + H_2O \longrightarrow (CH_3)_3COH + H^+ + Cl^- \qquad (20\text{-}41)$$

This reaction goes so rapidly in water at room temperature that it cannot be followed by ordinary methods. It can, however, be slowed down appreciably by dilution of the water with some organic solvent such as acetone, ethanol, or dioxane. The rate can then easily be followed by titration of the acid produced as the reaction proceeds.

The results for acetone-water mixtures shown in Fig. 20-5 are typical; the rate decreases by a factor of 10^5 when the water content drops by a factor of 9, from 90 to 10 vol percent water. A great variety of evidence indicates that this and similar reactions go through an ionic intermediate,

$$(CH_3)_3CCl \longrightarrow (CH_3)_3C^+ + Cl^- \qquad (20\text{-}42)$$

or more generally,

$$RCl \longrightarrow R^+ + Cl^- \qquad (20\text{-}43)$$

where in this case R^+ represents the *t*-butyl cation, $(CH_3)_3C^+$. The ions R^+ and Cl^- may recombine to form RCl, or either of them may react with other substances present. When R^+ is the *t*-butyl cation, its reaction with water,

$$R^+ + OH_2 \longrightarrow \left[RO{\overset{H}{\underset{H}{\diagdown}}} \right]^+ \longrightarrow ROH + H^+ \qquad (20\text{-}44)$$

is far faster than the formation of R^+ and Cl^- by (20-42) and is also faster than the reverse of (20-42). Thus in water the product formed is *t*-butyl alcohol, by (20-44), with (20-42) as the rate-determining step. The overall reaction is first order in *t*-butyl chloride and the rate is independent of the concentration of water, except insofar

---

[1] The interaction energy is given by $\epsilon = \mathbf{Q}_1 \mathbf{Q}_2 / \alpha D r$, where D is the *dielectric constant* of the medium. For a vacuum, $D = 1$ [see Eq. (3-1)]. The larger D, the smaller is the interaction between ions. For water, $D = 80$, a large value.

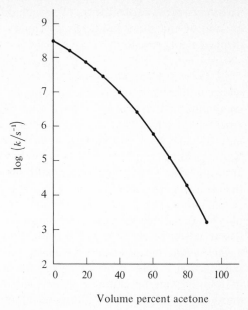

FIGURE 20-5 **Hydrolysis of t-butyl chloride at 25°C in acetone–H_2O mixtures.**

as the concentration of water affects the polarity of the medium. The stabilization of the ionic transition state by solvation is greatly reduced as the water is replaced by a less polar solvent, such as a mixture of water and acetone, and the energy of attraction of the ions for one another also rises as the dielectric constant of the medium falls. These two effects are, in fact, interrelated. They account qualitatively for the dramatic effect of solvent composition on the rate.

The high dielectric constant of water also makes some reactions possible that would be very unfavorable energetically in a medium of much lower polarity. These include, for example, reactions between ions of the same charge. Such ions may exchange electrons (oxidation-reduction) or ligands, or both. The transfer of an electron from a hydrated ferrous ion to a hydrated ferric ion is typical:

$$*Fe(H_2O)_6{}^{2+} + Fe(H_2O)_6{}^{3+} \longrightarrow *Fe(H_2O)_6{}^{3+} + Fe(H_2O)_6{}^{2+} \quad (20\text{-}45)$$

We have used the asterisk (*) to label the atom that was initially Fe(II). Reaction (20-45) produces no net chemical change. It can be followed, however, by using radioactive Fe(II) or Fe(III) initially and monitoring the distribution of radioactivity between the two oxidation states as the reaction progresses. This has been done, and the reaction has been found to be first order in each ion, with k about $1\ M^{-1}\,s^{-1}$ at $0°C$. This rate constant, and the observed activation energy of about 39 kJ mol^{-1}, are consistent with an approach of these ions so that their centers are about 7 Å apart, corresponding to approximate contact of their first hydration shells. At this distance, the electrostatic repulsion energy between them is only about 17 kJ mol^{-1} in water. In a nonpolar solvent, or in the gas phase, the repulsion energy would be about 10^3 kJ mol^{-1} and the ions could never get close enough to react.

Substitution reactions.   We conclude this section with some general comments on one of the most fundamental classes of solution processes—substitution (or displacement) reactions. These reactions can be symbolized in a general way as

$$X + RY \longrightarrow XR + Y \tag{20-46}$$

which is meant to imply that the group X is substituted for the group Y in a compound with the group R. An alternative viewpoint is that this is a *group-transfer reaction,* the group R having been transferred from X to Y. Any of the three groups may be charged. Commonly X and Y are, at least partially, negatively charged; X, for example, might be Cl^-, the oxygen atom of H_2O or CH_3OH, or the nitrogen atom of NH_3. The group R frequently carries a small positive charge; it may be the central metal atom of a coordination compound, with ligands other than Y that remain unaltered during the reaction, or a nonmetallic atom such as carbon or phosphorus joined to more electronegative atoms. When, as in these examples, X has an unshared electron pair and functions as a Lewis base (page 289), it is called a *nucleophile,* meaning that it donates this electron pair to a positive center, and the reaction is said to be a *nucleophilic substitution.* Most reactions of metal ions in solution can be characterized as nucleophilic substitutions of one ligand by another. Nucleophilic substitution reactions at carbon atoms are extremely common and thus play a key role in organic chemistry and biochemistry. Some substitution reactions can be classified as *electrophilic*—that is, the group being substituted is seeking, rather than bearing, an unshared electron pair—but these are less common. We shall restrict our attention to nucleophilic processes.

Two general classes of nucleophilic substitution reactions can be distinguished. In one, the reaction is first order in the substrate (the substance being acted upon, RY) and the rate is independent of the nucleophile X. An example is the hydrolysis of *t*-butyl chloride already discussed. Reactions of this kind are first order overall and the rate-determining step is unimolecular, as in (20-42).

The other class of nucleophilic substitutions consists of reactions that are second order overall—first order in the substrate and first order in the nucleophile. The rates of second-order substitution reactions depend on the natures of both X and Y, as well as R. This suggests that both the formation of the X-R bond and the breaking of the R-Y bond contribute to the energetics of the activated complex. It is not surprising that in some systems there is a strong correlation of increased reaction rate with increased basicity of the nucleophile X, for basicity reflects a tendency of an unshared electron pair to react with a center of high positive charge density, the proton. The availability of a low-lying empty orbital that can be used for bond formation by R—for example, an empty *d* orbital on a sulfur or phosphorus atom—also increases reactivity.

These lines of evidence, and others, suggest that the rate-determining step in these reactions is bimolecular and that in the activated complex there is at least a weak interaction of R with X and Y simultaneously:

$$X + RY \longrightarrow X \text{——} R \text{——} Y \longrightarrow XR + Y \tag{20-47}$$

Some reactions are believed to occur in a single step, in which the X-R bond is formed at the same tme the R-Y bond breaks—a *concerted mechanism.* In others, an intermediate that has (perhaps weak) bonds from R to both X and Y may persist for some time.

It may be difficult to distinguish between the unimolecular and bimolecular mechanisms, particularly when the nucleophile is the solvent and consequently the order of the reaction with respect to it is not easily established. The unimolecular reaction is suggested when it is observed that the rate increases in solvents that promote ionization, such as water. Studies involving optical isomers (Chap. 26) can also be helpful in determining the mechanism.

Not all substitution reactions can be regarded as strictly unimolecular or bimolecular. In some systems, the entering nucleophile plays a small role, but only a small role, in helping to break the R-Y bond. The effect of changes in X for a given RY depends, in part, on the strength of the R-Y bond; the stronger this bond, the smaller the effect. By systematic variations in the nature of X and Y, a smooth transition from the characteristics of one mechanism to those of the other can be observed.

Equilibrium control or kinetic control of product distribution.

In some reactions equilibrium is established so quickly that the relative amounts of reactants and products can be predicted solely on thermodynamic grounds, from equilibrium constants. The products of such a reaction are said to be *equilibrium-controlled*. Sometimes several different products are formed from a given reaction mixture by alternative pathways. If such a reaction is equilibrium-controlled, the proportions of the different products are entirely independent of the mechanisms and the relative rates of the reactions by which the various products are formed.

Most acid-base reactions and some others, including certain oxidation-reduction processes and acid-catalyzed reactions, are equilibrium-controlled. However, the great majority of the common reactions of organic, inorganic, and biological chemistry are usually not run under conditions in which equilibration of the possible products can occur. The relative amounts of the different products formed from a particular reaction mixture are therefore quite independent of their relative stabilities and depend only on their relative rates of formation. The distribution of products is said to be *kinetically controlled*.

For example, many of the reactions of the *t*-butyl halides—such as the hydrolysis of *t*-butyl chloride to *t*-butyl alcohol, Equation (20-41)—are kinetically controlled. In an alkaline solution, there are several possible reactions in addition to the formation of the alcohol. One of them is the elimination of the constituents of hydrogen chloride, with the resultant formation of a hydrocarbon containing a carbon-carbon double bond (an alkene):

$$\underset{\underset{Cl}{|}}{\overset{\overset{CH_3}{|}}{CH_3-C-CH_3}} + OH^- \longrightarrow \overset{\overset{CH_3}{|}}{CH_3-C=CH_2} + Cl^- + H_2O \qquad (20\text{-}48)$$

The analogous reactions for *t*-butyl bromide and *t*-butyl iodide also occur, but at quite different rates. Significantly, the proportion of the alkene is the same regardless of which halide is used. Thus, at room temperature in aqueous ethanol, about one-sixth of each halide is converted into the alkene. This is not the equilibrium proportion of the alkene, as can be shown from the known standard free energies of the alkene, the alcohol, and the other possible products.

The fact that the different halides react at different rates to give the alkene in the same proportions suggests strongly that the reactions proceed through an intermediate common to all three halides. The formation of this intermediate from the starting material is rate-determining and thus the rate depends on which halide is used. However, the proportion of products depends only on the ratio of the rates of the different possible reactions of this common intermediate and hence is independent of the starting material. As indicated earlier, the intermediate is believed to be the *t*-butyl cation, $(CH_3)_3C^+$. When it adds water and then loses a proton, as in (20-44), the product is *t*-butyl alcohol. When it loses a proton directly, the alkene is formed. Since all the protons in this cation are equivalent, only one alkene can be formed from it by loss of a proton. With some cations, however, loss of a proton can occur in several distinct ways to give isomeric alkenes. Under some conditions the proportions in which these are formed are kinetically controlled, depending only on the relative rates of loss of the different protons. Under other conditions the proportions may be equilibrium-controlled, the final ratio of products then being determined solely by the relative stabilities of the different alkenes that can be formed.

20-6   Reaction Mechanisms

We summarize here some ways of learning about the elusive details of what happens as reactants are changed into products.

Postulating a mechanism.   The first step in any study of kinetics is the determination of the rate law and its temperature dependence. If the rate law involves fractional orders or a sum of terms, the mechanism is certain to consist of several steps, no one of which is rate-determining. For simple rate laws there are some guidelines to aid in postulating a mechanism from the rate law and the known stoichiometry of the reaction. The key generalization is that the observed rate law must be explainable in terms of the reaction occurring in the rate-determining step. When the overall order is greater than 3, the mechanism very likely includes one or more equilibria and intermediates *prior* to the rate-determining step. No experimental evidence has ever been obtained for a molecularity greater than 3.

If the rate law contains an inverse dependence on the concentration of some species, that species must be a product of a rapid equilibrium prior to the rate-determining step and some other product of that same equilibrium must be a participant in the rate-determining step.

For example, iodide reacts with hypochlorite in alkaline solution to produce hypoiodite and chloride:

$$I^- + OCl^- \longrightarrow OI^- + Cl^- \tag{20-49}$$

The rate law is found to be

$$\frac{d[OI^-]}{dt} = k\frac{[I^-][OCl^-]}{[OH^-]} \tag{20-50}$$

The inverse dependence on $[OH^-]$ suggests that hydroxyl ion is the product of a rapid equilibrium prior to the rate-determining step. Since OCl^- appears in the numerator of the rate expression, its participation as a reactant in such an equilibrium is consistent with the experimental observations. Thus, a plausible rapid equilibrium is the hydrolysis of OCl^-:

$$OCl^- + H_2O \rightleftharpoons HOCl + OH^- \qquad \text{(rapid)} \tag{20-51}$$

The observed law is consistent with two additional steps in the mechanism, the sum of all the steps being the stoichiometric equation:

$$I^- + HOCl \longrightarrow HOI + Cl^- \qquad \text{(slow)} \tag{20-52}$$
$$OH^- + HOI \rightleftharpoons OI^- + H_2O \qquad \text{(rapid)} \tag{20-53}$$

The rate law is that of the slow step,

$$\frac{d[HOI]}{dt} = k'[I^-][HOCl] = k'K\frac{[I^-][OCl^-]}{[OH^-]} \tag{20-54}$$

with K the equilibrium constant for (20-51), the hydrolysis constant for OCl^-. Since each molecule of HOI is converted rapidly to one hypoiodite ion, OI^-, the rate of appearance of OI^- is given by the same expression, in agreement with the observed rate law (20-50), the experimental rate constant k being equal to $k'K$.

Testing a mechanism.   A mechanism can never be proved to be correct, but it can be proved wrong by one unambiguous piece of relevant experimental evidence. Sometimes several different mechanisms are consistent with the available evidence. It is then particularly desirable to design experiments that can distinguish among them. Among the useful kinds of experiments are:

1.   Careful analysis of reaction mixtures to establish the possible presence of small amounts of products other than the main products.

2.   Efforts to detect transient intermediates in reaction mixtures; spectroscopic methods are often valuable.

3.   The use of isotopes as *tracers* to permit following a reaction in which there is no net chemical change or to give information about the breaking of bonds during a reaction. Radioactive isotopes are often used, but stable isotopes, such as ^2H (deuterium) and ^{18}O, are also of great value. They can readily be detected and estimated precisely with a mass spectrometer (Fig. 1-10).

4.   The use of isotopes to provide mechanistic evidence by their influence on the rate of the reaction itself. This may be possible if an appropriate isotope can be substituted for one of the atoms that participates directly in bond breaking or bond formation during the rate-determining step of a reaction. The effect is discernible only for isotopes of relatively low atomic weight, no more than about 30, because it depends on the ratio of the masses of the normal isotope and that which is used in its place.

5.   Variation of the nature of the medium in which the reaction is carried out, which can provide important clues about the mechanism, e.g., comparing solvents of quite different polarity.

6.   Study of changes in the stereochemistry of the molecule that has been altered, that is, in the three-dimensional arrangement of its bonds. This method is particularly useful for molecules that are not superimposable upon their mirror images (Chaps. 25 to 27).

Reaction rates and equilibria.   For any reaction that occurs by a single elementary step, the rate of the forward reaction must, at equilibrium, be just equal to the rate of the reverse reaction. Thus, if in the elementary step A combines with B to form C,

$$A + B \longrightarrow C$$

then at equilibrium

$$k_f[A][B] = k_r[C] \tag{20-55}$$

with k_f and k_r the rate constants for the forward and reverse reactions, respectively. We know that at equilibrium $[C]/([A][B])$ is a constant, which we represent as K; thus, by (20-55),

$$K = \frac{[C]}{[A][B]} = \frac{k_f}{k_r} \tag{20-56}$$

The result in (20-56) can readily be generalized for reactions that proceed by a series of steps. The overall stoichiometric equation must be the sum of the equations for the individual steps, and we know from Chap. 13 that the equilibrium constant for a reaction that can be represented as the sum of other reactions is just equal to the product of the equilibrium constants for those other reactions. For a multistep reaction, each elementary step must proceed at the same rate as its reverse when the system is at equilibrium. Thus, if the overall reaction is the sum of three individual steps, denoted by the subscripts 1, 2, and 3, we apply (20-56) to each step and get for the overall equilibrium constant K

$$K = K_1 K_2 K_3 = \frac{k_{f,1}}{k_{r,1}} \frac{k_{f,2}}{k_{r,2}} \frac{k_{f,3}}{k_{r,3}} \tag{20-57}$$

20-7   Catalysis

Many reactions are speeded up by substances that are not themselves permanently changed during the reaction. These substances are called *catalysts,* a term introduced by Berzelius early in the nineteenth century. The effects are often dramatic. Reactions that proceed immeasurably slowly in the absence of catalysts may go too rapidly to be followed when a catalyst is present. Many catalysts are effective even when present in almost imperceptibly small amounts. Under certain conditions, for example, one molecule of the enzyme catalase can cause the decomposition of 10^5 molecules of hydrogen peroxide every second. In the absence of a catalyst, hydrogen peroxide and its solutions can be stored for years.

Since catalysts undergo no permanent change during the reaction whose rate they are accelerating, they do not appear in the stoichiometric equation for that reaction and can have no effect on the position of equilibrium. This implies that they must accelerate the reverse reaction by the same factor by which they accelerate the forward reaction, so that the equilibrium constant, the ratio of these rates at equilibrium, remains the same. For example, the hydrolysis of esters, a reaction exemplified by

$$CH_3C\overset{O}{\underset{OC_2H_5}{<}} + H_2O \longrightarrow C_2H_5OH + CH_3C\overset{O}{\underset{OH}{<}} \tag{20-58}$$

Ethyl acetate   Ethanol   Acetic acid
(an ester)   (an alcohol)   (a carboxylic acid)

is catalyzed by acids and by certain enzymes. Exactly the same catalysts have an equal effect on the rate of *formation* of esters and are used in the laboratory and in nature to speed up each of these reactions. Bases also speed up the hydrolysis of esters, but at the same time they react with the carboxylic acids formed to give salts such as sodium acetate. Thus their role is more than that of a catalyst.

Many catalysts are highly specific, being effective only for a certain reaction or a group of closely related reactions. This specificity is particularly characteristic of enzymes, a generic name for the thousands of different proteins that catalyze reactions in living systems.

The mode of action of catalysts. Catalysts are often classified according to whether they act homogeneously (in a single phase) or heterogeneously (at an interface between phases). Heterogeneous reactions are of great practical importance but are complex and not well understood; we discuss here only homogeneous reactions. Although by definition a catalyst is necessarily unaltered at the end of a reaction, every catalyst participates in the reaction it catalyzes. It interacts in some way with at least one of the reactants, and then in a subsequent step or steps, or perhaps as a continuation of the original interaction, it is regenerated in its original form while more or less simultaneously the final products are formed. Because catalysts are continually regenerated, they can act effectively when present in a very small molar proportion relative to the reactants—for example, 1:1000 or even smaller.

Some catalysts react chemically with one of the initial reactants, changing it to a substance that reacts more rapidly; the catalyst is then returned to its original state by a subsequent chemical change. Other catalysts form a rather loose complex with a reactant, perhaps involving only van der Waals forces and hydrogen bonds rather than any stronger chemical interactions. This complex reacts more rapidly than the original reactant and, as it is consumed, it leaves behind the unaltered catalyst, ready to act again.

In general, catalysts lower the activation energy of a reaction, thereby increasing its rate. Some lower the height of the barrier between reactants and products along a path close to that which the uncatalyzed reaction follows. Others open quite different pathways, which also have lower barriers than the original path but which are for some reason not accessible in the absence of the catalyst. In effect, such catalysts are like the scouts who directed bands of settlers in the Old West to mountain passes that were easier to cross than those first used, thus speeding up the migration to the lands beyond the mountains.

There are substances that slow down certain reactions, rather than accelerating them. These are called inhibitors or, sometimes, negative catalysts. The latter name is misleading, however, for they are not true catalysts since they are themselves permanently changed at the same time that they are inhibiting the reaction. True negative catalysts do not exist.[1]

We turn now to a few examples of different kinds of catalysis.

Oxidation-reduction reactions.   Many redox reactions, including oxidations by molecular oxygen, can be greatly accelerated by the addition of small amounts of ions of certain transition metals with two or more oxidation states that are accessible under the reaction conditions. Typical catalysts include Ag^+, Cu^+, Cu^{2+}, and Mn^{2+}. These catalysts, like many others, operate by what have sometimes been called compensating reactions, being alternately oxidized and reduced so that they are continually regenerated. For example, the oxidation of V^{3+} by Fe^{3+},

$$V^{3+} + Fe^{3+} \longrightarrow V^{4+} + Fe^{2+} \qquad (20\text{-}59)$$

is catalyzed by either Cu^+ or Cu^{2+}, the rate being proportional to the concentrations of V^{3+} and of either copper ion but independent of the concentration of Fe^{3+}. This result is explained by the two-step sequence of elementary processes

$$V^{3+} + Cu^{2+} \longrightarrow V^{4+} + Cu^+ \qquad \text{(rate-determining)} \qquad (20\text{-}59a)$$
$$Fe^{3+} + Cu^+ \longrightarrow Fe^{2+} + Cu^{2+} \qquad \text{(fast)} \qquad (20\text{-}59b)$$

whose sum is just (20-59). If Cu^+ is used as the catalyst, it is quickly changed to Cu^{2+} by reaction (20-59b) and the rate-determining step is still (20-59a). Although much has been learned during the last two decades about the mechanisms of many oxidation-reduction reactions, it is not yet clear why electron transfer from V^{3+} to Cu^{2+} and then from Cu^+ to Fe^{3+} should be so much faster than direct transfer from V^{3+} to Fe^{3+}.

Acid-base catalysis.   A great variety of reactions can be catalyzed by acids or bases, or by both acids and bases, some only by H^+ or OH^-, others by any acid or base present. We have already cited the formation and hydrolysis of esters; many

---

[1] There are substances that can interfere effectively and reversibly with the action of the catalysts, especially with enzymes. If they inhibit the action of the catalyst essentially completely, the net effect is to force the reaction to follow its normal uncatalyzed pathway. They do not, however, diminish the rate below that of the uncatalyzed reaction.

other reactions of organic compounds are similarly catalyzed. In general, the mechanism of acid catalysis involves transfer of a proton to a substrate (reactant) molecule, and catalysis by bases proceeds via proton transfer from the substrate to the base, resulting in a species of enhanced reactivity in each case.

An example of a sequence of elementary processes consistent with the available evidence about the acid-catalyzed hydrolysis of esters, with ethyl acetate,

$$CH_3C\overset{O}{\underset{OCH_2CH_3}{\diagup}} \text{, as an illustration, is}$$

$$CH_3C\overset{O}{\underset{OCH_2CH_3}{\diagup}} + H_3O^+ \rightleftharpoons CH_3C\overset{^+OH}{\underset{OCH_2CH_3}{\diagup}} + H_2O \tag{20-60a}$$

$$H_2O + CH_3C\overset{^+OH}{\underset{OCH_2CH_3}{\diagup}} \rightleftharpoons CH_3\overset{OH}{\underset{OCH_2CH_3}{\overset{|}{C}}}-OH_2{}^+ \rightleftharpoons CH_3\overset{OH}{\underset{\underset{H}{\overset{+}{\diagup}}OCH_2CH_3}{\overset{|}{C}}}-OH \tag{20-60b}$$

$$CH_3\overset{OH}{\underset{\underset{H}{\overset{+}{\diagup}}OCH_2CH_3}{\overset{|}{C}}}-OH \rightleftharpoons CH_3C\overset{^+OH}{\underset{OH}{\diagup}} + CH_3CH_2OH \tag{20-60c}$$

$$CH_3C\overset{^+OH}{\underset{OH}{\diagup}} + H_2O \rightleftharpoons CH_3C\overset{O}{\underset{OH}{\diagup}} + H_3O^+ \tag{20-60d}$$

The first and last of these steps are rapid proton transfers; the second and third involve the addition of a water molecule to form a tetrahedral intermediate and the elimination of a molecule of alcohol from this intermediate. Since each of the steps is reversible, the overall process is also reversible; the sum of all the steps is, of course, the overall hydrolysis equation (20-58).

Enzymes. The molecular weights of the proteins known as enzymes that catalyze reactions in living systems are usually in the range 10^4 to 10^6. A complex organism functions properly only because of a delicate balance of chemical reactions catalyzed by thousands of different enzymes, each with a distinct function and highly specific structure, essentially identical in all individuals of the same biological species and remarkably similar in different species. We shall defer consideration of any mechanistic details of enzyme action until the chemical nature and structure of proteins in general have been discussed (Chap. 27). However, a few of the facts and concepts relating to enzymes and their action that antedate, and are independent of, the knowledge of the structural details of any enzyme are worth discussing at this point.

The rates of many enzyme-catalyzed reactions have been found to be first order in the substrate at low substrate concentrations and, at a constant formality of the enzyme, to become essentially independent of the substrate concentration (zero order) as it becomes very high (Fig. 20-6). The rate is usually first order in the enzyme concentration as well, but this can be studied only at very low concentrations because the solubilities of proteins are low and their molecular weights are high.[1] Michaelis and Menten proposed in 1913 a mechanism that accounts for these facts—a mechanism that is still accepted today and is almost always associated with their names.

The catalytic activity of enzymes is attributed to particular regions of each molecule, called active sites. To simplify this discussion, we assume there is only one active site per enzyme molecule. A substrate molecule S can combine reversibly with the enzyme E, occupying the active site and forming an enzyme-substrate complex ES. The complex ES can then either dissociate to its original components, E and S, or it can decompose to give the products of the reaction that is being catalyzed as well as uncomplexed enzyme, ready to pick up more substrate. These reactions can be summarized as

$$E + S \rightleftharpoons ES \qquad (20\text{-}61a)$$
$$ES \longrightarrow E + \text{products} \qquad (20\text{-}61b)$$

This combination of reactions is far faster than the uncatalyzed reaction by which S is changed into the products. The rate-determining step is (20-61b) and this reaction is first order in the complex ES. It is not hard to show (see Prob. 20-17) that at constant formality of the enzyme this mechanism leads to just the relation between the reaction rate and the substrate concentration that is illustrated in Fig. 20-6. Although some enzymes have more than one active site, the principles of the foregoing discussion apply equally well to them.

---

[1]For a protein with molecular weight of 10^5, a 1 percent solution, which contains about 10 g protein per liter, has a concentration of 10^{-4} M.

FIGURE 20-6 **Rate of a typical enzyme-catalyzed reaction at constant formal concentration of enzyme.** The rate of reaction is proportional to the concentration of substrate at low values of [S] but becomes independent of it when [S] is high.

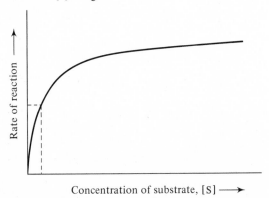

20-8   Concluding Remarks

Most of our present understanding of kinetics is derived from studies of the average behavior of large numbers of molecules with a range of energy states and modes of motion, and often in very complex environments. Because of the averaging, any resulting view is necessarily imprecise when it gets down to molecular details. However, there is every reason to believe that a revolution in the understanding of the dynamics of chemical change is now under way, a revolution that will eventually lead to extraordinarily detailed information about and control over the specific molecular interactions that occur as different kinds of reactions take place.

No real understanding of chemical dynamics could be contemplated until structural chemistry was on a firm foundation, both experimentally and theoretically, for chemical dynamics is essentially structural chemistry with the dimension of time added. Such a foundation has been provided in the last four or five decades. The rapid advances in electronics and instrumentation in the past decade or two have made it possible to design and carry out detailed experimental studies of elementary gaseous reactions by methods that employ intersecting beams of molecules, reactions in special mass spectrometers, ultrashort pulses of radiation, and related techniques. The effects of varying the mutual energies of the reactants, and of distributing this energy in varying proportions among translational, rotational, vibrational, and electronic modes, can be observed and analyzed, as can the ways in which energy and momentum are distributed among the products. These experiments are complex and expensive, but they should lead to valuable generalizations that can extend our understanding and control over the reactions of more complex molecules.

Problems and Questions

20-1   **Determination of order of reaction.**   Nitric oxide, NO, reacts with chlorine to form nitrosyl chloride, NOCl. At 22°C, with all reactants gaseous, the following data were obtained:

[NO]/M	[Cl$_2$]/M	$-\dfrac{d[Cl_2]}{dt}$ /$M\,s^{-1}$
0.100	0.100	8.0×10^{-3}
0.50	0.100	2.0×10^{-1}
0.100	0.50	4.0×10^{-2}

(a) What is the order of the reaction with respect to each reactant? (b) What is the rate constant for the reaction? (c) Suppose 0.30 mol NO and 0.60 mol Cl$_2$ are mixed in a 3.0-liter flask. How much chlorine disappears during the first 0.5 s?

Ans.   (a)   second order in NO, first order in Cl$_2$;   (b) $k = 8.0\ M^{-2}\,s^{-1}$;   (c)   about 0.024 mol

20-2   **Reaction rate.**   At constant hydrogen-ion concentration the rate expression for the reaction of bromide and bromate to give bromine,

$$5Br^- + BrO_3^- + 6H^+ \longrightarrow 3Br_2 + 3H_2O$$

can be written as

$$\frac{d[Br_2]}{dt} = k\,[Br^-][BrO_3^-]$$

The rate constant k is 2.1×10^{-4} liter mol^{-1} s^{-1}. A solution is prepared by dissolving 1.00×10^{-2} mol KBr in 100 ml water, which is mixed with 100 ml of a solution that is 1.00×10^{-2} F in KBrO$_3$ and 1 F in H$_2$SO$_4$. (a) At what rate is bromine appearing in the solution after exactly half the BrO$_3^-$ has reacted? (b) What is the molar concentration of Br$_2$ in the solution at this point? (c) State qualitatively the effect on the rate, at this point, of doubling the concentration of Br$_2$ by adding liquid bromine to the solution.

20-3   Mechanism and rate law. A reaction proceeds by the following mechanism:

$$mA + nB + pC \longrightarrow \text{products}$$

with m, n, and p all positive integers. Doubling the concentration of A, B, and C increases the overall rate of reaction by a factor of 16. Tripling the concentration of C has the same effect as tripling the concentration of A. An increase in the concentration of B has a larger effect than an increase in the concentration of A. What are m, n, and p? What is unrealistic about this mechanism?

Ans. 1, 2, 1

20-4   First-order kinetics. A substance X decomposes by a first-order reaction with half-life t_1. About how much time would be needed for the concentration of X to fall to 0.1 percent of its initial value?

20-5   Determination of order of reaction. A certain compound A decomposes in solution to form two other substances, B and C, under conditions such that the reverse reaction and competing reactions are negligible. The following kinetic data were measured:

Time/min	[A]/M	Time/min	[A]/M
0	0.100	58	0.037
10	0.084	92	0.020
20	0.071	140	0.009
36	0.054		

(a) What is the order of the reaction? (b) What is the rate constant?

Ans. (a) first order;   (b) $k = 0.017$ min^{-1}

20-6   Determination of order of reaction. A certain hydrocarbon A is unstable in the gas phase at temperatures above 300°C. Measurements of its concentration spectroscopically showed that, at 350°C and an initial pressure of 1.0 atm, the concentration of A fell to half its original value in 1.5 min; if the initial pressure was 0.10 atm at the same temperature, the half-life was 15 min. What can you deduce about (a) the order of the reaction by which A disappears, (b) the specific rate constant for this reaction, and (c) the activation energy of this reaction?

20-7   Quenching of a reaction. Some reactions in solution may be brought essentially to a halt by diluting the reaction mixture with an inert solvent at the same temperature. Does this possibility of quenching a reaction depend on the rate law? If so, how?

20-8 **Half-life of a fast reaction.** The rate constant for the transfer of a proton from H_3O^+ to OH^- is $1.4 \times 10^{11}\ M^{-1}\,s^{-1}$. If one assumes perfect mixing, how long would it take for half of a strong acid to be neutralized by a strong base if the initial concentration of each were $10^{-3}\ N$?

20-9 **First-order kinetics.** The gas-phase decomposition of sulfuryl chloride, SO_2Cl_2, into sulfur dioxide and chlorine is first order, with $k = 2.2 \times 10^{-5}\,s^{-1}$ at $320°C$. If a sample of sulfuryl chloride is heated for 10.0 hours at $320°C$, what fraction of it will still be present as SO_2Cl_2 at the end of this time? *Ans.* 0.45

20-10 **Activation energy.** The rate constant for the disappearance of chlorine in the third-order reaction of NO with Cl_2 to form NOCl is $4.5\ M^{-2}\,s^{-1}$ at $0°C$ and $8.0\ M^{-2}\,s^{-1}$ at $22°C$. What is the activation energy for this reaction?

20-11 **Activation energy.** At high temperatures nitrogen dioxide decomposes into NO and O_2 by the second-order rate law

$$\text{Rate} = -\frac{d[NO_2]}{dt} = k[NO_2]^2$$

At 592 K the rate constant is 4.98×10^{-1} liter $mol^{-1}\,s^{-1}$, and at 656 K it is 4.74 liter $mol^{-1}\,s^{-1}$. Calculate an activation energy that accounts for these values. *Ans.* 114 kJ mol^{-1}

20-12 **Activation energy.** Two second-order reactions have identical frequency factors. The activation energy of reaction 2 is 20 kJ mol^{-1} greater than that of reaction 1. Calculate the ratio of their rate constants (*a*) at $27°C$ and (*b*) at $327°C$.

20-13 **Implications of a mechanism.** The following observations have been made about a certain reacting system: (i) When A, B, and C are mixed at about equal concentrations in neutral solution, two different products are formed, D and E, with the amount of D about 10 times as great as the amount of E. (ii) If everything is done as in (i) except that a trace of acid is added to the reaction mixture, the same products are formed, except that now the amount of D produced is much smaller than (about 1 percent of) the amount of E. The acid is not consumed in the reaction. The following mechanism has been proposed to account for some of these observations and others about the order of the reactions:

1. $A + B \underset{k_{-1}}{\overset{k_1}{\rightleftharpoons}} F$ (rapid)

2. $C + F \xrightarrow{k_2} D$ (slow)

3. $C + F \xrightarrow{k_3} E$ (slow)

(*a*) Explain what this proposed scheme of reactions implies about the dependence (if any) of the rate of formation of D on the concentrations of A, of B, and of C. What about the dependence (if any) of the rate of formation of E on these same concentrations? (*b*) What can you say about the relative magnitudes of k_2 and k_3? (*c*) What explanation can you give for observation (ii) in view of your answer to (*b*)?

Ans. (*a*) $d[D]/dt = (k_1 k_2 / k_{-1})[A][B][C]$; $d[E]/dt = (k_1 k_3 / k_{-1})[A][B][C]$;

(*b*) k_2 is about $10 k_3$

20-14 Reaction mechanism. Ferrous ion is oxidized by chlorine in aqueous solution, the overall equation being

$$2Fe^{2+} + Cl_2 \longrightarrow 2Fe^{3+} + 2Cl^-$$

It is found experimentally that the rate of the overall reaction is decreased when either the ferric-ion or the chloride-ion concentration is increased. Which of the following possible mechanisms is consistent with the experimental observations?

(*a*) (1) $Fe^{2+} + Cl_2 \underset{k_{-1}}{\overset{k_1}{\rightleftharpoons}} Fe^{3+} + Cl^- + Cl$ (rapid)

  (2) $Fe^{2+} + Cl \xrightarrow{k_2} Fe^{3+} + Cl^-$ (slow)

(*b*) (3) $Fe^{2+} + Cl_2 \underset{k_{-3}}{\overset{k_3}{\rightleftharpoons}} Fe(IV) + 2Cl^-$ (rapid)

  (4) $Fe(IV) + Fe^{2+} \xrightarrow{k_4} 2Fe^{3+}$ (slow)

where Fe(IV) is Fe in the (+IV) oxidation state.

20-15 Rate of a proton-transfer reaction. Eigen and his coworkers found that the specific rate of proton transfer from a water molecule to an ammonia molecule in a dilute aqueous solution is $k_1 = 2 \times 10^5 \text{ s}^{-1}$. The equilibrium constant for the dissociation of "ammonium hydroxide" is $1.8 \times 10^{-5} M$. What, if anything, can be deduced from this information about the rate of transfer of a proton from NH_4^+ to a hydroxide ion? Write equations for any reactions you mention, making it clear to which reaction(s) any quoted constant(s) apply.

Ans.   rate constant $= 1.1 \times 10^{10} M^{-1} \text{s}^{-1}$

20-16 Rate and equilibrium. Consider the reaction

$$A + B \rightleftharpoons C + D$$

with all reactants and products gaseous (for simplicity) and an equilibrium constant K. (*a*) Assume that the elementary steps in the reaction are those indicated by the stoichiometric equation (in each direction), with specific rate constants for the forward reaction and the reverse reaction, respectively, k_f and k_r. Derive the relation between k_f, k_r, and K. Comment on the general validity of the assumptions made about the relation of elementary steps and the stoichiometric equation and also on the general validity of K. (*b*) Assume that the reaction as written is exothermic. Explain what this implies about the change of K with temperature. Explain also what it implies about the relation of the activation energies of the forward and reverse reactions and how this relation is consistent with your statement about the variation of K with temperature.

20-17 Rate of an enzyme-catalyzed reaction. Show that the Michaelis-Menten mechanism for enzyme action leads at constant formality of the enzyme to the relation depicted in Fig. 20-6, rate $= a[S]/(b + [S])$, with [S] the substrate concentration and a and b constants. {*Hint:* Use the formality of the enzyme—call it E_0—to express the concentration of free enzyme, $[E] = E_0 - [ES]$, so that [E] can be eliminated from the equilibrium expression for the reaction of (20-61*a*).}

"It strikes a sympathetic chord, I think, to learn that 700 years ago . . . the then Queen of England moved out of the city to Nottingham where she was residing because of the insufferable smoke; and that some 300 years later the brewers of Westminster offered to use wood instead of coal because of Queen Elizabeth's allergy to coal smoke. But it was only about the end of her reign that feeling began to lead to action; and then there was a prohibition—probably ineffective—of the use of coal in London while Parliament was sitting!"

Hugh E. C. Brewer,[1] 1955

21 Hydrogen, Oxygen, Nitrogen, and the Inert Gases

An enormous body of descriptive chemistry underlies the theoretical aspects of the science that have been discussed in the earlier chapters. Most often it is only after regularities in nature have been discovered by observation and experiment that a body of theory is developed to explain the experimental facts. Systematic studies that turn up unexpected deviations from "normal" behavior provide the impetus for modifications of existing theories and sometimes disprove theories entirely.

This chapter is the first of seven concerned with the descriptive chemistry of some of the elements, an essential part of the science of chemistry. Much current research is concerned with developing new materials and finding new pathways for the

[1]Hugh E. C. Brewer, in F. S. Malette (ed.), "Problems and Control of Air Pollution", Van Nostrand Reinhold Company, New York, 1955.

synthesis and modification of known substances. To accomplish this the chemist must have an intuition for the way substances react and how their physical properties depend on their structures. This intuition is developed by studying in a systematic way the body of chemical information that has been collected through the years.

We have not provided an encyclopedic coverage of descriptive chemistry. The aim of these chapters is rather to give the reader a feeling for the variety of chemical compounds and reactions and the importance of some of them in our lives and our society. This chapter deals with three important elements, hydrogen, oxygen, and nitrogen, as well as the inert gases. It includes a discussion of the chemistry of water and of the atmosphere.

21-1   Hydrogen

Occurrence and physical properties.   Hydrogen appears to be the most abundant element in the universe. In those parts of the earth that are most accessible—the crust, the oceans, and the atmosphere—hydrogen ranks ninth in abundance on a weight basis (Table 21-1), although there are more atoms of hydrogen than of any other element except oxygen and silicon. Virtually all the hydrogen is in combined form; any free hydrogen in the atmosphere is eventually lost because the earth's gravity is not sufficiently strong to retain this very light molecule.

The normal form of hydrogen is H_2 because of the large enthalpy change associated with the dissociation reaction

$$H_2 \longrightarrow 2H \qquad \Delta H^\oplus = 436 \text{ kJ} \qquad (21\text{-}1)$$

Atomic hydrogen can be formed at very high temperatures, for example in electric arcs, but even at 3000 K the degree of dissociation is only about 8 percent.

The van der Waals interactions between H_2 molecules are very weak. Hence the gas condenses to a liquid at extremely low temperature (20.4 K at 1 atm) and a very soft molecular solid forms at 14.1 K. Liquid hydrogen is the lightest of all liquids—its

TABLE 21-1   **Abundance of the elements on the earth's surface***

Oxygen	O	49.5		Chlorine	Cl	0.19	
Silicon	Si	25.7		Phosphorus	P	0.12	
Aluminum	Al	7.5		Manganese	Mn	0.09	
Iron	Fe	4.7		Carbon	C	0.08	
Calcium	Ca	3.4	99.2	Sulfur	S	0.06	0.7
Sodium	Na	2.6		Barium	Ba	0.04	
Potassium	K	2.4		Chromium	Cr	0.033	
Magnesium	Mg	1.9		Nitrogen	N	0.030	
Hydrogen	H	0.87		Fluorine	F	0.027	
Titanium	Ti	0.58		Zirconium	Zr	0.023	

All others < 0.1

*Percentages by weight in the earth's crust, the oceans, and the atmosphere.

density is 0.070 g cm^{-3}, only 7 percent the density of water—and solid hydrogen has the lowest density of all crystalline solids, 0.088 g cm^{-3}.

Deuterium, the hydrogen isotope of mass number 2, constitutes 1 part in 5000 of naturally occurring hydrogen; tritium, ^{3}H or T, occurs only to an extremely small extent in nature. The physical properties of isotopic molecules differ significantly only when there is a large percentage difference in the masses. Since the masses of H$_2$, HD, D$_2$, DT, and T$_2$ are in the ratio 2:3:4:5:6, the variation in the physical properties through the series is unusually large. For example, the boiling points of HD and D$_2$ are, respectively, 2.1 and 2.2 K higher than that of H$_2$, and at 20 K the molar volumes of the liquids are 9 and 17 percent smaller.

Isotopic differences are also observable in the spectra of molecules containing hydrogen. The frequency of vibration of a bonded pair of atoms X—D is lower than the X—H frequency. Thus, when D is substituted for H in a compound, there is a characteristic shift in the frequency of the spectral lines associated with the X—H bond vibrations and the lines can be readily identified. For example, a comparison between the spectra of CH$_3$OH and CH$_3$OD (Table 21-2) shows that the frequency shift is smallest for lines attributed to CH$_3$, which indeed is little affected by the substitution. The shift is only slightly larger for CO vibrations, but there is a large difference for OH and OD vibrations. Such changes in spectra can be very useful keys to the determination of structural features of molecules by the analysis of complex spectra.

Nuclear magnetic resonance spectroscopy. Another important tool for the determination of molecular structure is also based on a physical property of the hydrogen atom, its nuclear spin. Like the electron, the proton has a spin and an associated magnetic moment. In a magnetic field the proton can be in either of two energy states distinguished by the orientation of its magnetic moment with respect to the field direction (Fig. 21-1). The difference in energy between the states is directly proportional to H, the strength of the field:

$$\Delta\epsilon = \beta H \tag{21-2}$$

TABLE 21-2 Some infra-red lines of CH$_3$OH and CH$_3$OD

Group to which observed line is attributed	Wave numbers of lines observed, $\tilde{\nu}$/cm^{-1}	
	CH$_3$OH	CH$_3$OD
CO	1034	1040
OH or OD	1340	823
CH$_3$	1430	1427
CH$_3$	1455	1459
CH$_3$	1477	1480
CO	2053	2065
CH$_3$	2844	2849
CH$_3$	2977	2964
OH or OD	3682	2720

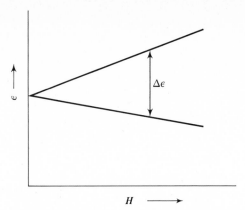

FIGURE 21-1   **Schematic energy levels for a proton in a magnetic field.** In the absence of a magnetic field the two states of the proton have equal energy. The energy difference between the states, $\Delta\epsilon$, is directly proportional to the field strength H.

A proton in the lower state can absorb radiation of frequency $\nu = \Delta\epsilon/h = \beta H/h$ and be excited to the upper state; protons in the upper state can drop to the lower state by emitting radiation of the same frequency. It is customary to describe these processes as *spin flips* between the two orientations with respect to the field.

Measurements of the spectral lines associated with nuclear spin flips can provide information about molecular structure because the magnetic field that determines the absorption or emission frequency for a given proton is the sum of the applied field, produced by the laboratory magnet, and the local field, produced by the presence of other atoms in the immediate surroundings of the proton. In effect, every proton in a molecule can act as a magnetic sensor of its atomic environment. Protons that have identical environments absorb or emit radiation at the same frequency; those in other environments are characterized by slightly different frequencies. This is the basis of nuclear magnetic resonance (nmr) spectroscopy.

A representative nmr spectrum is shown in Fig. 21-2. Even for very strong magnetic fields, the energy change associated with the flip of a proton spin is small, and the corresponding frequency is in the radio range. Nuclear magnetic resonance spectroscopy is therefore also known as radio-frequency spectroscopy. The sample spectrum is for ethanol, CH_3CH_2OH, which has protons in three different environments. Three protons are part of the $-CH_3$ group. They are magnetically equivalent; that is, they experience the same local field in their environment. The two protons in the $-CH_2$ group constitute another distinct group, and the third environment is that of the proton in $-OH$. Under high resolution (Fig. 21-2b) each of the three absorption lines shows fine structure that is characteristic of the group to which the protons belong. The area under each peak (or group of peaks in the high-resolution spectrum) is proportional to the number of structurally identical atoms. In principle, nmr spectra can be observed for all nuclei with spins. These include ^{19}F, ^{17}O, ^2H, and ^{14}N. Proton nmr is most commonly studied because hydrogen is present in so many compounds and because of the high signal strengths characteristic of proton spectra.

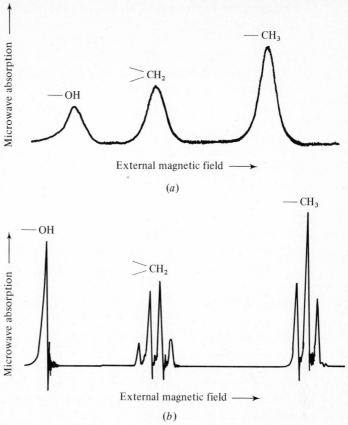

FIGURE 21-2 **Nmr spectrum of ethanol.** Graph (*a*) shows the nmr absorption spectrum of ethanol, CH_3CH_2OH, at relatively low resolution; (*b*) shows the "fine structure" at high resolution.

Preparation of hydrogen. Water is the raw material from which most hydrogen is produced industrially. Hydrogen of the highest purity is produced by the electrolysis of water, a relatively expensive process. There are various commercial methods of preparation, the most economical of which is the action of steam on coke at temperatures in the neighborhood of 1000°C:

$$H_2O + C \rightleftharpoons H_2 + CO \qquad (21\text{-}3)$$

The mixture of CO and H_2, known as *water gas,* finds wide use as a fuel since both components are combustible.

If reaction (21-3) is to be used as a source of hydrogen, the H_2 must be separated from the CO. This is accomplished by mixing the products with steam and passing them over a catalyst at 500°C to convert the CO to CO_2:

$$CO + H_2O \rightleftharpoons H_2 + CO_2 \qquad (21\text{-}4)$$

Carbon dioxide is very soluble in water; hydrogen is only slightly soluble. Hence the CO_2 is easily removed by passing the product mixture under pressure into water.

Bonding of hydrogen. Hydrogen forms compounds in which it has lost an electron, producing H^+, as well as many binary covalent molecules with elements of groups IV through VII, known as *covalent hydrides*. Boron, aluminum, and gallium, members of group III, form *complex hydrides,* ions with the general formula XH_4^-. A typical complex hydride is lithium aluminum hydride, which can be made by the reaction of lithium hydride and aluminum chloride in dry ether:

$$4LiH + AlCl_3 \longrightarrow LiAlH_4 + 3LiCl \qquad (21\text{-}5)$$

Complex hydrides find wide use as reducing agents.

Hydrogen can also gain an electron to form the hydride ion, H^-. This process occurs in the reaction of hydrogen with highly electropositive metals. An example is the reaction of molecular hydrogen with sodium at 700°C:

$$H_2(g) + 2Na \longrightarrow 2Na^+ + 2H^- \qquad (21\text{-}6)$$

The compounds formed are typical colorless crystalline salts known as *saltlike hydrides*. The hydride ion reacts vigorously with water:

$$H^- + H_2O \longrightarrow H_2 + OH^- \qquad (21\text{-}7)$$

Bridge hydrogens. Although a typical covalent bond involves the sharing of two electrons between a pair of atoms, there is a class of compounds, termed *electron deficient,* in which the number of electron pairs available for bonding is less than the number of covalent bonds. Many of the known electron-deficient molecules contain boron and hydrogen; the simplest is diborane, B_2H_6. It contains 12 electrons, 3 from each boron atom and 1 from each hydrogen atom. A conventional valence bond structure would require seven bonds, six between B and H atoms and one connecting the boron atoms.

Structural investigations have shown that there are two distinct types of H atoms in diborane (Fig. 21-3a). Four of them are terminal H atoms, each associated with a single B atom; the others are bridging atoms that are equidistant from the two B atoms. The bonding in B_2H_6 is depicted in Figs. 21-3b and c. The 2s and 2p orbitals of boron are hybridized to sp^3 orbitals and the bonds between the boron and the terminal hydrogen atoms can be thought of as arising from the overlap of the hybrid orbitals with the 1s orbitals of each hydrogen atom. These four bonds require four electron pairs. Each of the bridging hydrogen atoms is associated with *one* electron pair that bonds it to the two B atoms by what is called a three-center bond. This bond may be considered to result from the combination of one sp^3 orbital from each boron and the s orbital of a hydrogen atom.

21-2 Oxygen

Occurrence and physical properties. The most abundant of the chemical elements on earth is oxygen. Elemental oxygen constitutes 23 percent by weight (21 vol percent) of the atmosphere, and oxygen in a combined form comprises almost

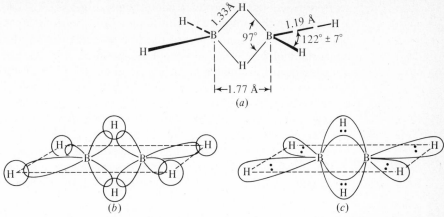

FIGURE 21-3   **Structure and bonding in diborane, B_2H_6.** (*a*) Structural formula. The wedge-shaped bonds imply that the H atoms are above the plane of the page; the dashed bonds imply that the H atoms are below the plane of the page. (*b*) Diagram depicting the 2*s* and 2*p* orbitals of the boron atoms, hybridized to sp^3 and overlapping with the 1*s* orbitals of the hydrogen atoms. (*c*) Diagram showing these overlapping orbitals schematically combined into bond orbitals, each occupied by an electron pair.

50 percent by weight of the crust of the earth and 89 percent of the oceans. It is a constituent of all living matter.

The low boiling and melting points of O_2 (90 K and 54 K, respectively) are evidence of the weakness of the van der Waals interactions between the molecules. The gas is odorless, tasteless, and colorless, but liquid oxygen is blue. Oxygen is slightly soluble in water; since its solubility is about twice that of nitrogen, the ratio of dissolved O_2 to N_2 in water is 1:2, rather than 1:4 as it is in air. The electronic structure and paramagnetism of O_2 were discussed in Chap. 12; the paramagnetism is observed in the gaseous, liquid, and solid states and vanishes only at very low temperatures.

Oxygen is obtained commercially from the liquefaction and subsequent distillation of air. Small quantities are also produced by electrolysis of water. Three isotopic species of oxygen occur in nature: ^{16}O (99.76 percent), ^{17}O (0.04 percent), and ^{18}O (0.20 percent). Both ^{17}O and ^{18}O can be concentrated by fractional distillation of water. Highly enriched $H_2^{17}O$ and $H_2^{18}O$ are commercially available, and their use as tracers, particularly in biochemical applications, is common.

Ozone.   Ozone, O_3, is a naturally occurring but less familiar molecular form of oxygen. At room temperature it is a pale blue gas with a characteristic pungent odor (Greek: *ozein,* smell) that liquefies at 162 K and solidifies to a dark violet solid at 80 K. Urban atmospheres contain ozone at concentrations as high as 0.3 ppm[1] on very smoggy days. It is an irritant and also plays an integral part in the complex

---

[1] ppm stands for *parts per million.*

series of reactions that produce photochemical smog (Sec. 21-6). Nevertheless ozone is essential to the presence of all life on earth; an ozone layer in the upper atmosphere, formed by photochemical reaction of oxygen at an altitude of about 25 km, filters the biologically lethal ultraviolet radiation from the sun. The concentration of ozone in this protective layer can be diminished significantly by reaction of O_3 with other reactive species, such as NO or halogen atoms. There is evidence that damagingly high concentrations of NO may be introduced into the upper atmosphere by large numbers of supersonic aircraft and that similarly damaging concentrations of Cl atoms may be produced photochemically in the upper atmosphere from the normally inert chlorofluorocarbon propellant gases used in some aerosol sprays.

The ozone molecule is similar to SO_2 in geometry and electronic structure. It is bent, with a bond angle of 117°. Its electronic structure can be represented by the

superposition of two resonance formulas, and, correspondingly, the bond distance lies between that of a single and a double O—O bond.

Ozone is generally prepared by passing oxygen through an electric discharge:

$$3O_2 \longrightarrow 2O_3 \qquad \Delta H^\ominus = 285 \text{ kJ} \qquad (21\text{-}8)$$

As one might expect from (21-8), energy-rich O_3 is a much more powerful oxidizing agent than O_2.

Bonding of oxygen.   Binary compounds of oxygen with all the elements save He, Ne, Ar, and Kr are known. Bonding in these oxides ranges from ionic to covalent.

There are three types of ionic oxides. The most important are those that contain the oxide ion, O^{2-}. Although considerable energy must be expended to form O^{2-} from molecular oxygen, the strong electrostatic interactions of the small O^{2-} ion with cations stabilize oxide crystal lattices. Oxide ion reacts vigorously in solution to form hydroxyl ion:

$$O^{2-} + H_2O \longrightarrow 2OH^- \qquad (21\text{-}9)$$

Compounds containing the O_2^{2-} ion are called peroxides. Examples are sodium peroxide, Na_2O_2, and barium peroxide, BaO_2, which react with water or dilute acid to form hydrogen peroxide, H_2O_2 (Sec. 21-3). A small class of ionic oxides contain O_2^-, the superoxide ion; an example is potassium superoxide, KO_2. Peroxides and superoxides find use as strong oxidizing agents.

The oxides of the nonmetals are all covalent. We discuss some of them in subsequent sections. The acidic, basic, and amphoteric properties of oxides have been treated in Sec. 7-4.

21-3 Compounds of Hydrogen and Oxygen

Hydrogen peroxide. This compound, H_2O_2, is a syrupy, faintly blue liquid that freezes at $-0.9°C$ and boils at $150°C$. The molecular structure of H—O—O—H is like that depicted in Fig. 11-2 for the atomic arrangement W—X—Y—Z, the torsion angle being about $100°$ and the bond angle O-O-H also about $100°$. The O-O distance of 1.46 Å is to be compared with those of 1.49 for O_2^{2-} in BaO_2, 1.26 for O_2^- in KO_2, 1.21 for O_2, and 1.12 for O_2^+ (observed in electric discharges in O_2).

Hydrogen peroxide has acid-base properties similar to those of H_2O. It has a somewhat larger acid constant than H_2O, and its more important properties are those of a strong oxidizing agent and somewhat weaker reducing agent. The pertinent standard potentials are given by the Latimer diagram shown on page 525.

Water. Water is doubtless the most thoroughly studied compound. Virtually every aspect of its chemical and physical behavior has been studied experimentally and treated by theory. One might expect, then, that there would be little more to be learned about this simple triatomic molecule, but this is not so. Interpretation of the special properties of water, particularly those relating to hydrogen bonding, is exceedingly complex and still presents significant challenges. We shall not review the chemistry of water here; it has been dealt with earlier in the discussions of equilibria, hydrogen bonding, hydration of ions, ionization, and acids and bases. The present focus is on physical properties.

Most liquids become more dense as their temperature is lowered. Water shows this behavior down to $4°C$, but below that temperature its density *decreases* as the temperature is lowered. The density decreases still further when water freezes, another unusual effect shown by comparatively few substances. These unusual properties are of profound significance to life in bodies of water in cold climates. Consider what happens when a lake, originally above $4°C$, cools in the winter. At first the coldest water sinks to the bottom because it is more dense than the warmer liquid. But once the temperature of the water reaches $4°C$, the coldest water begins to accumulate at the surface and eventually freezes. The ice, which is less dense than the liquid, floats on the surface and slows the rate of cooling of the liquid underneath. As a result, large lakes do not freeze solidly even in the coldest winters and aquatic life survives.

The unusual behavior of the density of water is caused by strong hydrogen bonding between the molecules. As can be seen in Fig. 21-4, the structure of ice is very open. Each water molecule is surrounded tetrahedrally by four nearest neighbors with which it forms hydrogen bonds. When ice melts, the tetrahedral structure is partially destroyed and the water molecules crowd somewhat more closely together, which accounts for the increase in density on melting. Some residual order due to hydrogen bonding persists in the liquid. As the temperature is raised a little, hydrogen bonds are broken and a more disordered, denser liquid is obtained. A rise in temperature also increases the thermal motion of the molecules, the origin

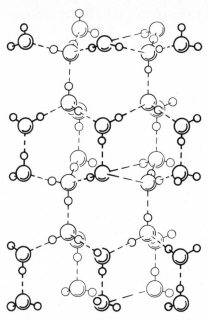

FIGURE 21-4   **The arrangement of water molecules in ice.**   Each H_2O molecule is surrounded tetrahedrally by four other H_2O molecules, with which it forms strong hydrogen bonds. Other crystalline forms of ice exist at high pressures; they are more dense than liquid water.

of the normal thermal expansion of liquids. At 4°C the effect of thermal motions overcomes the effect of loss of structure and above 4°C the temperature dependence of the density becomes normal.

The breaking of hydrogen bonds as the temperature is raised is also responsible for the abnormally high heat capacity of water. Energy must be spent to disrupt these bonds as well as to increase the average kinetic energy of the molecules. An important consequence of the high heat capacity of water is the stabilizing influence of large bodies of water on the climate; temperatures are always more moderate near the ocean than they are in the nearby interiors of continents. The oceans and lakes are efficient reservoirs of energy. They absorb energy from their surroundings during warm weather and release it to the surroundings in cold weather.

Heavy water.   Although there is very little D_2O (heavy water) present in natural water, high concentrations of D_2O can be obtained by electrolysis. Under suitable conditions ordinary hydrogen is liberated roughly six times more readily than D_2; the unelectrolyzed water is therefore enriched in D_2O.

The molecular weight of water is sufficiently low that isotopic substitution can have a marked effect on its physical properties. For example, D_2O freezes at 3.82°C and boils at 101.4°C. The dielectric constant of heavy water is lower than that of H_2O and its viscosity is higher. The chemical properties of the isotopic species are not identical. At 25°C the ion product of D_2O is 0.3×10^{-14} and its heat of formation is about 3 percent larger than that of ordinary water. The properties of HDO are intermediate between those of ordinary water and D_2O.

Water quality. Water is of paramount importance to life, not only life in the narrow biological sense but life as applied to the functioning of our society. When we speak of water in this sense, we are not restricting ourselves to the chemically pure substance H_2O. Instead, we refer to a complex chemical system of varying composition that we find in the oceans and rivers, in municipal water systems, in wastes, in the atmosphere—virtually everywhere in our environment.

Some of the constituents of natural waters are listed in Table 21-3. Even water that we might call "pure", such as the water in a mountain stream, is likely to contain many of these substances, at least in small amounts. The distinction between brackish water and the water in a mountain stream depends largely on the relative proportions of the contaminants. If Table 21-3 were not limited to natural waters but included

TABLE 21-3 Constituents of natural waters

| Source | Particle size classification | | | | |
| | Suspended | Colloidal | Dissolved | | |
			Molecules	Positive ions	Negative ions
Atmosphere	←——Dusts——→		CO_2 SO_2 O_2 N_2	H^+	HCO_3^- SO_4^{2-}
Mineral soil and rock	←—Sand → ←—Clays → ←—Mineral soil particles ——→		CO_2	Na^+ K^+ Ca^{2+} Mg^{2+} Fe^{2+} Mn^{2+}	Cl^- F^- SO_4^{2-} CO_3^{2-} HCO_3^- NO_3^- Various phosphates
Living organisms and their decomposition products	Algae Diatoms Bacteria ←—Organic soil (topsoil)——→ Fish and other organisms	Viruses Organic coloring matter	CO_2 O_2 N_2 H_2S CH_4 Various organic wastes, some of which produce odor and color	H^+ Na^+ NH_4^+	Cl^- HCO_3^- NO_3^-

industrial effluents and municipal sewage systems as well, the list of impurities and sources would have to be greatly expanded. In the sections that follow we consider the effect of some of the constituents of water on its properties and describe ways in which water can be treated to improve its quality.

Hard water.   Water that contains appreciable quantities of any of the ions Ca^{2+}, Mg^{2+}, and Fe^{2+} is said to be "hard". These ions, and others, are picked up by water percolating through even slightly soluble minerals or flowing over them in rivers and streams. For example, slightly acidic groundwater percolating through limestone ($CaCO_3$) or dolomite ($CaCO_3 \cdot MgCO_3$) reacts to form the soluble HCO_3^- salts $Ca(HCO_3)_2$ and $Mg(HCO_3)_2$. Some soils provide Fe^{2+} and Mg^{2+} in the form of the soluble salts $FeSO_4$ and $MgSO_4$.

When soaps are used in hard water, a grayish curdy scum is formed. It is a precipitate produced by the reaction of a metal ion with the active ingredient in soap, the sodium salt of a long-chain organic acid such as stearic acid, $C_{17}H_{35}COOH$:

$$M^{2+} + 2Na^+(C_{17}H_{35}COO^-) \longrightarrow 2Na^+ + M^{2+}(C_{17}H_{35}COO^-)_2 \quad (21\text{-}10)$$

The scum settles on the wash and is aesthetically unpleasing. More importantly, it reduces the cleaning power of the soap by removing the emulsifying agents, the organic acid ions, from solution.

Boiler scale, a hard deposit that coats the interior of boilers, hot water pipes, heat exchangers, and tea kettles, is formed when hard water containing appreciable quantities of HCO_3^- ion is heated:

$$M^{2+} + 2HCO_3^- \longrightarrow MCO_3 + CO_2 + H_2O \quad (21\text{-}11)$$

The scale consists of the carbonates of Ca, Mg, and Fe; $CaSO_4$ may be present as well if the water contains SO_4^{2-}. The deposits reduce the efficiency of heat transfer and constrict pipes so that the flow of water and steam is impeded.

The effects of hard water present a serious economic problem, and a number of methods have been devised to soften water or to inhibit the action of the metal ions. Precipitation of soap during laundering can be eliminated by the use of synthetic detergents. The active ingredient in these substances is an emulsifying organic ion that does not form an insoluble compound with divalent metal ions. Another component of detergents is a chelating agent such as sodium trimetaphosphate, $Na_3P_3O_9$, which forms soluble complexes with metal ions. Unfortunately, as we shall see later, the phosphates in many detergents have serious environmental effects in certain areas of the country. Washing soda, $Na_2CO_3 \cdot 10H_2O$, can be added to water prior to washing to precipitate the divalent metal ions in an acceptable granular form; it is a component of some low-phosphate or phosphate-free washing products. Care must be exercised in using washing soda, however, because it is highly alkaline.

An elegant way to soften water is to replace the objectionable divalent metal ions by Na^+. This may be accomplished with ion exchangers such as zeolites, which are complex aluminum silicates. These solids are built up of a rigid, negatively charged framework of AlO_4 and SiO_4 tetrahedra (Sec. 22-3) and contain channels into which positive ions can fit. If the zeolite is placed in a concentrated NaCl solution, the channels fill with Na^+ ions. Divalent metal ions bind more strongly to the zeolite than do the sodium ions. Thus, if hard water is passed through a vessel containing a sodium-filled zeolite, the sodium ions are displaced. If a singly charged site on the zeolite anion is represented by Z, the exchange reaction may be written

$$2NaZ + M^{2+} \rightleftharpoons 2Na^+ + MZ_2 \qquad (21\text{-}12)$$

When all the sodium ions have been exchanged, the zeolite can be regenerated (even though the equilibrium lies far to the right) by placing the zeolite in contact with a concentrated salt solution. The reaction is thereby driven backward in accordance with Le Châtelier's principle.

Zeolites are no longer used in most commercial water softeners; they have been replaced by more efficient organic ion-exchange resins, networks of polymer molecules containing groups that react with ions. Resins can be synthesized that exchange either positive or negative ions, and they can be made highly selective. Ion-exchange techniques are widely used for analysis as well as for purification. For example, the highly similar lanthanide ions can be separated from each other by ion-exchange chromatography.

21-4   Water Pollution

Although the divalent metal ions that cause hardness are undesirable constituents of water, they are not normally classified as pollutants. The word *pollutant* is usually reserved for contaminants present in high enough concentration to affect human beings adversely. This definition is obviously not precise, and even when agreement can be reached on definitions of the terms, it is usually very difficult to determine precisely how high the concentration of a substance must be in order to produce adverse effects. For example, do we mean adverse effects that occur immediately, or in weeks, months, years, or generations? Adverse effects may also be indirect. Our ecosystem (Greek: *oiko,* habitation) consists of many interrelated subsystems. An adverse effect on some small subsystem may be amplified as it moves to other subsystems, eventually emerging as a serious pollution problem.

Standards for drinking water. The federal standards, Table 21-4, for chemical substances in drinking water illustrate some of the complexities in the definition of pollution and some of the related economic, governmental, and scientific problems. There are two types of limits set on allowable concentrations of contaminants: (1) limits that may not be exceeded under any circumstances and (2) limits that should not be exceeded whenever in the judgment of the local water authority a more suitable supply is or can be made available at reasonable cost.

The economic side of pollution is immediately evident. What is "reasonable cost"? This depends on the benefits, which again cannot be sharply defined. Consider the standard for nitrate ion. A level of 45 mg liter^{-1} appears to be safe for adults

TABLE 21-4   **Federal standards for chemical substances in drinking water**

Substance	Concentration/mg liter^{-1}
1 The following chemical substances should not be present in a water supply in excess of the listed concentrations where, in the judgment of the reporting agency and the certifying authority, other more suitable supplies are or can be made available	
Alkyl benzene sulfonate (ABS)	0.5
Arsenic	0.01
Chloride	250.0
Copper	1.0
Carbon chloroform extract	0.2
Cyanide	0.01
Iron	0.3
Manganese	0.05
Nitrate*	45.0
Phenols	0.001
Sulfate	250.0
Total dissolved solids	500.0
Zinc	5.0
2 The presence of the following substances in excess of the concentrations listed shall constitute grounds for rejection of the supply	
Arsenic	0.05
Barium	1.0
Cadmium	0.01
Chromium	0.05
Cyanide	0.2
Lead	0.05
Selenium	0.01
Silver	0.05

*In areas in which the nitrate content of water is known to be in excess of the listed concentration, the public should be warned of the potential dangers of using the water for infant feeding.

(although there have been suggestions that further study might prove this too high). Infants, however, have bacteria in their digestive tracts that convert NO_3^- to NO_2^- (nitrite ion), and nitrite ions react with hemoglobin and thereby reduce the oxygen-carrying capacity of the blood. Nitrite poisoning is the cause of a potentially fatal disease, infantile methemoglobinemia, the blue-baby syndrome. In Minnesota between 1947 and 1950 there were 139 cases of this disease and 14 deaths, all associated with well water containing nitrates and in some cases with NO_3^- levels below the federal standard. Nitrates can be removed from water by a relatively expensive ion-exchange process. How does one measure the cost—with respect to the health of infants or with respect to the potential hazards to a larger percentage of the population?

Other sections of the federal drinking water standards deal with the level of bacterial contamination and physical characteristics such as taste, smell, color, and turbidity. Omitted from these standards are limits on the concentrations of viruses. There is evidence that disease-causing viruses do get into water supplies and may cause outbreaks of viral disease, but no routine tests for waterborne viruses have yet been developed. Only recently has the carcinogenic (cancer-causing) effect of asbestos fibers been discovered, and the possibility of widespread contamination of drinking water by asbestos is therefore just being recognized. It is clear, then, that there must be a continuing scientific, legal, and economic review of health standards.

Sources of water pollution.   Water is a recyclable resource. It is drawn from rivers, lakes, and wells, utilized in homes, on farms, and by industry, and then returned to the environment. Each time that water is used, its quality—its suitability for use—is altered. Sometimes the quality of water is significantly improved between uses, either by natural purification processes (such as filtration through soil layers) or by treatment in sewage plants, but often the intensity of water usage is so great that little purification takes place.

The volume of waste water from industry and sewage plants is staggering. Estimated yearly quantities for the United States are shown in Table 21-5; not included are the volumes of water used for agricultural purposes, mining, runoff from rainfall, and the sewage from the 37 percent of the population who were not served by sewers

TABLE 21-5   **Estimated annual volumes of industrial and domestic wastes before treatment, 1963**

	*Waste water** (10^6 m^3)
INDUSTRY	
Food and related products	2,800
Textile mill products	560
Paper and related products	7,600
Chemical and related products	15,000
Petroleum and coal	5,200
Rubber and plastics	640
Primary metals	17,000
Machinery	600
Electrical machinery	360
Transportation equipment	960
All other manufacturing	1,800
All manufacturing	52,000
DOMESTIC	
Served by sewers (120 million people)	20,000†

* For comparison the average annual flow of the Colorado River is $22,000 \times 10^6 \text{ m}^3$.

† Number of persons $\times$ 0.5 m³ per person per day $\times$ 365 days per year.

at the time the estimates were made. We consider here only a few of the major contaminants that arise from this intensive use of water.

Degradable substances.   Given sufficient time, natural water systems are able to cleanse themselves of organic impurities and certain nitrogen compounds by the action of bacteria and other microorganisms. These organisms consume dissolved oxygen as they function. A measure of the concentration of degradable substances in water is the weight of dissolved oxygen utilized in bacterial action during a given period, the *biological oxygen demand* (BOD). If the BOD is so high that the oxygen is used at a rate greater than the rate at which it can be replenished by aeration or furnished by plants growing under water, the character of the water changes. Aquatic life dies out, organic debris collects, and microorganisms that can survive without oxygen begin to multiply. They feed on organic matter and release carbon dioxide, methane, hydrogen sulfide, and foul-smelling organic sulfur compounds. The hydrogen sulfide reacts with metal ions to form black precipitates that float as a scum on the surface.

To maintain the BOD at acceptable levels it is necessary to control the discharge of organic materials into natural waters. This can be accomplished by proper treatment of sewage, not only in urban areas but also on farms and feedlots (waste production by farm animals in the United States has been estimated to be about 20 times that of the human population). This action is not sufficient, however. A high BOD can be produced by the decay of plants and algae whose growth has been encouraged by the presence of nitrogen- and phosphorus-containing plant nutrients present in waste water. Nitrates and phosphates in fertilizers are carried into groundwater during irrigation. Soluble phosphates come from human and animal wastes, and as much as three-fourths of the soluble phosphates in urban sewage can come from detergents. Effective control of plant nutrients in the water system requires a reformulation of detergents and removal of soluble phosphates in the treatment of sewage.

Nondegradable organic substances.   To an increasing extent new chemical substances have been developed to perform certain functions. Some of these persist for long times in the ecosystem because they are chemically very stable and are also not microbially degradable. Prime examples are the chlorinated hydrocarbon insecticides, such as DDT, dieldrin, and aldrin:

DDT   Dieldrin   Aldrin

Chlorinated hydrocarbons degrade very slowly in the soil and are washed into the water system. Although the levels of these compounds in the water are not very high, there is a natural amplification system that concentrates them. Fish scales are coated by a fatty layer. Since chlorinated hydrocarbons dissolve more readily in fats than in water, they are in effect extracted from the water by the fatty layer. When small aquatic animals are eaten, the insecticide is further concentrated in the body fat of larger animals. It works its way along the food chain and eventually reaches man.

Although DDT has been linked to metabolic disorders in birds, there has been no conclusive evidence that it constitutes a health hazard to human beings. As a precautionary measure, however, its use in the United States was banned in 1972. Since then, some scientists have argued that the ban on DDT is potentially more harmful than its continued use. They feel that careful, controlled usage is necessary to control insect-borne diseases such as malaria. All sales of dieldrin and aldrin were suspended in August 1974 because medical data indicated that they produce liver tumors in rats. This finding is not necessarily indicative that these substances can produce cancer in human beings: workers in plants that produce them apparently show no adverse effects even though they have been exposed to very high levels of the insecticides.

Acid mine drainage and coal wastes. Domestic coal has a sulfur content of 6 to 8 percent, and one-fifth to three-fifths of this may come from the mineral pyrite (iron disulfide, FeS_2). The iron disulfide remaining in the mines and left in heaps of coal wastes after coal has been processed oxidizes gradually in moist air to produce sulfuric acid, sulfates, iron oxides, and other compounds:

$$FeS_2 + O_2 + H_2O \longrightarrow H_2SO_4 + Fe_2O_3 + \text{other compounds} \quad (21\text{-}13)$$

Water that enters the mine or washes through the wastes dissolves the oxidation products and carries the resulting acid solution to surface waters.

In 1971 there were over 800 waste piles covering 12,000 acres in the anthracite region of Pennsylvania alone, and it is estimated that acid mine drainage pollutes 10,000 miles of streams in the United States. The best current solution to this problem is a program of prevention—sealing abandoned mines to keep out air and reducing water seepage. Attempts have also been made to control acid drainage by neutralization with CaO.

21-5 Nitrogen

About 78 vol percent of the earth's atmosphere consists of N_2, a colorless, odorless, and tasteless gas. Nitrogen in combined form is also present to a small extent in the earth's crust. Soluble nitrogen salts are important components of fertile soils, and nitrogen makes up about 16 percent by weight of all proteins. Liquid nitrogen boils at 77 K and freezes at 63 K. The element is obtained commercially by the distillation of liquid air; unless it is further purified it usually is contaminated with small amounts of oxygen.

Because of the strength of the N-N triple bond, N_2 is very unreactive at room temperature. The processes by which N_2 can be converted to more reactive and directly useful compounds are therefore important. The removal of nitrogen from the atmosphere and its conversion to soluble compounds is called *nitrogen fixation*. Some nitrogen is fixed when it reacts with oxygen in lightning discharges, and certain soil bacteria associated with the roots of legumes (alfalfa, clover, beans, peas) are also able to fix nitrogen. Commercially the most common first step for the utilization of N_2 is the preparation of ammonia.

Bonding of nitrogen.   Nitrogen is covalently bonded in most of its compounds, although it does form the nitride ion, N^{3-}, when it reacts at high temperatures with the most electropositive elements, such as lithium. Li_3N is a saltlike material, but most nitrides are covalent—for example, boron nitride (Fig. 4-4). Nitrogen normally forms three single covalent bonds, as in the pyramidal molecules NH_3 and NF_3, or multiple bonds, as in N_2 and HCN. The nitrogen octet can also be completed by the gain of an electron and the formation of only two covalent bonds. The amide ion, NH_2^-, which is isoelectronic with water, is an example of this mode of bonding. Amides can be formed by highly electropositive metals; the most common is sodium amide or sodamide, $NaNH_2$, used as a reagent in organic synthesis.

Loss of an electron leaves nitrogen isoelectronic with carbon, and thus N^+ forms four covalent bonds, as in ammonium ion (NH_4^+) and substituted ammonium ions (such as $CH_3NH_3^+$, methylammonium ion).

Ammonia.   The most important compound of nitrogen and hydrogen is ammonia. It can be synthesized by the direct reaction of gaseous N_2 and H_2:

$$N_2 + 3H_2 \rightleftharpoons 2NH_3 \qquad \Delta H^\oplus = -92 \text{ kJ} \qquad (21\text{-}14)$$

At $25°C$ the equilibrium constant for reaction (21-14) is quite large, $K = 7.1 \times 10^5$; the rate of reaction, however, is negligible. Raising the temperature increases the rate, but as indicated by the standard enthalpy of formation, the reaction is exothermic and an increase in temperature decreases K. Thus, at $450°C$, the reaction rate is only slightly more favorable and K has fallen to 6.5×10^{-3}—the equilibrium no longer favors the product.

In spite of these apparently mutually exclusive requirements for satisfactory yield and rate, reaction (21-14) is the basis of the most important industrial method for the production of ammonia. The German chemist Haber made the process viable through the use of a catalyst to promote the rate and by careful choice of the conditions. As shown in Fig. 21-5, even at temperatures as high as 400 to $600°C$ acceptable yields can be obtained if the reaction is run at elevated pressures. This result can be predicted by application of Le Châtelier's principle, since the number of moles in the reaction mixture is decreased when product is formed. A further enhancement in the yield is obtained by removing the ammonia as it is formed, thereby driving the reaction toward completion.

There are many similarities between ammonia and water. Like water, liquid ammonia is self-associated by hydrogen bonding. Its boiling point, $-33.4°C$, is abnormally high when compared with the other hydrides in its group, such as PH_3

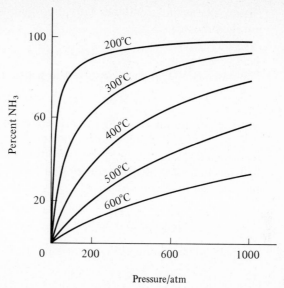

FIGURE 21-5   **Percentage conversion of a 3:1 mixture of N_2 and H_2 into NH_3 at equilibrium for different temperatures and pressures.**

and AsH_3, and it has a high dielectric constant. Analogies also exist between the chemistry of water and ammonia. Like water, ammonia can undergo a self-ionization equilibrium:

$$2NH_3 \rightleftharpoons NH_4^+ + NH_2^- \qquad K = [NH_4^+][NH_2^-] \approx 10^{-30} \text{ at } -50°C \quad (21\text{-}15)$$

Thus the ammonium ion can be thought of as the analog of the hydronium ion, while the amide ion can be regarded as the analog of the hydroxyl ion.

Ammonia is one of a group of compounds that are the chief materials of the chemical industry and are referred to collectively as *heavy chemicals*. The term *heavy* refers to the amounts of the substances produced and not to their densities. A sizable portion of our industrial capacity is connected with the production of heavy chemicals, but they draw little attention because they do not appear in finished products. Huge quantities of ammonia are produced—15 million tons in the United States alone during 1972. Much of this ammonia is used as a fertilizer, either directly or in the form of ammonium compounds.

Nitric acid.   The production of an important heavy chemical, nitric acid, requires large quantities of ammonia. Gaseous NH_3 is catalytically oxidized to nitric oxide at temperatures between 500 and 1000°C in the Ostwald process:

$$4NH_3 + 5O_2 \xrightarrow{\text{platinum catalyst}} 4NO + 6H_2O \qquad (21\text{-}16)$$

Without the presence of the platinum catalyst, the oxidation stops at elemental nitrogen:

$$4NH_3 + 3O_2 \longrightarrow 2N_2 + 6H_2O \qquad (21\text{-}17)$$

Nitric oxide readily combines with atmospheric oxygen at room temperature to give nitrogen dioxide:

$$2NO + O_2 \longrightarrow 2NO_2 \tag{21-18}$$

The gaseous NO_2 is then dissolved in water to form nitric acid,

$$3NO_2 + H_2O \longrightarrow 2HNO_3 + NO \tag{21-19}$$

and the NO produced in this step is collected and reoxidized to NO_2.

Nitric acid is a strong oxidizing agent; it will oxidize all metals except the noble metals gold, platinum, rhodium, and iridium. As indicated by two of the possible half-reactions that involve reduction of NO_3^-, the products of reduction depend on the hydrogen-ion concentration:

$$NO_3^- + 2H^+ + e^- \rightleftharpoons NO_2 + H_2O \qquad E^{\ominus} = 0.80 \text{ V} \tag{21-20}$$
$$NO_3^- + 4H^+ + 3e^- \rightleftharpoons NO + 2H_2O \qquad E^{\ominus} = 0.96 \text{ V} \tag{21-21}$$

Copper reduces dilute HNO_3 primarily to NO,

$$3Cu + 8H^+ + 2NO_3^- \longrightarrow 3Cu^{2+} + 2NO + 4H_2O \tag{21-22}$$

but gives NO_2 with concentrated acid:

$$Cu + 4H^+ + 2NO_3^- \longrightarrow Cu^{2+} + 2NO_2 + 2H_2O \tag{21-23}$$

Of course the products of reduction also depend on the strength of the reducing agent. For example, the powerful reducing agent zinc ($E^{\ominus}_{Zn^{2+}/Zn} = -0.76 \text{ V}$) reduces nitric acid all the way to ammonium ion:

$$4Zn + 10H^+ + NO_3^- \longrightarrow 4Zn^{2+} + NH_4^+ + 3H_2O \tag{21-24}$$

Nitric acid reacts with concentrated sulfuric acid to form nitronium ion, NO_2^+:

$$HNO_3 + H_2SO_4 \longrightarrow NO_2^+ + HSO_4^- + H_2O \tag{21-25}$$

This reaction is of great importance in organic chemistry because of the ease with which NO_2^+ reacts with aromatic molecules, e.g.,

$$\bigotimes + NO_2^+ \longrightarrow \bigcirc\!\!-NO_2 + H^+ \tag{21-26}$$

Hydrazine and hydroxylamine. Ammonia and the nitrides are compounds in which nitrogen is in a ($-$III) oxidation state. Nitrogen can also be found in two other negative oxidation states, ($-$I) and ($-$II), as in hydroxylamine, NH_2OH, and hydrazine, N_2H_4, respectively.

Hydroxylamine can be thought of as derived from NH_3 by the substitution of an OH group for one of the H atoms. The structure of hydrazine is related to that of H_2O_2. The two NH_2 groups are not in an eclipsed position (Sec. 11-2) but are rotated with respect to each other with a torsion angle φ reported to be 90 to 95° (Fig. 11-2). When free of water, hydrazine is a colorless liquid that boils at 144°C and freezes at 2°C. It is a weak base capable of accepting two protons:

$$H_2NNH_2 + H_2O \rightleftharpoons H_2NNH_3^+ + OH^- \qquad K_1 = 1.0 \times 10^{-6}$$
$$H_2NNH_3^+ + H_2O \rightleftharpoons H_3NNH_3^{2+} + OH^- \qquad K_2 = 9 \times 10^{-16}$$

Oxides of nitrogen. Two of the oxides of nitrogen have already been introduced, nitric oxide (NO) and nitrogen dioxide (NO_2). Four other oxides are known; their names, formulas, and properties are given in Table 21-6.

Nitrous oxide, N_2O, is the only nontoxic oxide of nitrogen. It causes hysteria when it is breathed for a short time and for this reason is sometimes called laughing gas. Prolonged inhalation produces unconsciousness and the gas is used as an anesthetic. Canisters of instant whipped cream contain cream and N_2O under pressure. The gas readily dissolves in the fat and when the pressure is released it forms many tiny bubbles that whip the cream.

NO is unusual in that it is an *odd molecule,* a molecule that contains an odd number of electrons. There is no unique Lewis structure for the molecule, but a resonance description based mainly on two formulas may be written

$$\left\{ :\ddot{N}=\ddot{O}: , :\dot{N}=\ddot{O}: \right\}$$

The presence of an unpaired electron makes nitric oxide paramagnetic.

The molecular orbitals for NO are shown in Fig. 21-6. There are 4 pairs of electrons in bonding orbitals and $1\frac{1}{2}$ electron pairs in antibonding orbitals, resulting in $2\frac{1}{2}$ bonds. The molecular orbital diagram is in agreement with the fact that the bond distance in NO is 1.15 Å, while that in the NO^+ ion, which has one less antibonding electron, is 1.06 Å. It explains the ease with which NO loses an electron to form NO^+ and the related fact that the bond energy in NO^+ is 1.6 times greater than that in NO.

Nitrogen dioxide, a red-brown paramagnetic gas, exists in equilibrium with the colorless, diamagnetic N_2O_4. Like NO, NO_2 is an odd molecule; the N_2O_4 dimer

TABLE 21-6   Oxides of nitrogen

Formula	Name	Color	Remarks
N_2O	Nitrous oxide	Colorless	Rather unreactive
NO	Nitric oxide	Gas, colorless; liquid and solid, blue	Moderately reactive
N_2O_3	Dinitrogen trioxide	Blue solid	Extensively dissociated as gas
NO_2	Nitrogen dioxide	Brown	Rather reactive
N_2O_4	Dinitrogen tetroxide	Colorless	Extensively dissociated to NO_2 as gas and partly dissociated as liquid
N_2O_5	Dinitrogen pentoxide	Colorless	Unstable as gas; ionic solid
NO_3; N_2O_6			Not well characterized and quite unstable

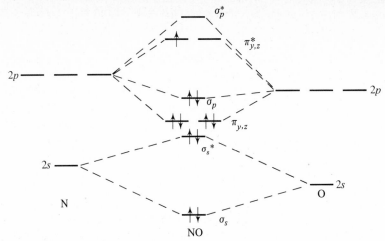

FIGURE 21-6   **Molecular orbitals of NO.**   The energies of the atomic orbitals of O are lower than those of N because O is more electronegative (see also Fig. 12-10).

necessarily has an even number of electrons. As predicted from the Latimer diagram[1]

$$NO_3^- \xrightarrow{\ 0.80\ } NO_2 \xrightarrow{\ 1.07\ } HNO_2$$

NO_2 and N_2O_4 disproportionate in water to give a mixture of nitrous acid (HNO_2) and nitric acid:

$$2NO_2 + H_2O \longrightarrow HNO_2 + H^+ + NO_3^- \qquad (21\text{-}27)$$

21-6   Air Pollution

The oxides of nitrogen have recently gained notoriety because of their involvement in air pollution. Standard equipment for automobiles in the United States now includes devices to limit the emissions of so-called NO_x, mixtures of nitrogen oxides. The oxides are formed in internal-combustion engines when the air drawn into the cylinders is subjected to high temperatures and pressures during the combustion process.

Originally the term *smog*, which is a combination of the words smoke and fog, was used to describe the air pollution caused primarily by the combustion of smoky and sulfur-containing fuels. In cities such as London and Pittsburgh, dense, dirty fogs containing fine particles of soot and ash were once very common. When fuels with high sulfur content are burned, the sulfur is oxidized to sulfur dioxide, a

---

[1] The acid/base species shown are those dominant when $[H^+] = 1$. They define the standard states with which the E° values given are associated. For nonstandard states the Nernst equation must be used with actual concentrations or partial pressures of the species shown, including the concentration of H^+. In Latimer diagrams applying to basic conditions, the species shown are those dominant when $[OH^-] = 1$ and the Nernst equation must be formulated accordingly.

poisonous gas with a characteristic irritating odor:

$$S + O_2 \longrightarrow SO_2$$

The SO_2 can be further oxidized to sulfur trioxide:

$$2SO_2 + O_2 \longrightarrow 2SO_3$$

This gas is also very poisonous and damaging to the lungs; it reacts with water to form sulfuric acid.

Smog laden with SO_2 and SO_3 can be literally deadly. In December 1952 heavy smog covered London for 4 days and was responsible for around 4000 deaths. The smog that hovered over Donora, Pennsylvania, in 1948 caused 20 deaths among the 14,000 inhabitants and made almost half the population ill.

Improved smoke-abatement measures and the use of low-sulfur fuels have helped to rid many cities of sulfur-laden and sooty smog. Unfortunately, a new kind of smog, more difficult to eliminate, has taken its place. It is called photochemical smog, a form of air pollution linked heavily but not exclusively to the automobile. Before discussing the chemistry of photochemical smog, we consider briefly the manner in which light can be involved in chemical reactions.

Photochemical reactions. Figure 21-7a shows schematically the changes in potential energy that may occur when a diatomic molecule AB in its ground electronic state absorbs a photon of frequency ν_1 (energy $h\nu_1$) and undergoes a transition to an excited electronic state AB*. The process can be described by the equation

$$AB + h\nu_1 \longrightarrow AB^* \tag{21-28}$$

FIGURE 21-7 **Schematic energy changes accompanying the absorption of a photon.** In (a) a molecule absorbs a photon of energy $\Delta\epsilon_1 = h\nu_1$ and undergoes a transition from the ground state AB to an excited electronic state AB*. In (b) the molecule absorbs a photon of different energy $\Delta\epsilon_2$ and is brought to an electronic state AB** that is unstable with respect to dissociation into A + B.

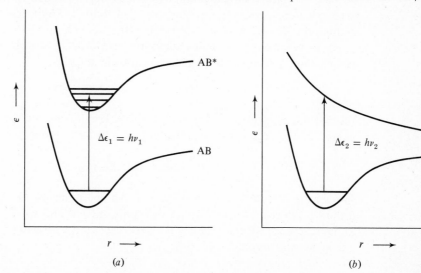

(a) (b)

The lower curve in Fig. 21-7*a* represents the way in which the potential energy of AB varies with internuclear separation in the ground state. The upper curve, which represents the potential energy curve for the excited molecule AB*, is similar but not identical because the equilibrium separation and the bond strength depend on the electronic structure. As long as the molecule remains in this excited state, its potential energy will be that represented by the upper curve.

Now consider the situation shown in Fig. 21-7*b*. Here again the molecule starts out in the ground state, but now it absorbs a photon of energy $h\nu_2$, which takes it to a *different* excited electronic state, AB**. Unlike the excited state AB*, this state is not stable to dissociation: there is no minimum in the graph of potential energy against internuclear separation. The potential energy of an A atom and a B atom in this state is lowest when the atoms are very far apart, and there is no bonding. Thus, the equation describing the light absorption process is

$$\text{AB} + h\nu_2 \longrightarrow [\text{AB**}] \longrightarrow \text{A} + \text{B} \tag{21-29}$$

The absorption of the photon causes the molecule to dissociate.

In this simple example, the photochemical reaction is produced by a photon of a very specific frequency, ν_2. Often in real systems molecules may be excited to a number of very closely spaced levels so that a photochemical reaction can be produced by light in a rather broad range of wavelengths. The extent of the reaction depends on the concentration of ground-state molecules and the intensity of the light, i.e., on the number of photons that strike the sample per second.

Reactions producing photochemical smog. There are three necessary ingredients for the production of photochemical smog: light, oxides of nitrogen, and hydrocarbons. The mechanism of smog formation is very complex and not completely understood. It involves many compounds present in the atmosphere at concentrations of only a few parts per million. In one attempt to describe the kinetics of smog formation, 60 elementary reactions were included in the analysis and 120 reactions were considered as possible contributors. Some reactions, however, are unquestionably essential to any mechanistic scheme.

Photochemical smog begins when nitrogen dioxide absorbs light of wavelength shorter than 430 nm and dissociates photochemically:

$$\text{NO}_2 + h\nu \longrightarrow \text{NO} + \text{O}$$

The atomic oxygen formed is very reactive. It can combine rapidly with molecular oxygen in the presence of a third molecule to form ozone:

$$\text{O} + \text{O}_2 \xrightarrow{\text{M}} \text{O}_3 \tag{21-30}$$

The third molecule, M, absorbs some of the energy in the O–O_2 collision. Without it the O_3 molecule formed would be highly excited and would immediately revert to the reactants. Atomic oxygen can also react with hydrocarbon vapors to produce aldehydes and oxyacyl radicals:

$$O + \text{hydrocarbons} \longrightarrow RCHO + RCO_2$$

Aldehyde   Oxyacyl
radical

The symbol R in these formulas represents a hydrocarbon part of the molecule, such as CH_3CH_2- (Chap. 26). Its identity depends on the particular hydrocarbon in the atmosphere encountered by the O atom.

Both ozone and aldehydes are components of smog. Ozone irritates mucous membranes and oxidizes double bonds in rubber and plastics, causing the materials to crack and weaken. Aldehydes sting the eyes and can attack the central nervous system.

Oxyacyl radicals are also very reactive. They can oxidize nitric oxide and produce another reactive species, acyl radicals:

$$RCO_2 + NO \longrightarrow NO_2 + RCO$$

Acyl
radical

The acyl radicals can react further to form the major eye irritants in smog, the peroxyacyl nitrates, known as PAN:

$$RCO + NO_2 + O_2 \longrightarrow RCO_3NO_2$$

Peroxyacyl
nitrate

Other reactions that contribute to smog involve other organic species, as well as CO and OH radicals.

The approach to the control of photochemical smog is based on limiting the emissions of the reactants to the atmosphere. Primary sources of oxides of nitrogen are internal-combustion engines and power plants. Hydrocarbons enter the atmosphere from incomplete combustion and from the evaporation of solvents and fuels. The automobile is a major source of both types of reactants. In an attempt to meet federal standards, manufacturers have changed the conditions of combustion to minimize the formation of nitrogen oxides and to reduce the concentration of unburned hydrocarbons in the exhaust. These design changes are not sufficiently effective to meet the new more stringent standards, and catalytic converters have been added to the exhaust systems in new autos to oxidize hydrocarbons that remain after combustion.

21-7   Group 0

The elements in group 0 have been known both as the rare gases and as the inert gases. Both these terms are in part misnomers. Argon is far from rare; it constitutes almost 1 percent of the atmosphere. Helium is found in concentrations of up to 7 percent in certain natural gases. Since 1962 it has been known that not all the elements in this group are completely inert. The story behind this discovery is an interesting episode in the history of science.

In 1894 Lord Rayleigh and Sir William Ramsay announced that a new element more inert than nitrogen had been isolated from air, and in 1895 Ramsay read a paper at the Royal Society in which he described in detail his unsuccessful efforts to form compounds of the new substance. The name *argon* (Greek: *argos,* inert, not working) seemed appropriate. The five other members of the group were all discovered within the next 5 years, and during the decades that followed no evidence of compound formation for any of the congeners was uncovered. During the twenties and early thirties it was suggested that the heavier inert gases might form compounds, and Pauling, on theoretical grounds, predicted that a compound of xenon and fluorine should be stable. An attempt to synthesize xenon fluoride and xenon chloride was reported in 1933; its lack of success stifled further experimentation.

It was not until 1962 that Neil Bartlett, after preparing an ionic compound of O_2^+ and the PtF_6^- ion, realized that the ionization energy of Xe is nearly identical to that of O_2. He mixed Xe with PtF_6 vapor and immediately obtained a solid product that was later identified as xenon hexafluoroplatinate, $Xe^+(PtF_6)^-$.

Once the first inert-gas compound was reported, a substantial number of others were discovered. These include XeF_4, XeF_2, KrF_2, XeO_3, XeO_4, XeO_6^{4-}, $XeCl_2$, and $FXeN(SO_2F)_2$, the first compound containing a Xe—N bond. No compounds of He, Ne, or Ar have been synthesized.

With the exception of helium and radon, the inert gases are obtained from the distillation of liquid air. The world's supply of helium comes mainly from natural gas supplies in the United States and the Soviet Union. Argon is used commercially to provide an inert atmosphere in incandescent light bulbs. It decreases the rate at which the tungsten filament evaporates and therefore allows the filament to be run at a higher temperature to produce a whiter light. Neon finds use in electric signs and other discharge tubes. Because of its very low boiling point (4.2 K), helium is important as a refrigerant for cryogenic studies. Plans to use superconducting metals in technical applications such as electrical transmission lines would require that large quantities of liquid helium be produced.

Problems and Questions

21-1   **Nmr and structure.** Give the number of peaks and the relative peak areas that should be observed in low-resolution proton magnetic resonance spectra of the following molecules: $CH_3CH_2CH_2CH_3$; CH_3OCH_3; CH_3NHCH_3.   *Ans.* 2, 3:2;   1;   2, 6:1

21-2   **Molecular weights of boron hydrides.** The empirical formulas and densities of two gaseous boron hydrides are as follows: BH_3 has a density of 0.629 g liter^{-1} at 17°C and 406 torr; B_2H_5 has a density of 0.1553 g liter^{-1} at 19°C and 52 torr. What are the respective molecular weights of these two compounds? What are their correct molecular formulas?

21-3   **Bond energies of O-O bonds.** Use data in Table D-4 to find the enthalpy change for the reaction $3O \longrightarrow O_3$, and calculate the "bond enthalpy" of the O-O bond in O_3. Repeat these calculations for the O-O bond in H_2O_2, using an $\widetilde{H}_f^{\ominus}$ value of -136 kJ mol^{-1} for $H_2O_2(g)$ as well as values for $H_2O(g)$, etc., from Table D-4. Compare these values with that given for the O-O single bond in Table 11-2 (which is actually an average "bond enthalpy" rather than a bond energy).   *Ans.* 300 kJ mol^{-1}, 142 kJ mol^{-1}

21-4 **Bonding in boron hydrides.** The structure of the B_4H_{10} molecule (Fig. 21-8) can be described in terms of two single-bonded BH groups and two BH_2 groups with single B-H bonds. The four B atoms are arranged in two triangles that share a side. The shared side consists of the BH groups, linked by a B-B single bond. Along the remaining four sides of the triangles are BHB three-center bonds. Give the total number of valence electrons in B_4H_{10} and show their distribution on a schematic diagram.

FIGURE 21-8

21-5 **Association of water molecules.** In the vapor phase water is believed to be dimerized to a small extent. Calculate the enthalpy change associated with the dimerization reaction $2H_2O(g) \rightleftharpoons H_4O_2(g)$ from the following data: at 40°C, mole fraction $H_4O_2 = 1.6 \times 10^{-3}$, $P_{H_2O} = 55$ torr; at 100°C, mole fraction $H_4O_2 = 5.0 \times 10^{-3}$, $P_{H_2O} = 760$ torr.   *Ans.* -24 kJ

21-6 **Hydrogen bonds.** Define the term *hydrogen bond* and explain the bearing of the geometry of hydrogen bonding on the relative densities of water and ice.

21-7 **Solid solutions of NH_4F in H_2O.** Ammonium fluoride tends to form solid solutions with ice and is the only known substance that does this. Explain this behavior in terms of the ice structure.

21-8 **Properties of a phosphate substitute.** A number of phosphate-free detergents use sodium metasilicate, Na_2SiO_3, as a replacement for phosphate compounds. Sodium metasilicate is the salt of metasilicic acid, H_2SiO_3, for which $K_1 = 2 \times 10^{-10}$ and $K_2 = 1 \times 10^{-12}$. When the recommended quantity of detergent is used, about 1.5 g Na_2SiO_3 is added to each liter of water. What is the pH of the resulting solution? Should any special precautions be taken when storing such products in the home?

21-9 **Water hardness.** The hardness of water is usually expressed in terms of milligrams per liter of equivalent $CaCO_3$, that is, the concentration of $CaCO_3$ that would by itself produce the hardness observed for that water. During 1974, the Los Angeles Water Department supplied 8.5×10^{10} gal water to residences, and the average hardness was 97 mg liter^{-1} of $CaCO_3$. (*a*) Assume that 10 percent of this water was used for washing and that soap can be represented as sodium stearate, $C_{17}H_{35}COONa$. What weight of soap would have been wasted by precipitation as calcium stearate if all washing had been done with soap? (*b*) If $Na_3P_3O_9$ was used to soften this water by complexing the Ca^{2+} as $CaP_3O_9^-$, what weight of softener would have been required? (1 gal = 3.79 liters.)   *Ans.* (*a*) 1.9×10^7 kg;   (*b*) 8.1×10^6 kg

21-10 **Ammonia synthesis.** The theoretically possible equilibrium yield of NH_3 by the Haber synthesis from N_2 and H_2 is higher at 1000 atm than at 1 atm total pressure. Why? It is lower at 500°C than at 25°C. How is this fact related to the sign of ΔH for the reaction? Is the reaction endothermic or exothermic?

21-11 Liquid ammonia. For solutions in liquid ammonia the role of the pH of aqueous solutions is played by $pNH_4 = -\log[NH_4^+]$. Estimate pNH_4 of pure NH_3 at $-50°C$ from Equation (21-15). Explain how it is possible to titrate a solution of NH_4NO_3 in liquid ammonia with a solution of known concentration of KNH_2 (potassium amide) in liquid ammonia; an indicator such as phenolphthalein may be used. How are acid and base constants of a conjugate acid/base pair related in liquid ammonia? *Ans.* 15; $K_aK_b = 1 \times 10^{-30}$

21-12 Oxidation of hydrazine. In an acid solution, hydrazine (H_2NNH_2) is oxidized quantitatively to N_2 by iodate, Cl_2, Br_2, and I_2. (a) Write a balanced equation for the reaction with IO_3^-, which goes to I^-. Note that hydrazine is a weak base and that the predominant species in an acid solution is hydrazinium ion, $N_2H_5^+$. (b) Suppose that 10.0 ml hydrazine solution yields 30.7 ml N_2 at STP. What is the concentration of hydrazine in moles per liter? What is the minimum weight of KIO_3 needed in this oxidation?

21-13 Oxidation and reduction of nitrous acid. Describe the following processes by balanced equations. (a) In an acid solution nitrous acid is oxidized to nitrate ion by Br_2, by MnO_4^-, or by $Cr_2O_7^{2-}$, the latter two species yielding Mn^{2+} and Cr^{3+}. (b) Nitrite ion is reduced to NO by I^- going to I_2 and is reduced to NH_3 by HSO_3^- going to SO_4^{2-}. (c) HNO_2 reacts with $NH_3(aq)$ to yield N_2. (d) Use data from Table D-5 to calculate the equilibrium constant for the oxidation of HNO_2 by $Br_2(l)$.

Ans. (d) $[H^+]^3[Br^-]^2[NO_3^-]/[HNO_2] = 1.7 \times 10^4 \, M^5$

21-14 Lewis formulas. Write possible Lewis formulas for (a) N_2O, NO^+, NO_2^+, NO_2^-, NO_3^-; (b) NH_3, N_2H_4, NH_2OH, CN^-. (c) Indicate the geometric configuration of each molecule or ion.

21-15 Solution of hydrazine. (a) From the data in Table D-3 calculate the pH of a 0.10 F hydrazine solution in water. (b) What is the concentration of $N_2H_5^+$ in a 0.10 F solution of hydrazine at pH 8.0? *Ans.* 10.5; 0.05 M

21-16 Photochemistry. In the photochemical decomposition of HI,

$$2HI(g) \longrightarrow H_2(g) + I_2(g)$$

the absorption of 500 J of energy in the form of light of a wavelength of 254 nm is found to decompose 2.12×10^{-3} mol HI. How many photons per molecule of HI does this correspond to?

21-17 Molecules of Ar_2? A. D. Buckingham has defined a molecule as "a grouping of atoms that can survive interactions with its environment". The potential energy curve associated with two Ar atoms has a well of depth $\epsilon/k = 100$ K (Fig. 3-13). Under what conditions might one be able to speak of Ar_2? *Ans.* at around 10 K

"I should be very glad to know what doctrine you teach now with regard to oxymuriatic acid [Cl_2]. Are you yet a convert to chlorine? I am impatient to see Lussac's paper on iodine, in particular to learn how far the facts respecting that substance go to confirm the new views of chlorine. Lussac appears to be a convert to Davy's sentiments, and certainly the acquisition of one who so strenuously opposed them must be accounted a very flattering occurrence."

Thomas Charles Hope, in a letter to William Allen, 1814

22 Some Families of Nonmetals

The preceding chapter was devoted primarily to the chemistry of three of the most common and important individual elements. We could continue to break down the fabric of chemistry into discussions of each of the elements in turn and, indeed, there are many monographs treating each element in depth. Fortunately, however, it is often possible and instructive to consider and contrast the properties of all the elements in one column at one time, rather than following a piecemeal approach. This is the method used in this chapter and Chap. 24.

We begin with a discussion of the halogens, a family of nonmetals that is especially regular in properties. Carbon and silicon, considered in Secs. 22-2 and 22-3, also show many similarities. The oxygen-family elements, sulfur, selenium, and tellurium, are discussed next, and the chapter concludes with an introduction to the chemistry of phosphorus and arsenic, congeners of nitrogen. The metallic elements in columns IVA and VA, tin, lead, antimony, and bismuth, are considered in Chap. 24.

22-1   The Halogens

F
Cl
Br
I
At

Occurrence and production.   No member of this highly reactive family is found as the element. Chlorine, the most common halogen, constitutes about 0.05 percent of the earth's solid crust as chloride ion. Concentrated deposits of NaCl (rock salt) are mined, and salt is also obtained from brines pumped out of wells and from the evaporation of seawater, in which it occurs to an extent of about 3 percent by weight. Most Cl_2 is prepared by the electrolysis of concentrated salt solutions, a procedure discussed in Sec. 24-1.

Electrolysis is also the common method of production of F_2. The element, occurring as fluoride ion, is about half as abundant in the earth's crust as chlorine; chief among its ores is fluorite, CaF_2. Fluoride-containing minerals are treated with H_2SO_4 to release HF, the gas is dissolved in molten KF, and F_2 is obtained by electrolysis of the mixture. Although KF itself melts at 846°C, the electrolysis can be performed at temperatures as low as 100°C because low-melting compounds are formed with formulas KF · HF (mp 250°C) and KF · 2HF (mp 72°C), in which the anions are the hydrogen-bonded species FHF^- and $FHFHF^-$. To combat the highly corrosive character of the melt, electrolysis cells are constructed only from metals such as nickel, copper, or monel (a Ni–Cu–Fe alloy), which become covered by protective fluoride layers, and nonreactive electrodes of graphite or nickel are used.

The concentration of bromide ion in seawater is about 70 ppm, sufficiently high to make the extraction of bromine economically feasible. In the industrial process, bromide ion in acidified seawater is oxidized by the relatively inexpensive Cl_2:

$$2Br^- + Cl_2 \longrightarrow Br_2 + 2Cl^- \tag{22-1}$$

Air is bubbled through the solution to sweep out the bromine. The bromine is then concentrated by passing the air through a solution of sodium carbonate in which the Br_2 is converted to bromide and bromate, BrO_3^-:

$$3Br_2 + 6CO_3^{2-} + 3H_2O \longrightarrow BrO_3^- + 5Br^- + 6HCO_3^- \tag{22-2}$$

The bromine is finally obtained by acidifying the solution, which in essence reverses reaction (22-2):

$$5Br^- + BrO_3^- + 6H^+ \longrightarrow 3Br_2 + 3H_2O \tag{22-3}$$

Iodides obtained from kelp (a seaweed) or oil-well brines are also oxidized with Cl_2 to produce I_2, but the major source of iodine is sodium iodate, $NaIO_3$, which is found as an impurity in deposits of sodium nitrate in Chile. The iodate is usually reduced to iodine with hydrogen sulfite ion:

$$2IO_3^- + 5HSO_3^- \longrightarrow I_2 + 5SO_4^{2-} + 3H^+ + H_2O \tag{22-4}$$

Physical properties of the halogens.   All the halogens exist as diatomic molecules that interact with each other by weak van der Waals forces. The variations in their physical properties (Table 22-1) show the regularity that parallels the increase in intermolecular forces accompanying increasing size and polarizability. For example, the melting points, boiling points, and critical temperatures all show a steady

TABLE 22-1 **Some properties of the halogens**

Property	F	Cl	Br	I
Atomic number	9	17	35	53
Stable isotopes, mass numbers	19	35, 37	79, 81	127
Covalent radius/Å	0.72	0.99	1.14	1.33
Ionic radius/Å	1.36	1.81	1.95	2.16
Melting point/°C	−223	−102	−7	113
Boiling point/°C	−188	−35	58	183
Color	Pale yellow	Yellow-green	Red-brown	Gray-black
E°/V for $X_2 + 2e^- \longrightarrow 2X^-$	2.87	1.36	1.06	0.54
ΔH°/kJ for $X_2 \longrightarrow 2X$	159	239	190	149

increase from fluorine to iodine. At room temperature F_2 and Cl_2 are gases, Br_2 is a liquid, and I_2 is a solid. As expected, the ionization energy and the electron affinity decrease as one goes down the group and the covalent and ionic radii increase. Closer inspection of the trends shows that the changes in properties between F_2 and Cl_2 are significantly larger than those between any other two members of the family. An important irregularity is the abnormally low heat of dissociation of F_2. It has been suggested that this low value is the result of electrostatic repulsions that occur at short range within the small F_2 molecule.

Bonding and reactions of the halogens. The principal modes of bond formation in the halogens involve the addition of one electron to the s^2p^5 configuration to complete the octet. The ease with which the halide ions, X^-, may be formed is indicated by the high values of the electron affinities of halogen atoms and the standard potentials for reduction of the halogen molecules (Table 22-1).

$$X_2 + 2e^- \longrightarrow 2X^- \tag{22-5}$$

The stability of the diatomic halogen molecules is also evidence of the tendency to achieve an inert-gas configuration.

The formation of positive ions is precluded by the very high ionization energies, but with the exception of fluorine, the halogens exhibit positive oxidation states in covalent compounds. Examples are the oxygen acids $HOCl$, $HClO_2$, $HClO_3$, and $HClO_4$, in which chlorine has oxidation states (+I), (+III), (+V), and (+VII), respectively. The formula for the first acid has been written $HOCl$ to show that in this compound the halogen forms only one covalent bond. The other acids are conventionally written as shown, but they also contain the HO— group and have, respectively, two, three, and four Cl-O bonds. Binary covalent compounds of the halogens with each other are known as interhalogen compounds. These include compounds with the general formula XX' (such as ICl, iodine monochloride), as well as those of formula XX'_3 (such as BrF_3, bromine trifluoride) and the compounds BrF_5, IF_5, and IF_7, in which the bonding involves *d* electrons.

The similarity in properties of the halogens makes it possible to describe much of their chemistry by general group reactions such as those given in Table 22-2. Differences between the congeners become evident when one considers the extent of reaction or the reaction rate. A case in point is the disproportionation reaction

$$X_2 + H_2O \rightleftharpoons H^+ + X^- + HOX \tag{22-6}$$

The equilibrium constant for (22-6) falls off drastically from chlorine to iodine. Fluorine also reacts according to (22-6), but the HOF is difficult to isolate because it reacts rapidly and vigorously with water. The overall reaction usually observed gives oxygen,

$$2F_2 + 2H_2O \longrightarrow 4H^+ + 4F^- + O_2 \tag{22-7}$$

as well as ozone and other oxidation products. When the equilibrium constant for reaction (22-7) is calculated from the standard electrode potentials for the half-reactions

$$F_2 + 2e^- \longrightarrow 2F^- \qquad\qquad E^\ominus = 2.87 \text{ V}$$
$$O_2 + 4H^+ + 4e^- \longrightarrow 2H_2O \qquad E^\ominus = 1.23 \text{ V}$$

a value of about 10^{111} is obtained. Similar calculations for Cl_2 lead to a smaller, but still large, value of K, 10^9, which is large enough to make HOCl highly unstable. Thus, the observed difference in behavior between F_2 and Cl_2 is a kinetic rather than a thermodynamic effect. Further evidence of this fact is the observation that solutions of Cl_2 release oxygen on standing.

Another reaction that illustrates differences between the halogens is the oxidation

$$X_2 + 2X'^- \rightleftharpoons 2X^- + X_2' \tag{22-8}$$

that is used to liberate Br_2 from seawater [Equation (22-1)]. It is apparent from the standard emfs in Table 22-1 that a halogen will displace from solution any congener that lies below it in the periodic table. Thus, F_2 can oxidize Cl^-, Br^-, and I^-, and Cl_2 can in turn oxidize Br^- and I^-.

Halides.   Many of the most familiar and important chemical compounds are halides. With the exception of He, Ne, and Ar, all the elements in the periodic system form halides.

TABLE 22-2   **Group reactions of the halogens**

$nX_2 + 2M \longrightarrow 2MX_n$	
$3X_2 + 2P \longrightarrow 2PX_3$	
$X_2 + PX_3 \longrightarrow PX_5$	not with I_2
$X_2 + H_2 \longrightarrow 2HX$	
$X_2 + H_2O \rightleftharpoons H^+ + X^- + HOX$	not with F_2
$X_2 + 2X'^- \longrightarrow X_2' + 2X^-$	X_2 a better oxidizing agent than X_2', $F_2 > Cl_2 > Br_2 > I_2$
$X_2 + H_2S \longrightarrow 2HX + S$	
$3X_2 + 8NH_3 \longrightarrow 6NH_4X + N_2$	not with I_2

Most metallic halides can be formed by direct interaction of the elements:

$$n\mathrm{X}_2 + 2\mathrm{M} \longrightarrow 2\mathrm{MX}_n \tag{22-9}$$

If the metal is highly electropositive and does not have a very small ionic radius, e.g., an alkali metal or the alkaline earths except Be, the halide is ionic. Metals with relatively high ionization energies and small atomic radii form covalent halides. These trends can be noted in the series of compounds KCl, CaCl_2, ScCl_3, and TiCl_4, in which there is a progression from a high-melting ionic solid to a molecular liquid. If a metal forms more than one halide with a particular halogen, the compound in which the metal has the lower oxidation state has a greater degree of ionic character. For example, SnCl_4 is a molecular liquid that freezes at $-30°\mathrm{C}$ whereas SnCl_2 is a crystalline solid that conducts electricity when molten.

The special character of fluorine is also observable in the halides. Melting points of metallic fluorides are markedly higher than those of other halides. Calcium fluoride melts above $1300°\mathrm{C}$; the melting points of the chloride, bromide, and iodide lie $600°\mathrm{C}$ lower.

Hydrogen halides. Although the hydrogen halides can be formed by a direct reaction between the elements, they are usually prepared by the reaction of a halide with a nonoxidizing acid, e.g.,

$$\mathrm{CaF}_2 + \mathrm{H}_2\mathrm{SO}_4 \longrightarrow 2\mathrm{HF} + \mathrm{CaSO}_4 \tag{22-10}$$

The term *nonoxidizing acid* is relative. Sulfuric acid, an inexpensive reagent, will serve for the production of HF and HCl, but it will oxidize bromides and iodides to Br_2 and I_2. To prepare HBr and HI, therefore, a poorer oxidizer, such as phosphoric acid, must be used:

$$\mathrm{NaBr} + \mathrm{H}_3\mathrm{PO}_4 \longrightarrow \mathrm{HBr} + \mathrm{NaH}_2\mathrm{PO}_4 \tag{22-11}$$

All the hydrogen halides are gases at room temperature. The physical properties of HCl, HBr, and HI show a smooth progression (Table 22-3) but HF stands alone. Its melting point, boiling point, enthalpy of fusion, and enthalpy of vaporization are abnormally high as a result of strong hydrogen bonding in the solid and liquid. Hydrogen fluoride is even associated in the gaseous state, where aggregates as large as $(\mathrm{HF})_5$ have been observed.

The acid strengths of HCl, HBr, and HI can be distinguished in nonaqueous solvents: HI is the strongest acid, HCl the weakest. All three acids are almost completely dissociated in water, however. Once again, HF is unique. It is a sur-

TABLE 22-3 **Properties of the hydrogen halides**

Property	HF	HCl	HBr	HI
Melting point/°C	-83	-115	-87	-51
Boiling point/°C	20	-85	-67	-35
$\Delta \widetilde{H}_{\mathrm{vap}}/\mathrm{kJ\ mol}^{-1}$	30.3	16.1	17.6	19.7
$\Delta \widetilde{H}_{\mathrm{fus}}/\mathrm{kJ\ mol}^{-1}$	4.58	2.11	2.40	2.87

prisingly weak acid ($K = 7 \times 10^{-4}$), the low acid strength being primarily a consequence of the strength of the H-F bond. In concentrated solutions (5 to 15 M), HF becomes a strong acid because of the formation of complexes, especially the very stable hydrogen difluoride ion HF_2^-:

$$F^- + HF \longrightarrow HF_2^- \qquad K = 5.1 \qquad (22\text{-}12)$$

Reaction (22-12) uses up F^- and thereby enhances the ionization of HF.

The well-known ability of hydrofluoric acid to etch glass is a result of a unique reaction between HF and silicates:

$$CaSiO_3 + 6HF \longrightarrow CaF_2 + SiF_4 + 3H_2O \qquad (22\text{-}13)$$

Oxyacids and their salts. Table 22-4 lists some of the oxyacids of the halogens and gives their names and the names of their salts. Hypofluorous acid, not given in the table, has also recently been isolated but is quite unstable. The strengths of the acids increase with the oxidation state of the halogen, and general rules for predicting their pK's were given at the end of Sec. 14-2. The solution chemistry of these substances is rather complex; the course of reaction depends sensitively on temperature and pH, and mixtures of products are often produced. As examples we consider some reactions involving the hypohalous acids and hypohalites.

In principle, hypohalous acids can be prepared by reaction (22-6), but the equilibrium yield is low and halide ion is also a product. A better procedure is to pass the halogen into an aqueous suspension of mercuric oxide, HgO:

$$2X_2 + 2HgO + H_2O \longrightarrow \{HgO\}HgX_2 + 2HOX \qquad (22\text{-}14)$$

The formula $\{HgO\}HgX_2$ is used here to indicate that the HgX_2 is removed from solution by adsorption on the solid HgO. This loss of HgX_2 drives the reaction to the right. Salts of hypohalous acids can be obtained by the disproportionation of a halogen in basic solution, e.g.,

$$Cl_2 + 2OH^- \longrightarrow Cl^- + ClO^- + H_2O \qquad (22\text{-}15)$$

The hypohalite ion, however, can itself disproportionate further,

$$3ClO^- \longrightarrow 2Cl^- + ClO_3^- \qquad (22\text{-}16)$$

TABLE 22-4 **Some oxyacids of the halogens**

Name of acid	Name of salt	Chlorine	Bromine	Iodine
Hypohalous	Hypohalite	HOCl	HOBr	HOI
Halous	Halite	HOClO		
Halic	Halate	$HOClO_2$	$HOBrO_2$	$HOIO_2$
Perhalic	Perhalate	$HOClO_3$	$HOBrO_3$*	$HOIO_3$

* Salts of perbromic acid have recently been prepared for the first time; the acid has not yet been isolated.

as can be seen from a Latimer diagram (Sec. 19-4) relating the various species in basic solution (at pOH = 0):

$$\text{ClO}_3^- \xrightarrow{\text{0.50}} \text{ClO}^- \xrightarrow{\text{0.40}} \text{Cl}_2 \xrightarrow{\text{1.36}} \text{Cl}^- \qquad (22\text{-}17)$$
$$\underset{\text{0.88}}{\underline{\qquad\qquad\qquad}}$$

EXAMPLE 22-1   **Disproportionation of halogens in basic solution.**   Predict on thermo-dynamic grounds the products of the disproportionation of Br_2 and I_2 in basic solution. The Latimer diagrams showing the standard reduction potentials for the various species are

$$\text{BrO}_3^- \xrightarrow{\text{0.54}} \text{BrO}^- \xrightarrow{\text{0.45}} \text{Br}_2 \xrightarrow{\text{1.06}} \text{Br}^- \qquad (22\text{-}18)$$

(with 0.61 spanning BrO_3^- to Br_2 and 0.76 spanning BrO^- to Br^-)

$$\text{IO}_3^- \xrightarrow{\text{0.14}} \text{IO}^- \xrightarrow{\text{0.45}} \text{I}_2 \xrightarrow{\text{0.53}} \text{I}^- \qquad (22\text{-}19)$$

(with 0.26 spanning IO_3^- to I_2)

Solution.   Since the potential for the Br_2/Br^- couple is more positive than the potential for the BrO^-/Br_2 couple, Br_2 will spontaneously disproportionate to BrO^- and Br^-. Examination of (22-18) shows that BrO^- has no tendency to disproportionate to Br_2 and BrO_3^-, but the potential of the BrO^-/Br^- couple is more positive than that of BrO_3^-/BrO^-; hence BrO^- will tend to disproportionate further to BrO_3^- and Br^-. No further disproportionation is possible. Similar reasoning shows that in basic solution I_2 is unstable with respect to I^- and IO^- and that IO^- will disproportionate further, forming IO_3^- and, eventually, I^-.

Even though ClO^- is thermodynamically unstable with respect to Cl^- and ClO_3^-, reaction (22-15) is a useful method for the production of hypochlorites because (22-16) is slow at or below room temperature. The analog of (22-16) for BrO^- is rapid at room temperature, so solutions containing hypobromite ion can only be prepared and kept at around $0°C$. Hypoiodites are not found because the disproportionation of IO^- is very fast at all practicable temperatures.

The magnitudes of the standard potentials for the reduction of the oxygen-containing halogen species in basic solution indicate that all these substances are strong oxidizers; this generalization is also true for the couples in acid solution. Solutions of sodium hypochlorite are used as a household bleach because of their oxidizing power and low cost. On a commercial scale the solutions are produced by electrolysis of concentrated NaCl solutions. The process is identical to that used to produce Cl_2 and NaOH, but the products obtained at the anode and cathode are allowed to mix and ClO^- is formed by (22-15). Bleach containers carry a warning not to mix the contents with household ammonia (to prepare a strong cleaning solution); the NH_3 reduces the OCl^- and poisonous Cl_2 gas is produced.

Potassium chlorate, the most common halate, is prepared by passing Cl_2 into hot KOH solution [Equations (22-15) and (22-16)]. When heated, chlorates decompose to oxygen and a chloride:

$$2KClO_3 \xrightarrow{\text{MnO}_2} 2KCl + 3O_2 \qquad (22\text{-}20)$$

Manganese dioxide catalyzes the reaction. When chlorates are placed in contact with easily oxidizable materials such as carbon, sugar, or phosphorus, they may react explosively. The reaction can be initiated by grinding, mixing, or shaking the materials together.

Perchloric acid is one of the strongest acids known. Unlike the other oxyacids of chlorine, $HClO_4$ can be isolated in pure form. The Latimer diagram (22-17) can be extended to show the relation of ClO_4^- to other oxygen-containing chlorine species:

$$ClO_4^- \xrightarrow{\quad 0.36 \quad} ClO_3^- \xrightarrow{\quad 0.50 \quad} ClO^- \qquad (22\text{-}21)$$

Chlorate ion is clearly unstable with respect to disproportionation to ClO^- and ClO_4^-, but the reaction occurs so slowly that it is not useful for the preparation of perchlorate salts. Instead, perchlorates are commonly prepared by the electrolytic oxidation of chlorates. Almost all perchlorate salts are very soluble in water, and the perchlorate ion forms almost no complexes; thus perchlorates are commonly used for experiments in which the properties of cations in solution are to be studied. Aqueous solutions containing perchlorates and organic substances can explode when concentrated by heating.

Pure crystals of periodic acid with the formula H_5IO_6 can be obtained. In solution, there are pH-dependent equilibria between several species: H_5IO_6 (paraperiodic acid), $H_4IO_6^-$, $H_3IO_6^{2-}$, and IO_4^- (metaperiodate ion). Perbromic acid and perbromates are predicted to be stable on thermodynamic grounds, but attempts to prepare them in pure form have been unsuccessful. Recently, however, solutions containing BrO_4^- have been obtained by oxidation of BrO_3^-, electrolytically and with XeF_2.

22-2 Carbon

Elemental carbon occurs in nature as diamond and graphite. Soot, lampblack, and charcoal consist primarily of microcrystals of graphite. Coal is a mineral rich in carbon; little of the carbon in soft (bituminous) coal is uncombined, but hard (anthracite) coal contains elemental carbon. Compounds of carbon are the basic structural materials of living matter.

The structures of graphite and diamond have already been discussed (Sec. 4-2). Diamond, the denser form, is less stable than graphite at low pressure, but its conversion to the more stable form does not occur under ordinary conditions. At very high pressures ($\sim 10^5$ atm), diamond is more stable than graphite. Thus, if the conversion is made rapid by raising the temperature and by the use of metal catalysts, diamonds can be produced by compressing graphite. The stones produced are small and not of gem quality, but synthetic diamonds, which are as hard as natural diamonds, are used industrially as abrasives.

Naturally occurring carbon contains 1.1 percent ^{13}C. Unlike ^{12}C, this isotope has a nuclear spin, and nmr spectroscopy (Sec. 21-1) with ^{13}C as a probe has become an important tool in structure determinations of complex molecules. Another isotope, ^{14}C, has a natural abundance of about 10^{-10} percent. Carbon 14 is radioactive and has a half-life (5730 years) that is short on a geological time scale. It is found in nature despite its continuous radioactive decay because it is continually formed in the upper atmosphere by the reaction of neutrons, produced by cosmic rays, with the very abundant ^{14}N. The pertinent equations for formation and decay are

$$^{14}_{7}N + ^{1}_{0}n \longrightarrow ^{14}_{6}C + ^{1}_{1}H \tag{22-22}$$

$$^{14}_{6}C \longrightarrow ^{14}_{7}N + ^{0}_{-1}e \tag{22-23}$$

The concentration of ^{14}C in natural carbon is a steady-state concentration—the number of ^{14}C nuclei disintegrating by (22-23) just matches the number formed from ^{14}N by (22-22).

^{14}C dating. The ^{14}C formed in the atmosphere by (22-22) is oxidized to CO_2 and mixes with ordinary CO_2, so that atmospheric CO_2 reaches a steady-state level of radioactivity: one of every 10^{12} molecules of CO_2 contains a radioactive ^{14}C atom. The atmospheric CO_2 may be transformed into cellulose and other carbon compounds by plants, and the plants may in turn serve as food for animals or human beings. The carbon compounds in living tissue thus reach the same steady-state level of one ^{14}C atom for every 10^{12} carbon atoms. Upon death of the plant or animal, however, the ^{14}C that decays by (22-23) is no longer replaced. The concentration of ^{14}C in dead wood, in cloth, and in other carbonaceous material derived from living matter diminishes slowly, falling to half the original steady-state level in 5730 years, to a quarter of this level in twice 5730 years, and so on. By measuring the ^{14}C content of ancient dead plant or animal matter, it is thus possible, if the measurements are sufficiently sensitive, to determine how long ago death occurred.

Radioactive decay is a classic example of a first-order rate process (Sec. 20-3). If the activity of the sample (the number of disintegrations per unit time) is A_0 at time $t = 0$ and is A at time t (for example, the present), then from (20-15) and (20-17), with $\tau_{1/2}$ the half-life of the radioactive atom,

$$\ln \frac{A}{A_0} = 2.30 \log \frac{A}{A_0} = -0.693 \frac{t}{\tau_{1/2}} \tag{22-24}$$

If the ratio of the initial activity to the present activity and the half-life are known, then the time t since the activity began to change can be measured.

EXAMPLE 22-2   **Dating of fossils.** Some fossil mammalian bones found in a certain stratified layer in the Olduvai Gorge in Africa have a ^{14}C activity about 0.285 times that existing in comparable living matter today. If it is assumed that the atmospheric level of ^{14}C was the same at the time of death of the animal as today, what is the approximate age of the sample?

Solution. The half-life of ^{14}C is 5730 years. Thus, given $A_0/A = \frac{1}{0.285} = 3.51$, we get, from (22-24),

$$\log 3.51 = \frac{0.693}{2.30} \frac{t}{5730 \text{ years}} = 0.545$$

or

$$t = \frac{0.545(5730)}{0.301} \text{ years} = \underline{10.4 \times 10^3 \text{ years}}$$

A basic assumption in the early applications of carbon dating was that the steady-state level of the ratio of ^{14}C to ordinary ^{12}C in the atmosphere is the same now as it was at the time of death of the plant or animal from which the material was taken. Comparisons of radiocarbon dates of historical and prehistorical objects with dates established by historical accounts and geological research suggested that there may have been significant variations in the level of atmospheric ^{14}C, and recent dating of wood samples by tree-ring counting has supported these results. It now appears that radiocarbon dates earlier than 500 B.C. are systematically in error. The deviation between the radiocarbon date and true date becomes progressively larger as the sample gets older, a radiocarbon date of 4700 B.C. corresponding to a true date of 5400 B.C. Smaller systematic deviations have also been observed for dates later than 500 B.C., but these errors are not greater than about 100 years.

Most of the chemistry of carbon is classified as organic chemistry, the subject of Chap. 26. A number of compounds of carbon, however, are considered inorganic. The classification is not clear-cut or necessarily logical—some "inorganic" compounds of carbon play an essential role in life processes—but the distinctions are commonly made. We discuss here the chemistry of some of the inorganic compounds of carbon.

Carbon oxides. When carbon is heated in a limited supply of oxygen, carbon monoxide is formed:

$$2C + O_2 \longrightarrow 2CO \tag{22-25}$$

Carbon monoxide is isoelectronic with and similar in physical properties to N_2. It is a colorless, odorless, and tasteless gas that is only very slightly soluble in water. Carbon monoxide is very toxic because it forms a complex with hemoglobin, the oxygen-carrying component of blood. The CO–hemoglobin complex is much stronger than the complex with O_2. When CO is inhaled, hemoglobin binds it in preference to oxygen and the transport of oxygen to body cells is diminished. Fortunately, the CO–hemoglobin reaction is reversible. If the concentration of CO in the lungs is decreased, CO–hemoglobin complexes will dissociate and the oxygen transport can return to normal if no major damage to the organism has yet occurred because of the lack of O_2.

Carbon monoxide can combine directly with halogens to form compounds with the formula COX_2. An example is the reaction with chlorine to produce the poisonous gas phosgene, $COCl_2$:

$$CO + Cl_2 \longrightarrow COCl_2 \tag{22-26}$$

Transition metals also react with CO to form metal carbonyls (Sec. 23-1). Most metal carbonyls are very hazardous to work with because they dissociate easily to form CO.

Although CO is quite inert at room temperature, it burns readily when heated, forming carbon dioxide:

$$2CO + O_2 \longrightarrow 2CO_2 \tag{22-27}$$

Because carbon dioxide is very stable and denser than air, it is commonly used as a fire extinguisher. Cylinders containing liquid CO_2 under pressure can be used to produce a blanket of inert gas to smother fires. In other extinguishers CO_2 generated by the reaction of bicarbonate and acid,

$$HCO_3^- + H^+ \longrightarrow CO_2 + H_2O \tag{22-28}$$

is used to spray liquid out of a reservoir. The reaction is initiated by inverting the extinguisher, which mixes an acid with a bicarbonate solution.

The acidic properties of solutions of CO_2, which is the cheapest and most abundant weak acid, were discussed in Sec. 14-6. This acid forms many salts, both carbonates and bicarbonates. Carbon dioxide constitutes about 0.03 percent of the atmosphere and about 0.014 percent of the oceans, where it is present chiefly as HCO_3^-. Because the mass of the oceans is much greater than that of the atmosphere, the amount of CO_2 in the oceans is considerably greater than that in the atmosphere, and large quantities are also present in combined form in various carbonate minerals, principally as $CaCO_3$ (Sec. 24-2). During the process of photosynthesis green plants remove CO_2 and H_2O from the atmosphere, form carbohydrates from them, and simultaneously return oxygen to the atmosphere.

When CO_2 reacts with ammonia under high pressure, urea is produced (see also Fig. 4-1):

$$CO_2 + 2NH_3 \longrightarrow OC(NH_2)_2 + H_2O \tag{22-29}$$
$$\text{Urea}$$

Urea is used as fertilizer. It is the principal end product of protein metabolism in the human body and is the chief nitrogen-containing component of urine. This association with life processes made the synthesis of urea from strictly inorganic substances by Wöhler in 1828 a milestone that removed any apparent boundary between inorganic and organic chemistry. It disproved the existence of a mysterious "vital force", a *vis viva*, once believed to be essential in the production of organic substances.

Other carbon compounds. The direct reaction of carbon and sulfur at high temperature gives carbon disulfide:

$$C + 2S \longrightarrow CS_2 \tag{22-30}$$

This pale yellow volatile liquid has an unpleasant odor; it is used extensively as a solvent. Carbon tetrachloride, another common solvent, is prepared by the reaction of CS_2 and Cl_2:

$$CS_2 + 3Cl_2 \longrightarrow CCl_4 + S_2Cl_2 \tag{22-31}$$

Carbon tetrachloride is toxic and can cause serious liver damage. It is used in some fire extinguishers because it has a dense vapor that can smother flames. It is not recommended for this purpose, however, because at high temperatures CCl_4 reacts with oxygen or water to produce phosgene.

Carbon forms binary compounds, called carbides, with many metals. The carbides of the most electropositive metals are saltlike and contain ions such as C^{4-}, C_2^{2-}, and C_3^{4-}. An important saltlike carbide is calcium carbide, CaC_2, which is made by the reduction of CaO at very high temperature:

$$CaO + 3C \longrightarrow CaC_2 + CO \qquad (22\text{-}32)$$

Carbides containing the C_2^{2-} ion are also called acetylides because they react with water to produce acetylene, C_2H_2, a compound important in organic synthesis:

$$CaC_2 + 2H_2O \longrightarrow Ca(OH)_2 + C_2H_2 \qquad (22\text{-}33)$$

The close-packed arrays of atoms in a metal contain octahedral holes (see Fig. 11-6 and Prob. 11-4). If the atoms have a radius of at least 1.3 Å (for example, Ti, Mo, W), the holes are sufficiently large to contain carbon atoms without distortion of the crystal structure. Carbides in which carbon occupies such sites are called *interstitial carbides*. The presence of the carbon does not fundamentally alter the characteristic metallic properties such as electrical conduction; however, melting points and hardness are enhanced because the C atoms are significantly involved in the bonding (Sec. 23-2).

An important compound that contains C and N is hydrogen cyanide, HCN, a highly poisonous gas used sometimes as a fumigant. In aqueous solution it is a very weak acid. Cyanide ion forms a large number of very stable complexes, such as $Ag(CN)_2^-$, $Fe(CN)_6^{3-}$, and $Fe(CN)_6^{4-}$. The poisonous action of cyanides is the result of complex formation: certain oxidative enzymes and other biological molecules are inactivated when the transition metals they contain are removed as cyanide complexes.

Two elements, silicon and boron, form completely covalent carbides. Silicon carbide, SiC, is known as carborundum. It exists in several crystalline forms, one of which has the diamond structure (Fig. 4-6). Carborundum is unusually hard and has a very high melting point. Boron carbide, B_4C, is one of the hardest materials known. Some metals of small radius form carbides intermediate in properties between the interstitial and covalent carbides. Important among these is iron; its carbide plays an essential role in steel making (Sec. 24-4).

22-3 Silicon

Our discussion of the chemistry of the halogens highlighted the strong similarities between the elements in this family. A comparison of the properties of silicon and carbon also shows many similarities. As we shall see, however, some of these similarities are superficial, and there are many significant differences between these elements.

Silicon is the second most abundant element in the earth's crust. Most of the rocks that constitute the crust are silicates, compounds of silicon with the most abundant element, oxygen. Pure silicon has the diamond structure: each Si is bonded to four other Si atoms in a tetrahedral arrangement. No silicon analog of graphite is known. Silicon is obtained by the reduction of quartz sand (silicon dioxide, SiO_2) with carbon in an electric furnace. Much of the demand for Si is for electronic components that require materials of very high purity. Silicon too impure for such use is refined by converting it to the tetrachloride,

$$Si + 2Cl_2 \longrightarrow SiCl_4 \qquad (22\text{-}34)$$

and purifying the liquid $SiCl_4$ by fractional distillation. The purified $SiCl_4$ is then reconverted to Si by reaction with hydrogen in a hot tube:

$$SiCl_4 + 2H_2 \longrightarrow Si + 4HCl \qquad (22\text{-}35)$$

The final stage in purification is zone refining (Fig. 22-1). In this process a short segment of a long rod of the element is heated until it melts. The impurities are more soluble in the molten Si than in the solid and therefore concentrate in the liquid. The heat source is slowly moved so that the narrow molten zone traverses the length of the rod, sweeping the impurities into the liquid. When the impure region is brought to the end of the rod, it is allowed to cool and is cut off. The process may be repeated. Zone refining makes it possible to produce silicon with only 1 part in 10^{11} of impurity.

Comparisons between Si and C. Like carbon, silicon forms four tetrahedral bonds. The tetrahalides, SiX_4, are analogous in structure to the carbon tetrahalides;

FIGURE 22-1   **Zone refining of silicon.** The silicon rod is drawn extremely slowly through the heated zone. Surface tension keeps the molten section of the rod intact.

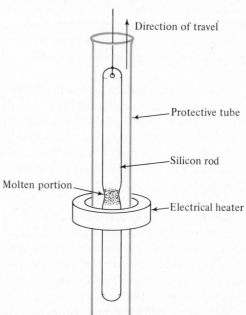

however, the silicon compounds are much more reactive. For example, $SiCl_4$ reacts vigorously with water:

$$SiCl_4 + 4H_2O \longrightarrow 4HCl + H_4SiO_4 \tag{22-36}$$

The product H_4SiO_4 is called silicic acid; its formula might better be written $Si(OH)_4$ because it is a mixture of $Si(OH)_4$ and the products of condensation reactions such as

$$2Si(OH)_4 \longrightarrow (HO)_3Si{-}O{-}Si(OH)_3 + H_2O \tag{22-37}$$

Very large molecules connected by Si-O-Si linkages can be formed.

Silicon also forms silanes, hydrides that are analogs of alkanes (Chap. 26): SiH_4, $Si_2H_6, \ldots, Si_6H_{14}$. These compounds are more reactive than their carbon counterparts; for example, they burn spontaneously in air:

$$Si_2H_6 + \tfrac{7}{2}O_2 \longrightarrow 2SiO_2 + 3H_2O \tag{22-38}$$

Compounds of silicon containing double Si-Si or Si-C bonds are known only as highly unstable intermediates.

Silicon dioxide, commonly known as silica, is the analog of CO_2; however, it bears little resemblance to the carbon compound. It is commonly found in nature in three crystalline forms: quartz, tridymite, and cristobalite. Quartz is the form that is stable at low temperature. At atmospheric pressure, quartz and tridymite are in equilibrium at 870°C and tridymite and cristobalite are in equilibrium at 1470°C. However, the transformation from one form to another is usually slow; hence crystals of tridymite and cristobalite, although rarer than quartz, exist at ordinary temperatures, presumably because they were cooled rapidly.

The basic structural unit of silica is an SiO_4 tetrahedron in each of the crystalline forms, as well as in glassy silica. The tetrahedra are linked together differently in the different forms, but a common feature of all is that every oxygen atom is common to two tetrahedra. Thus every silicon is bonded to four oxygen atoms and every oxygen to two silicon atoms, which accounts for the empirical formula SiO_2.

When molten silica is cooled it usually does not crystallize. Instead it becomes increasingly viscous until it hardens and can no longer flow. The solid is noncrystalline and can be described as a supercooled liquid or glass. On the microscopic scale it consists of linked SiO_4 tetrahedra that have local order but lack the long-range order of a crystal. Fused silica or quartz glass finds use in special laboratory equipment where its low thermal expansion and transparency to ultraviolet light may be of importance.

Silicates.   The silicates are silicon-oxygen compounds of various metals, chiefly the alkali and alkaline-earth elements, aluminum and iron. About 95 percent of the earth's solid crust consists of silicate minerals, and they are major components of meteorites and lunar rocks as well. The preponderance of silicates accounts for the fact that oxygen and silicon are the two most abundant elements on earth, together constituting three-fourths of the mass of the crust (Table 21-1).

More than 1000 different natural silicates have been characterized, and such synthetic materials as bricks, cement, glass, and ceramics are also silicates. All silicates are based on structures derived from SiO_4 tetrahedra. The simplest, such as the mineral zircon, $ZrSiO_4$, contain the SiO_4^{4-} ion. Much more commonly, however, the SiO_4 tetrahedra are linked together through Si-O-Si bonds to form large anions in one, two, or three dimensions. Three common chain and sheet structures built up from SiO_4 tetrahedra are illustrated in Fig. 22-2, and the formulas of some representative silicate minerals are given in Table 22-5.

The principles discussed in Chap. 4 concerning the formation of chains, two-dimensional networks, and three-dimensional frameworks apply to silicate structures as well. When each SiO_4 tetrahedron is linked to just two others, long chains (Fig. 22-2a) or isolated rings (not illustrated) are formed. The pyroxene minerals, of which diopside (Table 22-5) is typical, contain single chains of composition $(SiO_3)_n^{2n-}$. In the amphibole minerals, pairs of such chains are connected by Si-O-Si cross-links, Fig. 22-2b. Tremolite (Table 22-5), one form of asbestos, is a typical amphibole; some forms of it readily yield fibers that can be matted or woven into materials that are valuable for fireproof thermal insulation.

Layered structures are obtained when each SiO_4 tetrahedron is linked to three neighboring tetrahedra to form interconnected planar rings (Fig. 22-2c). Most minerals of this type also contain hydroxyl ions and Mg^{2+} or Al^{3+} that connect the sheets in rather complicated ways. The layered structures are characteristic of clays, talc (Table 22-5), and micas. In talc, for example, pairs of layers, like those in Fig. 22-2c, are held together by Mg^{2+} ions. The resulting double layers are electrically neutral and interact with each other only by relatively weak van der Waals forces and occasional hydrogen bonds. The layers therefore slip by one another readily, giving talc its soft, somewhat greasy, feel. In a typical mica, such as muscovite, one out of every four SiO_4 tetrahedra has been replaced by an AlO_4 tetrahedron. Because Al is in the (+III) oxidation state, whereas the Si it replaces is in the (+IV) state, an extra positive ion, usually K^+ or Na^+, must be incorporated into the structure for each AlO_4 tetrahedron. These extra ions are situated between the double layers, which are negatively charged, and serve to bind the layers together. Micas are consequently harder than talc; however, because of their layer structure, they can be cleaved readily into large sheets.

The extraordinary variety of silicates arises not only because SiO_4 tetrahedra

TABLE 22-5 **Some silicate minerals**

Name	Formula	Arrangement of SiO_4 tetrahedra
Zircon	$ZrSiO_4$	Individual SiO_4^{4-} groups
Diopside	$CaMg(SiO_3)_2$	Chains of tetrahedra (Fig. 22-2a), $(SiO_3)_n^{2n-}$
Tremolite	$Ca_2Mg_5(Si_4O_{11})_2(OH)_2$	Two parallel chains linked together (Fig. 22-2b), $(Si_4O_{11})_n^{6n-}$
Talc	$Mg_3Si_4O_{10}(OH)_2$	Double sheets of tetrahedra like those of Fig. 22-2c, $(Si_2O_5)_n^{2n-}$, with Mg^{2+} holding pairs of sheets together
Orthoclase	$KAlSi_3O_8$	Three-dimensional framework of linked tetrahedra; one-fourth of SiO_4 tetrahedra replaced by AlO_4 tetrahedra and K^+

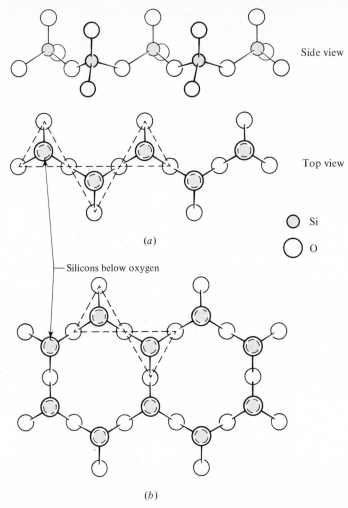

Side view

Top view

○ Si

◯ O

(a)

— Silicons below oxygen

(b)

FIGURE 22-2 **Structural units of silicates.** Each unit is composed of linked SiO_4 tetrahedra. Linear chains (a) are formed when an oxygen is shared by neighboring tetrahedra. In double chains (b), half the silicon atoms share three oxygens with other silicons and half share only two. Layers of planar rings (c) are obtained when each silicon shares three oxygens.

can be linked together in many different arrangements but also because of the ease with which ions of similar size can replace each other in the structures. For example, there is a continuous series of minerals of formula M_2SiO_4 containing $SiO_4{}^{4-}$, in which the metal ion M can be Mg^{2+} or Fe^{2+} or any mixture of the two. Olivine, a common green or yellow mineral in this group, contains about one Fe^{2+} for every ten Mg^{2+}. In some closely related mineral structures, two ions of charge +1 in one compound are replaced by one of charge +2 in another, or three of charge +2 substitute for two of charge +3. Because of differences in ion size and placement, substitution of one ion for another usually introduces some strain, so a given struc-

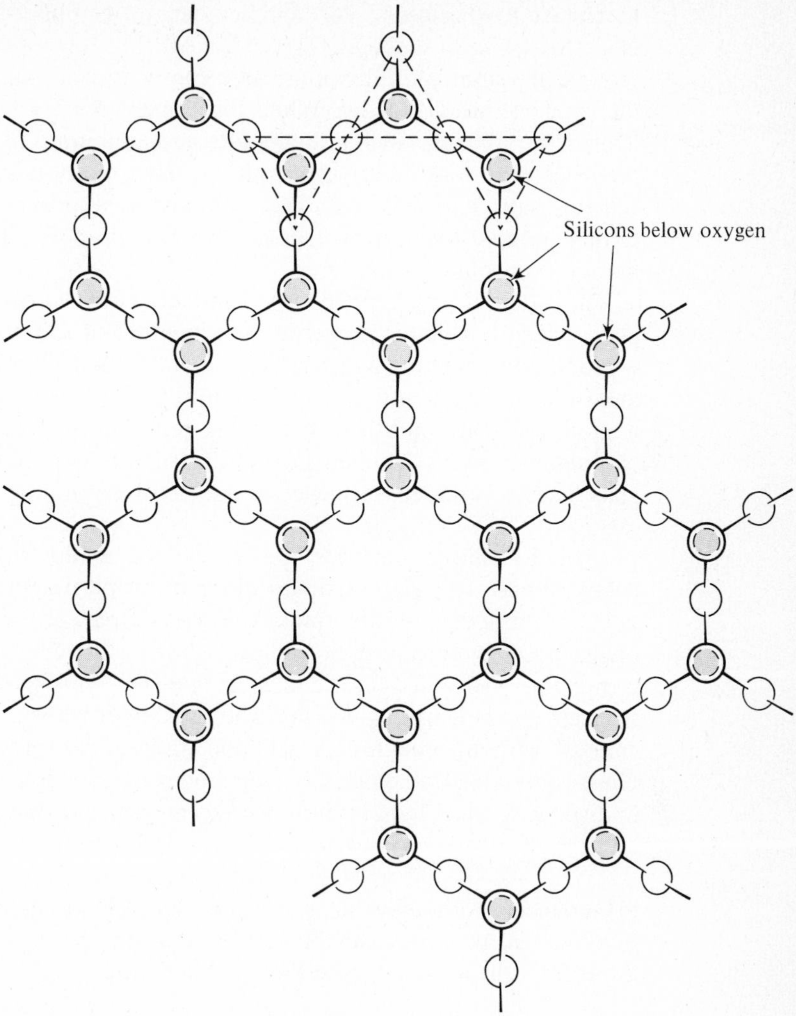

Silicons below oxygen

(c)

tural arrangement can tolerate only a limited number of substitutions and remain stable.

The most abundant silicates are the feldspars, a group of substances comprising about 60 percent of the earth's crust. They are aluminosilicates, with the general formula $M(Al,Si)_4O_8$, where M can be Na, K, Ca, or Ba, and the notation (Al,Si) stands for a ratio of Al to Si varying from $1:3$ to $2:2$. Note that the ratio of the number of oxygen atoms to the total of silicon and aluminum atoms is just 2, as it is in SiO_2. The feldspars have three-dimensional framework structures, with SiO_4 and AlO_4 tetrahedra linked by the sharing of all four oxygen atoms on each tetrahedron, just as the SiO_4 tetrahedra are linked in the various forms of SiO_2. Orthoclase (Table 22-5) and anorthite, $CaAl_2Si_2O_8$, are typical feldspars.

Cement and glass.   Portland cement is a complex mixture consisting mainly of Ca_2SiO_4, Ca_3SiO_5, $Ca_3Al_2O_6$, and Ca_2AlFeO_5. It is produced by "burning"[1] finely ground limestone and aluminosilicate clay with coal, oil, or gas fuel and crushing the product into a powder. When the cement is mixed with water the $Ca_3Al_2O_6$ hydrolyzes, forming calcium hydroxide and aluminum hydroxide, which react with the silicates to form intermeshed aluminosilicate crystals. Concrete is a mixture of cement, sand, and rock. Although one commonly thinks of cement as drying, the setting process does not in fact involve dehydration. Thus, cement and concrete harden even under water.

The fused silica glass previously described is brittle and difficult to work. Common glass, which has better properties, is a mixture of sodium and calcium silicates. It is produced by melting a mixture of sodium carbonate or sodium sulfate, limestone, and sand. The molten glass is poured into forms or flattened into sheets. Plate glass is ground flat and polished; it is being replaced by float glass, which is produced by floating a layer of molten glass on top of a molten metal, usually tin. A uniform glass surface is thereby achieved and expensive grinding and polishing is unnecessary.

Like the silicate minerals, glasses can be made with an enormous range of compositions. The glass varieties differ in properties such as color, transparency, index of refraction, and thermal expansion. If part of the SiO_2 is replaced by boric oxide, B_2O_3, the resulting borosilicate glass has a very low thermal expansion and is therefore well suited for use in cooking dishes. This glass was originally developed for train signal lamps, which must be capable of withstanding sudden temperature changes without breaking. A common trade name for borosilicate glass is Pyrex. Glass containing lead oxide has a high index of refraction that makes it very effective in dispersing white light into its colored components. Such glass is used for so-called crystal[2] glassware and jewelry.

Silicones.   Although silicon forms no long-chain compounds connected only by Si-Si bonds, polymers can be formed that are based on Si-O-Si linkages. They are called silicones and have the general formula

$$\begin{array}{ccccccc} & R & & R & & R & \\ & | & & | & & | & \\ -O-&Si&-O-&Si&-O-&Si&-O- \\ & | & & | & & | & \\ & R & & R & & R & \end{array}$$

where R represents a carbon-containing group such as $-CH_3$ or $-C_6H_5$. Silicone polymers have relatively high thermal and chemical stabilities because the Si-O bond energy is more than twice that of the Si-Si bond. Their properties depend on the chain length and the nature of the R groups. Short-chain methylsilicones (R = $-CH_3$) are oils and are used for lubrication and for hydraulic fluids; longer-

---

[1] Only the organic impurities in the feed mixture burn. Other processes occurring are the decomposition of the limestone and the reaction of CaO and SiO_2.

[2] The glassware known as crystal is not crystalline; i.e., it has no long-range structural order. Polished cut pieces of the glass sparkle much like crystals of high refractive index, such as diamonds.

chain-length methylsilicones find use as greases. Silicone oils have a very low temperature coefficient of viscosity and therefore remain more fluid at low temperatures and more viscous at high temperatures than typical hydrocarbon oils.

Solid, rubberlike materials can be produced when adjacent silicone chains are interconnected by cross-links, as in

$$
\begin{array}{ccccccccc}
 & R & & R & & R & & R & & R \\
 & | & & | & & | & & | & & | \\
-O-&Si&-O-&Si&-O-&Si&-O-&Si&-O-&Si&-O- \\
 & | & & | & & | & & | & & | \\
 & R & & O & & R & & R & & R \\
 & & & | & & & & & & \\
 & & & R & & R & & R & & R & & R \\
 & & & | & & | & & | & & | & & | \\
-O-&Si&-O-&Si&-O-&Si&-O-&Si&-O-&Si- \\
 & | & & | & & | & & | & & | & & | \\
 & R & & R & & R & & R & & R & & R \\
\end{array}
$$

Silicone rubber is used for electrical insulation and as a replacement for ordinary rubber in applications where chemical inertness or stability at high temperatures is required.

22-4   Sulfur, Selenium, and Tellurium

S
Se
Te

The elements in group VIA exhibit the general trends already noted in group VIIA. Sulfur, selenium, and tellurium show many similarities, and their chemical and physical properties change in a uniform and predictable manner with increasing atomic number. Oxygen, the first-period member of the family, has unique characteristics associated with its high electronegativity and the lack of d orbitals in its valence shell. The ($-$II) oxidation states of oxygen and the other members of the group show at least formal similarities; for example, all the analogs of H_2O—H_2S, H_2Se, and H_2Te—are known. Unlike oxygen, however, the other group VIA elements have an extensive chemistry in the ($+$IV) and ($+$VI) oxidation states as well. Their physical properties and elemental forms are also distinctly different from those of oxygen.

Occurrence and physical properties.   The name chalcogen (Greek: *khalkos,* copper; *genes,* born) is sometimes applied to the elements of group VIA because they can be obtained from copper ores, such as chalcocite (Cu_2S). Many other metal ores are also sulfides (for example, PbS and ZnS); they often contain traces of metal selenides and tellurides. About 30 percent of the sulfur produced in the United States is derived from these ores when the metal is extracted, and selenium and tellurium are also recovered as by-products. An increasingly important source of sulfur is the hydrogen sulfide component in some natural gas supplies.

Deposits of elemental sulfur also occur. Those in Texas and Louisiana are veins or pockets located in limestone at depths of 150 to 400 m. The sulfur is brought to the surface by the Frasch process, which avoids costly and dangerous mining operations. A hole is bored to the deposit and three concentric pipes are sunk. Water at a temperature of about 160°C (at which its vapor pressure is 6 atm) is pumped down the outer pipe and melts the sulfur, which collects in a pool at the base of the well. Compressed air is forced down the center pipe, mixes with the molten sulfur, and forms a froth that rises to the surface through the middle pipe. The mixture flows into enormous bins where the sulfur solidifies and the water runs off. When full each bin contains up to $\frac{1}{2}$ million tons of 99.5 percent pure sulfur. Molten sulfur is sometimes pumped directly into heated railroad tank cars for shipping to chemical plants.

Allotropy, the existence of an element in more than one form in the same physical state, is much more extensive in sulfur than in oxygen; at least 10 allotropic modifications of solid sulfur have been characterized, and almost 50 have been reported. The most important are the yellow crystalline forms called orthorhombic and monoclinic sulfur, in each of which the structural unit is an S_8 molecule in the form of a puckered ring (Fig. 4-2). They differ in the way the molecules are packed in the crystal.

Orthorhombic sulfur is the form stable at low temperature; above 95.6°C the monoclinic form, which melts at 119.5°C, is stable. The transitions between the crystalline forms occur very slowly; hence it is possible to heat orthorhombic sulfur to its melting point, 112.8°C, before any appreciable conversion to monoclinic sulfur occurs. Among the other solid allotropes of sulfur are a purple form that consists of S_2 molecules, an unstable form containing S_6 rings (Fig. 4-2), and a soft rubbery material, known as plastic sulfur, that contains long helical chains of sulfur atoms.

Liquid sulfur is a complex material. The orthorhombic and monoclinic solids melt to a mobile yellow liquid that for the most part consists of S_8 rings. When the liquid is heated above 160°C, it darkens and becomes increasingly viscous. By about 200°C it has become dark red-brown and is so thick that it cannot be poured from the container. These transformations are associated with the breaking of the S_8 rings and the subsequent reaction of the S_8 units to form long-chain molecules. If the viscous liquid is rapidly cooled in water, the chains persist and plastic sulfur is obtained. The viscosity of the liquid reaches a maximum at 200°C; further heating leads to an increase in fluidity because the chains break into smaller units. Sulfur vapor consists of a mixture of species, including S_8 and S_2 molecules.

Selenium also can exist in several allotropic modifications. A gray crystalline form, called metallic selenium because of its luster, contains infinite chains of selenium atoms (Fig. 1-2). A red form, which is less stable at room temperature, consists of Se_8 rings. Only one crystalline form of tellurium is known.

Metallic selenium conducts electricity poorly in the dark, but its conductivity in light is markedly higher and increases in proportion to the light intensity. This property of *photoconduction* is utilized in selenium photocells, found in some photographic light meters. It is also the basis of the dry copying process called xerography (Greek: *xeros,* dry). In this process a layer of photoconductive material deposited on an electrically conducting backing is electrostatically charged. When an image is focused on the layer, the illuminated areas become conductive and their charge flows away through the backing. Unilluminated areas, which remain charged, are allowed to pick up an ink powder and the powder is then transferred to paper, where it is set by heat.

Sulfides and hydrogen sulfide. Like oxygen, sulfur forms binary compounds by direct reaction with most of the elements. The alkali-metal and alkaline-earth sulfides (for example, Na_2S and BaS) are ionic and readily soluble in water. No transition-metal sulfide is predominantly ionic and, as one would predict from the relative electronegativities of oxygen and sulfur, the metal sulfides tend to have much less ionic character than the corresponding metal oxides. In some metal sulfides the distances between metal atoms are so short that there is very probably some direct metal-metal bonding. In fact, alloylike behavior—metallic luster, variable composition, and electrical conductivity—is found, among others, in the sulfides of iron, cobalt, and nickel. As a consequence of these differences between oxides and sulfides, the formulas of the sulfides of a metal frequently differ from those of the oxides. Compare, for example, FeO, Fe_2O_3, and Fe_3O_4 with FeS and FeS_2. Even when the formulas are analogous, the oxide and sulfide usually have different structures (for example, SnS_2 and SnO_2).

The insolubility of transition-metal sulfides is utilized to advantage in qualitative and quantitative analysis of metal ions. The hydrolysis of S^{2-},

$$S^{2-} + H_2O \rightleftharpoons HS^- + OH^- \qquad (22\text{-}39)$$

allows the concentration of sulfide ion in solution to be controlled by adjustment of the pH. Since K_2 for H_2S is about 10^{-15}, the equilibrium constant for (22-39), which is equal to K_2/K_w, has a value of about 10, so S^{2-} is extensively hydrolyzed unless the solution is highly dilute or very alkaline. The sulfide-ion concentration can be varied over a range of at least 10^{22} when the pH is varied between 14 and 0. Thus, in an appropriately buffered solution the concentration of S^{2-} can be set at a level at which a specific insoluble sulfide precipitates, leaving more soluble sulfides in solution.

Hydrogen sulfide, H_2S, is formed when metal sulfides are treated with dilute nonoxidizing acids such as HCl. Similar reactions of acids with selenides and tellurides produce H_2Se and H_2Te. All three hydrides are colorless and highly poisonous gases with extremely unpleasant odors; their properties are compared with those of H_2O in Table 22-6. Each molecule has an angular structure and, with the exception of water, a bond angle close to $90°$. The heats of formation become more positive with increasing size of the chalcogen. The formation of H_2Se and H_2Te from the elements requires *expenditure* of free energy, and these compounds slowly decompose

TABLE 22-6   **Properties of the hydrides of the group VIA elements**

Property	H_2O	H_2S	H_2Se	H_2Te
Boiling point/°C	100	-61	-42	-2
$\widetilde{H}_f^°$/kJ mol^{-1}	-286	-20	77	143
Bond angle/deg	104	92	91	90
Ionization constants, 25°C:				
K_1	1.8×10^{-16}*	9.1×10^{-8}	1.9×10^{-4}	2.3×10^{-3}
K_2	$\sim 10^{-24}$	1.2×10^{-15}	10^{-11}	10^{-11}

*See footnotes concerning H_2O in Table 14-2.

on standing. Acid strength increases in a parallel fashion from H_2O to H_2Te—the weaker the chalcogen-hydrogen bond, the stronger the acid. As noted before (Sec. 3-2), hydrogen bonding accounts for the anomalously high boiling point of H_2O while the steady increase in boiling point from H_2S to H_2Te correlates well with the increase in van der Waals interaction associated with increasing molecular size.

If solutions of sulfides are boiled with excess sulfur, polysulfide ions are formed:

$$S^{2-} + (n-1)S \longrightarrow S_n^{2-} \tag{22-40}$$

Ions containing chains of from two to six sulfur atoms are known. Examples of crystalline polysulfides are BaS_3, BaS_4, Cs_2S_6, and FeS_2 (pyrite, "fool's gold").

Selenides and tellurides closely resemble sulfides but tend to be more covalent. Polyselenide and polytelluride ions have also been characterized.

Oxides and oxyacids. Both the dioxides and the trioxides of sulfur, selenium, and tellurium are known, but only SO_2 and SO_3 are of importance. The dioxides can be prepared by burning the elements in air. Large quantities of SO_2 are produced when sulfides are heated in air or when sulfur-containing fuels are burned.

At room temperature SO_2 is a colorless gas with an unpleasant, irritating odor. SeO_2 is a white solid composed of infinite chains,

and TeO_2 exists in two forms, both of which are white solids with structures similar to those of metallic oxides.

Sulfur dioxide is quite soluble in water and forms acid solutions that contain sulfurous acid, H_2SO_3. Salts containing the SO_3^{2-} ion are called sulfites; those containing HSO_3^- are known as bisulfites or hydrogen sulfites. The alkali-metal sulfites and bisulfites are the most common. Sulfur dioxide is often used as a reducing agent and is also employed commercially as a preservative for prunes and other dried fruits because it destroys fungi and bacteria.

If solutions containing sulfites are boiled with sulfur, thiosulfate ion ($S_2O_3^{2-}$) is formed. The prefix *thio* is applied to species in which sulfur replaces oxygen. The thiosulfate ion can be thought of as derived from the tetrahedral sulfate ion by replacing one of the oxygen atoms by a sulfur atom. Sodium thiosulfate, $Na_2S_2O_3 \cdot 5H_2O$, is commonly known as *hypo* and is used in the photographic process (Sec. 25-1). After the film has been through the developer, undeveloped grains of silver bromide are removed by forming the soluble silver thiosulfate complex (Fig. 15-2)

$$AgBr + 2S_2O_3^{2-} \rightleftharpoons Ag(S_2O_3)_2^{3-} + Br^- \tag{22-41}$$

Thiosulfate is a mild reducing agent and can be oxidized by iodine to tetrathionate ion, $S_4O_6^{2-}$:

$$2\begin{bmatrix} & O & \\ & \| & \\ O - & S - & S \\ & \| & \\ & O & \end{bmatrix}^{2-} + I_2 \rightleftharpoons \begin{bmatrix} & O & & O & \\ & \| & & \| & \\ O - & S - & S - S - & S - & O \\ & \| & & \| & \\ & O & & O & \end{bmatrix}^{2-} + 2I^- \qquad (22\text{-}42)$$

a reaction analogous to the joining of two SO_4^{2-} tetrahedra to give peroxydisulfate ion, $S_2O_8^{2-}$ (Fig. 2-2). Reaction (22-42) is the basis of a procedure for the quantitative analysis of oxidizing agents. The oxidizer is allowed to react with an excess of iodide ion and the I_2 produced is titrated with a standard solution of thiosulfate, with a starch solution as indicator. Starch gives a characteristic blue color with iodine.

Almost 90 percent of the sulfur produced is converted to SO_2 and used in the production of sulfuric acid, for which it must be oxidized further to SO_3:

$$2SO_2 + O_2 \rightleftharpoons 2SO_3 \qquad (22\text{-}43)$$

This reaction occurs very slowly at low temperature. It is also appreciably exothermic; hence raising the temperature, which increases the reaction rate, decreases the equilibrium yield of product. A viable commercial process is made possible by the use of vanadium pentoxide or platinum catalysts.

The SO_2 is mixed with air and passed over the catalyst at a temperature of 400 to 450°C. The resulting SO_3 is dissolved in sulfuric acid to produce $H_2S_2O_7$, known as oleum or fuming sulfuric acid:

$$SO_3 + H_2SO_4 \longrightarrow H_2S_2O_7 \qquad (22\text{-}44)$$

Oleum is then diluted with water to make sulfuric acid of any desired concentration:

$$H_2S_2O_7 + H_2O \longrightarrow 2H_2SO_4 \qquad (22\text{-}45)$$

This roundabout procedure is used because the direct reaction of SO_3 and water produces a fine fog of acid that is difficult to condense.

At room temperature pure H_2SO_4 is a dense, oily liquid that is highly corrosive and hazardous. It is a relatively cheap strong acid and as such is one of the most important industrial chemicals. Sulfuric acid has a great affinity for water, a property that makes it useful as a dehydrating agent and as a solvent for reactions in which water must be released. A striking example of the dehydrating power of H_2SO_4 is its ability to break down sucrose, ordinary sugar, to carbon and water:

$$C_{12}H_{22}O_{11} \xrightarrow[H_2SO_4]{} 12C + 11H_2O \qquad (22\text{-}46)$$

Dilute sulfuric acid is not a good oxidizing agent, but hot concentrated acid is highly effective. For example, it can dissolve copper and it oxidizes carbon to carbon dioxide:

$$Cu + 2H_2SO_4 \longrightarrow CuSO_4 + 2H_2O + SO_2 \qquad (22\text{-}47)$$
$$C + 2H_2SO_4 \longrightarrow CO_2 + 2H_2O + 2SO_2 \qquad (22\text{-}48)$$

Selenic acid, H_2SeO_4, is a white crystalline solid that is similar in acid strength and molecular structure to H_2SO_4. It is a tetrahedral molecule in which the central selenium atom is surrounded by two oxygen atoms and two hydroxyl groups. The corresponding oxyacid of tellurium, telluric acid, has the formula H_6TeO_6 and is an octahedral molecule, containing six —OH groups. It is a weak diprotic acid with a pK_1 of 7.8.

22-5   Phosphorus and Arsenic

The group VA elements straddle the line between the metals and the nonmetals. Nitrogen and phosphorus are distinctly nonmetallic, arsenic is classed as a metalloid, and antimony and bismuth are primarily metallic. It is not surprising, then, that there are few strong family characteristics. One similarity that can be noted is the structure of the hydrides: all the group VA elements form gaseous hydrides with molecular formulas MH_3. All these hydride molecules are pyramidal; only NH_3 is appreciably basic. Another group characteristic is the existence of oxides with the *empirical* formula M_2O_5, all of which are acidic. We deal here with phosphorus and arsenic.

Occurrence and physical properties.   Phosphorus is derived from the relatively abundant phosphate minerals. A prime source is apatite, a group of minerals with compositions represented by the formula[1] $3Ca_3(PO_4)_2 \cdot CaX_2$, where X can be OH (hydroxyapatite) or can be either F or Cl or both, in any ratio (fluorapatite). The term *phosphate rock* is applied to ores that are rich in calcium phosphate and that may range in composition from pure $Ca_3(PO_4)_2$ to hydroxy-apatite.

Phosphorus is prepared commercially by heating a mixture of phosphate rock, silica, and coke in an electric furnace. The process is thought to occur in two stages: formation of P_4O_{10} vapor,

$$2Ca_3(PO_4)_2 + 6SiO_2 \longrightarrow P_4O_{10} + 6CaSiO_3 \qquad (22\text{-}49)$$

and its subsequent reduction by carbon,

$$P_4O_{10} + 10C \longrightarrow P_4 + 10CO \qquad (22\text{-}50)$$

The resulting gaseous mixture of P_4 and CO is passed through water, where the phosphorus condenses to a solid.

The soft, waxy, yellowish white crystals obtained in the commercial process contain tetrahedral P_4 molecules (Fig. 4-2) and are known as white phosphorus. This allotropic form of the element does not dissolve in water but is soluble in nonpolar solvents such as benzene and carbon disulfide. It is extremely poisonous and causes very painful and slow-healing burns. At temperatures higher than 40°C, white

---

[1] Formulas such as this for minerals are used to represent the composition and do not necessarily indicate structural features. Thus, for example, hydroxyapatite contains no distinct $Ca(OH)_2$ units and its formula might better be written as $Ca_5(OH)(PO_4)_3$.

phosphorus ignites spontaneously in air; it oxidizes slowly at room temperature and gives off a white light, the phenomenon from which phosphorus derives its name (Greek: *phosphoros,* light-bearing).

When white phosphorus is heated it changes to a more stable form, red phosphorus. The rate of conversion is increased by exposure to light or by the addition of small amounts of iodine. Unlike the white form, red phosphorus is relatively non-poisonous and unreactive; it will not ignite in air at temperatures below about 240°C and is practically insoluble in most liquids. Another crystalline form, black phosphorus, which is even more stable than the red, can be obtained by subjecting white phosphorus to high pressures. Commercial red phosphorus is amorphous; various crystalline red forms have been reported, but their structures are unknown. Black phosphorus has a layer structure consisting of sheets of puckered hexagons.

Most of the elemental phosphorus produced is oxidized to P_4O_{10} and used to prepare phosphoric acid. White phosphorus was once used for making matches but may no longer legally be used for this purpose because its toxicity led to serious health problems among workers in match factories. The head of a modern safety match is composed of a mixture of antimony sulfide, Sb_2S_3, which serves as a source of sulfur; potassium dichromate, which acts as an oxidizing agent; and glue as a binder. A mixture of red phosphorus, powdered glass, and glue comprises the strip on which the matches must be struck. When the match is rubbed against this strip, heat generated by friction raises the temperature of bits of phosphorus sufficiently that they are oxidized by the dichromate. The heat released by this reaction ignites the antimony sulfide.

Minerals containing arsenic have been known since early times. The bright yellow orpiment, As_2S_3, and the red realgar, AsS, were used by primitive people to decorate faces and bodies and to paint earthenware. Arsenic can be obtained from these ores by roasting to produce As_4O_6 and subsequent reduction by carbon. Most arsenic, however, comes from various iron, nickel, and cobalt arsenic sulfides (for example, $FeAsS$ and $CoAsS$), which contaminate metal ores.

The common form of the element is "metallic" arsenic, a steely-gray lustrous crystal that contains puckered sheets of As atoms and conducts electricity poorly. At 610°C gray arsenic sublimes at a pressure of 1 atm to tetrahedral As_4 molecules. If arsenic vapor is rapidly cooled, yellow arsenic, an unstable crystalline form consisting of As_4 molecules, is obtained. On standing, yellow arsenic soon reverts to the gray form.

Arsenic is a constituent of numerous alloys and its compounds are employed as insecticides and preservatives for hides. Some of the earliest drugs that were effective in combating microbial infections contained organic compounds of arsenic as the active ingredients.

Hydrides and halides. Phosphine, PH_3, is an extremely poisonous gas. It cannot be made by direct combination of the elements but can be produced by the reaction of white phosphorus with warm aqueous solutions of strong bases, which also yields hypophosphite ion, $H_2PO_2^-$:

$$4P + 3OH^- + 3H_2O \longrightarrow 3H_2PO_2^- + PH_3 \qquad (22\text{-}51)$$

Although pure phosphine is not spontaneously flammable in air, the gas produced by reaction (22-51) will often ignite because of the presence of impurities such as P_4 vapor or diphosphine, P_2H_4, which is very reactive.

Phosphine is only slightly soluble in water and does not yield basic solutions—its pK_b is estimated to be 25. When PH_3 is mixed with gaseous HCl, HBr, or HI, phosphonium halides, which contain the PH_4^+ ion, are formed. These salts are much less stable than the corresponding ammonium compounds and they decompose completely in solution, releasing PH_3.

All the trihalides of phosphorus and pentahalides other than PI_5 are known. With the exception of PF_3 they can be made by direct combination, e.g.,

$$2P + 3Br_2 \longrightarrow 2PBr_3 \tag{22-52}$$

The pentahalides can also be prepared from the trihalides,

$$PX_3 + X_2 \longrightarrow PX_5 \tag{22-53}$$

The trihalides are pyramidal molecules similar in structure to phosphine; in the vapor state the pentahalides have a trigonal bipyramidal structure. Crystals of the pentachloride and pentabromide are ionic: PCl_5 consists of tetrahedral $[PCl_4]^+$ and octahedral $[PCl_6]^-$ ions; PBr_5 consists of tetrahedral $[PBr_4]^+$ and Br^-.

All the halides react with water to give hydrogen halides and oxyacids of phosphorus. Trihalides give phosphorous acid, H_3PO_3,

$$PX_3 + 3H_2O \longrightarrow H_3PO_3 + 3H^+ + 3X^- \tag{22-54}$$

and the pentahalides undergo a two-step reaction giving first the oxyhalide, POX_3, and then phosphoric acid, H_3PO_4:

$$PX_5 + H_2O \longrightarrow POX_3 + 2H^+ + 2X^- \tag{22-55}$$
$$POX_3 + 3H_2O \longrightarrow H_3PO_4 + 3H^+ + 3X^- \tag{22-56}$$

Arsine can be formed by the reaction of zinc in acid solution with any soluble arsenic compound—for example, arsenite ion, AsO_2^-:

$$AsO_2^- + 3Zn + 7H^+ \longrightarrow AsH_3 + 3Zn^{2+} + 2H_2O \tag{22-57}$$

This reaction is the basis of the very sensitive Marsh test for the presence of arsenic. A substance suspected of containing arsenic is treated with zinc and acid; any arsine evolved is decomposed by heat to arsenic and hydrogen. The arsenic is detected by allowing it to condense as a shiny mirror on a cool smooth surface.

Arsenic forms all the trihalides, but only one pentahalide is known, AsF_5. The reactions of the trihalides with water are analogous to (22-54); unlike the phosphorus trihalides, however, the arsenic trihalides do not react completely and the reaction can be easily reversed by the addition of HX.

Oxides and oxyacids. The oxides and oxyacids of phosphorus all have structures involving tetrahedral bonding. When phosphorus is burned in an excess of air, the chief product is phosphorus(V) oxide,[1] P_4O_{10}, shown in Fig. 22-3a. With

---

[1] For historical reasons, the empirical formulas P_2O_5 and P_2O_3 are sometimes used for the phosphorus oxides, which are referred to as phosphorus pentoxide and phosphorus trioxide.

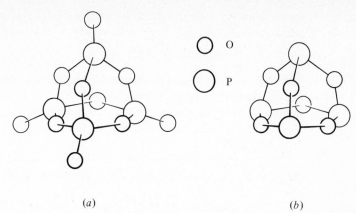

(a)   (b)

FIGURE 22-3   **Structures of phosphorus oxides.** (a) The phosphorus atoms in P_4O_{10} are situated at the corners of a tetrahedron, and six of the oxygens lie along the edges. The P-O-P bond angles are 124°; the O-P-O angles involving only bridging oxygens are 99°; those also involving the peripheral oxygens are 120°. (b) The four peripheral oxygen atoms of P_4O_{10} are missing in P_4O_6, and the P-O-P angle is slightly larger, 128°.

a limited supply of air, phosphorus(III) oxide, P_4O_6, is obtained (Fig. 22-3b). Both oxides are white crystalline solids. Phosphorus(III) oxide melts at about room temperature, 24°C, whereas the higher oxide sublimes at 1 atm at 258°C as discrete P_4O_{10} molecules.

Reaction of P_4O_6 with water produces phosphorous acid:

$$P_4O_6 + 6H_2O \longrightarrow 4H_3PO_3 \qquad (22\text{-}58)$$

As noted in the discussion of the strengths of the oxyacids of phosphorus in Sec. 14-2, hydrogen atoms directly bonded to phosphorus are not acidic. Phosphorous acid, which contains two —OH groups,

$$\begin{array}{c} \text{H} \\ | \\ \text{HO—P—OH} \\ | \\ \text{O} \end{array} \qquad (22\text{-}59)$$

is therefore diprotic.

Phosphorus(V) oxide has an extremely high affinity for water and therefore is much used as a drying agent. The reaction

$$P_4O_{10} + 6H_2O \longrightarrow 4H_3PO_4 \qquad (22\text{-}60)$$

gives orthophosphoric acid (frequently called simply phosphoric acid). Phosphoric acid is a poorly oxidizing triprotic acid.

Phosphate, hydrogen phosphate (HPO_4^{2-}), and dihydrogen phosphate ($H_2PO_4^{-}$) salts are known. In general, the dihydrogen phosphate salts are the most soluble. Among the phosphates, only the alkali-metal and ammonium salts are appreciably soluble in water; their solutions are strongly alkaline as a result of hydrolysis and they are widely used as cleaning agents. Fertilizers to supply the phosphate essential for all plant growth are made by treating insoluble phosphate rock with sulfuric acid to produce soluble calcium dihydrogen phosphate:

$$Ca_3(PO_4)_2 + 2H_2SO_4 \longrightarrow 2CaSO_4 + Ca(H_2PO_4)_2 \qquad (22\text{-}61)$$

Sufficient water is added to convert the calcium sulfate to gypsum, $CaSO_4 \cdot 2H_2O$, and the product mixture is marketed as "superphosphate of lime". Pure $Ca(H_2PO_4)_2$ is known as "triple phosphate" fertilizer because it contains three times as much phosphate per calcium as $Ca_3(PO_4)_2$. It is made by the reaction of H_3PO_4 with phosphate rock.

Other oxyacids of phosphorus include pyrophosphoric acid, $H_4P_2O_7$, which is obtained by loss of water from orthophosphoric acid,

$$2H_3PO_4 \longrightarrow H_4P_2O_7 + H_2O \qquad (22\text{-}62)$$

and the metaphosphoric acids, which have the composition $(HPO_3)_n$, with $n = 3$, 4, 5, . . . , and are produced when orthophosphoric acid is heated to 325–350°C:

$$nH_3PO_4 \longrightarrow (HPO_3)_n + nH_2O \qquad (22\text{-}63)$$

Organic phosphates play an essential role in the chemistry of life processes (Chap. 27), and hydroxyapatite is the main mineral constituent of bones and teeth.

The oxides of arsenic are similar in structure to those of phosphorus. Arsenic(III) oxide, As_4O_6, is obtained when arsenic is burned in air. Aqueous solutions of this oxide contain arsenious acid and are weakly acidic ($K = 6 \times 10^{-10}$). This oxide is somewhat amphoteric, with $K_b \approx 10^{-15}$. Arsenic(V) oxide, whose structure is not well established, cannot be prepared by direct reaction of arsenic with oxygen, but it can be made by boiling As_4O_6 with nitric acid. It is very soluble in water and gives solutions of arsenic acid (H_3AsO_4), a moderately strong oxidizing agent, comparable in acid strength to H_3PO_4. Many of the salts of arsenic acid resemble the corresponding phosphates in properties. Arsenates are toxic because arsenate can replace phosphates in some organic phosphates, but the resulting organic arsenates do not behave metabolically like the phosphates.

Problems and Questions

22-1 Reactions of halogens and halides. The following pairs of substances are mixed in aqueous solution: $NaCl + Br_2$; $NaBr + I_2$; $NaI + Cl_2$; $NaI + Br_2$. (*a*) Write balanced equations for any reactions that occur. (*b*) Use $E^\ominus$ values given in Table 22-1 to evaluate the equilibrium constant for the first mixture. *Ans.* (*b*) 7×10^{-11} atm

22-2 Inorganic reactions. Explain the following observations. (*a*) Chlorine is more soluble in cold NaOH solutions than in pure water at the same temperature. (*b*) Pure HBr(*aq*) can be prepared by distillation from a mixture of KBr and concentrated H_3PO_4 but not from KBr and concentrated H_2SO_4.

22-3 Halide solutions. Dilute aqueous solutions of NaCl are neutral but dilute solutions of NaF are slightly basic. Explain.

22-4 **Entropies of fusion and vaporization.** Use the data in Table 22-3 to calculate the entropies of fusion and vaporization of the hydrogen halides. What factors might account for the major trends in the entropies?

22-5 **Solutions of chlorine.** When gaseous chlorine is bubbled through water at 20°C, 2.26 liters of the gas (measured at STP) dissolve per liter of water. The Cl_2 reacts with H_2O according to (22-6) to give $HOCl$; the equilibrium constant is $K = 4.7 \times 10^{-4} M^2$. Calculate the concentration of $HOCl$ in a solution saturated with chlorine at 1 atm. *Ans.* 0.036 *M*

22-6 **Carbon dating and tree rings.** In 1951, wood from two sequoia trees was dated by the ^{14}C method. In one tree, clean borings located between the growth rings associated with the years A.D. 1057 and 1087 (i.e., wood known to have grown 880 ± 15 years prior to the date of measurement) had a ^{14}C activity about 0.892 of that of wood growing in 1951. A sample from a second tree had an activity about 0.838 of that of new wood and its age was established as 1377 ± 4 years by tree-ring counting. (a) What ages does carbon dating associate with the wood samples? (b) What values of $\tau_{1/2}$ can be deduced if the tree-ring dates given are used as starting point? (c) Discuss assumptions underlying the calculations in (a) and in (b), and indicate in what direction failures of these assumptions might affect the calculations.

22-7 **Solubility of limestone.** From data in Tables D-2 and D-3 calculate the solubility of limestone in water of a pH that is maintained at 7.00. How much limestone could be carried away in a year by a river with the average annual flow of the Colorado River ($22,000 \times 10^6 m^3$), if the pH is 7.00?
Ans. 0.36 g liter^{-1}, 8×10^9 kg year^{-1} (for pH = 6, 1.8 g liter^{-1}, 4×10^{10} kg year^{-1})

22-8 **Equilibrium between C, CO, and CO$_2$.** From Table D-4 find the values of $\Delta G^\ominus$ and $\Delta H^\ominus$ for the reaction $2CO(g) \longrightarrow C(graphite) + CO_2(g)$. (a) Which side would be favored at 298 K and a total pressure of 1 atm if equilibrium were established at a finite rate; which side at very high temperatures? (b) Assume $\Delta H^\ominus$ and $\Delta S^\ominus$ to be independent of T and estimate the temperature at which the partial pressures of CO and CO_2 are both 0.500 atm.

22-9 **Structures of silicates.** Using models or sketches, show how SiO_4 tetrahedra can be linked together to form linear, two-dimensional, and three-dimensional structures.

22-10 **Titration of a peroxide.** Peroxides such as Na_2O_2 and BaO_2 are highly reactive oxidizing agents. For example, they liberate I_2 from solutions of I^-, while dioxides such as PbO_2 and MnO_2 are more inert and do not oxidize I^-. A 0.250-g sample of impure Na_2O_2 was allowed to react with an I^- solution and the resulting I_2 was titrated with a 0.0637 *M* thiosulfate solution [Equation 22-42]. A volume of 47.8 ml thiosulfate solution was required. What is the purity of the Na_2O_2 sample? (Assume that there are no oxidizing impurities in the Na_2O_2.)

22-11 **Thiosulfates.** Thiosulfates are readily prepared by boiling elemental sulfur with a sulfite solution. The addition of acid to a thiosulfate solution essentially reverses this reaction, causing the formation of S and H_2SO_3. When radioactive elemental sulfur is used to prepare thiosulfate in this manner and the thiosulfate is subsequently decomposed, the H_2SO_3 recovered is not radioactive. Explain.

22-12 **Group VA halides.** Phosphorus forms both trihalides and pentahalides, but nitrogen forms only trihalides. Discuss a possible explanation for this difference.

22-13 **Separation by sulfide precipitation.** A solution is $1.0 \times 10^{-2} F$ in both $ZnSO_4$ and $CdCl_2$. It is desired to precipitate as much of the cadmium as possible, without precipitating the zinc, by bubbling H_2S at 1 atm through the solution at a pH controlled by a buffer. (*a*) What is the highest pH at which no ZnS would precipitate? (*b*) What fraction of the Cd would remain unprecipitated? Data: $[H_2S] = 0.10\ M$ in a solution saturated by H_2S at 1 atm. K_{sp} is $1.6 \times 10^{-24}\ M^2$ for ZnS and $7.9 \times 10^{-27}\ M^2$ for CdS.   *Ans.* (*a*) 0.6;   (*b*) 0.005

22-14 **Phosphorus acids.** (*a*) What is the oxidation state of phosphorus in each of the following acids: H_3PO_4, $H_4P_2O_7$, $(HPO_3)_3$, H_3PO_3 (one H bonded to P), H_3PO_2 (two H bonded to P). (*b*) Give structural formulas for these acids. (*c*) Name the acids. Use Appendix B or any other reference source.

22-15 **Double and triple bonds.** It is often stated that double or triple bonds involving elements beyond the second row of the periodic table are less stable than those involving second-row elements. Cite at least one specific example that supports or opposes this generalization, including references to the structures of at least two elements or compounds.

22-16 **pH of washing aids.** Many washing aids give an alkaline reaction in aqueous solution, the more alkaline the less "mild". Soap is the sodium salt of an organic acid for which $pK = 4.8$. Indicate the order of increasing alkalinity (increasing pH) of solutions of soap, sodium phosphate, sodium carbonate, sodium borate (which may be assumed for this purpose to be a salt of boric acid), and, for purposes of comparison, sodium sulfate and sodium chloride. Relevant acid constants may be found in Tables 14-2 and D-3. Assume the formality of all solutions to be the same.

22-17 **Disproportionation of I_2.** When iodine is dissolved in an aqueous solution of a strong base, it is completely converted into I^- and IO_3^-. If the solution is acidified, however, the iodide and iodate are converted completely into iodine again. Explain these facts with the help of any pertinent $E^{\ominus}$ values from Table D-5.

"Although those who burn, roast and calcine the ore, take from it something which is mixed or combined with the metals; and those who crush it with stamps take away much; and those who wash, screen and sort it, take away still more; yet they cannot remove all which conceals the metal from the eye and renders it crude and unformed. Wherefore smelting is necessary, for by this means earths, solidified juices, and stones are separated from the metals so that the metals obtain their proper color and become pure, and may be of great use to mankind in many ways."

Georgius Agricola, "De re metallica", 1556

23 Metals and the Metallic State

23-1 Characteristic Properties and Methods of Production

Some physical properties. The characteristic properties of metals (Chap. 4) include metallic luster, high thermal conductivity, and high electrical conductivity that decreases with increasing temperature. Pure metals normally have a silvery appearance, except for copper, which is red-brown, and gold, which is yellow; finely divided metals often appear black. It is not easy to distinguish between elemental metals and alloys by appearance or by superficial tests. Alloys that contain small amounts of nonmetals, such as carbon, silicon, or phosphorus, exhibit typical metallic properties.

Only three metals are liquid at or near room temperature: mercury melts at $-39°C$, cesium at $28°C$, and gallium at $30°C$. The highest melting points, boiling points, and densities are found near the middle of the periodic table, chiefly among the transition metals. "Heavy" metals are much denser than other common substances, representative values being, in grams per cubic centimeter, 7.9 (iron), 11.3 (lead), 19.3 (tungsten), and 21.4 (platinum); the most dense metal at 1 atm is osmium (22.5). In contrast, most common rocks have densities in the range 2 to 3.

Most metals can be deformed readily without breaking; they can be drawn into wires and hammered or rolled into sheets. A few, however, such as chromium and bismuth, are brittle or have brittle modifications, but they have a metallic appearance and are good conductors of electricity and heat.

Metals are widely used for structural purposes. Other important applications, particularly of copper, silver, and gold, are based on their exceptional electrical conductivity. Many uses are tied to more specific properties. For example, antimony and some of its alloys are used for casting type because they expand on solidification and therefore permit exact replication of the fine details of the mold. Tungsten, because of its very high melting point ($\sim$3400°C), is particularly suited for use as light-bulb filaments and electrodes in spark plugs. The high-temperature reaction vessels in chemical laboratories are frequently made from the relatively unreactive metal platinum.

Some chemical properties.   Metals are virtually insoluble in inorganic or organic solvents, unless there is a chemical reaction. They are soluble, however, in the melts of other metals. Trace quantities of mercury dissolve in water; alkali metals dissolve in liquid ammonia and, in trace amounts, in water. The dissolved alkali-metal atoms react almost immediately with the water, however, and the metal atoms eventually react with the ammonia (Sec. 24-1). Metals are monatomic in the vapor state and when dissolved in the melts of other metals (as can be shown, e.g., by the freezing-point depression).

The chemical reactivities of metals vary tremendously. Gold and the platinum metals (Ru, Rh, Pd, Os, Ir, and Pt) are chemically quite inert, whereas the alkali metals are extremely reactive. The most general manifestation of this reactivity is the tendency to form positive ions. This tendency increases toward the left and downward in the periodic table. It is strongest for cesium.

Production methods.   The natural sources of metals are ores, minerals from which a metallic constituent can be extracted. Some ores contain metals in the elemental state—for example, gold, platinum, silver, and copper. However, most metals occur naturally in positive oxidation states, often as oxides, sulfides, or binary compounds with other group VI elements, or as hydroxides or carbonates. The metal is then obtained from the ore by chemical or electrolytic reduction.

Metals in the zero-oxidation or *native* state can be separated from rocky contaminants by hand or other simple means. If the ore contains only a relatively small amount of the metal (a low-grade ore), chemical extraction methods are often used. Gold is sometimes extracted from ores with mercury. The mercury-gold solution, called an *amalgam,* is separated from the ore and the mercury is then distilled away. Alternatively, the ore may be treated with a cyanide solution in the presence of air. A soluble gold-cyanide complex is formed:

$$4Au + 8CN^- + O_2 + 2H_2O \longrightarrow 4Au(CN)_2^- + 4OH^- \qquad (23\text{-}1)$$

Although gold is not normally oxidized by oxygen, the strength of the cyanide complex helps to drive this reaction to the right. Similar methods are used with silver ores, including both those containing native silver and those containing its compounds, such as the mineral argentite, Ag_2S.

The least electropositive metals can sometimes be obtained from sulfide or oxide ores by heating the ore in air; for example:

$$HgS + O_2 \longrightarrow Hg + SO_2 \qquad (23\text{-}2)$$
$$Cu_2S + O_2 \longrightarrow 2Cu + SO_2 \qquad (23\text{-}3)$$
$$2HgO \longrightarrow 2Hg + O_2 \qquad (23\text{-}4)$$

Most other sulfide ores are converted into oxides when heated in air. The resulting oxides, as well as naturally occurring oxide ores, hydroxide ores, and carbonate ores, are reduced to the metal in various ways. One common method, used in the production of iron, nickel, copper, zinc, cadmium, and tin, is to heat the oxide with carbon. The carbon may reduce the oxide directly, with the formation of metal and either CO or CO_2, or some of the carbon may react first with oxygen to produce CO, which is itself an effective reducing agent. Since CO is gaseous, it can make much better contact with the solid oxide than can solid carbon. The reduction of iron oxide by CO in the blast furnace (Fig. 23-1) is illustrative.

Some oxides are reduced with aluminum or magnesium instead of with carbon, in part because of a tendency of certain metals, particularly transition metals, to react with carbon to form carbides.

Highly electropositive metals, such as the alkali and alkaline-earth elements and aluminum, are usually produced by the electrolysis of melts. Sodium is obtained in this way from $NaOH$ or $NaCl$, calcium from $CaCl_2$ (with some CaF_2), and aluminum from Al_2O_3 dissolved in molten cryolite, Na_3AlF_6 (Fig. 23-2).

Some ores are too low grade to be directly useful, but methods have been developed for concentrating the desired material. The most common of these is flotation, used especially with sulfide ores, which are wetted by oil but not by water. For example, 95 percent of the copper may be recovered from ores that contain only about 2 percent of Cu_2S. The finely crushed ore is first ground with a mixture of oil and water; then a foaming agent is added and air is bubbled through the mixture. The sulfide particles are concentrated, with the oil, in the foam that rises to the top, while the remainder of the ore, chiefly silicates, settles to the bottom.

Metals as first obtained from their ores must often be purified further for particular applications. Electrolytic refining is used for some, e.g., copper (Fig. 19-1). Many transition metals form volatile molecules on reaction with carbon monoxide, and these can be used to obtain very pure samples of the metal. In the Mond process for refining nickel, CO is passed over impure nickel; nickel carbonyl, $Ni(CO)_4$, a volatile and highly toxic liquid, is formed. It is purified by distillation and then decomposed into pure Ni and CO by heating. Iron can be similarly purified through the formation of the volatile $Fe(CO)_5$.

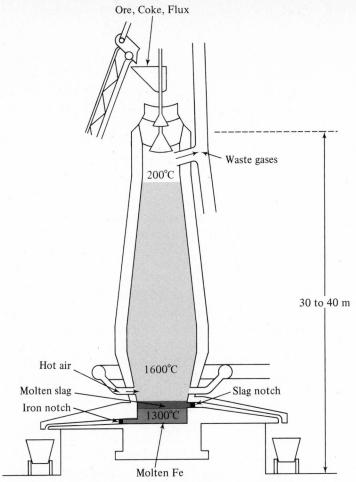

FIGURE 23-1   **Blast furnace for smelting iron ores.** The furnace is fed from the top, alternately with coke (chiefly carbon) and with a mixture of iron ore and a flux that produces liquid slag with ore components. With ores rich in silica (SiO_2), the flux is limestone; with basic ores, rich in $CaCO_3$ or $MgCO_3$, the flux is rich in SiO_2. Preheated air is blown in near the bottom. The solid materials are completely converted to gases and two liquids as they descend in the furnace. The reactions, somewhat simplified, are

$$2C + O_2 \longrightarrow 2CO$$
$$3CO + Fe_2O_3 \longrightarrow 2Fe(l) + 3CO_2$$
$$CaCO_3 \longrightarrow CaO + CO_2$$
$$CaO + SiO_2 \longrightarrow CaSiO_3 \text{ (molten slag)}$$

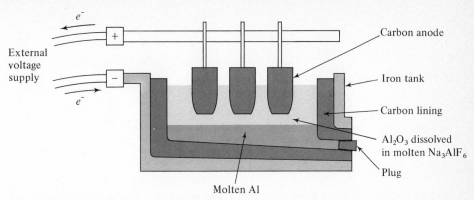

FIGURE 23-2   Electrolytic production of aluminum. In the Hall process, a melt of Al_2O_3 in Na_3AlF_6 is electrolyzed at a temperature near 1000°C. Sufficient heat to keep the electrolyte liquid is provided by the passage of current. Liquid aluminum is produced at the cathodic carbon lining by the reaction $Al^{3+} + 3e^- \longrightarrow Al$. It collects at the bottom and may be drawn off. The carbon anodes are consumed by the anode reaction $C + 2O^{2-} \longrightarrow CO_2 + 4e^-$.

23-2   Typical Crystal Structures of Metals

Close-packed structures. The atoms in crystals of about two-thirds of the metallic elements are packed as closely as possible in regular three-dimensional arrays. To describe this three-dimensional close packing of atoms, which we idealize as spheres, we begin with the close-packed layer illustrated in Fig. 23-3. Layers of this type are stacked together so that each sphere in a given layer is in contact with three spheres of the layer below and three spheres of the layer above. The spheres fit into niches of the adjoining layers. An ambiguity regarding the stacking of any *three* layers arises because the spheres of the outside layers may be fitted, from above and below, into the same or into different indentations of the layer in the middle (Fig. 23-4). A periodic repetition of the first mode of stacking is called *hexagonal close packing;* a periodic repetition of the second mode is termed *cubic close packing.*

In terms of the layer types described in Fig. 23-3, hexagonal close packing corresponds to the sequence $\cdots ABABAB \cdots$ and cubic close packing to the sequence $\cdots ABCABC \cdots$ (Fig. 23-5). The repeat in the first sequence is two layers; in the second it is three layers. Equivalent descriptions are $\cdots BCBCBC \cdots$ and $\cdots ACACAC \cdots$ for hexagonal and $\cdots ACBACBACB \cdots$ for cubic close packing. Stacking with other layer sequences (such as $\cdots ACABACABACAB \cdots$ with a four-layer repeat) and random stacking ($\cdots ABCACBCABACB \cdots$) are also possible, but the cubic and hexagonal modes are the most important ones.

The body-centered cubic structure and other structures. In another simple structure exhibited by metallic elements, atoms are located at the corners and the center of a cubic unit cell (Fig. 23-6); the structure is called *body-centered*

Layer

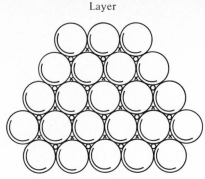

FIGURE 23-3   **Closest packing of equal spheres in a plane.**   Each sphere (atom) is in contact with six neighboring spheres in a hexagonal pattern and is surrounded by six interstices, three marked by crosses (x) and three by circles (o). Additional spheres can be placed on top of this layer by nesting them in the niches just above the interstices. An entire layer of spheres can be placed on top of the interstices marked by crosses, or those marked by circles, because the distances between the interstices in either set are exactly those between the centers of the original spheres. A new layer can also be nested against the original one from below. To describe the nesting of more than two layers (Figs. 23-4 and 23-5), it is convenient to designate the original layer and all layers whose spheres lie directly above or below those of the original layer as type *A*, all layers whose spheres lie above or below interstices marked with circles as type *B*, and those with spheres above or below crosses as type *C*.

FIGURE 23-4   **Hexagonal and cubic stacking of layers.**   In hexagonal close packing, the spheres of the first and the third layers are seated above and below the *same* interstices of the second layer; in cubic close packing, *different* interstices are utilized. Side views of the two ways of stacking are shown in Fig. 23-5. Cubic close packing corresponds to a face-centered cubic arrangement of atoms, although this is not apparent in these views.

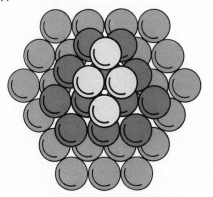

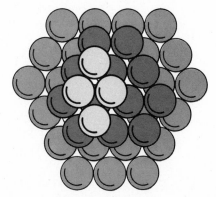

Hexagonal

Cubic

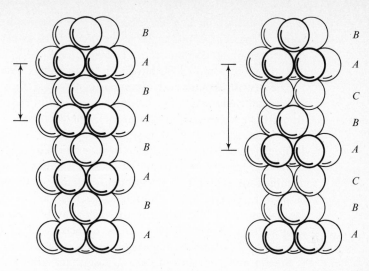

Hexagonal   Cubic

FIGURE 23-5 **Side views of hexagonal and cubic close packing.** The layers of hexagonal and cubic close packing shown are periodic repetitions of two layers A and B or of three layers A, B, and C, respectively. The double-headed arrows indicate the repeat distances.

FIGURE 23-6 **Body-centered cubic structure.** The atom at the center of the unit cube (shown in dashed outline) has eight nearest neighbors at the corners of the cube. They are at a distance $d_1 = a\sqrt{3}/2$, where a is the cube edge. It has six next-nearest neighbors at the centers of adjoining unit cubes, forming the corners of an octahedron. They are at the distance $d_2 = a = 2d_1/\sqrt{3} = 1.15d_1$, 15 percent larger than d_1. The environment of all atoms is the same.

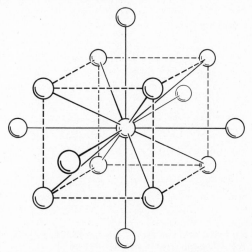

cubic. Atoms are packed less efficiently in this structure than in cubic close packing. Some 70 percent of all metallic elements have cubic or hexagonal close-packed structures,[1] about 25 percent are body-centered cubic, and 5 percent have structures that are more complex (Table 23-1). Many alloys also have complex structures.

The structures of tin and its congeners illustrate the relationship between structure and metallic character. The common form of tin, called white tin, is metallic. Each atom has six nearest neighbors in a deformed octahedral arrangement. A nonmetallic form of tin, called gray tin, is stable below 13°C. It has the diamond structure (Fig. 4-6), each atom being surrounded by four others in a regular tetrahedral arrangement at an average distance about 10 percent smaller than in the metallic phase. The metallic form is about 27 percent denser than gray tin because of the greater number of near neighbors, even though each neighbor is somewhat further away. The conversion from white to gray tin is slow at 13°C, but it becomes faster at lower temperatures. In very cold climates, objects made of tin suffer from the so-called tin pest or tin disease—they begin to disintegrate into a powder of the gray form.

Silicon and germanium, the elements just above tin in the periodic table, are metalloids. They have the diamond structure under normal conditions, but each can be transformed into a metallic form with the white tin structure at pressures of a few thousand atmospheres. It may be possible to transform diamond into a metallic form also, but it has been estimated that a pressure of some 2 million atm would be needed.

[1] Crystals of the inert-gas elements also contain atoms packed in hexagonal and cubic close-packed arrays.

TABLE 23-1   The crystal structures of some metals and metalloids

Li	Be	B											C
bcc; ccp; hcp	hcp; bcc	comp.											graph-ite; diam.
Na	Mg	Al											Si
bcc; hcp	hcp	ccp											diam.
K	Ca	Sc	Ti	V	Cr	Mn	Fe	Co	Ni	Cu	Zn	Ga	Ge
bcc	ccp; bcc	hcp	hcp; bcc	bcc	bcc	dist. ccp; comp.	bcc;	ccp; hcp	ccp	ccp	hcp	comp.	diam.
Rb	Sr	Y	Zr	Nb	Mo	Tc	Ru	Rh	Pd	Ag	Cd	In	Sn
bcc	ccp; hcp; bcc	hcp; bcc	hcp	bcc	bcc	hcp	hcp	ccp	ccp	ccp	hcp	dist. ccp	diam.; comp.
Cs	Ba	La	Hf	Ta	W	Re	Os	Ir	Pt	Au	Hg	Tl	Pb
bcc	bcc	ccp; bcc	hcp; bcc	bcc	bcc	bcc	hcp	ccp	ccp	ccp	dist. ccp	hcp; bcc	ccp

ccp = cubic close packing; hcp = hexagonal close packing; bcc = body-centered cubic; diam. = diamond structure; dist. = distorted; comp. = complex structure

Alloys. As discussed briefly in Sec. 4-4, alloys are metallic substances that are intimate mixtures of two or more elements, at least one of which is a metal. The mixture may be homogeneous, or there may be an intermingling of crystallites with different compositions and atomic arrangements. In other words, a solid alloy may consist of one phase or it may be heterogeneous and contain several solid phases. Many alloys have desirable properties that differ substantially from those of the parent elements. For example, addition of tin to copper, both of which are soft, produces hard and tough bronzes. Small amounts of carbon change soft, pure iron to steel.

For many alloy phases the crystal structure is that of the more abundant parent element, with an appropriate number of original atoms replaced by atoms of the admixed element. Other alloys have different and unique structures. For example, in alpha brass (a Cu alloy with less than about 35 percent Zn) some of the atoms in the copper structure are replaced by Zn atoms, while eta brass (consisting of more than about 95 percent Zn) is based on the structure of zinc, with some of the Zn atoms replaced by Cu atoms. There are three other intermediate structures among the brasses, each associated with a specific composition: $CuZn$ (beta brass), Cu_5Zn_8 (gamma brass), and $CuZn_3$ (epsilon brass). In brasses that contain somewhat more Cu than corresponds to one of these formulas, Zn atoms in the appropriate structure are replaced by Cu atoms. When Zn is in excess the opposite substitution takes place.

Interstitial compounds. Even in the close-packed structures of Fig. 23-4, about 26 percent of the space is unfilled, corresponding to the interstices (spaces) between the spheres. Atoms such as H, B, C, N, and even O are small enough to fit into the interstices between the atoms in crystals of some transition metals without causing considerable distortion. Such substances are called interstitial compounds.[1] The small atom does not simply fill a hole in the structure; there is strong chemical bonding between metal and nonmetal atoms. This bonding is evidenced, for example, by the fact that the compounds are harder and higher melting than the pure metal. Interstitial compounds are often nonstoichiometric, as might be expected from the foregoing description.

The bonding in interstitial compounds is of the electron-deficient kind; i.e., there are more bonds than there are pairs of bonding electrons. It is similar to the bonding in the boranes (Sec. 21-1) and in metals (Sec. 23-3). Interstitial compounds usually have metallic properties, such as good electrical conductivity and silvery sheen, but they are often brittle.

One unusual interstitial compound is palladium hydride. When there is less than one hydrogen atom for every two palladium atoms, which corresponds to less than 0.5 percent hydrogen by weight, the material is a good electrical conductor. With more hydrogen, conductivity is poor. Gaseous H_2 can diffuse through a hot membrane of palladium. The rate of diffusion is proportional to $(P_{H_2})^{1/2}$, so that the diffusing species is atomic hydrogen. Since nothing else diffuses so rapidly through palladium, this procedure can be used to obtain extremely pure hydrogen.

---

[1] Not all transition-metal compounds of the nonmetallic elements mentioned are of the interstitial type, particularly not all oxides.

Some substances that are referred to as interstitial compounds have a structure different from that of the metal to which they are related, a structure that cannot be explained simply in terms of filling interstices with small atoms. An example is cementite, Fe_3C, grains of which are in part responsible for the toughness of some steels.

Dislocations. The regularity of the sequence of atoms in crystals is usually not perfect. Among the different kinds of imperfections are dislocations, one type of which is shown schematically in Fig. 23-7. In the neighborhood of such an edge dislocation, the bonding between the atoms above and below the plane through A and B is weakened because of the distortion in the structure—compression above the plane and dilation below. As a result, when a stress is applied to the crystal in the manner symbolized by the arrows, the atoms below the AB plane may slip, as a unit, to the right with respect to the atoms above this plane. In effect the dislocation moves to the left under stress.[1] Dislocations therefore facilitate slippage and allow a metal to be deformed. Another type of dislocation, the screw dislocation (Fig. 23-8), is important for crystal growth as well as for mechanical properties. For a screw dislocation, slippage occurs easily in a direction parallel to the dislocation axis.

There are other more complex sorts of dislocations. Edge and screw dislocations may interact to form complicated patterns. Moving dislocations eventually pile up at regions where impurity atoms are concentrated, or at the boundaries of crystal

---

[1] Charles Kittel gives an analogy for the motion of an edge dislocation: "The motion of an edge dislocation through a crystal is analogous to the passage of a wrinkle across a rug: the wrinkle moves more easily than the whole rug, but passage of the wrinkle across the rug does amount to sliding the rug on the floor."

FIGURE 23-7 **An edge dislocation.** An edge dislocation can be described as the presence of an extra plane of atoms (indicated by the hatching) in part of the crystal. The edge of this plane (just above E) is called the dislocation line. There is maximum distortion of the structure near this line—compression above it and dilation below.

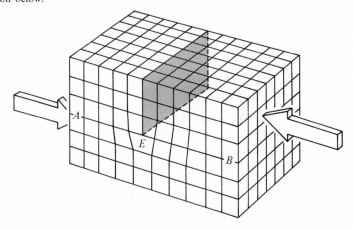

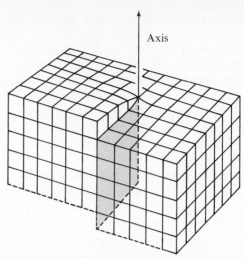

Axis

FIGURE 23-8   **A screw dislocation.** In a screw dislocation a single layer of atoms winds like a helical staircase around a line, the dislocation axis. There is maximum strain in the immediate environment of the axis.

grains. Such pile-ups often mark the locations and the beginnings of mechanical failure. This is true for crystals of nonmetals as well as of metals.

Repeated deformation of a metal, e.g., by hammering or bending, is called *cold working* and tends to produce dislocations. While a few dislocations promote slippage of crystal planes, a high concentration of dislocations hinders motion of planes because the dislocations disrupt the structural regularity. The result is a mechanical strengthening and toughening of the metal called *work hardening*. The majority of the dislocations can be removed and the metal can be made soft again by annealing, that is, by heating to about half the melting temperature. For example, an annealed piece of copper tubing is easy to bend, but bending becomes progressively more difficult as it is continued. The original softness can be restored by heating the copper red hot.

Crystals may also be strengthened by the addition of a small concentration of impurity atoms that inhibit the movement of dislocations. This is often why alloys have better mechanical properties than the component metals, which may be soft in the pure state. For example, pure iron is relatively soft, whereas mild carbon steels (containing less than about 1 mol percent carbon) are strong, in part because carbon atoms distort the structure and reduce the mobility of dislocations.

Whiskers.   Calculations show that metals should be hundreds of times stronger than they are ordinarily observed to be. The reason is the presence of dislocations: the calculations are made for perfect crystals. It has been found that theoretical strength is approached by *whiskers,* extremely thin threads of metals that sprout from metal surfaces under certain conditions. The whiskers are single crystals and are virtually free of dislocations, excepting the screw dislocations responsible for their growth. Iron whiskers with a strength more than 400 times that of ordinary iron have been made.

23-3   Electronic Structure of Metals

There are two distinct ways of describing the electronic structure of metals: (1) the resonance picture, which is most effective in describing the chemical bonding that holds the atoms of a metal together; (2) the molecular orbital picture, which accounts in a natural way for the electrical properties of metals as well as of electrical insulators and of materials called semiconductors. The two theories are complementary and together provide a deeper insight into the metallic state than does either alone.

The resonance approach.   Pauling has described the bonding in metals as a superposition of or a resonance among many formulas that involve the sharing and transfer of electrons. Consider the alkali metals, where there is one valence electron per atom so that on the average only one bond can be formed for every two atoms. The two-dimensional resonance structures shown in Fig. 23-9 are indicative of the possible three-dimensional resonance structures that can be drawn to represent the bonding. The bonding in other metals can be represented by similar formulas, but the average number of electrons involved in bonding per atom, the so-called metallic valence, will be different. An important requirement for formulas in which some of the atoms show extra bonds—such as formulas (*c*) through (*f*) in Fig. 23-9—is the availability of one low-energy unoccupied orbital per atom, enabling the atom to accept the extra bonds. Such extra orbitals are indeed generally present for metal atoms, as Pauling has shown.

The resonance description of metallic bonding is consistent with the fact that most metals can readily be deformed without breaking. New bonds are easily formed during the deformation, when atoms torn from their original neighbors find themselves moved close to new atoms. In a qualitative manner this picture can, for example, explain the relative brittleness of transition metals, lanthanides, and actinides as being caused by the high directionality of the *d* and *f* orbitals involved

FIGURE 23-9   **Resonance formulas for four alkali-metal atoms.**   In formulas (*c*), (*d*), (*e*), and (*f*) of this schematic two-dimensional representation, the positively charged atom has lost its valence electron while the atom with two covalent bonds has gained an extra electron. The bonding between the four atoms can be described as a superposition of the formulas shown.

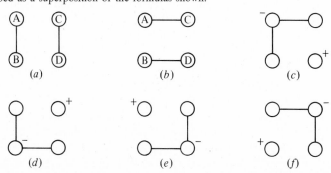

in the bonding. In contrast, alkali metals and metals on the right side of the periodic system are soft because the bonding involves nondirectional *s* orbitals (or *p* orbitals, which are less highly directed than *d* and *f* orbitals).

Metallic valence and properties of metals. As a starting point for the discussion of metallic valence we consider the packing of atoms in the metals in the first long period, exemplified by the graph in Fig. 23-10, which shows densities that have been adjusted for differences in atomic weights. For the first six metals on the left the adjusted density increases with the atomic number, indicating closer packing of the atoms and suggesting an increase in the strength of the interatomic bonds. The adjusted density stays relatively constant for the elements chromium through copper and decreases again for still higher atomic numbers, paralleled presumably by a similar behavior of the bond strength. Other properties of the metals shown, such as their melting points, hardness, and tensile strengths, vary similarly.

According to Pauling the variation in these properties is an indication of the metallic valence. In potassium each atom has one valence electron available for bonding, while in calcium there are two, in scandium three, and so on, until chromium, with six valence electrons, has been reached. It is this rise in metallic valence that accounts for the increase in the adjusted density, the melting point, and other properties in the sequence of metals discussed. From chromium through nickel there are no substantial changes in these properties, which is interpreted by Pauling to mean that the metallic valence remains at about 6 even though the number of electrons beyond the filled argon shell rises to 10. Beyond nickel the metallic valence

FIGURE 23-10   **Adjusted densities of some metals.** The adjusted densities of the metals of the first long period are shown, put on the same basis by multiplying the actual density by the ratio 54.0/atomic weight, where 54.0 is the average atomic weight of the metals shown. The adjusted (or idealized) densities reflect the closeness of the packing of the atoms.

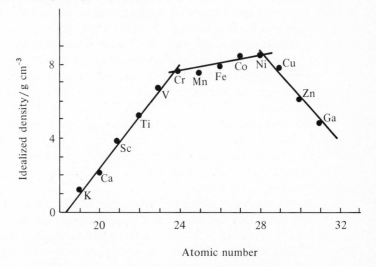

decreases again, in steps of one, accompanied by a decrease in the adjusted density, in hardness, and in other properties. The metallic valences in the sequence discussed are thus, at least approximately, the following:

K	Ca	Sc	Ti	V	Cr	Mn	Fe	Co	Ni	Cu	Zn	Ga	Ge
1	2	3	4	5	6	6	6	6	6	5	4	3	2

The congeners of the metals shown exhibit similar metallic valences.

The leveling off and subsequent fall in the metallic valence is related to the fact that it is difficult to form more than six good bonding hybrids with s, p, and d orbitals. This also explains why there are at most six ligands in complexes formed by transition metals in the first long period. As an illustration of the reduced metallic valence for elements beyond nickel, consider zinc. It has 12 valence electrons. There are nine orbitals in its valence shell (one $4s$, six $3d$, and three $4p$). One of the orbitals must be reserved to allow resonance formulas such as (c) to (f) in Fig. 23-9, a feature that is postulated to be essential in metallic bonding. Four of the eight remaining orbitals must then contain pairs of electrons, and there are four unpaired electrons available for bonding. The result is a metallic valence of 4.

The transition metals in the higher long periods and the lanthanides and actinides have f orbitals available in addition to s, p, and d orbitals, and more than six good bonding orbitals can now be formed. Complexes of such metals do indeed sometimes have more than six ligands, and metallic valences greater than 6 are also found.

The molecular orbital approach. Another useful model pictures metals as composed of electrically neutral aggregates of positive ions held in relatively fixed positions and highly mobile electrons. The electrons are free to roam within the confines of the crystal and are thus able, for example, to act as carriers of electric current. This is the *electron gas* or *electron-in-a-box* model, which was largely conceived and developed by Sommerfeld in the late twenties. Refinement of this approach by Bloch and others shows that the electrons can assume energy values in certain ranges only, a result that can be understood from molecular orbital theory.

Consider a sequence of molecules, such as Li_2, Li_3, Li_4, . . . , Li_N, where N is a very large number, so that Li_N represents a metallic crystal. In each of these molecules the atomic $2s$ orbitals of the individual atoms may be combined into molecular orbitals and the resulting energy levels worked out. Figure 23-11 shows the situation for short chains containing two, three, and four lithium atoms.

Two features of the figure remain correct when the number of atoms is increased to macroscopic proportions and when three-dimensional rather than linear aggregates of atoms are considered. The number of states remains equal to the original number of atomic orbitals involved, and the spacing of the new states becomes closer as the number of atoms increases (Fig. 23-12). For macroscopic crystals the spacing is so dense that the levels virtually coalesce into continuous *energy bands* of permitted states, each band containing as many states as there are contributing atomic orbitals.

Bands may result from any atomic states that interact to form molecular orbitals (i.e., from overlap of suitable orbitals, such as s, p, d, or even hybridized orbitals

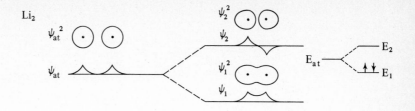

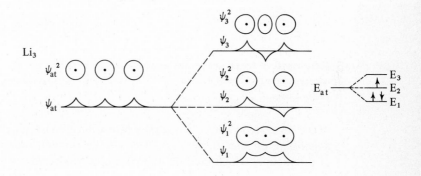

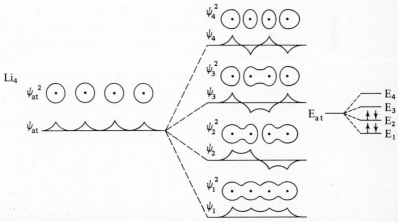

FIGURE 23-11 **Schematic representation of orbitals and energies for aggregates of lithium atoms.** On the left are schematic side views of the atomic wave functions for two, three, and four separate lithium atoms (ψ_{at}), as well as top views of the corresponding electron distributions (ψ_{at}^2). No details about the shapes of these functions are implied by the diagrams. To the right of the diagrams of individual atoms are schematic views of the molecular orbitals (ψ_i) obtained from the atomic orbitals; also shown are the resulting electron distributions (ψ_i^2), again viewed from above. On the far right are the atomic and molecular energy states, the lowest states being filled with the valence electrons of the atoms. The number of molecular states is equal to the number of contributing atomic orbitals; in each of the three cases shown, the energy increases with the number of nodes in the corresponding molecular orbital.

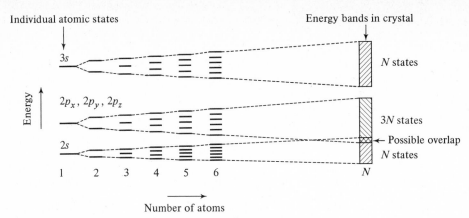

Individual atomic states

Energy bands in crystal

Energy

3s   N states

$2p_x, 2p_y, 2p_z$ 3N states

←— Possible overlap

2s   N states

1   2   3   4   5   6   N

Number of atoms

FIGURE 23-12   **Schematic energy states in aggregates of atoms.** Each of the interacting states of the *N* original atoms gives rise to a band of *N* states with energies that are, for all practical purposes, spaced continuously.

of the separated atoms). There may or may not be overlap between different bands, depending on the energy separation of the original atomic states and the degree of interaction, which controls the widths of the bands.

Metals, semiconductors, and insulators. The details of the band structure of the energy states available to the valence electrons of a solid determine whether the solid has metallic properties. Figure 23-13 illustrates in an idealized way the energy level structure for three different types of electrical behavior. In all three cases the electrons fill the lowest states possible, except for a small fraction of thermally excited electrons that may be in higher states. In the absence of an external electric field the distribution of electron velocities in the three types of solids is isotropic, and no current flows. For every electron moving in a specified direction there is always an electron moving in the opposite direction. Application of an electric field causes preferential flow in one direction, *provided that levels are available to accommodate the increased energy of these electrons.* This is how the three cases differ.

The good electrical conductivity of metals is explained by the availability of easily accessible states in the incompletely filled valence band, the band produced by the interactions of valence orbitals. Each state in an energy band may, by Pauli's principle, hold two electrons with antiparallel spins. Thus, for example, the alkali metals, with one valence electron, have half-filled valence bands (one state for each atom and two electrons per state). For the alkaline earths, the *N* states of the lowest band would be exactly filled by the $2N$ valence electrons available (two per atom) were it not for the fact that the band produced by the interaction of the *p* orbitals overlaps the band originating from the *s* orbitals. The decrease in electrical conductivity when the temperature is raised can be related to the increased thermal motion of the atoms and the increase of the average interatomic distances, which interfere with the overlaps between the orbitals and hence with the conductivity.

The availability of low-energy unfilled states also favors good thermal conductivity. Electrons that are readily mobile absorb energy from vibrating atoms at sites

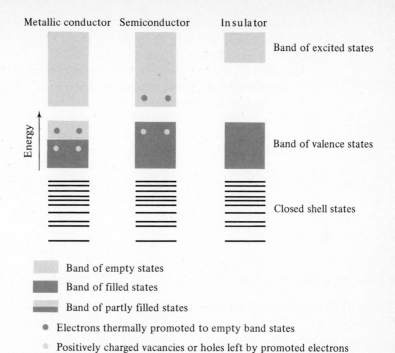

Metallic conductor   Semiconductor   Insulator

Band of excited states

Energy

Band of valence states

Closed shell states

Band of empty states
Band of filled states
Band of partly filled states
● Electrons thermally promoted to empty band states
○ Positively charged vacancies or holes left by promoted electrons

FIGURE 23-13   **Schematic energy states for solids of different electrical behavior.** The low-lying closed-shell states of the individual atoms are relatively unaffected by the presence of other atoms and thus remain separated, while there is orbital overlap and band formation for the higher states. The band containing the valence electrons is only partly filled in metals, but it is completely filled in semiconductors and insulators. In semiconductors there is a gap of the order of a few kT between the valence band and the empty band lying above it; in insulators the gap is large compared to kT. Electrons that have been thermally promoted from one band to another are shown schematically, together with the positively charged vacancies that are created by this promotion.

of vigorous atomic motion, transport it to places where the atoms are moving less vigorously, and there transfer the energy to these atoms, thereby increasing their motion.

In *semiconductors* the electrons in the valence band are unable to act as carriers because the band is completely filled. However, a few electrons are thermally excited to the next higher band, which is otherwise unoccupied. These electrons can act as carriers of current because close-lying energy states are available to them, and the band they are in is therefore called the *conduction band*. Additional conduction is provided by the vacancies left in the valence band by the thermally promoted electrons. These vacancies are called *holes* and behave like mobile positive charges, because an electron in a nearby filled state may move to fill one hole, thereby creating another hole elsewhere. The electrical conductivity of semiconductors is much poorer than that of metals because many fewer carriers participate. It increases with temperature, in contrast to the conductivity of metals, as a result of the promotion of additional electrons to the conduction band.

In *insulators* (also called *dielectrics*) the gap between the valence band and the conduction band is so large that valence electrons cannot be boosted to the higher states by thermal energy. These substances are therefore not conductors.

The difference between semiconductors and insulators is one of degree. At elevated temperatures an insulator may become a semiconductor; at very low temperatures a semiconductor ceases to conduct and becomes an insulator. Thermal conductivity is notably poor in both semiconductors and insulators because of the absence of freely movable electrons.

When light strikes some semiconductors, such as selenium (Sec. 22-4) or heat-treated zinc sulfide, electrons may be promoted to the conduction band with a resulting increase in the electrical conductivity. This effect is called photoconduction. The inverse effect also exists, in which electrons are promoted to excited states by an electric field and return to the valence band with the emission of light.

Problems and Questions

23-1   Reduction of metal oxides by carbon. Consider the reduction by carbon of (*a*) Fe_2O_3, (*b*) Fe_3O_4, and (*c*) MnO_2. Find the values of ΔH, ΔG, $T\Delta S$, and ΔS for the reactions at 298 K of these oxides with C (graphite) to produce CO and to produce CO_2. (*d*) Comment on the effect on ΔS that one of the products is a gas. (*e*) Give rough estimates of the temperatures above which the reductions considered might be feasible. (*f*) In what way would your considerations be modified by the fact that at temperatures actually used for reductions the metals are liquid while the metal oxides and carbon are mainly solid?

Ans. In sequence $\Delta \widetilde{S}/(J\ mol^{-1}\ K^{-1})$ and T_{estim}^{equil}/K, (*a*)   CO: 542, 905; CO_2: 277, 837;
(*b*)   CO: 705, 958; CO_2: 351, 940;   (*c*)   CO: 364, 820; CO_2: 187, 674

23-2   Discovery of metals. In the biblical book Numbers (chapter 31, verse 22), the priest Eleazar classifies the following as substances that can be purified by fire: "the gold, and the silver, the brass, the iron, the tin, and the lead." (The word *brass* is more properly translated as *bronze*, a copper-tin alloy.) What characteristics of these six metals (Au, Ag, Cu, Sn, Fe, Pb) might have led to their discovery and use in such early times? Lithium, sodium, and potassium were not recognized as elements until the early years of the nineteenth century; rubidium and cesium were first isolated in 1860 and 1861. To what properties might the relative lateness of the discovery of the alkali metals be attributed?

23-3   Stabilization of Au(I). (*a*) Use the Latimer diagram relating the (0), (I), and (III) oxidation states of gold

$$Au^{3+} \xrightarrow{\ 1.5\ } Au^+ \xrightarrow{\ 1.7\ } Au$$

to explain why Au(I) is not found in aqueous solution. (*b*) The half-reaction

$$Au(CN)_2^- + e^- \rightleftharpoons Au + 2CN^-$$

has an electrode potential $E^\ominus = -0.60$ V. By combining this reaction with the reduction

$$Au^+ + e^- \rightleftharpoons Au$$

show quantitatively how the formation of the $Au(CN)_2^-$ complex stabilizes Au(I) in solution.

23-4 **Aluminum production.** (*a*) How much electricity is required to produce 1 metric ton (1000 kg) of aluminum by electrolysis of Al_2O_3 dissolved in cryolite? (*b*) Estimate the energy needed, assuming a voltage requirement of 5 V. (*c*) What is the cost of the energy, assuming that 1 industrial kilowatt-hour costs 0.7¢?

23-5 **Purification of copper.** The electrolytic purification of copper is described in Fig. 19-1. Assume that among the impurities are $Ag(s)$ and $Fe(s)$ and that $[Cu^{2+}] = 1.00$ in the electrolyte. (*a*) What is the equilibrium ratio of the concentrations of Fe^{3+} and Fe^{2+} during the electrolysis? (*b*) What is the maximum Ag^+ concentration that can be achieved by the dissolving of $Ag(s)$ under the conditions given? (*c*) What is the theoretical value of $[Fe^{2+}]$ required for $Fe(s)$ to be plated out? Use the emf values given in Appendix D.

Ans. (*a*) 4.7×10^{-8}; (*b*) $1.6 \times 10^{-8} M$; (*c*) $1.8 \times 10^{26} M$

23-6 **Impurity atoms.** A cubic crystal contains a mole fraction of 1×10^{-5} impurity atoms. Estimate the average distance between impurity atoms in terms of d', the average nearest-neighbor distance characteristic of the crystal. (Note that each cube of 100,000 atoms contains on the average one impurity atom.)

23-7 **Common alloys.** Look up (e.g., in a good encyclopedia or a chemical handbook) and report the compositions of the following alloys: brass, phosphor bronze, type metal, plumber's solder, pewter, white gold.

23-8 **Mechanical properties.** Discuss structural interpretations of the characteristic ductility and malleability of many metals. Why are alloys frequently harder than pure metals?

23-9 **Metallic valence.** Consider the electrons available for bonding, and predict the relative melting points and the relative hardnesses of the metals rubidium, strontium, and yttrium.

23-10 **Group I metals.** Why are group IA metals, such as potassium and rubidium, much softer than group IB metals, such as copper and silver?

23-11 **Defects in copper crystals.** Metallic crystals can contain vacancies, that is, some sites that should be occupied by atoms but are not. In copper crystals at 900 K the ratio of such vacancies to atoms under equilibrium conditions is about 10^{-5}. Use this information and the Boltzmann expression (5-42) to estimate the energy involved in forming a vacancy, assuming that only energetic considerations are of importance. *Ans.* $\Delta E \approx 86$ kJ mol^{-1}

23-12 **Electrical conductivity.** Give definitions of the terms *insulator, semiconductor,* and *metallic conductor*. Describe how the band structure of these materials influences their electrical conductivities.

"Went to the Royal Institution to see Davy. . . . Pepys went with me. He showed us his new experiments on the decomposition of potash [K_2CO_3] and soda [Na_2CO_3]. From the oxygen, or zinc end of a combination of troughs, pure potash was decomposed . . . and a new substance produced in little globules, which has the properties of a metal. . . . The globule explodes and ignites in contact with water. . . . Pepys and I concluded we would have cheerfully walked fifty miles to see the experiment. Here is another grand discovery in chemistry."

From the diary of William Allen, November 16, 1807

24 Metallic Elements and Their Compounds

This chapter includes a discussion of both representative metals and transition metals and, in Sec. 24-4, a consideration of radioactivity and nuclear reactions.

24-1 The Alkali Metals

Li
Na
K
Rb
Cs
Fr

The highly reactive alkali metals (column IA) can be prepared by electrolysis of their molten hydroxides or chlorides. They are comparatively soft and low-melting (Table 24-1) and their densities are low: lithium, sodium, and potassium are less dense than water. The electrical and thermal conductivities of the alkali metals are high; liquid sodium is used as a coolant in nuclear reactors, and the feasibility of using sodium for underground power transmission, sheathed in suitable cables, is being considered.

The chemistry of the alkali metals is dominated by their having a single outer electron and a low ionization energy. Lithium is the best reducing agent in an

TABLE 24-1   **Some properties of alkali metals**

	Li	Na	K	Rb	Cs	Fr
Density/g cm^{-3}	0.53	0.97	0.86	1.53	1.87	
Melting point/°C	179	98	64	39	28	27
Boiling point/°C	1317	883	758	700	670	
E°/V for M$^+$(aq) + e$^-$ $\longrightarrow$ M	−3.03	−2.70	−2.92	−2.99	−2.95	

aqueous medium (Table 24-1), even though cesium has a lower ionization energy (Fig. 10-2). The ionization energy refers to a change in state involving gaseous atoms and ions, not a change from the metal to a hydrated ion, and the hydration energy of the small lithium ion is appreciably greater than that of Cs$^+$ (Prob. 24-1).

The alkali metals are highly electropositive and most of their compounds are ionic, although covalent bonds can be formed, as in methyl sodium, $NaCH_3$. The metals are usually stored in an inert medium, such as kerosene, to protect them from reaction with oxygen, water vapor, and carbon dioxide in air.[1] These reactions would lead, respectively, to the formation of peroxides, hydroxides, and bicarbonates or carbonates:

$$2Na + O_2 \longrightarrow Na_2O_2 \tag{24-1}$$
$$2Na + 2H_2O \longrightarrow 2NaOH + H_2 \tag{24-2}$$
$$NaOH + CO_2 \longrightarrow NaHCO_3 \tag{24-3}$$

Reaction of the alkali metals with liquid water can be extremely hazardous. Even with a tiny piece of metal, the heat generated may be sufficient to ignite the hydrogen produced; under some circumstances the hydrogen-air mixtures that are formed explode when ignited. Metallic sodium is sometimes used as a reducing agent in preparative chemistry.

The alkali metals dissolve in liquid ammonia (normal boiling point, −33°C) to form solutions that are very good conductors of electricity and that appear to contain solvated electrons, that is, free electrons associated loosely with one or more ammonia molecules. The solutions are blue when dilute and copper-colored at higher concentrations of the alkali metal; the more concentrated solutions have high conductivity and metallic luster and resemble liquid metals. The solutions decompose slowly, with the evolution of hydrogen and the formation of (dissolved) amides (for example, $Na^+NH_2^-$).

The alkali metals and their compounds impart characteristic colors to flames and have characteristic line spectra (Fig. 8-11). The sodium-vapor lamps commonly used for highway illumination have a yellow-orange color that is produced by the same electronic transition in sodium atoms responsible for the color in flames.

Almost all salts of the alkali metals are water-soluble. The most important is sodium chloride. Seawater now contains about 3 wt percent of NaCl, a figure that has gradually increased over the eons as rivers have carried soluble salts into the

---

[1] A fresh lithium surface will darken rapidly in nitrogen as a result of the formation of lithium nitride, Li_3N. The other alkali metals do not react directly with N_2.

ocean basins. The importance of sodium chloride in animal diets is doubtless related to the fact that much of early evolution took place in the oceans.

Two of the most widely used basic compounds, in both industry and the laboratory, are $NaOH$ and Na_2CO_3. Sodium hydroxide is prepared by electrolysis of an aqueous solution of $NaCl$ with a mercury cathode (Fig. 24-1). The sodium that is generated dissolves in the cathode, forming an amalgam. Reaction of the dissolved sodium with water yields $NaOH$ and hydrogen; chlorine is liberated at the anode.

Sodium carbonate is used in the manufacture of soaps, laundering agents, and glass. It is made by the *Solvay process,* in which a solution of $NaCl$ is saturated with NH_3 and CO_2. The carbon dioxide reacts with the weakly alkaline solution to form the bicarbonate ion, HCO_3^-,

$$NH_3 + H_2O + CO_2 \longrightarrow NH_4^+ + HCO_3^- \qquad (24\text{-}4)$$

and the relatively insoluble salt $NaHCO_3$ (baking soda) precipitates. The sodium carbonate is formed by heating the dried bicarbonate,

$$2NaHCO_3 \longrightarrow Na_2CO_3 + H_2O + CO_2 \qquad (24\text{-}5)$$

and the CO_2 produced in this step is recycled to form additional bicarbonate ion. Calcium hydroxide, $Ca(OH)_2$, obtained from the $CaCO_3$ that is the original source of the CO_2 (Sec. 24-2), is added to the remaining solution so that the ammonia can also be recovered and recycled. Calcium chloride is produced as a by-product:

$$Ca(OH)_2 + 2NH_4Cl \longrightarrow CaCl_2 + 2NH_3 + 2H_2O \qquad (24\text{-}6)$$

Sodium peroxide, the principal product formed by the reaction of sodium with (dry) air, is the active ingredient of some bleaching powders. Its effectiveness is a result of a reaction with water in which oxygen is produced:

$$2O_2^{2-} + 2H_2O \longrightarrow 4OH^- + O_2 \qquad (24\text{-}7)$$

When potassium, rubidium, and cesium react with air, the chief product is the corresponding superoxide, containing the ion O_2^-. Both potassium superoxide (KO_2)

FIGURE 24-1 **An electrolysis cell for the production of chlorine and sodium hydroxide.**

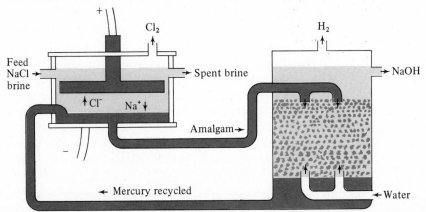

and Na_2O_2 have been used in space capsules to regenerate oxygen from water vapor by reactions analogous to (24-7) and from reaction with CO_2 as well:

$$4KO_2 + 2CO_2 \longrightarrow 2K_2CO_3 + 3O_2 \qquad (24\text{-}8)$$

The properties of potassium compounds are generally quite similar to those of the corresponding sodium compounds, but they are more expensive and much less widely used. Potassium nitrate is found in some desert regions, where extensive deposits of it were left by evaporation of earlier seas. It is used in fertilizers and as an oxidizing agent in gunpowder.

Potassium and sodium ions have essential roles in some biological processes, including the transmission of signals by nerves. An important part of a nerve cell is a slender extension, the axon, along which the nerve impulses are propagated from the main body of the cell to the synapse, which is the junction of the axon with another nerve cell or a muscle cell. At the synapse the impulse is transmitted by release of a molecule called acetylcholine; however, transmission along the axon involves changes in the concentrations of Na^+ and of K^+ on the inside and the outside of the axon. In the resting state of a nerve, the potassium-ion concentration is higher inside an axon than outside; conversely, the sodium-ion concentration is higher outside than inside. The concentration differences across the axon membrane produce a potential difference, the inside of the axon being 60 to 95 mV more negative than the outside. These concentration differences are maintained by a mechanism referred to as the *sodium pump*, which is believed to involve large chelating molecules that carry Na^+ to one side and K^+ to the other side of the cell membrane enclosing an axon. During transmission of a nerve impulse, temporary local breakdowns of this mechanism occur at a given location, starting at the main body of the nerve cell and moving progressively toward the synapse at a speed of the order of 100 m s^{-1}. As a consequence of ion movement, the potential difference across the membrane is altered, the inside of the axon becoming about 50 mV more *positive* than the outside. It is this electrical phenomenon that corresponds to the nerve impulse.

The principal use of rubidium and cesium is in the manufacture of photoelectric cells that respond efficiently to light in the visible region of the spectrum.

24-2 The Alkaline-Earth Metals

Be
Mg
Ca
Sr
Ba
Ra

The elements of column IIA are prepared chiefly by electrolysis of molten halides. They are less reactive than the alkali metals, having greater ionization energy, and also are denser, harder, higher melting, and higher boiling (Table 24-2). Since there are two electrons in the valence shell, the ions of these metals have a charge of $+2$. The metals and their compounds give characteristic colors to flames, a property made use of in the manufacture of fireworks.

The alkaline-earth elements react with water to form hydroxides, the reaction at room temperature being slow for beryllium and magnesium and rapid for the others. They all form stable oxides, and calcium, strontium, and barium form ionic peroxides.

The chemistry of beryllium is quite different from that of the other alkaline earths.

TABLE 24-2 **Some properties of alkaline-earth metals**

	Be	Mg	Ca	Sr	Ba	Ra
Density/g cm^{-3}	1.85	1.74	1.55	2.54	3.5	5.0
Melting point/°C	1280	650	850	770	725	700
Boiling point/°C	2970	1107	1487	1384	1140	
$E^{\ominus}$/V for $M^{2+}(aq) + 2e^{-} \longrightarrow M$	-1.85	-2.37	-2.87	-2.89	-2.90	-2.92

Beryllium is less electropositive and has a greater tendency than its congeners to form covalent bonds, e.g., in $BeCl_2$. Some of its compounds contain three-center bonds, as in the polymeric solid $(BeH_2)_n$, which can be represented schematically as

$$\begin{array}{ccccc} & H & H & H & H & H \\ Be & Be & Be & Be & Be \\ & H & H & H & H & H \end{array} \tag{24-9}$$

Beryllium hydroxide is amphoteric whereas the hydroxides of the other alkaline earths are strong bases, although their solubility is limited.

Magnesium is widely used in low-density structural alloys. It is a valuable reducing agent in the production of other strongly electropositive metals and is used to make the important Grignard reagent in organic chemistry (Chap. 26). Because it burns with a brilliant and very hot flame, the metal is used, in powdered or ribbon form, in incendiary devices and photoflash bulbs. The magnesium ion has a number of vital biochemical roles. It is present in chlorophyll, the green pigment in plants that initiates photosynthesis, the process by which plants convert carbon dioxide and water into carbohydrates and oxygen. Magnesium ion is also involved in the activation of smooth muscle and in the reactions by which energy is provided for many biological processes, reactions involving the breakdown of the molecule adenosine triphosphate (ATP) (Chap. 27).

Calcium is widespread in nature, chiefly as $CaCO_3$, which is essentially the sole component of limestone and marble and is the principal ingredient of chalk, of coral reefs, and of many shells. When calcium carbonate is roasted at high temperature, it decomposes to carbon dioxide and calcium oxide (quicklime) (Sec. 13-4, Figs. 13-2 and 13-3):

$$CaCO_3 \longrightarrow CaO + CO_2 \tag{24-10}$$

Calcium hydroxide, *slaked lime,* is the least expensive strong base, so it is widely used in industry. It is produced from the oxide by the addition of water. Slaking of lime is a rather dramatic procedure, since part of the water added is turned to steam by the high temperature generated by the reaction. Common mortar is a 3:1 mixture of sand and slaked lime, which hardens by absorption of CO_2 from air and evaporation of water:

$$Ca(OH)_2 + CO_2 \longrightarrow CaCO_3 + H_2O \tag{24-11}$$

Calcium ion participates in many biochemical reactions, including those that lead to muscle contraction and to the clotting of blood. Calcium phosphate is the major component of teeth and bones.

The chemical similarity between strontium and calcium makes the radioactive ^{90}Sr from the fallout of nuclear weapons tests in the atmosphere especially dangerous. The ^{90}Sr replaces calcium in bones, where the radiation it produces can damage the red blood corpuscles being formed in the bone marrow. Barium sulfate is highly insoluble and is used for quantitative analysis of barium and of sulfate. Because the high atomic number of barium makes it relatively opaque to X rays, suspensions of barium sulfate are widely used in medical X-ray diagnosis to make features of the intestinal tract more visible. Soluble barium compounds are poisonous, so that the insolubility of the sulfate is an important factor in this technique. Barium sulfate is used as a whitening agent for pigments and leather and also to add weight to drilling mud to help maintain pressure at the bottom of holes drilled for oil.

24-3 Some Other Representative Metals: Groups IIIA, IVA, and VA

B	C	N
Al	Si	P
Ga	Ge	As
In	Sn	Sb
Tl	Pb	Bi

Aluminum and its congeners. Aluminum is the most abundant metallic element in the earth's crust, occurring widely in aluminosilicate minerals. The other metals of column IIIA, gallium, indium, and thallium, are less common and less important. All four of these metals may be prepared by electrolysis of suitable melts, as described for aluminum in Fig. 23-2, and all but aluminum may also be obtained by electrolysis of aqueous solutions, because of high overvoltages for hydrogen evolution on the metals. Aluminum and gallium are comparatively hard; indium and thallium are softer. Gallium and indium are unusual in being liquid over a range of about 2000 K, the widest ranges for any known substances; gallium melts just above room temperature.

Aluminum is the most reactive of these metals and would react rapidly with the oxygen or moisture in air if it were not protected by a thin, transparent, and strongly adherent layer of aluminum oxide. All four elements form ions of charge $+3$; thallium also forms stable compounds containing the thallous ion, Tl^+, many of whose salts resemble the corresponding salts of the alkali metals. The salts of Tl(I) are, however, extremely poisonous.

Aluminum has low density for a metal, 2.7 g cm^{-3}, and high tensile strength. It is easily malleable, can be rolled into thin foils, and is an excellent electrical conductor. Its conductivity is only a little more than half those of copper, silver, and gold for wires of equal diameter, but because the density of aluminum is less than one-third that of any of these other elements, the conductivity of wires of equal mass per unit length is greatest for aluminum. Since it is by far the least expensive of any of these excellent conductors, aluminum is widely used in high-voltage transmission lines.

Aluminum and its alloys find broad application in aircraft and space vehicle manufacture, in the construction industry, and in household utensils. It is the metallic ingredient of "silver" paint. A mixture of powdered aluminum with iron oxide is used in the *thermite process* for welding iron:

$$2Al + Fe_2O_3 \longrightarrow 2Fe + Al_2O_3 \tag{24-12}$$

Molten iron is produced by this highly exothermic reaction, which is often started with a mixture of barium peroxide and powdered aluminum. A temperature of about 3000°C is reached and the white-hot liquid iron formed by the reduction can be directed into the space between the pieces to be welded.

Despite its strong reducing power ($E^\diamond = -1.66$ V), aluminum does not normally react with even hot water because it is protected by the tough oxide film. However, amalgamated aluminum reacts readily:

$$2Al(amalgam) + 6H_2O \longrightarrow 2Al(OH)_3 + 3H_2 \tag{24-13}$$

Since aluminum oxide is amphoteric [Secs. 7-4 and 14-6, Equation (14-51)], it is soluble in strong acids and strong bases, and thus these materials will normally[1] dissolve metallic aluminum readily:

$$2Al + 2OH^- + 6H_2O \longrightarrow 2Al(OH)_4^- + 3H_2 \tag{24-14}$$
$$2Al + 6H^+ \longrightarrow 2Al^{3+} + 3H_2 \tag{24-15}$$

Aluminum oxide, Al_2O_3, occurs naturally as the very hard mineral corundum (used as an abrasive) and as various gemstones that have the same structure as corundum but contain traces of transition-metal ions. These ions impart characteristic colors; for example, traces of Cr(III) color aluminum oxide a deep red and traces of Ti(IV) or Fe(II) and Fe(III) color it blue, producing the gems ruby and sapphire, respectively. Synthetic corundum and synthetic rubies and sapphires are made by melting precipitated aluminum hydroxide, with appropriate additives, in an electric furnace and then cooling the melt slowly under carefully controlled conditions.

The fact that $Al(H_2O)_6^{3+}$ is acidic ($K \approx 10^{-5}$) and thus readily "hydrolyzes" is the basis for several important commercial uses of the alums, which are salts containing this ion. They have the general formula $MAl(SO_4)_2 \cdot 12H_2O$, with M representing an alkali-metal cation (other than Li^+), NH_4^+, Ag^+, or Tl^+. Ammonium and potassium alums, the most important ones, are widely used both in baking powders, in which they furnish the acid that causes the release of CO_2 from sodium bicarbonate, and as mordants in dyeing. Mordants bind the dye to the material. They are substances that attach themselves to fibers of fabrics and also can easily adsorb dyes. In the dyeing process the fabric is immersed in a solution of the alum, and aluminum hydroxide is precipitated on the fibers by the addition of sodium carbonate or calcium hydroxide. The dye is added later. Alums are also used to clarify water. The gelatinous $Al(OH)_3$ that is formed by hydrolysis carries down suspended material as it slowly precipitates.

Aluminum chloride, $AlCl_3$, forms ionic crystals in which each aluminum atom is octahedrally coordinated. On the other hand, when the chloride is dissolved in nonpolar solvents, or is melted or vaporized, it has a quite different structure, consisting of discrete nonionic molecules of formula Al_2Cl_6. These molecules have

[1]Nitric acid, however, renders aluminum passive (Sec. 19-3) and thus can be shipped in aluminum containers.

a bridged structure, with each aluminum atom surrounded by four halogens in an approximately tetrahedral arrangement:[1]

$$\text{(structure: Cl atoms bridging two Al atoms)} \tag{24-16}$$

Aluminum bromide and iodide have the same structure, even in the crystalline state.

Aluminum, like its congener boron, also forms electron-deficient compounds. An example is $Al_2(CH_3)_6$, the geometric structure of which resembles that of Al_2Cl_6, the carbon atoms of the methyl groups being situated where the chlorine atoms are in (24-16). In each of these molecules the peripheral atoms are covalently bonded to the aluminum atoms. In the methyl compound, however, there is only one electron pair available for bonding each bridging methyl group to the two aluminum atoms, one bonding electron being provided by the methyl group and one by an aluminum atom. There are thus two three-center bonds involved in the bridging, each using one electron pair.

$$\text{(structure of } Al_2(CH_3)_6 \text{)} \tag{24-17}$$

In the halides, on the other hand, each bridging halide atom has three unshared electron pairs, one of which can be used to bond to the second aluminum atom. Each bridging atom is thus held by two covalent bonds:

$$\text{(structure with bridging Cl atoms)} \tag{24-18}$$

Tin and lead.   These two metals of group IVA are congeners of carbon, silicon, and the metalloid germanium, and indeed tin behaves as a metalloid at low temperatures (p. 636). Both tin and lead are relatively low-melting and are soft and malleable. Copper and tin were known thousands of years before metallic iron was discovered; their ores are much more easily reduced than those of the more electropositive iron. Before 3000 B.C. it was discovered in the Middle East that copper could be hardened by alloying it with tin. The resulting bronzes could be forged

---

[1] The wedge-shaped bonds imply that the Cl atoms are above the plane of the page; the dashed bonds imply that the Cl atoms are below the plane of the page.

into utensils, tools, and weapons so superior to others then available that they eventually produced a major revolution in technology, ushering in the Bronze Age.

The chief ores of tin and lead are cassiterite, SnO_2, and galena, PbS. The metals, which are readily made by reduction of these ores, are good reducing agents in alkaline or strongly acid solutions. When pure they do not readily liberate hydrogen from dilute acids because of high overvoltages. Both metals form compounds of oxidation states (II) and (IV), including covalent as well as ionic substances. Stannous ion, Sn^{2+}, is a fairly good reducing agent and is also quite acidic in water because of extensive hydrolysis,

$$Sn^{2+} + H_2O \longrightarrow SnOH^+ + H^+ \qquad K \approx 10^{-2} \qquad (24\text{-}19)$$

Stannic oxide, SnO_2, is amphoteric, dissolving in hot NaOH to form the octahedral $Sn(OH)_6^{2-}$ ion and in concentrated HCl to form $SnCl_6^{2-}$. The chief uses of tin are in the manufacture of bronzes and other alloys—including type metal (Pb–Sb–Sn), pewter (Sn–Sb–Cu), and solders (low-melting Sn–Pb alloys)—and as the coating on tin cans.

Pb(II) salts are more stable than Pb(IV) compounds; both are important in the lead storage battery (Chap. 18). Considerable quantities of lead are consumed in the manufacture of storage batteries and in the production of the gasoline additive tetraethyllead, $Pb(CH_2CH_3)_4$, a covalent compound. When this compound is heated, it liberates ethyl radicals, CH_3CH_2, which help to keep the heated gasoline-air mixture from igniting before the proper time in the engine cycle and thus to minimize "knocking". To prevent lead deposits from forming in the engine, dibromoethane ($C_2H_4Br_2$) is also added to the fuel. It reacts with the lead to form lead bromide, a volatile compound that becomes part of the exhaust gases. Other important uses of lead are in paint pigments, of which "white lead" [essentially $2PbCO_3 \cdot Pb(OH)_2$], "chrome yellow" ($PbCrO_4$), and "red lead" (Pb_3O_4) are especially important.

Lead compounds are poisonous, chiefly because Pb^{2+} reacts with the essential —SH groups of proteins to form insoluble precipitates. Metallic lead was used for pipes and for wine vessels as far back as Roman times, and some historians claim that widespread lead poisoning played a significant role in the decline of the Roman empire. More recently paints containing lead have sometimes been a source of chronic poisoning. They are now banned by law from being used on objects such as toys or baby furniture that are likely to be chewed by infants. Much effort is being made to find suitable substitutes for tetraethyllead as an antiknock fluid, in order to reduce atmospheric lead pollution.

Antimony and bismuth.   The congeners of these elements of group VA are nitrogen, phosphorus, and arsenic, all nonmetals. Although arsenic forms steel-gray lustrous crystals, they are brittle and are poor electrical conductors. Antimony is a metalloid, occurring in a silvery brittle metallic form and in several nonmetallic modifications; bismuth exists only as a metal. Antimony and bismuth may both be prepared by reduction of their oxides, obtained by roasting sulfide ores in air. The elements are purified electrolytically.

Each element forms ionic as well as covalent compounds, with oxidation states (III) and (V). Antimony is stable in air at ordinary temperatures, but it burns when heated, forming Sb_2O_3. At high temperatures this compound consists of molecules of formula Sb_4O_6, containing a tetrahedron of antimony atoms sharing oxygen atoms along each edge, like the phosphorus and arsenic analogs (see Fig. 22-3). Bismuth is also stable in air; at high temperature it reacts to form Bi_2O_3. Both antimony and bismuth react readily with chlorine, forming $SbCl_5$ and $BiCl_5$, respectively. Antimony pentachloride is a powerful reagent for introducing chlorine atoms into other molecules; BiF_5 is an equally effective fluorinating agent.

Many of the alloys of antimony and bismuth, as well as the pure metals themselves, expand upon freezing and thus are useful for castings. A number of bismuth alloys melt below the boiling point of water—e.g., Woods metal (mp 71°C), containing Bi, Pb, Sn, and Cd in proportions $4:2:1:1$. Such alloys are used in automatic fire sprinklers, in safety plugs in steam boilers to guard against overheating, and in various kinds of fuses.

24-4 Some Transition Metals and Nuclear Reactions

The transition metals include many of the most important metals of industry and commerce. Iron is the central metal of modern civilization—we are still living in the Iron Age that began about 3000 years ago. Many other transition metals are used to impart desired qualities to iron alloys (steels). The coinage metals, copper, silver, and gold, are important not only commercially but also because of special properties, such as their high electrical conductivity, comparative inertness to attack by the atmosphere, and the sensitivity of the silver halides to light, the basis of the photographic process. The unique properties of many transition metals lead to other important applications, e.g., tungsten in light-bulb filaments, platinum and palladium as catalysts for hydrogenation and other reactions, and iron, cobalt, nickel, and their alloys in magnets. Transition-metal compounds, sometimes only in trace quantities, are essential for almost all known forms of life (Chap. 27).

In this section we discuss a few selected and representative transition metals and some of their compounds, as well as radioactivity and nuclear reactions.

Sc
Y
La Ce $\cdots$ Lu
Ac Th $\cdots$ Lr

Group IIIB elements, the lanthanides, and the actinides. Lanthanum and actinium, the congeners of scandium and yttrium in column IIIB, give their names to the series of 14 elements referred to as the lanthanides and actinides, which correspond, respectively, to the filling of the $4f$ and $5f$ subshells. The chemistry of the lanthanides is very similar to that of Sc and Y, oxidation state (III) predominating. Just as Sc and Y readily lose three electrons to form ions of charge $+3$ (lose the electrons in the $4s$ and $3d$, or $5s$ and $4d$, orbitals, respectively), so each of the lanthanides forms a tripositive ion with no electrons in the $6s$ and $5d$ orbitals. The $4f$ subshell in these ions is progressively increased in population, from no electrons in La^{3+} to 14 in Lu^{3+}. The only other common oxidation states found for any of these elements can be explained in terms of the special stability of filled or half-filled subshells: Ce^{4+} (xenon configuration), Tb^{4+} and Eu^{2+} (half-filled $4f$ subshell), and Yb^{2+} (filled $4f$ subshell).

The lanthanides are difficult to separate from one another because of their nearly identical chemical properties, but ion-exchange chromatography of appropriate complexes provides efficient separations of almost all of them. These elements were formerly called the *rare-earths* (the term *earth* implying oxide) but they are not unusually scarce. Some have important applications. Yttrium vanadates, with an added trace of Eu(III), serve as red phosphors for color television. (The blue and green phosphors are usually ZnS with small quantities of added silver and copper, respectively.) Ceric oxide, CeO_2, mixed with a large proportion of ThO_2, is used in the white-glowing mantles of gasoline and propane lanterns.

The radii of the tripositive lanthanide ions decrease steadily with increasing atomic number, from 1.06 Å for La^{3+} to 0.85 Å for Lu^{3+}. This *lanthanide contraction* arises because of poor screening of the nuclear charge by the $4f$ electrons, so that there is a progressive increase in the effective nuclear charge experienced by all electrons as the atomic number increases. A similar contraction occurs in the actinide series, for the same reason. An important consequence of the lanthanide contraction is that it approximately offsets the increase in ionic radius that might have been expected within a given column of the periodic table because of the increase in the number of electrons and the higher principal quantum numbers of the valence shell. Such an increase is found between the first and second transition series, which involve no f electrons. In group IVB, for example, the radii of the $+4$ ions are 0.68 Å for Ti^{4+} (first transition series), 0.79 Å for Zr^{4+} (second series), and 0.78 Å for Hf^{4+} (third series). The radii of Zr and Hf are nearly identical in other oxidation states as well, and atoms of these elements readily substitute for one another in minerals and other substances. Zirconium and hafnium are very difficult to separate; their chemical properties are more similar than those of any other pair of congeners.

The actinides do not resemble each other as much as the lanthanides do, because the $5f$ electrons are less effectively shielded from neighboring atoms by the $6s$ and $6p$ electrons than are the $4f$ by the $5s$ and $5p$ electrons. Although the actinides resemble the lanthanides in some ways, their chemistry is more complex. Many actinides have several stable oxidation states, ranging from (III) to (VI). The most stable oxidation state of the most important actinide, uranium, is (VI)—exemplified in the uranyl ion, UO_2^{2+}, which forms a number of salts, and in UF_6, a highly reactive molecule that forms volatile yellow crystals with a vapor pressure of 1 atm at 57°C. The volatility of UF_6 made possible the gaseous-diffusion separation of uranium isotopes at Oak Ridge, Tennessee, during World War II (Example 5-16).

All isotopes of the actinides are radioactive.[1] Actinium, thorium, protactinium, and uranium occur naturally; the other actinides occur on earth only because they are made by nuclear reactions. The chemical properties of the elements beyond uranium, which were unknown before 1940, were deduced initially from experiments with microgram or even smaller quantities.

To clarify why nuclei of high atomic number are unstable and to explain the nature of the changes that accompany radioactivity and other nuclear reactions, we digress briefly from the principal topic of this chapter.

---

[1] All isotopes of seven other elements that have lower atomic numbers than that of Ac (89) are also radioactive. Four of these (Tc, 43; Pm, 61; At, 85; and Fr, 87) exist at most in trace amounts on the earth's surface and their properties are known only from experiments with synthetic samples. The other three (Po, 84; Rn, 86; and Ra, 88) are decay products of long-lived radioactive isotopes of U or Th.

Nuclear stability, radioactivity, and nuclear reactions. Radioactivity has been mentioned briefly in Secs. 1-5 and 1-6, and radiocarbon dating was discussed in Sec. 22-2. Radioactivity is a manifestation of nuclear instability. Many radioactive nuclei decay by emission of charged particles, most commonly electrons or helium nuclei (alpha particles). When a nucleus emits an electron, the atomic number Z increases by 1 but the mass number A does not change. In effect, one neutron in the nucleus has been converted into a proton and an electron. In representing such a nuclear reaction by an equation, our concern is only with the nuclei involved, not with any accompanying changes involving extranuclear electrons:

$$\ce{_{Z}^{A}X} \longrightarrow \ce{_{Z+1}^{A}Y} + \ce{_{-1}^{0}}e \qquad (24\text{-}20)$$

When a nucleus decays by emission of an alpha particle, A must decrease by 4 and Z by 2:

$$\ce{_{Z}^{A}X} \longrightarrow \ce{_{Z-2}^{A-4}Q} + \ce{_{2}^{4}He} \qquad (24\text{-}21)$$

Some nuclei decay only by emission of a photon (a gamma ray); neither A nor Z changes. The original nucleus is in an excited state, corresponding to some unstable configuration of its nucleons. As a more stable arrangement is achieved, a photon is emitted corresponding to the difference in energy of the two nuclear states, just as a photon is emitted by an atom in an excited electronic state that decays to some more stable arrangement of the electrons. Nuclear energy levels are typically separated by energies around 10^6 eV (1 Mev), about 1 million times the separation of the outer electronic energy levels in atoms. Gamma rays have wavelengths of the order of 10^{-2} Å.

Unstable nuclei can decay in other ways than those just described, and stable nuclei can be altered artificially by bombardment with sufficiently energetic particles. All nuclear reactions conform to the same general principles: there is conservation of mass-energy, conservation of charge, and conservation of nucleons (in the absence of antiprotons and antineutrons, the usual situation). Other quantities are conserved as well, but they are not of concern here. The only difference from the conservation laws for ordinary chemical reactions is that atoms are conserved in chemical reactions but not in nuclear reactions. In ordinary chemical reactions it is possible to speak of conservation of mass by itself, because the energy changes involved are so small that the mass differences corresponding to them (through the relation $E = mc^2$) are of the order of 1 part in 10^{10} or even smaller, and are thus normally undetectable. In nuclear reactions the energy changes are often so large that mass changes are readily detectable, but the sum of Δm and $\Delta E/c^2$ is always zero, which is what is implied by the condition that mass-energy is conserved.

Because nuclear energy levels are so widely spaced relative to ordinary thermal and chemical-bond energies, the rates of radioactive disintegrations and other nuclear reactions are generally independent of temperature, physical state, and chemical form.

Consideration of the properties of stable nuclei shows that the presence of neutrons in the nucleus is associated with nuclear stability. Every stable nucleus containing more than two protons has at least as many neutrons as protons. Furthermore, the ratio of the number of neutrons, N, to the number of protons, Z, in stable nuclei increases as Z increases. Figure 24-2 shows graphically the relation between Z and $N \, (= A - Z)$ for stable nuclei. Up to $Z = 20$ there are numerous stable isotopes with $N = Z$, but beyond that N gradually becomes greater than Z. For the heaviest known stable isotope, ^{209}Bi, $N = 1.52Z$; there are no stable isotopes with Z greater than 83.

Nuclear radii vary from about 1×10^{-5} to 8×10^{-5} Å. At these distances there are very strong coulombic repulsive forces between pairs of protons as a consequence of their positive charge; at a distance of 1×10^{-5} Å the proton-proton coulombic repulsive energy amounts to more than 10^8 kJ mol^{-1}. Very strong short-range attractive forces between nucleons (neutron-neutron, neutron-proton, and proton-proton), arising through the exchange of subatomic particles called π mesons, can provide enough binding energy to offset these repulsions for nuclei of moderate size. However, these attractive forces are of much shorter range than the coulomb repulsions, and as the number of protons continues to increase, the repulsions tend to dominate and the nucleus becomes unstable. Addition of more neutrons cannot compensate for this because nucleons must obey the Pauli principle and the added neutrons would have to occupy high energy levels and thus could not confer added stability.

FIGURE 24-2 **Relation between the number of protons and the number of neutrons in stable nuclei.** This broad curve, which tapers at each end, represents the general trend of the relation between N and Z for the stable nuclei. There are, however, many unstable nuclei within this "band of stability", in particular most of those with both N and Z odd, and some with either N or Z odd. The majority of the approximately 280 known stable nuclei have both N and Z even. Only five stable nuclei have odd Z and odd N (^{2}H, ^{6}Li, ^{10}B, ^{14}N, and ^{180}Ta; in addition ^{50}V is almost stable, with a half-life of 4×10^{14} years, so that it has scarcely disintegrated during the lifetime of the earth). Some elements of even atomic number have many stable isotopes—for example, Sn (10) and Xe (9). Many odd-Z elements have no more than one stable isotope, although many unstable isotopes are known for some (for example, 19 for iodine).

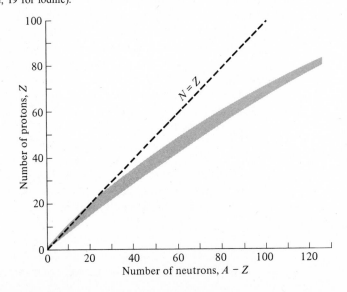

When an unstable nucleus is created that is outside the band of stability indicated in Fig. 24-2, it will change to a more stable nucleus in such a way as to return to the band of stability. If the N/Z ratio is too high, for example, emission of an electron from the nucleus will decrease N and increase Z, thus lowering the N/Z ratio. Conversely, if that ratio is too low, the nucleus may capture one of the atom's extranuclear electrons or emit a positron (a positive electron); in either case N increases by 1 and Z decreases by 1, raising the N/Z ratio. Alternatively, a nucleus of high atomic number for which N/Z is too low may emit an alpha particle, thereby changing into a more stable nucleus with a higher N/Z ratio.

Table 24-3 illustrates some of these nuclear transformations. It shows the series of disintegration products arising from the long-lived isotope ^{238}U, whose half-life, 4.5 billion years, is comparable to the age of the earth. Successive radioactive disintegrations occur by either electron or alpha-particle emission until the stable nucleus ^{206}Pb is formed. All mass numbers in this series differ from that of the parent nucleus, ^{238}U, by an integral multiple of 4. There are three other similar series, two of which originate with naturally occurring long-lived radioactive isotopes, ^{235}U and

TABLE 24-3   **The uranium-radium radioactive decay series**   The arrows diagonally to the left correspond to the emission of an alpha particle by a nucleus; the shorter arrows pointing to the right correspond to nuclear emission of an electron. The figures below the symbols for the elements represent the half-lives of the corresponding isotopes; abbreviations are as follows: y, years; d, days; h, hours; m, minutes; s, seconds. A few of the nuclei shown have alternative modes of decay, but they are of minor importance, occurring only about once in a thousand disintegrations.

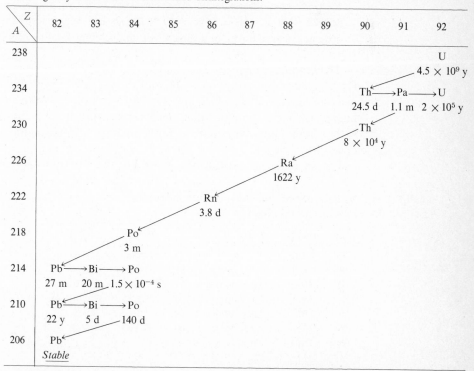

^{232}Th; the third was discovered during World War II and is named for its longest-lived member, ^{237}Np, with a half-life of 2.2 million years.

Even the most stable isotopes of the transuranium elements (i.e., elements with $Z > 92$) have half-lives (Table 24-4) sufficiently short that any of these elements created at the time the elements of the earth were formed would have long since become undetectable. The transuranium elements have been manufactured principally by bombardment of more stable isotopes—including some that are themselves radioactive—with high-energy protons, deuterons, and other low-mass nuclei that have been accelerated to very high energies in cyclotrons, bevatrons, and other particle accelerators.

The stability of different nuclei can be compared by calculation of the *binding energy* for each. This quantity is defined for a given nucleus as the energy equivalent to the difference between the sum of the masses of the individual nucleons present and the mass of the nucleus itself.

EXAMPLE 24-1 **Binding energy.** The nucleus of an ordinary helium atom contains 2 neutrons and 2 protons. Its mass, and those of a neutron and a proton, are, respectively, 6.64462×10^{-27} kg, 1.67495×10^{-27} kg, and 1.67265×10^{-27} kg. Calculate the binding energy for ^{4}He.

Solution. The difference between the sum of the masses of the nucleons constituting the helium nucleus and the mass of the nucleus itself is $(2 \times 1.67495 + 2 \times 1.67265 - 6.64462) \times 10^{-27}$ kg $= 5.058 \times 10^{-29}$ kg. Hence by the Einstein relation, $E = mc^2$, the difference in mass corresponds to an energy of 5.058×10^{-29} kg $(2.9979 \times 10^8 \text{ m s}^{-1})^2 = \underline{4.546 \times 10^{-12} \text{ J}}$. This is a huge energy for the formation of just one nucleus; for 1 g of ordinary helium nuclei the binding energy is 6.84×10^8 kJ.

The binding energy per nucleon rises rapidly with increasing A up to a maximum for A values in the range 55 to 65, and then it falls slowly. Thus energy is released in a nuclear reaction in which low-A nuclides combine to form a nucleus of somewhat higher A value (a process called nuclear *fusion*) or in a reaction in which a nucleus with a high value of A splits to form nuclides of intermediate A values (a *fission* reaction).

Fusion reactions provide the energy of the sun and other stars. The principal

TABLE 24-4 **Half-lives* of the most stable isotopes of some actinides**

Z	Isotope	Half-life	Z	Isotope	Half-life
89	^{227}Ac	22 y	95	^{244}Am	8×10^3 y
90	^{232}Th	1.4×10^{10} y	96	^{248}Cm	4×10^5 y
91	^{231}Pa	3.4×10^4 y	97	^{247}Bk	7×10^3 y
92	^{235}U	7.1×10^8 y	98	^{251}Cf	7×10^2 y
	^{238}U	4.5×10^9 y	99	^{254}Es	1 y
93	^{237}Np	2.2×10^6 y	100	^{253}Fm	3 d
94	^{244}Pu	8×10^7 y			

*Abbreviations: y, years; d, days.

reaction in the sun is the fusion of four protons to form a single ^{4}He nucleus.[1] This reaction predominates at temperatures below about 20 million deg. At considerably higher temperatures further fusion reactions occur, producing carbon and heavier elements.

When certain very heavy nuclei, notably ^{235}U and ^{239}Pu, absorb slow neutrons, fission occurs. The products are nuclei of intermediate mass numbers, ranging from about 75 to 155; on the average, two to three neutrons are produced as well for each nucleus undergoing fission. A representative fission reaction for ^{235}U is

$$^{235}_{92}\text{U} + {}^1_0n \longrightarrow [{}^{236}_{92}\text{U}] \longrightarrow 3{}^1_0n + {}^{94}_{36}\text{Kr} + {}^{139}_{56}\text{Ba} \qquad (24\text{-}22)$$

Because the N/Z ratios of the product nuclei are higher than those for stable nuclei in this range of A values, fission products are usually intensely radioactive, emitting highly energetic electrons and gamma rays (and occasionally further neutrons) before they decay into stable nuclides.

The energy released in a fission reaction of ^{235}U is about 2×10^{10} kJ mol^{-1}, about 10^7 times the energy released in a typical chemical reaction of this same quantity of uranium (its oxidation or fluorination, for example). Because neutrons are emitted during each fission, they can (if slowed down) be absorbed by other ^{235}U nuclei, causing them in turn to undergo fission. Thus it is possible to carry out fission in such a way that it will be a self-sustaining chain reaction, the products of one step in the chain initiating further steps. If this chain reaction is allowed to proceed so that each fission causes several more fissions, the entire sample will react in a very small fraction of a second—and if the sample is of even modest size, a powerful explosion will result. The first nuclear weapons were fission bombs.

Under controlled conditions, with each fission producing exactly one more fission so that there is no sudden growth in the number of nuclei splitting per unit time, nuclear fission can be an enormously valuable source of energy. It provides power at a steady rate under easy and precise control for a considerable length of time before more fuel is needed. A nuclear reactor (Fig. 24-3) is a device for carrying out fission under such controlled conditions. Confinement and disposal of the extremely hazardous radioactive fission products in a secure and permanently safe way is an ever-present problem that is becoming increasingly critical as reactors become more commonplace and fossil fuels scarcer and more expensive.

Reactors are not only efficient and reliable energy sources but are also sources of isotopes of almost every element, a few directly from the fission products and most indirectly by neutron irradiation of stable isotopes. In addition, intense beams of neutrons are available from reactors and are of great value in many kinds of experiments, including precise qualitative and quantitative analysis of even trace quantities of materials by neutron activation analysis[2] and determination of molecu-

---

[1] This process proceeds through several stages and involves the emission of two positrons (and two neutrinos, uncharged particles of negligible mass and enormous penetrating power). In effect, two of the protons are converted into two neutrons; the remaining two protons and the two neutrons constitute the ^{4}He formed.

[2] In neutron activation analysis, a sample is irradiated with neutrons, causing radioactive isotopes to be formed. The characteristic half-lives of the products and the energies of the radiation are used to identify the elements in the sample.

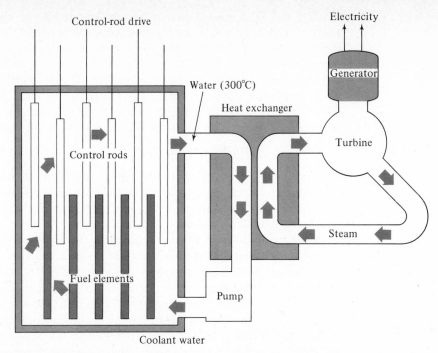

FIGURE 24-3 **Schematic diagram of a nuclear reactor.** The reactor is designed to permit carrying out a self-sustaining fission reaction as a source of power. The fuel elements contain some nuclide such as ^{235}U that undergoes fission after absorption of a slow neutron, with the emission of additional neutrons. These neutrons may cause other nuclei to undergo fission or may be absorbed by the control rods, which are made of cadmium or some other material that strongly absorbs neutrons. By moving the control rods into or out of the matrix of fuel elements, the number of nuclei undergoing fission per unit time, and thus the power output of the reactor, can be controlled.

lar structure and molecular motion in solids (and even some liquids) by diffraction and scattering of neutrons.

Power from fusion reactions is not yet a reality on earth, as it is in the sun and stars. Fusion reactions are effected in an explosive way in thermonuclear weapons (hydrogen bombs). In these awesomely frightening devices, deuterium and tritium are fused, with ^6Li present also, to form ^4He and a neutron. The reaction is initiated by a fission bomb, which raises the temperature to around 10^7 K in a matter of microseconds. Highly exothermic fusion reactions then begin, further increasing the temperature and, thereby, the rate of fusion. The energy released is of the order of 10^3 or more times that of a fission bomb.

Controlled fusion would have the advantage over present nuclear power sources that the primary products are not radioactive and that the needed fuel, hydrogen, is available in virtually unlimited supply. The technical problems in achieving the temperatures necessary for fusion, confining the reactants adequately, and carrying out the process in a safe and controlled manner are enormous. They are the subject of intense research all over the world.

Cr   Mn
Mo   Tc
W   Re

Chromium and manganese. The transition elements in columns IVB to VIIB (Ti, V, Cr, Mn, and their congeners) exhibit many oxidation states, exemplified in Table 24-5. The highest oxidation state of each element is identical with the number characterizing its column and is usually found in compounds or anions containing oxygen or fluorine. We discuss here only chromium and manganese.

The Latimer diagrams for the different oxidation states of chromium and manganese are typical of the relationships among oxyanions and hydrated cations. When the equilibria involve H^+ or OH^- it is necessary to consider acidic and basic solutions separately. The Latimer diagrams for Cr are

Acidic solution:

$$\overset{\displaystyle \overset{-0.74}{\overbrace{\hspace{6cm}}}}{Cr_2O_7{}^{2-} \xrightarrow{+1.33} Cr^{3+} \xrightarrow{-0.41} Cr^{2+} \xrightarrow{-0.91} Cr(s)} \tag{24-23}$$

Basic solution:

$$CrO_4{}^{2-} \xrightarrow{-0.13} Cr(OH)_3(s) \xrightarrow{-1.1} Cr(OH)_2(s) \xrightarrow{-1.4} Cr(s) \tag{24-24}$$

The corresponding diagrams for Mn are

Acidic solution:

$$MnO_4{}^- \xrightarrow{+0.56} MnO_4{}^{2-} \xrightarrow{+2.26} MnO_2(s) \xrightarrow{+0.95} Mn^{3+} \xrightarrow{+1.51} Mn^{2+} \xrightarrow{-1.18} Mn(s)$$

$$\underset{+1.69}{\underbrace{\hspace{3cm}}} \qquad \underset{+1.23}{\underbrace{\hspace{3cm}}} \tag{24-25}$$

Basic solution:

$$MnO_4{}^- \xrightarrow{+0.56} MnO_4{}^{2-} \xrightarrow{+0.60} MnO_2(s) \xrightarrow{-0.2} Mn(OH)_3(s) \xrightarrow{+0.1} Mn(OH)_2(s) \xrightarrow{-1.55} Mn(s)$$

$$\underset{+0.59}{\underbrace{\hspace{3cm}}} \qquad \underset{-0.05}{\underbrace{\hspace{4cm}}}$$

$$\tag{24-26}$$

The diagrams in acidic and basic solution differ because the preponderant species are not always the same and because reaction quotients involving H^+ or OH^- are written with $[H^+]$ in acidic solutions and with $[OH^-]$ in basic solutions. The differences in the corresponding values of $\mathbf{E}^\ominus$ depend on K_w and other relevant equilibrium constants, as discussed in Chap. 19. Some of the values given are known only with

TABLE 24-5 Oxidation states in some binary compounds

	Ti	V	Cr	Mn
F	III, IV	III, IV, V	II, III, IV, V, VI	II, III, IV
Cl	II, III, IV	II, III, IV	II, III, IV	II
Br	II, III, IV	II, III, IV	II, III, IV	II
I	II, III, IV	II, III	II, III	II
O	II, III, IV	II, III, IV, V	II, III, IV, VI	II, III, IV, VII

low precision because some of the measurements are experimentally difficult, e.g., those involving gelatinous, slightly soluble hydroxides such as $Cr(OH)_2$ or $Mn(OH)_3$.

Chromium is a rather hard, high-melting, silvery metal. It is comparable to zinc in reducing power, as implied by its $\mathbf{E}^{\ominus}$ values, and dissolves readily in nonoxidizing strong acids such as HCl and H_2SO_4. It does not, however, dissolve in oxidizing media such as HNO_3 because it is rendered passive (Sec. 19-3). The metal is also extremely resistant to atmospheric attack, because of formation of an adherent oxide film. Thus it is extensively used for corrosion protection, especially of iron, which is first electroplated with copper and then with chromium. Chromium is also an important component of many steels: the most common stainless steel contains about 18 percent Cr and 6 percent Ni, and a typical tool steel contains about 1 percent each of Cr, Mn, W, and C.

The three common oxidation states of chromium, other than that of the element itself, are Cr(II), which is basic, Cr(III), which is amphoteric, and Cr(VI), which is acidic. The chromous ion, Cr^{2+}, is a strong reducing agent, easily oxidized by atmospheric oxygen and even slowly liberating hydrogen from water. Green chromic oxide, Cr_2O_3, which can be obtained by heating the metal to high temperatures in air, is used as a pigment. Cr(III) has a great tendency to form hexacoordinate complexes, and thousands of these are known. They include complexes with water, ammonia, halide ions, and many other anions and neutral molecules.

The aqueous chemistry of Cr(VI) is influenced strongly by equilibria involving the yellow chromate ion, CrO_4^{2-}, and the orange dichromate ion, $Cr_2O_7^{2-}$:

$$HCrO_4^- \rightleftharpoons H^+ + CrO_4^{2-} \qquad K = 3.2 \times 10^{-7} \qquad (24\text{-}27)$$
$$2HCrO_4^- \rightleftharpoons H_2O + Cr_2O_7^{2-} \qquad K = 40 \qquad (24\text{-}28)$$

In alkaline solutions, most Cr(VI) exists as CrO_4^{2-}; in strongly acidic solutions, $Cr_2O_7^{2-}$ predominates. Many Cr(VI) compounds are generally good oxidizing agents; aqueous solutions of sodium dichromate, $Na_2Cr_2O_7$, are used to retard rust formation in automotive cooling systems, and a solution of this same salt in concentrated sulfuric acid is a potent cleaning agent for laboratory glassware. This solution contains dichromic acid, $H_2Cr_2O_7$, and sometimes deposits red needlelike crystals of chromium trioxide, CrO_3.

Manganese is one of the more abundant transition metals in the earth's crust, exceeded in quantity only by iron and titanium. Although it is stable in air unless heated to high temperatures, it is readily soluble in dilute acids, forming salts of Mn(II), and even dissolves slowly in cold water, producing $Mn(OH)_2$. The metal is important principally as a component of steels, but it is also used in many alloys of aluminum and magnesium to improve corrosion resistance and mechanical properties.

Manganese has five common positive oxidation states: Mn(II) and Mn(III) are basic, Mn(IV) is amphoteric, and Mn(VI) and Mn(VII) are acidic. As implied by the Latimer diagram (24-25), Mn^{3+} is unstable in aqueous solution, disproportionating even at low concentrations to form Mn^{2+} and insoluble MnO_2:

$$2Mn^{3+} + 2H_2O \rightleftharpoons Mn^{2+} + MnO_2(s) + 4H^+ \qquad (24\text{-}29)$$

Manganese dioxide, MnO_2, which is found naturally as the lustrous black mineral pyrolusite, the principal ore of manganese, is extensively used in dry cells (Fig. 18-4). The green manganate ion, MnO_4^{2-}, is stable in aqueous solution only if the solution is basic; in acidic or neutral solution it disproportionates to MnO_2 and permanganate ion, MnO_4^-. The latter ion is intensely purple and is a widely used oxidizing agent, chiefly in the form of the potassium salt. With concentrated sulfuric acid, it forms Mn_2O_7, a volatile oily liquid that explodes when warmed, decomposing to MnO_2 and oxygen. In the formation of this unstable heptoxide, Mn(VII) resembles Cl(VII), which forms a similar volatile compound on dehydration of perchloric acid, $HClO_4$. The heptoxide, Cl_2O_7, is a liquid that boils at 85°C; it is somewhat more stable than Mn_2O_7, but is also liable to explode if heated excessively or subjected to shock. There are other similarities in the chemistry of Mn(VII) and Cl(VII); for example, the tetrahedral permanganate and perchlorate ions are almost the same size and can substitute for one another in many crystals.

Fe	Co	Ni
Ru	Rh	Pd
Os	Ir	Pt

Iron, cobalt, and nickel. These elements, with their congeners, constitute the three columns of the periodic table jointly labeled VIII. Iron commonly assumes oxidation states (II) and (III), and rarely (VI), as in K_2FeO_4, a powerful oxidizing agent. Oxidation states (II) and (III) are also common for cobalt; for nickel, oxidation state (II) is the most important, with (IV) occurring only in a few compounds. The congeners of these elements are all noble metals, not readily attacked chemically. They form many complexes, and higher oxidation states than those observed for Fe, Co, and Ni are common. Thus, for ruthenium and osmium, the congeners of iron, states (VI) and (VIII) are well known; OsO_4, a volatile and poisonous compound, is a useful oxidizing agent in organic chemistry. Palladium and platinum, the congeners of nickel, form many stable complexes of oxidation state (IV) as well as (II); some of these have been mentioned in Chap. 15.

Iron is the second most abundant metallic element in the earth's crust (after aluminum) and is believed to form a significant fraction of the earth's core. It is also common in meteorites. It is a high-melting silvery metal that easily corrodes in air containing water vapor (Sec. 19-3) and dissolves readily in dilute acids, but it is rendered passive by strong oxidizing agents. Cobalt and nickel are somewhat more inert; polished nickel retains its luster in air, although the powdered metal reacts with the oxygen in air. Both cobalt and nickel dissolve slowly in dilute acids but not in concentrated nitric acid, which makes them passive also.

All three of these elements form stable ionic compounds, especially in oxidation state (II), and a large number of complexes in each of their common oxidation states. Two of the most important complexes biologically are that of iron with the large organic heme group in the oxygen-carrying proteins hemoglobin and myoglobin (Chaps. 25 and 27) and that of cobalt with a related large ring system in vitamin B_{12}, which has various biochemical roles and is essential for the treatment and prevention of pernicious anemia. The radioactive isotope ^{60}Co, produced by irradiation of stable ^{59}Co with neutrons, has a half-life of 5.3 years and is widely used in cancer therapy as a source of highly penetrating gamma rays.

Pure iron is quite soft, but its mechanical properties are considerably altered by the presence of small amounts of carbon and other elements. The molten iron from a blast furnace (Fig. 23-1) contains 3 to 4 percent carbon from the coke with which it has been in contact, and also smaller amounts of silicon, manganese, phosphorus, and sulfur. When this melt is cooled quickly, as in making castings, the resulting *cast iron* is hard and brittle. It contains large amounts of the compound Fe_3C, cementite. Cast iron, the most inexpensive form of iron, can be converted by heat treatment into a more malleable form, and if refined so that the content of carbon and other impurities is only a few tenths of a percent, the product is called *wrought iron*. This material, which is much stronger and tougher than cast iron, can readily be welded and forged and was widely used until this century. It has now largely been replaced by mild steel.

Steels are purified alloys of iron that contain carefully specified amounts of carbon and other elements. More than 600 million tons of steel are made in the world each year, by a complex and varied technology. Impure liquid iron is first refined to obtain the desired carbon content and remove other impurities; then the desired alloying elements are added and the product is handled under carefully controlled conditions to obtain the proper composition and texture for a particular application.

There are two broad commercial classes of steels: carbon steels and alloy steels. Carbon steels, which constitute about 90 percent of United States steel production, contain from about 0.03 to 1.5 percent carbon, sometimes as much as 1.5 percent manganese, and very small quantities of other elements. They are further classified according to increasing carbon content (which is also the order of increasing hardness and strength) as low-carbon, mild, medium, or high-carbon steels. Carbon steels are used extensively for appliances, containers, automobile bodies, ships, machinery, rails, and the structural framework of buildings.

The most important alloying elements other than carbon are nickel, chromium, and manganese, but molybdenum, tungsten, vanadium, titanium, and others are also used. Low-alloy steels, containing a total of no more than about 5 percent of elements other than iron, are used when special properties such as stress resistance or corrosion resistance are needed but cost is important. Applications include aircraft landing gears and other highly stressed parts of machines, such as axles, shafts, and gear wheels. High-alloy steels contain more than 5 percent of one alloying element and are used for their special properties: toughness and heat resistance for cutting tools, which contain tungsten or molybdenum, or corrosion resistance and appearance in stainless steels, used for tableware and kitchen utensils, chemical equipment, and jet engines.

Iron has very pronounced and almost unique magnetic properties, properties classified as ferromagnetic. Compounds with unpaired electrons are paramagnetic, and when samples of such compounds are placed in a nonuniform magnetic field they are pulled into the region of stronger field. In ferromagnetic substances there is very strong interaction between the unpaired electron spins on neighboring atoms, which causes the spins to be oriented parallel to each other throughout large regions of the sample, called magnetic domains. The result is an enormous reinforcement of the magnetism inherent in the individual atoms containing unpaired electrons. Ferromagnetic substances are pulled very strongly into magnetic fields and also remain magnetized even after they are removed from magnetic fields.

Ferromagnetism is restricted to solid substances and requires special conditions on the atomic level, so that few substances are ferromagnetic. The effect is not, however, restricted to metals; for example, the mineral magnetite, Fe_3O_4, one of the chief ores of iron, is ferromagnetic. It was probably the first known magnetic substance, referred to in Greek writings as early as 800 B.C. and known to the English as the lodestone. Metallic cobalt and nickel are also ferromagnetic, and some Fe–Co–Ni alloys are important technically in the manufacture of magnets for industrial and scientific applications—e.g., alnico, which contains about 51 percent Fe, 24 percent Co, 14 percent Ni, 8 percent Al, and 3 percent Cu.

Cu
Ag
Au
Copper, silver, and gold.   These metals of group IB are all quite soft, malleable, and ductile, and are excellent conductors. Their handsome appearance, resistance to corrosion, and comparative scarcity has made them valued for use in jewelry and other ornamentation and as money. They are often called the coinage metals. Silver coins manufactured in the United States until the mid-1960s were made of an alloy containing 90 percent silver and 10 percent copper, and a copper-nickel alloy containing 75 percent copper has long been used for United States coins. Copper is a component of many important alloys, notably brass and bronze.

The valence shells of atoms of copper, silver, and gold all have ten *d* electrons and one *s* electron, and thus a predominance of the (+I) oxidation state in their compounds might be expected. This is found, however, only for silver. Most copper compounds are those of Cu(II), with many fewer known for Cu(I). Au(III) is more stable than Au(I), although both are found in solution only in complexes. A few compounds of Ag(II), Ag(III), and Cu(III) have been reported, but they are rare and unusual.

When copper is heated in air it first forms Cu_2O and then CuO. Long exposure to moist air at ordinary temperatures gives copper and bronze a green patina, consisting of basic copper carbonates, mixtures of $Cu(OH)_2$ and $CuCO_3$. Silver is attacked chiefly by traces of H_2S in air, which produce a tarnish, a thin layer of black Ag_2S. None of these group IB metals is attacked by dilute nonoxidizing acids, but both copper and silver dissolve in nitric acid, which is reduced to NO and NO_2. Hot concentrated sulfuric acid is a good enough oxidizing agent to dissolve copper, being itself in part reduced to SO_2:

$$Cu + 2H_2SO_4 \longrightarrow Cu^{2+} + SO_4^{2-} + SO_2 + 2H_2O \qquad (24\text{-}30)$$

Gold can be dissolved by a mixture of concentrated HNO_3 and concentrated HCl (aqua regia), as can platinum and some other noble metals. This action results from a combination of the strongly oxidizing properties of the nitric acid and the strong complexing properties of the chloride ion present. The stable complex $AuCl_4^-$ is formed (and $PtCl_6^{2-}$ for platinum):

$$Au + 4HCl + HNO_3 \longrightarrow AuCl_4^- + H^+ + NO + 2H_2O \qquad (24\text{-}31)$$

Many other strong complexes are formed by each of these elements, such as deep blue-violet $Cu(NH_3)_4^{2+}$, blue $Cu(H_2O)_4^{2+}$, colorless $Ag(NH_3)_2^+$, and the complexes of Ag^+ with thiosulfate ion that are used in photography (Secs. 22-4 and 25-1).

The most important salt of copper is the sulfate, which is white when anhydrous but blue as the pentahydrate, $CuSO_4 \cdot 5H_2O$ (Fig. 4-8). Copper sulfate is used in copper plating, in dye and pigment manufacture, as an insecticide in vineyards [mixed with $Ca(OH)_2$ and water], in tanning, in engraving, and as an algicide in water treatment.

Silver nitrate is the most widely used soluble silver compound, important for the manufacture of silver halides for photography, as an antiseptic, in silver plating and the manufacture of mirrors, in hair dyes, and as a laboratory reagent. The silver halides, except for the fluoride, are insoluble in water. They are all photosensitive, darkening by decomposition into the elements when exposed to light of sufficiently short wavelength.

24-5   The Posttransition Metals: Zn, Cd, and Hg

Zn
Cd
Hg

Zinc is the most reactive of the metals of group IIB; it burns in air and dissolves readily in acidic and basic solutions. It is, however, protected from atmospheric attack by an adherent oxide-carbonate film. Cadmium also burns in air, but is dissolved only with difficulty by nonoxidizing acids and not at all by strong bases. Mercury is quite noble and is stable in air, although it reacts with chlorine; it dissolves in nitric acid, forming mercuric nitrate, $Hg(NO_3)_2$.

All three elements form compounds in oxidation state (II). Mercury also forms compounds of Hg(I), containing two covalently bonded mercury atoms, in such molecules as Hg_2Cl_2 (which is linear) and in the mercurous ion, Hg_2^{2+}. Zinc hydroxide is amphoteric, dissolving readily in both acids and bases. Cadmium hydroxide is basic, as is mercuric oxide, HgO; mercury forms no hydroxides.

Both zinc and cadmium are used for plating to protect against corrosion; galvanized iron is iron coated with zinc (Sec. 19-3). All three metals form many alloys. Brass is a hard, yellow, comparatively inert alloy containing copper and zinc in proportions that may vary rather widely. Alloys of mercury with other metals are called amalgams; some of these have definite composition, corresponding to the existence of specific compounds. Mercury reacts vigorously with some metals, such as sodium and potassium, forming compounds like Hg_2Na and Hg_2K. However, it has no effect on others, such as iron, and iron containers are used to store and ship mercury.

Mercury and its compounds are highly toxic. Despite its extremely low vapor pressure, about 2×10^{-10} atm at 20°C, the metal can be particularly harmful in the presence of decaying organic matter because of the formation of CH_3Hg^+, which is easily ingested and highly poisonous. The toxicity of mercury compounds is due to their irreversible combination with proteins. Very dilute solutions of $HgCl_2$, which is little dissociated, are toxic to bacteria and are used as disinfectants.

Problems and Questions

24-1   Reducing power of alkali metals.   The enthalpy change for the half-reaction

$$M(s) \longrightarrow M^+(aq) + e^-$$

is a measure of the reducing power of an alkali metal M. Since enthalpy is a state function, the enthalpy change for this reaction can be obtained by calculating the enthalpy change for the following equivalent three-step process:

(I)   Sublimation of the solid

$$M(s) \longrightarrow M(g)$$

(II)   Ionization of the monatomic vapor

$$M(g) \longrightarrow M(g)^+ + e^-$$

(III)   Hydration of the gaseous ion

$$M(g)^+ \longrightarrow M^+(aq)$$

(Both the overall reaction and step III involve H_2O implicitly.) Use this thermodynamic process to compare the reducing powers of lithium and cesium. What is the effect of hydration energy on the relative reducing powers? (Enthalpies of sublimation: Li, 160 kJ mol^{-1}; Cs, 78 kJ mol^{-1}. First ionization energies: Li, 520 kJ mol^{-1}; Cs, 376 kJ mol^{-1}. Enthalpies of hydration: Li, -506 kJ mol^{-1}; Cs, -259 kJ mol^{-1}.)   *Ans.* The overall enthalpy change for Li is 21 kJ less positive than that for Cs.

24-2   **Balanced chemical equations.**   Write balanced equations for the following reactions: (*a*) a blue solution of sodium in liquid ammonia becomes colorless by the formation of sodium amide ($NaNH_2$) and hydrogen; (*b*) permanganate ion, MnO_4^-, is reduced to Mn^{2+} in acidic solution by ethanol, CH_3CH_2OH, which is oxidized to acetic acid, CH_3COOH; (*c*) ethanol is oxidized in acidic solution to acetaldehyde, CH_3CHO, by dichromate ion, $Cr_2O_7^{2-}$, which is reduced to Cr^{3+}; (*d*) a melt of MnO_2 in KOH is oxidized by air to potassium manganate, K_2MnO_4; (*e*) manganate ion is oxidized to permanganate ion by chlorine in alkaline solution.

24-3   **Acidic properties of metal oxides.**   Arrange the metal oxides in each of the following groups in order of decreasing acidity and explain the reasons for your ranking: (*a*) Al_2O_3, Na_2O, MgO; (*b*) Cr_2O_3, CrO_3, CrO.   *Ans.* (*a*) Al_2O_3, MgO, Na_2O;   (*b*) CrO_3, Cr_2O_3, CrO

24-4   **Oxidations in alkaline and acidic solution.**   It is often asserted that it is much easier to oxidize an element to an oxyanion of high oxidation number in an alkaline solution than in an acidic solution. Analyze this assertion by calculating the fraction of Mn^{2+} oxidized to MnO_4^- by gaseous oxygen ($P = 1$ atm) at pH $= 0$ and at pH $= 14$. Use $\mathbf{E}^\oplus$ values from Table D-5. If the assertion seems generally true, what is its explanation? (*Hint:* Consider why and how the half-cell potential for oxidation to form a typical oxyanion depends on the pH.)

24-5   **Lanthanides and actinides.**   Explain the predominance of the ($+$III) oxidation state in lanthanide compounds. Why are higher oxidation states more common in actinide compounds? Why are actinides easier to separate from each other than lanthanides?

24-6   **Nuclear reactions.**   Give the chemical symbols, atomic numbers, and mass numbers of X, Y, and Z in the following nuclear reactions: (*a*) a neutron reacts with ^{6}Li to form an α particle plus a particle X; (*b*) an α particle combines with ^{14}N to yield a proton and a particle Y; (*c*) a proton combines with ^{12}C to form a particle Z and a photon.

24-7   **Survival of a nuclide.**   Suppose the elements of the earth were created 5×10^9 years ago (a very conservative estimate). What fraction of a nuclide with a half-life of 1×10^8 years would still remain today?   *Ans.* $(\frac{1}{2})^{50} \approx (10^{-3})^5 \approx 10^{-15}$

24-8   **Disintegration of ^{50}V.**   Ordinary vanadium contains about 0.25 percent ^{50}V, which has a half-life of 4×10^{14} years. Suppose that you have a 1.00-g sample of vanadium that contains no other radioactive material. What will be the average number of nuclei disintegrating per minute?

24-9   Stability of manganate.   (*a*) Discuss the relative stability of MnO_4^{2-} in acidic and in alkaline solutions by using the Latimer diagrams given in the text. (*b*) What is the equilibrium constant for the equilibrium $3MnO_4^{2-} + 2H_2O \rightleftharpoons MnO_2(s) + 2MnO_4^- + 4OH^-$? What is $[MnO_4^{2-}]$ at equilibrium with $[MnO_4^-] = 0.0100$ and $[OH^-] = 1.00$?

Ans.   (*b*) $K = 2 \times 10^1\ M^3$; $[MnO_4^{2-}] = 0.02\ M$

24-10   Precipitation of chromate.   In an acid solution Cr(+VI) exists predominantly as $Cr_2O_7^{2-}$. When Ba^{2+} is added to such a solution, practically all the chromium is precipitated as $BaCrO_4$. Explain in terms of equilibria, making whatever reasonable assumptions are needed.

24-11   Manganese in steel.   In order to determine the manganese content in a sample of steel, a 0.200-g sample of the steel is dissolved in an acidic solution and the manganese is oxidized to permanganate ion. The permanganate is then reduced to Mn^{2+} by addition of an excess of ferrous ion, 50.0 ml of 0.0625 N ferrous sulfate being used. The excess of ferrous ion is determined by titration with 0.059 N dichromate, 18.7 ml being required. What is the percentage of Mn in the steel?

Ans. 11.1

24-12   Behavior of $FeSO_4$ solutions.   When air is bubbled through aqueous $FeSO_4$, both the pH and the color of the solution change. In which direction does the pH change? Explain, giving equation(s) for any reaction(s) mentioned.

24-13   Relative stability of oxidation states.   Consider the disproportionation of a 1 M solution of Fe^{2+} into Fe(*s*) and Fe^{3+}. (*a*) Use the values of $\tilde{G}_f^{\oplus}$ given in Table D-4 to decide whether the reaction is spontaneous. (*b*) Calculate (at 298 K) the equilibrium constant involved.

Ans.   (*b*) 1.1×10^{-41}

24-14   Steel manufacture.   In the Bessemer process for making steel, oxygen is bubbled through impure molten iron. What is the effect of this procedure on the equilibrium $Fe_3C \rightleftharpoons 3Fe + C$ involving the cementite (Fe_3C) component of the impure iron?

24-15   Use of carbonates in steel making.   In steel making, small quantities of nonmetallic impurities such as phosphorus, sulfur, and silicon are oxidized and removed from molten iron. This process is often carried out in furnaces lined with $MgCO_3$ and $CaCO_3$, which decompose at high temperatures to MgO and CaO. Explain with equations how the presence of this lining can help in the removal of the oxides of nonmetallic impurities from the iron.

24-16   Balanced chemical equations.   Write balanced equations for the following reactions: (*a*) copper dissolves in dilute sulfuric acid in the presence of air; (*b*) hot concentrated sulfuric acid is reduced to SO_2 by copper; (*c*) gold dissolves in aqua regia to give $AuCl_4^-$ and NO_2; (*d*) zinc dissolves in a KOH solution to give $Zn(OH)_4^{2-}$.

24-17   Stability of Cu^+ ion in aqueous solution.   (*a*) Use $E^{\oplus}$ values from Table D-5 as an aid in predicting whether Cu^+ in aqueous solution might disproportionate to Cu^{2+} and Cu. (*b*) Estimate the equilibrium constant for the reaction $2Cu^+ \rightleftharpoons Cu^{2+} + Cu(s)$.

Ans.   (*b*) $1.7 \times 10^6\ M^{-1}$

"With due acknowledgment to G. B. Shaw for the inspiration, it might be said that theories of chemical bonding—neglecting quite a few which are entirely valueless—fall into one of two categories: those which are too good to be true and those which are too true to be good. 'True' in this context is intended to mean 'having physical validity' and 'good' to mean 'providing useful results, especially quantitative ones, with a relatively small amount of computational effort'. The proper, rigorous wave equation for any molecular situation represents a theory of that situation which is too true to be good. Most theories which are too good to be true are those in which the real problem per se is not treated, but rather an artificial analogue to the real problem, contrived so as to make the mathematics tractable, is set up and solved.

"The electrostatic crystal field theory, using the point charge approximation, is such a theory—at its best, or, perhaps, at its worst."

F. Albert Cotton[1]

25 Transition-Metal Complexes

Much of the recent research in inorganic chemistry has been concerned with transition-metal complexes. These compounds are of great practical importance in analytical chemistry, industrial processes, agriculture, and biochemistry. They offer challenges to the chemist interested in preparing new and unusual compounds, and the significant progress that has been made in understanding their properties is an example of the successful application of quantum mechanics to complex multiatomic systems.

[1]*Journal of Chemical Education,* vol. 41, p. 466, 1964.

Some important features of the chemistry of transition-metal complexes were introduced in the discussion of ionic equilibria (Sec. 15-3); that material is a necessary prerequisite to this chapter and should be reviewed. The focus here will be on understanding the reasons for the qualitative distinctions between the properties of the transition metals and those of the representative metals, such as the alkali and alkaline-earth elements, aluminum, and tin. Representative metals usually have only one, or at most two, stable oxidation states and form colorless diamagnetic compounds. In contrast, transition metals usually have several stable oxidation states and their compounds are frequently colored and paramagnetic. The reactions of representative metal ions in solution occur virtually instantaneously, while transition-metal ions often react so slowly that it takes minutes or even hours for their reactions to go to completion.

We begin with examples of applications of transition-metal complexes in order to show their technological importance. The stereochemistry of complex ions is discussed next, and the chapter concludes with sections on the crystal-field and molecular orbital theories, which provide explanations for the properties of complexes.

25-1   Some Applications of Metal Complexes

Complex ions are frequently employed in analytical chemistry, as described in Chap. 15. They also have many nonscientific applications, most of which involve control of the solubility of metal ions. Laundry detergents usually contain complexing agents to prevent the precipitation of calcium and magnesium ions (Sec. 21-3). The iron in plant foods is often in the form of its EDTA complex [Formula (15-21) and Fig. 15-4]; large quantities of iron can damage plants and the slow dissociation of this soluble complex provides a controlled supply of trace amounts of the metal.

Control of the concentration of metal ions by complexing is a technique also used in electroplating. When a metal is plated with a more noble metal, e.g., when copper is silver-plated, the plating reaction occurs spontaneously, i.e., with no applied voltage. As a consequence, the rate of formation of the metal layer cannot be controlled by adjusting the voltage and current, and the metal finish produced is of poor quality. If the plating is carried out in a solution containing a suitable ligand such as cyanide ion, however, the concentration of free metal ion can be so drastically reduced that, as predicted from the Nernst equation (Sec. 18-3), the plating process is no longer spontaneous. Plating will occur only if electrical work is performed, and the plating can therefore be controlled.

Complex formation is also an essential part of the photographic process. The light-sensitive emulsion that coats photographic film consists of a thin layer of gelatin in which sulfur and very fine grains of silver bromide are suspended. When the film is exposed, the AgBr grains that have been struck by light undergo chemical reaction, probably the formation of a small amount of Ag_2S on their surfaces. Such grains can be attacked and reduced to silver by treatment in the dark with a developer, typically an alkaline solution of an organic reducing agent. Grains that have not been exposed to light do not react with the developer and are removed from the emulsion with a solution of "hypo", sodium thiosulfate. Chemically, this part of the photographic process, which is called *fixing*, is the formation of soluble, very stable thiosulfate complexes such as $[Ag(S_2O_3)]^-$ and $[Ag(S_2O_3)_2]^{3-}$. The resulting film is a negative of the original image—black in the regions where silver has been deposited because of exposure to light and transparent in unexposed regions, from which AgBr grains have been removed.

Some recent applications of metal complexes involve their reactions with molecular oxygen and nitrogen. Complexes of several transition metals combine reversibly with O_2 and may be useful for storing oxygen in spacecraft and submarines. It may be possible to develop complexes to which nitrogen can be bound, reduced to ammonia, and then released. Such compounds could function like the nitrogen-fixing soil bacteria that are associated with nodules on the roots of legumes.

25-2 Stereochemistry of Complexes

The effect of the ligands on the central metal atom in a complex depends on their spatial arrangement—the stereochemistry of the complex. We have already discussed some features of the stereochemistry of complexes (Sec. 7-6, Table 7-3) and the coordination geometries in ionic crystals (Sec. 11-3, Fig. 11-6).

Coordination numbers 2 and 4. A number of transition metals form complexes with just two ligands. Typical examples are the linear silver (I) complexes $Ag(NH_3)_2^+$ (Sec. 15-3), $Ag(CN)_2^-$, and $AgCl_2^-$. Those with coordination number 4 are more common. Some have tetrahedral structures, as in $FeCl_4^-$, $CoCl_4^{2-}$, $Ni(CO)_4$, and HgI_4^{2-}; others, such as $Ni(CN)_4^{2-}$, $PdCl_4^{2-}$, $PtCl_4^{2-}$, and $Pt(NH_3)_4^{2+}$, are square planar. Figure 25-1 shows some examples of isomers for square planar complexes with two different ligands.

Octahedral complexes. Many transition-metal complexes have sixfold coordination and an octahedral geometry. An important feature of this spatial arrangement is the existence of isomers for complexes with certain general formulas, such as MA_4B_2 or MA_3B_3, where M represents the central metal atom and A and B are

FIGURE 25-1   **Examples of square planar complexes.** The two compounds on the left have the same composition as the compound on the right, but half the formula weight. The first two molecules are isomers. Their relationship with the third molecule is not strictly one of isomerism, although it has been considered as such historically.

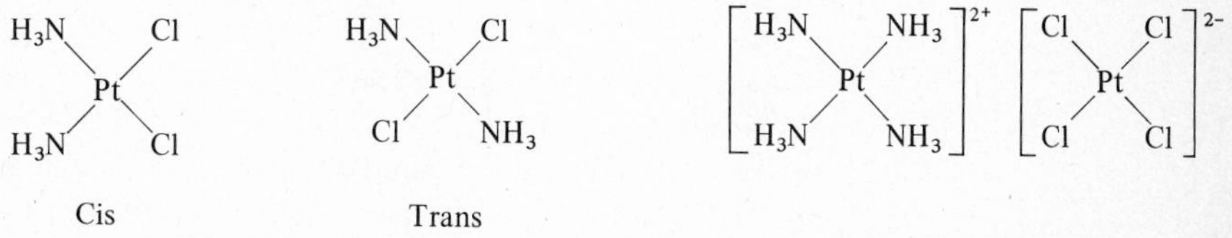

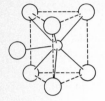

TaF$_7^{2-}$, NbF$_7^{2-}$

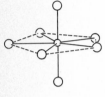

ZrF$_7^{3-}$, UF$_7^{3-}$

two kinds of *monodentate* ligands (i.e., ligands capable of forming one bond to the central atom) (Fig. 25-2). The proofs of the existence of such isomers, and their isolation, were important experimental verifications of Werner's hypothesis of the octahedral structure of coordination compounds, as discussed in Sec. 15-3.

Structural arguments lead to the conclusion that there must also exist pairs of isomers that are mirror images of each other and cannot be superimposed, as shown in Fig. 25-3. They are related as are a right hand and a left hand. Such isomers can be distinguished by their interaction with polarized light (Sec. 26-4), but they are identical in most other respects and cannot be distinguished on the basis of physical properties such as boiling point and melting point.

Complexes with coordination numbers other than 4 and 6 are relatively rare.

Polynuclear complexes.   Although all the complexes described so far contain only one central metal atom, it is possible to prepare coordination compounds called polynuclear complexes that contain several metal atoms (Sec. 7-6). An example is the condensation of the yellow-colored chromate ion to orange dichromate ion.

FIGURE 25-2   **Isomers expected for an octahedral ion or molecule.**   In MA$_6$ all six ligands occupy the corners of a regular octahedron, with M at the center. Since the six corners of the octahedron are equivalent, there is only one species MA$_5$B. For the formula MA$_4$B$_2$ there are two isomers, one with the two B ligands at adjoining corners of the octahedron (cis) and the other with the two B's at diagonally opposite corners (trans). The complex of formula MA$_3$B$_3$ also has two isomers, in which all three B's either are at the corners of the same face or define a plane that also goes through M. The arrows show from which precursor a given isomer can be made by substituting B for A.

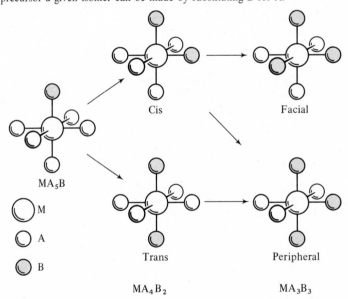

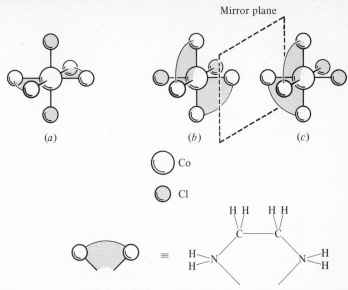

FIGURE 25-3 **Isomers of Co(en)$_2$Cl$_2^+$.** There are three isomers of this complex, in which en represents ethylene-diamine, H$_2$NCH$_2$CH$_2$NH$_2$. While the mirror image of isomer (*a*), in which the two Cl atoms are in trans position, is superimposable on the original ion, this is not true for isomers (*b*) and (*c*), in which the Cl atoms are in cis position. These two isomers are mirror images of each other and are not superimposable.

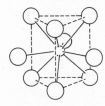

Nd(H$_2$O)$_9^{3+}$, ReH$_9^{2-}$

Chromate ion is stable in aqueous solution at high pH. If acid is added, however, the binuclear dichromate ion is formed (Sec. 24-4):

$$2 \left[\begin{array}{c} O \\ | \\ O-Cr-O \\ | \\ O \end{array} \right]^{2-} + 2H^+ \longrightarrow \left[\begin{array}{c} O \quad\quad O \\ | \quad\quad | \\ O-Cr-O-Cr-O \\ | \quad\quad | \\ O \quad\quad O \end{array} \right]^{2-} + H_2O \quad (25\text{-}1)$$

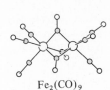

Fe$_2$(CO)$_9$

If the solution is sufficiently acidic, Cr$_2$O$_7^{2-}$ reacts with another CrO$_4^{2-}$ to form trichromate ion, Cr$_3$O$_{10}^{2-}$. The complexes can grow still larger, a process that leads eventually to a form of chromic oxide, (CrO$_3$)$_n$O^{2-} with n very large, a red poly-nuclear complex that consists of essentially infinite chains of oxygen-bridged chromium atoms similar in structure to the SO$_3$ chains shown in Fig. 4-2*e*. The same sequence of reactions also occurs with OH as the ligand, as discussed in Sec. 7-6.

Many polynuclear carbon monoxide complexes are known; in some the structural units are connected only by metal-metal bonds and in others the metal atoms are bridged by CO ligands as well. Heavy transition metals such as molybdenum and tantalum form ions containing clusters of many metal atoms. (Cluster compounds of representative metals are also known. For example, bismuth forms an ion with the formula Bi$_9^{5+}$.)

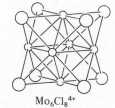

Mo$_6$Cl$_8^{4+}$

Sandwich compounds.   The preparation of the compound $(C_5H_5)_2Fe$, commonly known as ferrocene, was reported in 1951. It was first assumed to have the structure

However, chemical and physical studies showed and X-ray crystallographic studies confirmed that this structure is incorrect. The iron atom is sandwiched between the hydrocarbon rings as shown in Fig. 25-4a. The molecule may be considered to be composed of two five-membered ring anions (cyclopentadienyl ions, $C_5H_5^-$, each of which contains six π electrons) and Fe^{2+}. Many other sandwich compounds have been synthesized, including complexes of other metals such as chromium and ruthenium with cyclopentadienyl anions, as well as complexes involving other hydrocarbon rings such as benzene and cyclooctatetrene, C_8H_8.

The bonding in sandwich compounds involves the π electrons of the hydrocarbon ring and is best described in terms of molecular orbitals that are combinations of the π orbitals of the two rings and the valence-shell orbitals of the metal atom. Most of these complexes satisfy the rule that the total number of electrons involved in the bonding is 18, so that the valence shell of the metal atom can in a sense be regarded as filled. In dibenzene chromium (Fig. 25-4b), for example, each of the two benzene molecules has 6 π electrons and the chromium atom has 6 valence electrons, a total of 18.

Complexes involving π bonding need not be of the symmetric sandwich form. As shown by the example in Fig. 25-4d, they can contain a single hydrocarbon in conjunction with other ligands. Nonaromatic hydrocarbons with π electrons, such as ethylene, also form complexes with some transition metals (Fig. 25-4e).

Complexes based on porphin.   Several biologically important complexes are based on derivatives of porphin, a planar molecule made up of four 5-membered rings of carbon and nitrogen atoms. The four nitrogen atoms near the center of the molecule are positioned so that they are capable of chelating a metal atom.

FIGURE 25-4   **Sandwich compounds and related molecules.**   The metal atoms in compounds (a) to (c) are sandwiched between parallel rings. The CO ligands in (e) occupy five of the six corners of an approximate octahedron, the sixth corner of which is at the center of the C-C double bond.

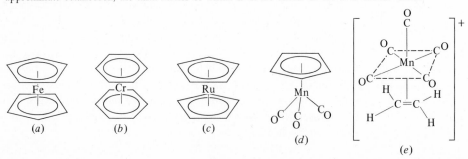

Naturally occurring molecules with the porphin framework are called porphyrins. The green pigments in plants are chlorophylls A (Fig. 25-5a) and B, which are porphyrin chelates of magnesium. These pigments are found in the chloroplasts, the primary sites of photosynthesis. Another important porphyrin chelate is the heme group, shown in Fig. 25-5b. This group is a structural unit in hemoglobin and myoglobin, the biological molecules that are involved in the transport and storage of oxygen in many mammals, including humans.

25-3 Crystal-Field Theory

Octahedral surroundings of an ion. The crystal-field theory[1] of bonding in metal complexes focuses on the five d orbitals of the central metal atom or ion. When an atom is undisturbed by an external field these orbitals have the same energy—there is fivefold degeneracy. If the atom is in a spherically symmetric electrostatic field this degeneracy is maintained, but in a field of lower symmetry the degeneracy may be partly or completely broken.

Consider a specific case in which an electrostatic field of octahedral symmetry is created by six identical negative ions placed at the corners of an octahedron surrounding an atom or ion. For example, we may picture the hexafluorocobaltate(III) ion, CoF_6^{3-}, as a Co^{3+} ion situated in the field produced by six F^- ions (Fig. 25-6a).

---

[1] So called because it was first applied by H. A. Bethe to ions in crystals, but the theory can be applied to isolated complexes as well.

FIGURE 25-5 **Some derivatives of porphin.** Chlorophyll B has the same structure as chlorophyll A, except that the —CH₃ group at the upper right is replaced by a —CHO group. The heme group is found in hemoglobin, myoglobin, and certain oxidation enzymes such as the cytochromes. Just one of the possible resonance formulas is shown for each molecule.

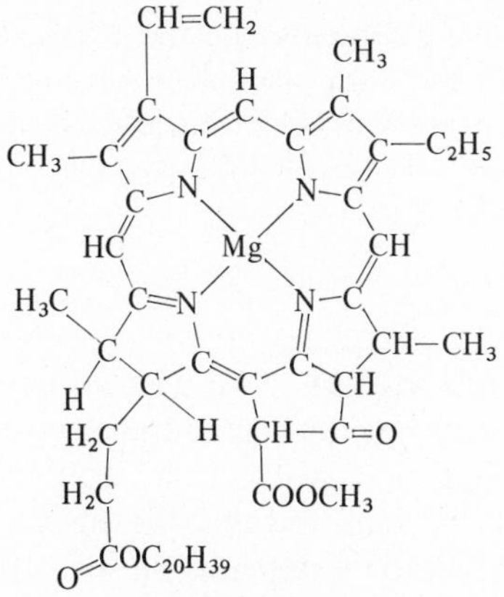

Chlorophyll A
(a)

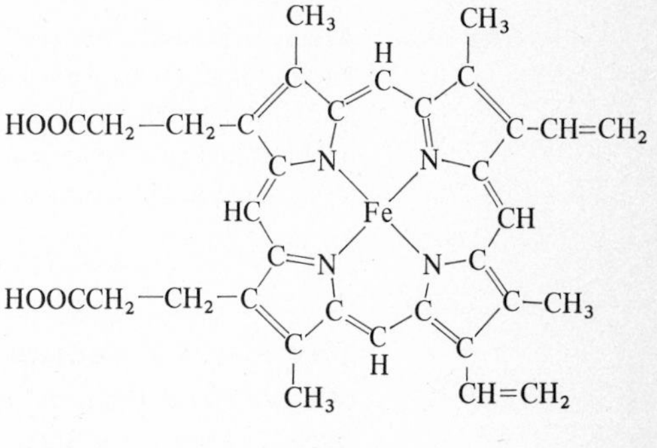

Heme group
(b)

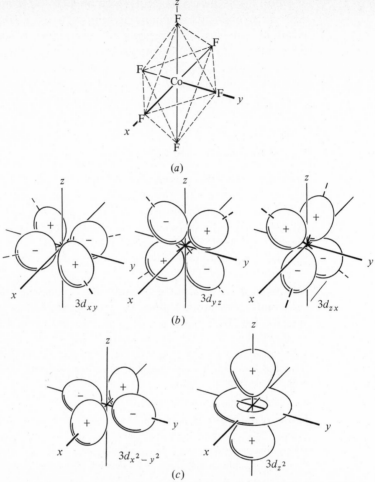

FIGURE 25-6   **An octahedral species and the five *d* orbitals.**   (*a*) The environment of Co(III) in CoF$_6{}^{3-}$. The Co atom is surrounded by six F ions that occupy the corners of an octahedron. (*b*) and (*c*) The five *d* orbitals. The signs on the lobes are associated with the amplitudes of the corresponding wave functions and have nothing to do with electric charges. Electrons occupying these orbitals are most likely to be found in the directions of the lobes.

The energy changes of the *d* states.   We need to know how the energies of electrons in the *d* orbitals are altered by the octahedral electrostatic field. To assess these changes qualitatively, we note that an electron occupying any of the five *d* orbitals will be repelled by the six F$^-$ ions and that the energy associated with this repulsion—the quantity we wish to know—depends on the orbital being occupied. Consider, for example, the $3d_{xy}$, $3d_{yz}$, and $3d_{zx}$ orbitals (Fig. 25-6*b*). The lobes of the $3d_{xy}$ orbital are directed at a point halfway between the F$^-$ ions in the *x,y* plane and those of the $3d_{yz}$ point halfway between the ions in the *y,z* plane; those of the

$3d_{zx}$ orbital have a corresponding orientation. This similarity in orientation of the lobes with respect to the surrounding ions requires that electrons in any of these orbitals experience the same energy change (an increase because they approach the F^-). Next, consider an electron in the $3d_{x^2-y^2}$ orbital, the lobes of which point toward the F^- ions in the x,y plane (Fig. 25-6c). On the average such an electron will be closer to some of the ions than electrons in any of the three orbitals just discussed, so its energy will be even higher. Turning to the $3d_{z^2}$ orbital, we see from its shape that there is a high probability that an electron occupying it will be near one of the F^- ions in the $+z$ or $-z$ directions and that the probability of finding the electron near one of the ions in the x,y plane is also high. Thus, the increase in energy caused by electrostatic repulsion is also higher for this electron than it is for electrons in the $3d_{xy}$, $3d_{yz}$, or $3d_{zx}$ orbitals. More detailed considerations show that the energy increase for an electron in a $3d_{z^2}$ orbital is exactly the same as that for an electron in a $3d_{x^2-y^2}$ orbital.

The foregoing geometric arguments apply quite generally to all octahedral complexes. If the ligands were uncharged species such as NH_3 molecules, the electrostatic field could be thought of as originating from the partial negative charge on the nitrogen resulting from the polar N—H bonds. The electrostatic field produced by these ligands would differ in strength from the field generated by the fluoride ions, but it would have the same geometry. The *pattern* of splitting of the d orbitals of the central metal ion would therefore be identical.

In summary, the interaction of an octahedral electrostatic field with the d orbitals of the metal results in an overall increase in the energy of the five d states and in a splitting of the states into two sets: three states of relatively lower energy (d_{xy}, d_{yz}, d_{zx}) and two of relatively higher energy ($d_{x^2-y^2}$ and d_{z^2}). The lower energy states are usually referred to collectively as t_{2g} states and the upper ones as e_g states. Chemical bonding between the ligands and the central metal ion produces a decrease in the energy of the orbitals with respect to the isolated metal atom, an attraction called the *stabilization energy* of the complex.

A schematic energy-level diagram showing the changes in electron energy caused by the interactions in an octahedral complex is shown in Fig. 25-7. The energy difference between the t_{2g} and the e_g states is called the *crystal-field splitting* and is often designated by Δ. The level splitting is not symmetric: the three t_{2g} states lie 0.4Δ below the average energy of the d orbitals and the two e_g states lie 0.6Δ above the average. The values of Δ and the stabilization energy depend on the nature of the complex. They can be calculated in some special cases but they are usually obtained from experiments, commonly from spectroscopic measurements.

Electronic structure of octahedral complexes. The schematic energy-level diagram in Fig. 25-7 can be used to predict the d electron structure of octahedral complexes by the same principles used to predict the electronic structures of atoms (Sec. 10-3) and molecules (Sec. 12-3). To begin, one must know the total number of d electrons in the complex. Spectroscopic observations show that all the valence electrons in transition-metal *ions* are d electrons. In other words, while the $4s$ state, for example, lies below the $3d$ states in *atoms* of the transition elements (Table 10-3), the sequence of states is reversed in the corresponding ions—the d states are filled

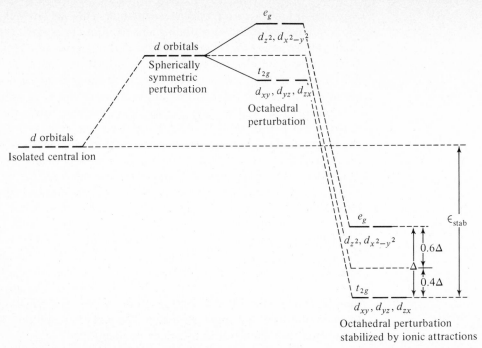

FIGURE 25-7   **Energies of *d* orbitals.**   The change in the energy of an electron in a *d* orbital by the interactions between six octahedral ligand anions and the positive central metal ion may be considered to be the sum of three effects: (1) the average repulsion of the electron by a negative charge equal to the charge on the ligands but smeared evenly over the surface of a sphere; (2) the change in repulsions when the negative charge is apportioned equally to the corners of an octahedron; (3) the overall stabilization caused by the attraction between ligands and central ion, an attraction that constitutes the bonding between ligands and metal ion.

before the *s* states. Thus, once the charge on a transition-metal ion is known, it is a simple matter to determine the number of *d* electrons from the atomic number, as shown in Example 25-1.

EXAMPLE 25-1   **Electron configurations of ions.**   What is the *d* electron configuration of the ions Fe^{2+}, Fe^{3+}, and Pt^{4+}?

Solution.   The atomic number of iron is 26 and the next lower inert gas is argon, for which $Z = 18$, so iron has eight valence electrons, Fe^{2+} has six, and Fe^{3+} has five. The *d* electron configurations of the ions are therefore d^6 and d^5, respectively. The easiest way to arrive at the answer for Pt^{4+} is to note that platinum is a congener of nickel (element 28), which has 10 valence electrons. The loss of four electrons to form the Pt^{4+} ion leaves a d^6 configuration. The same result can be obtained from the atomic number of platinum, 78, which indicates that the atom has 24 more electrons than krypton, the closest inert gas. Of these electrons, 14 are in the $4f$ orbitals, leaving 10 *d* electrons in the neutral atom.

Now consider the sequence in which the d states in an octahedral complex are filled by the available electrons. The first three electrons go into the three t_{2g} orbitals. Each of the electrons goes into a separate orbital because this is the configuration in which the electrostatic repulsion between the electrons is at a minimum (Hund's rule, Sec. 10-3, Example 10-1). There are two alternatives for configurations involving four electrons (Fig. 25-8). The fourth might be paired, with opposite spin, with one of the electrons in the t_{2g} orbitals; or it might go, without change of spin, into one of the e_g orbitals, a configuration that minimizes electron repulsion but also requires the expenditure of energy because of the use of the higher-energy e_g orbital. Both d orbital configurations are found. The first (pairing in lower-energy orbitals before any higher-energy orbitals are occupied) occurs when the ligand field splitting Δ is large compared to the energy of repulsion between the electrons; this is called

FIGURE 25-8 **Patterns of filling d orbitals in octahedral complexes.** The pattern of filling d orbitals is shown for values of the ligand-field splitting, Δ, that are large and small compared to the electron repulsion energy. Also shown for each configuration is N, the number of unpaired electrons. The total spin S is $N/2$ because each electron has spin $\frac{1}{2}$. Only for the d^4 through d^7 configurations (inside box) is there a difference between the weak-field and strong-field cases. For these configurations large S is associated with small Δ and small S with large Δ.

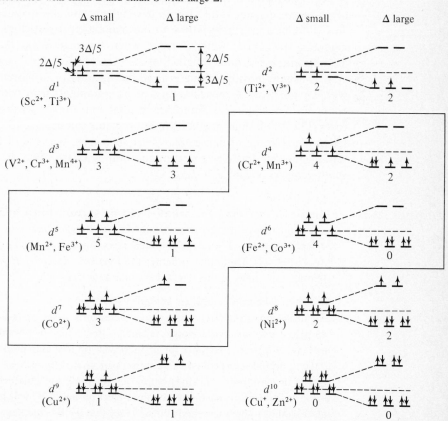

the *strong-field* case. The second is found when Δ is small compared to the repulsion energy (the *weak-field* case).

Similar considerations apply to five-electron configurations. If Δ is small, each electron occupies a different orbital; if Δ is large, two of the t_{2g} levels contain paired electrons and one is singly occupied. Diagrams showing the configurations for from 1 to 10 d electrons are sketched in Fig. 25-8. The strong-field and weak-field configurations differ when there are four, five, six, or seven electrons. As we shall see in the following sections, these differences have a significant effect on the properties of complexes.

Although the discussion of crystal-field splitting has been confined to octahedral complexes, similar considerations apply to other geometries as well. For ions with tetrahedral configuration, for instance, it is found that there are again three t_{2g} levels and two e_g levels. For such ions, however, the e_g levels are more stable than the t_{2g}.

Complexes that contain unpaired electrons are *paramagnetic*. As mentioned in Sec. 12-3, paramagnetic substances are pulled into nonuniform magnetic fields. In contrast, *diamagnetic* substances, which do not contain unpaired electrons, are forced out of such fields. A relatively simple technique for measuring the magnetic properties of a substance is the Gouy method, in which a tube containing the sample is suspended from a balance and placed partway into the field between the poles of an electromagnet. The sample is weighed with the magnet off and with it on. Any difference in weight is due to the magnetic properties of the sample. A paramagnetic substance "weighs" more when the field is on than when it is off; a diamagnetic substance "weighs" less.

It is usually possible to determine the number of unpaired electrons associated with an ion from a measurement of its paramagnetism. For the d^4 through d^7 configurations there is a difference in total spin for the weak-field and strong-field cases (Fig. 25-8); thus, magnetic measurements can be used to determine the character of the splitting of energy states for such ions. Because of this relation between paramagnetism and electron configuration, ions are often described as being *high spin* (small Δ) or *low spin* (large Δ).

EXAMPLE 25-2   **Magnetic criteria for weak-field or strong-field behavior.**   The magnetic properties of the following octahedral ions suggest that they contain the number of unpaired electrons indicated in parentheses: $Fe(CN)_6^{3-}$ (1); FeF_6^{3-} (5); $Co(NO_2)_6^{3-}$ (0); CoF_6^{3-} (4). Determine whether the ions are examples of weak- or strong-field splitting, and indicate the electron configuration for each.

Solution.   The iron complexes contain Fe(III), which has five d electrons; the cobalt complexes contain Co(III), which has six d electrons. In the weak-field case no electrons are paired until all five orbitals are singly filled. Thus, in a weak field a d^5 ion will have five unpaired electrons; for d^6 there will be four unpaired electrons, since one of the t_{2g} orbitals must contain an electron pair. In the strong-field case there is electron pairing in the three t_{2g} orbitals before any of the e_g orbitals becomes populated. A d^5 ion will therefore have one unpaired electron and a d^6 will have none. Applying this reasoning to the ions listed above, we see that the F⁻ complexes must be weak-field cases and the others must be strong-field. We might also have noted that for a given metal ion the high-spin case always corresponds to weak-field splitting and the low-spin case to strong-field splitting.

25-4   The Spectra of Complexes

The spectrochemical series.   While it is possible to make theoretical estimates of the crystal-field splitting, in practice Δ is adjusted to fit experimental data. This procedure is less satisfying than direct calculations, but it is meaningful because the values of Δ obtained from different experiments are in substantial agreement and form a general pattern for the different transition-metal ions.

Typical values of Δ lie between 80 and 400 kJ mol^{-1}. They depend on the metal and ligand atoms that form the complex and on the oxidation state of the metal atom, and are much larger for high oxidation states than for low ones. This dependence on oxidation state is shown, for instance, by complexes of cobalt and ammonia: the weak-field case applies for Co(II) and the strong-field case for Co(III).

$$
\begin{array}{cc}
\uparrow \quad \uparrow & \underline{\quad}\ \ \underline{\quad} \\[2pt]
\uparrow\downarrow \quad \uparrow\downarrow \quad \uparrow & \uparrow\downarrow \quad \uparrow\downarrow \quad \uparrow\downarrow \\[2pt]
(t_{2g})^5(e_g)^2 & (t_{2g})^6 \\
\mathrm{Co(NH_3)_6^{2+}} & \mathrm{Co(NH_3)_6^{3+}}
\end{array}
$$

This effect may be thought of as related to the decrease in size of the central ion as its charge is increased, so that ligands can approach the metal atom more closely and interact with it more strongly.

Many common ligands can be arranged in a sequence called the *spectrochemical series* that indicates the relative magnitude of the splitting they produce. A partial listing, in order of decreasing power to split d levels, is

$$ CN^- \gg NH_3 > O^{2-} \approx OH^- \approx H_2O > F^- > Cl^- > Br^- > I^- $$

For example, CoF_6^{3-} and $Co(H_2O)_6^{3+}$ are weak-field complexes and have four unpaired electrons whereas $Co(NH_3)_6^{3+}$ and $Co(CN)_6^{3-}$ are strong-field complexes and have no unpaired electrons:

$$
\begin{array}{cc}
\uparrow \quad \uparrow & \underline{\quad}\ \ \underline{\quad} \\[2pt]
\uparrow\downarrow \quad \uparrow \quad \uparrow & \uparrow\downarrow \quad \uparrow\downarrow \quad \uparrow\downarrow \\[2pt]
(t_{2g})^4(e_g)^2 & (t_{2g})^6 \\
\mathrm{Co(H_2O)_6^{3+},\ CoF_6^{3-}} & \mathrm{Co(NH_3)_6^{3+},\ Co(CN)_6^{3-}}
\end{array}
$$

The crystal-field splitting is similar for the transition-metal atoms in a given column of the periodic table. However, the repulsion energies between electrons of antiparallel spin are much larger for the elements in the first long period (Sc through Cu) than in the next two periods (Y through Ag and Hf through Au), in part because $4d$ and $5d$ orbitals are spatially much more extended than $3d$ orbitals. Low-spin complexes are therefore quite common for metals in the second and third long periods.

The order of ligands in the spectrochemical series clearly demonstrates one of the weaknesses of crystal-field theory. If the interaction between the metal ion and the ligands were purely electrostatic, then surely a ligand of high charge density such as fluoride ion would produce a larger splitting than neutral ammonia or water. This inconsistency between experiment and theory is only one of a number of observations that lead to the conclusion that in many complexes there is overlap between the *d* orbitals of the metal and orbitals associated with the ligands.

Ligand-field theory is a modification of crystal-field theory that allows for some degree of covalent bonding between metal and ligand in addition to the purely electrostatic interaction. In many respects it is merely a method for patching up the deficiencies of the electrostatic theory, and it is therefore of limited utility. A molecular orbital theory of metal complexes (Sec. 25-5) can in principle provide a more realistic picture of metal-ligand bonding. Both the molecular orbital and ligand-field treatments lead to level-splitting schemes that agree qualitatively with the predictions of the simpler electrostatic model. Qualitative arguments based on crystal-field theory are thus little changed by the use of more realistic theories.

Absorption spectra. The characteristic colors of many transition-metal complexes result from electronic transitions involving *d* electrons. The absorption spectrum in the visible region of a pale red-purple solution containing the ion $Ti(H_2O)_6^{3+}$ is shown in Fig. 8-10. It consists of a band with a maximum at about 20,000 cm^{-1}, corresponding to a wavelength of about 5000 Å. The titanium ion in the complex has a d^1 configuration, and the absorption of light is thought to result in the transition of the *d* electron from a t_{2g} to an e_g state:

The value of Δ necessary to account for the wavelength of this transition is approximately 240 kJ mol^{-1} $[= 20 \times 10^3$ cm^{-1} $(1.2 \times 10^{-2}$ kJ/cm$^{-1})]$. Visible light that passes through the solution with little absorption falls into two wavelength ranges, one around 7000 Å (red light) and the other around 4000 Å (purple light), which accounts for the red-purple color.

When a complex contains several *d* electrons, the number of possible electronic transitions increases and the corresponding absorption spectrum has several bands. While the analysis of complex spectra is not always unambiguous, a great deal of information about crystal-field splitting has been obtained from spectroscopic studies. It has also been possible to explain many general observations about colors of ions in terms of the behavior and number of *d* electrons. For example, the absence of color of the alkali-metal and alkaline-earth ions and of ions such as Sc^{3+}, Y^{3+}, Ti^{4+}, and Zr^{4+} is readily associated with the absence of *d* electrons in the valence shell. Similarly, Zn^{2+}, Cd^{2+}, Hg^{2+}, and Ag^+ are colorless because they have a complete *d* shell and no readily accessible states of appropriate energy into which *d* electrons can be promoted.

The intensity of the color associated with an ion depends on the concentration of the ion and also on the intrinsic strength of the absorption. Solutions of $Mn(H_2O)_6^{2+}$ are faintly pink, rather than strongly colored, because the electronic transition associated with the absorption of radiation has a low probability of occurring. Manganese(II) has five d electrons and the octahedral environment of the six water molecules is known to produce weak-field splitting, so the electron configuration of the complex is $(t_{2g})^3(e_g)^2$. Transitions to low-lying excited states involve the promotion of electrons from the t_{2g} to the e_g orbitals; for example,

Every conceivable transition involves a change in the total number of unpaired electrons and, for reasons explained by quantum mechanics, transitions that require a change in the total spin occur only rarely.

If cyanide ions are added to a solution containing Mn(II), the deep violet $Mn(CN)_6^{4-}$ complex is formed. Cyanide ion is at the top of the spectrochemical series, indicating that it forms strong-field complexes. The five d electrons in this complex are therefore in a $(t_{2g})^5$ configuration and it is possible to have transitions to the e_g state that do not require a change in electron spin; for example,

Not all the transitions that produce color in transition-metal complexes are between d levels, so-called d–d transitions. Another mechanism involves the transfer of an electron to the metal ion from one of the ligands. This type of electronic excitation, called charge transfer, produces very intense color.

25-5 The Molecular Orbital Approach

While our discussion of crystal-field theory has centered mainly on octahedral ions, the behavior of ions of other geometries is also well understood. There are many other applications than those discussed; in particular the theory has shed much light on details of the crystal structures of transition-metal compounds. Nevertheless, as we have already pointed out, crystal-field theory is not entirely satisfactory because of its undue emphasis on purely electrostatic interactions.

Like the crystal-field theory, the molecular orbital theory of bonding in transition-metal complexes predicts that the d states of the metal ion are grouped in different sets, such as the t_{2g} and e_g orbitals. In addition it gives a more realistic picture of the bonding. It is sufficiently flexible to deal with a variety of metal-ligand interactions: electrostatic, covalent, and even multiple bonding. The theory also yields at least semiquantitative values for the observed splittings and therefore explains the spectrochemical series. However, it is quite complicated, and our discussion of it will be concerned only with a qualitative energy-level diagram predicted by the theory for the $Co(NH_3)_6^{3+}$ complex (Fig. 25-9).

We know from crystal-field theory that $Co(NH_3)_6^{3+}$ is a strong-field complex, in which the configuration of the six d electrons is $(t_{2g})^6$. The molecular orbital treatment leads to an equivalent picture in which three molecular states with t_{2g} charac-

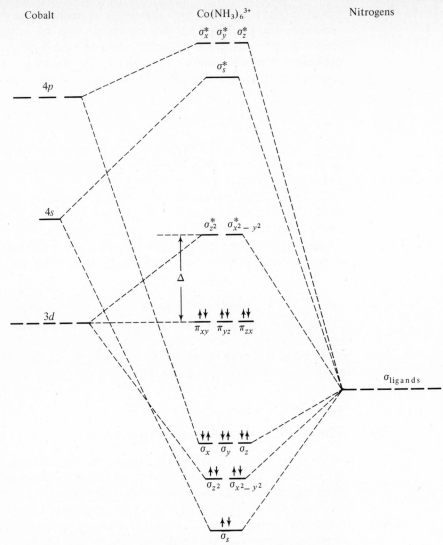

FIGURE 25-9 **Schematic energy levels of cobalt, nitrogen, and Co(NH₃)₆³⁺.**

teristics are filled before two e_g states when the 18 valence electrons in the complex are accommodated in the lowest-lying orbitals. There are 18 valence electrons to be considered because the Co^{3+} ion contributes six electrons and each of the six ammonia molecules contributes its unshared pair.

The energy level diagram is similar to those shown for molecular orbitals in Sec. 12-3. Approximate energies of the valence states of a cobalt atom are shown at the left, and the $2s$ states of six nitrogen atoms are on the right. The nitrogen states lie below those of cobalt because nitrogen is more electronegative than cobalt. The center of the diagram shows the approximate relative energies of the molecular orbitals obtained by suitable combinations of the cobalt and nitrogen atomic orbitals, as shown by the dashed lines.

The symbols used to designate the molecular orbitals are similar to those used in Chap. 12. The subscripts of the states π_{xy}, π_{yz}, and π_{zx} are meant to imply that these orbitals have the character of the cobalt d_{xy}, d_{yz}, and d_{zx} orbitals; those on the $\sigma^*_{x^2-y^2}$ and $\sigma^*_{z^2}$ orbitals refer to the $d_{x^2-y^2}$ and d_{z^2} orbitals. These π and σ^* orbitals correspond to the t_{2g} and e_g orbitals of the crystal-field theory, and the energy difference between the molecular states is identical to Δ. In this case, however, Δ can be obtained from theory and need not be chosen to fit experiments.

Problems and Questions

25-1 **Isomers of cobalt complexes.** Two compounds with the formula $Co(NH_3)_5ClBr$ can be prepared. Use structural formulas to show how they differ.

25-2 **Isomers of cobalt complexes.** Draw structural formulas for the isomers of the octahedral complex CoA_4BD^+, with $A = NH_3$, $B = OH^-$, and $D = Cl^-$.

25-3 **Isomers.** (*a*) Two isomers with the molecular formula $Pt(NH_3)_2Cl_2$ are known. Is this number of isomers consistent with an arrangement of the ligands (1) at the corners of a planar square? (2) At the corners of a tetrahedron? (*b*) Three isomers with the molecular formula PtABCD are known, where A, B, C, and D symbolize four different atoms. Discuss the possibilities for isomerism with a square planar configuration and with a tetrahedral configuration about the Pt atom.

25-4 **Isomers.** Draw all possible isomers of a compound of the formula MA_2XYZ, where M is located in the center of a symmetric square pyramid and the atoms A, A, X, Y, Z are at the five corners.

25-5 **Isomers.** Draw structures for the six possible isomers with formula $CrCl_3(H_2O)_6$ (not all of them are known). [*Hint:* Cr(III) is octahedrally coordinated in all the complexes.]

25-6 **Color change.** A solution of $Cr(NO_3)_3$ is violet. Addition of HCl changes the color to green. Give a possible explanation in terms of complex ions.

25-7 **Inorganic reaction.** Explain the following observation: When cupric hydroxide is shaken with pure water, the solvent remains practically colorless; however, if the water contains dissolved ammonium chloride, a blue solution results.

25-8 **Redox couples.** Explain the following experimental observations. (*a*) The reducing power of the Fe^{3+}/Fe^{2+} couple is considerably increased by the addition of phosphate ions, known to form strong complexes with Fe^{3+}. (*b*) Ba^{2+} combines with MnO_4^{2-} to form sparingly soluble $BaMnO_4$; in the presence of Ba^{2+} the alkaline couple MnO_4^-/MnO_4^{2-} becomes much more strongly oxidizing.

25-9 **Stable carbonyls.** Count the number of electrons in the valence shell of the central atom, including those contributed by the ligands, in each of the following known species, and compare this number with the number associated with a filled valence shell of the central atom: $V(CO)_6^-$, $Cr(CO)_6$, $Mo(CO)_6$, $W(CO)_6$, $Ru(CO)_5$, $Os(CO)_5$, $MnH(CO)_5$, $TcH(CO)_5$, $ReH(CO)_5$, $Fe(CO)_5$, $CoH(CO)_4$. *Ans.* 18 for all

25-10 Pi complexes. Establish the number of electrons involved in the bonding to the central atom of the following species: $V(C_5H_5)(CO)_4$, $Mn(C_5H_5)(CO)_3$, $Co(C_5H_5)(CO)_2$.

25-11 Energies of d orbitals. Draw two diagrams showing how the five $3d$ energy states of a transition-metal ion are split in an electric field of octahedral symmetry. In the first diagram, show where the $3d$ electrons of a Fe^{2+} ion would be found if the electric field induced by the ligands were very weak. In the second, show the disposition of these same $3d$ electrons if the ligand field were very strong. If the electronic structures you show in the two diagrams are different, what sort of experiment will distinguish between them?

25-12 Occupancies of d levels. Consider, separately, octahedral complexes of V(III), Fe(II), and Fe(III). Show the splitting of the $3d$ levels of the central metal atom and the occupancies of these levels in the weak-field and the strong-field situations. If the two situations differ, indicate which is the low-spin and which the high-spin complex.

25-13 Color of ions of transition elements. Which of the following hydrated ions would you expect to be colorless: Ti^{3+}, V^{4+}, Mn^{2+}, Sc^{3+}, Cu^+, Hf^{4+}? Explain your answer.
Ans. Mn^{2+}, Sc^{3+}, Cu^+, Hf^{4+}

25-14 Paramagnetism. Explain why the species CoF_6^{3-} is paramagnetic while $Co(NH_3)_6^{3+}$ is not. Similarly, why does the paramagnetism of $Fe(H_2O)_6^{3+}$ correspond to five unpaired electrons and that of $Fe(CN)_6^{3-}$ to only one?

25-15 High-spin and low-spin cases. Predict the numbers of unpaired electrons in the strong-field and weak-field situations for octahedral species containing Mn(III), V(III), Cr(II), or Rh(III).
Ans. strong field: 2,2,2,0; weak field: 4,2,4,4

25-16 Crystal-field theory. (*a*) Why is the crystal-field theory not concerned with s states? (*b*) Predict the effect of the ligands on the p_x, p_y, and p_z states of the Co atom of Fig. 25-6a. Draw a diagram similar to that of Fig. 25-7. Is the degeneracy of p states broken by an octahedral field? (*c*) Consider the effect of a tetrahedral field on the p states.

25-17 Crystal-field splitting for an elongated octahedron. Consider an atom surrounded octahedrally by ligands, two of which, at opposite corners of the (distorted) octahedron,

FIGURE 25-10

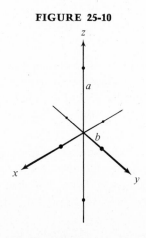

are more distant than the other four. Simulate the effect of the ligands on the central atom by negative charges, two of which are located along the $+z$ and $-z$ axes at a distance a and the other four along the $+x$, $-x$, $+y$, and $-y$ axes at a distance b, with a larger then b (Fig. 25-10). Consider the effect of this charge distribution on the energy of each of the five d states of the atom, and draw a diagram that shows qualitatively the relative values of the energies of these states.

Ans. The e_g states are split into d_{z^2} and $d_{x^2-y^2}$, of which the second has the higher energy. The t_{2g} states are split into d_{xy} and the doubly degenerate d_{zx} and d_{yz}, where d_{xy} has the higher energy.

"Organic chemistry just now is enough to drive one mad. It gives me an impression of a primeval tropical forest, full of the most remarkable things, a monstrous and boundless thicket, with no way to escape, into which one may well dread to enter."

F. Wöhler, in letter to J. J. Berzelius, 1835

"The world has been in an 'Organic Chemical Age' for almost thirty years, and the era is just reaching maturity. Multimillion-dollar industries based on organic compounds have taken their places alongside the other commercial giants of our civilization. Farm products, wood, coal, petroleum, and natural gas have been chemically tailored to provide organic products that vitally affect our environment. Fabrics, dyes, paints, coatings, and structural materials literally surround us. Rubber takes the violence out of movement, gasoline fuels our travel, gas both warms and cools our houses, plastics fashion our implements, and drugs guard our health and prolong our lives. Unprecedented opportunities to exercise the imagination are available in organic research, and much still remains to be done."

D. J. Cram and G. S. Hammond,[1] 1958

26 Organic Chemistry

Organic chemistry is the chemistry of the compounds of carbon.[2] More than a million organic compounds are now known, almost all of them containing hydrogen as well as carbon and the great majority containing other elements as well, most commonly

[1] "Organic Chemistry", p. 3, McGraw-Hill Book Company, New York, 1959.

[2] For historical reasons, CO and CO_2 and substances derived from CO_2, such as carbonates and CS_2, are not considered organic, nor are HCN and other cyanides and related compounds.

oxygen and nitrogen. No other element forms as many stable compounds as carbon. Its uniqueness arises from the fact that it forms stable covalent bonds with itself as well as with many other elements and that each carbon atom forms four bonds, which makes possible extended three-dimensional structures (Chap. 4) and almost innumerable possibilities for isomerism.

Many organic compounds are found, as the name implies, in living organisms, but a significant fraction of the organic substances now known have been created in the laboratory. The quotations at the start of this chapter suggest the enormous progress and changes that have occurred in organic chemistry in a little more than a century. In 1835 the concepts of chemical bonding and structure were unknown and only a handful of organic substances had been characterized. Now the chemist can determine the detailed three-dimensional molecular architecture of compounds containing thousands of atoms and devise laboratory syntheses for complex molecules, including many that are designed and created to serve a specific purpose. In this chapter we introduce some of the simplest classes of organic compounds and reactions and suggest the kinds of considerations that govern the relation between structure and properties. We begin by examining some of the important types of hydrocarbons. We then consider organic compounds containing other elements and investigate their structural relationship to the hydrocarbons. The following chapter is concerned with a few of the essential compounds and reactions that are found in living organisms, the realm of biochemistry.

26-1 Hydrocarbons

The simplest organic substances are the hydrocarbons, compounds containing only C and H. Representative hydrocarbons include ethane (CH_3CH_3), ethylene ($CH_2{=}CH_2$), and benzene (C_6H_6), the structures of which were discussed in Chaps. 11 and 12. These three molecules differ in the nature of their carbon-carbon bonds; ethane is the simplest molecule containing a C-C single bond, ethylene is the simplest containing a double bond, and benzene is the prototype aromatic hydrocarbon, with a completely delocalized π-bond system (Sec. 12-3). The hydrocarbons provide a logical starting point for a discussion of organic chemistry because they contain only two elements, their chemistry is comparatively simple, and they serve as the basis for the systematic naming of all classes of organic compounds. A systematic scheme of nomenclature is essential because of the structural complexities possible with organic compounds; some of the simpler rules for naming organic substances are given in Appendix B and in Sec. 26-3 of this chapter.

Some of the molecular formulas of hydrocarbons and other types of organic compounds on the following pages may appear bewildering at first. However, a little practice and attention to a few fundamental principles established in earlier chapters will quickly reveal the simple and regular patterns followed by the formulas and structures. The most important points to keep in mind are these:

1. The number of covalent bonds normally formed by an atom of each element and the number of unshared pairs remaining on that bonded atom (Table 26-1).

TABLE 26-1 **Number of covalent bonds and unshared electron pairs for representative bonded atoms***†

	C	H	N	N$^+$	O	O$^+$	F, Cl, Br, I
Number of covalent bonds	4	1	3	4	2	3	1
Number of unshared pairs on bonded atom	0	0	1	0	2	1	3
Examples	CH_4	CH_4	NH_3	NH_4^+	H_2O	H_3O^+	CH_3F
	$H_2C{=}CH_2$				$(CH_3)_2C{=}O$		CBr_4

* Sulfur and phosphorus in some organic compounds conform to the pattern indicated here for their congeners O and N, but they may also form more bonds, e.g., in the molecules SF_6 and PCl_5 and in the groups $-SO_2OH$, $-OSO_2OH$, and $-OPO(OH)_2$.

† Note that N$^+$ is isoelectronic with C and O$^+$ is isoelectronic with N.

A common error of beginners is to draw organic structural formulas with three or five bonds to a carbon atom, rather than four, or with two bonds to a hydrogen atom rather than one.

2. The geometric arrangement of the bonded neighbors around an atom that has no unshared pairs and has four, three, or two bonded neighbors (Table 12-1), as does carbon in different common bonding situations. These considerations are important in the determination of the number and the nature of possible isomers.

3. The fact that at ordinary temperatures there is normally almost free rotation about single bonds and no rotation about double bonds. These facts are important in the determination of the possibilities of isomerism.

Alkanes. Methane, ethane, and propane are the three simplest representatives of this class of compounds:

$$
\begin{array}{ccc}
\underset{\text{Methane}}{\overset{\displaystyle H}{\underset{\displaystyle H}{H-\overset{\textstyle |}{\underset{\textstyle |}{C}}-H}}} \text{ or } CH_4
&
\underset{\text{Ethane}}{\overset{\displaystyle H\ \ H}{\underset{\displaystyle H\ \ H}{H-\overset{\textstyle |}{\underset{\textstyle |}{C}}-\overset{\textstyle |}{\underset{\textstyle |}{C}}-H}}} \text{ or } CH_3CH_3
&
\underset{\text{Propane}}{\overset{\displaystyle H\ \ H\ \ H}{\underset{\displaystyle H\ \ H\ \ H}{H-\overset{\textstyle |}{\underset{\textstyle |}{C}}-\overset{\textstyle |}{\underset{\textstyle |}{C}}-\overset{\textstyle |}{\underset{\textstyle |}{C}}-H}}} \text{ or } CH_3CH_2CH_3
\end{array}
$$

The two ways of writing molecular formulas shown here were discussed briefly in Sec. 2-3. Although those shown on the right do not display each bond explicitly, they are much more convenient typographically and hence more common. Their implications with regard to the number of C-C and C-H bonds for each carbon atom must be carefully noted. In these *condensed* formulas, the hydrogen atoms are usually written immediately to the right of the carbon atom to which they are bonded; C-C bonds are shown explicitly only if they are in a vertical direction on the page (as in later examples).

Each of these three hydrocarbons has only single bonds, and their molecular formulas differ successively by CH_2 (CH_4, C_2H_6, and C_3H_8). These molecules are the first three members of a series with general formula C_mH_{2m+2}, where m is a positive integer. Hydrocarbons with this general formula are called *alkanes* or saturated hydrocarbons, the term *saturated* implying that they contain no multiple bonds.

Ethane may be regarded as derived from methane by (mentally) replacing one of the four equivalent hydrogen atoms of methane by a CH_3 (methyl) group. Similarly, propane can be considered to be derived from ethane by the replacement of one of the six equivalent[1] hydrogen atoms of ethane by CH_3. The replacement of one H of propane by CH_3 can lead to either of two possible molecules because there are two nonequivalent kinds of bonding environments for hydrogen atoms in propane. The two hydrogen atoms attached to the central carbon atom are equivalent to one another but are quite distinct from the six attached to the end carbon atoms. The latter six are in turn equivalent to one another because there is essentially free rotation about the C-C bonds. Thus there are two isomeric molecules of formula C_4H_{10} (butanes):

$$
\begin{array}{cccc}
 & H & H & H & H \\
 & | & | & | & | \\
H- & C- & C- & C- & C-H \\
 & | & | & | & | \\
 & H & H & H & H
\end{array}
\quad \text{or} \quad CH_3CH_2CH_2CH_3
\qquad\qquad
\begin{array}{ccc}
 & H & H & H \\
 & | & | & | \\
H- & C- & C- & C-H \\
 & | & | & | \\
 & H & C & H \\
 & & H \quad H \\
 & & H
\end{array}
\quad \text{or} \quad CH_3CHCH_3 \\ \qquad\qquad CH_3
$$

n-Butane   Isobutane (2-methylpropane)

The letter *n* signifies *normal* and implies an unbranched chain of carbon atoms. Isobutane is the simplest branched alkane, the term *branched* implying that at least one carbon atom is bonded directly to three or four other carbons. The systematic name of isobutane is 2-methylpropane, a name derived by choosing the longest chain of carbon atoms in the molecule, numbering the atoms consecutively in this chain starting at one end, and indicating the point of attachment of the substituent as shown (see Appendix B).

The physical and chemical properties of isobutane are similar to but nonetheless distinct from those of *n*-butane. For example, its melting temperature is 10 K lower, it boils 11 K lower, and it has a slightly different density in the liquid state.

The alkanes with more than four carbon atoms are named by combining the ending -*ane* with a Greek root that signifies the number of carbon atoms in the molecule (*pent* for 5, *hex* for 6, and so on), as indicated in Table 26-2. There are three isomeric pentanes, with significantly different physical properties:

$$
CH_3CH_2CH_2CH_2CH_3 \qquad CH_3\underset{\underset{\displaystyle CH_3}{|}}{C}HCH_2CH_3 \qquad CH_3\underset{\underset{\displaystyle CH_3}{|}}{\overset{\overset{\displaystyle CH_3}{|}}{C}}CH_3
$$

n-Pentane   2-Methylbutane   2,2-Dimethylpropane
(bp 36°C)   (isopentane)   (neopentane)
      (bp 28°C)   (bp 10°C)

---

[1] Inspection of Fig. 11-3 or a model of ethane will show that the six hydrogen atoms are equivalent to one another, although they may not at first seem to be in the two-dimensional formulas used above to represent the structure of this molecule.

TABLE 26-2 **Properties of some *n*-alkanes**

Name	Molecular formula	Molecular weight	mp/°C	bp/°C	Number of isomers
Methane	CH_4	16	−183	−162	1
Ethane	C_2H_6	30	−172	−89	1
Propane	C_3H_8	44	−187	−42	1
n-Butane	C_4H_{10}	58	−135	0	2
n-Pentane	C_5H_{12}	72	−130	36	3
n-Hexane	C_6H_{14}	86	−94	69	5
n-Heptane	C_7H_{16}	100	−91	98	9
n-Octane	C_8H_{18}	114	−57	126	18
n-Nonane	C_9H_{20}	128	−54	151	35
n-Decane	$C_{10}H_{22}$	142	−30	174	75
n-Eicosane	$C_{20}H_{42}$	282	36	*	3.66×10^5

* n-Eicosane boils at about 205°C at a pressure of 15 torr.

The number of possible isomers rises rapidly for the higher alkanes, as indicated in Table 26-2. The lower alkanes are colorless, odorless gases under ordinary conditions, and those from C_5 through C_{18} are colorless liquids. The higher alkanes are waxy solids. All alkanes are nonpolar and water-insoluble. The liquids and solids have densities between about 0.6 and 0.8 g ml^{-1} and thus float on water.

EXAMPLE 26-1   **Implications of formulas.** Which of the following formulas represent the same molecules and which are incorrect in their implications about bonding?

(a) $(CH_3)_2CHCH_2CH_3$

(b) CH$_3$CHCH$_2$
         | \
       CH$_3$ CH$_3$

(c) $CH_3CH_2CH(CH_3)_3$

(d) CH$_3$CH$_2$CHCH$_3$
                  |
                 CH$_3$

(e) $(CH_3)_2CH(CH_3)_2$

Solution. Formulas (a), (b), and (d) all represent the same molecule, 2-methylbutane, written in different ways. This is apparent if one locates the longest chain of carbon atoms and finds the position of the methyl substituent on this chain. A formula such as (b), with a bend in the chain, is unconventional; since there is freedom of rotation about a C-C single bond, the methyl group written at the lower right might just as well have been written on the same line as the upper three carbon atoms. Formulas (c) and (e) are incorrect; in (c) the third carbon from the left has five attached groups (CH_2, H, and three CH_3) and in (e) the central carbon atom has five attached groups.

EXAMPLE 26-2   **Isomers of hexane.** Write the formulas for the isomeric hexanes, C_6H_{14}, and name the molecules.

Solution. A systematic approach is to start with the three pentanes, substitute CH_3 for H at each unique position of each of the pentanes, and then eliminate the duplicate molecules. For simplicity we show here merely the carbon skeletons, with the newly introduced CH_3 in boldface type.

From *n*-pentane, we get three hexanes:

$$C-C-C-C-C-C \qquad \text{\textit{n}-hexane}$$

$$C-C-C-C-C \longrightarrow \begin{array}{c} C \\ | \\ C-C-C-C-C \end{array} \qquad \text{2-methylpentane}$$

$$\begin{array}{c} C \\ | \\ C-C-C-C-C \end{array} \qquad \text{3-methylpentane}$$

Substitution on the fourth carbon of pentane is equivalent to substitution on the second; substitution on the fifth is equivalent to that on the first.

Substitution of CH_3 for H on each of the carbon atoms of 2-methylbutane gives

$$\begin{array}{c} C \\ | \\ C-C-C-C-C \end{array} \qquad \text{3-methylpentane (again)}$$

$$\begin{array}{c} C \\ | \\ C \\ | \\ C-C-C-C \end{array} \qquad \text{3-methylpentane (again)}$$

$$\begin{array}{c} C \\ | \\ C-C-C-C \\ | \\ C \end{array} \qquad \text{2,2-dimethylbutane}$$

$$\begin{array}{c} C \quad C \\ | \quad | \\ C-C-C-C \end{array} \qquad \text{2,3-dimethylbutane}$$

$$\begin{array}{c} C \\ | \\ C-C-C-C-C \end{array} \qquad \text{2-methylpentane (again)}$$

The fact that the first two of these are the same reflects the equivalence of the two methyl groups on the second carbon atom of 2-methylbutane; do not be confused by the writing of the five-carbon chain "around a corner" in the second formula.

Since all 12 hydrogen atoms of neopentane (2,2-dimethylpropane) are equivalent to each other, there is only one possible substitution product:

$$\begin{array}{c} C \\ | \\ C-C-C \\ | \\ C \end{array} \longrightarrow \begin{array}{c} C \\ | \\ C-C-C-C \\ | \\ C \end{array} \qquad \text{2,2-dimethylbutane (again)}$$

Inspection of these formulas and names shows that there are five unique isomeric hexanes, in agreement with Table 26-2.

Natural gas and petroleum are composed principally of alkanes and constitute their chief natural sources. A typical natural gas contains about 80 percent methane, with about 10 percent ethane and smaller amounts of propane, butanes, and pentanes. Methane is also produced by the bacterial degradation of cellulose in vegetable matter in the absence of oxygen and thus is a primary constituent of the gas formed under water in marshes and below the surface of garbage dumps. It is found as well in significant quantities in some coal mines, where it can form dangerously explosive mixtures with air.

Petroleum is a naturally occurring mixture of hydrocarbons. Some crude oils consist chiefly of alkanes; others may have as much as 40 percent of cycloalkanes and aromatic hydrocarbons (see below). Some sources of petroleum contain up to about 1 percent of sulfur compounds, which are ecologically objectionable because, when burned, they produce sulfur dioxide, an irritating and noxious gas. Petroleum is refined principally by a series of fractional distillations (Sec. 6-6). Gasoline is the fraction distilling in the temperature range from about 35 to about 210°C; it contains hydrocarbons with from 5 to about 12 carbon atoms. The fraction distilling from about 200 to 275°C is kerosene; less volatile fractions find use as lubricating oils, petroleum jelly, and paraffin. These higher fractions can be *cracked* by various methods to break them down to hydrocarbons of lower molecular weight, suitable for inclusion in gasoline. These methods include both pyrolysis (decomposition by heating) at temperatures up to 700°C, often at high pressure, and treatment with special catalysts at somewhat lower but still elevated temperatures. The development of successful cracking methods has more than doubled the yield of gasoline from petroleum.

Some alkanes are also found in small amounts in plant materials. *n*-Heptane occurs in the turpentine from certain pines, and the waxy coatings on pears, apples, and cabbage leaves contain the C_{29} *n*-alkane. Beeswax contains alkanes in the C_{27} to C_{31} range.

Alkanes are normally quite inert chemically, being unaffected by hot acids, bases, metals, and most oxidizing and reducing agents. They do, however, react with oxygen when heated, forming CO_2 and H_2O with an excess of air or oxygen and forming CO and H_2O when the quantity of O_2 is limited. They also react with chlorine and bromine under the influence of light and with fluorine even in the dark. For example, methane reacts with chlorine to form HCl and a mixture of products with from one to four chlorine atoms substituted for the hydrogen atoms of methane: CH_3Cl, CH_2Cl_2, $CHCl_3$, and CCl_4.

Cycloalkanes. These hydrocarbons contain at least one ring of carbon atoms and no double or triple bonds. Those with only one ring have the general formula C_mH_{2m}; cyclopropane (C_3H_6) is the simplest representative of this class. Cyclopentane and cyclohexane are found in some samples of petroleum, as are several other alkanes containing five- and six-membered rings:

$$
\begin{array}{cc}
\quad\ \text{CH}_2 & \qquad\ \text{CHCH}_3 \\
\text{CH}_2\quad\text{CH}_2 & \text{CH}_2\quad\text{CHCH}_3 \\
\text{CH}_2\!-\!\text{C(CH}_3)_2 & \text{CH}_2\quad\text{CH}_2 \\
 & \quad\ \text{CHCH}_3
\end{array}
$$

1,1-Dimethylcyclopentane	1,2,4-Trimethylcyclohexane

The chemical and physical properties of the cycloalkanes are very similar to those of the alkanes.

Alkenes. Alkenes contain one carbon-carbon double bond and have the general formula C_mH_{2m}, with $m = 2$ or more. Those with m greater than 3 are isomeric with cycloalkanes containing only one ring; e.g., propene, $CH_3CH{=}CH_2$, is isomeric with cyclopropane. The simplest alkene is ethylene, $CH_2{=}CH_2$. There is only one propene but there are four different butenes:

$$CH_3CH_2CH{=}CH_2$$

$$\underset{\substack{H\\CH_3}}{}C{=}C\underset{\substack{H\\CH_3}}{}$$

$$\underset{\substack{H\\CH_3}}{}C{=}C\underset{\substack{CH_3\\H}}{}$$

$$H_2C{=}C\underset{\substack{CH_3\\CH_3}}{}$$

1-Butene	*cis*-2-Butene	*trans*-2-Butene	Isobutene (2-methylpropene)

The two forms of 2-butene illustrate the phenomenon of *geometrical isomerism*, alluded to on page 326. It arises because of the impossibility of rotation about the double bond at ordinary temperatures. The term *cis* is Latin for "on this side" while *trans* is Latin for "on the other side". The naming of alkenes is discussed briefly in Appendix B.

The most important reactions of alkenes are those in which the double bond is converted to a single bond. For example, hydrogen reacts with an alkene (an unsaturated hydrocarbon) in the presence of an appropriate catalyst to form the corresponding alkane (a saturated hydrocarbon):

$$CH_3CH{=}CHCH_3 + H_2 \xrightarrow{\text{Pt or Ni}} CH_3CH_2CH_2CH_3 \qquad (26\text{-}1)$$

Bromine reacts with alkenes to form the dibromoalkane with one bromine atom attached to each carbon atom originally involved in the double bond, a reaction often used to test for the presence of a C-C double bond:

$$(CH_3)_2C{=}CH_2 + Br_2 \longrightarrow (CH_3)_2CBrCH_2Br \qquad (26\text{-}2)$$

$$\text{2-Methylpropene} \qquad\qquad \text{1,2-Dibromo-2-methylpropane}$$

Each of these reactions may be considered to involve an attack on and opening of the π bond in the alkene (Sec. 12-3). The electrons originally paired in this bond are thereby made available for the formation of new covalent bonds.

Alkynes. These hydrocarbons contain a carbon-carbon triple bond and have the general formula C_mH_{2m-2}. Acetylene, $HC{\equiv}CH$, is the prototype of this series. Alkynes undergo most of the reactions characteristic of alkenes. Those with the triple bond at the end of a chain, that is, with the grouping $-C{\equiv}CH$, are very weakly acidic; the terminal hydrogen atom reacts with metallic sodium to form H_2 and with Ag^+ to produce an insoluble silver salt of the alkyne. For example,

$$CH_3C{\equiv}CH + Na \longrightarrow CH_3C{\equiv}C^-Na^+ + \tfrac{1}{2}H_2 \qquad (26\text{-}3)$$

$$CH_3C{\equiv}CH + Ag^+ \longrightarrow CH_3C{\equiv}CAg + H^+ \qquad (26\text{-}4)$$

Naphthalene,
$C_{10}H_8$

Aromatic hydrocarbons.   This is an extensive and important group of compounds, of which benzene, C_6H_6, is the simplest example, the π-electron system extending over all 6 carbon atoms.[1] In naphthalene, it extends over all 10 carbon atoms. There are many other aromatic systems, with even more rings "fused" together as in naphthalene, and also numerous hydrocarbons derived from benzene, napthalene, and the other ring systems. An example is toluene (methylbenzene), $C_6H_5CH_3$, whose trinitro derivative is the common high explosive TNT:

$$CH_3$$

$$O_2N \qquad NO_2$$

$$NO_2$$

2,4,6-Trinitrotoluene
(TNT)

Aromatic compounds react much less readily than alkenes and alkynes, despite their unsaturation, but nonetheless do undergo many reactions (not discussed here).

26-2   Alkyl Groups and Functional Groups

Most organic compounds contain not only carbon and hydrogen but other elements as well, in various common groupings such as —OH, —NH₂, —Br, and —COOH. The unconnected line emanating from each group is *not* a negative charge; it merely denotes the potential for forming a covalent bond. Simple examples of compounds with such groups are

$$CH_3CH_2OH \qquad CH_3COOH \qquad CH_3CH_2CH_2CH_2NH_2 \qquad CH_3CHFCH_3$$

Ethanol   Acetic acid   1-Aminobutane   2-Fluoropropane
                                    (*n*-butylamine)   (isopropyl fluoride)

Each of these molecules may be regarded as derived from an alkane by substitution of an atom or group of atoms for one of the hydrogen atoms of the alkane. Because of the chemical inertness of the alkanes, the reactions of the derivative molecules depend to a great extent upon the nature of the substituted groups, which are therefore referred to as *functional groups*. The hydrocarbon fragment formed from an alkane by substituting for one of the hydrogen atoms is called an *alkyl group*—for example, methyl (CH_3—) or $CH_3CH_2CH_2$— (*n*-propyl).

Alkyl groups are necessarily uncharged (since they correspond to the removal of a neutral atom from a neutral molecule), and they contain an unpaired electron (since they contain an odd number of electrons). Species that contain unpaired

[1] In the formulas used conventionally for aromatic rings, the corners of the rings denote carbon atoms. A hydrogen atom is assumed to be attached to each such corner if no other atom or group is shown, and a circle inside the hexagon indicates the presence of a delocalized π-electron system.

electrons are called *radicals*.[1] If we symbolize an alkyl radical as R— and a functional group as Y—, then we can represent compounds like those above as RY. The simplest alkyl groups, together with two other common hydrocarbon radicals, are depicted in Table 26-3; the common functional groups are indicated, in representative compounds, in Table 26-4. The names of the hydrocarbon radicals in Table 26-3 are used not only in the common names of many compounds with comparatively few carbon atoms but also in the systematic names of branched-chain higher hydrocarbons and other more complex molecules.

The fact that the properties of organic compounds are, to a first approximation, a weighted sum of the properties attributable separately to their hydrocarbon portions and their functional groups provides a great simplifying feature in learning and understanding organic chemistry. It is not necessary to treat each new compound as a new entity. Rather, it is possible to predict the kinds of reactions a molecule will undergo by considering its component parts. The reactivity is governed largely by the nature of the functional groups present.

26-3 Properties and Reactions of Common Types of Organic Compounds

In this section some of the characteristic properties and reactions of typical compounds containing common functional groups are considered in the order in which they appear in Table 26-4, which should be used as a guide in reading the section.

Most organic compounds are not ionic, in contrast to many of the common inorganic materials discussed in earlier chapters. Thus, most of the organic reactions that we shall consider involve changes in bonds that are predominantly covalent.

---

[1] Although most radicals are extremely reactive and have very short lifetimes under normal conditions, some are more stable. Among the common molecules that contain unpaired electrons, and thus are radicals, are NO and NO_2 and, in its most stable state, O_2.

TABLE 26-3   Some common hydrocarbon radicals

CH_3— Methyl	CH_3CH_2— Ethyl	$CH_3CH_2CH_2$— *n*-Propyl	$CH_3CH_2CH_2CH_2$— *n*-Butyl
$\begin{matrix} H_3C \\ \\ H_3C \end{matrix}$CH— Isopropyl	$\begin{matrix} H_3C \\ \\ H_3C \end{matrix}$CH—CH_2— Isobutyl	CH_3—CH—CH_2—CH_3 *sec**-Butyl	$\begin{matrix} H_3C \\ H_3C \\ H_3C \end{matrix}$C— *tert**-Butyl
$CH_2{=}CH$— Vinyl	C_6H_5— or ⬡ Phenyl		

*The abbreviations *sec* and *tert* (or *s* and *t*) stand for secondary and tertiary. They relate to the fact that a carbon atom with four bonded neighbors is referred to as primary, secondary, or tertiary, depending on whether it is directly linked to one, two, or three other carbon atoms.

TABLE 26-4 **Some common classes of organic compounds***

Compound class	General formula†		Comment	Specific example
Alkyl halides	R—X	(RX)	X may be F, Cl, Br, or I	CH_3I Methyl iodide
Alcohols	R—O—H	(ROH)		$CH_3CH_2CHCH_3$ $\mid$ OH 2-Butanol
Ethers	R—O—R′	(ROR′)		$CH_3CH_2OCH(CH_3)_2$ Ethyl isopropyl ether
Amines	R—N⟨H H	(RNH₂)	One or both of the H's on the N may be replaced by an alkyl radical	$CH_3CH_2CH_2CH_2NH_2$ *n*-Butylamine $CH_3CH_2N(CH_3)_2$ *N,N*-Dimethyl ethylamine
Aldehydes	R—C⟨H O	(RCHO)		CH_3CHO Acetaldehyde
Ketones	R—C⟨R′ O	(RCOR′)	Note that there is no O—R′ bond	CH_3COCH_3 Acetone $CH_3CH_2COCH_3$ Butanone
Carboxylic acids	R—C⟨O—H O	(RCOOH)		$CH_3CH_2CH_2CH_2COOH$ Pentanoic acid
Esters	R—C⟨O—R′ O	(RCOOR′)		$CH_3COOCH(CH_3)_2$ Isopropyl acetate
Amides	R—C⟨N⟨H H / O	(RCONH₂)	One or both of the H's on the N may be replaced by an alkyl radical	$CH_3CH_2CONH_2$ Propanamide $CH_3CONHCH_2CH_3$ *N*-Ethyl acetamide

*Not including hydrocarbons. The functional group characteristic of each class is shown in **boldface** in the general formula.

†The formula given in parentheses is that normally used for typographical convenience. R and R′ signify alkyl radicals (Table 26-3) and may be the same or different.

This does not imply, however, that organic substances are nonpolar. Because fluorine, oxygen, nitrogen, and chlorine are appreciably more electronegative than carbon, bonds between carbon and these elements have significant ionic character, with carbon the more positive atom. This bond polarity plays an important role in many organic reactions.

Organic reactions are usually much slower than most familiar ionic reactions in aqueous solutions. Furthermore, most organic reactions are under kinetic control rather than equilibrium control. Careful specification of such reaction conditions as solvent and temperature is frequently necessary in order to maximize the yield of desired product, which is often far less than 100 percent because a number of different reactions compete with each other.

Alkyl halides.   The simpler alkyl halides are usually named in terms of the alkyl radical present, but more complicated ones are named systematically in a fashion parallel to that used for hydrocarbons (Appendix B):

$$\begin{array}{cc} & CH_3 \\ & | \\ (CH_3)_2CHBr & CH_3CHICHCH_2CH_3 \end{array}$$

$$\begin{array}{cc} \text{Isopropyl bromide} & \text{2-Iodo-3-methylpentane} \end{array}$$

They may be prepared by direct substitution of a halogen atom for a hydrogen atom in an alkane, as mentioned earlier, or by addition of HX to an appropriate alkene:

$$(CH_3)_2C{=}CH_2 + HCl \longrightarrow (CH_3)_3CCl \qquad (26\text{-}5)$$

$$\begin{array}{cc} \text{2-Methylpropene} & t\text{-Butyl chloride} \end{array}$$

Although isobutyl chloride, $(CH_3)_2CHCH_2Cl$, is another possible product of (26-5), it is formed in much smaller quantity. The general rule is that in addition of HX to a double bond, the hydrogen atom normally becomes attached to the doubly bonded carbon atom already richer in hydrogen. Alkyl halides may also be formed by reaction of the corresponding alcohol with the appropriate hydrohalic acid; for some alcohols, a catalyst such as the corresponding zinc halide, ZnX_2, must also be present:

$$ROH + HX \xrightarrow[ZnX_2]{} RX + H_2O \qquad (26\text{-}6)$$

The chief importance of the alkyl halides, especially the bromides, lies in their use in the preparation of the extremely useful *Grignard reagents* by reaction with metallic magnesium in dry diethyl ether:

$$RBr + Mg \longrightarrow RMgBr \qquad (26\text{-}7)$$

These reagents are among the most valuable in organic chemistry; a few applications are mentioned below.

Another important reaction of alkyl halides is removal of the elements of HX (dehydrohalogenation) from adjacent carbon atoms under the influence of KOH dissolved in alcohol, with the resultant formation of an alkene, as exemplified by

$$CH_3CHICH_3 \xrightarrow[KOH]{\text{alcoholic}} CH_3CH{=}CH_2 \qquad (26\text{-}8)$$

$$\begin{array}{cc} \text{Isopropyl iodide} & \text{Propene} \end{array}$$

Alcohols.   Alcohols contain the hydroxyl group, —OH, joined to a (nonaromatic) carbon atom to which no other oxygen atoms are bonded. They are named systematically as derivatives of the longest carbon-atom chain to which the —OH group is attached, with that group given the lower possible number if there is a choice. The suffix *-ol* denotes the —OH group:

$$CH_2$$

Cyclopentanol

$$CH_3CH_2CHCHCH_2CH_3$$
$$OH$$

4-Methyl-3-hexanol
(not 3-methyl-4-hexanol)

Alcohols with three or fewer carbon atoms are completely miscible with water because they can readily form hydrogen bonds with the water, both as donors and acceptors, without great disruption of the water structure. As the size of the hydrocarbon portion increases further, however, the solubility in water decreases.

CH_3OH
Methanol

Methanol and ethanol, the simplest and most important alcohols, are volatile liquids that boil at 65 and 78°C, respectively. Methanol (methyl alcohol, wood alcohol) is extremely poisonous, causing damage to brain tissue and being oxidized readily in the body to formic acid, which attacks the optic nerves and causes blindness. Methanol was first made by heating dried wood at high temperatures, but it is now produced industrially by the direct combination of CO and H_2 at 350 to 400°C and about 300 atm in the presence of a chromic oxide–zinc oxide catalyst. It is one of the crucial starting materials for the synthetic organic chemical industry.

CH_3CH_2OH
Ethanol

Ethanol (ethyl alcohol, grain alcohol, "alcohol") has been made since antiquity by fermentation of various grains, molasses, and other materials containing carbohydrates (Sec. 27-1). In moderate quantities ethanol is intoxicating, but slightly larger doses can be fatal through depression of the respiratory center in the brain. About 10^9 kg ethanol is now produced annually in the United States, the bulk of it for industrial purposes, by the hydration of ethylene with an acid catalyst at elevated temperatures and pressures.

It is important to note that alcohols do not ionize significantly in water; they are nonelectrolytes. They may be considered as alkyl derivatives of water, with one of the hydrogen atoms of water replaced by an alkyl group. They are at once extremely weak bases and extremely weak acids, much weaker than water. They accept protons only from the strongest acids to form ROH_2^+ (analogous to H_3O^+) and they react with the most reactive metals (e.g., sodium) to produce hydrogen and a salt:

$$CH_3OH + Na \longrightarrow CH_3O^- + Na^+ + \tfrac{1}{2}H_2 \qquad (26\text{-}9)$$

The resulting alkoxide ions, RO^-, are stronger bases than OH^-. They therefore cannot exist in appreciable concentrations in aqueous solutions, because they react with water to form the alcohol ROH and OH^-.

The reactions of alcohols are those of the —OH group or of its hydrogen atom (as in 26-9). Dehydration of an alcohol results in an alkene by removal of the —OH from one carbon atom and of a hydrogen atom from an adjoining carbon atom:

$$(CH_3)_3COH \xrightarrow{\text{catalyst}} (CH_3)_2C{=}CH_2 + H_2O \qquad (26\text{-}10)$$

The catalyst may be sulfuric acid or aluminum oxide. The hydroxyl group may be replaced by a halogen atom [Equation (26-6)], and the alcohol may react with a carboxylic acid to form an ester:

$$CH_3CH_2CH_2C\begin{matrix} OH \\ \\ O \end{matrix} + CH_3CH_2OH \xrightarrow[\text{catalyst}]{\text{acid}} CH_3CH_2CH_2C\begin{matrix} OCH_2CH_3 \\ \\ O \end{matrix} + H_2O$$

| Butanoic acid (butyric acid) | Ethanol | | Ethyl butanoate (ethyl butyrate) | (26-11) |

Alcohols can also react with inorganic acids containing —OH groups to form esters:

$$CH_3OH + (HO)_3PO \longrightarrow CH_3OPO(OH)_2 + H_2O \qquad (26\text{-}12)$$

| Phosphoric acid | Methyl dihydrogen phosphate (or just methyl phosphate) | |

$$\begin{matrix} CH_2ONO_2 \\ | \\ CHONO_2 \\ | \\ CH_2ONO_2 \end{matrix}$$
Nitroglycerine

Phosphate esters are essential components of many biochemical systems (Chap. 27). The sodium salts of sulfate esters of long-chain alcohols are important synthetic detergents, and nitrate esters such as glycerol trinitrate (nitroglycerin) are ingredients of some explosives.

Ethers. These compounds contain an oxygen atom bonded to two carbon atoms. They are comparatively inert chemically and are used chiefly as solvents. Ordinary "ether", the first substance used as a general anesthetic for operations (in the 1840s), is diethyl ether. It has a high vapor pressure, boiling at about 35°C, and since it is flammable and its mixtures with air are explosive, it has generally been superseded as an anesthetic. It is, however, still a common, and somewhat hazardous, laboratory solvent. Ethers are somewhat less soluble in water than are alcohols with the same number of carbon atoms (which are isomeric with them); because ethers have no —OH group, they can only act as acceptors, not donors, in hydrogen-bond formation.

Amines. Amines are derivatives of ammonia, with one, two, or all three of the hydrogen atoms of ammonia replaced by hydrocarbon radicals. The group —NH_2 is called the *amino* group. Amines are customarily named by attaching the names of the radicals to the word *amine*, as in isopropylamine, $(CH_3)_2CHNH_2$. When two or more different radicals are bonded to the nitrogen atom, the compound is named as a derivative of the largest radical, with the prefix *N* used to designate other radicals substituted on the amino group, as in *N,N*-dimethyl ethylamine (Table 26-4).

The most characteristic property of amines is their basicity, a consequence of the availability of the unshared pair of electrons on the nitrogen atom. Alkyl amines are comparable to ammonia in strength as bases, reacting slightly with water. For example,

$$(CH_3CH_2)_2NH + H_2O \rightleftharpoons (CH_3CH_2)_2NH_2^+ + OH^- \qquad (26\text{-}13)$$

| Diethylamine | Diethylammonium ion |

Similarly, amines form salts with acids stronger than water; for example, trimethylamine, $(CH_3)_3N$, reacts with acetic acid to give trimethylammonium ion $[(CH_3)_3NH^+]$ and acetate ion.

Amines can be prepared by treating an appropriate alkyl halide with ammonia, which produces an alkylammonium salt, and then liberating the amine by treating the salt with a strong base such as NaOH:

$$RX + NH_3 \longrightarrow RNH_3^+X^- \xrightarrow{\text{OH}^-} RNH_2 \qquad (26\text{-}14)$$

Ammonia reacts with alkyl halides because the C-X bond polarity makes the carbon atom slightly positive, and ammonia is a good nucleophilic reagent (Sec. 20-5) because of its unshared electron pair. Detailed kinetic and mechanistic studies indicate that in reactions of this kind the ammonia molecule attacks the halo-substituted carbon atom from the side opposite the C-X bond, displacing the halogen atom with its bonding pair as a halide ion. For example,[1]

$$H_3N: + \quad \underset{\underset{CH_3CH_2}{H}}{\overset{CH_3}{C}}{-}Cl \longrightarrow \left[H_3N{-}\underset{\underset{CH_2CH_3}{H}}{\overset{CH_3}{C}} \right]^+ + Cl^- \qquad (26\text{-}15)$$

The reactions are typical nucleophilic displacements (or substitutions). Amines with more than one alkyl substituent on the nitrogen can be prepared in similar fashion by treating an alkyl halide with an appropriate amine. For example,

$$CH_3NH_2 + CH_3CH_2CH_2Br \longrightarrow [CH_3NH_2CH_2CH_2CH_3]^+Br^-$$
$$\xrightarrow{\text{OH}^-} CH_3NHCH_2CH_2CH_3 \qquad (26\text{-}16)$$

Amines with few carbon atoms are water-soluble and volatile, but as the hydrocarbon portion of the molecule becomes larger, the volatility and water solubility decrease. Methylamine, dimethylamine, and trimethylamine are all found in brine in which herring have been soaked, and amines generally tend to have a somewhat fishy odor.

Amines and related compounds play a major role in many biochemical systems, and we consider them again in later parts of this chapter and in Chap. 27.

Aldehydes and ketones. Aldehydes and ketones have very similar reactions; the chemistry of these compounds is that of the carbonyl group, C=O. In aldehydes, the carbonyl group is joined to at least one hydrogen atom (and in formaldehyde, $H_2C{=}O$, to two hydrogen atoms); in ketones, it is joined to two other carbon atoms. Aldehydes are named systematically with the aid of the suffix *-al,* but the simplest ones are usually given common names based on the name of the corresponding acid (especially formaldehyde and acetaldehyde). Ketones are named systematically with the help of the suffix *-one* and numbers designating the position of the carbonyl group in the longest carbon chain. Alternatively, the names of the radicals attached to the carbonyl group and the word *ketone* may be used.

---

[1] The wedge-shaped and dashed bonds imply that the CH_3CH_2 group is above and the H atom below the plane of the page.

$$CH_3CH_2CHO \qquad CH_3CH_2COCH_2CH_3 \qquad \overset{\overset{\displaystyle Cl}{|}}{CH_3CHCOCH_3}$$

Propanal   3-Pentanone
                (diethyl ketone)   3-Chloro-2-butanone

Note that a too casual inspection of these formulas for ketones might suggest that they are ethers; that this is not so is evident when one counts the bonds to the carbon atom written to the left of the oxygen atom.

The double bond from carbon to oxygen is quite polar, with the oxygen atom the more negative, and the reactions of the carbonyl group reflect this polarity. One of the most useful reactions is that with a Grignard reagent, RMgX [Equation (26-7)], for this provides a means of lengthening a carbon chain or adding side chains. Because of the highly electropositive character of magnesium, the C-Mg bond in RMgX is polarized:

$$\overset{|}{\underset{|}{-C}}\!\!-\!\!\overset{\delta-}{} \ \overset{\delta+}{Mg}-$$

Consequently the carbon atom attached to the magnesium atom is nucleophilic and becomes bonded to the more positive atom of the carbonyl group, the carbon atom, while the —MgX group becomes attached to the oxygen atom. The product is a magnesium salt of an alcohol, and treatment with water readily frees the alcohol. For example, with acetaldehyde,

$$RMgBr + \overset{CH_3}{\underset{H}{\diagdown}}C\!\!=\!\!O \longrightarrow H\!-\!\overset{CH_3}{\underset{R}{\overset{|}{C}}}\!-\!OMgBr \xrightarrow{H_2O} H\!-\!\overset{CH_3}{\underset{R}{\overset{|}{C}}}OH \qquad (26\text{-}17)$$

The product is a secondary alcohol, that is, an alcohol with two other carbon atoms joined to the carbon bearing the —OH. One of these atoms is in the methyl group from acetaldehyde; the other is in the R— group from the Grignard reagent. With formaldehyde instead of acetaldehyde, the product would be a primary alcohol, RCH_2OH. With a ketone, the product is a tertiary alcohol:

$$RMgBr + \overset{R'}{\underset{R''}{\diagdown}}C\!\!=\!\!O \longrightarrow \xrightarrow{H_2O} R\!-\!\overset{R'}{\underset{R''}{\overset{|}{C}}}OH \qquad (26\text{-}18)$$

Aldehydes and ketones can easily be reduced to give alcohols with the same number of carbon atoms. This can be done with hydrogen and a suitable catalyst, such as Pt or Ni, or with a reducing agent like lithium aluminum hydride ($LiAlH_4$) or sodium borohydride ($NaBH_4$). An aldehyde gives a primary alcohol, and a ketone gives a secondary alcohol:

$$(CH_3)_2CHCHO \xrightarrow{LiAlH_4} (CH_3)_2CHCH_2OH \qquad (26\text{-}19)$$

2-Methylpropanal   2-Methylpropanol
(isobutyraldehyde)   (isobutyl alcohol)

$$(CH_3)_2C{=}O \xrightarrow{\text{LiAlH}_4} (CH_3)_2CHOH \qquad (26\text{-}20)$$

Acetone   2-Propanol
(isopropyl alcohol)

Conversely, primary and secondary alcohols can be oxidized under appropriately mild conditions to produce the corresponding aldehydes and ketones. A suitable reagent is aqueous potassium dichromate, $K_2Cr_2O_7$. Further oxidation of the aldehyde will give the corresponding carboxylic acid,

$$RCHO \xrightarrow{\text{ox}} RCOOH \qquad (26\text{-}21)$$

and indeed it is sometimes difficult to stop the oxidation of a primary alcohol at the stage of the aldehyde. Most ketones are resistant to mild oxidation; like most organic compounds, they can be oxidized to CO_2 and H_2O under more severe conditions.

EXAMPLE 26-3

Isomers of C_3H_8O. (*a*) Write the structural formulas for the isomeric molecules with molecular formula C_3H_8O. (*b*) Mild oxidation of one of these isomers gives a carboxylic acid. What is the isomer?

Solution. (*a*) The formula indicates that the molecules must be alcohols or ethers without a ring or a double bond. The only possibilities with three carbon atoms are 1-propanol, 2-propanol, and methyl ethyl ether:

$CH_3CH_2CH_2OH$ $CH_3\underset{\underset{\displaystyle OH}{|}}{C}HCH$ $CH_3OCH_2CH_3$

(*b*) Of these three molecules, only 1-propanol would form a carboxylic acid (propanoic acid) on mild oxidation. Acetone would be formed from 2-propanol, and the ether is inert under mild oxidizing conditions.

EXAMPLE 26-4

Synthesis of 3-hexanol from propanol. Devise a scheme for synthesizing 3-hexanol from propanol and any needed inorganic substances.

Solution. In devising a synthetic scheme, it is usually simplest to start at the end and work backward. The desired product, 3-hexanol, $CH_3CH_2\underset{\underset{\displaystyle OH}{|}}{C}HCH_2CH_2CH_3$, is a secondary alcohol that could be made by the addition of a Grignard reagent to an aldehyde, reaction (26-17). The corresponding aldehyde and alkyl halide could be propanal and *n*-propyl bromide:

$$CH_3CH_2CH_2MgBr + CH_3CH_2CHO \longrightarrow \xrightarrow{\text{H}_2\text{O}} CH_3CH_2\underset{\underset{\displaystyle OH}{|}}{C}HCH_2CH_2CH_3$$

The available starting material, propanol, could be converted into propanal by mild oxidation and into *n*-propyl bromide by appropriate treatment with HBr [Equation (26-6)]:

$$CH_3CH_2CH_2OH \xrightarrow{ox} CH_3CH_2CHO$$

$$CH_3CH_2CH_2OH \xrightarrow[ZnBr_2]{HBr} CH_3CH_2CH_2Br \xrightarrow[\substack{dry \\ ether}]{Mg} CH_3CH_2CH_2MgBr$$

The carbonyl group is common in various natural products, including sugars, which are polyhydroxy aldehydes or polyhydroxy ketones (Sec. 27-1). The pungent odor of formaldehyde is familiar to zoology students since its aqueous solution (formalin) is used in preserving specimens. The simplest ketone, acetone, is a common solvent in the laboratory, in industry, and in some common household glues and similar products.

Carboxylic acids.[1]   The simpler carboxylic acids have common names that reflect their origin: formic acid is the active ingredient in the sting of red ants (Latin: *formica,* ant), acetic acid is the acidic component of vinegar (Latin: *acetum,* vinegar), and butyric (or butanoic) acid is present in rancid butter (Latin: *butyrum,* butter). The systematic names, now usually used for acids with three or more carbon atoms, involve the root hydrocarbon name with the final *e* changed to *oic* and the word *acid* added. The carbon atom of the —COOH group is included in counting the number of carbon atoms for the purpose of naming the longest chain.

$$HCOOH \qquad\qquad BrCH_2CH_2\overset{\displaystyle |}{\underset{\displaystyle CH_3}{C}}HCOOH$$

Formic acid   2-Methyl-4-bromobutanoic acid

Most carboxylic acids are weakly acidic, like the familiar acetic acid. Those with four or more carbon atoms have limited solubility in water, but they are soluble even in as weakly basic a solution as one containing sodium bicarbonate because they are converted to salts:

$$RCOOH + HCO_3^- \longrightarrow \underset{\text{A carboxylate ion}}{RCO_2^-} + H_2O + CO_2 \qquad (26\text{-}22)$$

Carboxylic acids can be prepared by controlled oxidation of primary alcohols, RCH_2OH, or aldehydes, $RCHO$, with the same number of carbon atoms as the acid, as just discussed. The carboxyl group may also be introduced into a molecule by adding CO_2 to a Grignard reagent. For example,

$$\underset{\substack{\text{Ethylmagnesium} \\ \text{bromide}}}{CH_3CH_2MgBr} + CO_2 \longrightarrow CH_3CH_2COOMgBr \xrightarrow{H_2O} \underset{\substack{\text{Propanoic} \\ \text{acid}}}{CH_3CH_2COOH} \quad (26\text{-}23)$$

Finally, acids may be produced by the reaction of water with (hydrolysis of) an ester or an amide:

$$RCOOR' + H_2O \longrightarrow RCOOH + R'OH \qquad (26\text{-}24)$$
$$RCONH_2 + H_2O \longrightarrow RCOOH + NH_3 \qquad (26\text{-}25)$$

[1] The typographic convenience of the representation RCOOH should not obscure the fact that there is no —O—O—H (hydroperoxy) group in these molecules (see Table 26-4).

Esters are usually hydrolyzed in the presence of a strong base, which not only acts as a catalyst but also increases the extent of hydrolysis by removing the product acid as RCO_2^-. Amides are normally hydrolyzed in an acidic medium, which catalyzes the reaction and converts the ammonia or amine that is produced into the corresponding ammonium salt.

An important type of compound related to an acid is the corresponding acid chloride, $RC\begin{smallmatrix} \nearrow Cl \\ \searrow O \end{smallmatrix}$. Acid chlorides are valuable intermediates for the ready preparation of esters, amides, and other acid derivatives.

Esters.   The formation of carboxylate esters from alcohols and carboxylic acids was considered earlier in the discussion of alcohols [Equation (26-11)]. Equation (26-24) represents the reverse of this reaction, the hydrolysis of an ester, the mechanism of which was considered in Sec. 20-7. Esters are named by combining the name of the radical corresponding to the alcohol with the name of the acid, altered by replacing the ending *-ic* by *-ate:*

$$HCOOCH_2\underset{\underset{CH_3}{|}}{C}HCH_2CH_3 \qquad (CH_3)_2CHCOOCH_2CH_2CH_3$$

2-Methylbutyl formate   *n*-Propyl isobutyrate
                                          (*n*-propyl 2-methylpropanoate)

Esters are important as laboratory and commercial solvents. The most common is ethyl acetate; about 200 million pounds of it are used in the United States annually, 90 percent of this as a solvent, chiefly for coatings, from synthetic rubbers to nail polishes. The ester linkage is common in many naturally occurring and biologically important compounds. For example, fats are triesters of the trihydroxy alcohol glycerol (1,2,3-trihydroxypropane) with primarily long-chain carboxylic acids, C_{16} and C_{18} acids predominating. It is interesting that when these acids are unsaturated, the fats are normally liquid at ordinary temperatures (rather than solid, like beef tallow). The presence of double bonds leads to bent molecules that do not pack together readily. As a result, interactions between adjacent side chains are weak and the solid has a low melting point. These unsaturated fats predominate in the fatty tissue of aquatic mammals that live in polar regions, such as arctic seals and polar bears, permitting this tissue to remain mobile at low temperatures. Unsaturated fats are also the major constituents of plant oils, such as olive, corn, peanut, and safflower oils.

CH_2OH
$|$
$CHOH$
$|$
CH_2OH
Glycerol

Natural waxes, such as beeswax and the protective waxy coating on leaves, feathers, and wool, consist in part of esters of long-chain acids with long-chain alcohols. Many fruits owe their odor and flavor to simple esters. For example, bananas contain 3-methylbutyl-3-methylbutyrate and raspberry flavoring consists of a mixture of up to seven different esters, with from three to nine carbon atoms.

Amides.   The naming of amides is illustrated in Table 26-4. Although these molecules contain an amino group, they have no significant basicity, because of delocalization of the unshared pair formally on the nitrogen atom by resonance interaction with the adjacent carbonyl group:

$$\left\{ \begin{array}{c} \\ R-C \end{array} \overset{\displaystyle NH_2}{\underset{\displaystyle O}{\diagdown}} \;,\; R-C \overset{\displaystyle NH_2{}^+}{\underset{\displaystyle O^-}{\diagup}} \right\} \tag{26-26}$$

Amides are most readily prepared in the laboratory by interaction of an acid chloride, RCOCl, with ammonia or with an appropriate amine that has at least one hydrogen atom bonded to the nitrogen:

$$CH_3COCl + NH_3 \longrightarrow CH_3CONH_2 + HCl \tag{26-27}$$
Acetyl   Acetamide
chloride

$$CH_3CH_2COCl + HN(CH_3)_2 \longrightarrow CH_3CH_2CON(CH_3)_2 + HCl \tag{26-28}$$
Propanoyl   Dimethyl-   N,N-Dimethyl-
chloride   amine   propanamide

The amide linkage, —CONH—, is an essential structural feature of all protein molecules (Sec. 27-2) and also of synthetic polymers like nylon, discussed in a later section of this chapter.

Other kinds of organic substances. Although alkyl, or occasionally cyclo-alkyl, radicals have been used exclusively in the examples in this section so far, aromatic or alkenyl radicals such as phenyl or vinyl (Table 26-3) might also have been used. In most instances the properties of the resulting compounds would be very similar to those described. The greatest difference in properties occurs for compounds in which a hydroxyl group is substituted directly on an aromatic ring. The simplest of these substances is C_6H_5OH, which has the common name *phenol;* the general class of compounds with hydroxyl groups attached to aromatic rings is called phenols. Phenols are weak acids, which will lose their proton to a base such as $CO_3{}^{2-}$, although not usually to $HCO_3{}^-$. Thus phenols will dissolve in a solution of sodium carbonate:

$$C_6H_5OH + CO_3{}^{2-} \longrightarrow C_6H_5O^- + HCO_3{}^- \tag{26-29}$$
Phenol   Phenolate
ion

Aromatic amines are much weaker bases than ammonia or alkylamines. For example, K_b for aniline, $C_6H_5NH_2$, is smaller by a factor of 10^5 than K_b for ammonia because of partial delocalization of the unshared electron pair from the nitrogen atom to the aromatic ring.

We have mentioned so far only ring systems containing nothing but carbon atoms. However, there are many important *heterocyclic* compounds, that is, compounds with other atoms in the ring in addition to carbon, most notably nitrogen and oxygen (and occasionally sulfur). Some of these will be encountered in Chap. 27.

Only a few molecules containing more than one functional group have been mentioned; they include CCl_4, TNT, glycerol and its esters, and sugars. One very important class of molecules with more than one functional group is the amino acids, which contain an amino group and a carboxyl group. Just 20 of these naturally occurring molecules are used, in highly patterned ways, to construct most protein molecules in every known organism. These 20 amino acids (Table 27-3) all have the carboxyl group and the amino group joined to the same carbon atom; they can be represented generally as

$$\underset{\underset{\displaystyle \text{RCHCOOH}}{|}}{\text{NH}_2} \qquad\qquad (26\text{-}30)$$

where R now may stand for H or for a radical derived from various possible substituted alkanes, aromatic molecules, or heterocyclic molecules. Actually, because an ammonium ion is a much weaker acid than a carboxylic acid (that is, ammonia is a stronger base than a carboxylate ion), these molecules are more correctly represented as

$$\underset{\underset{\displaystyle \text{RCHCOO}^-}{|}}{\text{NH}_3{}^+} \qquad\qquad (26\text{-}31)$$

Thus they are "internally ionized" and exist in a dipolar form. Amino acids other than these 20 occur in various natural sources and have a variety of biochemical roles, although they are not incorporated into most proteins.

Physical properties of organic compounds. Most organic compounds of relatively low molecular weight are gaseous or liquid for the reasons discussed in Chaps. 3 and 4; the intermolecular attractions are not strong. Substances in which there is considerable opportunity for hydrogen bonding are at once less volatile and more viscous; the trihydroxy compound glycerol, for example, is a syrupy high-boiling liquid, although its molecular weight is about the same as that of hexane, which boils at 69°C. Amino acids of even low molecular weight are relatively high-melting solids, because there are strong forces between the dipolar molecules.

The solubility of organic substances in different solvents varies widely. A general rule is that compounds tend to dissolve in substances of similar structure and composition and not to dissolve in substances of very different structure and composition. Thus, compounds in which the proportion of oxygen atoms (and especially of —OH groups) or nitrogen atoms is high tend to be soluble in water and insoluble in hydrocarbons. Substances that consist primarily of a hydrocarbon chain or other nonpolar portion tend to dissolve in hydrocarbons and not in water.

A classification of organic reactions. There are hundreds of known reactions and the mechanisms of many have been studied carefully. In many of these, a bond is broken in such a way that one of the fragments formed retains both the bonding electrons. Such reactions may be classified in one of the following four categories: displacement (or substitution), elimination, addition, and rearrangement.

In a displacement reaction, a species with an unshared pair of electrons (a nucleophile) becomes attached to an alkyl group by means of the electron pair and another atom or group is displaced from the alkyl group, taking with it the electrons that formerly bound it to the carbon atom. Examples include reactions (26-6), (26-15) and (26-16), and (26-27) and (26-28). In an elimination reaction, a small fragment is ejected from a molecule and multiple bonding is introduced. Examples include reactions (26-8) and (26-10). An addition reaction involves, in essence, the reverse of an elimination, the degree of multiple bonding being reduced. Examples include reactions (26-1), (26-2), and (26-5). Reactions (26-17) and (26-18) involve a sequence of processes, addition and then displacement. In rearrangement reactions, not considered here, the atoms of a molecule are rearranged into a new bonding pattern without the substitution, elimination, or addition of any atoms.

Some reactions proceed by means of a bond cleavage in which one electron is left with each of the two fragments. This leads to radicals, and hence such reactions are called radical reactions. Examples include the halogenation of alkanes and, under some circumstances, the polymerization of unsaturated molecules, as discussed in Sec. 26-5.

26-4 Optical Isomerism

A number of kinds of isomerism have been discussed in Chap. 2 and in earlier sections of this chapter. One additional widespread type of isomerism, of particular significance in biochemistry, is that of two molecules that are mirror images of one another but are not superimposable, even in thought, like a right hand and a left hand. Such isomers are called optical isomers or enantiomers; the property of being not superimposable upon (congruent with) one's own mirror image is called *chirality* (Greek: *cheir,* hand).

Molecules that are not congruent with their own mirror images are said to be optically active because their solutions will rotate the plane of plane-polarized light (Fig. 26-1). Two of the isomers of $Co(en)_2Cl_2{}^+$ illustrated in Fig. 25-3 have this property. Any molecule containing a tetrahedrally substituted carbon atom joined to four different atoms or groups of atoms, such as bromochlorofluoromethane (Fig. 26-2), is chiral and therefore optically active. In fact, it was the optical activity of certain naturally occurring compounds containing a carbon atom bonded to four different groups[1] that led van't Hoff and Le Bel (quite independently) to postulate in 1874 that the four single bonds at a carbon atom were arranged tetrahedrally. Their brilliant insight was not confirmed by direct structural evidence until the arrangement of the carbon atoms in diamond crystals was determined more than 40 years later.

The physical properties of optical isomers are identical when they are measured by methods that have either no directional implications (e.g., density, melting point, enthalpy) or no chirality associated with a directional property (such as dipole moment). Their chemical properties are also identical, except when they involve interaction with other chiral species. Thus the rates of esterification of two optically isomeric acids with a nonchiral alcohol such as ethanol would necessarily be identical, but their rates of reaction with a chiral molecule, such as one of the optical isomers

---

[1] Such a carbon atom is referred to as an *asymmetric carbon atom.*

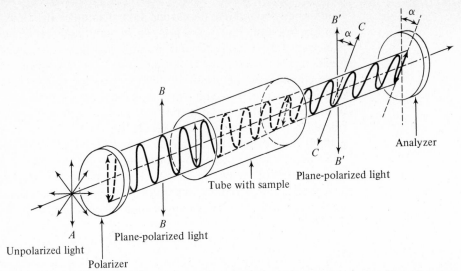

FIGURE 26-1 **Rotation of the plane of plane-polarized light by an optically active sample.** Light is electromagnetic radiation and as such is associated with transverse oscillations of electric and magnetic fields (i.e., oscillations perpendicular to the direction of propagation). In this figure unpolarized light enters from the left; this is light in which transverse oscillations occur in all directions. A vertically oriented polarizer permits the passage of the vertical component of the electrical oscillation only, producing plane-polarized light. The optically active sample rotates the plane of polarization clockwise from the orientation BB and $B'B'$ to the orientation CC by an angle α that is proportional to the length of the tube and to the concentration of the active substance in it. The surface described by the oscillation of the electric field of the light in the sample looks like a twisted ribbon. The orientation of the analyzer shown permits maximum passage of the light with the new orientation of its plane of polarization.

If all the molecules of the optically active sample were replaced by their mirror images, the figure would also have to be replaced by its mirror image. In this new situation, the polarization plane of the light passing through the sample would be rotated by $-\alpha$. Molecules that are superimposable on their mirror images are not optically active (Fig. 26-2).

of 2-butanol, would be quite different. Because all organisms contain numerous chiral molecules, the interactions of optical isomers with organisms almost invariably differ. For example, a particular chiral molecule may be essential for the proper functioning of an organism whereas its mirror image molecule is not utilized at all by that species.

A molecule whose internal symmetry is such that one-half of it is the mirror image of the other half is congruent with its own mirror image and hence cannot be optically active. Such a molecule may have a plane of symmetry. An example is bromochloromethane (Fig. 26-2c), whose symmetry plane passes through the centers of the Br, C, and Cl atoms and bisects the H–C–H angle. Molecules without internal mirror planes and without certain other symmetry elements[1] are necessarily chiral. Since

---

[1] These include a center of symmetry and a fourfold rotation-reflection axis, but molecules that possess one of these elements and lack a plane of symmetry are uncommon.

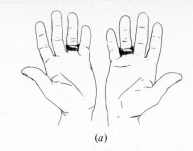

(a)

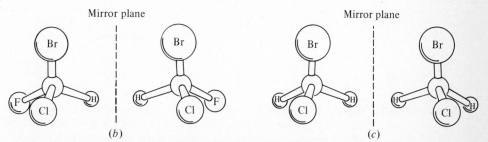

(b)   (c)

FIGURE 26-2   **(a) A left hand and its mirror image, a right hand; (b) an optically active molecule and its mirror image; (c) an optically inactive molecule and its mirror image.** A mirror plane, viewed here edge on, has the effect that a mirror appears to have—(a) changing a right hand held a certain distance from the mirror into a left hand that is equally distant on the opposite side of the mirror. The two molecules in (b) are both bromochlorofluoromethane; they are nonsuperimposable mirror images of one another, or optical isomers. Study of the drawings will show that if the Br, C, and Cl atoms are superimposed, the F atom of one molecule will fall on top of the H atom of the other, and conversely. These molecules are chiral; that is, they represent "right-handed" and "left-handed" forms. A solution of the right-handed form will rotate the plane of polarized light by the same amount, but in the *opposite direction,* as a solution of the left-handed form of the same concentration in the same solvent (Fig. 26-1).

The two molecules of bromochloromethane in (c) are also mirror images of each other, but they are congruent, that is, superimposable and thus indistinguishable. This substance is not optically active.

a helix is an inherently chiral object (Fig. 26-3), all helical molecules are chiral. When they are synthesized in an environment that is itself chiral, as are such biologically important helical molecules as DNA and proteins containing helical regions (Chap. 27), then there is normally a preference for one sense of the helix, either "right-handed" or "left-handed". Otherwise, both kinds of helices are created in equal numbers.

26-5   Organic Polymers

The term *polymer* (Greek: *poly,* many; *meros,* part) refers to any very large molecule, with molecular weight of the order of 10^4 or more, that is made by combining many identical or very similar small molecules (monomers). Naturally occurring polymers

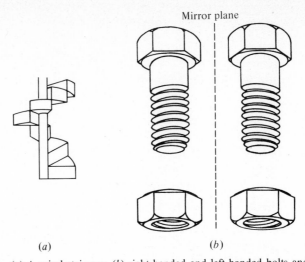

(a)   (b)

FIGURE 26-3   **Some helices.** (*a*) A spiral staircase; (*b*) right-handed and left-handed bolts and nuts.   A helix is the curve described by a point rotating about a straight line and simultaneously advancing parallel to the line through a distance proportional to the angle of rotation. Helices have a chirality associated with the sense of the rotation; the two shown in (*b*) are mirror images of one another. It is impossible to thread the right-handed nut onto the left-handed bolt, and vice versa. Turning either of the nuts (or bolts) end over end does not change its chirality; each nut still fits the same bolt that it did before. Some important helical structures in large molecules are shown in Figs. 1-2, 26-4, 27-2*b*, and 27-7*a*.

are not only the principal ingredients of most biological structures but also play central roles in metabolism and in genetics. Synthetic polymers have been tailor-made by chemists during the last half century to fill specific needs in industry, commerce, and the home, often as substitutes for less durable or more expensive materials.

The preparation and properties of polymers involve the same types of reactions and the same general principles of structure and reactivity that are applicable to smaller molecules. Nevertheless, there was little real progress in understanding the detailed chemical nature or structures of most polymeric materials, especially those that are naturally occurring, until the middle third of this century because of the great difficulty in characterizing most of them.

Vinyl polymers.   Many alkenes can be polymerized, and since most of these compounds contain the vinyl group, $CH_2{=}CH{-}$, the products are known generally as vinyl polymers. For example, ethylene polymerizes to form polyethylene, and the related molecule tetrafluoroethylene undergoes a similar reaction to form the polymer polytetrafluoroethylene (Teflon):

$$\text{Many } CH_2{=}CH_2 \longrightarrow \sim\!\!\sim CH_2CH_2CH_2CH_2CH_2CH_2CH_2CH_2 \sim\!\!\sim$$
$$\text{Many } CF_2{=}CF_2 \longrightarrow \sim\!\!\sim CF_2CF_2CF_2CF_2CF_2CF_2CF_2CF_2 \sim\!\!\sim$$

Each of these reactions involves an attack on and opening of the π bond, as with other reactions of alkenes [(26-1) and (26-2)].

Although a few monomers are so reactive that in the absence of refrigeration and inhibitors they may polymerize spontaneously, polymerization is usually initiated catalytically or by addition of substances that form radicals. Radical-initiated polymerization is normally done with a small quantity of an organic peroxide, which readily splits to form radicals:[1]

$$ROOR \longrightarrow 2RO\cdot \tag{26-32}$$

One of the radicals so formed then attacks the π bond of a monomer molecule, $CH_2{=}CHY$, producing a new radical:

$$RO\cdot + CH_2{=}CHY \longrightarrow ROCH_2\overset{\cdot}{C}HY \tag{26-33}$$

This radical can then in turn attack another monomer molecule:

$$ROCH_2\overset{\cdot}{C}HY + CH_2{=}CHY \longrightarrow ROCH_2CHYCH_2\overset{\cdot}{C}HY \tag{26-34}$$

This process continues in a chain reaction and the polymer grows rapidly; polymer growth is terminated only when two radicals combine, which happens on the average after 10^3 or 10^4 steps. Thus most of the resulting polymer molecules are extremely large and the effects of the RO— end groups are negligible. The resulting polymer molecules usually have a spread of molecular weights in the range 10^4 to 10^5. Hydrocarbon polymers are chemically inert, and polytetrafluoroethylene is especially unreactive because of the great stability of the C-F bond.

Polymerization of vinyl monomers can also be effected with a variety of catalysts,[2] many of them heterogeneous; typical is a mixture of crystalline $TiCl_4$ with triethyl aluminum, $Al(C_2H_5)_3$. Such catalysts bring about polymerization through ionic intermediates, although the mechanisms are not yet clearly established.

Polymerization of a vinyl monomer, $CH_2{=}CHY$, can give rise to structural variations in the product not possible with polyethylene, because in the polymer the groups Y may be arranged either randomly or in various ordered ways along the newly formed chain. For example,

Isotactic (Greek: *iso*, same; *taxis*, arrangement)

$$-CH_2\underset{\underset{Y}{|}}{C}HCH_2\underset{\underset{Y}{|}}{C}HCH_2\underset{\underset{Y}{|}}{C}HCH_2\underset{\underset{Y}{|}}{C}HCH_2\underset{\underset{Y}{|}}{C}HCH_2\underset{\underset{Y}{|}}{C}HCH_2\underset{\underset{Y}{|}}{C}HCH_2\underset{\underset{Y}{|}}{C}H- \tag{26-35}$$

Syndiotactic (alternating)

$$-CH_2\overset{\overset{Y}{|}}{C}HCH_2\underset{\underset{Y}{|}}{C}HCH_2\overset{\overset{Y}{|}}{C}HCH_2\underset{\underset{Y}{|}}{C}HCH_2\overset{\overset{Y}{|}}{C}HCH_2\underset{\underset{Y}{|}}{C}HCH_2\overset{\overset{Y}{|}}{C}HCH_2\underset{\underset{Y}{|}}{C}H- \tag{26-36}$$

[1] In the following equations, the unpaired electron in each radical is symbolized by a dot.

[2] The organic peroxide discussed earlier is often termed a catalyst, even though its mode of action is not strictly that of a catalyst. The substances mentioned here may also not be catalysts by strict definition.

Atactic (random, disordered)

$$-CH_2\overset{\overset{\displaystyle Y}{|}}{CH}CH_2\overset{\overset{\displaystyle Y}{|}}{CH}CH_2\overset{\overset{\displaystyle Y}{|}}{CH}CH_2\overset{\underset{\displaystyle Y}{|}}{CH}CH_2\overset{\overset{\displaystyle Y}{|}}{CH}CH_2\overset{\underset{\displaystyle Y}{|}}{CH}CH_2\overset{\underset{\displaystyle Y}{|}}{CH}CH_2\overset{\overset{\displaystyle Y}{|}}{CH}- \qquad (26\text{-}37)$$

Radical-initiated polymerization invariably leads to random or atactic polymers. Catalytic polymerization can lead to stereoregular polymers—e.g., the isotactic form exemplified by (26-35) or the alternating arrangement of (26-36).

The difference in physical properties between polymers of random structure and those of ordered structure can be enormous. Comparison of models of atactic and isotactic polypropylene indicates the reason. The regularity of isotactic polymers permits efficient packing of the chains and thus gives rise to "crystalline" (highly ordered) regions in the solid polymer and hence to stronger interchain forces and a consequent greater rigidity and higher melting or softening temperature. We consider this topic further below.

In addition to ethylene, propene, and tetrafluoroethylene, many other unsaturated molecules serve as monomers from which common vinyl polymers are made. Some of these are listed in Table 26-5. The polymers can be represented by formulas similar to (26-35) to (26-37). Sometimes two different monomers are mixed before polymerization. The resulting product is called a copolymer and usually has properties significantly different from those of the polymer formed from either monomer alone. Saran and similar kitchen wraps are copolymers of vinyl chloride or acrylonitrile with vinylidene chloride, $CH_2{=}CCl_2$.

Natural rubber and most synthetic rubbers are closely related to vinyl polymers. Natural rubber is a stereoregular polymer with the repeating unit

$-CH_2\overset{\overset{\displaystyle CH_3}{|}}{C}{=}CHCH_2-$. It can be made from the monomer 2-methyl-1,3-butadiene

(isoprene), $CH_2{=}\overset{\overset{\displaystyle CH_3}{|}}{C}CH{=}CH_2$. When this diene polymerizes, the two double bonds are converted to single bonds, with introduction of a new double bond in the center of the molecule and, in effect, an unpaired electron at each end that permits the isoprene units to be joined together by single bonds. The reaction of the first two

TABLE 26-5   **Some other vinyl polymers**

	Monomer		
Name		*Formula*	*Trade names of polymer*
Styrene (vinyl benzene)		$CH_2{=}CHC_6H_5$	Polystyrene, Styron, Lustron
Vinyl chloride*		$CH_2{=}CHCl$	PVC, Koroseal, Geon
Acrylonitrile (vinyl cyanide)		$CH_2{=}CHCN$	Orlon, Acrilan
Methyl methacrylate		$CH_2{=}\overset{\overset{\displaystyle CH_3}{|}}{C}COOCH_3$	Lucite, Plexiglas, Perspex

* Long exposure to vinyl chloride monomer has been correlated with above-normal incidence of certain kinds of cancer.

molecules can be represented as

$$2CH_2=\overset{\overset{\displaystyle CH_3}{|}}{C}CH=CH_2 \longrightarrow 2-CH_2\overset{\overset{\displaystyle CH_3}{|}}{C}=CHCH_2- \longrightarrow$$

$$-CH_2\overset{\overset{\displaystyle CH_3}{|}}{C}=CHCH_2CH_2\overset{\overset{\displaystyle CH_3}{|}}{C}=CHCH_2- \quad (26\text{-}38)$$

$$CH_2=\overset{\overset{\displaystyle Cl}{|}}{C}CH=CH_2$$
2-Chloro-1,3-buta-
diene

Neoprene, a common synthetic rubber, is a polymer of 2-chloro-1,3-butadiene and is more polar than natural rubber. Natural rubber has a tendency to swell in and be dissolved by hydrocarbons; neoprene does not. Consequently, neoprene is substituted for natural rubber whenever the material may be in contact with hydrocarbons, e.g., in hoses for gasoline and oils.

Condensation polymers. As the preceding examples show, the polymerization of unsaturated molecules proceeds by the addition of one molecule to another, all atoms of the monomers being retained in the polymer. In condensation polymerization there is simultaneously a formation of covalent bonds linking monomers together and an elimination of a low-molecular-weight compound, such as water, an alcohol, or a halide. The monomers must contain two functional groups, and in the process of polymerization functional groups on different monomers react with one another. For example, the common synthetic fiber known as Dacron or Terylene is a polyester made by heating ethylene glycol with terephthalic acid; water is eliminated as ester bonds are formed:

$$HOCH_2CH_2OH + HOOC-\bigcirc-COOH \longrightarrow$$

$$H\left(-OCH_2CH_2O\underset{\underset{\displaystyle O}{\|}}{C}-\bigcirc-\underset{\underset{\displaystyle O}{\|}}{C}-\right)_n OH \qquad (26\text{-}39)$$

The formula of the product is intended to imply that the portion in parentheses is repeated n times, where n is some large integer. Nylon, one of the first and most successful of the synthetic fibers, is a polyamide made by reacting a six-carbon diamine with a six-carbon dicarboxylic acid. The initial product is an ammonium salt, as when any amine reacts with an acid, but when heated above 200°C this forms the amide[1] and water:

$$HOOCCH_2CH_2CH_2CH_2COOH + H_2NCH_2CH_2CH_2CH_2CH_2CH_2NH_2 \longrightarrow$$
$$HOOC(CH_2)_4CONH(CH_2)_6NHCO(CH_2)_4CONH(CH_2)_6NHCO\cdots \quad \text{or}$$
$$HO[CO(CH_2)_4CONH(CH_2)_6NH]_n H \quad (26\text{-}40)$$

---

[1] For convenience, we represent the $-\overset{\overset{\displaystyle O}{\|}}{\underset{\underset{\displaystyle H}{|}}{C}}-N-$ group as $-CONH-$.

Proteins are naturally occurring polyamides. However, each of the "monomers" from which they are made has one amino group and one carboxyl group rather than two similar groups. They are the 20 amino acids referred to earlier, whose general structure may be represented as RCHCOOH, so that a segment of a protein chain may be represented as

$$\underset{NH_2}{|}$$

$$\overset{R}{\underset{|}{}} \quad \overset{R'}{\underset{|}{}} \quad \overset{R''}{\underset{|}{}} \quad \overset{R'''}{\underset{|}{}}$$

$$H_2NCHCONHCHCONHCHCO\cdots NHCHCOOH \qquad (26\text{-}41)$$

Each of the thousands of different proteins is characterized by a highly specific sequence of amino acids, corresponding to the different R groups in (26-41), as discussed in the following chapter.

Physical properties of polymers. The physical properties of common polymeric substances vary enormously. Some are soft but somewhat elastic; others are hard and brittle. Some of those that are hard at ordinary temperatures soften or melt on heating; others do not. Some are soluble in solvents such as benzene, acetone, or ethyl acetate, and some in water; others are unaffected by solvents. These differences in properties can be explained by considering the chemical nature of the materials and the general principles of structure and interatomic interactions discussed in earlier chapters, especially Chaps. 3, 4, and 11. There are no new kinds of forces in polymers. Different polymer molecules and different regions of the same molecule interact with each other by van der Waals forces and, when appropriate groups are present, by hydrogen bonding. Different molecules may even be cross-linked by covalent bonds, or by ionic interactions if oppositely charged groups are present.

As the temperature of a sample of polymer is raised, the component atoms and groups of atoms begin to vibrate increasingly. When the forces between the chains are weak, as with hydrocarbon chains and other comparatively nonpolar portions of molecules, not much thermal energy is needed to overcome intermolecular attractions. However, the energy required does depend significantly on how closely the chains fit together, since van der Waals forces are extremely sensitive to interatomic distance and are thus very dependent on the fit of one molecule to its neighbors. Atactic polymers, which have no ordered arrangement of side groups, pack together much less effectively than do the highly regular isotactic polymers. Consequently, appreciably more thermal energy is needed to overcome the interchain attractions in isotactic materials and they have higher softening and melting points.

The flexibility of polymer chains, that is, their ability to adopt different conformations by rotation of adjoining segments about the bonds between them, also plays an important role. A chain in which the barrier to rotation about the bonds is comparatively low will be flexible at ordinary temperatures. On the other hand, double bonds will stiffen a chain, and if the substituted groups are relatively bulky, rotation even about single bonds may be highly hindered except at elevated temperatures. This effect is manifested in the much greater flexibility of polyethylene objects, e.g., tubing, than of those made from Teflon. Teflon is very stiff. Its polymer chains are comparatively inflexible because the rotation of one CF_2 group past that adjacent to it requires appreciable energy, much more than does the rotation of one CH_2 relative to the next in a polyethylene chain. The van der Waals radius (packing radius) of a fluorine atom is only 15 to 20 percent greater than that of a hydrogen atom, but this difference has an appreciable effect (Fig. 26-4).

The packing of chains in polyamides such as nylon is stabilized by the formation of strong hydrogen bonds between the $>$NH groups of one chain and the $>$C$=$O groups of an adjacent one, $>$NH---O$=$C$<$. When nylon is first made, the individual chains are randomly oriented. The polymer is melted and extruded into filaments, which are then stretched to four or five times their original length. This orients the molecules more or less parallel to the filament axis. The tensile strength of these fibers is greater than that of the original unoriented polymer, because of the hydrogen bonding and the more efficient packing of the oriented chains. Hydrogen bonds between $>$NH and $>$C$=$O groups play a crucial role in the structure of the fibers of silk and wool as well; these are both natural protein fibers.

We pointed out in Chap. 4 that elements whose atoms can bond to only two neighbors form long-chain molecules (or rings) (Fig. 4-2) but not two- or three-dimensional networks. This same principle applies to monomers that can form two bonds. They produce long-chain polymers, either by opening a double bond or by condensation reactions involving two functional groups. In elements whose atoms

FIGURE 26-4 **Packing models of chains in Teflon and extended polyethylene.** The increased bulkiness of the fluorine atoms in the Teflon chain (*a*), relative to that of the hydrogen atoms in polyethylene (*b*), stiffens the Teflon because of the barrier to rotation of one CF_2 group relative to the next. It also imparts a twist to the Teflon chain, making it necessarily helical. Polyethylene is much more flexible. It can adopt the extended zigzag configuration illustrated here, with 180° torsion angles around the C-C bonds, or can coil up, with the most favorable torsion angle then being 60° (Sec. 11-2).

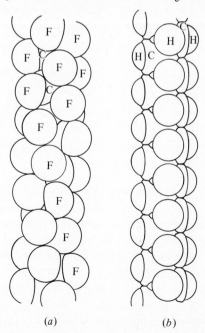

(*a*)   (*b*)

can bond to three or more neighbors, two-dimensional or three-dimensional structures can be built by linking chains together. Similarly, when a monomer can form more than two bonds, it can bind chains together to give extended networks or three-dimensional structures. If these bonds are strong, the resulting polymer will consist of only one or a few giant molecules, rather than many separate chains, and will be hard and insoluble. Only a few such cross-links can produce an appreciable change in the properties of a polymer. For example, addition of less than 0.1 percent of divinylbenzene to styrene before polymerization will produce a polymer that is insoluble in benzene, whereas ordinary polystyrene readily dissolves. The divinylbenzene provides cross-links between chains; one of the vinyl groups forms a segment of one chain, the second a component of a second chain, and the benzene ring forms the bridge linking the chains together.

The vulcanization of rubber by heating it with sulfur, which converts the rubber from a soft gummy material into a product of varying hardness depending on the amount of sulfur used, involves the creation of cross-links that consist of —S—S—

groups. This same disulfide linkage binds together different chains, or different regions of one chain, of some proteins and thus helps to maintain the highly specific chain conformations that are essential to the proper biological functioning of the protein. The disulfide links in proteins are brought about naturally by oxidation of —SH groups in nearby segments of the protein chains.

Synthetic polymers with extensive cross-linking are classified as *thermosetting* polymers, the term implying that heat is needed to bring about the polymerization but that once the polymer is finally formed it remains hard and rigid. This class includes such materials as Bakelite, a polymer made from phenol and formaldehyde.

Thermoplastic polymers, more often called just plastics (Greek: *plastikos,* that can be molded), soften upon heating and can then be shaped into any desired form. They will retain that form upon being cooled sufficiently and will soften again when reheated. Thus scraps or broken objects can be reused by reheating and molding them into any desired shape.

At a sufficiently low temperature, any plastic material is, like glass, hard and brittle. When heated it becomes first somewhat flexible and pliable and then may reach a state in which it can be deformed but will return to its original shape if the deformation is not maintained for too long a period—that is, it shows some elasticity. Finally, on continued heating it softens and then melts to a viscous liquid.

Different plastics exhibit these various properties at quite different temperatures because of variations in the nature and arrangements of their chains and the side groups on the chains. In the glassy state there are often highly ordered microcrystalline regions of the polymer where some chain segments are packed together in a regular array, although the length and the flexibility of the chains prevents the formation of large regions in which three-dimensional order persists. As the temperature is raised, the increasing mobility of chain segments leads to a more flexible material. Elasticity arises when the chains, normally twisted and coiled, are sufficiently flexible that they can be straightened to some degree when tension is applied but are not free to slip past one another. When the tension is relaxed, the chains coil up again and the material contracts once more to its original shape—provided that the tension has not been applied for so long a time that the chains have been able to slip by one another, a phenomenon that becomes increasingly likely as the temperature is raised. At still higher temperatures, the chains can move more or less independently of one another and the polymer softens and melts, although because of the length of the chains, which ensures their remaining to some degree tangled, the resulting liquid is always viscous.

At ordinary temperatures, thin sheets of polyethylene are sufficiently flexible to be used for raincoats, packaging materials, and the like whereas a polymer such as Lucite is in the glassy state. Polyethylene softens below the temperature of boiling water; hence polyethylene kitchenware put in a dishwasher often sags and loses its shape. Polypropylene softens at around 120°C, Lucite near 160°C or higher, and polystyrene at around 200°C.

Variations in the chemical behavior of polymers, such as their stability in the presence of acids and bases or on heating, can also be understood in terms of the chemical nature of the polymer. Since both amides and esters are readily hydrolyzed by acids and bases, it is not surprising that a drop of acid will make a hole in a silk or nylon blouse or that sodium hydroxide will disintegrate a sweater made of Dacron. Polyethylene containers are understandably inert to acids and bases, but they will burn readily, as other hydrocarbons do. Teflon, on the other hand, is stable at relatively high temperatures, because of the great stability of the C-F bond.

By varying the nature of the polymeric mix—the structures of the monomer and the catalyst, the possible presence of cross-linking agents or of a second monomer that can form a copolymer—chemists have succeeded in tailor-making polymers with an enormous range of characteristics. More than half of all chemists employed in industry work on various facets of polymer chemistry.

26-6   Conclusion

We have only given a glimpse of the kinds of compounds and reactions with which organic chemists are concerned, and hardly any hint at all about what they do. Refined synthetic schemes, taking into account the preferred conformations of the starting materials and of the desired products, as well as what is known about the mechanisms of organic reactions under different circumstances, permit the creation of hitherto unknown molecules. Some of these might be new polymers of potential value. Others may be synthesized that differ in minor ways from materials of known biological activity, in the hope that the new compounds may be even more desirable than the natural materials—perhaps less toxic, less expensive, more selective, more stable, or more soluble. The product might be a new insecticide, a dye, an antibiotic, a contraceptive, or a flavoring agent. It might be a molecule with no potential application but of quite novel structure, dreamed up first by an imaginative chemist and then synthesized to permit a test of some point of theory.

Proof of the structure of a molecule isolated from some natural source or from a reaction mixture is essential. The separation methods discussed in Chap. 6, most especially the various forms of chromatography, distillation, and crystallization, are used for isolation and purification. Then a combination of chemical tests for various functional groups, molecular-weight determination, and spectroscopic methods, including nmr, infra-red, and ultraviolet as well as high-resolution mass spectroscopy, will usually provide clues to all or most of the structure unless the molecule is very complex. If suitable crystals (a fraction of a millimeter on an edge) can be obtained, diffraction of X rays by these crystals can be used to give detailed information about the molecular geometry, even of protein molecules with thousands of atoms.

Problems and Questions

26-1   Structural formulas. Write formulas for the following compounds: *n*-heptane; 2-amino-ethanol; 1,1,2-trichloroethane; 2,3-hexanediol; 4-methyl-2-pentanone; 3-ethyl-2,3,5-trimethyl-hexane; 3-methylbutanal; isopropyl *t*-butyl ether; ethyl butanoate; 2-pentyne; 3-bromocyclo-butanol.

Ans. $CH_3CH_2CH_2CH_2CH_2CH_2CH_3$; $H_2NCH_2CH_2OH$; Cl_2CHCH_2Cl;

$$CH_3\underset{\underset{\text{OH}}{|}}{CH}\underset{\underset{\text{OH}}{|}}{CH}CH_2CH_2CH_3;$$

$$CH_3COCH_2\underset{\underset{\text{CH}_3}{|}}{CH}CH_3; \quad CH_3\underset{\underset{\text{CH}_2CH_3}{|}}{\overset{\overset{H_3C \quad CH_3}{|\quad\,\,\,|}}{CH}}\overset{\overset{CH_3}{|}}{C}CH_2CHCH_3; \quad CH_3\overset{\overset{CH_3}{|}}{CH}CH_2CHO;$$

$$(CH_3)_2CHOC(CH_3)_3; \quad CH_3CH_2OOCCH_2CH_2CH_3; \quad CH_3C\equiv CCH_2CH_3;$$

$$\begin{array}{c} BrCH—CH_2 \\ |\qquad\quad| \\ CH_2—CHOH \end{array}$$

26-2 **Names and formulas of isomers.** Write formulas and give systematic names for the eight unique compounds of formula $C_5H_{11}Cl$, not considering optical isomers.

26-3 **Isomeric dibromopropanes.** Write structural formulas and names for all possible dibromopropanes, $C_3H_6Br_2$. Do not repeat identical structures. *Ans.* four structures

26-4 **Names and formulas of isomers.** Write structural formulas and give names for all compounds having the general formula (a) $C_4H_{10}O$; (b) $C_2H_2Cl_2$; (c) C_4H_8.

26-5 **Intermolecular interactions.** Explain why the boiling point of ethanol (78°C) is much higher than that of its isomer dimethyl ether (−25°C) and why the boiling point of CH_2F_2 (−52°C) is far above that of CF_4 (−128°C).

26-6 **Formula of an amine.** A compound $C_4H_{11}N$ is known from its reactivity and spectroscopic properties to have no hydrogen atoms attached directly to the nitrogen atom. Write all structural formulas consistent with this information.

26-7 **Formula of an unknown compound.** A compound $C_4H_{10}O$ does not react with sodium. When the compound reacts with chlorine in light, three, and only three, monochloro derivatives, C_4H_9OCl, are formed. What structural formulas for the original compound are consistent with this information? *Ans.* only one formula is consistent: $(CH_3)_2CHOCH_3$

26-8 **Formula of an unknown compound.** A compound C_3H_6O has a hydroxyl group but no double bonds. Write a structural formula consistent with this information.

26-9 **Formulas of benzene derivatives.** Indicate with structural formulas the number of different isomers of (a) chlorobenzene, C_6H_5Cl; (b) dichlorobenzene, $C_6H_4Cl_2$; (c) bromochlorobenzene, C_6H_4BrCl; (d) trichlorobenzene, $C_6H_3Cl_3$; and (e) bromodichlorobenzene, $C_6H_3BrCl_2$.
Ans. (a) one; (b) three; (c) three; (d) three; (e) six

26-10 **Synthesis of a ketone.** Devise a reaction sequence for the synthesis of 2-butanone from acetaldehyde and ethyl bromide. Assume the availability of any inorganic reagents and of organic solvents.

26-11 **Esterification.** Acetyl chloride, CH_3COCl, reacts with the hydroxyl groups of alcohols to form ester groups with the elimination of HCl. When an unknown compound X with formula $C_4H_8O_3$ reacted with acetyl chloride, a new compound Y with formula $C_8H_{12}O_5$ was formed. (a) How many hydroxyl groups were there in X? (b) Assume that X is an aldehyde and write a possible structure for X and a possible structure for Y consistent with your structure for X.
Ans. (a) two; (b) one possible structure for X is $HOCH_2CHCH_2CHO$
$$\qquad\qquad\qquad\qquad\overset{|}{O}H$$

26-12 **Geometric structure.** Which of the following molecules are linear, which are planar, and which are neither linear nor planar: CH_2O, ClCN, BrCCH, N_2H_4, C_6H_6, H_2CCCH_2?

26-13 **Cis-trans isomerism.** For which of the following compounds are isomeric cis and trans forms possible: (a) propene; (b) 2-pentene; (c) 1-butene; (d) 2-methyl-2-butene; (e) 3-methyl-2-pentene; (f) 1,2-dichloropropene; (g) 2-butyne? *Ans.* (b), (e), (f)

26-14 **Reaction products.** Write structures for the products of the following reactions: (a) bromine reacts with cyclopentene in the dark; (b) HBr reacts with 1-butene; (c) isopropyl alcohol is dehydrated; (d) 2-methyl-2-butanol reacts with acetic acid; (e) trimethylamine reacts with

HI; (*f*) 2-pentanone is reduced with $LiAlH_4$; (*g*) 2-methylpropanal is oxidized mildly; (*h*) an amide is formed from diethylamine and the acid chloride of formic acid; (*i*) vinyl chloride is polymerized.

26-15 Syntheses. Devise a sequence of reactions for synthesis of each of the following molecules, starting with ethyl alcohol and any necessary inorganic materials and organic solvents: 2-butanol; 2-methylbutanoic acid; 3-methyl-3-bromopentane.

26-16 Optical isomerism. Which of the following molecules are chiral?

$$(a) \quad CH_3CH_2\underset{\underset{\displaystyle OH}{|}}{C}HCH_3 \qquad (b) \quad CH_3CH_2\underset{\underset{\displaystyle CH_3}{|}}{C}HCH_3$$

$$(c) \quad \underset{Br}{\overset{Cl}{\diagdown}}C{=}CHCH_3 \qquad (d) \quad H_2C\underset{\underset{\displaystyle CHBr}{|}}{\overset{\overset{\displaystyle CHCl}{\diagup}}{\diagdown}}$$

26-17 Polymers. (*a*) Define the terms vinyl monomer, isotactic polymer, atactic polymer, copolymer, condensation polymer, and thermoplastic polymer. (*b*) Which of the following materials is likely to be damaged by spilled acid or base? Teflon, Lucite, polyvinyl chloride, polystyrene, nylon, polyester, Orlon. Explain.

"It has become clear that the basic metabolic processes of all living cells are very similar. A number of identical compounds, mechanisms, structures, and reaction pathways are found in all living things so far observed, including such diverse cells as those of bacteria, begonias, bees, birds, and biochemists.

"The biopolymers, the synthetic pathways, and the small molecules that are common to all living things have been preserved almost unchanged for more than three billion years. The proto-organism would have had a degree of complexity sufficient to include all these common components shared by present living things. . . . The inferred biochemical structure of this primitive cell provides evidence that it was itself the product of many evolutionary steps, very similar to ones which have occurred in its descendants."

M. O. Dayhoff and R. V. Eck,[1] 1972

27 Biochemistry

Biochemistry has flowered remarkably in the last three decades. New tools and new methods have provided detailed information at the molecular and atomic level about the structures of the various components of and the reactions that occur in living organisms. As with any science, the earlier stages were chiefly descriptive. They involved the accumulation of information about the chemical nature and the distribution of the thousands of molecules and ions present in living systems, a task that is still incomplete because of the complexity of some of the molecules and molecular interactions in cells and because of their sensitivity to changes in conditions. However, as even partial information became available, it began to be possible

[1]M. O. Dayhoff (ed.), "Atlas of Protein Sequence and Structure", vol. 5, National Biomedical Research Foundation, Bethesda, Md., 1972.

to understand in some detail the chemical changes that accompany and indeed are responsible for many of the processes and activities that, taken together, characterize life—digestion, energy utilization, reproduction, growth, muscle contraction, nerve-impulse transmission, immunity, response to various drugs and other stimuli, and so on. Much is now known about some of these topics, little about others. Only a few of them are considered in this chapter. Our aim is to indicate the ways in which chemistry is able to provide some explanations for and insight into biological phenomena.

27-1   Biochemical Uniformity

There is a quite remarkable uniformity, indeed a near universality, in the general chemical makeup of all living matter on the earth and a striking similarity in many of the sequences of metabolic reactions in the most diverse organisms. Even the specific details of some structures and reactions are nearly universal, although many of them vary in at least minor ways from one living system to another. In this section, we examine some of the universal biochemical themes, devoting only minor attention to the variations on them.

Living matter is composed primarily of 4 elements: carbon, hydrogen, oxygen, and nitrogen. Another 15 or 20 also play essential roles, although together they compose only 1 or 2 percent of the atoms present. Some of these minor but essential elements are indicated in Table 27-1. A few others are of occasional importance; of these only molybdenum has atomic number above 30. It is noteworthy that silicon, a congener of carbon, is of very minor significance—it is present only in the skeletons of certain algae called *diatoms* and possibly in some other plants—despite the fact that silicon atoms are 100 times more abundant than carbon atoms on the earth's surface. The lack of silicon compounds in living systems is related to the weakness of the Si-Si bond as compared with the C-C bond.

Common molecules of biological systems.   There are four general classes of relatively small molecules that constitute the building blocks of biological macro-molecules, as well as being important in their own right. One class is the amino acids, described briefly in the previous chapter; they are the "monomers" from which proteins are built. Amino acids and proteins are discussed in Sec. 27-2. The other three groups are sugars and the large molecules made from them, which together are called carbohydrates; nucleotides and their polymers, the nucleic acids; and lipids.

Carbohydrates.   This class of compounds includes various simple sugars (also called monosaccharides), which are polyhydroxyaldehydes or ketones containing from three to seven carbon atoms, and macromolecules that yield sugars as products of hydrolysis. Starch, cellulose, and glycogen are typical macromolecules of this sort (polysaccharides). The names of most carbohydrates end in *-ose*; the sugars with five and six carbon atoms, to which we give most attention, are called pentoses and hexoses.

TABLE 27-1 Some minor elements in living systems*

Element	Predominant form and role
Fluorine	F^-. In teeth and some similar hard bony structures
Sodium	Na^+. Principal extracellular cation
Magnesium	Mg^{2+}. Essential for the activity of many enzymes; in chlorophyll
Phosphorus	Phosphate. Component of nucleic acids and some lipids; essential for all biosynthetic and energy-transfer processes
Sulfur	Usually S(-II) or S(-I), sometimes as sulfate. In many proteins and other molecules
Chlorine	Cl^-. Principal mobile anion
Potassium	K^+. Principal intracellular cation
Calcium	Ca^{2+}. Principal cation in bone and in shells; role in muscle contraction
Manganese	Mn(II). In some enzymes
Iron	Fe(II), Fe(III). Oxygen transport and storage in hemoglobin and myoglobin; in key enzymes involved in oxidation-reduction reactions
Cobalt	Co(III). In vitamin B_{12}
Copper	Cu(II). In certain oxidative enzymes; in oxygen-transport protein in some marine organisms
Zinc	Zn(II). In several enzymes and in insulin
Iodine	I($-$I). Essential for thyroid activity

*Not including C, H, N, and O. The elements are listed in order of increasing atomic number. The cations of the transition elements are often present only in trace amounts (in coordination complexes attached to large molecules).

Sugar molecules contain unbranched chains of carbon atoms, and although their structures are often represented in a straight-chain or "open-chain" form, most sugars occur predominantly with a more stable cyclic structure. The conversion of the open chain to the cyclic form results from the reaction of the sugar carbonyl group with an alcoholic hydroxyl group in the same molecule to form a product called a hemiacetal:

$$RC\!\!\underset{O}{\overset{H}{\diagdown}} + R'OH \longrightarrow RC\!\!\underset{OH}{\overset{H}{\diagup}}\!\!-OR' \tag{27-1}$$

A hemiacetal

For example, the open-chain form of the most common hexose, glucose, has an aldehyde group at one end of a six-carbon chain, with a hydroxyl group on each of the other carbon atoms. The hydroxyl group on the fifth carbon atom reacts with the aldehyde group, forming a hemiacetal with a six-membered ring that includes one oxygen atom (the circled numbers identify corresponding carbon atoms in the two forms).

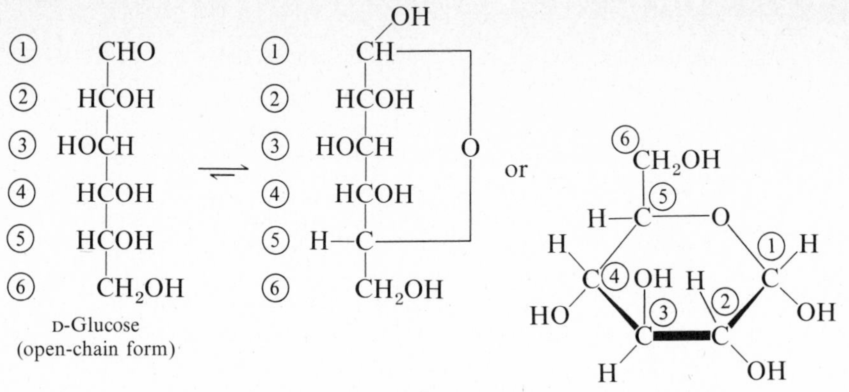

(27-2)

α-D-Glucose
(one cyclic form;
a hemiacetal)

CHO
HOCH
HCOH
HOCH
HOCH
CH₂OH

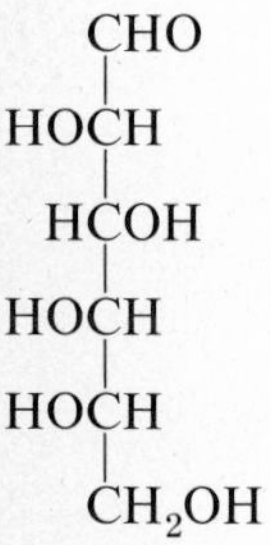

L-Glucose

CHO
HOCH
HCOH
HCOH
HOCH
CH₂OH

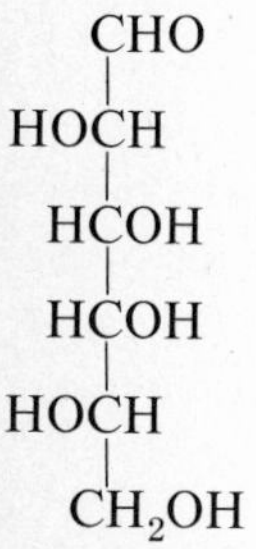

L-Galactose
(open-chain
form)

The six-membered ring is so much more stable than the noncyclic form that under most circumstances the equilibrium between them strongly favors the cyclic form. This is true as well of most other hexoses and pentoses, and we usually represent them in the cyclic form.

There are four asymmetric carbon atoms in the open-chain form of glucose—the four (C-2, 3, 4, and 5) between the —CHO and the —CH$_2$OH groups. D-Glucose has a specific configuration about each of these asymmetric centers; its mirror image, called L-glucose,[1] has the opposite configuration about *all four* asymmetric carbon atoms. A molecule with a different configuration about only some, but not all, of these atoms is a molecule of a different sugar, with different properties. For example, if the positions of the H and the OH are interchanged on the fourth carbon atom, the structure is that of a molecule of the sugar galactose, which also exists in both D and L forms, both cyclic and noncyclic. These hexoses contain four asymmetric carbon atoms, each of which can have two possible configurations. Thus, there are $2^4 = 16$ possible aldehyde hexoses, or eight enantiomeric pairs. The two optical isomers of glucose and the two of galactose are two of these eight pairs.

Cyclization of one of these hexoses, as in (27-2), introduces a fifth asymmetric center at the carbon atom that was formerly in the aldehyde group, C-1 in (27-2). Hence two possible forms of each cyclic molecule exist; they are distinguished by the prefixes α- and β-, rather than by separate names. Only the α form of cyclic D-glucose is illustrated in (27-2); in the β form, the positions of the H and the OH on C-1 would be interchanged.

The cyclic forms of sugars are found in many larger molecules—for example, combined with other sugar molecules in polysaccharides. These compounds are formed by reactions parallel to (27-1); the hemiacetal produced in that reaction might combine, for example, with another molecule of an alcohol (or a phenol) to form an acetal:

[1]The prefix D- (Latin: *dexter,* right) is often used to designate one of the two enantiomeric forms of an optically active molecule. The prefix L- (Latin: *laevus,* left) denotes its mirror image.

$$\underset{\text{A hemiacetal}}{RC\overset{\displaystyle H}{\underset{\displaystyle OH}{-}}OR'} + R''OH \longrightarrow \underset{\text{An acetal}}{RC\overset{\displaystyle H}{\underset{\displaystyle OR''}{-}}OR'} + H_2O \qquad (27\text{-}3)$$

An acetal made from the cyclic hemiacetal form of a sugar is called a glycoside; the specific name for a particular glycoside is derived from the names of the alcohol and the sugar involved. For example,

$$CH_3CH_2OH + \alpha\text{-}D\text{-glucose} \longrightarrow \quad + H_2O \qquad (27\text{-}4)$$

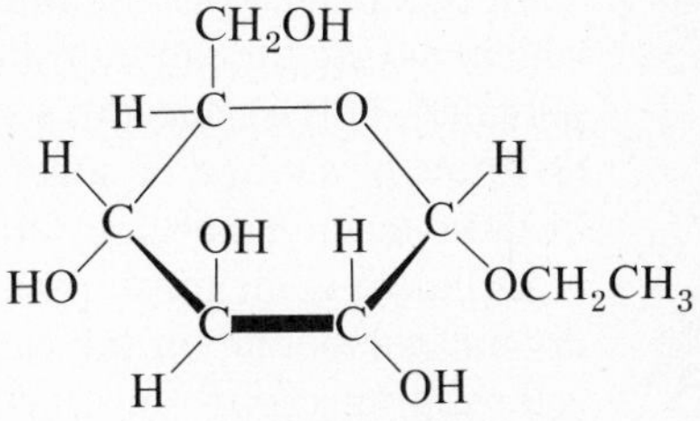

Ethyl α-D-glucoside
(an acetal)

Amines with at least one hydrogen atom on the nitrogen atom can react with sugars to form similar products; the nucleotides, discussed below, have this linkage. Many naturally occurring alcohols, phenols, and amines are found as glycosides, in combination with various sugars.

A disaccharide is a glycoside that is formed by combining two monosaccharides, the hemiacetal group of one combining with a hydroxyl group of the second as in (27-3). Sucrose and lactose are representative disaccharides:

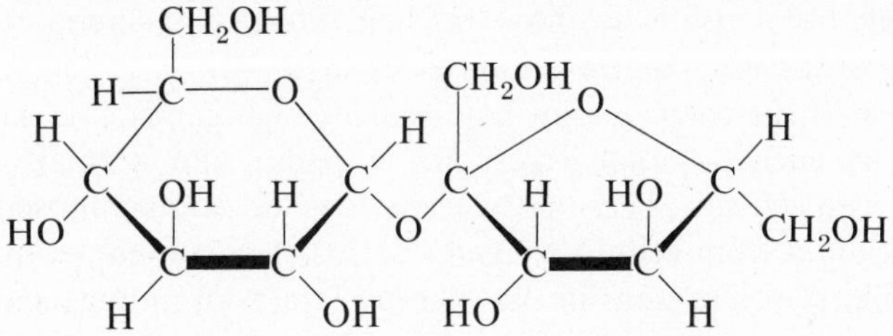

Glucose   Fructose

Sucrose

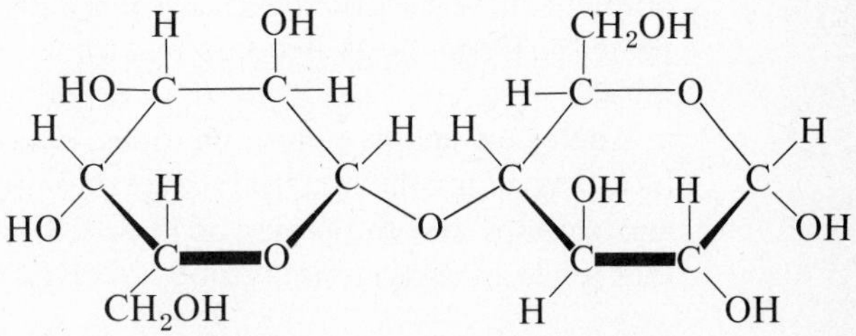

$$(27\text{-}5)$$

Galactose   Glucose

Lactose

$$CH_2OH$$
$$|$$
$$C{=}O$$
$$|$$
$$HOCH$$
$$|$$
$$HCOH$$
$$|$$
$$HCOH$$
$$|$$
$$CH_2OH$$

D-Fructose
(open-chain
form)

Sucrose is the common sugar of the home and of commerce; it is widely distributed in all photosynthetic plants, although it is produced commercially only from sugarcane and sugar beets. It contains one glucose unit joined by an α-acetal linkage from its first carbon atom to the second carbon of fructose, a hexose that has a ketone group at C-2 in its open-chain form. Lactose is a compound of galactose and glucose joined by a β-acetal linkage from C-1 of galactose to C-4 of glucose—a so-called β-1,4 linkage. This disaccharide, sometimes called "milk sugar", constitutes as much as 5 percent of the milk of most mammals.

In polysaccharides, hundreds or even many thousands of monosaccharide units, which may be the same or different, are joined together to form linear or branched polymers. Polysaccharides serve both as structural components of living matter and as forms of storage of sugars. Glucose is the most common monosaccharide unit in these macromolecules, but galactose and others are found as well.

Cellulose is the most abundant polysaccharide and indeed is the most abundant organic compound on the earth. It is the major structural element of plants, whose cell walls must provide structural strength since plants lack a bony skeleton. Wood consists of about half cellulose and half a noncarbohydrate heterogeneous polymer called lignin. Such plant fibers as cotton and flax are more than 90 percent cellulose. Cellulose is a linear polymer of glucose units, joined by β-1,4 linkages; the molecular weight is of the order of 10^6.

It might seem that a substance that consists of nothing but linked glucose units should be an excellent foodstuff, since glucose itself is a prime fuel for organisms. However, cellulose is insoluble and must be broken down into small units before it can be utilized as food. The enzymes present in the saliva and intestines of many animals, including human beings, cannot catalyze the hydrolysis of the β-1,4 linkage in cellulose, so cellulose is not a practical foodstuff[1] for these species. Grazing animals and insects like termites do have in their intestines bacteria that have the proper enzymes, so they can digest cellulose readily.

About half the total energy requirement of organisms is supplied by carbohydrates. The energy-storage reservoirs in plants and animals contain starch and glycogen, respectively. These polysaccharides are also composed entirely of glucose units, but differ from cellulose chiefly in that they contain primarily α-1,4 linkages. Since the enzymes present in the digestive tracts of animals can readily hydrolyze these α linkages, the plant polysaccharide starch is a valuable animal food. There are several forms of starch, some unbranched and some with occasional linkages involving other carbon atoms that provide branches and cross-links. Glycogen, the animal counterpart of starch, is even more highly branched and is often bound to proteins. Deposits of it occur in various tissues and organs, especially liver and muscle.

There are many polysaccharides besides cellulose that play structural roles, including chitin, the material that makes up the shells of crustaceans (such as lobsters and shrimp) and the scales of insects. Chitin is rather similar to cellulose, except that it has an acetylamino group, $-NHCOCH_3$, on C-2 of the glucose units instead

---

[1] A few animals, such as dogs, have so high a concentration of HCl in their stomachs that they can hydrolyze cellulose at a significant rate even though they lack the proper enzymes.

of an —OH group. Complex polysaccharides containing this *N*-acetylglucosamine group are also present in bone, skin, and various connective tissues, as well as in the cell walls of bacteria (where they are cross-linked by short chains of amino acids) and in the coatings of many animal cells. Many other polysaccharides are found in both plants and animals.

Nucleotides and nucleic acids. Mononucleotides are molecules that consist of three portions: (1) a weakly basic nitrogen-containing ring compound, joined by a glycosidic N-C bond to (2) a five-carbon sugar (ribose or deoxyribose), one hydroxyl group of which is esterified by (3) phosphoric acid. Base-sugar-phosphate molecules such as those in Fig. 27-1*a* and *b* polymerize in highly patterned sequences to form the various kinds of nucleic acids. Simple mono- and dinucleotides, sometimes with additional phosphate groups attached, participate in all energy-transfer processes in organisms, act as catalysts for metabolic oxidation-reduction reactions (Fig. 27-1*c*), and play other vital biochemical roles.

FIGURE 27-1 **Formulas of three nucleotides.** (*a*) Adenosine 5′-phosphate; (*b*) thymidine 3′-phosphate; (*c*) nicotinamide adenine dinucleotide. Adenosine 5′-phosphate, also known as adenosine monophosphate (AMP), is a ribonucleotide containing the sugar ribose. Thymidine 3′-phosphate is a deoxyribonucleotide, its sugar being 2-deoxyribose, i.e., ribose without the hydroxyl group on the second carbon atom. Nicotinamide adenine dinucleotide (NAD$^+$), which has a positive charge localized primarily in the ring of the nicotinamide residue, contains two base-ribose-phosphate combinations, joined through their phosphate groups. This dinucleotide is an important oxidizing agent in many biochemical reactions; its reduced form, NADH, is readily oxidized again. The nicotinamide portion of the molecule is the part that is reversibly reduced and oxidized.

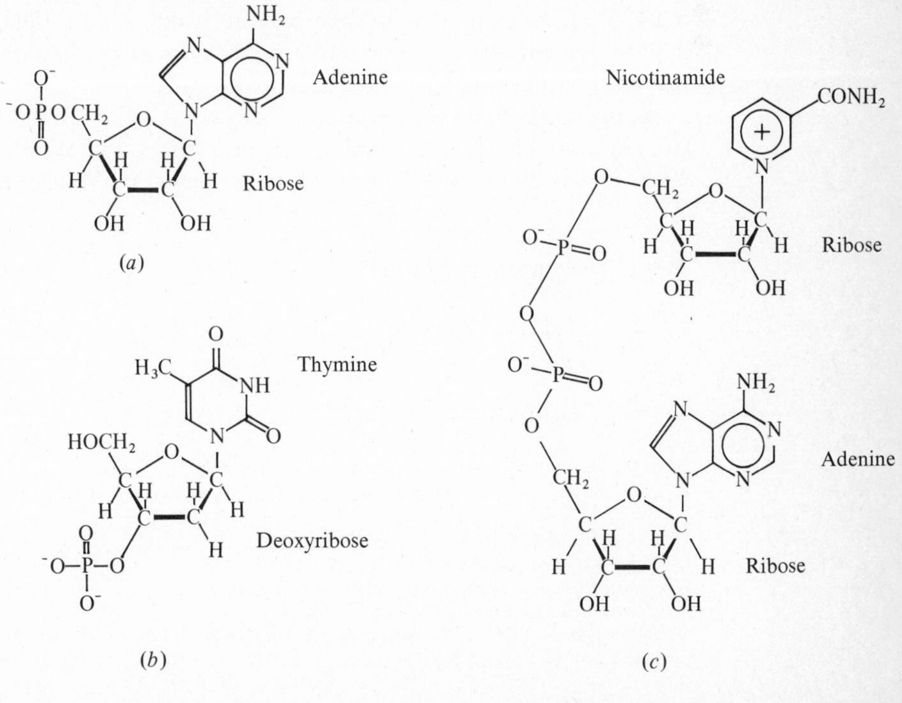

CHO
|
HCOH
|
HCOH
|
HCOH
|
CH$_2$OH

D-Ribose

Adenosine 5′-phosphate (Fig. 27-1*a*) is one of the four nucleotides that are combined together in various sequences and proportions to make up molecules of the different ribonucleic acids (RNA). It consists of the base adenine, the sugar ribose, and a phosphate group attached to the fifth carbon atom (numbered 5′) of ribose; the combination of adenine and ribose is called adenosine. Adenosine triphosphate, or ATP, has two additional phosphate groups joined to that already present:

$$\text{Adenine-ribose}-\text{O}-\overset{\overset{\displaystyle O}{\|}}{\underset{\underset{\displaystyle O^-}{|}}{P}}-\text{O}-\overset{\overset{\displaystyle O}{\|}}{\underset{\underset{\displaystyle O^-}{|}}{P}}-\text{O}-\overset{\overset{\displaystyle O}{\|}}{\underset{\underset{\displaystyle O^-}{|}}{P}}-\text{O}^- \qquad (27\text{-}6)$$

ATP is remarkable in its universality as the common currency for the exchange or transformation of energy in all living cells. No biosynthesis occurs, no nerve impulse travels, no heart beats, no eye opens, no firefly glows without the participation of ATP, by reactions considered further below.

Thymidine 3′-phosphate (Fig. 27-1*b*) is one of the four nucleotide components of deoxyribonucleic acid (DNA). It consists of the base thymine; the sugar deoxyribose, which differs from ribose only in that the hydroxyl group on the second carbon atom (numbered 2′) is missing; and a phosphate group, joined to the third carbon atom, 3′. The four bases present in most DNA are adenine, thymine, guanine, and cytosine, abbreviated A, T, G, and C, respectively (Table 27-2). Three of these four bases, A, G, and C, are also major components of RNA; the fourth base in RNA is usually uracil, abbreviated U (Table 27-2).

The polymerization of mononucleotides to form the nucleic acids occurs by linking of the sugar molecules through phosphate groups, joining the C-5′ of one sugar to the C-3′ of the next. This linkage scheme is the same in DNA, in which the only pentose present is deoxyribose, and in RNA, in which the only sugar molecules are ribose. It is illustrated schematically in Fig. 27-2*a*.

As Figure 27-2*b* shows, molecules of DNA consist of two individual polynucleotide strands wound around a common axis in a spiral, with an outer diameter of about 20 Å. This is the famous double helix discovered by Watson and Crick in 1953. At

TABLE 27-2   **Major bases present in nucleic acids***

Name	Common abbreviation	Present in
Adenine	A	DNA, RNA
Guanine	G	DNA, RNA
Cytosine	C	DNA, RNA
Thymine	T	DNA only
Uracil*	U	RNA only

*The structures of A, G, C, and T are illustrated in Fig. 27-3. Uracil has the same structure as thymine (Fig. 27-3*a*) except that it has a hydrogen atom in place of the methyl group of thymine. Its hydrogen-bonding potentiality is almost identical to that of thymine, so T and U are interchangeable in base pairs, such as that in Fig. 27-3*a*.

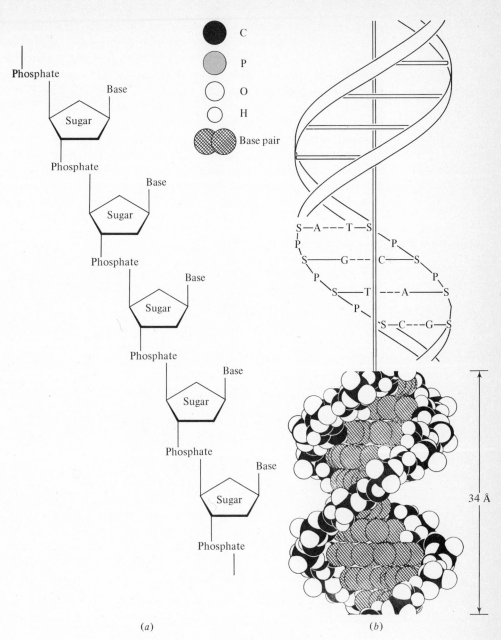

(a)   (b)

FIGURE 27-2 **(a) Schematic representation of a portion of a single chain of a nucleic acid molecule; (b) the double helix of DNA, represented in different ways.** At the top of (b), the base pairs are represented by the crossbars and the phosphate-sugar chains by the spiraling ribbons. In the center is a somewhat more detailed representation. The P and S represent the phosphate and sugar groups; the bases are denoted by A, G, T, and C. A space-filling model is depicted at the bottom.

the perimeter are the two sugar-phosphate chains; the bases lie inside these chains, with their planes perpendicular to the axis of the helix. Each base on one chain is paired by hydrogen bonding with a base on the other chain (Fig. 27-3).

Although there are four bases in DNA—A, T, G, and C—only two kinds of hydrogen-bonded base pairs are found: A-T and G-C. These are called complementary base pairs and are the only pairs bound by strong hydrogen bonds that fit properly into the DNA helix; they are of the right length, and the glycosidic N-C bonds (the bonds between the bases and the sugars) project from them at angles appropriate for the sugar residues in the sugar-phosphate chains. Other possible base-pair combinations (for example, A-G, C-T, A-C, and so on) are too bulky, or not large enough, or otherwise inappropriate for the steric requirements of this remarkably precisely engineered helix.

FIGURE 27-3   **Base pairs in DNA.** (*a*) Adenine-thymine (A-T); (*b*) guanine-cytosine (G-C). In each of these pairs, the hydrogen bonds are nearly linear and of the appropriate length for strong interaction. Although other pairings of the four bases in DNA with reasonable hydrogen bonding and mutual disposition of the glycosidic N-C' bonds are conceivable, they are less favorable than the two shown, which are, as far as is known, the only pairs that normally occur in DNA.

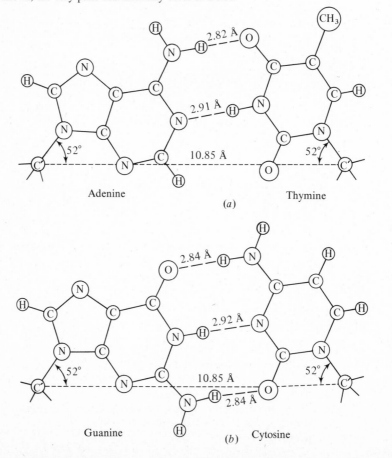

The fact that only two of the many possible base pairs are actually present has profound implications: it means that every time the base A occurs in one strand, the base T must occur at the same level in the second strand, and every time G occurs in one, C must be present in the second. Thus the sequence in one strand is determined completely by that in the other. The two strands are complementary; if the sequence is $\cdots$ A-G-C-A $\cdots$ in one, it must be $\cdots$ T-C-G-T $\cdots$ at the corresponding place in the other.

In *replication* of DNA during cell reproduction, the paired strands first separate. Then free deoxyribonucleotides, present in the cell as their triphosphates, are polymerized by special enzymes present and two new strands are created, each one complementary to and paired with one of the original strands. Thus there are now two DNA molecules for every one present initially, each identical to the original one.

Protein synthesis is effected by *transcription* of DNA. Molecules of DNA may contain from about 10^4 to more than 10^6 base pairs, depending on the nature of the organism. The particular sequence of bases in each strand constitutes the unique genetic information repository, the specific coded instructions that determine protein structure and synthesis and, thereby, all the characteristics of each unique organism. The marvelous mechanism by which this code is interpreted and protein synthesis effected begins with the creation of many special RNA molecules, called *messenger RNA* (m-RNA), each complementary to an appropriate section of the DNA. In RNA, the base complementary to A is U rather than T. Thus the base pairs formed as RNA is synthesized on DNA are A-U, T-A, G-C, and C-G, where the bases associated with DNA are given first and those associated with RNA given second in each pair. These messenger RNA molecules, patterned uniquely by the particular DNA present during their formation, then initiate protein synthesis (considered briefly at the end of Sec. 27-2).

Lipids.   The lipids constitute a heterogeneous group of compounds, the only common feature of which is that they are all soluble in benzene and ether and are relatively insoluble in water. The most abundant subclass of lipids includes esters (and some amides) of long-chain carboxylic acids with glycerol and with other alcohols and amino alcohols. Some of these compounds contain phosphate groups as well. Long-chain acids are often called *fatty acids* because their triesters with glycerol (triglycerides) constitute common animal and plant fats and oils (Chap. 26). Among the other materials classed as lipids are long-chain hydrocarbon waxes, certain kinds of animal and plant pigments, and the steroids, a group of compounds containing four rings that includes a number of hormones and related substances.

We restrict this brief discussion to lipids containing fatty acids. Most of these acids have an even number of carbon atoms; the great majority have unbranched chains, with C_{16} and C_{18} acids predominating. Some of these acids have one or more double bonds, as mentioned in the previous chapter; they are found especially in seeds and in cell membranes, chiefly as esters.

Lipids have two principal natural functions, serving for energy storage and as components of macroscopic structures. Fats (triglycerides) are the main source of stored energy in animals and in the seeds of plants. At least 10 percent of the body weight of most mammals is normally in the form of triglycerides, which can be utilized rapidly to meet demands for energy. No other form of food can be stored in comparable quantities; for example, the capacity of an animal to store carbohydrate in the form of glycogen is usually less than a tenth as great. Furthermore, oxidation of the hydrocarbon chains of lipids produces a higher energy yield than is available from other foods.

Lipids also have an important structural role—particularly in cell membranes and at other interfaces, where they are often bound to proteins. Phospholipids are especially abundant in membranes and in nerve and brain tissue. The vital role of lipids in various membranes is attributable to the fact that they contain both non-polar and polar regions (Fig. 27-4). The long hydrocarbon chains are *hydrophobic* (Greek: *hydor,* water; *phobos,* fear, flight)—i.e., they tend to avoid water, showing a preference for hydrocarbons or ether. In contrast, the polar portions of lipid molecules, the ester and phosphate groups, show an affinity for an aqueous medium; they are *hydrophilic.*

FIGURE 27-4 **Drawings of models of two lipids.** The black dots represent carbon atoms, with the appropriate number of attached hydrogen atoms. The lines between the dots denote single or double bonds, and the curved contours outline the overall shape of the molecule. In both structures, the saturated fatty acid chains are more regular in shape than the unsaturated ones, which have all double bonds in the cis configuration. Lipid (*a*) is a triglyceride, with three fatty acid chains. Lipid (*b*) is also a triester of glycerol, but one of the hydroxyl groups of glycerol has been esterified by phosphoric acid, which is esterified as well by the amino acid serine, $HOCH(NH_2)COOH$.

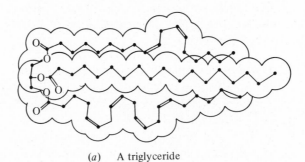

(*a*)   A triglyceride

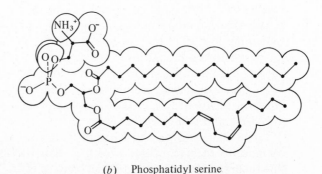

(*b*)   Phosphatidyl serine

Thus the lipid molecules are easily arrayed in layers or double layers (Fig. 27-5), which may be given additional stability if the molecules are somehow bound together, as by interaction with the nonpolar portions of proteins. A schematic model of a membrane is depicted in Fig. 27-5. Membranes are not passive, static structures. They are flexible and somewhat elastic and perhaps even at times somewhat fluid, and there is usually a continual and highly selective transport of ions and molecules through them as the metabolism of the cell proceeds. Lipid structures play a key role in these processes. Much still remains unknown about the organization, structure, and properties of membranes, which are the subject of a great deal of current research.

Lipids serve in part as thermal insulators—for example, in fat deposits beneath the skin in aquatic mammals—and they also function as buffers against mechanical shock by forming deposits around sensitive organs.

27-2 Amino Acids and Proteins

As mentioned in the previous chapter, the naturally occurring amino acids from which proteins are formed can be written generally as

$$R-\underset{\underset{H}{|}}{\overset{\overset{NH_3^+}{|}}{C}}-COO^- \tag{27-7}$$

The amide linkages joining amino acids together in proteins, illustrated in formula (26-41), are called *peptide bonds* because they are broken during digestion (Greek:

FIGURE 27-5   **A highly schematic representation of a portion of a membrane.** According to this model, some membranes consist of two layers of phospholipid molecules, with their polar layers outward and their hydrophobic tails in contact. The different molecules in this double-layer sandwich are held together in part by proteins, presumably through interaction at their hydrophilic ends as well as by nonpolar interactions.

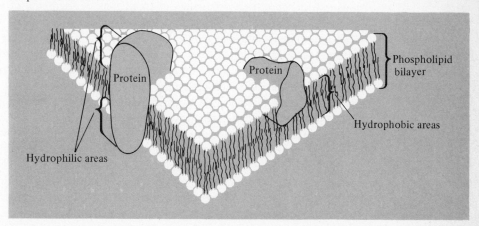

peptein, to digest). The general name *polypeptides* is given to molecules containing many amino acids so linked, while terms like dipeptide and tripeptide are used for molecules with just two or three amino acids. The word *protein* is reserved for polypeptides containing more than about 50 amino acids.

It is a remarkable fact that every molecule of a pure, biologically active protein or other polypeptide from a given source is identical—the same amino acids, joined together in precisely the same sequence to form a chain of the same length, or sometimes several distinct chains, assembled together to make one protein molecule. When some variation occurs—e.g., the substitution of one amino acid for another—it occurs because there has been a change in the code that dictates the specific amino acids to be used, in sequence, to build that polypeptide or because there has been an error in reading the code.

In this section we first describe the formulas of representative amino acids and a few simple peptides and examine some typical three-dimensional structural features of proteins. We then consider some of the many different roles played by proteins and discuss their synthesis briefly. Their catalytic function as enzymes is the topic of Sec. 27-3.

Amino acids and simple peptides.   The 20 amino acids that are commonly found in proteins and for which genetic code words are known to exist vary in the nature of the side chain R in (27-7). Table 27-3 gives the common names and accepted abbreviations for 13 representative amino acids and the formulas of their side chains. It also lists the names of the other protein amino acids that are chemically similar to those whose formulas are given; their structures are not of concern here.

As Table 27-3 shows, some side chains are nonpolar aliphatic or aromatic radicals. Others contain —OH, —SH, amide, or acidic or basic groups, all of which have hydrogen-bonding potentialities. Furthermore, at physiological pH, somewhat above 7, glutamic and aspartic acids are negatively charged because of ionization of the second carboxyl group; lysine and arginine bear a net positive charge because functional groups in their side chains are sufficiently strong bases to pick up a proton even at higher pH values. When these acidic and basic amino acids are incorporated into proteins, the side chains retain their charges under normal physiological conditions and hence attract oppositely charged ions or segments of proteins.

The 20 amino acids in Table 27-3 are found in a wide variety of proteins. A few other amino acids occur in particular proteins; for example, 4-hydroxyproline constitutes about 12 percent of collagen, an important protein in connective tissue in animals. All the amino acids found in proteins except glycine are optically active and all have the same configuration at the asymmetric carbon atom bearing the amino and carboxyl groups, arbitrarily designated the L configuration. No D-amino acids, those with opposite chirality, are found in any protein, although some do occur in a few smaller peptides. For example, D-phenylalanine is a component of gramicidin-S, a cyclic decapeptide that is an effective antibiotic. Other naturally occurring peptides include the tripeptide glutathione (a widely distributed biological reducing agent that has the sequence Glu-Cys-Gly) and a number of hormones—ranging from the relatively simple nonapeptide oxytocin, which stimulates lactation and uterine

CH$_2$-CHCOO$^-$

HOCH   NH$_2$$^+$

CH$_2$

4-Hydroxyproline

TABLE 27-3 Representative amino acids of proteins* General formula: $R\overset{\overset{\displaystyle NH_3^+}{|}}{C}COO^-$
$\underset{\displaystyle H}{|}$

R—	Name of amino acid	Common abbreviation	Other amino acids with similar chemical nature
H—	Glycine	Gly	
CH_3—	Alanine	Ala	
$(CH_3)_2CHCH_2$—	Leucine	Leu	Valine (Val) Isoleucine (Ile, Ilu)
†	Proline	Pro	
$C_6H_5CH_2$—	Phenylalanine	Phe	Tryptophan (Try)
$CH_3SCH_2CH_2$—	Methionine	Met	
$HSCH_2$—	Cysteine	Cys	
$HOCH_2$—	Serine	Ser	Threonine (Thr)
$HO\!-\!\bigcirc\!-\!CH_2$—	Tyrosine	Tyr	
$H_2NCOCH_2CH_2$—	Glutamine	Gln	Asparagine (Asn)
$HOOCCH_2CH_2$—	Glutamic acid	Glu	Aspartic acid (Asp)
$H_2NCH_2CH_2CH_2CH_2$—	Lysine	Lys	Arginine (Arg)
$HC\!=\!CCH_2-$ $\;\;\;\;N\;\;\;NH$ $\;\;\;\;\;\;\;C$ $\;\;\;\;\;\;\;H$	Histidine	His	

---

*Arranged with the less polar side chains near the top of the list and the more polar ones near the bottom.

†Proline contains a ring that incorporates the amino-nitrogen atom and so cannot be represented by the general formula above. It is $H_2C\overset{\displaystyle CH_2}{\underset{\displaystyle CH_2}{\diagup\diagdown}}\overset{+}{N}H_2$ ... $C\!-\!COO^-$

contraction, to the adrenocorticotropic hormone (ACTH), containing 39 amino acid residues, which stimulates the surface layers or cortex of the adrenal gland.

The metabolic systems of plants are able to manufacture all the amino acids needed to build the essential plant proteins. However, human beings can synthesize in adequate quantity only 12 of the amino acids they require and must therefore get the other 8 (Val, Leu, Ile, Phe, Try, Met, Thr, Lys) in the diet, either in proteins or in other sources from which the amino acids can readily be obtained. If these eight indispensable amino acids are not present in adequate quantity in the diet, the body is unable to manufacture essential enzymes and other proteins containing these amino acids. This malnutrition leads first to general debility and eventually to death. Other animals must also have certain amino acids in their diets.

Proteins. Proteins are such complex molecules, with molecular weights usually in the tens or hundreds of thousands, that until the 1950s no one was sure it would ever be possible to learn the details of the chemical structure and three-dimensional architecture of any of them. There was serious debate about whether all molecules of a particular protein from a given source, e.g., hemoglobin isolated from the blood of a particular individual, were identical in chemical composition and structure or whether there might be some variations. It is now known that they are indeed all the same if they are programmed by identical genes during their biosynthesis.

With the improvement in methods of purification of proteins and the development of chromatographic methods for the separation and quantitative estimation of amino acids and simple peptides in the 1940s and 1950s, precise analyses became possible and the exact amino acid composition could be determined even for a protein of high molecular weight. Techniques were then worked out for discovering the exact sequence of the amino acids in each protein chain. This sequencing is now done chiefly by breaking specific bonds by enzyme action and identifying the fragments present. The amino acids at the ends of the original chains, or the ends of the chains in each fragment, can be identified because their free amino or free carboxyl groups react with special reagents. The process of deducing from these data the original sequence of amino acids in the intact protein is in some ways like putting together a giant jigsaw puzzle—with a unique solution possible if and when all the pieces have been properly found and identified. In the last two decades, the precise sequence of amino acids has been determined for each of around a thousand pure proteins. This sequence, including the location of any disulfide bridges that may link cysteine residues within one chain or between nearby chains, is called the *primary* structure of the protein.

Proteins play two quite distinct roles: some serve only as structural materials; others are biologically active as catalysts or regulators of metabolic reactions, as antibodies to protect the organism from the effects of foreign materials, or in other ways. Some structural proteins have protective functions (such as the proteins in skin), some are parts of connective tissue, and still others are part of the motive machinery that constitutes muscle. These structural proteins are almost always fibrous in nature, composed of long-chain molecules joined together by hydrogen bonding or other cross-linking into extended filaments, which may in turn be coiled together to make insoluble fibers and other macroscopic structures.

Biologically active proteins are usually ellipsoidal in shape. Their polypeptide chains are folded and twisted to form a globular mass rather than an extended one. The biological activity of most globular proteins is carried on in aqueous biological fluids. The folding of the polypeptide chains is such that the amino acids with polar side chains tend to be on the outside of the molecule, where these polar side chains make the molecule water-soluble, while the nonpolar amino acids, especially those with large side chains, are usually in the interior of the molecule (Fig. 27-6).

Certain specific hydrogen-bonded arrangements of the polypeptide chains are characteristic of fibrous proteins—for example, the α helix and several sheetlike structures (Fig. 27-7). The α helix is the predominant structural feature of the protein α-keratin, which is present in wool, feathers, claws, and similar materials. This stable helical arrangement is also found to varying extents in many globular proteins. About

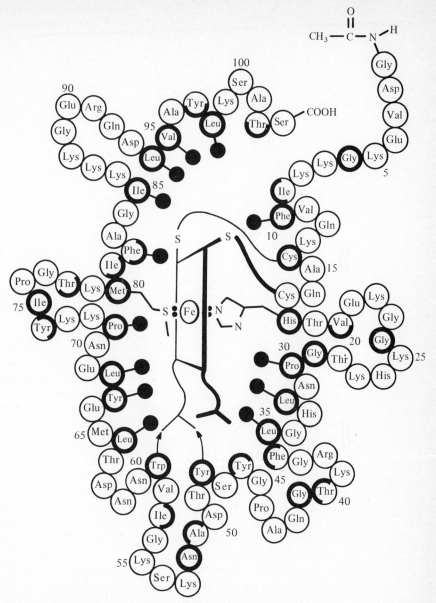

FIGURE 27-6 **Amino acid sequence in cytochrome c.** This schematic diagram shows the sequence of the 103 amino acid residues in tuna reduced cytochrome c, an oxidation-reduction enzyme. A circle ringed in black represents an amino acid in the interior of the molecule; those without such rings indicate amino acids on the outside. Note that the nonpolar amino acids (Table 27-3) tend to be in the interior whereas the polar ones congregate on the exterior.

The square group in the middle of the figure represents a heme group (Fig. 25-5), a group of four heterocyclic rings joined together in such a way that their four nitrogen atoms can coordinate to a central iron atom. Octahedral coordination of the iron atom is completed in this molecule, as indicated, by one of the nitrogen atoms of a His residue and a sulfur atom in a Met residue. The heme group is bound to the protein by covalent bonds and hydrogen bonds. The iron atom in this complex is oxidized and reduced when cytochrome c participates in the electron-transport chain by which ATP is synthesized. (*Reproduced by courtesy of R. G. Dickerson.*)

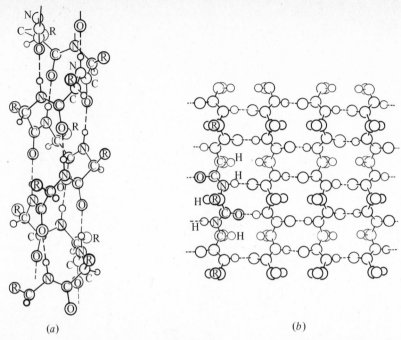

FIGURE 27-7   **(a) The right-handed α helix, found in many proteins; (b) a portion of a sheet structure, found in silk and other proteins.** The side chains, denoted by R, extend out from the helix and above and below the sheet. Hydrogen bonds between NH and CO groups in successive turns of the helix help to stabilize this structure. Similar N-H---O hydrogen bonds between different chains hold the sheet structure together.

80 percent of each polypeptide chain in the oxygen-transport blood protein hemoglobin and the related muscle protein myoglobin (Fig. 27-8) is in α-helical form, in eight separate segments of varying length; on the other hand, less than 10 percent of the electron-transport agent cytochrome c is helical. Silk fibroin has the sheet structure illustrated in Fig. 27-7b, which is also found to a limited extent in a wide variety of globular proteins.

These helix and sheet structures, and other similar specific hydrogen-bonded arrangements involving the polypeptide backbone, are referred to as *secondary* structures in proteins. The intricate folding and convoluting of the polypeptide chain of each globular protein that is brought about by interactions involving the side chains and that gives the protein its characteristic overall shape (for example, that shown in Fig. 27-8) are called the *tertiary* structure. The secondary and tertiary structural features of different proteins seem to be determined entirely by the primary structure. In other words, the formation of helical regions and hydrogen-bonded sheets and the folding and twisting of the chains in globular proteins are dictated by the sequence of amino acids, which is in turn determined by the sequence of nucleotide base pairs in the DNA of the organism. When a genetic mutation occurs, it produces a change in one of the base pairs and hence a corresponding substitution of one amino acid by another. If this change merely replaces one amino acid by

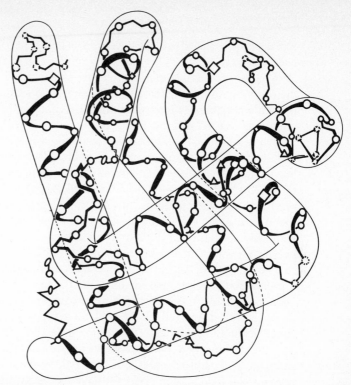

FIGURE 27-8 **The tertiary structure of myoglobin and many other globins.** The different circles, diamonds, and triangles represent the —CHR— groups of different amino acids in the myoglobin chain. The links joining them represent the peptide bonds, —NHCO—. The extension of side chains outside the helix is not indicated. Note the distinct helical regions, joined by randomly coiled segments of the polypeptide chains. Myoglobin, an oxygen-storage protein in muscle, is closely related to the blood protein hemoglobin and other globins in a great variety of organisms. Although their amino acid sequences are in some cases quite distinct, certain key positions in the sequences are invariant in all these molecules and the other substitutions do not affect the tertiary structure in more than minor ways. Doubtless all these molecules have a common ancestor. The preservation of this intricate molecular shape indicates that the structural features found in these molecules are those essential to their biological functioning. *(From M. O. Dayhoff (ed.), "Atlas of Protein Sequence and Structure", p. 18, National Biomedical Research Foundation, Bethesda, Md., 1967–68.)*

another of similar chemical structure, the secondary and tertiary structure of the protein may be unaltered and its biological functioning remain unchanged. If, on the other hand, the mutation replaces a nonpolar amino acid by a polar one, the secondary and tertiary structure of the protein may be drastically altered and hence its biological activity may be changed in a major way, since this activity depends on the three-dimensional architecture of the molecule.

The effect of a change in a single amino acid in a protein chain on the properties of a globular protein is dramatically illustrated by the contrast in properties between hemoglobin molecules (hemoglobin A) in the blood of normal adult humans and those (hemoglobin S) in the blood of people who suffer from sickle-cell anemia. Each hemoglobin molecule has four polypeptide chains. The only difference between hemoglobins A and S is that in two of the four chains a glutamic acid residue in

A has been replaced by a valine in S. There are 574 amino acid residues in the hemoglobin molecule. This alteration of just two of them, from acidic to nonpolar, changes the properties of the molecule sufficiently in the deoxygenated state that the solubility of S is decreased by a factor of 25 relative to that of A. The abnormal molecule polymerizes into microfilaments that combine to form a rodlike insoluble mass that distorts the red blood cell into a sickle shape. The sickled cells cannot easily pass through the capillaries, so vital organs are deprived of their normal blood supply and gradually deteriorate. Individuals who have only hemoglobin S seldom live beyond the age of 40, and about half of them die in infancy. On the other hand, those who have half normal hemoglobin A and half hemoglobin S (because they received from their parents one A-forming gene and one S-forming gene, rather than two A genes or two S genes) usually do not have the serious problems of those with sickle-cell anemia, although sickling of their red cells may occur at very low oxygen pressure (for example, at high altitudes).

The survival of any protein molecules as potentially lethal as hemoglobin S in a population suggests that in some environment they must confer an evolutionary advantage. This turns out indeed to be true. When the parasite that causes malaria is developing in the red blood cells, the oxygen pressure is reduced sufficiently to cause sickling of many cells containing hemoglobin S. Such cells are then attacked by other cells called phagocytes, whose function is to engulf and destroy alien particles. The malarial parasite is destroyed at the same time. Individuals who have only normal hemoglobin A do not have this protective mechanism against malaria, so that the population in malarial regions tends to be enriched in individuals with one A gene and one S gene. Mating of two of these individuals can produce in the next generation individuals with gene combinations SS and AA as well as AS. This explains why the incidence of sickle-cell anemia in the United States is highest in the black population descended from individuals who came from malarial regions of Africa.

As mentioned in Sec. 27-1, protein synthesis is initiated by messenger RNA molecules that are complementary to the DNA in the appropriate genes. An ingenious series of experiments has shown that each *codon,* the unit that specifies which amino acid is to be added to the growing chain at a particular point, is a sequence of three bases in the messenger RNA molecule. For example, glutamic acid is coded either by the triplet G-A-A or by the triplet G-A-G; if the central base in either of these triplets is changed to U, giving G-U-A or G-U-G, the amino acid specified by the codon is valine. Thus the genetic change responsible for the production of hemoglobin S in place of hemoglobin A must have been a change at one point in the DNA such as to change the central base in this triplet at one point in the complementary RNA from A to U, with the result that a particular glutamic acid in each of two chains of each hemoglobin molecule is replaced by valine.

Since there are four different bases in RNA, there are $4^3 = 64$ different possible triplets of bases to be used to code for the 20 different amino acids that occur in most proteins. Most of these amino acids can be specified by several different codons, as in the examples just cited. The complete code has been deciphered. Sixty-one of the sixty-four possible triplets are used for encoding amino acids; the other three are used for terminating or interrupting protein synthesis.

The details of the way in which information in DNA is translated into a synthesized protein are exceedingly complex and far from completely understood. They involve the orderly interactions of more than 100 different macromolecules (including various kinds of RNA and proteins) and many small molecules and ions (including nucleoside triphosphates and Mg^{2+}).

27-3 Enzymes

Without the catalytic effects of enzymes, life would be impossible because almost all the reactions that occur in organisms would proceed at negligible rates under the mild conditions of temperature, pressure, and pH characteristic of normal cells. Indeed, most common poisons act by interfering with just one, or a few, of the thousands of enzymes in the body, diminishing the catalytic activity sufficiently that the corresponding reaction is effectively stopped. Some aspects of the kinetics of enzyme-catalyzed reactions were discussed in Sec. 20-7, with emphasis on the essential role of the complex formed by the reacting (substrate) molecule with the enzyme at the so-called active site. In this section we consider briefly the nature of enzymes, their often remarkable specificities, and ways in which their activity can be inhibited.

The nature of enzymes.   More than 1000 enzymes have been characterized and more than 100 of them have been studied in considerable detail. All are proteins; many require for effective catalytic activity the presence of one or more additional components, called *cofactors*. Some cofactors are firmly bound to the protein—for example, a heme group in some enzymes or a dinucleotide fragment in others. Other cofactors are small organic molecules, easily separable from the protein and referred to as *coenzymes*. Many coenzymes are closely related to vitamins; if they cannot be synthesized by the body, they must be in the food supply, although the amounts need only be small since they act catalytically in concert with the enzyme. Still other cofactors are metallic ions, such as K^+, Ca^{2+}, or Mn^{2+}, which may be bound to the protein through carboxyl, hydroxyl, amino, or phosphate groups.

Enzymes may be divided into three general categories on the basis of their size and complexity. The simplest, which constitute a rather small group, consist of only a single polypeptide chain, with molecular weight typically in the range 10^4 to 4×10^4. All have hydrolytic functions; some break down carbohydrates, others proteins, and still others nucleic acids. A second, larger group consists of *multisubunit enzymes*—that is, enzymes containing several independent polypeptide chains, some of which may be identical. The enzyme activity is associated only with the assembled entity of subunits; the separated chains are catalytically inactive. Finally, there are in many cells *multienzyme complexes*—highly structured groups of enzymes arranged so as to facilitate easy transfer of substrates from one enzyme to another during a sequence of reactions within a cell.

The catalytic activity of enzymes is quite remarkable in a number of ways. As indicated, they act under mild conditions, often thousands of times more effectively than the best nonenzyme catalysts for the same reactions. Most enzymes will cause the reaction of around 100 to 1000 substrate molecules every second for each molecule of enzyme present, but some can process (one at a time) as many as 10^6 or even 10^7 molecules per second. Among the fastest are catalase, which catalyzes the decomposition of hydrogen peroxide to water and oxygen and prevents the accumulation of H_2O_2 in cells, and carbonic anhydrase, which promotes the equilibration of H_2CO_3, CO_2, and water, a process important in respiration.

Like all catalysts, enzymes are needed only in small amounts because they are continually regenerated as the reaction they catalyze proceeds. An enzyme or any other protein that loses its biological activity without breaking of the polypeptide chain is said to be *denatured*. Denaturation is likely to occur upon change of pH or temperature because such changes easily disrupt the tertiary (and sometimes the secondary) structure of the protein (e.g., by breaking intramolecular hydrogen bonds) and the activity is sensitively dependent on the details of the tertiary structure. Most mammalian enzymes are optimally active at physiological pH and at temperatures near 37°C. Enzymes in organisms adapted to environmental extremes sometimes have optimal temperatures for their catalytic activities admirably suited to their environment. For example, algae in hot springs contain enzymes that function well at temperatures that would destroy most mammalian enzymes whereas certain enzymes in bacteria that grow in the arctic are most active near 0°C.

Enzyme specificity.   Unlike many catalysts, enzymes are often highly specific. Some have what is called absolute specificity: they will catalyze the reaction of only one particular molecule, and not others, no matter how closely those others resemble the usual substrate in structure. Urease, for example, will decompose only urea, and glucokinase catalyzes only the transfer of a phosphate group from ATP to D-glucose to form glucose-6-phosphate. Other enzymes will catalyze the reactions of a group of compounds more or less closely related in structure. Hexokinase promotes the transfer of phosphate from ATP to glucose and to certain other hexoses as well, but not to any other sugars less closely similar to glucose. One of the digestive enzymes will break down almost any protein, although it will not degrade other amides; other digestive enzymes are more selective, breaking peptide bonds only if they occur adjacent to particular amino acid residues.

The specificity of an enzyme is a consequence of its highly intricate tertiary structure. The folding of the polypeptide chain brings together amino acid residues from quite different regions of the chain, thus providing at once an environment that can activate the substrate for reaction and a uniquely shaped template into which only a molecule of the proper shape and proper polar character can fit snugly. Because enzymes are themselves chiral objects, composed of L-amino acids, their specificity extends to what has been termed *chiral recognition*. They distinguish clearly between optical isomers, favoring those found in the natural environment with which they have been evolved to deal. Glucokinase will not catalyze the transfer of phosphate to L-glucose, and no digestive enzyme will attack a polypeptide made of D-amino acids.

Enzyme inhibition.   Substances that interfere with the catalytic effects of enzymes are termed inhibitors. Some are molecules that do not react like the normal substrate but resemble it so closely in shape that they can attach themselves to the enzyme molecules at the active sites and thus compete with the normal substrate for these sites. As the concentration of such an inhibitor is increased, it competes more and more effectively for the active sites on the enzyme and thus reduces more and more the rate of reaction of the substrate. Increase of substrate concentration will counteract this trend by Le Châtelier's principle. Sulfanilamide, one of the first of the sulfa drugs that were so important in treating bacterial infections in the 1930s and 1940s, competes with a molecule of similar shape and hydrogen-bonding properties—4-aminobenzoic acid—that bacteria normally need for biosynthesis of essential amino acids and nucleic acid bases.

Many inhibitors are noncompetitive. They alter some enzyme, often irreversibly, in such a way that its biological activity is destroyed. These include such toxic materials as CN^-, which combines essentially irreversibly with transition metals in various enzymes, and heavy-metal ions such as Ag^+ or Pb^{2+}, which inactivate the —SH groups of cysteine residues in proteins. Nerve gases, which are alkylfluorophosphates, $(RO)_2POF$, react with the hydroxyl groups of serine residues in the enzyme acetylcholine esterase (among others), thereby preventing the regeneration of the resting state of nerve fibers after a nerve has transmitted an impulse. The consequence is that all muscular control is lost, including that over such vital functions as breathing and heart action.

Although noncompetitive inhibition is often more dramatic because of the spectacular effects of poisons, competitive inhibition of enzyme activity is of much greater significance since it plays a key role in the regulation of metabolism in all organisms. One of the most important regulatory mechanisms is feedback inhibition: a product of a long sequence of reactions is a competitive inhibitor for an enzyme that catalyzes one of the initial reactions in the sequence. Thus, as this product accumulates in the organism, its rate of synthesis is diminished; as it is depleted, its synthesis is again enhanced. The availability of the product in the cells can thereby be maintained at a level appropriate for whatever biochemical function the substance is to serve.

The nature of the active site and the mechanism of enzyme action. Within the last decade, the three-dimensional molecular structures of more than 20 enzymes and complexes of enzymes with molecules that resemble their usual substrates have been worked out by X-ray diffraction studies of single crystals, and many more such studies are in progress. Because the experimental data are collected over periods of weeks or months, the resulting structural models are of a time-averaged, essentially static, situation. They show the details of the structure of the enzyme and, in some cases, of its complex with a molecule much like its usual substrate but with which the enzyme does not react during the experiment. The mechanism of interaction with the proper substrate must be inferred, but the evidence on which to make this inference, at least for the initial stage of the reaction, is concrete and highly detailed. We are thus in the paradoxical situation of knowing more about the structures of the initial reaction intermediates for some of these hitherto mysterious reactions involving molecules of enormous complexity than about the activated complexes or intermediates for most of the simplest reactions not involving enzymes.

For example, it has been known for some time from studies of chemical reactivity that the closely related digestive enzymes chymotrypsin and trypsin are inactivated if any alteration is made in two particular amino acids that occur in the polypeptide chain of each, a histidine residue that is the 57th amino acid from one end and a serine that is 195th from that end. When the structures of these enzymes were determined in detail, it was found not only that the two molecules are remarkably similar in overall shape and tertiary structure, but that the molecular conformation of each is such that these two critical amino acid residues, although widely separated along the chain, are brought close together by the folding of the chain. A third residue, an aspartic acid at position 102, is also in this same region of space. These three residues are intimately involved in the mechanism by which these enzymes, and some others closely similar to them, hydrolyze proteins. They thus constitute a portion of the active site of each enzyme.

The region around the active site of an enzyme has two functions: recognizing the appropriate substrate or substrates specific for that enzyme and accelerating the appropriate reaction of that substrate. The specificities of trypsin and chymotrypsin are somewhat different, and the difference can be explained in terms of the shape and lining of a cleft surrounding the active site in each molecule. In chymotrypsin, which is specific for splitting peptide bonds adjacent to large nonpolar side chains, this cleft is fairly large and is lined with hydrophobic (nonpolar) amino acids. In trypsin, on the other hand, the cleft has a negatively charged aspartic acid residue in its interior. This is consistent with the fact that trypsin will hydrolyze peptide chains only at positions adjacent to positively charged side groups. Other digestive enzymes closely related to these two are also known, and their specificities can be correlated with their molecular shapes and structural details as well.

While no quantitative explanation has yet been made of the rate enhancement characteristic of an enzyme-catalyzed reaction, qualitative interpretations are now available for a few that have been studied in sufficient detail. They are provided by schemes developed from a careful analysis of the structural details of each molecule and the relative positions of the enzyme and its substrate in the activated complex or intermediate, as well as from other chemical information about the specificity of the enzyme and its inhibitors.

27-4   Bioenergetics and the Role of ATP

In this section we consider some of the principles, compounds, and reactions involved in the flow of energy through organisms. Living cells are open systems. They exchange matter and energy with their environment and are not at equilibrium, although to a good approximation they are in a steady-state condition. A cell is a very efficient engine, operating at nearly constant temperature, pressure, and volume, and is thus unlike most engines contrived by man. Nevertheless, the thermodynamic principles that were developed originally from consideration of heat engines apply to living systems.

Light from the sun is the ultimate energy source upon which all life depends.

Photosynthetic plant cells and some bacteria use this energy directly for the synthesis of essential organic molecules by reduction of carbon dioxide, with water or another inorganic compound as the reducing agent. All higher animals, most microorganisms, and even the nonphotosynthetic cells of plants utilize oxidation-reduction reactions to supply the energy they need, chiefly through the oxidation of reduced carbon compounds such as glucose by atmospheric oxygen, producing ultimately CO_2. They are dependent on the photosynthetic cells for the supplies of glucose and other reduced carbon compounds and for returning oxygen to the atmosphere. In turn, the photosynthetic cells are dependent on the others for resupplying carbon dioxide to the atmosphere. This symbiotic (Greek: *symbiosis,* state of living together) inter-dependence channels solar energy indirectly into the energetic demands of animals—for biosynthesis, for transport of chemical substances throughout the organism, for muscle action, and for nerve-impulse transmission—through the gradual release of the very considerable energy available by combustion of glucose and similar reduced substances.

The fact that this chemical energy is made available gradually is crucial, just as it is for any other useful energy source, such as a battery. The energy from a battery is most useful if it is released under controlled conditions when and where it is needed, rather than by short-circuiting the battery or direct combination of the oxidizing agent and the reducing agent. Similarly, in a normal organism energy is made available in comparatively small amounts at times and places best suited to the needs of the organism. Many of the essential processes in an organism, e.g., the biosyntheses of various molecules, occur with an increase of free energy and thus are not thermodynamically possible unless they are coupled with some process that simultaneously provides sufficient free energy that there is an overall decrease in free energy. Many different molecules, most of them phosphate esters, are used to supply free energy sufficient to drive reactions in living cells essentially to completion. The most widespread of these "high-energy molecules" is ATP, whose formula was given in (27-6), page 734.

Adenosine triphosphate (ATP). This molecule appears to be present in every cell of every organism. It is continually being transformed to ADP (adenosine diphosphate) by loss of its terminal phosphate group, either as some inorganic phosphate-ion species (symbolized here as P_i) appropriate to the pH,

$$\text{ATP} + \text{H}_2\text{O} \longrightarrow \text{ADP} + \text{P}_i \qquad (27\text{-}8)$$

or through transfer of the phosphate to some other molecule.[1] At physiological pH, near 7, both ATP and ADP exist as anions with charges -4 and -3, respectively; the predominant species of each molecule in cells is its $1:1$ complex with Mg^{2+}, an ion present in appreciable concentration in intracellular fluid. It has been estimated that each person synthesizes and breaks down about his own body weight of ATP every day; its half-life in typical cells is of the order of 1 to 10 s.

---

[1] In (27-8) and subsequent equations, charges on ATP, ADP, and the inorganic phosphate species P_i are omitted.

Two properties of ATP are critical to its central role in the exchange of energy in organisms. (1) It has a relatively high standard[1] free energy of hydrolysis, $\Delta G^{\oplus\prime}$, about -31 kJ mol^{-1}. (2) Despite its thermodynamic instability, it is kinetically stable—that is, it is not broken down to ADP until an appropriate catalyst is present, so that it can be stored almost indefinitely for use when needed.

Since the breakdown of ATP in (27-8) occurs with a significant decrease in free energy, the reverse process (the synthesis of ATP from ADP and phosphate) requires free energy. Some ATP is made directly during photosynthesis in green leaves, with sunlight supplying the energy; we shall not discuss here the reactions involved. Most of the ATP in both plant and animal cells is made by coupling the reverse of reaction (27-8) with an oxidation reaction that provides sufficient free energy or with the splitting of a phosphate compound that has an even more negative free energy of hydrolysis than does ATP (Sec. 27-5). A typical phosphate compound of this kind, synthesized during the oxidative breakdown of glucose in cells (glycolysis), is 1,3-diphosphoglycerate:

$$\underset{\underset{O}{\overset{\displaystyle\|}{}}}{{}^{-2}O_3POCH_2\overset{\overset{\displaystyle OH}{\overset{\displaystyle |}{}}}{CH}COPO_3{}^{2-}} \qquad (27\text{-}9)$$

In the presence of an appropriate enzyme, this anion transfers a phosphate group directly to ADP to form ATP, with a significant decrease in standard free energy. The overall reaction is just the sum of the reverse of (27-8), for which $\Delta G^{\oplus\prime} = +31$ kJ mol^{-1}, and the hydrolysis reaction for the acyl[2] phosphate group involved, for which $\Delta G^{\oplus\prime} = -49$ kJ mol^{-1}:

$$\text{1,3-Diphosphoglycerate} + \text{ADP} \longrightarrow \text{3-phosphoglycerate} + \text{ATP} \quad (27\text{-}10)$$

The corresponding standard free-energy change, $\Delta G^{\oplus\prime}_{10}$, is

$$\Delta G^{\oplus\prime}_{10} = (-49 + 31)\ \text{kJ mol}^{-1} = -18\ \text{kJ mol}^{-1}$$

and thus ATP can be produced in good yield by this reaction. There are numerous other "high-energy" molecules used for storage of energy in chemical systems. They share with ATP the property of being kinetically stable in the absence of enzymes specific for the reactions in which their energy contribution is needed.

Biological oxidations of carbohydrates, fats, and other reduced molecules that serve as energy sources involve sequences of a dozen or so individual reactions. In each successive oxidizing stage, the oxidizing agent used is one whose oxidation potential is sufficient for the reaction involved but not much greater than needed. Were it much greater, the efficiency of recovery of the free energy liberated in the oxidation of the reduced molecules would be much smaller than it is.

[1] Standard conditions for hydrolytic reactions of phosphate esters in organisms have been defined as pH $= 7.0$, $t = 37°$C, the presence of an excess of Mg^{2+}, and all reactants and products at 1.0 F concentration. Corresponding values of the standard free-energy change are designated $\Delta G^{\oplus\prime}$. The actual free energy of hydrolysis of ATP under physiological conditions differs from the standard value because of variations in the concentrations of the reactants. It is usually of the order of -50 kJ mol^{-1}.

[2] An acyl group has the general formula $RC\overset{\overset{\displaystyle O}{\displaystyle\|}}{}$—; an example is the acetyl group, for which $R = -CH_3$.

27-5   Metabolic Pathways

The term *metabolism* (Greek: *metabole,* change) refers to the chemical reactions occurring as cells exchange matter and energy with their surroundings. These reactions serve a number of specific purposes, which can be described broadly as (1) the extraction of energy from food or sunlight and its storage in chemical form, as discussed in the previous section; (2) the conversion of nutrients provided to the cell into small molecules or portions of molecules, the precursors for making proteins, lipids, nucleic acids, and other specialized molecules needed by the cell; (3) the synthesis of the macromolecular components of the cell and the other molecules necessary for its proper functioning; (4) the disposal of undesirable materials; and (5) providing for various specialized functions of the cell and the organism.

Metabolic changes occur in a highly regulated manner and are remarkably organized in both space and time. As mentioned in Sec. 27-3, many multienzyme systems exist in cells, assembled in such a way that the product released by one enzyme becomes the substrate for a neighboring enzyme that is specific for the next reaction in a sequence. Many of these reaction sequences or pathways have been studied in great detail and have been found to be closely similar in all forms of life. They are often displayed in intricate metabolic flowcharts that show the reactants and products of each change as well as the enzyme specific for it. Our concern is not with the details of any of these pathways but rather with some general comments about them and about metabolic reactions.

Catabolism and anabolism.   Cells can both degrade or break down macromolecules and molecules of intermediate sizes and synthesize molecules from smaller components, as they are needed. These processes are called respectively *catabolism* and *anabolism* (Greek: *kata,* down; *ana,* up). It might seem logical that cells would utilize the same sequences of reactions, traversed in opposite directions, for both the degradation and the synthesis of macromolecules and other necessary cell components. However, there are always some differences in catabolic and anabolic pathways, although some of the individual steps may be catalyzed by the same enzymes and thus involve the identical reaction in opposite directions. The reason for these differences can be understood in terms of the thermodynamics of spontaneous reactions.

Any spontaneous process, whether it be a biosynthesis or a degradation, must occur with a decrease in free energy. Although some of the individual steps in the pathway in either direction may involve such small free-energy changes that they can be effectively reversed by variations in the concentrations of the reactants and products, others involve large decreases in free energy. The products are strongly favored in the latter reactions, which can be reversed only by supplying a large amount of free energy—for example, by coupling with some other reaction or by providing the free energy photochemically. Thus reactions in steps involving large free-energy changes invariably differ in catabolism and anabolism.

Degradative and synthetic pathways differ in other respects as well. They are usually situated in different parts of the cell and their rates are regulated independently. This permits parallel catabolic and anabolic pathways to operate simultaneously and independently of one another, with their rates individually dependent on the supplies of various molecules and of energy available and on the needs of the cell.

As indicated in the schematic diagram in Fig. 27-9, both catabolism and anabolism can be considered to take place in three general stages. In the first stage of catabolism, stage I in the diagram, the major nutrients of cells are broken down into their major components—proteins into their constituent amino acids, polysaccharides into sugars, and lipids into fatty acids, glycerol, and other molecules. In stage II, the products of stage I are degraded still further; the ultimate product of stage II can be a two-carbon acetyl group, which becomes attached by esterification to an —SH group of a molecule called *coenzyme* A, an adenine nucleotide with additional groups attached. Acetyl coenzyme A acts as an acetyl-transfer agent. It can, for example, convert a four-carbon acid to a six-carbon acid, as indicated in stage III. Stage III is a cyclic stage, in which entering acetyl groups are oxidized to two molecules of CO_2 and simultaneously the nucleotide NAD^+ (Fig. 27-1c) and another similar nucleotide are reduced. A series of coupled oxidations of these nucleotides (not shown in Fig. 27-9) then results in the production of a dozen molecules of ATP for each acetyl group oxidized, so that much of the energy released in the oxidation of the acetyl groups is eventually stored as the energy of ATP molecules.

FIGURE 27-9   **The three stages of metabolism.** Catabolic pathways are indicated with solid arrows, anabolic pathways with dashed arrows. Stage II and stage III are both catabolic and anabolic. For example, stage III, the citric acid cycle, is the final stage in the breakdown of food to CO_2 and can also furnish small molecules and fragments of molecules for biosynthesis. At physiological pH, citric acid and the other acids of this cycle are present principally in the form of their anions.

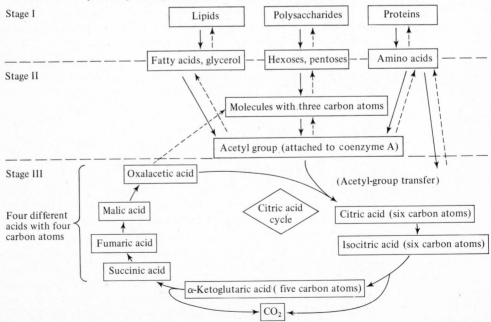

The citric acid cycle of stage III, sometimes called the Krebs cycle after Hans Krebs, the biochemist who proposed it in 1937, is catalytic with respect to the degradation of acetyl groups to CO_2 and liberation of hydrogen to NAD^+. Each acid in the cycle is regenerated in each complete turn of the cycle. These organic acids can play noncatalytic roles as well, serving as the starting materials for the biosyntheses of various essential molecules. For example, as indicated in Fig. 27-9, oxalacetic acid can be converted through three-carbon intermediates into sugars. The five-carbon dicarboxylic acid, α-ketoglutaric acid, is converted by enzymatic reduction in the presence of ammonia into L-glutamic acid (Table 27-3). The latter reaction is an essential one for the biosynthesis of all amino acids in all species, since it is the chief (if not the only) mode of formation of the amino group of an amino acid directly from ammonia. Most other such amino groups are introduced by transfer of the amino group from glutamic acid. This phenomenon of group transfer is common to many biosynthetic reactions. We have already mentioned examples of the transfer of phosphate and acetyl groups and the transfer of hydrogen atoms or electrons in redox reactions. Other commonly transferred groups are methyl, carboxyl, and formyl (—CHO).

The biosynthesis of large molecules involves the addition of small units to molecules that already exist. The units may be, for example, one- or two-carbon groups (such as formyl or acetyl), which are carried by specialized molecules like coenzyme A that act with the enzymes specific for the reaction in question. The added units may also be complete molecules that are, in essence, monomers for the macromolecule—for example, glucose in the buildup of glycogen or amino acids in the synthesis of proteins.

One of the three-carbon intermediates in both the degradation and the synthesis of carbohydrates is 1,3-diphosphoglycerate [Formula (27-9)], mentioned in the previous section as one of the molecules with enough energy to convert ADP to ATP efficiently [Equation (27-10)]. How is this high-energy diphosphate itself synthesized? By oxidative phosphorylation of glyceraldehyde-3-phosphate,

$$\overset{\displaystyle OH}{\underset{\displaystyle |}{^{-2}O_3POCH_2CHCHO}} \qquad (27\text{-}11)$$

that is, by coupling the oxidation of the aldehyde group in (27-11) with a reaction that transfers a phosphate to the resulting carboxylate group. The oxidizing agent is the ubiquitous NAD^+. The appreciable free energy released in the oxidation is sufficient to provide the energy needed to form the thermodynamically unstable acyl phosphate [Formula (27-9)]:

$$\text{Glyceraldehyde-3-phosphate} + NAD^+ + H_2O \longrightarrow$$
$$\text{3-phosphoglycerate} + NADH + 2H^+ \qquad (27\text{-}12)$$

with $\Delta G^{\oplus\prime}_{12} = -43 \text{ kJ mol}^{-1}$, and

$$\text{3-Phosphoglycerate} + P_i \longrightarrow \text{1,3-diphosphoglycerate} \qquad (27\text{-}13)$$

for which $\Delta G^{\oplus\prime}_{13} = 49 \text{ kJ mol}^{-1}$. Thus, the overall standard free-energy change for the oxidative phosphorylation, the sum of (27-12) and (27-13),

Glyceraldehyde-3-phosphate + NAD$^+$ + P$_i$ + H$_2$O $\longrightarrow$

$$1,3\text{-diphosphoglycerate} + NADH + 2H^+ \quad (27\text{-}14)$$

is $\Delta G^{\circ\prime}_{14} = +6 \text{ kJ mol}^{-1}$. This positive standard free-energy change is sufficiently small that the corresponding equilibrium constant is not far from unity and the reaction can be made to proceed in either direction with good yield, depending on the concentrations in the cell.

Regulation. The regulation of metabolic pathways, that is, the control of the rates at which the myriad catabolic reactions and biosynthetic reactions in the cell form their products, can occur in several different ways. At the simplest level, the rate of a particular metabolic sequence is a function of the pH and of the concentration within the cells of each enzyme of the pathway, of each substrate of these enzymes, and of all essential coenzymes and metal ions.

Many metabolic pathways, especially those involving biosyntheses, are controlled as well by particular regulatory enzymes, which normally catalyze a reaction at the start of a multienzyme sequence and are inhibited by an end product of the sequence. Thus the cell makes no more of each product than it needs at any time. Some regulatory enzymes are *multivalent,* that is, are stimulated or inhibited by two or more molecules, which might be the products of different pathways. The rate of one sequence can thus be made responsive to the integrated effect of several sequences.

The rates of many catabolic reactions are controlled by the immediate needs of the cell for energy in the form of ATP. The enzymes catalyzing these reactions are inhibited by excess ATP and are stimulated by ADP or AMP[1]. In other words, cells degrade food no more rapidly than necessary to provide the energy they need for their moment-to-moment activities. This is not to imply that an organism always lives "from hand to mouth". Metabolic regulation is such that cells tend to store reduced carbon as glycogen and fats in times when the supply of energy is plentiful and the need for carbon for biosynthesis is minimal and to call on these stored reserves as conditions change.

Another method of regulation of metabolic pathways is through control of the rate of enzymatic synthesis. Although many enzymes are present in essentially constant amounts in cells, others are synthesized only in response to need. The genes that specify their synthesis are normally repressed or inactive; certain molecules can, however, activate these genes—that is, start the synthesis of the enzymes of some sequence. This mode of control makes it possible for cells to operate economically in expenditure of energy and nutrients, utilizing metabolic pathways only when they are needed. For example, certain bacteria can synthesize all the nitrogenous compounds they need—amino acids, nucleotide bases, and others—from ammonia as the only source of nitrogen if that is all that is provided. However, if these bacteria are fed a mixture of different amino acids and other nitrogenous molecules, the enzyme system that can synthesize these essential compounds directly from ammonia is not needed and the manufacture of the enzymes is suppressed. If the concentrations

---

[1]Adenosine monophosphate.

of the essential compounds in the environment fall below critical levels, the manufacture of the enzymes is resumed.

Finally, in higher animals, hormones secreted internally by various endocrine or ductless glands can either stimulate or inhibit metabolic reactions in other cells. There are many such hormones; they control a great variety of activities, such as growth of bone and muscle, levels of blood sugar, muscle relaxation, secretion of salt and water, and sexual activity. The mechanism of hormone action is still little understood, but hormones clearly operate through regulation of the chemical activities of the cell.

Problems and Questions

27-1   Molecular weight of proteins.   Hemoglobin and myoglobin contain, respectively, 0.344 and 0.330 percent iron. (*a*) Calculate the minimum molecular weight of each protein. (*b*) Each hemoglobin molecule contains four Fe atoms and each myoglobin molecule contains one Fe atom. Calculate the actual molecular weight of each protein.

Ans. (*a*) 1.62×10^4, 1.69×10^4; (*b*) 6.5×10^4, 1.69×10^4

27-2   Acid-base properties of amino acids.   A representative protein-forming amino acid that has one amino group and one carboxyl group, such as alanine (Table 27-3), is a diprotic acid; that is, it has two acid dissociation constants. The usual form of the amino acid, represented in Table 27-3, picks up a proton at very low pH and loses one at very high pH. The pK's for alanine are about 2.3 and 9.6. Write structural formulas for the low-pH and high-pH forms of this amino acid, and suggest why the values of pK_1 and pK_2 are, respectively, in the ranges 2–3 and 9–10. (*Hint:* Compare the forms before and after dissociation of each proton with related species, such as acetic acid and ammonia.)

27-3   Important atoms in living systems.   Table 21-1 lists the 20 most abundant elements on the earth's surface in the order of their abundance. Nine of them are nonmetals, all of which, except silicon, are found in Table 26-1. Show how this striking difference in the behavior of silicon and carbon and the unimportance of Si—Si, Si—C, and Si—H linkages in living systems is consistent with bond energies (Table 11-2). In particular, contrast the energies of C—C and C—H bonds with those of C—O, C—N, C—F, and C—Cl. Then contrast the energies of Si—C and Si—H bonds with those of Si—O, Si—N, Si—F, and Si—Cl. (The bond energy for a Si—N bond is not well established, but it is probably within 10 percent of 440 kJ mol^{-1}.) What do the differences suggest?

27-4   Properties of water.   Water is the most important compound in all known organisms, typically constituting 60 to 90 percent of the gross weight of the organism. It is also important in the environment of most organisms. Three physical properties of water, among others, are important for many forms of life: the high heat of vaporization, the high heat capacity, and the fact that water has its maximum density at about 4°C. Discuss reasons for the importance of these properties, and consider the bearing of the hydrogen-bonding tendencies of water on these properties.

27-5   Acidity of nucleic acids.   Nucleic acids are diesters of phosphoric acid. Esters are not usually acidic. Why are nucleic acids acidic, or is this name inappropriate?

27-6 **Genetic code.** The codons in the genetic code are sequences of three bases in the messenger RNA molecules. Explain why sequences of only two bases could not be used to code for the 20 different amino acids commonly found in proteins.

27-7 **DNA for hemoglobin S.** What base change must have occurred in the corresponding DNA to produce the messenger RNA described in the text as responsible for the genetic change from hemoglobin A to hemoglobin S? *Ans.* thymine to adenine

27-8 **Anions of citric acid.** Citric acid is a triprotic acid, $C_3H_4(OH)(COOH)_3$, with pK values 3.13, 4.77, and 6.40. Verify the assertion in the legend of Fig. 27-9 that the acids of the citric acid cycle exist principally in the form of their anions by calculating the fraction of citric acid in the form of the -3 ion and the fraction in the form of the -2 ion at pH $= 6.9$, the pH of a typical liver cell.

27-9 **Biochemical molecules.** Indicate the important structural features and a principal biological role of the following kinds of molecules: monosaccharides, amino acids, fats, disaccharides, ribonucleotides, enzymes, polysaccharides, fibrous proteins.

27-10 **Metal ions.** Discuss the biochemical function of various metal ions of both representative metals and transition metals. Indicate why many heavy-metal ions are poisons with cumulative effects (that is, the ions are not effectively eliminated from the system even over comparatively long periods of time).

27-11 **Length of a DNA molecule.** The average distance between base pairs measured parallel to the axis of a double-helical DNA molecule is 3.4 Å. The average molecular weight of a pair of nucleotides is about 650. What is the approximate length in millimeters of a single DNA molecule of molecular weight 2.8×10^9 (a value typical for the DNA of some bacteria)? About how many base pairs does this DNA contain? *Ans.* 1.5 mm; 4.3 million

27-12 **Hydrogen bonding.** Indicate the number and type of hydrogen bonds formed per amino acid unit in the alpha-helical and the sheet-structure portions of proteins. Indicate also the number and type of hydrogen bonds formed between the complementary base pairs in DNA.

"Methanol has a heating power about half that of gasoline. However, it is denser than gasoline, so that 2 gallons of methanol do not require quite as much storage space as 2 gallons of gasoline."

Newspaper account

"As the result of a great and continuing international effort there now exists an internationally agreed language of physical science."

M. L. McGlashan, 1967

Appendix A
Units and Dimensions

The following related topics are discussed in this appendix: SI units, equivalent energies, dimensionless quantities, electrostatic energy and force, units in intermediate steps.

SI units. SI units (Système Internationale d'Unités) were adopted in 1960 by the General Conference on Weights and Measures, the international organization that defines metric units. These units have been in use at the U.S. National Bureau of Standards since 1964. The system is an outgrowth of the mks (meter-kilogram-second) system and is based on seven units considered to be independent. Among them are the meter (m), kilogram (kg), second (s), kelvin (K), mole (mol), and ampere (A).[1] Units related to these by division or multiplication by powers of 10 are denoted by the prefixes in Table A-1.[2] The kilogram is anomalous in that it includes the prefix *kilo-* but has nevertheless been chosen as a basic unit. Prefixes that do not

---

[1] The remaining unit is the candela (cd), a unit of luminous intensity, which is not used in this book.

[2] Tables A-1, A-4, and A-5 are also given inside the back cover.

TABLE A-1   **Prefixes denoting powers of 10**

Factor	Prefix	Symbol	Factor	Prefix	Symbol
10^{-1}	deci	d	10	deka	da
10^{-2}	centi	c	10^2	hecto	h
10^{-3}	milli	m	10^3	kilo	k
10^{-6}	micro	μ	10^6	mega	M
10^{-9}	nano	n	10^9	giga	G
10^{-12}	pico	p	10^{12}	tera	T
10^{-15}	femto	f			
10^{-18}	atto	a			

represent powers of 10^3 are to be used sparingly, an example being *centi-*, meaning 10^{-2}: cm $= 10^{-2}$ m. Units may be raised to powers, the exponent applying to the prefix also, so that km^2 = (1000 m)^2 = 10^6 m^2 and μm^3 = $(10^{-6}$ m$)^3$ = 10^{-18} m^3.

Table A-2 presents various units derived by combining basic units. Several of the derived SI units have special names and symbols. A few of the units used in this text are not SI units. The most common are noted in Table A-3.

Equivalent energies.   Although the SI unit of energy is the joule, many other energy units are commonly used in chemistry. Some important interrelations of these units are given in Tables A-4 and A-5; values given on the same line are equivalent. For example, by Table A-4, 1 J is equivalent to 0.239 cal, so that multiplication by 0.239 converts an energy in joules to one in calories [see also Equation (A-1)].

The energy units in Table A-5 require additional comment. The first two columns concern quantities that apply to *one mole* of some substance rather than to an isolated atomic or molecular event. On the other hand, energies in electronvolts (eV) usually refer to individual atomic and molecular events, 1 eV being the kinetic energy acquired by one electron accelerated by an electric field of 1 V. The fifth column

TABLE A-2   **Examples of derived SI units***

Physical quantity	Unit	Special symbol and name
Speed	m s^{-1}	
Acceleration	m s^{-2}	
Force	kg m s^{-2} = J m^{-1}	N, newton
Pressure	kg s^{-2} m^{-1} = N m^{-2}	
Energy	kg m^2 s^{-2} = N m	J, joule
Power	kg m^2 s^{-3} = J s^{-1}	W, watt
Electric charge	A s	C, coulomb
Electric potential	kg m^2 A^{-1} s^{-3}	V, volt
difference	= J A^{-1} s^{-1}	
	= W A^{-1}	
	= J C^{-1}	
Electric resistance	V A^{-1}	Ω, ohm

*For further details on the physical quantities involved, see the Study Guide.

TABLE A-3   Additional units

Physical quantity	Unit	Special name
Volume	liter $= 1000 \text{ cm}^3 = 10^{-3} \text{ m}^3$	liter
	ml $= \text{cm}^3 = 10^{-6} \text{ m}^3$	milliliter
Density	$\text{g cm}^{-3} = 10^3 \text{ kg m}^{-3}$	
Molar concentration	$M = \text{mol liter}^{-1} = 10^{-3} \text{ mol m}^{-3}$	molarity
Length	$\text{Å} = 10^{-10} \text{ m} = 0.1 \text{ nm}$	angstrom
Pressure	atm $= 760 \text{ torr} = 101.325 \times 10^3 \text{ N m}^{-2}$	atmosphere
	torr $= 1.316 \times 10^{-3} \text{ atm}$	torr
	$= 133.3 \text{ N m}^{-2}$	

TABLE A-4   Some equivalent energies

calorie (cal)	joule (J)	liter atmosphere (liter atm)
1	4.1840	0.041293
0.23901	1	9.8692×10^{-3}
24.217	101.325	1

in Table A-5 shows the temperature T associated with the energy RT per mole (where R is the gas constant, discussed on pages 102 and 124) or kT per atom or molecule (where k is Boltzmann's constant, $R/N_A = 1.3807 \times 10^{-23} \text{ J K}^{-1}$). The significance of these quantities of energy is discussed in Sec. 5-4.

The final column is the wave number equivalent to the energy by the relation $\tilde{\nu} = 1/\lambda = E/hc$, where h is Planck's constant and c is the speed of light.

A table of physical constants is also presented inside the back cover. The values given are based on a report by the U.S. National Bureau of Standards (*Journal of Physical and Chemical Reference Data,* vol. 2, no. 4, p. 741, 1973).

Dimensionless numerical values of physical quantities. It is customary to regard the symbol for a physical quantity, such as absolute temperature T or energy E, as representing the quantity including its physical dimensions. In other words, the symbol represents the product

$$\text{(Physical quantity)} = \text{(numerical value)} \times \text{(unit)}$$

TABLE A-5   Additional equivalent energies

kJ mol^{-1}	kcal mol^{-1}	eV	J	K	cm^{-1}
1	0.23901	0.010364	1.6606×10^{-21}	120.27	83.60
4.1840	1	4.3364×10^{-2}	6.948×10^{-21}	503.2	349.8
96.48	23.060	1	1.6022×10^{-19}	11.605×10^3	8066
6.022×10^{20}	1.4393×10^{20}	6.241×10^{18}	1	7.243×10^{22}	5.034×10^{22}
8.314×10^{-3}	1.9872×10^{-3}	8.617×10^{-5}	1.3807×10^{-23}	1	0.6949
1.196×10^{-2}	2.859×10^{-3}	1.240×10^{-4}	1.986×10^{-23}	1.439	1

The symbol used for the physical quantity does not and should not imply any *particular* choice of units. On the other hand, if the symbol for the physical quantity is divided by an appropriate unit, the result is a pure number that represents the value of the quantity in the units considered. For example, $T/K = 50$ implies a temperature of 50 K and $E/J = 40$ implies an energy of 40 J. The coordinate axes of graphs are often labeled by such pure numbers, as are the columns in tables of numerical values. For example, the labels T/mK or $1000T/K$ imply that the numbers given refer to temperatures in millikelvin (an entry of 5 implies that $1000T/K = 5$ so $T = 5 \times 10^{-3}$ K = 5 mK).

It is instructive to see how the manipulation of such symbols leads to the energy conversion considered earlier, in which multiplication by 0.239 changed an energy in joules to one in calories. Writing the relationship 1 J = 0.239 cal in the form J/cal = 0.239, we find[1] that

$$E/\text{cal} = (E/J)(J/\text{cal}) = (E/J)0.239 \qquad \text{(A-1)}$$

where E/cal and E/J are the respective numbers of calories and joules.

Electrostatic energy and force.   By the laws of electrostatics the potential energy ϵ_p (coulomb energy) associated with two point charges Q_1 and Q_2 at a distance r in a vacuum is

$$\epsilon_p = \frac{Q_1 Q_2}{\alpha r} \qquad \text{(A-2)}$$

For historical reasons the proportionality constant α is sometimes written as $\alpha = 4\pi\epsilon_0$, where ϵ_0 is called the permittivity constant. The force f between the two charges (coulomb force) is given by

$$f = \frac{-Q_1 Q_2}{\alpha r^2} \qquad \text{(A-3)}$$

When SI units are used (Q_1 and Q_2 in C = s A; r in m; ϵ_p in J = kg m^2 s^{-2}; and f in N = kg m s^{-2}), the proportionality constant α has the value[2]

$$\alpha = 1.11265 \times 10^{-10} \text{ C}^2 \text{ J}^{-1} \text{ m}^{-1} \qquad \text{(A-4)}$$

The units of α are seen to be those required by (A-2) or (A-3); reduced to basic units, $\text{C}^2 \text{ J}^{-1} \text{ m}^{-1} = \text{kg}^{-1} \text{ m}^{-3} \text{ s}^4 \text{ A}^2$.

---

[1] An alternative approach is to write the equation 1 J = 0.239 cal in the form 0.239 cal/J = 1. Since multiplication by unity causes no change, it follows, for example, that 40 J = (40 J)(0.239 cal/J) = (0.239 × 40) cal. Some discretion is required with units that are not strictly energy units. For example, the energy kT characteristic of a temperature of 1 K corresponds also to 8.617×10^{-5} eV, which is sometimes expressed by saying that 1 K "equals" 8.617×10^{-5} eV. Only when both units are interpreted as energy units does 8.617×10^{-5} eV/K represent unity.

[2] The numerical value of α is related to c, the velocity of light (in a vacuum):

$$\frac{\alpha}{\text{C}^2 \text{ J}^{-1} \text{ m}^{-1}} = 10^7 \left(\frac{c}{\text{m s}^{-1}}\right)^{-2}$$

Units in intermediate steps. It is essential to formulate a problem in terms of all units involved, but it is often cumbersome to express the units at each of the intermediate steps. Proper units must, however, be affixed to all final results.

A widespread habit of deleting units in intermediate steps is illustrated by the relation

$$x = 0.10 - [H^+] \tag{A-5}$$

which is more properly written as

$$x = 0.10 - \frac{[H^+]}{M} \tag{A-6}$$

However, the meaning of (A-5) is unambiguous and division of $[H^+]$ by M (meaning mol liter^{-1}) is generally omitted.

A related procedure concerns equilibrium constants, such as

$$K_{N_2O_4} = \frac{(P_{NO_2})^2}{P_{N_2O_4}} = 0.141 \text{ atm} \tag{A-7}$$

for the reaction

$$N_2O_4(g) \rightleftharpoons 2NO_2(g) \tag{A-8}$$

and
$$K_w = [H^+][OH^-] = 1.0 \times 10^{-14} M^2 \tag{A-9}$$

for the reaction

$$H_2O \rightleftharpoons H^+ + OH^- \tag{A-10}$$

(both values at 25°C). As discussed in Sec. 13-3, equilibrium constants are closely associated with the corresponding balanced chemical equations. Hence, when the chemical equation is explicitly stated or known from the context, the units of equilibrium constants, such as atm and M^2 (in A-7) and (A-9), are often dropped. We usually follow this practice.

Finally, consider logarithmic expressions such as

$$pH = -\log \frac{[H^+]}{M}$$

where $[H^+]$ is divided by M because only logarithms of dimensionless (and nonnegative) quantities have simple definitions. Other examples are $\log (K_w/M^2)$ and $\log (T/K)$. It is often customary to omit the division by the units; that is, one writes $-\log [H^+]$, $\log K_w$, and $\log T$, with the understanding that the units are simply dropped when taking logarithms. Caution must be exercised when there may be uncertainty about the appropriate units, as in $\log P$, which could imply, for example, $\log (P/\text{atm})$ or $\log (P/\text{torr})$. The units *must* be specified in ambiguous cases.

"The Chemical signs ought to be letters, for the greater facility of writing, and not to disfigure a printed book. . . . I shall take, therefore, for the chemical sign, the initial letter of the Latin name of each elementary substance: but as several have the same initial letter, I shall distinguish them in the following manner: 1. In the class which I call metalloids, I shall employ the initial letter only, even when this letter is common to the metalloid and some metal. 2. In the class of metals, I shall distinguish those that have the same initials with another metal, or metalloid, by writing the first two letters of the word. 3. If the first two letters be common to two metals, I shall, in that case, add to the initial letter the first consonant which they have not in common."

J. J. Berzelius, 1814*

Appendix B

Names and Formulas of Compounds

This appendix has two sections. Section B-1 gives the names and formulas of many common inorganic ions and compounds, and it discusses the simpler rules of inorganic chemical nomenclature. A beginning student must become familiar with these names, formulas, and rules. Section B-2 contains some of the rules for naming organic compounds.

An important aspect of the systematic naming of compounds is that knowledge of the name must permit the reconstruction of the correct molecular or ionic formula. Names of this kind are called *chemical* or *systematic* names and stand in contrast

---

*Quoted by Leicester and Klickstein, "Source Book in Chemistry, 1400–1900", McGraw-Hill Book Company, New York, 1952; *Annals of Philosophy*, vol. 3, p. 51, 1814.

to trivial names, which are often used for important compounds, such as "aspirin", "vitamin A", "indigo" (a dye), or "muriatic acid" (HCl). Trivial names have little or no relationship to the formula.

The names and rules given here suffice to illustrate the general principles of systematic nomenclature and should be adequate for most of the compounds discussed in this text—particularly in the derivation of a formula from a given name, which is easier than the reverse procedure. An incidental but important point in chemical nomenclature is that a given compound may have a number of acceptable chemical names. All these must, of course, relate to the same formula.

B-1 Inorganic Ions and Compounds

Some of the rules of inorganic nomenclature are based on chemical knowledge, in particular on oxidation states (Chap. 7 and Appendix C). If you are unfamiliar with the concept of oxidation states, simply learn the names and formulas.

An essential part of the formula of an ion is its charge. For example, there is an enormous difference in properties between hydrogen atoms (H) and hydrogen ions (H^+), between sulfur trioxide (SO_3) and sulfite ion (SO_3^{2-}), and between the hydroxyl radical (OH) and the hydroxide ion (OH^-). The charge of an ion can usually be understood in terms of the electronic structure of the ion, but for a beginning student it is best first to study and remember the charge of an ion as part of its formula. Differences such as that between ammonia (NH_3) and ammonium ion (NH_4^+) must also be noted and remembered.

The following subsections also indicate the geometric structure of some ions. These structures need not be memorized; they can be deduced from the electronic structure.

Cations

1. The following metals form only one stable positive ion (cation). The names of these ions are the same as those of the metals, such as potassium ion and magnesium ion.

Alkali metals: Li^+, Na^+, K^+, Rb^+, Cs^+, Fr^+
Alkaline-earth metals: Be^{2+}, Mg^{2+}, Ca^{2+}, Sr^{2+}, Ba^{2+}, Ra^{2+}
Others: Al^{3+}, Zn^{2+}, Cd^{2+}, Ni^{2+}, Pb^{2+}, Ag^+

2. Many metallic elements form two or more positive ions with different charges. For example,

Cr^{2+}	chromous	Cr^{3+}	chromic
Mn^{2+}	manganous	Mn^{3+}	manganic
Fe^{2+}	ferrous	Fe^{3+}	ferric
Co^{2+}	cobaltous	Co^{3+}	cobaltic
Cu^+	cuprous	Cu^{2+}	cupric
Sn^{2+}	stannous	Sn^{4+}	stannic
Hg_2^{2+}	mercurous[1]	Hg^{2+}	mercuric

[1]Note the diatomic nature of the Hg_2^{2+} (mercurous) ion.

3. *Some polyatomic cations*

H_3O^+ hydronium or oxonium (pyramidal)
NH_4^+ ammonium (tetrahedral)
PH_4^+ phosphonium (tetrahedral)
NO_2^+ nitryl (linear)

Anions

1. *Halides*

F^- fluoride Br^- bromide
Cl^- chloride I^- iodide

2. *Chalcogenides*

O^{2-} oxide S^{2-} sulfide
O_2^{2-} peroxide S_2^{2-} disulfide
OH^- hydroxide HS^- hydrogen sulfide

3. *Planar trigonal oxyanions and related species*

CO_3^{2-} carbonate NO_3^- nitrate
HCO_3^- bicarbonate or
         hydrogen carbonate

4. *Tetrahedral oxyanions and related species*

SiO_4^{4-} silicate ClO_4^- perchlorate
         (orthosilicate)
PO_4^{3-} phosphate CrO_4^{2-} chromate
         (orthophosphate)
HPO_4^{2-} hydrogen $HCrO_4^-$ hydrogen chromate
         phosphate
$H_2PO_4^-$ dihydrogen MnO_4^- permanganate
         phosphate
SO_4^{2-} sulfate MnO_4^{2-} manganate
HSO_4^- hydrogen sulfate
         or bisulfate
$S_2O_3^{2-}$ thiosulfate[1]

5. *Condensed polyanions linked through oxygen atoms:* Each of the individual units linked is tetrahedral (for example, $O_3SOSO_3^{2-}$ for $S_2O_7^{2-}$):

$S_2O_7^{2-}$ disulfate or pyrosulfate
$Cr_2O_7^{2-}$ dichromate
$P_2O_7^{4-}$ diphosphate or pyrophosphate
$(PO_3)_n^{n-}$ polyphosphate ring, also called metaphosphate
$Si_2O_7^{6-}$ disilicate
$(SiO_3)_n^{2n-}$ polysilicate ring or (if n is very large) chain, also called metasilicate

---

[1]The relationship between SO_4^{2-} and $S_2O_3^{2-}$ is that one oxygen atom in sulfate has been replaced by a sulfur atom in thiosulfate.

6. *Some other important anions*

H^-	hydride	OCN^-	cyanate (linear)
CN^-	cyanide	SCN^-	thiocyanate (linear)
BH_4^-	borohydride (tetrahedral)	N_3^-	azide (linear)
ClO^-	hypochlorite	SO_3^{2-}	sulfite (pyramidal)
ClO_2^-	chlorite (bent)	NO_2^-	nitrite (bent)
ClO_3^-	chlorate (pyramidal)	NH_2^-	amide (bent)

Compounds formed from ions. These are named by suitable combinations of the names of the ions, with the cation(s) named first—e.g., ferric chloride ($FeCl_3$), calcium hydride (CaH_2), sodium hydroxide (NaOH), and lead dichromate ($PbCr_2O_7$).

Acids usually have special names (see "Rules and comments" and "Oxyacids and their anions" below), although there are a few exceptions.

Almost all positively charged ions are hydrated in aqueous solution, even though this is not usually indicated in their formulas (see Secs. 4-5 and 7-5). Negatively charged ions (anions) are hydrated also, but the structures are less well defined than those for most cations. Water of crystallization (or hydration) in solids is indicated explicitly in formulas, usually with a dot separating it from the formula for the remainder of the compound, as in $CuSO_4 \cdot 5H_2O$ (Fig. 4-8).

Rules and comments. An important feature of inorganic nomenclature is the way in which oxidation states are specified for each part of the compound, unless there is only one possibility, as with cations of the alkali and alkaline-earth metals.

The system used in the names just listed is one of endings and prefixes. In this system the lower and the higher of two oxidation states of a cation are distinguished by the endings *-ous* and *-ic*, respectively, as in the examples under "Cations", item 2, above. The specific oxidation states indicated by these suffixes vary from one element to the next and thus an alternative and unambiguous means of denoting oxidation states has been developed. They are shown by positive and negative roman numerals enclosed in parentheses after the name of the element or by (0) for oxidation state zero. The system is used most commonly for metallic elements that exhibit several stable oxidation states. Thus $FeCl_2$ is called iron(II) chloride as well as ferrous chloride, and SnS_2 may be referred to as tin(IV) sulfide as well as stannic sulfide.

Anions involving only one atom (such as F^- and S^{2-}) are given the ending *-ide*. The corresponding hydrogen compounds have names such as hydrogen fluoride (HF) and hydrogen sulfide (H_2S), except that the hydrogen halides (HF, HCl, HBr, HI) are often referred to (especially when in solution) by names such as hydrochloric acid to emphasize their acid character. A few polyatomic anions also have the ending *-ide,* as indicated in the lists of names given above.

Oxyacids and their anions. The oxidation state of the central atom in anions of oxygen acids is indicated by the endings *-ite* and *-ate* (*-ous* and *-ic* for the acids themselves) and by the prefixes *hypo-* and *per-*, in a way best illustrated by examples:

Ionic species	Oxidation state of central atom	Name	Corresponding acid	Name
ClO^-	I	Hypochlorite ion	$HClO$	Hypochlorous acid
ClO_2^-	III	Chlorite ion	$HClO_2$	Chlorous acid
ClO_3^-	V	Chlorate ion	$HClO_3$	Chloric acid
ClO_4^-	VII	Perchlorate ion	$HClO_4$	Perchloric acid
MnO_4^{2-}	VI	Manganate ion	H_2MnO_4	Manganic acid
MnO_4^-	VII	Permanganate ion	$HMnO_4$	Permanganic acid
SO_3^{2-}	IV	Sulfite ion	H_2SO_3	Sulfurous acid
SO_4^{2-}	VI	Sulfate ion	H_2SO_4	Sulfuric acid

Again, the prefixes and suffixes give no information about the oxidation states themselves; they only indicate that there may also be lower or higher oxidation states.

Other important prefixes are *di-* and *bi-*. The first of these indicates that the key element is represented by two atoms, as in disulfide ion (S_2^{2-}) and dichromate ion ($Cr_2O_7^{2-}$). The second prefix (*bi-*) is used as an alternative in names such as bicarbonate ion (HCO_3^-) and bisulfate ion (HSO_4^-), for which the more modern names are hydrogen carbonate ion and hydrogen sulfate ion. The rationale for the older name is that, for example, for a given quantity of sodium ion the salt sodium bicarbonate ($NaHCO_3$) contains twice as much carbonate as does sodium carbonate (Na_2CO_3).

Coordination Compounds   (compounds containing complex ions or similar neutral species). The following additional rules apply to coordination compounds:

1.   The oxidation state of the central atom is indicated by a roman numeral or zero in parentheses, unless it is unambiguous.

2.   Names of complex anions end in *-ate,* while for complex cations and neutral molecules no special ending is given.

3.   Neutral ligands are named as the molecule, negative ligands end in *-o* (common endings are *-ido, -ito,* and *-ato*), and positive ligands (rare) end in *-ium.* Examples are

$NH_2CH_2CH_2NH_2$	ethylenediamine	ONO^-	nitrito
Cl^-	chloro	NO^-	nitroso
O^{2-}	oxo	CH_3COO^-	acetato
OH^-	hydroxo	SO_4^{2-}	sulfato
CN^-	cyano	NH_2^-	amido
NO_2^-	nitro		

Exceptions to these rules are H_2O (aquo) and NH_3 (ammine; note the double *m*).

The different names given to the NO_2^- group reflect different possibilities in linkage: with the nitro group the nitrogen atom is bonded to the central atom of the coordination compound, whereas for the nitrito group one of the oxygen atoms forms the bond. Such *linkage isomerism* can also occur for other ligands. In the list above, the coordination bond normally involves the atom given first in the formula, except for $SO_4{}^{2-}$ and acetate, which are written in the conventional way and for each of which an oxygen atom is bonded to the central atom of the coordination compound.

4. The sequence of the ligands in naming a complex is negative, neutral, and positive; no hyphens are used. Within each category the order is that of increasing complexity.

5. The prefixes *di-*, *tri-*, *tetra-*, *penta-*, *hexa-*, . . . are used before such simple names as bromo, nitrito, amido. The prefixes *bis-*, *tris-*, *tetrakis-*, . . . (Greek: *kis*, times) are used in front of more complex names, such as ethylenediamine, particularly if these names contain prefixes such as *mono-*, *di-*, *tri-*, ... Examples are

$Li[AlH_4]$	lithium tetrahydridoaluminate[1]
$K_2[PtCl_4]$	potassium tetrachloroplatinate(II)
$K_3[Fe(CN)_6]$	potassium hexacyanoferrate(III)
$Na_4[Ni(CN)_4]$	sodium tetracyanonickelate(0)
$CoCl_2(en)_2$	dichlorobis(ethylenediamine)cobalt(II)
$Co(NH_3)_3(NO_2)_3$	trinitrotriammine cobalt(III)
$[Pt(NH_3)_4(ONO)Cl]SO_4$	chloronitritotetrammineplatinum(IV) sulfate

B-2 Modern Organic Nomenclature

The complexity of many organic compounds necessitates an elaborate set of rules so that any compound may be given a name from which the structural formula of that compound can be reconstructed. Furthermore, the rules must, ideally at least, be sufficiently definite that a given compound cannot be named in several different ways, because the names of organic substances are often so complex that the different names would not always be recognized as applying to the same compound. We do not deal with these complexities here, but rather present only a few rules and examples to show how the system works. Examples of the naming of compounds containing the common functional groups are given in Chap. 26.

*n-***Alkanes and alkyl groups** (Secs. 26-1 and 26-2). Alkanes are hydrocarbons with the general formula C_mH_{2m+2}. They contain no multiple bonds and no rings; every carbon atom forms four bonds and every hydrogen atom one. In molecules of the "straight-chain"[2] or normal alkanes (Table 26-2), designated by the prefix *n-*, no carbon atom has more than two other carbon atoms bonded to it. The simplest alkyl groups, corresponding formally to alkanes from which one hydrogen atom has been removed, are listed in Table 26-3.

Branched alkanes. Branched alkanes are named as follows: (1) Find and name the longest carbon chain in the molecule; (2) mentally attach the numbers 1, 2, 3,

---

[1] This compound, a powerful and frequently used reagent, is commonly called lithium aluminum hydride.

[2] The C-C-C bond angle in these chains is about 112°, so the chains are far from linear, but topologically they are "straight", which in this context implies "without branches".

. . . to successive carbon atoms in this chain, the direction being chosen so as to label the side chains by the lowest numbers possible; (3) name the alkyl groups that represent the side chains; (4) put these pieces of information together as shown by the examples below:

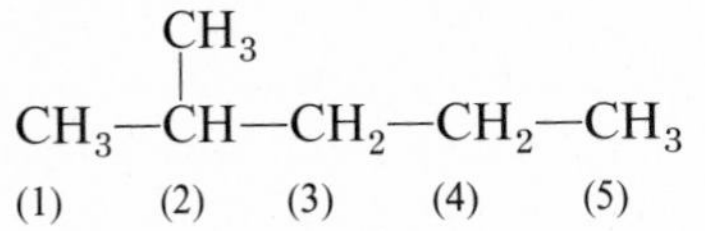

$$CH_3-\underset{(2)}{\overset{\overset{\displaystyle CH_3}{|}}{CH}}-\underset{(3)}{CH_2}-\underset{(4)}{CH_2}-\underset{(5)}{CH_3}$$
(1)

2-Methylpentane

$$CH_3-\underset{(2)}{\overset{\overset{\displaystyle CH_3}{|}}{\underset{\underset{\displaystyle CH_3}{|}}{C}}}-\underset{(3)}{\overset{\overset{\displaystyle \overset{\displaystyle CH_3 \quad CH_3}{\diagdown \diagup}}{CH}}{\underset{|}{CH}}}-\underset{(4)}{CH_2}-\underset{(5)}{CH_2}-\underset{(6)}{CH_2}-\underset{(7)}{CH_3}$$
(1)

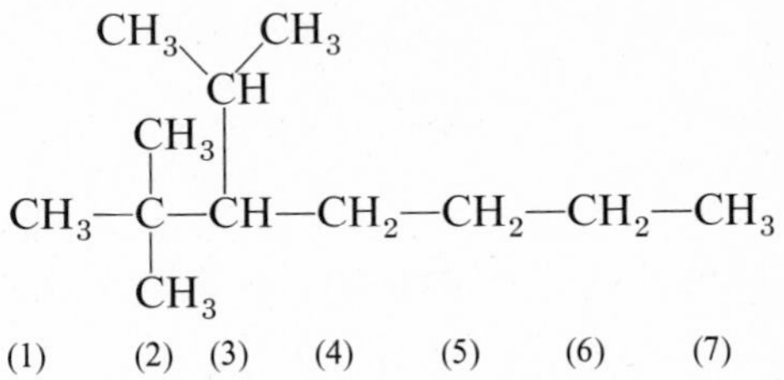

2,2-Dimethyl-3-isopropyl heptane

Alkenes and alkynes. Hydrocarbons with double bonds are called alkenes or olefins. They are named by the suffix *-ene*. The numbering system introduced earlier is used to show the position of the double bond when this is necessary:

$$CH_2{=}CH_2 \qquad CH_2{=}CHCH_3 \qquad \underset{(1)}{CH_2}{=}\underset{(2)}{CH}\underset{(3)}{CH_2}\underset{(4)}{CH_3}$$

Ethylene Propene 1-Butene

$$\underset{(1)}{CH_3}\underset{(2)}{CH}{=}\underset{(3)}{CH}\underset{(4)}{CH_3} \qquad \underset{(1)}{CH_2}{=}\underset{(2)}{CH}{-}\underset{(3)}{CH}{=}\underset{(4)}{CH_2}$$

2-Butene 1,3-Butadiene

Hydrocarbons with triple bonds are called alkynes. Examples are

$$HC{\equiv}CH \qquad \underset{(1)}{CH_3}\underset{(2)}{C}{\equiv}\underset{(3)(4)}{CCH_3}$$

Acetylene 2-Butyne

Ring compounds. Cyclic hydrocarbons are denoted systematically by the prefix *cyclo-*:

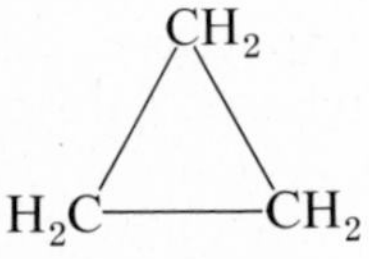

Cyclopropane 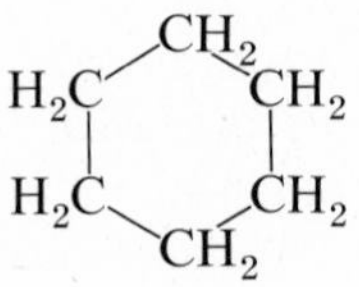 Cyclohexane 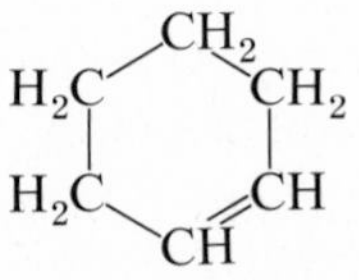 Cyclohexene 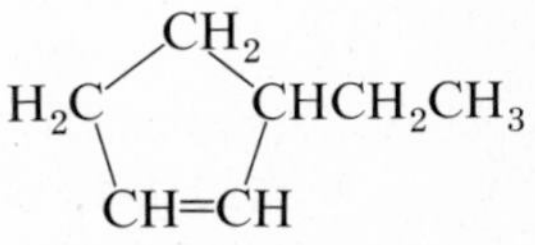 3-Ethylcyclopentene

Note in the last example that the double bond is assumed to be between C-1 and C-2; the substituent is then given the lowest number consistent with this assumption.

Polyfunctional compounds. Many common polyfunctional compounds have trivial names, but they can be named systematically also. Examples are

$$\begin{array}{cc} CH_2\text{---}CH_2 \\ | \quad\quad | \\ OH \quad\; OH \end{array}$$

1,2-Ethanediol
(ethylene glycol)

$$\begin{array}{c} CH_3 \\ | \\ CH_3\overset{|}{C}\text{==}CHCH_2OH \end{array}$$

3-Methyl-2-butene-1-ol

3-Chloronitrobenzene

3-Aminocyclopentanone

Problems and Questions

B-1 Formulas of inorganic compounds and ions. Give the correct formula, including charges if any, for the following compounds and ions: sulfuric acid; sodium sulfate; potassium bromide; ammonium ion; ammonium nitrate; nitric acid; carbonate ion; calcium bicarbonate; calcium phosphate; silver iodide.

   Ans. H_2SO_4; Na_2SO_4; KBr; NH_4^+; NH_4NO_3; HNO_3; CO_3^{2-}; $Ca(HCO_3)_2$; $Ca_3(PO_4)_2$; AgI

B-2 Formulas of inorganic compounds and ions. Give the correct formula, including charges if any, for the following: ammonium bicarbonate; barium perchlorate; iron(III) nitrate hexahydrate; calcium fluoride; aluminum sulfate; hydroxide ion; strontium bromide; barium phosphate; sulfite ion; copper(II) sulfate pentahydrate.

B-3 Names of inorganic compounds and ions. Name the following compounds and ions: BaO_2; $CaHPO_4$; $Cr(OH)_3$; HCN; $MgSO_3$; LiSCN; S^{2-}; $FeCl_3$; SrI_2; $NiSO_4$; HBr; OCl^-; NH_3.

   Ans. barium peroxide; calcium hydrogen phosphate; chromic hydroxide or chromium(III) hydroxide; hydrogen cyanide or hydrocyanic acid; magnesium sulfite; lithium thiocyanate; sulfide ion; ferric chloride or iron(III) chloride; strontium iodide; nickel(II) sulfate; hydrogen bromide or hydrobromic acid; hypochlorite ion; ammonia

B-4 Names of inorganic compounds and ions. Name the following compounds and ions: $LiHSO_4$; HS^-; $FePO_4$; H_2CO_3; $BaCrO_4$; $HClO_4$; $Na_2S_2O_3$; $Be(CN)_2$; NH_4ClO_4; Cu_2S; K_2SO_4; HSO_3^-.

B-5 Formulas of inorganic complexes. Give chemical formulas for the following compounds and ions: sodium bis(thiosulfato)argentate(I); ammonium tetrathiocyanatodiammine chromate(III); sodium hexanitrocobaltate(III); hydroxopentaaquoaluminum ion; tris(ethylenediamine)cobalt(III) sulfate; dihydroxotetraaquochromium(III) chloride; chlorotriamminecobalt(II) chloride; potassium tetrachloroaurate(III); tetrakis(trichlorophosphine)nickel(0); dichlorotetramminecobalt(III) chloride.

   Ans. $Na_3[Ag(S_2O_3)_2]$; $NH_4[Cr(NH_3)_2(SCN)_4]$; $Na_3[Co(NO_2)_6]$; $Al(H_2O)_5OH^{2+}$; $[Co(en)_3]_2(SO_4)_3$; $[Cr(H_2O)_4(OH)_2]Cl$; $[Co(NH_3)_3Cl]Cl$; $K[AuCl_4]$; $Ni(PCl_3)_4$; $[Co(NH_3)_4Cl_2]Cl$

B-6 Formulas of hydrocarbons. Give formulas for the following hydrocarbons: *n*-pentane; 2,3-dimethyl-1-hexene; 3-heptyne; 3-ethyloctane; cyclobutane; 3-methylcyclohexene; 1,3-dimethylbenzene.

Appendix C

Oxidation Numbers and the Balancing of Redox Equations

Although it is sometimes easy to balance chemical equations by trial and error, a systematic approach is often needed, especially with equations for redox reactions. We present here two methods for balancing such equations. The first is based on oxidation numbers, the second on the separate balancing of half-reactions. Familiarity with the contents of Sec. 7-7 is assumed.

Oxidation numbers and oxidation states. We begin with rules for assigning oxidation numbers for the purpose of balancing redox equations. We designate *oxidation numbers* by arabic numerals or fractions (for example, 5, -3, $\frac{1}{2}$) to avoid their confusion with *oxidation states,* which we have (in accord with common practice) designated by roman numerals [for example, Fe(III), S(VI), O($-$II)]. Numerals designating oxidation *numbers* for atoms in compounds or ions will very frequently be identical in value with those designating the oxidation *states* of these same atoms, but the distinction is made because, for purposes of balancing equations, oxidation numbers may be assigned quite arbitrarily provided only that they are consistent with one essential rule (rule 1 below). Additional rules have been adopted for the assignment of oxidation states; these rules are also to some degree arbitrary. The oxidation state of an atom in a particular structure is, of course, important in inorganic chemical nomenclature (Appendix B), but it should not be taken to imply the actual charge associated with that atom unless the atom happens to be a monatomic species or the situation is equally unambiguous (for example, O in peroxide ion O_2^{2-}). Rules 2 to 5 below are really rules for oxidation states, but they are often used for oxidation numbers as well. Rules 6 and 7 may give different oxidation numbers than would rules 2 to 5, but they may be adopted when convenient, as shown in several examples.

Rules for assigning oxidation numbers. There is only one essential rule: *Rule 1. The sum of the oxidation numbers of all the atoms in a molecule or ion must equal the charge on that molecule or ion.* A corollary to this rule is that the oxidation number of a monatomic ion, such as S^{2-} or Al^{3+}, must be the charge on that ion.

Any set of oxidation numbers consistent with rule 1 can be used successfully for balancing redox equations, as we shall see shortly. We use the following additional rules:

Rule 2. Fluorine in its compounds has oxidation number -1.

Rule 3. Hydrogen in its compounds is assigned oxidation number $+1$ except in metallic hydrides (such as NaH), where it is given oxidation number -1.

Rule 4. Oxygen in its compounds is assigned oxidation number -2 except

(*a*) In compounds containing O-O bonds: peroxides (such as H_2O_2) and superoxides (such as KO_2), with oxidation numbers -1 and $-\frac{1}{2}$, respectively.

(*b*) In compounds with fluorine, where rule 2 takes precedence. Thus in OF_2 the oxidation number of oxygen is $+2$.

Rule 5. Oxidation numbers may be assigned by analogy.

Consider the thiosulfate and sulfate ions, $S_2O_3^{2-}$ and SO_4^{2-},

$$\begin{bmatrix} & O & \\ & \| & \\ O{-}S{-}O \\ & | & \\ & S & \end{bmatrix}^{2-} \qquad \begin{bmatrix} & O & \\ & \| & \\ O{-}S{-}O \\ & | & \\ & O & \end{bmatrix}^{2-}$$

In $S_2O_3^{2-}$, a sulfur atom is seen to take the place of one of the oxygen atoms in SO_4^{2-}, as indicated by the prefix *thio-*. In the sulfate ion, each oxygen atom has the oxidation number -2; the oxidation number of the central sulfur atom is thus $+6$. By analogy, the peripheral and central sulfur atoms in $S_2O_3^{2-}$ are given the oxidation numbers -2 and $+6$, respectively.

Rule 6. If there are several atoms of a given kind in a molecule or ion, they *may* all be assigned the same oxidation number.

Thus in O_3 all atoms are given oxidation number zero, and in S_2^{2-} each sulfur atom is assigned oxidation number -1. This rule can lead to fractional oxidation numbers; for example, in I_3^- each I may be given oxidation number $-\frac{1}{3}$. In thiosulfate, $S_2O_3^{2-}$, each S atom might be given the oxidation number $+2$, which is the average of the values $+6$ and -2 assigned to the two S atoms according to rule 5. By either of these assignments, the sum of the oxidation numbers of the two sulfur atoms in the thiosulfate ion is $+4$, and this is all that is needed for balancing a redox equation involving $S_2O_3^{2-}$.

Rule 7. Oxidation numbers may for convenience be assigned to entire groups of atoms that remain together during the redox reaction—for example, a cyano group, CN, or a hydrocarbon radical, such as CH_3 or C_6H_5.

Suppose HCN is oxidized to HOCN. The cyano group is assigned oxidation number -1 in the first compound and $+1$ in the second, in accord with rules 3 and 4. There is no need to consider the carbon and nitrogen atoms individually.

Balancing equations with the aid of oxidation numbers. The equation for a redox reaction may readily be balanced by the method of oxidation numbers. When ions are involved, the reaction is assumed to take place in a medium favoring their existence and this medium is usually understood to be an aqueous solution. The reaction may then be defined solely by giving the oxidized and reduced forms of the couples present. *No other redox couples may be introduced in the balancing procedure,* although H^+, OH^-, and H_2O may participate in the reaction in the aqueous medium and thus may be used in balancing the equation.

EXAMPLE C-1   **Reduction of nitrate ion by aluminum.** Write a balanced equation for the reaction between Al and NO_3^- in a basic solution. The products are the $Al(OH)_4^-$ ion and ammonia.

Solution.

Step 1. Balancing of changes in oxidation states: The oxidation numbers of the reagents are

	Start	End
Al	Zero in Al	Al(3) in $Al(OH)_4^-$
N	N(5) in NO_3^-	N(-3) in NH_3

Thus the changes in oxidation state are $+3$ for Al and -8 for N, corresponding to a loss of $3e^-$ by each Al atom and a gain of $8e^-$ by each N atom. Since the number of electrons lost by one redox couple must be the same as that gained by the other, the coefficients must be in the ratio $8:3$,

$$8Al + 3NO_3^- \longrightarrow 8Al(OH)_4^- + 3NH_3 \qquad (C\text{-}1)$$

This expression is not yet balanced with regard to all atoms nor with regard to charges.

Step 2. Balancing of charges: We add either H^+ or OH^- ions as needed to balance the charges. Since this reaction occurs in a basic solution, in which OH^- ions far outnumber H^+, we use OH^-. The charge on the left of (C-1) is -3; that on the right side is -8. Thus we add $5OH^-$ to the left side:

$$8Al + 3NO_3^- + 5OH^- \longrightarrow 8Al(OH)_4^- + 3NH_3 \qquad \text{(C-2)}$$

This expression is still not completely balanced.

Step 3. Balancing of atoms: Since the charges and the atoms whose oxidation states change are now balanced, it remains only to balance the other atoms present, H and O, without any change in charge or oxidation state. The electrically neutral species H_2O may be added as needed. We note that there are $3 \times 3 + 5 = 14$ oxygen atoms on the left and $8 \times 4 = 32$ oxygen atoms on the right; thus to balance these atoms we add $32 - 14 = 18$ water molecules on the left:

$$8Al + 3NO_3^- + 5OH^- + 18H_2O \longrightarrow 8Al(OH)_4^- + 3NH_3 \qquad \text{(C-3)}$$

There are now $5 + 18 \times 2 = 41$ hydrogen atoms on the left and $8 \times 4 + 3 \times 3 = 41$ hydrogen atoms on the right. Thus these atoms are also balanced, and (C-3) is completely balanced. It is always true that *the last kind of atom considered is necessarily found to be balanced if no errors have been made.* This serves as a check that all steps have been properly followed.

A final check on the balancing of all atoms and charges should be made after the complete equation has been written.

EXAMPLE C-2 **Oxidation of methanol by dichromate.** The products of the reaction in acid solution between dichromate ion, $Cr_2O_7^{2-}$, and methanol, CH_3OH, are Cr^{3+} ions and formic acid, $HCOOH$. Write a balanced equation for this reaction.

Solution.

Step 1: The oxidation numbers of the reagents are

	Start	End
Cr	Cr(6) in $Cr_2O_7^{2-}$	Cr(3) in Cr^{3+}
C	C(-2) in CH_3OH	C(2) in $HCOOH$

Thus the changes in oxidation numbers are

Cr: -3 per Cr atom -6 per $Cr_2O_7^{2-}$ ion
C: $+4$ per C atom $+4$ per CH_3OH molecule

These changes correspond to a gain of $6e^-$ by each $Cr_2O_7^{2-}$ ion and a loss of $4e^-$ by each methanol molecule. The coefficients must therefore be in the ratio $2:3$,

$$2Cr_2O_7^{2-} + 3CH_3OH \longrightarrow 4Cr^{3+} + 3HCOOH \qquad \text{(C-4)}$$

Step 2: Since the solution is acidic, we use H^+ to balance the charge. The total charge on the left of (C-4) is -4; that on the right is $+12$. Thus $16H^+$ must be added to the left:

$$16H^+ + 2Cr_2O_7^{2-} + 3CH_3OH \longrightarrow 4Cr^{3+} + 3HCOOH \qquad \text{(C-5)}$$

Step 3: We note that there are $2 \times 7 + 3 = 17$ oxygen atoms on the left of (C-5) and $3 \times 2 = 6$ oxygen atoms on the right of this same expression. Thus we add $17 - 6 = 11$ water molecules on the right:

$$16H^+ + 2Cr_2O_7^{2-} + 3CH_3OH \longrightarrow 11H_2O + 4Cr^{3+} + 3HCOOH \qquad (C\text{-}6)$$

There are now seen to be $16 + 3 \times 4 = 28$ hydrogen atoms on the left and $11 \times 2 + 3 \times 2 = 28$ hydrogen atoms on the right. Thus these atoms have automatically been balanced, and (C-6) is completely balanced.

One step must be added to the procedure in Examples C-1 and C-2 when there are atoms or groups other than O and H that do not change oxidation number during the reaction. These atoms or groups should be balanced as soon as the oxidation numbers have been balanced, that is, before the charges are balanced (as in step 2 above). This step is usually a simple one, as indicated in the next example, which also illustrates the use of group oxidation numbers.

EXAMPLE C-3   **Balancing by assigning oxidation numbers to groups.** Write a balanced equation for the reaction

$$SCN^- + MnO_4^- \longrightarrow HCN + SO_4^{2-} + Mn^{2+} \qquad (C\text{-}7)$$

occurring in an acidic solution.

Solution. The thiocyanate ion is the source of both HCN and SO_4^{2-}. The CN group in HCN is assigned the oxidation number -1, and for convenience we assign the same number to the CN in SCN^-, which implies then that S in SCN^- has oxidation number zero. Sulfur ends up at $+6$ in SO_4^{2-}, an overall loss of $6e^-$. The Mn originally in MnO_4^- goes from $+7$ to $+2$, a gain of $5e^-$. The coefficients for balancing the changes in oxidation numbers are thus 5 and 6:

$$5SCN^- \text{ go to } 5SO_4^{2-}, \text{ with loss of } 30e^-$$
$$6MnO_4^- \text{ go to } 6Mn^{2+}, \text{ with gain of } 30e^-$$

Although the CN group does not change oxidation number, this group comes only from the SCN^- and thus the HCN that appears as a product in (C-7) must be equal in number to the SCN^- on the left. At this point we have

$$5SCN^- + 6MnO_4^- \longrightarrow 5SO_4^{2-} + 6Mn^{2+} + 5HCN$$

The remaining steps are similar to those used in the previous example. We add $13H^+$ on the left to balance charges and then $4H_2O$ on the right to balance oxygen atoms. Hydrogen atoms then balance, confirming the correctness of the earlier steps:

$$13H^+ + 5SCN^- + 6MnO_4^- \longrightarrow 4H_2O + 5SO_4^{2-} + 6Mn^{2+} + 5HCN$$

We use two different sets of oxidation numbers in the next example to illustrate that such assignments are arbitrary as long as they are consistent with the essential rule 1.

EXAMPLE C-4   **Balancing with sets of arbitrary oxidation numbers.** Find a balanced equation for the reaction between iron pyrite (FeS_2) and nitric acid,

$$FeS_2 + H^+ + NO_3^- \longrightarrow Fe^{3+} + SO_4^{2-} + NO_2 \qquad \text{(C-8)}$$

Solution. The iron in FeS_2 is transformed into Fe^{3+} with oxidation number $+3$, and the sulfur is changed to S(6) in sulfate ion. The nitrogen is reduced from N(5) in nitrate ion to N(4) in NO_2, a gain of $1e^-$ for each nitrate ion. We now illustrate the balancing of the equation for (C-8) with several sets of oxidation numbers for Fe and S, each consistent with the condition that the sum of the oxidation numbers for the atoms in FeS_2 must be zero, since this species is uncharged.

Set 1: Since iron ends up as Fe^{3+}, let us assume that this element has the oxidation number $+3$ at the start as well. This means that the S_2 group in FeS_2 has the oxidation number -3; on the average, each S atom is at -1.5 and changes to $+6$, a net change of $+7.5$ for each S atom or $+15$ for the S_2 group. Each S_2 group thus loses $15e^-$; since each NO_3^- gains $1e^-$, the coefficients for balancing changes in oxidation state must be in the ratio $1:15$,

$$FeS_2 + 15NO_3^- + H^+ \longrightarrow Fe^{3+} + 2SO_4^{2-} + 15NO_2 \qquad \text{(C-9)}$$

As (C-9) now stands, the charge on the left is $1 - 15 = -14$ and that on the right is $3 - 4 = -1$. This means that 13 more H^+ are needed on the left, making a total of $14\,H^+$ there.

$$FeS_2 + 15NO_3^- + 14H^+ \longrightarrow Fe^{3+} + 2SO_4^{2-} + 15NO_2$$

At this stage Fe, S, and N are balanced and charges are balanced. We need only balance either H or O; it is simplest to note that there are 14 hydrogen atoms on the left and none on the right, so that 7 H_2O molecules must be added to the right. Oxygen atoms are then automatically balanced: there are $15 \times 3 = 45$ on the left and $2 \times 4 + 15 \times 2 + 7 = 45$ on the right.

$$FeS_2 + 15NO_3^- + 14H^+ \longrightarrow Fe^{3+} + 2SO_4^{2-} + 15NO_2 + 7H_2O \qquad \text{(C-10)}$$

Other sets: Any other set of oxidation numbers for S and Fe in FeS_2, consistent with the condition that their sum must be zero, is equally good. This compound is in some ways analogous to a peroxide; thus a reasonable oxidation number for the S_2 group is -2 and that for Fe is $+2$. The change in oxidation numbers for one formula unit of FeS_2 is then $+1$ for each Fe atom (from $+2$ to $+3$) and $+7$ for each of the two S atoms (from -1 to $+6$), or a total change of $(1 + 2 \times 7) = 15$ for the entire FeS_2 unit. This result is identical to that obtained with set 1, as it must be, so all further steps in balancing must also be the same.

Suppose we had decided that for convenience we did not want the sulfur to change oxidation state and had assigned the oxidation number $+6$ to it in FeS_2; the two sulfur atoms would then be at $+12$, and the Fe would necessarily be at -12. The change in oxidation numbers as each FeS_2 was oxidized by nitric acid would then be zero for the sulfur atoms and $+15$ for the iron (from -12 to $+3$), again a total change of $+15$, as with the earlier sets.

Example C-4 illustrates a very important point. Although the change in oxidation number for any individual atom in a polyatomic species that gains or loses electrons is arbitrary, the change in oxidation number of the species as a whole is by no means arbitrary. The exchange of electrons between discrete entities is a very real phenomenon, as electrochemistry shows (Chaps. 18 and 19). The stoichiometry is highly precise; the electrons exchanged can be counted accurately. It is only the assignment of electrons to different components within the same polyatomic ion or molecule that is arbitrary.

Balancing redox equations with the aid of half-reactions. Instead of operating with oxidation numbers, we may write balanced equations for the half-reactions of each redox couple involved in the overall reaction and combine them in such a way that the electrons cancel. This method is usually more cumbersome than that of oxidation numbers, but it is free from the arbitrary rules used in assigning oxidation numbers and it may be easier to apply in complex situations. The method is best illustrated by example. Note that the sequence of steps is different from that used in the method of oxidation numbers.

EXAMPLE C-5   **Balancing by half-reactions.** Balance the reaction between *n*-propyl alcohol (C_3H_7OH) and permanganate in an acid solution to yield propionic acid (C_2H_5COOH) and Mn^{2+}:

$$MnO_4^- + CH_3CH_2CH_2OH \longrightarrow Mn^{2+} + CH_3CH_2COOH \qquad (C\text{-}11)$$

Solution.

Step 1.   Separation into half-reactions: Two couples are involved in the reaction: MnO_4^-/Mn^{2+} and C_3H_7OH/C_2H_5COOH. (No implication is attached to the sequence of the symbols in each couple, because we have not as yet established which substance is oxidized and which reduced.)

Step 2.   Balancing of atoms and of charges: Since the solution is acidic we have H^+ and H_2O available. For the couple MnO_4^-/Mn^{2+}, the four O atoms on the left must appear on the right, as $4H_2O$. Thus $8H^+$ are required on the left to furnish the hydrogen. With charges as yet unbalanced, we have

$$MnO_4^- + 8H^+ \longrightarrow Mn^{2+} + 4H_2O \qquad (C\text{-}12)$$

The charges are now balanced by inserting the appropriate number of electrons, leading to the balanced half-reaction

$$5e^- + MnO_4^- + 8H^+ \longrightarrow Mn^{2+} + 4H_2O \qquad (C\text{-}13)$$

This is a familiar reduction reaction; it is consistent with our earlier use of oxidation number $+7$ for MnO_4^- and $+2$ for Mn^{2+} but was derived with no consideration of oxidation numbers.

With the C_3H_7OH/C_2H_5COOH couple, one oxygen is needed on the C_3H_7OH side (the left) to balance oxygen atoms; it is supplied as H_2O. This gives an excess of four hydrogen atoms on the left, which must then be balanced on the right by $4H^+$ (since this reaction is in an acidic solution):

$$C_3H_7OH + H_2O \longrightarrow C_2H_5COOH + 4H^+ \tag{C-14}$$

To balance the charges, $4e^-$ are required on the right:

$$C_3H_7OH + H_2O \longrightarrow C_2H_5COOH + 4H^+ + 4e^- \tag{C-15}$$

Step 3. Combination of half-reactions: For the electrons to cancel in the overall equation, the half-reactions must be multiplied by the coefficients 4 and 5 before they are combined:

$$32H^+ + 4MnO_4^- + 5C_3H_7OH + 5H_2O \longrightarrow$$
$$4Mn^{2+} + 16H_2O + 5C_2H_5COOH + 20H^+ \tag{C-16}$$

The final equation is obtained by canceling the superfluous H^+ and H_2O:

$$12H^+ + 4MnO_4^- + 5C_3H_7OH \longrightarrow 11H_2O + 4Mn^{2+} + 5C_2H_5COOH \tag{C-17}$$

Problems and Questions

C-1 Reactions in acidic solutions. Balance equations for the following reactions, supplying missing ingredients (if any) on each side, on the assumption that the reactions occur in acidic aqueous solutions.

(a) $MnO_4^- + Cl^- \longrightarrow Mn^{2+} + OCl^-$
(b) $H_2O_2 + CH_3OH \longrightarrow CO_2 + H_2O$
(c) $Ti^{2+} + NO_3^- \longrightarrow N_2H_5^+ + TiO^{2+}$
(d) $Fe^{2+} + I_2 \longrightarrow Fe^{3+} + I^-$
(e) $V^{2+} + H_2O_2 \longrightarrow V(OH)_4^+$
(f) $Sn^{2+} + O_2 \longrightarrow Sn^{4+}$
(g) $PH_4^+ + Cr_2O_7^{2-} \longrightarrow P_4 + Cr^{3+}$
(h) $Fe^{3+} + I^- \longrightarrow Fe^{2+} + I_3^-$

C-2 Reactions in alkaline solutions. Balance equations for the following reactions, supplying missing ingredients (if any) on each side, on the assumption that the reactions occur in basic aqueous solutions.

(a) $MnO_4^- + Fe_3O_4 \longrightarrow Fe_2O_3 + MnO_2$
(b) $CuO + NH_3 \longrightarrow Cu + N_2$
(c) $As_2S_3 + NO_3^- \longrightarrow HAsO_4^{2-} + S + NO_2$
(d) $CrO_4^{2-} + CN^- \longrightarrow CNO^- + Cr(OH)_3$
(e) $Au + CN^- + O_2 \longrightarrow Au(CN)_2^-$
(f) $AsO_2^- + O_2 \longrightarrow AsO_4^{3-}$

C-3 Conventional oxidation numbers. Listed below are a number of compounds. For each give the oxidation number obtained by rules 1, 2, 3, 4, and 6 for the element indicated.

(a) P in H_3PO_3, P_4O_6, P_4O_{10}, P_4, H_3PO_4, H_3PO_2, HPO_4^{2-}, HPO_3, PH_4^+
(b) S in H_2S, SO_2, HSO_3^-, $S_2O_3^{2-}$, $S_4O_6^{2-}$, S_8, S_2^{2-}
(c) Mn in $Mn(OH)_2$, Mn^{3+}, Mn_3O_4, MnO_4^-, Mn_2O_7, MnO_4^{2-}, MnO_2
(d) Cl in Cl^-, $HClO$, ClO_2^-, $HClO_3$, Cl_2O, ClF
(e) N in NO, NO_2, HNO_3, NO_2^-, NH_3, N_2H_4, H_2NOH

Ans. (a) 3, 3, 5, 0, 5, 1, 5, 5, -3;   (c) 2, 3, $\tfrac{8}{3}$, 7, 7, 6, 4

C-4   **Balancing of reaction equations; equivalents of reactants.** Balance the following equations, adding, if necessary, water and hydrogen ions.

(a) $Cu^{2+} + CN^- \longrightarrow Cu(CN)_4{}^{3-} + (CN)_2$

(b) $ClO_3{}^- + H_2C_2O_4 \longrightarrow ClO_2 + CO_2$

(c) $NO_3{}^- + I_2 \longrightarrow IO_3{}^- + NO_2$

(d) $ClO_3{}^- + H_2S \longrightarrow SO_4{}^{2-} + Cl^-$

(e) $CuO + NH_3 \longrightarrow Cu + N_2$

Identify the oxidizing agent in (a), (c), and (e), and indicate the number of equivalents per mole of species. In (b) and (d) identify the reducing agent and calculate its equivalent weight.

C-5   **Balancing of reaction equations; equivalents of reactants.** Balance the following equations, adding, if necessary, water and hydrogen ions.

(a) $IO_3{}^- + N_2H_4 \longrightarrow N_2 + I^-$

(b) $HBiO_3 + Mn^{2+} \longrightarrow MnO_4{}^- + BiO^+$

(c) $BrO_2{}^- + NO_2{}^- \longrightarrow Br_2 + NO_3{}^-$

(d) $NH_4{}^+ + MnO_4{}^- \longrightarrow NO_3{}^- + Mn^{2+}$

In reactions (a) and (c) identify the reducing agent and indicate for each how many equivalents correspond to 1 mol. In (b) and (d) indicate the oxidizing agent. Give the equivalent weight of KIO_3 for reaction (a) and that of NH_4Br for reaction (d).

> *Ans.* Reducing agents: (a) N_2H_4, 4 eq mol^{-1}; (c) $NO_2{}^-$, 2 eq mol^{-1}.
> Oxidizing agents: (b) $HBiO_3$, 2 eq mol^{-1}; (d) $MnO_4{}^-$, 5 eq mol^{-1}.
> Equivalent weights: (a) KIO_3, 35.7; (d) NH_4Br, 12.24

C-6   **Redox titration.** Vanadic ion, V^{3+}, forms green salts and is a good reducing agent, being itself changed in neutral solutions to the nearly colorless ion $V(OH)_4{}^+$. A 0.200 F solution of vanadic sulfate was used to reduce a 0.500-g sample of an unknown substance X; exactly 15.0 ml of the solution was needed. What is the equivalent weight of X? What are possible values of its molecular weight?

C-7   **Redox titration.** Molybdenum forms a +3 ion that is green and is a strong reducing agent, being itself oxidized to molybdate, $MoO_4{}^{2-}$. A 0.100 F solution of Mo^{3+} was used to reduce an unknown substance Y; exactly 20.0 ml of the solution was needed to reduce a 0.360-g sample of Y. (a) What is the equivalent weight of Y? (b) What are possible values of its molecular weight?   *Ans.* (a) 60;   (b) 60n, with n a positive integer

C-8   **Equivalent weight and molecular weight.** A new antibiotic A, which is an acid, can readily be oxidized by hot aqueous permanganate; the latter is reduced to manganous ion, Mn^{2+}. The following experiments have been carried out with A: (a) 0.293 g A consumes just 18.3 ml of 0.080 F $KMnO_4$; (b) 0.385 g A is just neutralized by 15.7 ml of 0.490 N NaOH. What is the equivalent weight of A in reaction (a) and what is it in reaction (b)? (c) What can you conclude about the molecular weight of A?

"There is measure in all things."

Horace, 35 B.C.

Appendix D

Tables

This appendix contains the following tables:

D-1. Vapor pressure of water at various temperatures

D-2. Solubility-product constants

D-3. Acid constants K_a

D-4. Values of $\widetilde{H}_f^{\ominus}$ and $\widetilde{G}_f^{\ominus}$

D-5. Standard electrode potentials

TABLE D-1 **Vapor pressure of water at various temperatures**

$t/°C$	$P_{\text{vap,H}_2\text{O}}/\text{torr}$	$t/°C$	$P_{\text{vap,H}_2\text{O}}/\text{torr}$
−78 (ice)	0.00056	30	31.8
−20 (ice)	0.78	35	42.2
−10 (ice)	1.95	40	55
0	4.58	50	93
5	6.54	60	149
10	9.2	70	234
15	12.8	80	355
18	15.5	90	526
20	17.5	100	760
22	19.8	110	1075
24	22.4	120	1489
26	25.2	150	3570
28	28.3	200	11,659

TABLE D-2 **Solubility-product constants**

$Ag \cdot Ag(CN)_2$*	5×10^{-12}	$Fe(OH)_3$	10^{-36}
$AgBr$	5.2×10^{-13}	FeS	5×10^{-18}
$AgBrO_3$	5.2×10^{-5}	Hg_2Br_2†	1×10^{-21}
$AgCl$	1.8×10^{-10}	Hg_2Cl_2†	1.3×10^{-18}
Ag_2CrO_4	2.0×10^{-12}	Hg_2I_2†	7×10^{-29}
AgI	8.3×10^{-17}	Hg_2SO_4†	6.8×10^{-7}
$AgIO_3$	3.0×10^{-8}	$La(IO_3)_3$	6.2×10^{-12}
$AgOH$	2×10^{-8}	$MgCO_3$	5.6×10^{-6}
$AgSCN$	1.0×10^{-12}	MgF_2	6.6×10^{-9}
Ag_2SO_4	1.6×10^{-5}	$Mg(OH)_2$	1.8×10^{-11}
$Al(OH)_3$	2.0×10^{-33}	$Mn(OH)_2$	1.6×10^{-13}
$BaCO_3$	5.5×10^{-10}	$Ni(OH)_2$	6.3×10^{-18}
$BaCrO_4$	3×10^{-10}	NiS	2.0×10^{-21}
BaF_2	1.7×10^{-6}	$PbCO_3$	3×10^{-14}
$BaSO_4$	1.1×10^{-10}	$PbCl_2$	1.7×10^{-5}
$CaCO_3$	5×10^{-9}	$PbCrO_4$	2×10^{-14}
CaC_2O_4 (oxalate)	2.6×10^{-9}	$Pb(IO_3)_2$	2.6×10^{-13}
CaF_2	3.4×10^{-11}	$PbSO_4$	1.7×10^{-8}
$Ca(IO_3)_2$	3.3×10^{-7}	$SrCO_3$	1.1×10^{-10}
$CaSO_4$	2×10^{-5}	$SrCrO_4$	3×10^{-5}
$Ce(IO_3)_3$	3.2×10^{-10}	SrF_2	2.9×10^{-9}
$Cr(OH)_3$	10^{-30}	$SrSO_4$	2.8×10^{-7}
$CuBr$	4×10^{-8}	$TlCl$	1.9×10^{-4}
$CuCl$	1×10^{-6}	$ZnCO_3$	3×10^{-8}
CuI	4×10^{-12}	$Zn(OH)_2$	7×10^{-18}
$Cu(IO_3)_2$	7.4×10^{-8}	$ZnS(\alpha)$	2.5×10^{-22}
CuS	8×10^{-36}	$ZnS(\beta)$	1.6×10^{-24}

*When $AgCN(s)$ is shaken with water, the preponderant species that go into solution are Ag^+ and $Ag(CN)_2^-$, because of the great stability of the complex $Ag(CN)_2^-$. For this reason $AgCN(s)$ is often formulated as $Ag \cdot Ag(CN)_2(s)$, even though all the silver atoms in the crystal are equivalent. The solubility product of silver cyanide is usually given the form

$$[Ag^+][Ag(CN)_2^-] = K_{sp} = 5 \times 10^{-12}$$

rather than

$$[Ag^+][CN^-] = const\ (= 7 \times 10^{-17})$$

†Solutions contain predominantly the species Hg_2^{2+} rather than Hg^+, and the solubility products are formulated as $[Hg_2^{2+}][SO_4^{2-}]$ and $[Hg_2^{2+}][Br^-]^2$.

TABLE D-3   **Acid constants*** K_a (see also Table 14-2)

Acetic	HOAc (or CH_3COOH)	1.8×10^{-5}
Ammonium ion	NH_4^+	5.5×10^{-10}
		$(K_b = 1.8 \times 10^{-5})$
Arsenic	H_3AsO_4	$K_1 = 5.6 \times 10^{-3}$
		$K_2 = 1.7 \times 10^{-7}$
		$K_3 = 3 \times 10^{-12}$
Arsenious	H_3AsO_3	$K_1 = 6 \times 10^{-10}$
		$K_2 = 3 \times 10^{-14}$
Benzoic	C_6H_5COOH	6×10^{-5}
Boric	H_3BO_3	$K_1 = 6.4 \times 10^{-10}$
Carbonic	H_2CO_3	$K_1 = 4.4 \times 10^{-7}$
		$K_2 = 4.8 \times 10^{-11}$
Chlorous	$HClO_2$	1.0×10^{-2}
Chromic	H_2CrO_4	$K_1 = 1.2$
		$K_2 = 3.2 \times 10^{-7}$
Dichromic	$H_2Cr_2O_7$	K_1 large
		$K_2 = 8.5 \times 10^{-1}$
$2HCrO_4^- \rightleftharpoons Cr_2O_7^{2-} + H_2O$		$K = 40$
Isocyanic	HNCO	2.0×10^{-4}
Formic	HCOOH	1.7×10^{-4}
Hydrazinium ion	$^+H_3NNH_2$	1.0×10^{-8}
		$(K_b = 1.0 \times 10^{-6})$
Hydrazoic	HN_3	1.2×10^{-5}
Hydrocyanic	HCN	2×10^{-9}
Hydrofluoric	HF	6.7×10^{-4}
Hydrogen sulfide	H_2S	$K_1 = 9.1 \times 10^{-8}$
		$K_2 = 1.2 \times 10^{-15}$
Hydroxylammonium ion	^+H_3NOH	8.2×10^{-7}
		$(K_b = 1.2 \times 10^{-8})$
Hypobromous	HBrO	2.0×10^{-9}
Hypochlorous	HClO	1.1×10^{-8}
Hypoiodous	HIO	3×10^{-11}
Iodic	HIO_3	2×10^{-1}
Nitrous	HNO_2	4.5×10^{-4}
Oxalic	HOOCCOOH	$K_1 = 5.6 \times 10^{-2}$
		$K_2 = 7.2 \times 10^{-5}$
Phosphoric	H_3PO_4	$K_1 = 7.1 \times 10^{-3}$
		$K_2 = 6.2 \times 10^{-8}$
		$K_3 = 4.4 \times 10^{-13}$
Phosphorous	H_3PO_3	$K_1 = 1.0 \times 10^{-2}$
		$K_2 = 2.6 \times 10^{-7}$
Sulfuric	H_2SO_4	K_1 large
		$K_2 = 1.2 \times 10^{-2}$
Sulfurous	H_2SO_3	$K_1 = 1.0 \times 10^{-2}$
		$K_2 = 5.0 \times 10^{-8}$
Isothiocyanic	HNCS	1.4×10^{-1}

* Some K_b values are given for the corresponding conjugate bases.

TABLE D-4 Values* of $\tilde{H}_f^\circ$ and $\tilde{G}_f^\circ$

	$\tilde{H}_f^\circ$/kJ mol^{-1}	$\tilde{G}_f^\circ$/kJ mol^{-1}		$\tilde{H}_f^\circ$/kJ mol^{-1}	$\tilde{G}_f^\circ$/kJ mol^{-1}
Ag(s)	0.00	0.00	HI(g)	25.9	1.3
Ag$^+$(aq)	105.9	77.1	H$_2$(g)	0.00	0.00
AgCl(s)	−127.0	−109.7	H$_2$O(g)	−241.8	−228.6
Ba(s)	0.00	0.00	H$_2$O(l)	−285.8	−237.2
Ba^{2+}(aq)	−538.4	−560	H$_2$O$_2$(l)	−187.6	−114.0
Br(g)	111.8	82.4	H$_2$O$_2$(aq)	−191.1	−131.7
Br$^-$(aq)	−120.9	−102.8	H$_2$S(g)	−20.2	−33.0
Br$_2$(g)	30.7	3.1	HS$^-$(aq)	−17.7	12.6
Br$_2$(l)	0.00	0.00	S^{2-}(aq)	41.8	83.7
C(g)	716.7	671.3	Hg(g)	60.8	31.8
C (diamond)	1.9	2.8	Hg(l)	0.00	0.00
C (graphite)	0.00	0.00	Hg$_2$Cl$_2$(s)	−264.9	−210.7
CCl$_4$(g)	−102.9	−60.6	I(g)	106.6	70.2
CCl$_4$(l)	−135.4	−65.3	I$^-$(aq)	−55.9	−51.7
CHCl$_3$(l)	−134.5	−73.7	I$_2$(g)	62.3	19.4
CH$_4$(g)	−74.9	−50.8	I$_2$(s)	0.00	0.00
CO(g)	−110.5	−137.3	K(s)	0.00	0.00
CO$_2$(g)	−393.5	−394.4	K$^+$(aq)	−251.2	−282.3
CO$_3^{2-}$(aq)	−676.3	−528.1	KCl(s)	−435.9	−408.3
H$_2$CO$_3$(aq)	−698.7	−623.4	KNO$_3$(s)	−492.7	−393.1
HCO$_3^-$(aq)	−691.1	−587.1	Mg(s)	0.00	0.00
C$_2$H$_2$(g)	226.7	209.2	Mg^{2+}(aq)	−462.0	−457.9
C$_2$H$_4$(g)	52.3	68.1	MgCl$_2$(s)	−641.8	−592.3
C$_2$H$_6$(g)	−84.7	−32.9	Mn(s)	0.00	0.00
C$_3$H$_8$(g)	−103.8	−23.5	Mn^{2+}(aq)	−223.0	−227.6
CH$_3$OH(l)	−238.6	−166.2	MnO$_2$(s)	−519.7	−464.8
C$_2$H$_5$OH(l)	−277.7	−174.8	MnO$_4^-$(aq)	−542.7	−449.4
CH$_3$COOH(l)	−484.5	−389.9	N(g)	472.7	455.6
CH$_3$COOH(aq)	−485.8	−396.6	NH$_3$(g)	−46.2	−16.7
CH$_3$COO$^-$(aq)	−486.0	−369.4	NH$_4^+$(aq)	−132.8	−79.5
C$_6$H$_6$(l)	49.0	124.5	NO(g)	90.4	86.7
C$_6$H$_6$(g)	82.9	129.7	NO$_2$(g)	33.8	51.8
Ca(s)	0.00	0.00	N$_2$(g)	0.00	0.00
Ca^{2+}(aq)	−543.0	−553.0	N$_2$O(g)	81.5	103.6
CaCO$_3$ (calcite)	−1206.9	−1128.8	N$_2$O$_4$(g)	9.7	98.3
CaCO$_3$ (aragonite)	−1207.0	−1127.7	Na(s)	0.00	0.00
CaO(s)	−635.5	−604.2	Na$^+$(aq)	−239.7	−261.9
Ca(OH)$_2$(s)	−986.6	−896.8	NaCl(s)	−411.0	−384.0
Cl(g)	121.4	105.4	NaHCO$_3$(s)	−947.7	−851.9
Cl$^-$(aq)	−167.4	−131.2	Na$_2$CO$_3$(s)	−1130.9	−1047.7
Cl$_2$(g)	0.00	0.00	ONCl(g)	52.6	66.4
Cr(s)	0.00	0.00	O(g)	247.5	230.1
Cr^{3+}(aq)	−256.1	−215.5	OH$^-$(aq)	−230.0	−157.3
Cr$_2$O$_7^{2-}$(aq)	−1523.0	−1319.6	O$_2$(g)	0.00	0.00
Cu(s)	0.00	0.00	O$_3$(g)	142.3	163.4
Cu^{2+}(aq)	64.4	65.0	Pb(s)	0.00	0.00
CuCl(s)	−134.7	−118.8	PbCl$_2$(s)	−359.2	−314.0
Fe(s)	0.00	0.00	S (s, rhombic)	0.00	0.00
Fe^{2+}(aq)	−87.9	−84.9	S (s, monoclinic)	0.30	0.10
Fe^{3+}(aq)	−47.7	−10.5	SO$_2$(g)	−296.9	−300.4
Fe$_2$O$_3$(s)	−822.2	−741.0	SO$_3$(g)	−395.2	−370.4
Fe$_3$O$_4$(s)	−1117.1	−1014.2	SO$_4^{2-}$(aq)	−907.5	−742.0
H(g)	217.9	203.3	Zn(s)	0.00	0.00
H$^+$(aq)	0.00	0.00	Zn^{2+}(aq)	−152.4	−147.2
HBr(g)	−36.2	−53.2	ZnCl$_2$(s)	−415.9	−369.3
HCl(g)	−92.3	−95.3	ZnO(s)	−348.0	−318.2

*The standard states associated with these values are described on pages 446 and 473. The values for ions are based on the convention that $\tilde{H}_f^\circ$ and $\tilde{G}_f^\circ$ are zero for H$^+$(aq) in its standard state (1 M, 1 atm, and 25°C, unless another temperature is explicitly specified).

TABLE D-5 **Standard electrode potentials/volts**

Reaction	Potential
$Ag^+ + e^- \rightleftharpoons Ag(s)$	0.799
$Ag_2O(s) + H_2O + 2e^- \rightleftharpoons 2Ag(s) + 2OH^-$	0.342
$AgCl(s) + e^- \rightleftharpoons Ag(s) + Cl^-$	0.222
$AgBr(s) + e^- \rightleftharpoons Ag(s) + Br^-$	0.071
$AgI(s) + e^- \rightleftharpoons Ag(s) + I^-$	-0.152
$Al^{3+} + 3e^- \rightleftharpoons Al(s)$	-1.66
$Al(OH)_4^- + 3e^- \rightleftharpoons Al(s) + 4OH^-$	-2.35
$H_3AsO_4 + 2H^+ + 2e^- \rightleftharpoons H_3AsO_3 + H_2O$	0.56
$Ba^{2+} + 2e^- \rightleftharpoons Ba(s)$	-2.90
$Be^{2+} + 2e^- \rightleftharpoons Be(s)$	-1.85
$Br_2(l) + 2e^- \rightleftharpoons 2Br^-$	1.065
$Ca^{2+} + 2e^- \rightleftharpoons Ca(s)$	-2.87
$Cd^{2+} + 2e^- \rightleftharpoons Cd(s)$	-0.402
$Cl_2(g) + 2e^- \rightleftharpoons 2Cl^-$	1.359
$2HClO + 2H^+ + 2e^- \rightleftharpoons Cl_2(g) + 2H_2O$	1.63
$HClO_2 + 2H^+ + 2e^- \rightleftharpoons HClO + H_2O$	1.64
$ClO_2(g) + H^+ + e^- \rightleftharpoons HClO_2$	1.27
$ClO_3^- + 2H^+ + e^- \rightleftharpoons ClO_2(g) + H_2O$	1.15
$ClO_4^- + 2H^+ + 2e^- \rightleftharpoons ClO_3^- + H_2O$	1.19
$Co^{2+} + 2e^- \rightleftharpoons Co$	-0.28
$Co^{3+} + e^- \rightleftharpoons Co^{2+}$	1.82
$Cr^{3+} + 3e^- \rightleftharpoons Cr(s)$	-0.74
$Cr_2O_7^{2-} + 14H^+ + 6e^- \rightleftharpoons 2Cr^{3+} + 7H_2O$	1.33
$Cs^+ + e^- \rightleftharpoons Cs(s)$	-2.952
$Cu^+ + e^- \rightleftharpoons Cu(s)$	0.521
$Cu^{2+} + 2e^- \rightleftharpoons Cu(s)$	0.337
$Cu^{2+} + e^- \rightleftharpoons Cu^+$	0.153
$Cu^{2+} + I^- + e^- \rightleftharpoons CuI(s)$	0.85
$F_2(g) + 2e^- \rightleftharpoons 2F^-$	2.87
$Fe^{2+} + 2e^- \rightleftharpoons Fe(s)$	-0.440
$Fe^{3+} + e^- \rightleftharpoons Fe^{2+}$	0.771
$2H^+ + 2e^- \rightleftharpoons H_2(g)$	0.0000
$Hg_2^{2+} + 2e^- \rightleftharpoons 2Hg(l)$	0.792
$Hg_2Cl_2(s) + 2e^- \rightleftharpoons 2Hg(l) + 2Cl^-$	0.268
$2Hg^{2+} + 2e^- \rightleftharpoons Hg_2^{2+}$	0.907
$I_2(s) + 2e^- \rightleftharpoons 2I^-$	0.534
$2IO_3^- + 12H^+ + 10e^- \rightleftharpoons I_2(s) + 6H_2O$	1.19
$K^+ + e^- \rightleftharpoons K(s)$	-2.925
$Li^+ + e^- \rightleftharpoons Li(s)$	-3.03
$Mg^{2+} + 2e^- \rightleftharpoons Mg(s)$	-2.37
$Mn^{2+} + 2e^- \rightleftharpoons Mn(s)$	-1.190
$MnO_2(s) + 4H^+ + 2e^- \rightleftharpoons Mn^{2+} + 2H_2O$	1.23
$MnO_4^- + 8H^+ + 5e^- \rightleftharpoons Mn^{2+} + 4H_2O$	1.51
$HNO_2 + H^+ + e^- \rightleftharpoons NO(g) + H_2O$	0.99
$NO_3^- + 3H^+ + 2e^- \rightleftharpoons HNO_2 + H_2O$	0.94
$Na^+ + e^- \rightleftharpoons Na(s)$	-2.698
$Ni^{2+} + 2e^- \rightleftharpoons Ni(s)$	-0.23
$H_2O_2 + 2H^+ + 2e^- \rightleftharpoons 2H_2O$	1.77

TABLE D-5 (Continued)

$O_2(g) + 4H^+ + 4e^- \rightleftharpoons 2H_2O$	1.229
$O_2(g) + 2H^+ + 2e^- \rightleftharpoons H_2O_2$	0.69
$H_3PO_4 + 2H^+ + 2e^- \rightleftharpoons H_3PO_3 + H_2O$	-0.276
$Pb^{2+} + 2e^- \rightleftharpoons Pb(s)$	-0.126
$PbSO_4(s) + 2e^- \rightleftharpoons Pb(s) + SO_4{}^{2-}$	-0.356
$PbO_2(s) + 4H^+ + 2e^- \rightleftharpoons Pb^{2+} + 2H_2O$	1.47
$PbO_2(s) + SO_4{}^{2-} + 4H^+ + 2e^- \rightleftharpoons PbSO_4(s) + 2H_2O$	1.685
$Rb^+ + e^- \rightleftharpoons Rb(s)$	-2.93
$S(s) + 2H^+ + 2e^- \rightleftharpoons H_2S(g)$	0.141
$HSO_4{}^- + 9H^+ + 8e^- \rightleftharpoons H_2S(g) + 4H_2O$	0.316
$HSO_4{}^- + 3H^+ + 2e^- \rightleftharpoons SO_2(g) + 2H_2O$	0.14
$S_4O_6{}^{2-} + 2e^- \rightleftharpoons 2S_2O_3{}^{2-}$	0.09
$Sn^{2+} + 2e^- \rightleftharpoons Sn(s)$	-0.140
$Sn(OH)_6{}^{2-} + 2e^- \rightleftharpoons HSnO_2{}^- + H_2O + 3OH^-$	-0.90
$Sn(IV) + 2e^- \rightleftharpoons Sn(II)$	0.14 (in 1 F HCl)
$Sr^{2+} + 2e^- \rightleftharpoons Sr(s)$	-2.89
$Tl^+ + e^- \rightleftharpoons Tl(s)$	-0.336
$Tl^{3+} + 2e^- \rightleftharpoons Tl^+$	1.28
$Zn^{2+} + 2e^- \rightleftharpoons Zn(s)$	-0.763

Index

Index